Materials for Advanced Packaging

Daniel Lu · C.P. Wong
Editors

Materials for Advanced Packaging

Editors

Daniel Lu
Henkel Locite Co., Ltd.
Yantai, Shandong, P.R.
China
Daniel.Lu@cn.henkel.com

C.P. Wong
Georgia Institute of Technology
Atlanta, GA, USA
cp.wong@mse.gatech.edu

ISBN: 978-0-387-78218-8 e-ISBN: 978-0-387-78219-5
DOI: 10.1007/978-0-387-78219-5

Library of Congress Control Number: 2008932162

© Springer Science+Business Media, LLC 2009
All rights reserved. This work may not be translated or copied in whole or in part without the written permission of the publisher (Springer Science+Business Media, LLC, 233 Spring Street, New York, NY 10013, USA), except for brief excerpts in connection with reviews or scholarly analysis. Use in connection with any form of information storage and retrieval, electronic adaptation, computer software, or by similar or dissimilar methodology now known or hereafter developed is forbidden.
The use in this publication of trade names, trademarks, service marks, and similar terms, even if they are not identified as such, is not to be taken as an expression of opinion as to whether or not they are subject to proprietary rights.

Printed on acid-free paper

springer.com

Preface

With consistently active involvement in the electronic packaging conferences such as the IEEE Electronic Components and Technology Conference (ECTC) in the last several years, we have witnessed many advances in advanced electronic packaging technology, especially in materials and processing aspects. We have come to the decision to prepare this book so that readers can learn these recent advances in electronic packaging.

This book provides a comprehensive review on the most recent developments in advanced packaging technologies including emerging technologies such as 3 dimensional (3D), nanopackaging, and biomedical packaging with a focus on materials and processing aspects.

This book consists of 19 chapters which are written by well recognized experts in each field. Chapter 1 reviews various 3D package architectures, and processes and materials to enable these 3D packages. Chapter 2 provides an overview on several new bonding and joining techniques to make large area void-free bonding interface for electrical and/or mechanical interconnections. Chapter 3 reviews some novel approaches to make electrical interconnects between integrated circuit (IC) and substrates to improve both electrical and mechanical performance. Most recent developments in wire bonding are covered extensively in Chapter 4. Various wafer thinning techniques and associated materials and processing are reviewed in Chapter 6. Latest advances in several key packaging materials including lead-free solders, flip chip underfills, epoxy molding compounds, conductive adhesives, die attach adhesives/films, and Thermal Interface Materials (TIMs) are reviewed in great detail in Chapters 5, 9, 10, 11, 12, and 13, respectively. Advances on organic substrate and printed circuit boards are covered in Chapters 7 and 8, respectively. Chapter 14 reviews the materials advent on embedded passives including capacitors, inductors and resistors. Chapters 16 and 17 review the advent in materials and processing aspects on MicroElectroMechanical System (MEMS) and wafer level chip scale packaging, respectively. Emerging technologies such as nanopackaging, Light Emitting Diode (LED) and optical packaging, and biomedical packaging are covered in Chapters 15, 18 and 19, respectively.

We greatly thank all the contributors for their efforts to bring this wonderful book to the readers.

Contents

Preface v

Contributors.. ix

1   **3D Integration Technologies – An Overview** 1
    Rajen Chanchani

2   **Advanced Bonding/Joining Techniques** 51
    Chin C. Lee, Pin J. Wang and Jong S. Kim

3   **Advanced Chip-to-Substrate Connections**...................... 77
    Paul A. Kohl, Tyler Osborn and Ate He

4   **Advanced Wire Bonding Technology: Materials, Methods, and Testing**... 113
    Harry K. Charles

5   **Lead-Free Soldering**.. 181
    Ning - Cheng Lee

6   **Thin Die Production**....................................... 219
    Werner Kroeninger

7   **Advanced Substrates: A Materials and Processing Perspective**.. 243
    Bernd Appelt

8   **Advanced Print Circuit Board Materials** 273
    Gary Brist and Gary Long

9   **Flip-Chip Underfill: Materials, Process and Reliability** 307
    Zhuqing Zhang and C. P. Wong

10	Development Trend of Epoxy Molding Compound for Encapsulating Semiconductor Chips 339
	Shinji Komori and Yushi Sakamoto

11	Electrically Conductive Adhesives (ECAs) 365
	Daoqiang Daniel Lu and C.P. Wong

12	Die Attach Adhesives and Films 407
	Shinji Takeda and Takashi Masuko

13	Thermal Interface Materials 437
	Ravi Prasher and Chia-Pin Chiu

14	Embedded Passives. 459
	Dok Won Lee, Liangliang Li, Shan X. Wang, Jiongxin Lu, C. P. Wong, Swapan K. Bhattacharya and John Papapolymerou

15	Nanomaterials and Nanopackaging 503
	X.D. Wang, Z.L. Wang, H.J. Jiang, L. Zhu, C.P. Wong and J.E. Morris

16	Wafer Level Chip Scale Packaging 547
	Michael Töpper

17	Microelectromechanical Systems and Packaging 601
	Y. C. Lee

18	LED and Optical Device Packaging and Materials 629
	Yuan-Chang Lin, Yan Zhou, Nguyen T. Tran and Frank G. Shi

19	Digital Health and Bio-Medical Packaging 681
	Lei Mercado, James K. Carney, Michael J. Ebert, Scott A. Hareland and Rashid Bashir

Subject Index. 713

About the Editors. 717

Contributors

Bernd K Appelt, Ph.D. Marketing and Sales, ASE (U.S.) Inc., 3590 Peterson Way, Santa Clara, CA 95054
bernd.appelt@aseus.com

Rashid Bashir, Ph.D. Micro and Nanotechnology Laboratory, Department of Electrical and Computer Engineering and Department of BioEngineering, University of Illinois at Urbana-Champaign, 2000 Micro and Nanotechnology Laboratory, MC-249, University of Illinois at Urbana-Champaign, 208 North Wright Street, Urbana, Illinois 61801 USA

Swapan K. Bhattacharya Senior Research Scientist, School of Electrical and Computer Engineering, Georgia Institute of Technology, TSRB, 85N 5th Street, Atlanta, GA 30308

Gary Brist, Sr. PCB Technologist, Intel Corp. 5200 NE Elam Young Parkway, Hillsboro, OR 97124
gary.a.brist@intel.com

James K. Carney, Ph.D. Cardiac Rhythm and Disease Management, Medtronic, Inc., 8200 Coral Sea St. NE, Mounds View, MN 55112

Rajen Chanchani, Ph.D. Sandia National Laboratories, Sandia National Laboratories, Mail stop: 0352, Albuquerque, NM 87185
chanchr@sandia.gov

Harry K. Charles, Professor The Johns Hopkins University, The Johns Hopkins University, Applied Physics Laboratory, 11100 Johns Hopkins Road, Laurel, MD 20723-6099
harry.charles@jhuapl.edu

Chia-Pin Chiu, Ph.D. Intel Corporation, 5000 W. Chandler Blvd., AZ 85226, USA

Michael J. Ebert, Ph.D. Cardiac Rhythm and Disease Management, Medtronic, Inc., 8200 Coral Sea St. NE, Mounds View, MN 55112

Scott A. Hareland, Ph.D. Cardiac Rhythm and Disease Management, Medtronic, Inc 8200 Coral Sea St. NE, Mounds View, MN 55112

Ate He, Ph.D. Chemical and Biomolecular Engineering, Georgia Institute of Technology, 311 Ferst Dr., Atlanta, GA, 30332-0100

Hong Jin Jiang, Ph.D. Materials Science and Engineering, Georgia Institute of Technology, 500 Northside Circle, Apt. I3, Atlanta, GA, 30309

Jong S. Kim, Ph.D. Optical Platform Division, Intel Corporation, 8674 Thornton Ave, Newark, CA, 94560

Paul A. Kohl, Ph.D./Regents' Professor Chemical and Biomolecular Engineering, Georgia Institute of Technology, 311 Ferst Dr., Atlanta, GA, 30332-0100
paul.kohl@chbe.gatech.edu

Shinji Komori, M.S. Electronic Device Materials Research Laboratory 1, Sumitomo Bakelite Co., Ltd., 20-7, Kiyohara Industrial Park Utsunomiya-city Tochigi prefecture, 321-3231 Japan

Werner Kröninger Dipl. Phys. Univ., IFAG OP FEP T UPD 5 Infineon Technologies AG, P.O. Box 10 09 44, D-93009 Regensburg, Germany
werner.kroninger@infineon.com

Chin C. Lee, Ph.D./Professor Electrical Engineering and Computer Science, University of California, Irvine, 2226 Engineering Gateway Building, Irvine, CA 92697-2660
cclee@uci.edu

Ning-Cheng Lee, Ph.D. VP of Technology, Indium Corporation of America, 1676 Lincoln Ave., Utica, NY 13502
nclee@indium.com

Y.C. Lee, Ph.D./Professor Department of Mechanical Engineering, University of Colorado, Department of Mechanical Engineering, University of Colorado, Boulder, CO 80309-0427
leeyc@colorado.edu

Dok Won Lee Stanford University, Stanford, CA, USA

Yuan-Chang Lin, Ph.D. H.S. School of Engineering, University of California, Optoelectronics Packaging & Materials Labs, 916 Engineering Tower, University of California, Irvine, CA 92697-2575

Liangliang Li Stanford University, Stanford, CA, USA

Gary Long, Sr. PCB Technologist, Intel Corp., 5200 NE Elam Young Parkway, Hillsboro, OR 97124

Contributors

Daoqiang Daniel Lu, Ph.D. Henkel Corporation, Henkel Loctite (China) Co., Ltd., No. 90 Zhujiang Road, Yantai ETDZ, Shandong, China 264006
daniel.lu@cn.henkel.com

Jiongxin Lu, Ph.D. School of Materials Science and Engineering, Georgia Institute of Technology, 771 Ferst Drive, N.W., Atlanta, GA 30332, USA

Takashi MASUKO, Staff Researcher Ph.D. Electronic Materials R&D Center, Hitachi Chemical Co., Ltd., 48 Wadai, Tsukuba, Ibaraki, 300-4247, Japan

Lei Mercado, Ph.D. Neuromodulation, Medtronic, Inc., 4000 Lexington Ave N, Shoreview, MN, 55126
lei.l.mercado@medtronic.com

James E. Morris, Ph.D., Professor Electrical & Computer Engineering - Engineering & Computer Science, Portland State University, ECE, PO Box 751, Portland, OR 97207

Tyler Osborn, B.S. Chemical and Biomolecular Engineering, Georgia Institute of Technology, 311 Ferst Dr., Atlanta, GA, 30332-0100

John Papapolymerou, Professor School of Electrical and Computer Engineering, Georgia Institute of Technology, TSRB, 85N 5th Street, Atlanta, GA 30308

Ravi Prasher, Ph.D. Intel Corporation, 5000 W. Chandler Blvd., AZ 85226, USA
ravi.s.prasher@intel.com

Yushi Sakamoto, M.S. Electronic Device Materials Research Laboratory 1, Sumitomo Bakelite Co., Ltd., 20-7, Kiyohara Industrial Park Utsunomiya-city Tochigi prefecture, 321-3231 Japan
Yushis@sumibe.co.jp

Frank G. Shi, Ph.D./Professor H.S. School of Engineering, University of California, Optoelectronics Packaging & Materials Labs, 916 Engineering Tower, University of California, Irvine, CA 92697-2575
fgshi@uci.edu

Shinji TAKEDA, Ph.D. Research & Development Division, Hitachi Chemical Co., Ltd., 4-13-1 Higashi, Hitachi, Ibaraki, 317-8555, Japan
shin-takeda@hitachi-chem.co.jp

Michael Töpper, Ph.D. Fraunhofer IZM, Fraunhofer IZM, Gustav-Meyer-Allee 25, D-13355 Berlin, Germany
michael.toepper@izm.fraunhofer.de

Nguyen T. Tran, Ph.D. H.S. School of Engineering, University of California, Optoelectronics Packaging & Materials Labs, 916 Engineering Tower, University of California, Irvine, CA 92697-2575

Pin J. Wang, Ph.D. candidate, Materials Manufacturing Technology, University of California, Irvine, 2226 Engineering Gateway Building, Irvine, CA 92697-2660

Zhong Lin Wang, Ph.D./Regents' Professor School of Materials Science and Engineering, GA Tech, 771 Ferst Dr. NW, Atlanta, GA 30332
zhong.wang@mse.gatech.edu

Xudong Wang, PhD/Research Scientist School of Materials Science and Engineering GA Tech, 771 Ferst Dr. NW, Atlanta, GA 30332

Shan X. Wang, Ph.D./Professor Stanford University, Stanford, CA, USA
sxwang@ee.stanford.edu

C.P. Wong, Ph.D./Regents' Professor School of Materials Science and Engineering, Georgia Institute of Technology, 771 Ferst Drive, N.W., Atlanta, GA 30332, USA

Lingbo Zhu, Ph.D. New Products Group, Dow Chemical Company, 2616 Abbott Road, Apt. I7, Midland, MI 48642

Zhuqing Zhang, Ph.D. Imaging and Printing Group, Hewlett-Packard Company, 1000 NE Circle Blvd, Corvallis, OR 97333, USA
zhuqing.zhang@hp.com

Yan Zhou, Ph.D. Student, H.S. School of Engineering, University of California, Optoelectronics Packaging & Materials Labs, 916 Engineering Tower, University of California, Irvine, CA 92697-2575

Lingbo Zhu, Ph.D. New Products Group, Dow Chemical Company, 2616 Abbott Road, Apt. I7, Midland, MI 48642

Chapter 1
3D Integration Technologies – An Overview

Rajen Chanchani

Abstract The next generation of integrated micro-system technologies can only keep up with increased functionality and performance demands by using the 3rd dimension. The primary drivers for 3D integration are miniaturization, integration of different technologies in a small form-factor, and performance. 3D integration technologies can be grouped into 3 main categories, namely 3D On-chip integration, 3D IC-stacking, and 3D-packaging. This chapter provides a detailed review of each of these categories.

Keywords Micro-system · 3D integration · Die stacking · Thru-silicon vias (TSVs) · Wafer bonding

1.1 Introduction

Figure 1.1 illustrates a schematic example of an integrated micro-system showing five different individual micro-system functional blocks. Traditionally, integration of these different functional blocks is done in 2-Dimension (2D) on a package or a Printed Wiring Board (PWB) [1]. In 3-Dimension (3D) architecture, these functional blocks can be stacked vertically, each built on a separate layer in the stack. Each layer is connected together using vertical interlayer interconnects. When integration is done in 3D, micro-system form-factor significantly shrinks i.e. the X and Y dimensions are significantly reduced, with a very small to negligible increase in Z-dimension. This size reduction also results in reduction in interconnect length between the functional blocks. Reduced interconnect length results in improved system performance. This will be explained in more detail in a later section.

R. Chanchani (✉)
Sandia National Laboratories, Sandia National Laboratories, Mail stop: 0352, Albuquerque, NM 87185
e-mail: chanchr@sandia.gov

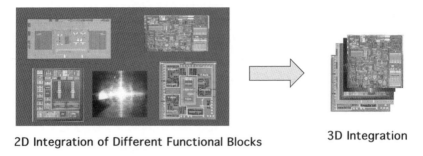

Fig. 1.1 Integrated microsystem in 2D and in 3D

1.1.1 Categories of 3D Integration Technologies

Although 3D integration is a relatively new concept, there are many different technologies that are either being investigated or implemented. All of these technologies can be classified into three broad categories [2].

1.1.1.1 3D On-Chip Integration

3D on-chip integration is a vertical extension of IC technology. Active semiconductor device layers are sequentially built-up on the first IC layer as shown in Fig. 1.2. This technology is truly a homogeneous, 3D system-on-chip (SOC) technology and it is still at R&D stage, mostly in the universities [3].

1.1.1.2 3D IC-Stacking with Thru-Silicon Vias (Thru-Si Vias or TSVs)

In this category of 3D integration, individual wafers are first fabricated and then wafers or ICs are stacked in 3D with thru-Si Vias which provide the

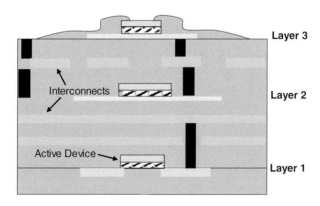

Fig. 1.2 On-chip 3D integration of IC

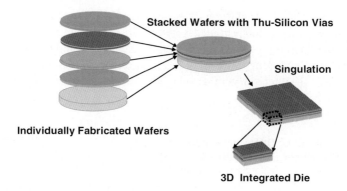

Fig. 1.3 3D IC-stacking technology with Thru-Si Vias

interconnections between dice. IC-stacking can either be done at wafer-level or at die-level. The wafer-level version of this technology is shown in Fig. 1.3. In this category, vias can go though either bulk Si or through SiO_2. The term 'thru-Si Via' refers to both these cases. In recent years, this technology has been the focus of world-wide research & development activities in industry, university and research institutes.

1.1.1.3 3D-Packaging

3D-packaging provides the least disruptive way to make integrated microsystems by stacking packaged ICs in 3D. There are many versions of this category of 3D integration. Two versions, wire-bonded die-stack and Ball-Grid-Array (BGA)-stack, are illustrated in Figs. 1.4a and 1.4b. Currently, 3D packages are widely used in many consumer applications. Some of the advanced concepts of

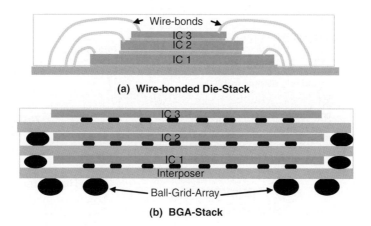

Fig. 1.4 Examples of 3D-packaging

3D packages are difficult to differentiate from 3D IC-stacking (second category). In this chapter, I have differentiated between the second and third categories, based on how die-to-die interconnections are made. If the interconnections are made with thru-Si vias, the technology is categorized as 3D IC-Stacking (second category) and if the interconnections are made outside of the die, then it has been categorized as 3D-Packaging.

1.1.2 Motivation for 3D Integration

The electronic industry had been successfully coping with increasing demand for micro-systems with higher functional density and performance by: (i) scaling, i.e. increasing IC functional density by reducing the feature size; and (ii) the use of advanced IC packaging and integration techniques like flip-chip, Ball grid Array (BGA), Chip-Scale-Package (CSP), multi-chip module (MCM), System-in-package (SIP), and embedded passives. Both IC functional density and advanced packaging/integration techniques have reached their practical limits in 2D. The next generation of technologies can only keep up with increased functionality and performance demands by using the 3rd dimension.

Figure 1.5 illustrates the historical progression into 3D integration. One measure of circuit density is Silicon Packaging Efficiency (SPE), which is the ratio of total silicon area to total area of circuitry. Figure 1.5 shows a traditional circuit board, which typically has a SPE of 10–15%. In the next generation of technologies, multichip modules, the SPE is in higher range of 50–70%.

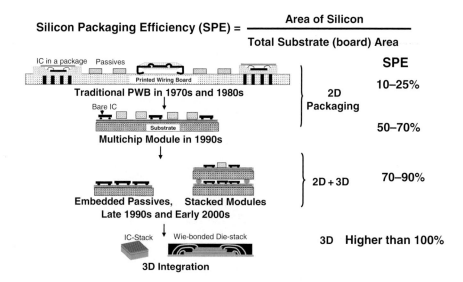

Fig. 1.5 Historical progression of integration technologies from 2D to 3D

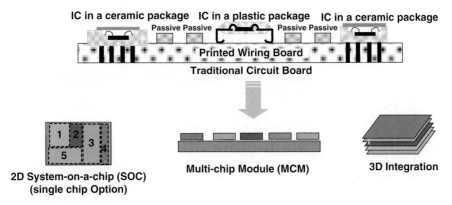

Fig. 1.6 Technology options for miniaturization of future systems

A higher SPE is obtained when 2D integration is combined with 3D integration, e.g. embedded passives and stacked module technologies. A fully 3D integrated ICs provides a SPE of over 100%. 3D integration not only allows a higher density of integration, but also offers several other practical advantages over 2D technologies as discussed next.

The motivation to use 3D integration technologies becomes apparent when they are compared to 2D options. As shown in Fig. 1.6, traditional systems have been built using packaged devices on Printed wiring Boards (PWBs). If these systems have to be further miniaturized for smaller size and better performance, we have three options. The first option is the ideal situation of a 2D system-on-a-chip (SOC), where all functional blocks are integrated in 2D on a single IC. The second option is also a 2D solution where different functional blocks are integrated on a substrate (MCM) or in a package (SIP). Third option is to integrate in 3D. The main motivating factors in choosing the best technology option are form-factor, cost, technology integration and performance. Table 1.1 shows the relative ranking of the three technology options for each of these factors. These rankings capture the summary of the detailed discussion below.

1.1.2.1 Form-Factor

It is obvious that 3D IC-stacking and 2D SOC offers the smallest form factor. Although MCMs have a smaller form-factor than the traditional PWB assemblies, their form-factor will be significantly larger than that of 3D IC-stacks.

1.1.2.2 Cost

In order to get full-scale integration on a single IC (2D SOC), the chip size has to be relatively large. According to the semiconductor industry roadmap, the

Table 1.1 Ranking of different technology options

	Technology Options ▶		
	2D SOC	2D MCM	3D
Form factor	5	2	5
Cost	1	4	3
Technology Integration	1	5	4
Performance	3	1	5

Ranking: 1 is poorest, 5 is best

chip size will not grow much bigger than 400 mm^2 because of the higher cost associated with making larger chips. As chip size grows, both lower manufacturing yields and lower number of dice available per wafer will contribute to its higher cost [4]. Thus a 2D SOC is not a cost-effective option for full integration. This leaves both MCM and 3D integration as feasible options as shown by ranking in Table 1.1. Currently, 3D integration is more costly than the MCM option, but many groups are developing 3D manufacturing processes, and believe that with technology maturation and increased production volumes, the cost will significantly reduce.

1.1.2.3 Integration of Different Heterogeneous Technologies

Full integration of different technologies on a 2D SOC is technically very difficult. Even if the technical feasibility exists, prohibitive production cost and the capital investment required will discourage such integration on a single IC (SOC). Both MCM and 3D integration allow integration of different technologies by first making individual dice, and then combining them either on a substrate (MCM) or stacking them in 3D. Thus from integration perspective, MCM and 3D options are more adaptable to a wide variety of microsystem applications.

1.1.2.4 Performance

System performance is by far much better in 3D integration than either of the two 2D technologies, SOC and MCM [3, 5, 6–8]. One of the main contributing factor for better performance of 3D technologies is that the longer interconnects in 2D are replaced with much shorter vertical interconnects in 3D as illustrated

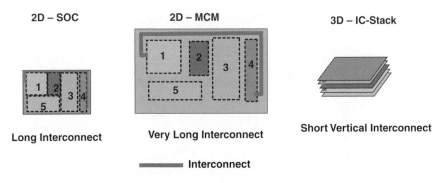

Fig. 1.7 Interconnect lengths in different technology options

in Fig. 1.7. In circuit layouts, there are three kinds of interconnects, namely local, semi-global and global. Local interconnects are the shortest interconnects between the elements within a functional block. Semi-global interconnects, which are of intermediate length, are used in interconnecting adjacent blocks. Global interconnects are of longer length, spanning the entire circuitry. An example of global interconnects will be timing interconnects that connect the clock with functional blocks. Global and semi-global interconnects are shorter in the case of 3D circuits providing the maximum impact on the circuit performance. Shorter interconnects results in better performance because they lower interconnect time delays, cross-talk and power dissipation.

Interconnect time delays The IC industry has continued to meet the demand for high performing, low power, and cost-effective miniaturized electronic products by packing more transistors in a single chip. This has been done primarily by scaling or reducing the feature size, with every generation of IC manufacturing. As the feature size goes down, the gate length reduces resulting in reduced gate delay. However, scaling increases interconnect length to accommodate increasing transistor densities [3, 9]. Higher interconnect length results in higher interconnect delay as shown in Fig. 1.8 [3, 9–11]. In an effort to improve the interconnect delay, Copper (Cu) /low k (Dielectric constant) technology was developed in 1990s. Even with using Cu/Low k technology in the future ICs, the interconnect delay will be too long and thus adversely affect the performance. In Fig. 1.9, the effects of scaling on interconnect and gate delay for traditional Al/SiO_2 and Cu/Low k technologies are shown [2, 9, 12]. The gate delays in both technologies have direct dependence on scaling. Cu/Low k interconnect delay is significantly lower than in Al/SiO_2 technology. However interconnect delay increases with lower feature size in both technologies. This figure illustrates that Cu/Low K will not give the desirable performance of lower signal delay in future circuits with < 0.1 µm feature size. This will be where 3D technologies (specifically IC-stacking with Thru-Si vias) will bridge the performance gap as illustrated in Fig. 1.10 [3]. When we convert from 2D design to 3D IC design

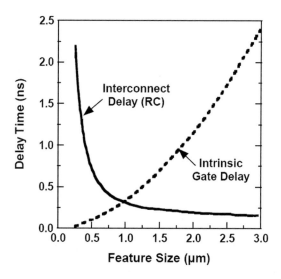

Fig. 1.8 Intrinsic gate and interconnect delays as a function of minimum feature size [10]

with stacked functional blocks, the interconnect delay reduces because long interconnects are now replaced with very short vertical interconnects. In addition, there is increased space available for interconnect routing in 3D, which can result in increased interconnect pitch and cross-sectional area. Shorter interconnect lengths, higher pitch, and cross-sectional areas result in lower time delays and reduction in parasitic R, L, C losses associated with interconnects. Thus we will not be able to continue to increase circuit densities in the future on 2D ICs, and the 3D IC-stacking option offers a better solution to these performance issues.

Cross-talk Cross-talk is directly related to the length of coupled interconnects and how closely they are spaced. In 3D designs, shorter interconnect lengths will reduce interconnect coupling and thus decrease cross–talk associated with these interconnects [13]. Due to the limited routing space available in 2D circuit, interconnect lines have to be spaced closer with finer pitch. In 3D designs, since the vertical dimension is also available for routing, there is ample interconnect routing area to allow increased pitch that will further reduce cross-talk.

Power Dissipation Interconnects consume a large portion of the power budget in a chip due to high parasitic losses. Since interconnect parasitics are proportional to their length, 3D designs will have lower parasitics and lower power consumption than in 2D circuits [6, 7, 14]. The effect of interconnect length on power consumption has been shown earlier with MCMs [6]. It was shown that when a traditional PWB assembly is implemented in a MCM, the power dissipation associated with interconnects is reduced by 80%, mainly due to the reduction

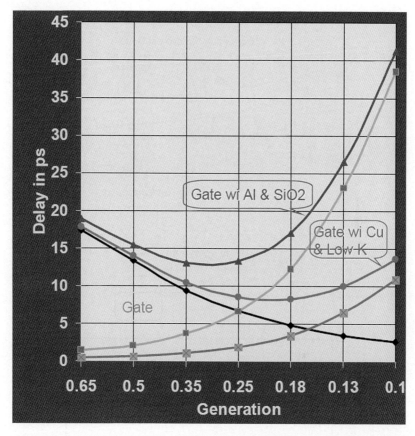

Fig. 1.9 Interconnect and gate Delay as a function of feature size in Al/SiO2 and Cu/Low K technologies [12]

in interconnect length. Similarly, with 3D technologies, the power consumption will decrease significantly, in proportion to the decrease in parasitic losses due to shorter interconnect length, higher interconnect pitch and cross-sectional area. This is illustrated in Fig. 1.11 [7], which shows the power consumption in a Field-Programmable Gate Array (FPGA) as a function of the number of layers (strata) in the stack. With 3D integration (2–4 strata), the decrease in power dissipation is 35–55%. The maximum effect is seen for the transition from a 2D to 2-layer stack, where most of interconnects will be vertical. As the number of strata increases, a higher percentage of interconnects will be in-plane, reducing the effect on reduction in power consumption.

The reasons for using 3D integration are very compelling, prompting universities, industry and the research institutes in the USA, Europe and Asia, to pursue the development of 3D technologies.

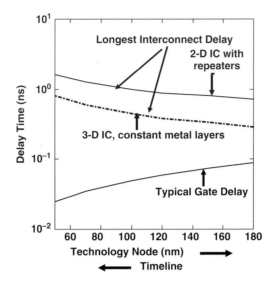

Fig. 1.10 Interconnect and gate delays as a function of feature size for 2D and 3D Circuits [3]

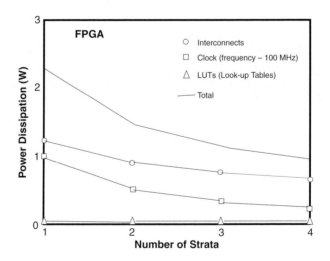

Fig. 1.11 Effect of number of 3D layers on power dissipation [7]

1.2 Description of Technologies

Three different categories of technologies, which were briefly introduced in Section 1.1.1 will be described next in more detail. All these three are in different stages of maturity. The first category (see Fig. 1.2), on-chip 3D integration, is still in the early stages of research & development. The second category, 3D IC-Stacking with TSV (see Fig. 1.3), is in the advanced stages of development and is currently being considered for application in some specialized advanced products. The third category, 3D-packaging (See Fig. 1.4), is the most mature category, which is already used in many high-end consumer products.

1.2.1 3D on Chip Integration

This is a 'bottom-up' approach for 3D integration, where active silicon layers are built-up sequentially and separated from each other with interlayer dielectric (refer to Fig. 1.2). This category is truly a homogeneous 3D SOC, offering the most efficient way to integrate. However, major technical challenges and R & D issues are still being addressed at the leading universities. Studies are currently pursuing three techniques to achieve 3D on-chip integration, including beam recrystallization, silicon epitaxial growth, and solid phase crystallization.

1.2.1.1 Beam Recrystallization

In this technique (shown in Fig. 1.12a) a second active layer on an existing substrate is deposited by first depositing a polysilicon layer followed by the fabrication of thin film transistors (TFTs) on top [3, 15]. However, transistors on polysilion perform poorly, because they exhibit very low surface mobility and high threshold voltages. To improve the performance, an intense laser beam is directed toward the polysilicon layer to recrystallize and eliminate grain boundaries. However this is not practical for 3D devices, because of the high temperatures, 1000°C, involved during melt. The high temperature will adversely affect the devices in the lower layers. There has been on-going work to recrystallize polysilicon at lower temperatures, which can be tolerated better by lower layer transistors.

1.2.1.2 Silicon Epitaxial Growth

In this technique (shown in Fig. 1.12b), 3D silicon layers are built-up by etching a hole in the passivation layer to form a window followed by epitaxially growing single crystal Si seeded from this window [3, 16]. The silicon crystal grows vertically and then laterally. The process starts by first depositing and patterning silicon dioxide. Next windows are created in the oxide layer for epitaxial growth. Next, a single crystal Si epitaxial layer is grown vertically and then laterally. This newly grown silicon is planarized to the oxide layer by using Chemical-Mechanical Polishing (CMP). Metal Oxide Semiconductor Field-Effect Transistor (MOSFET) devices are fabricated in this grown Si. This whole fabrication process cycle is repeated for the subsequent layers. The major concern is that the high temperatures ($\sim 1000°C$) involved in the epitaxial

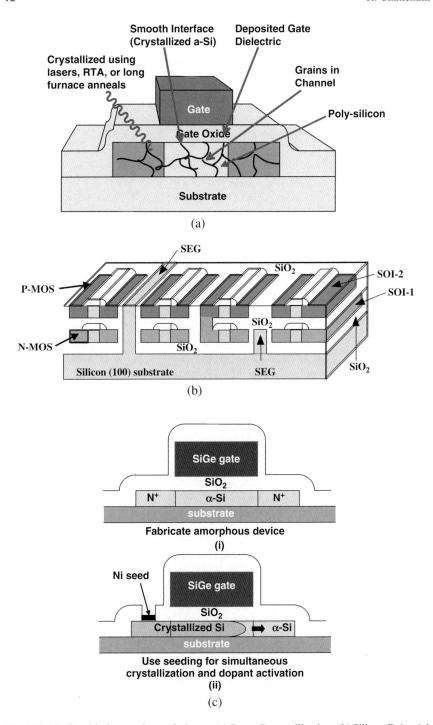

Fig. 1.12 3D On-chip integration techniques: (**a**) Beam Recrystllization, (**b**) Silicon Epitaxial Growth, and (**c**) Solid Phase Crystallization [5]

process will degrade the devices below. Low temperature processes using ultra-high vacuum Chemical Vapor Deposition (UHV-CVD) are being developed [17] to resolve the issues with high temperature processes.

1.2.1.3 Solid Phase Crystallization

This is a relatively lower temperature technique to fabricate devices in 3D. The technique is schematically shown in Fig. 1.12c [3, 18]. Amorphous silicon (alpha-Si) is first deposited at a low temperature on the first layer of the active devices. The second active layer of devices is built on this amorphous layer. Then the amorphous silicon is crystallized to form polysilicon. Crystallization is induced by Ni seed implanted in a small patterned window at temperature <500°C. This step is repeated for the multiple layers of the active devices. Some recent results show the feasibility of making better performing devices at lower temperatures. However, the electrical characteristics of these devices are inferior to those made on single crystal Si.

On-chip 3D integration technologies are still in research. The processing difficulty to build circuitry in 3D and its ability to integrate only a limited number of devices will probably enable its use to only highly specialized applications.

1.2.2 3D IC-stacking Using Thru-Si Vias (TSVs)

3D IC-stacking with Thru-Si vias (See Fig. 1.3) offers a very desirable 3D integration solution. This technology eliminates many problems associated with 3D on-chip integration. In 3D IC-stacking, integration of different technologies can be cost-effectively achieved because each IC is first made individually and then stacked together. Each of the layers in the stack can now contain circuits with different voltage, performance and fabrication process requirements. IC-stacking will also give a better performance because 3D silicon layers are connected with very short vertical interconnects as discussed in the earlier section.

1.2.2.1 Typical Process Steps

3D IC-stacking is a 'top down' approach. Processing of IC-stacks can be done either at wafer-to-wafer [19], chip-to-wafer [20, 21] or chip-to-chip [22, 23] levels as illustrated in Fig. 1.13. Wafer-level processing offers the most cost advantage, whereas chip-to-chip processing is high yielding because it allows the use of known-good-dice. Most world-wide R & D is currently concentrated on wafer-level because of the potential for cost reduction.

The process sequence will mainly depend on the approach taken for fabricating vias – 'vias first' and 'vias last'. The 'vias first' process involves fabricating vias before ICs are bonded and the 'via last' process involves fabricating vias after wafers/ICs are bonded. Typical process steps involved in the 'via first' and

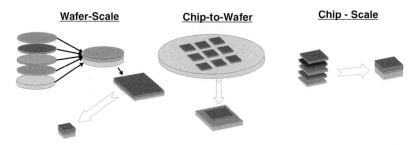

Fig. 1.13 Schematic illustration of different options for IC-stacking

the 'via last' approaches are shown in Figs. 1.14 [24] and 1.15 [13], respectively. The major process steps common to either of these processes are preparation of top and bottom wafers, wafer thinning, fabrication of thru-Si vias, alignment and bonding.

The ICs can be stacked Face-to-Back or Face-to-Face [20, 25]. The examples for processing these two stacking approaches are shown in Fig. 1.16 [20]. In the Face-to-Back, the face of the bottom wafer is bonded to the back of the top wafer. The top wafer is first prepared with thru-Si vias and then thinned from the backside using a handling wafer attached to the front side. The appropriate dielectric and metal are deposited on the front side of the bottom wafer before it is bonded with the back of the top wafer. In the Face-to-Face approach, thru-Si vias and metal pads are fabricated on the Front side of the top wafer. The top wafer is then bonded Face-to-Face with bottom wafer. In this case, the top wafer is thinned without a carrier wafer since it is already supported by the bottom wafer. In these examples, thru-Si vias are made before the wafers are bonded. There are other approaches where the vias are made after bonding [13].

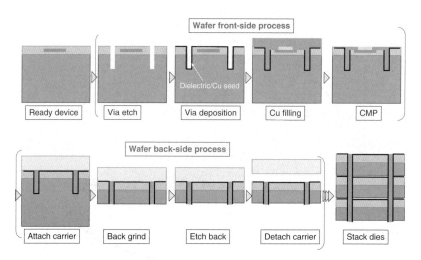

Fig. 1.14 3D IC-stacking using the 'via first' process steps [24]

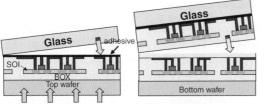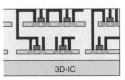

1. Attach circuit to a glass carrier
2. Thin wafer from back-side to expose burried oxide layer
3. Align top wafer to bottom wafer
4. Bond two wafer by annealing at 300°C
5. Remove glass carrier
6. Make thru-wafer via interconnects

Fig. 1.15 3D IC-stacking using the 'via last' process steps [13]

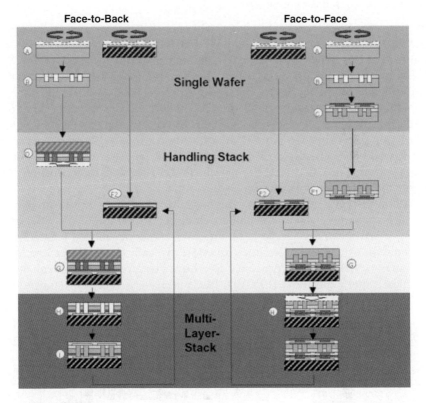

Fig. 1.16 Process steps for face-to-face and face-to-back wafer stacking [20]

1.2.2.2 3D IC-stacking Technologies

There are many different versions of IC-stacking technologies being developed by different groups around the world. The major difference between the various technologies is how the wafers/ICs are bonded. The other major steps like wafer thinning, fabrication of thru-Si vias, and precision alignment do not vary much between different versions of the technology. There are two kinds of wafer bonding techniques, namely direct bonding (without any intermediate layers) and indirect bonding (with intermediate layers). In the latter case, the two most common intermediate materials used are metals and polymers. In this chapter, I have divided IC-stacking technologies into three groups based on the bonding techniques. As illustrated in Fig. 1.17, these groups are direct oxide (SiO_2) bonding, metal-to-metal bonding and adhesive bonding.

Technologies Based on Direct (Fusion) Bonding

In Direct (Fusion) Bonding, two extremely smooth wafer surfaces are brought into very close contact so that intermolecular Van der Waals attractive forces and surface OH bonds create a weak bonding between the wafers [19, 26–29]. This bonding occurs at Room temperature without any external forces. With further annealing at higher temperatures, covalent bonds are formed, which strengthens and secures the bond. Fusion bonding occurs between many Si-based materials namely SiO_2-Si, Si_3N_4-Si, Si_3N_4-SiO_2 and SiO_2-SiO_2. SiO_2-SiO_2 mating surfaces are most commonly used in wafer bonding because (i) the final bond is made-up of the standard semiconductor material as shown in Fig. 1.18, and (ii) oxide is a good etch stop for both wet and dry etch processing during the wafer thinning process.

In order to get a good bond, the mating surfaces have to be well prepared. The surfaces should be pore-free (higher oxide density), smooth, flat, clean and with surface reactivity [27, 28]. In order to avoid the higher temperature degradation of active devices, the oxides are typically deposited at low temperatures (<420°C) using processes like Low Pressure or Plasma-enhanced chemical vapor deposition. (LPCVD, PECVD) [27–29]. The low temperature

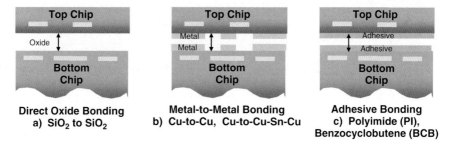

Fig. 1.17 3D IC-stacking technologies grouped based on bonding mechanism used

1   3D Integration Technologies – An Overview

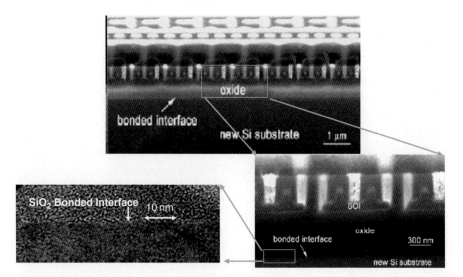

Fig. 1.18 SEM micrographs of SiO_2-SiO_2 bond line [13]

techniques deposit oxides with fine pores which could contain trapped gas molecules and absorbed OH ions. Evolution of gases and water molecules during post-bond annealing could cause bond-line voids and defects. To avoid these defects, the oxides are densified by annealing in a N_2 atmosphere at 350°C for 5 h prior to bonding.

Since Direct Bonding is based on short range intermolecular forces, it is very critical to have very smooth surfaces. To achieve high quality bonding, the root mean square (rms) roughness has to be below 1 nm. Table 1.2 [27] summarizes the mean and rms roughness measured by Atomic Force Microscopy (AFM) for wafers with various surface preparations. Generally, a 5000 angstrom thick thermal oxide having a rms roughness of 0.273 µm is acceptable for bonding. The rms roughness of the as-deposited PECVD (low temperature oxide deposition) is 9.757 nm. When the PECVD oxide on the wafer is densified at 350°C for 16 h in N_2, the roughness improves to 8.501 nm and the porosity in the oxide

Table 1.2 Mean and root mean square (rms) roughness of wafers with different surface preparations. (Surface roughness measured by Atomic Force Microscope, AFM) [27]

Wafer description	Mean roughness (nm)	rms roughness (nm)
Bare Si wafer	0.097	0.143
5000 Å SiO_2/Si	0.202	0.273
4 µm PECVD SiO_2/SOI (as prepared)	7.929	9.757
4 µm PECVD SiO_2/SOI (annealed at 350°C for 16 h)	6.877	8.501
4 µm PECVD SiO_2/SOI (annealed at 350°C for 16 h + 3 min. CMP)	0.312	0.394

layer decreases. After subsequent CMP for 3 min, the rms roughness decreases significantly to 0.394 nm, which is an acceptable smoothness for bonding. A combination of densification by annealing and CMP is an effective way to achieve very smooth surfaces.

Deviation from a surface flatness of <25 μm over 6" wafers also adversely affects bonding. In addition to the bow caused by materials deposited and processed during semiconductor processing, oxide deposited for wafer bonding can also contribute additional 10 μm to wafer bow [27]. Any force applied during bonding and post-annealing will help in reducing the bow, and typically a force of 1 kN is sufficient.

The mating surfaces have to be very clean. Dry cleaning with oxygen plasma followed by wet cleaning with piranha ($H_2O_2:H_2SO_4$) and a final rinse with deionized water was found to be effective to get a good bond [26]. The cleaning step not only removes any foreign contaminant, but also activates the mating surfaces. The cleaning step terminates the oxide surface with OH groups to initiate wafer bonding. The initial bonds at room temperature are weak forces based on Van der Waals attractive force and hydrogen bond. Subsequent annealing at higher temperatures will make this bond stronger. The reaction at higher anneal temperature converts the weak Si-OH-Si bond to stronger Si-O-Si bond,

Si-OH + OH-Si → Si-O-Si + H_2O (reaction is enhanced at higher temperature).

Figure 1.19 shows the correlation between annealing temperatures, times and bonding strength [27]. The bonds made at room temperature have strength of 170 mJ/m^2. After a 6 h anneal at 300°C, the bond strength increases to 432 mJ/m^2. In an attempt to further lower the bonding temperatures, to near room temperature and to improve the quality of the bond, surface chemistry modifications of the oxide surfaces have been tried. One such successful attempt

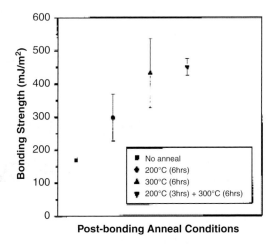

Fig. 1.19 Bonding strength of bonded wafer pairs with different post bond annealing [27]

is the patented process for ZiBond [30, 31]. The wafers are dipped in ammonia prior to bonding to change the surface chemistry to Si-NH$_2$. The reaction that occurs at room temperature is:

$$\text{Si-NH}_2 + \text{Si-NH}_2 \rightarrow \text{Si-N-N-Si} + \text{H}_2.$$

The hydrogen by-product that replaces water by-product can be removed from the bond interface at lower temperature. A good bond at room temperature has been reported with this technique [30, 32].

Some of the key examples of IC-stacking technology based on Direct Oxide Bonding are discussed next. IBM researchers have reported on 3D IC-stacking using oxide bonding [19, 28, 33, 34]. The devices used were 0.13 μm Cu/Low K SOI CMOS technology with intrinsic devices, ring oscillator circuits and interconnects. The process used for IC-stacking is shown earlier in Fig. 1.14. The process used was for 'via last', front-to-back bonding. The SEM micrograph of the bonded wafer is shown in Fig. 1.18. The investigators found that process steps do not adversely affect the device performance. MIT Lincoln Lab has developed 3D IC-stack using Direct Oxide Bonding for applications like the Laser Radar Imager. Figure 1.20 [35] shows the cross-section of a 3-wafer stack and the low temperature bonding process parameters used. Unlike IBM and MIT-LL who use 'via last' approach, Ziptronix uses 'via first' approach. Ziptronix has a patented Direct Bond Interconnect (DBI) and ZiBond technologies to bond oxide and metal interconnects [30, 31, 36,–32]. The detailed process and a cross-section SEM micrograph are shown in Fig. 1.21.

Technologies Based on Metal-to-Metal Bonding

Metal-to-metal bonding, illustrated in Fig. 1.22, is the most commonly used approach for IC-stacking. The advantages of using metal-to-metal bonding are

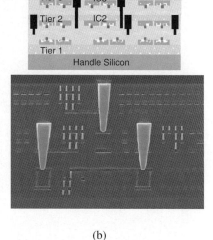

Parameter	Process
Film type	Low temperature oxide deposited by LPCVD
Surface preparation	CMP with Megasonic clean
Surface activation and clean	H$_2$O$_2$, 80°C, 10 min
Bond temperature	275°C
Bond time	10 hours
Wafer bow	< 30 μm

(a)   (b)

Fig. 1.20 MIT Lincoln Lab 3-wafer IC-stack using Direct oxide Bonding: (**a**) the bonding process parameters used and (**b**) cross-sectional image of a 3-wafer stack [35]

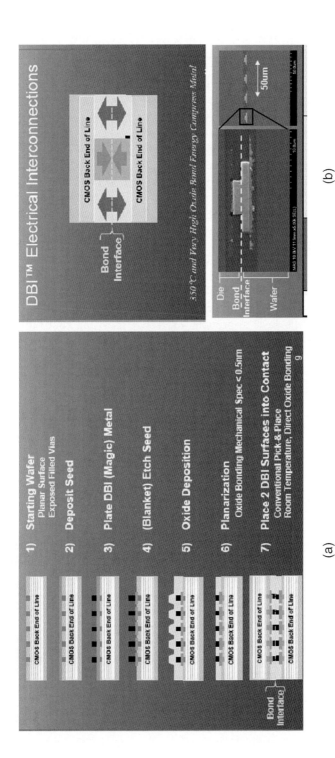

Fig. 1.21 Ziptronix's Wafer bonding technology using Direct Bond Interconnect (DBI): (**a**) detailed process and (**b**) a cross-sectional SEM micrograph [32]

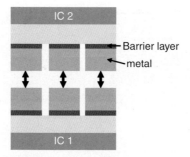

Fig. 1.22 3D IC-stacking using metal-to-metal bonding

(i) the metal bond line is thermally conductive, allowing heat to dissipate more easily to the side of the die or transfer vertically thru vias, and (ii) metal bonding can be used for the dual function of both mechanical support as well as electrical interconnection between the ICs. Many different metals have been investigated; some of the successful metals used are Cu-Cu [22, 23, 26, 37–42], Cu-Sn-Cu, [20, 21, 25, 43–41,43–48] Au-Au [49], Ti-Si [50] and In [51, 52]. The most widely accepted choices of metals are Cu-Cu and Cu-Sn-Cu, which will be discussed next.

Cu-Cu Bonding Prior to bonding, both mating surfaces should have Cu deposited over typically 50 nm of Ta. Ta acts as a diffusion barrier to prevent Cu diffusion into the device layer. Cu-Cu bonding is achieved by thermocompression, which requires temperature ($<400°C$) and Force (~ 4 kN) in vacuum. This is followed by an annealing step in N_2 at a temperature in the range of 300–400°C for 1 h [26] to achieve higher bond strength by allowing Cu inter-diffusion and grain growth. Figure 1.23 shows cross-sectional TEM images of the bond-line of a Cu-Cu bond and the Cu grains before bonding, after bonding and after annealing [53–55]. After bonding, the grain size ranges from 300 to 700 nm and there is a distinct bond line. However after annealing, the grain structure is well developed with a grain size of 800 nm and the bond line is completely gone. Table 1.3 shows the effect of bonding temperature on bond strength, determined by bond failures caused by stresses during dicing [56, 57]. The die bonded at 400°C and 350°C did not fail. A very high failure rate occurs for dice bonded at 200°C or lower. It was further determined that excellent bond quality is obtained by bonding at 350°C for 30 min followed by annealing at 350°C for 60 min.

It has been shown that an increased Cu pattern density on the bonding surface improves the bond yield [37]. The pad-to-pad bonding with proper isolation has to be made for electrical interconnection. In addition, mechanical pads should fill in the areas between electrical pads to provide strength and stability to the bond line. The SiO_2 around Cu pads should be recessed in order to make a good contact. Various surface preparation methods were tested [37] to get a higher Cu pad height above the SiO_2 surface. It was determined that

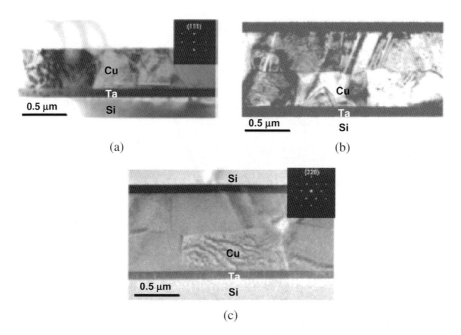

Fig. 1.23 Evolution of grain morphologies during Cu-Cu bonding [53]: (**a**) TEM image of evaporated blanket film (average grain size 300 nm); (**b**) TEM image of Cu-Cu bond after bonding (clear interface); and (**c**) TEM image of Cu-Cu bond after annealing (no visible interface)

Table 1.3 Percentage of dice failed at die sawing after different annealing times [56]

Bonding Temperature (°C)	Bonding Duration		
	30 min. Bonding	30 min Bonding + 30 min Annealing	30 min Bonding + 600 min Annealing
400	0% failed	0% failed	0% failed
350	0% failed	0% failed	0% failed
300	0% failed	5% failed	4% failed
250	1% failed	21% failed	22% failed
200	18% failed	86% failed	75% failed
150	37% failed	90% failed	96% failed

SiO$_2$ CMP followed by 3 min etch in HF gave the best results of Cu pad height of 100 nm above the SiO$_2$ surface. This surface treatment also produces a "dome" shaped Cu pad, which is very desirable during bonding because it will enable bonding to occur from center of the pad to its edge. Having a seal-ring of Cu around the edge of the die or the edge of the wafer is preferred, because it provides a "mechanical wall" protection from any damage during downstream processing. All of these attributes of good Cu pad design and fabrication are captured in an example in Fig. 1.24.

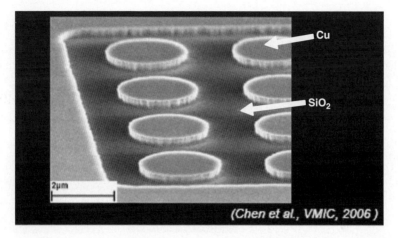

Fig. 1.24 Recessed SiO$_2$ around the Cu pads and seal ring [37]

Surface cleanliness prior to bonding is very critical. A surface roughness of rms 1.1 nm is recommended [58]. Oxide removal from the Cu surface is also very crucial. Soon after deposition, the Cu surface oxidizes in ambient atmosphere. Typically, HCl has been used to remove this oxide layer. A small amount of oxide that forms in the interval time between HCl treatment and bonding does not appear to affect the bond quality. The optimum bonding parameters, including those discussed above, are shown in Table 1.4.

Some of the key examples of IC-stacking technology based on Cu-Cu Bonding are discussed next. Using the 'via first' process steps shown in Fig. 1.12, the Japanese Consortium, Association of Super Advanced Electronic Technology (ASET), has developed Cu-Cu based 3D IC technology. Figure 1.25 shows the

Table 1.4 The optimum parameters to get good Cu-Cu- bonding [58]

Bonding Parameter	Condition
N2 anneal	When bonding above 300°C
Bonding temp	Above 300°C
Duration	30 min
Bonding pressure	4000 mbar
Chamber ambient	10–3 torr
Surface roughness	1.1 nm
Surface cleaning	HCl cleaning for 30 s

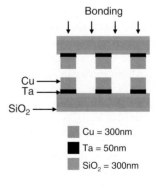

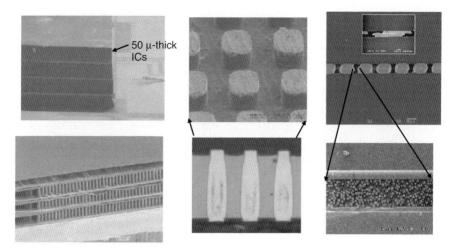

Fig. 1.25 Cu-Cu bonding used to stack 4 dice by Super Advanced Electronic Technology, ASET [24]

picture of a 4-die stack. In this version of the technology, the gaps between the Cu pads are filled with underfill material [24]. Figure 1.26 shows the Research Triangle Institute's version of a Cu-Cu bond based technology [59].

Cu-Tin (Sn)-Cu Bonding This technology, illustrated in Fig. 1.27, is similar to Cu-Cu bonding, except that Cu pads are topped with a thin layer of Sn [20, 21, 25, 43–46]. Sn can be on one or both surfaces. When pressure (5 bar) and temperature (<300°C) are applied, Sn melts and forms an alloy, Cu_3Sn. This alloy melts at 600°C. Thus, we have a technique where bonding is done at <300°C, but the bond, when done, is stable up to 600°C. This technique was initially developed at IZM Fraunhofer in Munich, Germany under the trade name of ICV-SLID Technology [20, 25, 46]. A 3-die stack using this technology is shown in Fig. 1.28.

Cu-Sn-Cu technology has been getting more popular in the last few years. Two major companies, Samsung and Intel, have announced prototype IC-stacks using this technology. Figure 1.29 shows 16 Gb memory from Samsung made from an 8-die stack using Cu-Sn-Cu bonding technology [47]. Figure 1.30 shows a 7-die stack from Intel using this technology [39, 48]. Oki, NEC, and Toshiba in Japan have implemented the slight variation of using Sn 2.5% Ag instead of pure Sn. Their 9-die stack is shown in Fig. 1.31 [41, 44].

Technologies Based on Adhesive Bonding

In this set of technologies, adhesives are used to bond stacked ICs as shown in Fig. 1.32. The most commonly used adhesives used are polymer dielectrics, namely Polyimide (PI) or Benzocyclobutene (BCB). Liquid polymers dissolved in a solvent are spin coated on either one of the mating surfaces or both. After the polymers are spin-coated, the wafers are aligned and bonded. Then, the

1   3D Integration Technologies – An Overview

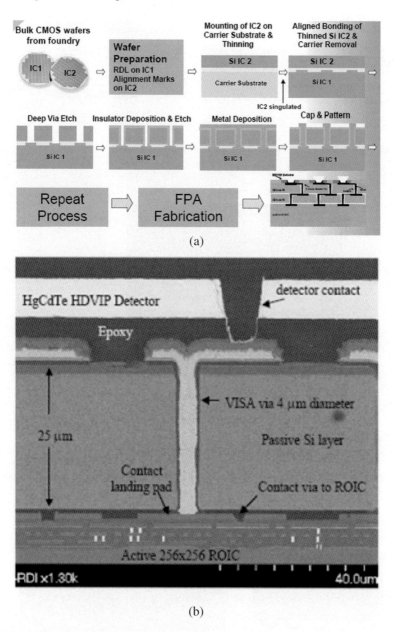

Fig. 1.26 3-Die stack using Cu-Cu bonding at Research triangle Institute [59] (**a**) Process flow diagram, (**b**) SEM Cross-section micrograph

polymer in the bonded wafers is cured in N_2 at 250°C in the case of BCB or at 300–400°C in case of PI. The use of adhesives does not require ultra-smooth surfaces as in the case of Direct Oxide and Cu-Cu bonding because the adhesive coating can smooth out any microscopic unevenness in the surface. If the

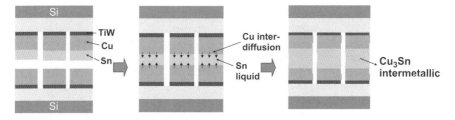

Fig. 1.27 Process steps for the Cu-Sn-Cu bonding [20]

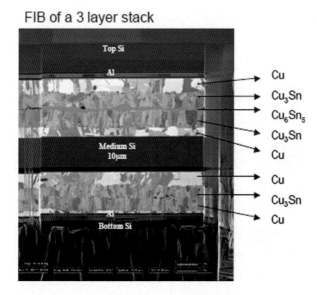

Fig. 1.28 Field Ion Beam Cross-section of a 3-die Stack using Cu-Sn-Cu technology [20]

Fig. 1.29 Samsung's 16 Gb memory module made with an 8-die stack using Cu-Sn-Cu technology [47]

1 3D Integration Technologies – An Overview

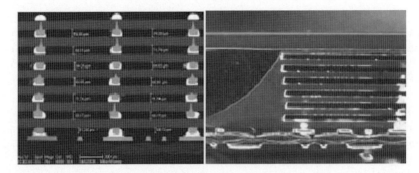

Fig. 1.30 Intel's 7-die stack prototype using Cu-Sn-Cu technology [39]

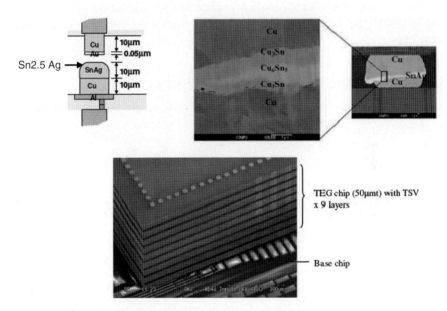

Fig. 1.31 Oki's 9-die stack using Cu-Sn2.5Ag-Cu technology [41]

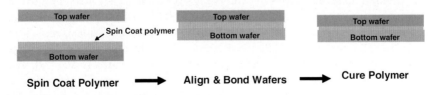

Fig. 1.32 Illustration of adhesive use in wafer/IC bonding

bonding step is not done correctly, adhesive bonding will tend to have entrapped voids [60, 61]. The sources of the voids are entrapped contaminants, out-gassing solvents and air. Thus care has to be taken in cleaning the wafers and drying them, using adhesion promoters, pre-curing the polymer to remove any out-gassing solvents, bonding in vacuum, and bonding so that joining occurs from the center towards the outer edge.

The typical process steps used are shown in Fig. 1.33 [20, 62]. Examples of completed modules are shown in Fig. 1.34 [14, 20, 63]. In recent years, polymer adhesives are also used in conjunction with Cu-Cu bonding [64]. An example is shown in the Rensselaer Polytechnic Institute (RPI) work in Fig. 1.35.

The 3D IC-stacking technologies discussed so far do not provide a way to redistribute the interconnects between the ICs and to integrate passive components. Recently, a US patent [65] was issued for a novel concept, shown in Fig. 1.36. In this concept, the glue layer between the ICs in a stack contain BCB/Cu interconnect layers, which could provide a way to redistribute interconnects and to embed passives. The basic concept has been demonstrated in an earlier study [66].

1.2.2.3 Key Enabling Technologies for 3D IC-Stacking

3D IC-stacking requires the following key common enabling technologies.

- Wafer-Thinning
- Thru-Vias in Silicon
    - Etching
    - Via Isolation
    - Metallization
- Wafer Alignment

Wafer Thinning

Wafer thinning is a very critical enabling technology because the size (diameter) of the thru-Si vias and via yields is determined by the thickness of ICs. Thin ICs can allow vias of smaller diameter and of shorter depth. Thin ICs also enable a thin profile of the stack. When wafers are thinned to <50 μm, the silicon wafer becomes very flexible as shown in Fig. 1.37 [67]. Figure 1.38 shows a typical wafer thinning process. Since thin wafers are very fragile, they are very hard to handle. Therefore the front face of the wafer is first to be mounted on a handling or carrier wafer, and then the wafer is thinned from the backside as shown in Fig. 1.38a. The mounted handling wafers should withstand the thinning processes. When thinning is completed and the thinned wafer is stacked, the handling wafer must be cleanly detached and removed. The thinning process involves many stages as shown in Fig. 1.38b. The initial few steps, coarse and fine grinding, help to remove bulk Si quickly and cost-effectively. The stresses

1 3D Integration Technologies – An Overview

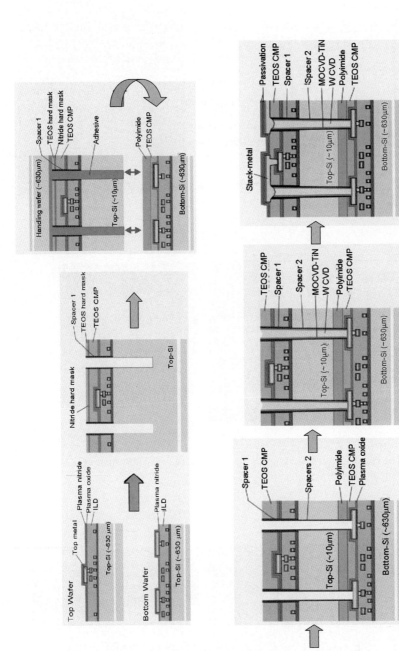

Fig. 1.33 Typical process steps for 3D IC-stacking using adhesive bonding [20]

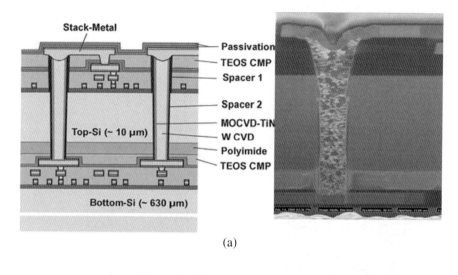

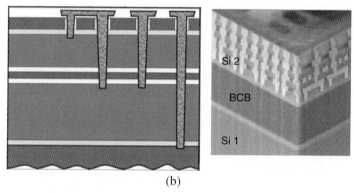

Fig. 1.34 The use of adhesive bonding in a 3D die-stack, (**a**) using polyimide at IZM [20], and (**b**) using BCB at RPI [14]

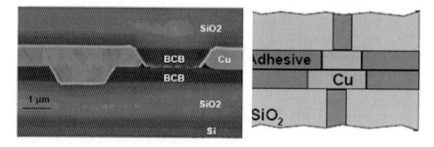

Fig. 1.35 A die-Stack using a combination of Cu-Cu and BCB adhesive bonding at RPI [14]

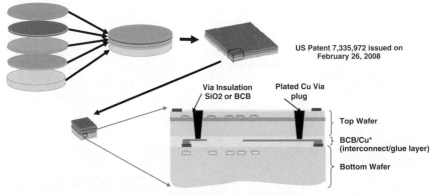

Fig. 1.36 3D IC-stacking technology with BCB/Cu interconnect layers embedded in the 'glue' layer [65, 66]

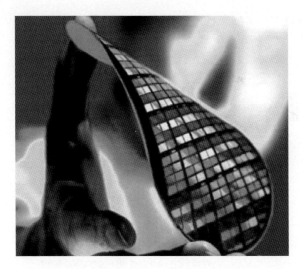

Fig. 1.37 Picture of a thinned Si wafer showing its flexibility [67]

from grinding leaves a damage layer on silicon, approximately 10–20 μm deep. The damage layer contains micro-cracks, which make these wafers mechanically very weak [67, 68]. After grinding, the damage layer is removed by wet etching. Since the wet etching makes the surface rougher, the finishing step is usually a CMP step. After removal of the damage layer by wet etching, the wafer strength increases as shown in Fig. 1.39 [69]. The strength reaches a plateau after the removal of 20 μm, which indicates that the damage layer is approximately 20 μm deep. Table 1.5 compares the removal rate, total thickness variation achieved, process temperature and application of different wafer thinning techniques [68].

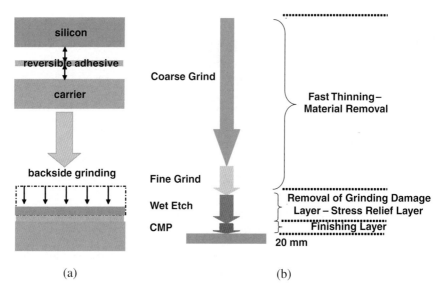

Fig. 1.38 Typical Process for Wafer Thinning. (**a**) Use of carrier wafer for handling during wafer thinning. (**b**) Sequence of thinning steps

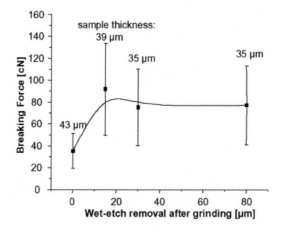

Fig. 1.39 Si Wafer strength as a function of the amount of Si removed by wet etching after grinding [69]

Die Singulation by sawing is not desirable because it causes rough edges and high stress points, which reduce the die strength. An alternate singulation method is to dry etch the streets, which do not show degrade the die strength. Figure 1.40 shows a top and side view of the rough edges formed in Si by dicing and the smooth edges formed by dry-etching. The preferred singulation method for thinned wafer is 'dicing by thinning' [69, 70] as illustrated in Fig. 1.41. The streets are first partially dry etched. Then when the thinning is done from backside past the bottom of the dry etch line, the dice get singulated.

Table 1.5 Comparison of thinning processes [68]

	Grinding	Spin-etching	Dry-etching	Polishing
Type of process	Mechanical abrasion	Wet-chemical etching	Plasma, Reactive ions	Chemical mechanical
Process medium	Diamonds in ceramic wheel	$HF + HNO_3$ + additives	SF_6, NF_3, XeF_2	Slurry: SiO_2 grains in soft etchant
Removal rate	300 μm/min	10…40 μm/min	3…30 μm/min	<2 μm/min
TTV: total thickness variation	0.5…3 μm/min	5..10% of removal	n.a.	<1 μm
Process temperature	cool	30…40°C	50…300°C	30…40°C
application	Thinning	Stress-relief	Stress-relief, MEMS thinning	Surface finish planarization

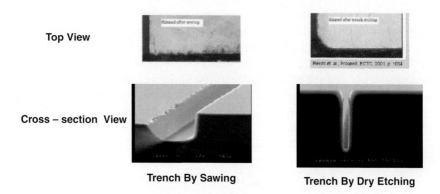

Fig. 1.40 A top and side view of the edges formed by sawing and dry etching [69]

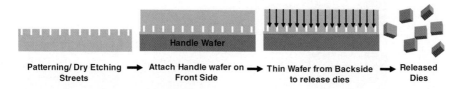

Fig. 1.41 The "Dicing by Thinning" process

Fabrication of Thru-Si Vias

Thru-Si vias are fabricated in two steps: (1) via etching, and (2) via metallization or filling.

Via etching There are two methods for etching vias, namely wet and dry etching. Wet etch is a fast and cost-effective etching technique; however, its application is limited because the etching occurs only on certain crystalline axis. Thus, the most common via etching technique is the Deep Reactive Ion Etching (DRIE). In DRIE, a high etch rate is achieved because highly reactive SF_6 chemistry is used. The etch reaction is:

$$Si + 4F- \rightarrow SiF_4 + heat$$

High anisotropy for fabricating large aspect ratio vias is achieved by short steps of isotropic etching with SF_6 and depositing the polymer C_4F_8 on the side walls of the via. The etch mechanism is illustrated in Fig. 1.42a [71]. Typically, a scalloping effect is seen on the side walls of the vias with this technique and this scalloping could adversely affect getting reliable metallization of the vias. As shown in Fig. 1.42b, with some of the recent development in controlled DRIE etching, this scalloping effect is minimized or eliminated. DRIE etching allows the fabrication of a wide range of via shapes, via diameters and aspect ratios as illustrated in Fig. 1.43. Vias having an aspect ratio of 30–40 are routinely fabricated with this technique.

Via metallization Metallization of the thru-Si vias in IC-stacks is the most critical step in the fabrication process. The most commonly used via fill metals are Tungsten (W) and Cu. The metal-filled vias have to be electrically isolated from each other and the rest of the circuitry. If the vias are through SiO_2, then an insulation layer between metal and oxide is not needed. But if the vias are through bulk Si, then a conformal insulation layer has to be deposited first on via wall. The two most commonly used insulation layers are Silicon Oxide and polymers like BCB or PI. The process shown in Fig. 1.14 shows vias thru-bulk Si, where the conformal insulation layer of low temperature oxide using TEOS

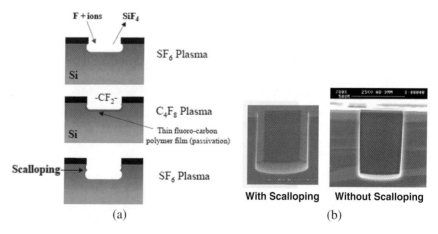

Fig. 1.42 Deep reactive Ion Etching (DRIE) etching – (a) Illustration of etch mechanism, (b) Examples of vias made with DRIE [71]

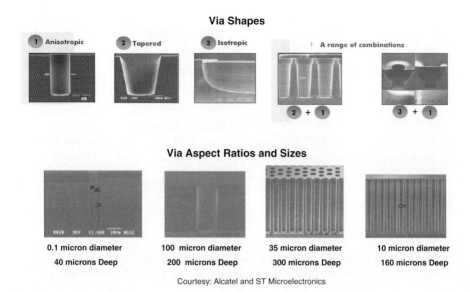

Fig. 1.43 A wide range of via shapes, diameter and aspect ratios made by DRIE [71]

(tetraethylorthosilicate) process was deposited [24]. On the other hand, polymer insulation was used in RTI's process shown in Fig. 1.26.

In the case of a W plug, the most commonly used adhesion layer is Ti or Ti-W and the barrier layer is TiN. The metallization process steps are shown in Fig. 1.44a and Fig. 1.44b for via plug thru SiO_2 and thru bulk Si, respectively. The only difference between the two processes is the additional insulation layer deposition for vias thru bulk-Si. Ti and TiN are deposited by sputtering or evaporation and the W plug is deposited by Chemical Vapor Deposition (CVD). A final CMP step is needed to remove the excess metals from the top surface [72, 73].

Figure 1.45 shows the Damascene process used for depositing Cu via plug [10, 74–81]. In Fig. 1.45a shows the process steps thru oxide. First vias are etched in the oxide layer and then the barrier layer is deposited, followed by deposition of the Cu plug. The excess Cu and barrier layer on the top surface are then removed by CMP. Figure 1.45b shows the process for vias thru-bulk Si. This process has an additional step for the deposition of conformal oxide layer for insulation. The most critical aspect Cu plug technology is the choice of the barrier layer. The requirements for the barrier layer [78, 79] are that it should have a good adhesion to oxide and be a low stress metal. Also the texture and roughness of the barrier layer determines the microstructure characteristics of Cu-fill. Ti, W, Ta and their nitrides have all been evaluated. A Ti, Ta or TaN barrier are the most widely accepted barriers. The vias are filled by electroplating Cu over a seed Cu layer. The Cu seed layer is deposited by sputtering. Figure 1.46 shows a schematic diagram of the electroplating bath. In the

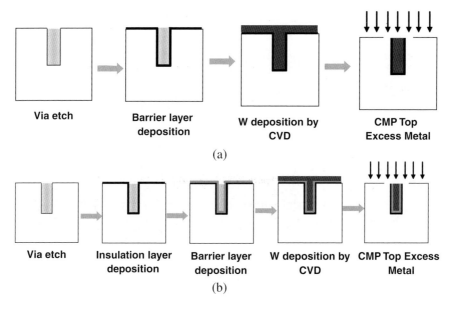

Fig. 1.44 Tungsten Via Plug process steps through SiO$_2$ (**a**) and through bulk Si (**b**)

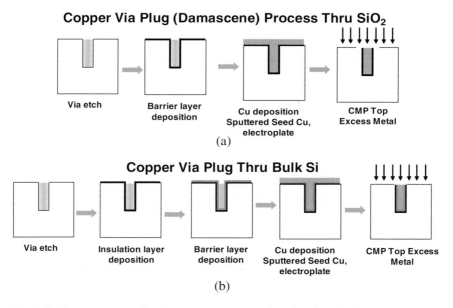

Fig. 1.45 The process steps for Damascene process used to deposit a Cu via plug

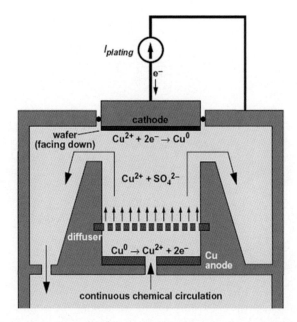

Fig. 1.46 A schematic diagram of the electroplating bath for Cu deposition [10]

electroplating process, Cu seed layer is the cathode; Cu plate is the anode and Cu Sulfate solution is the electrolyte. In the electrochemical process, as the Cu ions get deposited on the seed layer, the anode oxidizes to replenish the lost Cu ions. For better manufacturability of void-free and high aspect-ratio vias, the pulsed plating waveform and bath chemistry are optimized. In pulsed plating, the magnitude and direction of current is modulated [10].

Precision Alignment

In a 3D IC-stack, the precision alignment of dice is very critical. The alignment accuracy required will depend on the feature sizes being stacked. In a very dense circuitry, the alignment tolerance of $\sim +/- 1$ μm is needed. There are two types of alignment techniques, namely direct and indirect [82, 83]. In direct alignment, the alignment marks on the mating wafers are observed simultaneously. This is possible only if one of the wafers is transparent (see Fig. 1.47a) or by using an Infra-red microscope. The use of this technique is limited to only a very few applications. In the majority of the cases, the indirect alignment technique is used. In this technique (See Fig. 1.47b), the first alignment mark on one of the mating surface is captured and digitally stored. This wafer is then moved, so the alignment mark on the second surface can be seen. Then the second wafer is aligned to the digitized image of the first wafer. Using a precision positioning system, the first wafer's position is restored back to its original position, at which time both wafers

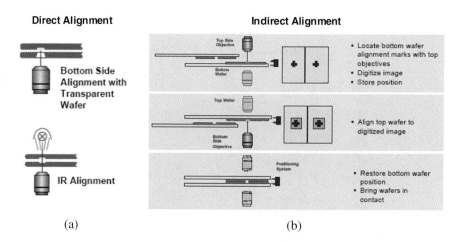

Fig. 1.47 7 Precision alignment techniques. (**a**) direct alignment method, and (**b**) indirect alignment method [82]

are aligned with accuracy. In state-of-the art alignment/bonding equipment, an accuracy of ± 1.3 μm has been demonstrated [82].

1.2.3 3D-Packaging

3D-packaging, which is the most mature integration category, includes many different technologies, most of which are an extension to 3D of already existing single-chip packaging technologies. 3D packaging allows the use of known-good-dice and thus the yield and reliability issues are minimized. From system and manufacturing perspective, 3D-packaging allows significant reduction in number of discrete components to be assembled on a board. Due to all these factors, the implementation of 3D packages has been easier, more cost-effective and less disruptive. As a result, it can be found in many of the latest consumer products like smart cell phones, cameras, MP3 players, and laptop computers. 3D-packaging technologies meet the requirements for the current generation of products; however, the level of performance and miniaturization needed in many future systems cannot be adequately met by these technologies.

Different 3D-packaging technology can be sub-divided in four major types, as illustrated in Fig. 1.48. These are wire-bonded die-stack, BGA-stack, folded–stack using chip-on-flex, and ultra-thin package stack. The first two types, wire-bonded die-stack and BGA-stack, are the most commonly used 3D-packaging technologies.

Figure 1.49 shows some of the advanced wire-bonded die-stacks [47, 84, 85]. The dice are first stacked using die-attach adhesives, and then

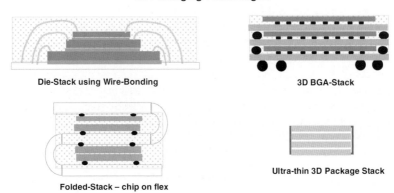

Fig. 1.48 Illustration of different 3D-packaging technologies

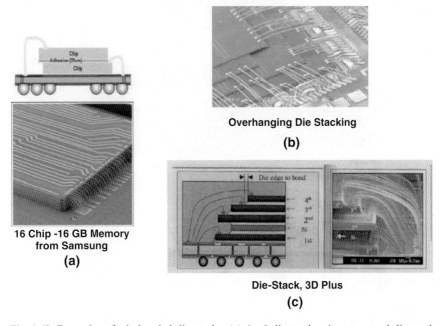

Fig. 1.49 Examples of wirebonded die-stacks. (**a**) An 8-die stack using staggered die-stack approach [47], (**b**) A 4-die stack using spacers between dice [84], and (**c**) A 4-die stack using very thin, over-hanging dice [85]

wire-bonded to a package. Such die-stacks require special core capabilities like higher pitch wire bonds, low loop-height wire-bonding (<75 μm), die thinning to 50–75 μm and thin die handling [86]. In addition, thin spacer technology has to be developed to maintain space between similar size dice for wire-bonds (See Fig. 1.49c).

Figure 1.50 shows an example of a BGA-Stack from Tessera [87]. In BGA-stack technology, single die is first packaged and completely tested before BGA packages are stacked. The pros and cons of wire-bonded die-stacks are compared with BGA-stacks in Table 1.6 [86]. The advantages of wire-bonded stacks are that they are more cost-effective, and they provide a low package profile. The main advantages of BGA stacks are that each die package can be independently manufactured by different vendors and completely pre-tested before stacking. Furthermore, it is easier to accommodate design changes in BGA-stack, because each of the single packages can be swapped without affecting the rest of the design.

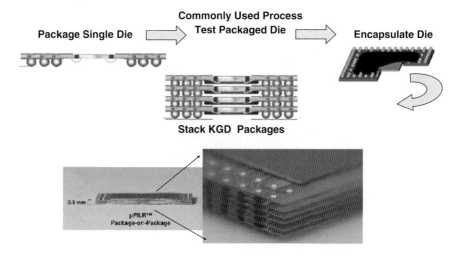

Fig. 1.50 Example of a BGA-stack using micro-PILR technology from Tessera [87]

Table 1.6 Comparison of wire-bonded die-stack and BGA-stack technologies [86]

Die-Stack by Wire-bonding	BGA-Stack
Advantages – Low cost packaging with low package substrate cost – Low package profile available by advanced wafer thinning technology	**Advantages** – Multi-sourced packages, stacked through conventional logistics – Known Good Package (KGP) testing prior to stacking – Easy device swap with qualified packages
Disadvantages – Single-sourced "assembled" product – KGD essential for high product yield – Design/Development required to change stacked device	**Disadvantages** – High Package profile due to multiple packages in stack – Packaging cost

1   3D Integration Technologies – An Overview

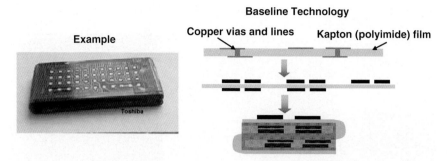

Fig. 1.51 3D-folded packaging technology

A 3D Folded-Stack is shown in Fig. 1.51. It uses a Flex PWB (e.g. Kapton) as an interconnect substrate. After the components are assembled on it, the circuit is folded in 3D. This technology is used only in very specialized applications.

The final group of 3D-packaging is advanced ultra-thin 3D packages. There are many different versions in this group. Three examples from Vertical Circuits, Irvine Sensors, and 3D Plus are shown in Fig. 1.52. A typical process

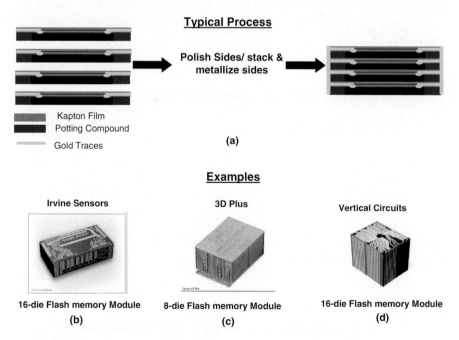

Fig. 1.52 Examples of Ultra-thin 3D packages, (**a**) Typical Process – Neo-Stacking technology from Irvine Sensors [88], (**b**) 16-die Memory Module from Irvine Sensors [88], (**c**) 8-die Flash memory from vertical Circuits [89], (**d**) 16-die Memory module from 3D-Plus [90]

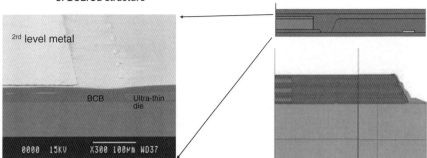

Fig. 1.53 Embedded die-stack technology [91]

involves first fabricating thin-film interconnects on individual die assembly so that all interconnects are routed to the edge of the dice. Then these individual die with interconnects are stacked. The die-to-die connections are made on the side of the stack [88–90]. One other noteworthy technology in this category is embedded IC technology developed at IMEC in Belgium. As shown in Fig. 1.53, very thin dice (~10 thick) are embedded within the multi-layers of BCB. Interconnect routing is also provided within these BCB layers [91].

1.3 Main Issues in 3D Integration Technologies

1.3.1 Issues in 3D IC-Stacking

There are many issues that still need to be fully resolved before 3D IC-stacking technologies can be fully commercialized. Many of the issues and their solutions are strongly dependent on the application and the technology used. Therefore each application is being individually evaluated to understand the application-specific issues. Some of the common issues that have been raised are thermal management, IC-stack yields, uncommon die size, and inadequate infrastructure for design, equipment and processing.

1.3.1.1 Thermal Management

All of the initial applications envisioned for 3D IC-stacking are for microsystems, where power dissipation is low, for example in memory modules (See Fig. 1.29) [40], logic-memory stacks [32] and image sensor read-out modules [59]. The extent of the thermal problem for applications using next generation

1 3D Integration Technologies – An Overview

of IC technologies is currently not well understood and is still being evaluated [3, 5, 14, 41, 92–95]. Some of the initial thoughts and examples are:

1. Many applications may require the use of more efficient advanced cooling concepts like micro-gap cooling as illustrated by Toshiba as illustrated in Fig. 1.54 [41].
2. Layout of the architecture and design of the functional blocks significantly affects the maximum temperature in a 3D IC-stack. This has been illustrated in the case of Logic-memory stack as illustrated in Fig. 1.55 [5]. The optimum layout involves avoiding having two heat-generating functional blocks in close proximity of each other.
3. The design layout of the vias also plays a role in minimizing the temperature of the IC-stack [93]. Smaller via pitch and additional thermal vias will allow higher heat dissipation.

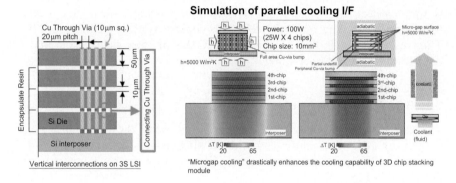

Fig. 1.54 Toshiba's Micro-gap cooling of 3D IC-stack [41]

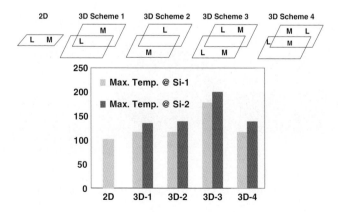

Fig. 1.55 The chart shows the effect of design layout of the functional blocks on maximum temperatures in a Logic-memory stack [5]

4. Minimize the use of low thermal conductivity materials and maximize the use of high thermal conductivity materials within the stack. This is where metal-to-metal bonding like Cu-Cu and Cu-Sn-Cu has an additional advantage in helping with thermal management.

1.3.1.2 IC-Stack Yield

The yield for a wafer-level IC-Stack, Y_{stack}, is Y_{die}^n, where Y_{die} is the yield of individual die on a wafer and n is the number of dice in the stack. Thus if die yield is 80%, the 3-die stack yield will be 51%. If the die yield is 99%, the 3-die stack yield will be 97%. Thus only high yielding dice will give high stack yield. The chip-to-chip and chip-to-wafer stacking is not affected by wafer yields because only known-good-die (KGD) are used. One way to resolve the issue with low yielding dice in wafer-level processes is to pre-select known-good-die (KGD) and rearrange them in a wafer-format on a handling wafer [20, 25]. This concept is illustrated in Fig. 1.56. This rearranged handling wafer with KGD can then be stacked like a regular wafer on a target wafer.

1.3.1.3 Uncommon Die Sizes

The wafer-level IC-stacking processing cannot be used if the die sizes in a stack are not the same. Thus in this case, the stacking technology is limited to 2-die or 3-die stack in pyramid-like structure. The chip-to-wafer stacking process is illustrated in Fig. 1.57 [20].

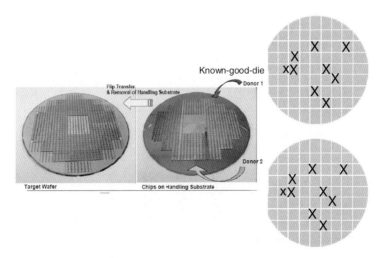

Fig. 1.56 Rearranging known-good-die (KGD) from donor wafers to a handling wafer. Rearranged handling wafer is then used to stack KGD on a target wafer [20]

1 3D Integration Technologies – An Overview

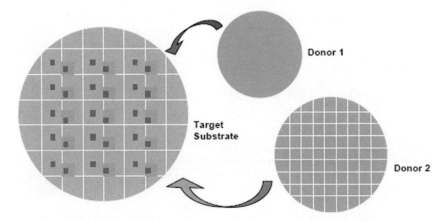

Fig. 1.57 IC-stacking with uncommon die sizes using the chip-to-wafer process [20]

1.3.1.4 Inadequate Infrastructure

Before IC-stacking processes can be commercialized, they requires an infrastructure for design, equipment and processing. Infrastructure vendors like EVG, Karl-Suss and Cadence have been increasingly putting more resources in developing 3D IC-stacking capability. Having large corporations like IBM, Intel, Infineon, Toshiba and NEC being seriously interested in the capability will further motivate some of these vendors.

1.3.1.5 Cost

3D IC-stacking technology using thru-Si via technology is an expensive technology because of high capital investment as well as high direct production cost. However, like many other technologies in the past, per unit cost will come down with technology maturation and higher production volumes. Use of high yielding ICs and wafer-level processes will further help in keeping the cost down.

1.3.2 Issues in 3D-packaging

Since 3D-packaging is already used in many applications, many issues have already been resolved. However, two main issues still remains, namely thermal management and cost.

1.3.2.1 Thermal Management

Each specific 3D-packaging technology has to be individually evaluated for thermal issues and solutions. Use of heat spreaders and heat sinks can be

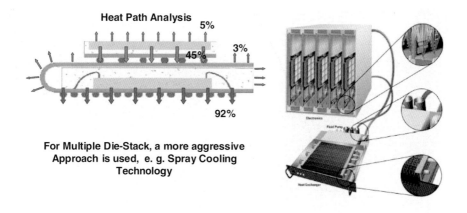

Fig. 1.58 Heat path analysis and spray cooling in multiple die-stack [96]

extended to 3D in a few cases. Some systems may require exotic heat dissipation apparatus as shown in Fig. 58.

1.3.2.2 Cost

Although 3D-packaging is the lowest cost 3D integration category, there is still a push to further lower the cost in the consumer product applications market. In some analysis, it has been shown [86] that 3D-packaging can be even more cost-effective than traditional approach. In cost analysis, the higher cost of assembling 3d-packages can be countered by the following cost advantages:

1. Fewer discrete components have to be assembled on a board.
2. A reduction in overall packaging cost i.e. in wire-bonded die-stacks.
3. Cost savings due to the reduction in area of printed wiring board assemblies.

1.4 Conclusions

The next generation of integrated micro-system technologies can only keep up with increased functionality and performance demands by using the 3rd dimension. The main driving factors for 3D integration are miniaturization, integration of different technologies in a small form-factor, and performance. There are many different versions of 3D integration technologies, which can be grouped into 3 main categories, namely 3D On-chip integration, 3D IC-stacking, and 3D-packaging.

3D On-chip integration is truly a homogeneous 3D IC technology to fabricate a system-on-a-chip. This technology is in early stages of R &D and still faces many technical challenges.

The second category is 3D IC-stacking, in which individual wafers are fabricated and then integrated in 3D either at the wafer-level or the chip-level. In this category, each IC wafer design is first fabricated independently. Then the wafers or ICs are bonded and electrical interconnection between ICs is made using thru-Si vias. There are many different types of 3D IC-stacking technologies, the major difference between them being how they are bonded. Three main bonding technologies used are direct oxide bonding, metal-to-metal bonding, and adhesive bonding. The other enabling technologies needed for 3D IC-stacking are wafer/ IC thinning, fabrication of thru-Si vias and precision alignment. Many major companies around the world like IBM, Intel, Toshiba and Infineon are pursuing 3D IC-stacking.

The last category, 3D-packaging, is an extension of single-chip packages into 3D. These technologies are matured and are currently used in many consumer products like smart cell phones, cameras, MP3 players and laptop computers. There are also many versions of this category; the two main technologies being wire-bonded die-stack and BGA-stack.

3D IC-stacking technologies have many unresolved issues, some of which are thermal management, low module yields, inadequate infrastructure, and higher cost. 3D-packaging technologies, being already in use in many applications, have relatively fewer unresolved issues, two of which are thermal management and cost. 3D integration will play a major role to enable many future products in consumer, medical, defense, and security applications.

References

1. Chanchani, Rajen, "An Overview of 3D Integration Technologies – Motivation, Options and Status," Workshop on 3D Integration of Semiconductor Devices, in conjunction with Advanced Metallization Conference, sponsored by University of California, Berkeley, San Diego, October 18, 2004
2. Chanchani, Rajen, "3D Integration Technologies – An Overview", Short course presented at Polytronics 2007 Conference, Tokyo, January16–18, 2007. 56th Electronic Component and Technology Conference, San Diego, CA, May 30–June 3, 2006
3. Saraswat, Krishna C. et al., Proceedings of The IEEE, Vol. 89, No. 5, May 2001, pp. 602–633
4. Davidson, E., Transactions IEEE-CPMT-B, Vol. 20, No. 4, 1967, pp. 361–374
5. Saraswat, Krishna C., "3-Dimensional ICs: Motivation, Performance Analysis and Technology," Conference on 3D Architecture for Semiconductors and Packaging (sponsored by Research Triangle Institute), Burlingame, CA, April 14–15, 2004
6. Franzon, Paul D. et al., Transactions IEEE-CPMT-B, Vol. 21, No. 1, February, 1998, pp. 2–14
7. Reif, Rafael et al., Transactions IEEE-VLSI Systems, Vol. 11, No. 1, February, 2003, pp. 44–54
8. Saraswat, Krishna C. et al., Transactions IEEE-Electron Devices, Vol. ED-29, No. 4, April, 1982, pp. 645–650
9. Bohr, Mark T., IEDM Tech. Dig.,1995, pp. 241–244
10. Loke, Alvin L. S., "Process Integration Issues of Low Permittivity Dielectric with Copper for High Performance Interconnects," Ph. D. Dissertation, Stanford University, March, 1999. www.ewh.ieee.org/r5/denver/sscs/Presentations/Loke_PhD_Thesis.pdf

11. Reif, Rafael et al., Transactions IEEE-VLSI Systems, Vol. 8, No. 6, December, 2000, pp. 671–678
12. Vitkavage, Susan C.,"3D Interconnects and the ITRS Roadmap," Conference on 3D Architecture for Semiconductors and Packaging (sponsored by Research Triangle Institute), Burlingame, CA, April 14–15, 2004
13. Guarini, K. W. et al., "3D IC Technology: Capabilities and Applications," Conference on 3D Architecture for Semiconductors and Packaging (sponsored by Research Triangle Institute), Burlingame, CA, April 14–15, 2004
14. Lu, James Q., "Wafer-level Hyper-Integration for 3D IC and packaging," Conference on 3D Architecture for Semiconductors and Packaging (sponsored by Research Triangle Institute), Tempe, AZ, June 13–15, 2005
15. Saraswat, Krishna C. et al., Transactions IEEE-Electron Devices, Vol. 47, 2000, pp. 1035–1043
16. Neudeck, G. W. et al., J. Vac. Sci. Technol. – B, Vol. 17, No. 3, 1999, pp 994–998
17. Lin, H-Y et al., Japanese J. App. Phys., Part 1, Vol. 36, July, 1997, pp. 4278–4282
18. Saraswat, Krishna C. et al., Proc. 196th Meeting Electro-chemical Soc., Honolulu, HI, 1999
19. Topol, A. W. et al., IBM J. Res. & Dev., Vol. 50, No. 4/5, July/September, 2006, pp. 491–506
20. Krupp, Armin et al., "3D Integration with ICV-Solid Technology," Conference on 3D Architecture for Semiconductors and Packaging (sponsored by Research Triangle Institute), Tempe, AZ, June 13–15, 2005
21. Weber, Werner, Proc. Mat. Res. Soc. Symp., Vol. 970, 0970-Y03-01, 2007
22. Schaper, L. W, Proc. of 53rd Electronic Components and Technology Conference, 2003, pp. 631–633
23. Schaper, L. W. et al., Trans. IEEE Adv. Packaging, Vol. 28, No. 3, August 2005, pp. 356–366
24. Bonkhara, Manabu, "3D Stacked LSI Interconnection by Cu-Vias & 3D System Integration," Conference on 3D Architecture for Semiconductors and Packaging (sponsored by Research Triangle Institute), Burlingame, CA, April 14–15, 2004
25. Ramm, Peter, "Vertical Integration technologies," Workshop on 3D Integration of Semiconductor Devices, in conjunction with Advanced Metallization Conference, sponsored by University of California, Berkeley, San Diego, October 18, 2004
26. Tan, C. S. et al., Proc. Materials Res. Soc. Symposium, Vol. 970, 0970-Y04-01, 2007
27. C. S. Tan et al., Applied Phys. Letters, Vol. 82, No. 16, April, 2003, pp. 2649–2651
28. Topol, A. W. et al., Proc. of 54th Electronic components and Technology Conference, May, 2004, pp. 931–938
29. Young, Albert et al., "perspectives on 3D-IC Technology," Conference on 3D Architecture for Semiconductors and Packaging (sponsored by Research Triangle Institute), Tempe, AZ, June 13–15, 2005
30. Enquist, Paul, Proc. Mat. Res. Soc. Symp., Vol. 970, 0970-Y01-04, 2007
31. US patent 6,902,987, June 7, 2005
36. www.ziptronix.com
32. Enquist, Paul, "Direct Bond Interconnect Technology for Scalable 3D SOCs," Conference on 3D Architecture for Semiconductors and Packaging (sponsored by Research Triangle Institute), Burlingame, CA, Oct. 31–Nov. 2, 2006
33. Topol, A. W. et al., IEDM Tech. Dig., 2005, pp. 363–366
34. Guarini, K. W. et al., IEDM tech. Dig. , 2002, pp. 943–944
35. Keast, Craig, "3D Integration Program at MIT LL," Conference on 3D Architecture for Semiconductors and Packaging (sponsored by Research Triangle Institute), Tempe, AZ, June 13–15, 2005
37. Chen, Kuan-Neng, "Science, Materials and Process Technology of Cu Bonding for 3D Integration," IMAPS International Conf. on Device Packaging, Scottsdale, AZ, March 2007
38. Chen, K. N. et al., Joun. Of Materials Science, Vol. 37, 2002, pp. 3441–3446

39. Morrow, Patrick et al., Proc. Mat. Res. Soc. Symp., Vol. 970, 0970-Y03-02, 2007
40. Lee, Kangwook, "The Next generation Packaging Technology for Higher performance and Smaller System," Conference on 3D Architecture for Semiconductors and Packaging (sponsored by Research Triangle Institute), Burlingame, CA, Oct. 31–Nov. 2, 2006
41. Takahashi, Kenji, "3D Chip Stacking," Conference on 3D Architecture for Semiconductors and Packaging (sponsored by Research Triangle Institute), Tempe, AZ, June 13–15, 2005
42. Beyne, Eric et al., Proc. Mat. Res. Soc. Symp., Vol. 970, 0970-Y01-02, 2007
43. Ramm, Peter et al., Proc. Mat. Res. Soc. Symp., Vol. 970, 0970-Y02-04, 2007
44. Mitsuhashi, Toshiro et al., Proc. Mat. Res. Soc. Symp., Vol. 970, 0970-Y03-06, 2007
45. Jang, Dong Min et al., Proc. Mat. Res. Soc. Symp., Vol. 970, 0970-Y05-06, 2007
46. Ramm, Peter et al., Workshop on Thin Semiconductor Devices – Manufacturing and Applications, Munich, Germany, Nov. 25, 2003
47. Lee, Kangwook, "The Next Generation Package Technology for Higher Performance and Smaller Systems" Conference on 3D Architecture for Semiconductors and Packaging (sponsored by Research Triangle Institute), Burlingame, CA, Oct. 31–Nov. 2, 2006
48. Newman, Michael et al., Proc. of 56th Electronic Components an Technology Conference, May, 2006, pp. 394–398
49. Naito, T. et al., Proc. of 55th Electronic Components and Technology Conference, 2005, pp. 788
50. Yu, Jian et al., Proc. Mat. Res. Soc. Symp., Vol. 863, B10.7.1, 2005
51. Motoyoshi, Makoto et al., "3D LSI and its key Supporting technologies," Conference on 3D Architecture for Semiconductors and Packaging (sponsored by Research Triangle Institute), Burlingame, CA, Oct. 31–Nov. 2, 2006
52. Bonkohara, Manabu et al., Proc. Mat. Res. Soc. Symp., Vol. 970, 0970-Y03-03, 2007
53. Chen, K. N. et al., Appl. Physics letters, Vol. 81, No. 20, Nov. 2002, pp. 3774–3776
54. Chen, K. N. et al., Journal of Elec. Materials, Vol. 30, No. 4, 2001, pp. 331–335
55. Chen, K. N. et al., Journal of Elec. Materials, Vol. 32, No. 12, 2003, pp. 1371–1374
56. Chen, K. N. et al., Electro-Chem. And Solid-State Letters, Vol. 7, No. 1, January, 2004, pp. G14–G16
57. Chen, K. N. et al., Journal of Elec. Materials, Vol. 34, No. 12, 2005, pp. 1464–1467
58. Chen, K. N., J of Elec. Materials, Vol. 35, No. 2, 2005, pp. 230–234
59. Williams, Ken, "Pixelated Architectures: Drives for 3D Integration Techniques", Conference on 3D Architecture for Semiconductors and Packaging (sponsored by Research Triangle Institute), Tempe, AZ, June 13–15, 2005.
60. Niklaus, Frank et al., Journ. Micromech. Microeng. Vol. 11, 2001, pp. 100–107
61. Niklaus, Frank et al., Proc. Mat. Res. Soc. Symp., Vol. 863, B10.8.1, 2005.
62. Weiland et al., 2nd International Workshop on Thin semiconductor Devices – Manufacturing and Applications, Munich, Germany, Dec. 3–4, 2001
63. Lu, J-Q et al., " 3D Integration Using Wafer Bonding," Advanced Metallization Conference, Oct. 3–5, 2000, San Diego, CA
64. Lu, J-Q et al., Proc. Mat. Res. Soc. Symp., Vol. 970, 0970-Y04-02, 2007
65. Chanchani, Rajen, US Patent 7,335,972 issued on Feb 26, 2008
66. Chanchani, Rajen, Transactions IEEE Components and Packaging Technologies, Vol. 30, No. 3, September 2007, pp. 478–485
67. Landesberger, Christof et al., 2nd International Workshop on Thin semiconductor Devices – Manufacturing and Applications, Munich, Germany, Dec. 3–4, 2001
68. Landesberger, Christof, Book – "Foldable Flex and Thinned Silicon Multichip Packaging technology," Chapter 5, edited by Jack balde, Kluwer Academic Publishers, ISBN 0-7923-7676-5, 2003
69. Landesberger, Christof, " New Dicing and Thinning Concept Improves Mechanical Reliability of Ultra-thin Silicon," Proc. of Advanced Packaging materials, processes, Properties and Interfaces, ISBN 0-930815-64-5, pp. 92–97

70. Reichel, H. et al., Proc. of 51st Electronic Components and Technology Conference, 2001, p 1034
71. Puech, Michel, "Fabrication of 3D Packaging TSV Using DRIE," IMAPS International Conf. on Device Packaging, Scottsdale, AZ, March 2007
72. Licata, T. J. et al., IBM Jorn. Of Res. & Dev., Vol. 39, No. 4, 1995, pp. 419–435
73. Mann, R. W. et al., IBM Jorn. Of Res. & Dev., Vol. 39, No. 4, 1995, pp. 403–416
74. Venkatraman, R. et al., Proc. of Mat. Res. Soc. Symp., Vol. 514, April, 1998, pp. 41–52
75. Ryun, C. et al., Symp. On VLSI Technology Tech. Dig. , June, 1988, pp. 156–157
76. Dubin, V. M. et al., Proc. of Mat. Res. Soc. Symp., Vol. 514, April, 1998, pp. 275–280
77. Taylor, T. et al., Solid State Technology, Vol. 41, No. 11, pp. 47–57, Nov. 1998
78. Wang, S. Q., MRS Bulletin, Vol. 19, No. 8, August 1994, pp. 30–40
79. Singer, P., Semiconductor International, Vol. 21, No. 6, June, 1998, pp. 90–98
80. Edelstein, D. et al., International Electron Device Meeting Digest, Dec. 1997, pp. 773–776
81. Kim, Bioh, Proc. Mat. Res. Soc. Symp., Vol. 970, 0970-Y06-02, 2007.
82. Mathias, Thorsten, "Processes and Equipment for Volume Manufacture of 3D Integrated Devices," Conference on 3D Architecture for Semiconductors and Packaging (sponsored by Research Triangle Institute), Tempe, AZ, June 13–15, 2005
83. Mathias, Thorsten, Proc. Mat. Res. Soc. Symp., Vol. 970, 0970-Y04-08, 2007
84. Walker, Jim, "3D Packaging: A Market Opportunity or Interim Solution?" Conference on 3D Architecture for Semiconductors and Packaging (sponsored by Research Triangle Institute), Burlingame, CA, April 14–15, 2004
85. Val, Christian, Proc. of IMAPS Annual Conference, boston, Nov. 2003
86. St. Amand, Roger, "Advances in Assembly Technology for 3D CSP Packaging," Conference on 3D Architecture for Semiconductors and Packaging (sponsored by Research Triangle Institute), Burlingame, CA, April 14–15, 2004
87. Haba, Belgacom, "Wafer-level Stacking: Novel Approach to stacking Very thin dies," Conference on 3D Architecture for Semiconductors and Packaging (sponsored by Research Triangle Institute), Burlingame, CA, October 22–24, 2007
88. Gann, Keith, "Neo-Stacking Technology," HDI Magazine, December, 1999, Miller-Freeman, Inc.. www.irvine-sensors.com
89. Val Christian, "Very High Speed 3D 'System-in-Package," HDI, Vol. No. 5, pp. 22–29, May, 2001. www.3D-plus.com
90. Robinson, Marc, "A High-Performance CSP Die Stacking Technology," Conference on 3D Architecture for Semiconductors and Packaging (sponsored by Research Triangle Institute), Burlingame, CA, April 14–15, 2004
91. Beyne, Eric et al., Proc. of Electronic Components and technology Conference, May, 2001
92. Banerjee, Kaustav et al., IEDM Tech. Dig. 2000, pp. 727–730
93. Banerjee, Kaustav et al., IEDM Tech. Dig. 2000, pp. 261–264
94. Joshi, R. V., "Thermal Modeling of Bonded SOI/3D ICs," Workshop on 3D Integration of Semiconductor Devices, in conjunction with Advanced Metallization Conference, sponsored by University of California, Berkeley, San Diego, October 18, 2004
95. Joshi, R. V. et al., Proc. of International Conference on Semiconductor Processes and Devices (SISPAD), 2001 pp. 242–245
96. Tuckerman, David, "3D packaging of Electronic Systems: Current trends and Future Challenges," Conference on 3D Architecture for Semiconductors and Packaging (sponsored by Research Triangle Institute), Burlingame, CA, April 14–15, 2004

Chapter 2
Advanced Bonding/Joining Techniques

Chin C. Lee, Pin J. Wang, and Jong S. Kim

Abstract In this chapter, three advanced bonding/joining techniques, adhesive bonding, direct bonding, and lead-free soldering, are presented. For each technique, we first review the bonding principles and applications in electronic industries, followed by novel bonding materials and processes.

For adhesive bonding, four popular adhesives, epoxy resins, silicon resins, polymides, and acrylics, are reviewed. Two new adhesives, liquid crystal polymer (LCP) and SU8, are covered too. LCP has the properties of both polymers and liquid crystals. It, thus, can be bonded to silicon, metal, and glass, and used as flexible circuit board. SU 8, an epoxy-based negative type photoresist, has been applied to zero-level-packaging technology for low-cost wafer-level MEMS packaging.

For direct bonding, three popular methods, anodic bonding, diffusion bonding, and surface-activated bonding, are discussed. Anodic bonding process has extensive applications in silicon-glass bonding and glass-glass bonding. Diffusion bonding process forms chemical bonds by inter-diffusion of two different atoms over the bond line. Surface-activated bonding is valuable in bonding objects with large difference in coefficients of thermal expansion because of low process temperature, usually room temperature. A novel Ag-to-Cu direct bonding technique at bonding temperature of 250°C is reported.

In lead-free soldering, fundamental soldering principle is presented. To eliminate the use of fluxes, oxidation-free fluxless soldering technology has been developed. It has been applied to developing numerous soldering processes based on systems such as Sn-Au, Sn-Cu, Sn-Ag, In-Au, In-Cu, and In-Ag. Two fluxless processes are reported. One is bonding between Si/Cr/Au/Sn/Ag and Si/Cr/Au. The other is between Si/Cr/Au/Ag and Cu/Ag/In/Ag. In either process, high bonding quality is achieved without using any flux. Fluxless process has also been demonstrated in flip-chip configuration using Sn-rich solder joints.

C.C. Lee (✉)
Electrical Engineering and Computer Science, University of California, Irvine, 2226 Engineering Gateway Building, Irvine, CA 92697-2660
e-mail: cclee@uci.edu

Keywords Bonding · Soldering · Fluxless soldering · Direct bonding · Anodic bonding · Adhesives · Epoxies

2.1 Adhesive Bonding Techniques

2.1.1 Adhesives in the Electronic Industries

The chemistry of polymeric materials used for adhesives in the electronics industry does not differ from that of polymers for other applications. Commonly used and important adhesives are based on epoxy resins, epoxy-phenolic resins, epoxy-silicone compositions, silicone resins, acrylic resins, and polyimides. In this chapter, four most popular adhesives, epoxies, silicones, acrylics, and polyimides, are briefly reviewed. Their applications are then presented. Conductive adhesives and die-attach adhesives are not included in this chapter. They are discussed in separate chapters.

2.1.1.1 Epoxy Resins

Epoxy resin formulations are important with many applications in electronic packaging industry [1]. Theses resins have a common feature: the three-membered oxygen containing epoxy (oxirane) rings that are incorporated into organic molecules by using either condensation or oxidation reactions. The resin formulations are used in numerous steps in the manufacturing of electronic packages including conductive adhesives, flip chip encapsulation, die coatings, encapsulation, surface mount placement, and lead bonding adhesives. Epoxy resins are excellent electrical insulator and can protect electrical components from short circuiting, dust, and moisture. Due to their excellent electrical properties, mechanical strength and processability, epoxy resins are widely adapted in the field of electronics [2, 3]. Epoxy resins can be aromatic, cycloaliphatic or aliphatic, monofunctional or polyfunctional, physically ranging from low-viscosity liquids to high-melting solids [4]. Commercial epoxy adhesives are primarily comprised of an epoxy resin and a curing agent. They are available in different forms of liquids, gels, pastes, and films.

2.1.1.2 Silicone Resins

Silicone resins are a type of silicone materials which is formed by branched, cage-like oligosiloxanes with the general formula of RnSiXmOy, where R is a non-reactive substituent, usually Me or Ph, and X is a functional group H, OH, Cl or OR. These groups are further condensed in many applications, to give highly crosslinked, insoluble polysiloxane networks. Silicone resins have been used in many applications. A thermosetting resin is used for developing a waveguide material with a low birefringence, low propagation loss, and good environmental stability [5]. Thermally curable silicone resin was also used for high-speed optical coating with specific feature of coating thickness controllability [6].

2.1.1.3 Polyimides

Polyimide (PI) is a polymer of imide monomers. Polyimides have a glass transition temperature at least 200°F greater than epoxy resins [7]. As a result, polyimide adhesives can operate at higher temperature than epoxies or phenolics, and semiconductor industry uses polyimides as a high-temperature adhesive. They are often used in the electronic industry in the form of films for flexible circuits and cables, deposited films for interlayer dielectrics, passivation and buffer coating, substrates for multichip modules, adhesive pastes or tapes with some unique features of low dielectric constant, high thermal stability, and excellent mechanical properties [8]. A good example is the cable in a laptop computer, which connects the main logic board to the display. It is often made of polyimide with copper conductors. Polyimides in use can either be cured by condensation reaction or addition reaction mechanism. Three most important categories are polyimide recursors, self-standing polyimide films, and polyimide adhesives [4].

2.1.1.4 Acrylics

Acrylic resins (polymethyl methacrylate) are known to have exceptional optical clarity and good weather resistance, strength, electrical properties, and chemical resistance with low water absorption characteristics [7]. Acrylic resins as adhesives in electronic industry are formed through radical or anionic polymerization [9]. Radical polymerization can be initiated by UV radiation as well as by heat. Cyanoacrylates are of special interest for systems with very high reaction rates. Their reaction follows an anionic polymerization mechanism [10]. Since the cyanoacrylates have very high polarity, water is able to act as an initiator.

2.1.2 Applications of Adhesives in Electronics

2.1.2.1 Integrated Circuits

Polyimides are commonly used in the fabrication of integrated circuits. Polyimides, particularly as dielectric, passivation, and protective layers, have advantages over other types of inorganic materials used for the same process [11]. Photosensitive polyimides can produce direct patterns using common photolithography process without using any toxic chemicals. In typical process for three-level metal interconnect design, polyimide films are inserted (coated) between metal 1 & 2 and metal 2 & 3. After all integrated circuits are implemented on the silicon wafer, a thicker outermost layer of polyimide is coated as final buffer coating to absorb the interfacial stresses and prevent passivation crack and electrode displacement during the pressure cooker test [7]. An efficient fabrication process was recently developed for superconducting integrated circuits using new fine resolution photosensitive polyimide. It is synthesized using aliphatic material as the KrF photoresist [12, 13]. This specific photosensitive polyimide is synthesized directly by block copolymerization using a catalyst in solvent at 180°C. The fabrication is simplified because the photosensitive polyimide insulation layer can be patterned by conventional photolithography process without etching.

2.1.2.2 Flexible Circuit

Flexible circuit is a technology for building electronic circuits by depositing electronic devices on flexible substrates. Flex circuits have traditionally been made with polyimide or polyester films. Many techniques have been made using flexible circuits to overcome the limitations of the rigid multilayer board technology [14]. Basically, tapes intended to the flexible circuit market are fabricated by using two- or three-layer construction schemes. For two-layer tapes, they are usually made by coating solutions of polyimide precursors over copper foil or plating copper onto polyimide films. For three-layer tapes, they are formed of polyimide films coated with organic adhesive and laminated to 35 μm copper foil [4]. The entire circuit board can be fabricated on a flexible substrate, and then folded and stacked in an organized pattern to achieve the desired compactness. Common applications of flex circuits are in cameras, cell phones, computer keyboard, printers, and medical applications.

2.1.2.3 Liquid Crystal Display

Liquid crystal displays (LCDs) are becoming ever more popular due to low power consumption, compactness, flatness, lightweight, and high compatibility with large scale integrated circuits [15]. Not only is it very important to have high resolution and large capacity but also critical to have cost reduction by choosing competitive materials in display applications. Many different types of adhesives are used to connect driver chips to LCDs, including thermoplastic adhesives and thermo-setting adhesives [4, 16]. There are two different mechanisms used for curing the adhesives, which are heat curing and UV curing. Heat curing process has been more common. On the other hand, the use of the transparent glass substrate in the LCD makes UV curing an interesting alternative because the bonding process can be done rapidly by UV irradiating at room temperature. The low temperature curing is critical for LCDs because liquid crystals are particularly heat sensitive and cannot withstand normal soldering temperatures [17, 18].

2.1.3 New Adhesives

2.1.3.1 Liquid Crystal Polymer (LCP)

Liquid crystal polymers (LCPs) are a unique class of wholly aromatic thermoplastic polyester polymers that provide previously unavailable high performance properties. LCPs combine the properties of polymers with those of liquid crystals. While LCPs show the same mesophases characteristic of ordinary liquid crystals, they retain many of the useful properties of polymers. These polymers were synthesized by linkage of rod-like or disk-like mesogenic side groups with flexible spacers to the polymer main chain [19]. Mesogens must be incorporated into the chains for flexible polymers to display liquid crystal effect. LCPs are known to be inert to organic solvents and acids and mechanically flexible. It has several unique properties such as low dielectric constant, low loss tangent, low water absorption coefficient, and low cost. Due to their excellent properties, LCP has been frequently used in microwave application and become commercially available as high performance flexible circuit boards [20, 21]. Not only could LCP be used as the substrate for microwave frequency application but also RF MEMS packaging. The temperature to bond LCPs is around 280°~310°C, which is acceptable for most RF MEMS switches. LCPs can be patterned by microfabrication or laser cutting. They can be directly bonded to metal, silicon, and glass without using adhesive [22]. It can thus be called as adhesive-less bonding technique. The coefficient of thermal expansion (CTE) for LCP can be adjusted through thermal treatments. It can facilitate integration of integrated circuits in SOP modules. Being flexible, LCP can lead to deployment of antennas in space [23]. Due to its unique features, large sheets of LCP containing antenna arrays can be flexed, rolled up, and easily deployed.

2.1.3.2 SU 8 Adhesive Bonding

Su-8 is a high contrast, epoxy-based photoresist designed for micromachining and other microelectronic applications, where a thick chemically and thermally stable image is desired. It is a negative type photoresist, and thus the exposed portion is cross-linked while unexposed portion is soluble to liquid developer. A normal process is listed in sequence.

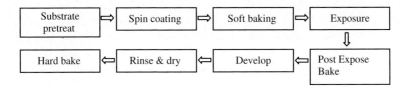

Process guidelines give us the duration of the soft bake as well as the post exposure bake depending on film thickness of the resist and different kinds of SU-8 resist. The advantages of SU-8 are its flexibility of layer thickness up to several hundreds of micrometers, its high chemical and thermal stability as well as its good mechanical properties [24]. SU-8 is known to be well suited for permanent applications where it is imaged, cured, and left in place. An adhesive bonding method has been developed at wafer level using SU-8 photoresist as intermediate layer [25]. Adhesive bonding is good for joining silicon or glass wafers at lower temperature (usually below 200°C), and this technique is known to be less dependent on the substrate material, particles, and surface roughness of the bonding surfaces. In their development, the layer was selectively deposited on one of the bonding surface by contact imprinting method. It is a good approach for some applications where the adhesive layer cannot be applied directly using classical spinning method. A cover die for a pressure sensor was bonded using SU-8. Zero-level-packaging (ZLP) technology has been developed for high aspect ratio microstructures with selective adhesive bonding using SU-8 photoresist [26]. In the processing, they developed three partial steps as the basis of the ZLP technology: (1) coating and patterning of the SU-8 photoresist, (2) adhesive bonding of the MEMS and the cap wafer, and (3) etching of the wafer stack using deep reactive ion etching (DRIE) method. The techniques developed are very promising for low-cost wafer-level MEMS packaging for monolithic integration of microelectronics.

2.2 Direct Bonding Methods

Direct bonding means a bonding process of object A to object B without any thing such as adhesive or solder in between. Only one bonding interface exists. The bonding mechanism relies on attractive force between two flat and smooth surfaces. In conventional direct bonding processes, the initial bonds rely on the van der Waals dispersion force between two extremely flat and clean surfaces. In principle, if the atoms on contact surfaces are close enough and can see each other, strong van der Walls force can exist along the interface of almost any two materials and make strong bond even at room temperature. In practice, it is not possible to produce two surfaces, one on each object, and bring them to contact within a few atomic distance over an extensive area. Thus, bonding two objects directly at room temperature seems nearly impossible. Over the past several decades, techniques have been developed to provide direct intimate contact condition. They can be categorized into three methods: anodic bonding, diffusion bonding, and surface-activated bonding. In what follows, these three methods are reviewed. A new and novel Ag-to-Cu direct bonding is then presented.

2.2.1 Anodic Bonding

The anodic bonding was initially developed for bonding metal to glass by Wallis and Pommerants in 1969 [27]. It was subsequently extended to silicon-to-glass and glass-to-glass bonding. Typical bonding is performed between a sodium-baring glass wafer and a silicon wafer under a high voltage around 200–1000 V at temperature range of 300-400°C [28]. Compared to glass softening temperature, anodic bonding temperature is relatively low.

Anodic bonding process can be conducted in atmosphere or vacuum environment. Figure 2.1 depicts the bonding apparatus [28]. The glass wafer is biased as the cathode, and the silicon wafer is the anode. After cleaning process, one mirror finish wafer is placed over the other one. Two surfaces have direct contact only at limited locations, as exhibited in Fig. 2.2 [27]. When high voltage is applied at elevated temperature, mobile Na^+ in the glass migrate toward the silicon through the contact locations, leaving behind negative charge on the glass (depletion region). An electric field is thus established across the bonding interface. Since the gap is very small, this electric field can be very high. Stronger electrostatic force is thus produced, which pulls the glass surface and silicon surface together. On the contact locations, Si-O or Si-Si covalent bonds form. The bonding action spreads out from contact locations over the entire wafers. With proper bonding conditions, the anodic bonding strength can reach 10–25 MPa and bonding efficiency (ratio of

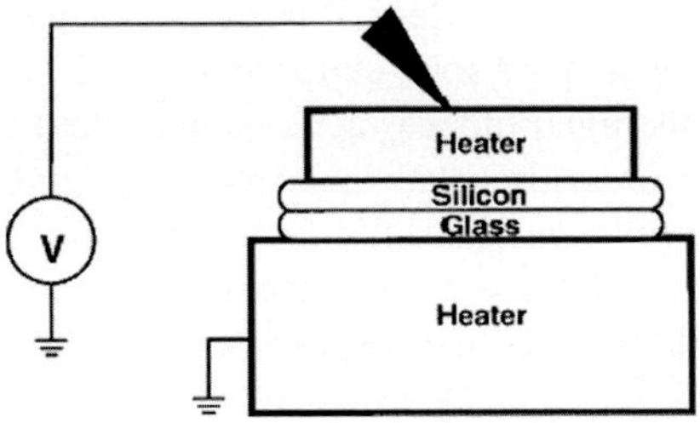

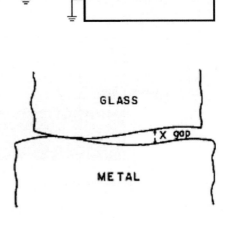

Fig. 2.1 The anodic bonding apparatus [28]

Fig. 2.2 Cross section of glass-metal interface before anodic bonding [27]

bonded area to whole wafer area) ranges from 94 to 99.9% [29]. As we can see, for this technique to work, one of the two wafers needs to be relatively flexible so that it can be pulled towards by the electrostatic force and conform to the surface of the other wafer. This means that one of the two wafers needs to be relatively thin.

Anodic bonding has been applied to electronic packaging applications, including hermetic sealing, encapsulation, and device fabrication. In packaging MEMS (microelectromechamical system) devices, laser diodes, photonic and fiber optical modules, and medical components, anodic bonding can provide high quality hermetic sealing [30–32]. Sealing process using epoxy can be performed at low temperature and low cost. However, epoxy out-gasses, making hermeticity impossible. Soldering process, on the other hand, needs flux to remove oxides. Flux and flux residues are easily trapped inside the package, causing the out-gassing problem. Thus, anodic bonding is a possibility to achieve high quality hermetic sealing. Anodic bonding has been applied to encapsulating silicon chips with three-dimensional microfluidic structures [33–35], as illustrated in Fig. 2.3 [33]. Without using any glue, fluidic interconnects are built robustly. Glass transparency at optical wavelengths makes alignment of glass and silicon wafers simple and accurate. This technique is compatible to wafer-level packaging.

2.2.2 Diffusion Bonding

Diffusion bonding process involves diffusion of atoms of object A and object B along the interface. It is often referred to as pressure joining, thermo-compression welding, or solid-state welding. This technique can join dissimilar materials, i.e. dissimilar metals, metal-to-glass, and metal-to-ceramic, etc. [36–38]. The bonding process almost always needs high temperature to activate interdiffusion of atoms. For materials difficult to bond, the process can be performed in vacuum or an inert gas environment to suppress detrimental surface oxides. It is often thought, even by experts, that bonding will occur as long as there is interdiffusion.

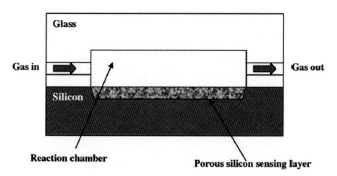

Fig. 2.3 The hybrid device made by anodic bonding [33]

Fundamentally, this is only partly correct. Bonding is not possible if atoms of object A and atoms of object B do not attract each other after the interdiffusion.

To achieve bonding, two objects should contact intimately for a period of a few minutes to a few hours to ensure sufficient atomic diffusion. Figure 2.4 displays the bonding mechanism [39]. When the two objects are held together, only a few locations have physical contact due to surface asperities. Thin surface oxide layers often exist that block inter-diffusion. Plastic deformation of surface asperities occurs when the applied pressure increases to the yield strength of the objects. As pressure increases, more plastic deformation occurs. This leads to reducing voids and dispersing oxides. When brittle oxides disrupt, fresh surfaces are exposed and contact each other in atomic scale. In the meantime, diffusion and creep behavior take place at the contact area, and the bond coalesces. Finally, all voids diminish and bond is formed along the entire interface.

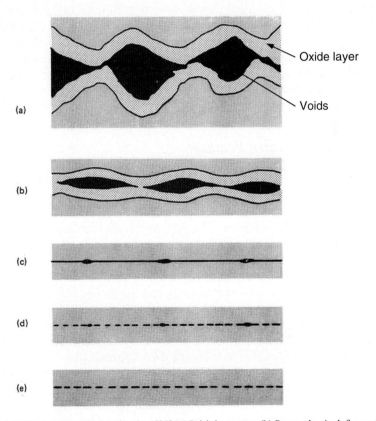

Fig. 2.4 Diffusion bonding mechanism [29] **(a)** Initial contact, **(b)** Some plastic deformation of surface asperities, **(c)** Diffusion and creep behavior, **(d)** Atoms continue diffusion and oxide layers are eliminated, **(e)** Final joint

Diffusion bonding generally has three key parameters: temperature, pressure, and dwell time. The bonding temperature should be approximately 50–80% of the melting point (in Kevin) of one of the objects. Elevated temperature prompts interdiffusion and aids plastic deformation. Pressure is applied to ensure intimate contact. It also helps surface deformation and improves bonding efficiency. Dwell time is important for bonding quality as well. Dwell time must be sufficient for atoms to diffuse throughout. Intermetallic layers might grow if excessive dwell time is employed.

Although diffusion bonding is one of the original techniques for electronic packaging, it is now used in some specific applications. In early days, wire bonding is completed by diffusion bonding [40]. To reduce the temperature required, thermal-sonic and ultrasonic methods are developed, where acoustic energy is used to break up surface oxides. In ceramic packages, the metal leads (pins) are bonded to embedded electrodes for electrical connection. Other than brazing and welding techniques, diffusion bonding can also be used [41]. Diffusion bonding has also been used in flip chip assembly, called Chip-on-Dot [42]. In this technique, aluminum pads on test chips are bonded directly to Gold Dot^{TM} on the flexible substrate at 300°C. Final joints meet the reliability required by most low–cost consumer electronics and telecommunications applications. The most important breakthrough of this technique is elimination of solder and under bump metallization on test chips. The assembly thus becomes easier, faster, and cheaper.

2.2.3 Surface-Activated Bonding

To alleviate surface smoothness requirement, one method is surface-activated bonding (SAB). The SAB not only cleans the surface but also creates incomplete chemical bonds, which are active and highly desirable to react with other atoms to form bonds. Formation of bonds on activated surfaces, thus, relieves surface asperity problems. The surfaces are usually activated by fast ion beam or plasma irradiation. The activation process induces bond defects, which are responsible for increasing chemical reactivity. Therefore, activated surfaces show significantly enhanced surface energy and like to form bonds with other chemicals even at room temperature. After the activation process, the surfaces should be kept active and clean until joining. Thus, the activation process and the bonding process are completed in a high vacuum system. The high vacuum system is expensive and it also restricts the size of parent components.

The surface-activated bonding has been applied in extensive fields and in bonding various materials, including silicon-to-silicon, silicon-to-ceramic, metal-to-metal, metal-to-ceramic [43–45]. One of the main advantages of this technique is that it can be conducted at room or low temperature. It is also an attraction to fabricate high frequency electro-optic devices, which is bonding piezoelectric materials, such as lithium niobate ($LiNbO_3$) and lithium tantalite

(LiTaO$_3$), to Si. However, the mismatch in coefficients of thermal expansion (CTE) of piezoelectric materials and silicon is quite large. LiNbO$_3$ and LiTaO$_3$ have CTE of 14.4 (*a* axis)-7.5 (*c* axis) ppm/°C and 16 (*a* axis)-4(*c* axis) ppm/°C, respectively, whereas CTE of Si is 3 ppm/°C. Shear stress, thus, is developed in the bonded structure when cooling down to room temperature. Bonding of LiNbO$_3$ on Si is demonstrated at room temperature using argon-beam surface activation [45, 46]. Figure 2.5 exhibits a bonding interface [46]. The bonding strength is equivalent to bulk materials.

As electronic devices are moving quickly towards high-speed and miniaturization, interconnect density in flip-chip technology has continued to increase. In conventional soldering processes, it is difficult to fabricate fine pitch flip-chip interconnect with high reliability due to CTE mismatch between chips and substrates [47, 48]. Accurate alignment is also a key issue to implement fine pitch bonding [49]. Several works have been done to develop high-density packaging using SAB method [50–52]. For example, bumpless interconnect with pitch of 10 μm and the diameter of 3 μm has been demonstrated [51]. Using SAB method, the flip-chip joints are made without intermetallic growth, resulting in reliability and the performance improvement.

2.2.4 Novel Ag-to-Cu Direct Bonding

A novel direct bonding of thick Ag foils (280 μm) on Cu substrates has been successfully developed recently to produce Ag-Cu dual layer substrate structure [53]. Among metallic and non-metallic substrates, Cu has been widely used in nearly all electronic packaging due to its high electrical and thermal

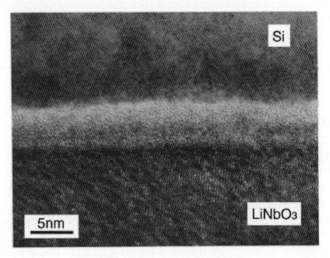

Fig. 2.5 High-resolution TEM images of the LiNbO3-Si bonding interface [46]

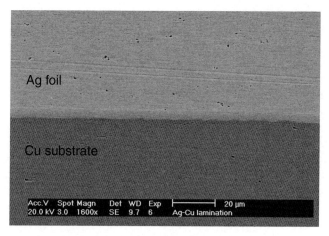

Fig. 2.6 Cross-section SEM image of Ag-Cu bonded structure [53]

conductivities, high strength, adequate rigidity, and low cost [54]. However, bonding semiconductor chips to Cu is always a challenge because of severe CTE mismatch, i.e. Si of 3 ppm/°C versus Cu of 17 ppm/°C. To relieve shear stress caused by CTE mismatch, a thick Ag foil is used as a buffer in this study. Ag is chosen because the yield strength of Ag is only 28% of Sn3.5Ag solder and 10% of Cu. Compared to Sn3.5Ag, the electrical conductivity of Ag is 7.7 times higher and thermal conductivity is 5 times higher. Thus, Ag layer is the optimal choice. The question is the following: how do you bond Ag to Cu without anything in between at conditions compatible to electronic packaging? In this study, mirror finish Ag foil is bonded directly onto Cu substrate at 250°C in 50 millitorr vacuum with a static pressure of 1,000 psi to ensure intimate contact. Compared to other direct bonding techniques, the bonding process is relatively easy. Figure 2.6 displays the Ag-Cu bonded interface. The Ag foil actually deforms to mate the Cu surface. The bonding strength is very strong. It is impossible to break the joint. The fundamental bonding mechanism is still under investigation.

2.3 Lead-Free Soldering and Bonding Processes

2.3.1 Basic Soldering Processes

Here, we briefly review the basic soldering process that has been used in industries for several decades. At the soldering temperature, the solder turns into molten phase. To have low soldering temperature, solders need to have low melting temperature. Thus, solder alloys always contain an element that has low melting temperature such as tin (Sn), indium (In), lead (Pb), and bismuth (Bi). In the soldering process, the molten solder reacts to a base metal to ignite the bonding. The bonding action is initiated by intermetallic compound formation, which is a chemical reaction. Consider Sn-based Pb-free solder on copper as an example. During the soldering process, the solder melts and contacts copper. The Sn in the molten solder reacts with copper to form Cu_6Sn_5 intermetallic compound (IMC), often known as wetting action, on the interface as portrayed in Fig. 2.7. Cu_6Sn_5 does not melt until temperature reaches 415°C. At typical soldering temperature of 250°C, it remains in solid state. It is this interfacial layer, Cu_6Sn_5 in this example, which links the solder and copper together. This IMC formation occurs on all known soldering systems. Without it, a soldering process could not be successful. However, since 1986, we have developed many fluxless soldering processes that do not use IMC formation as the fundamental bonding mechanism. This fluxless bonding technology will be presented in the next section.

Since the fundamental requirement of solder bonding is the chemical reaction that forms an IMC, the soldering environment must provide the condition that favors this chemical reaction. However, both the solder and the base metal have oxides on their surfaces. These oxide layers have very high melting temperature and do not melt at the soldering temperature [55]. For example, the melting temperature of SnO and SnO_2 is 1,080 and 1,630°C, respectively. The oxides are also lighter than the solder. They thus form barriers on solder surfaces and prevent the molten solder from having intimate contact with the base metal to initiate chemical reaction. Bonding thus cannot be achieved without first dealing with the oxide. This is where fluxes come into the picture. The purpose of fluxes is to reduce the oxide and to shield both solder and base metal against further oxidation.

There are many flux formations [56]. The key ingredients are resin acids such as abietic, neoabietic, dehydroabietic, palustric, pimaric, and isopimaric acids. Resin acids can react with metal oxides such CuO and SnO as follows:

$$2\ R - COOH + CuO \rightarrow (R - COO)_2\ Cu + H_2O$$
$$2\ R - COOH + SnO \rightarrow (R - COO)_2\ Sn + H_2O$$

where R represents the carboxyl residue. For the case of abietic acid, $R = C_{19}H_{29}$. In above equations, the copper salt is green and the tin salt is tan. The salt can be dissolved in the molten flux. At room temperature, both appear as a soapy film and are usually embedded in and mixed with the bulk of flux. As

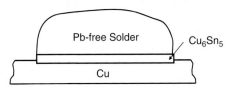

Fig. 2.7 For all conventional soldering process, intermetallic formation is necessary to produce a joint because the initial action is a chemical reaction

Fig. 2.8 Molten flux converts oxides into salts to expose fresh solder and fresh base metal, and to shield them from further oxidation

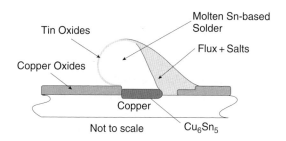

exhibited in Fig. 2.8, the oxide is now removed to expose the fresh molten solder and fresh base metal. The molten solder flows on the base metal and readily react with it to form an IMC. When this occurs, a bond is essentially produced. On the surface of molten solder that is covered with molten flux, further solder oxidation is prevented. We see that the flux must remain in molten state and be able to flow freely during the entire soldering process. Thus, the physical properties of a flux formulation are tightly controlled. Soldering temperature is the most important consideration.

2.3.2 The Fluxless Processes Dealing with Tin Oxides

The key requirement for successful soldering is to remove or convert the oxides. The most common technique is to use fluxes as explained in the section above. The soldering process that can be achieved without the use of fluxes is called fluxless or flux-free process. For the benefits of the readers, we briefly review the fluxless processes that deal with oxides that already exist. We try to review all the fluxless processes reported. Prior to 1980, scrubbing action was often applied in die-attach operation to break up the oxide layer. Many chemicals such as H_2, formic acid vapor, CO, and silane have been tried by various research groups to reduce SnO back to Sn, but with little success. In 1990, a process using fluorine treatment was developed, that is called Plasma Assisted Dry Soldering (PADS) process [57, 58]. In this technique, RF generated plasma is used to disassociate an innocuous fluorine containing source gas such as CF_4 or SF_6 to produce the atomic fluorine, a very reactive radical species. When solder is treated with atomic fluorine, the following reaction takes place:

$$SnO_x + yF \rightarrow SnO_xF_y$$

The resulting compound SnO_xF_y can be readily dissolved in the molten solder and the oxide is thus removed. This process has worked well on Pb-Sn solders of various compositions. After the treatment, the solders can store in air for several days before losing the treatment effectiveness. Potential problems are (a) fluorine is known to etch SiO_2 and SiN [59] and (b) the RF power used may damage IC chips. Recently, Ar + 10% H_2 plasma produced by 100–500 W of RF power is used as the dry cleaning agent to etch away the oxide layer on Sn3.5Ag and Sn37Pb solders [59, 60]. This process appears to be successful. One concern is that the high RF power may damage IC chips or sensitive devices. Chemicals other than fluxes, noticeably formic acid vapor, have been employed to treat the oxide layer [61, 62]. Fundamentally, these processes are similar to the traditional processes that use fluxes except that chemicals other than fluxes are used. The most dominating chemical is formic acid vapor [61]. The chemical reaction between the oxides and the acids was not experimentally confirmed. Effect of the resulting residues is also unclear. Most recently, an electron attachment technique is successfully developed to produce atomic hydrogen anions in low concentration H_2 environment at 200–300°C temperature range. It was shown that SnO and In_2O_3 oxides on solder perform pellets can be reduced (de-oxidized) by the atomic hydrogen anions in the reflow process to make nice solder bumps that bonded to copper substrate without using any flux [63].

To make this section complete, we include the fluxless processes reported using Au80Sn20 eutectic alloy as the bonding medium [64–66]. Among popular solders, this is the only solder that does not have oxidation problem. Here is the reason. At thermal equilibrium, this alloy is a mixture of AuSn and Au_5Sn intermetallic compounds [67, 68]. Oxidation is unlikely if the bonding process is performed in inert environment. As long as the AuSn eutectic solder is what ought to be at thermal equilibrium, fluxless bonding is achievable. However, commercial AuSn eutectic performs may have significant tin oxides on the surface [64]. Thus, the quality of performs is critical in achieving fluxless capability. Since early 1970, AuSn eutectic has been used in laser diode industries to attach laser diode chips on a package without using flux. An interesting fluxless process is to bond Au bumps onto Sn pads (90 μm × 90 μm) or SnPb solder pads (100 μm × 100 μm) [65]. The bonding process was carried out using Karl Suss model 950 flip chip bonder at 300°C with a bond force of 7–10 cN/pad. No information was given as to how the oxide layers on Sn and SnPb were overcome.

2.3.3 Oxidation-Free Fluxless Soldering Technology

Fundamentally, this technology provides oxidation-free environment from the beginning to the end, i.e., from solder manufacture to solder joint formation. There are four basic requirements: (a) oxidation prevention measure during solder manufacture, (b) capping layer to block oxygen penetration into the solder afterwards, (c) capping layer being dissolved into and becoming a part of the solder joint, and (d) proper environment to inhibit oxidation during the bonding process. Since we reported this technology in 1991 [55], it has been applied to developing various fluxless processes based on Sn-Au, Sn-Cu, Sn-Ag, Sn-Bi, Sn-In, In-Au, In-Cu, In-Ag binary systems and In-Pb-Au ternary system [69–77].

We now present our recent bonding processes based on Sn-Ag and In-Ag systems, respectively, to illustrate these fundamental requirements [77, 78]. The process using Sn-Ag system is chosen as a representative one because Sn-rich Sn-Ag alloy gets oxidized easily. Thus, it is particularly difficult to achieve fluxless feature. The process based on In-Ag system is interesting as it involves transient liquid phase bonding effect where the molten phase solidifies even at the bonding temperature due to solid liquid reaction.

To begin, a thin 0.03 μm Cr layer and 0.1 μm Au layer are deposited on a Si wafer in a high vacuum E-beam evaporator (2×10^{-6} torr). The Cr layer acts as an adhesion layer and the Au layer prevents the Cr from oxidation. The Cr/Au dual layer is used as a seed layer of electroplating as well as the underbump metallurgy (UBM). A 10 μm layer of Sn is then electroplated in a stannous Sn-based bath at 21.5 mA/cm^2 in 25 min. The plating bath temperature and pH value are 46°C and 1, respectively. Then, Ag film is plated over Sn layer for 1 min expecting 0.2 μm thickness. Prior to the Ag plating process, the sample is chemically treated to reduce the possibility of an oxide layer over the Sn. The Ag plating bath is a cyanide-free, mildly alkaline plating solution at pH 10.5. The current density and process temperature are 4 mA/cm^2 and room temperature, respectively. Expected composition of the joint is 96.9 at.% Sn and 3.1 at.% Ag, which is near the eutectic composition of the Sn-Ag system. The Ag layer over the Sn prevents the inner Sn layer from oxidation. This Si wafer with Cr/Au/Sn/Ag structure is diced into 4.5 mm × 4.5 mm chips. Another Si wafer is deposited with 0.03 μm Cr, followed by 0.1 μm Au in one vacuum cycle again, and diced into 6.5 mm × 6.5 mm substrates. The Si chip and substrate are held together with a static pressure of 50 psi (0.35 MPa) in a graphite fixture to ensure intimate contact. The assembly is mounted on a graphite heating platform inside a small vacuum chamber that is pumped down to 100 millitorrs. The graphite platform is heated using a temperature controller/driver. The fixture temperature is monitored by a thermocouple and controlled by the temperature controller. Optimal bonding temperature appears to be 240°C with a dwell time at peak temperature of 1 min. Reflow time is about 6 min. The heating platform is turned off and the assembly is allowed to cool naturally to room temperature in the same vacuum ambient.

Figure 2.9 depicts the schematics of the bonding principle. As mentioned earlier, the electroplated Si chip (with Cr/Au/Sn/Ag) and Si substrate deposited with Cr/Au are placed together and mounted on the heating platform inside a vacuum chamber that is pumped to 100 millitorrs. In the solder structure on the Si chip, thin Ag layer covers the inner Sn as illustrated in Fig. 2.9(a). The melting temperature of the designed composition (96.9 at.% Sn and 3.1 at.% Ag) is slightly higher than Sn-Ag eutectic point of 221°C. As temperature increases towards the bonding temperature of 240°C, the thick Sn layer melts at 232°C and the molten Sn starts to react with the Ag capping layer to initially form Ag$_3$Sn intermetallic compound and subsequently dissolve this layer to turn into Sn-rich (L) phase. The molten (L) phase would wet the Au layer on the substrate to form AuSn$_4$ intermetallic compound, as shown in Fig. 2.9(b). As temperature goes up to 240°C bonding temperature, the (L) phase dissolves the Ag$_3$Sn and AuSn$_4$ intermetallics completely, depicted in Fig. 2.9(c). When this

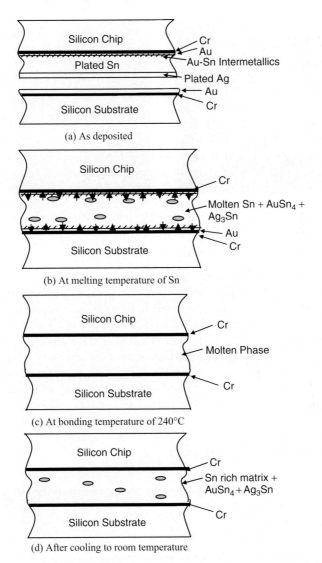

Fig. 2.9 Principle of the fluxless bonding using Sn-Ag multilayer structure [77]

happens, the essential condition of producing a joint is achieved. Upon cooling down to room temperature, the joint solidifies and is expected to consist of small $AuSn_4$ and Ag_3Sn intermetallic grains in a Sn rich matrix, indicated in Fig. 2.9(d).

The samples fabricated are examined by a reflection mode SAM (C-SAM) to evaluate the quality of fluxless Sn-Ag joints. Figure 2.10 shows C-SAM images of two samples. In the reflection mode SAM, the voids show up as bright spots

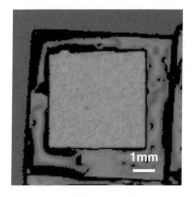

Fig. 2.10 Reflection scanning acoustic microscope (C-SAM) image of two samples bonded using fluxless Sn-Ag joint. The joints are virtually void free [77]

on a gray background. The joints are virtually void free. To study the microstructure of the joint, several samples are cut in cross-section and polished. SEM and EDX analysis are used to examine the cross-section of these samples. Figure 2.11 exhibits the secondary electron (SE) image of a joint cross section. The bonding layer is quite uniform with a thickness of 2.5 μm, which is less than expected. The reason is that significant amount of molten Sn is squeezed out in the bonding process. This molten Sn wets, reacts with, and stays on the area of Cr/Au coated substrate not covered by the chip. The surface of the cross section is not as smooth as desired, caused by the polishing process that still needs to be refined. The SEM image shows a homogeneous phase with about 97 at.% of Sn. To our surprise, Ag_3Sn and $AuSn_4$ intermetallic grains are not observed. One possible reason is that the SEM image does not pick up very small $AuSn_4$ or

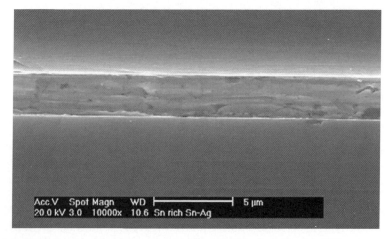

Fig. 2.11 SEM images of a eutectic Sn-Ag joint at 10,000x magnification [77]

Ag$_3$Sn grains with adequate contact. Another possibility is that the small amount of Au and Ag is not enough for intermetallic nucleation in the Sn matrix. In the latter case, the Au and Ag atoms are just dissolved in Sn to form a solid solution. The solder joints are very strong. We tried to break the joint with a hand tool but the silicon chip always break first. A de-bonding test is performed on several samples to measure the melting temperature. The melting temperature ranges from 219 to 226°C, which is close to the expected solidus temperature.

We next move onto the In-Ag process [78]. Figure 2.12 depicts the bonding design. Layer of Ag is electroplated on the silicon chip. The substrate chosen is Cu laminated with 280 μm thick of Ag foil by a new direct bonding process developed recently. In (Indium) is electroplated on the Ag foil, followed by a cap layer of Ag. We have found that electroplated In over Ag foil does not exist as pure In, but forms AgIn$_2$ IMC layer by reaction with Ag atoms during the plating process. AgIn$_2$ has an interesting characteristic; at 166°C, it turns into a mixture of In-rich molten phase (L) and Ag$_2$In solid grains. This situation continues until temperature reaches 205°C. At and above 205°C, Ag$_2$In grains convert to γ-phase grains. Thus, between 166 and 205°C, the (L) phase exists and can react with the Ag layer on the Si chip. In experiment, the Si chip is placed over Ag-laminated Cu substrate and held with static pressure to ensure intimate contact. The bonding is performed at 205°C for 3 min in 50 millitorr vacuum. Many samples were fabricated. When we tried to break the samples with a hand tool, the silicon chip always broke first. It means that the joint is really strong. To study the microstructure, several samples were cut in cross section and polished. SEM and EDX analysis were used to examine these samples. Figure 2.13 (a) and (b) are the secondary electron images of a joint cross section. Three distinct layers are identified as Ag/Ag$_2$In/Ag. While 6 μm of Ag still remains on the chip side connected to silicon, the Ag$_2$In layer has grown to 18 μm, resulting from rapid solid liquid reaction between Ag and the molten phase (L). Based on the observations, we present bonding mechanism. As temperature increases towards the bonding temperature of 205°C, the AgIn$_2$ layer starts melting at 166°C and converts to a mixture of molten phase (L) and

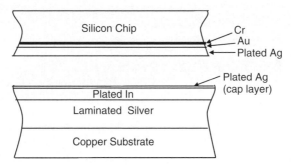

Fig. 2.12 Design of bonding process between Si/Cr/Au/Ag and Cu/Ag/In/Ag [78]

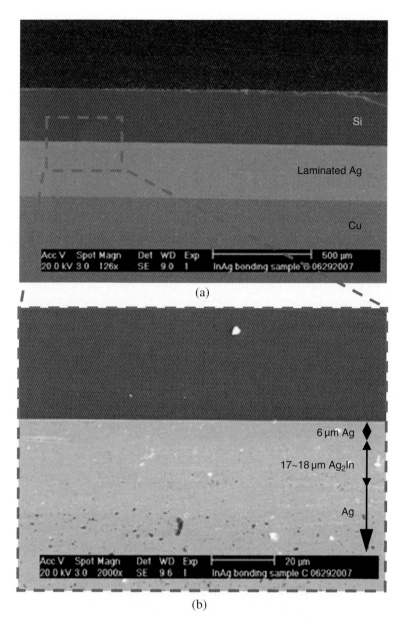

Fig. 2.13 Secondary images of the joint cross-section. The joint consists of three distinct layers of Ag, Ag2In, and Ag layers, **(a)** low magnification **(b)** high magnification [78]

Ag_2In solid grains. The molten phase now reacts with upper and bottom Ag layers and dissolves some of the Ag layers and a joint is formed. After cooling down to room temperature, the resulting joint would consist of Ag, Ag_2In, and $AgIn_2$ layers. The SEM image in Fig. 2.13 shows that the joint consists of only Ag and Ag_2In without $AgIn_2$. The absence of $AgIn_2$ indicates that, during the bonding process, the molten phase (L) dissolves enough Ag and turns into Ag_2In completely. The Ag_2In compound is in solid phase at the bonding temperature. Therefore, the joint solidifies during the bonding process and before cooling to room temperature. This effect is usually referred to as transient liquid phase bonding. At and beyond 300°C, Ag_2In turns into γ phase which remains solid until temperature reaches 630°C. Thus, the joint produced has very high melting temperature even though it is made at 205°C.

2.3.4 Fluxless Flip Chip Interconnect Technology

As solder joints shrink in flip chip interconnect, the gap between the Si chip and the package substrate also decreases. Eventually, the gap will become so small that it not realistic to clean the flux residues trapped in the gap. We thus envision the need of fluxless flip chip soldering processes. An initial process is reported using electroplated Sn-rich Sn-Au bumps [79].

To fabricate the Sn-Au solder bumps, the electroplating process is performed on Si wafers having a thick negative photoresist (Electrochem SU-8)

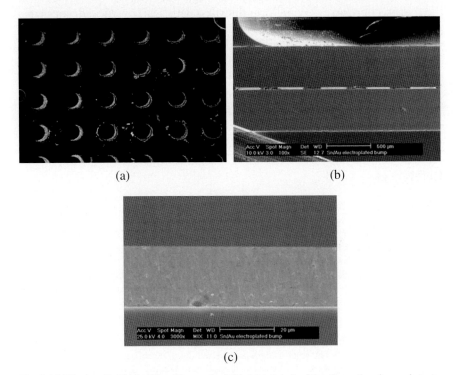

Fig. 2.14 Fluxless Sn2%Au flip chip results [79], (**a**) joints looking from the glass substrate, (**b**) Cross section of a sample showing even joints with nearly vertical walls, (**c**) Backscattered electron image of a single joint with 98 wt.% Sn

pattern. In the first step, 0.03 μm of Cr and 0.1 μm of Au are deposited in a vacuum chamber as a blanket UBM and a plating seed layer. To define the solder bumps, negative SU-8 resist is coated and photolithographically patterned to produce cavities with nearly vertical sidewalls. The pattern has 10×10 circular cavities each with a diameter of 250 μm and pitch of 500 μm. Sn is electroplated in the cavities, followed immediately by a thin Au capping layer. The Au and Sn layers are plated in disc shape with composition of 98 wt.% Sn and 2 wt.% Au. The Si wafer is precisely diced into 7 mm × 7 mm chips. Borosilicate glass is used as the substrate because it is transparent and makes alignment easier. On the glass wafer, Cr (0.03 μm) and Au (0.1 μm) are deposited. Lithographic and etching processes are performed to define a 10×10 array of 200 μm circular bond pads. The glass wafer is diced into 10 mm × 10 mm substrates. Both the silicon chip and the glass substrate are aligned in a special fixture and held together with a static pressure of 413 kPa (60 psi). The flip chip bonding is carried out in a tube furnace in hydrogen environment. The furnace is heated until the fixture reaches 250°C and the dwell time is 3 min.

Figure 2.14(a) exhibits the optical microscope image looking into glass substrate. It shows that all the bumps are joined to Cr/Au pads with good alignment. To confirm the quality of the flip chip joints, SEM image on cross section of a bonded sample is displayed in Figure 2.14(b). The solder joints on the cross-section have very uniform thickness with nearly vertical drum shape, which indicates that solder bumps are well aligned to the Cr/Au pads on the glass substrate. The good solder joint alignment probably implies that there is some self-alignment effect during the fluxless bonding process. Figure 2.14(c) shows the backscattered electron (BSE) image on the cross section of a typical solder joint. EDX data gave an average composition of 98 at.% of Sn. The small bright spots are believed to be $AuSn_4$ grains.

References

1. Luo S, Wong C (2001) Fundamental study on moisture absorption in epoxy for electronic application. Int Symp Adv Packag Mater, pp 293–298
2. Teh P, Mariatti M, Beh K et al (2007) The properties of epoxy resin coated silica fillers composites. Mater Lett 61:2156–2158
3. Chiu C, Lin J, Hsu K et al (2003) Thermally cleavable epoxy resins for electronic and optoelectronic applications. IEEE Electron Packag Technol Conf, pp 425–428
4. Cognard P, editor (2005) Adhesives and Sealants: Basic concepts and high tech bonding. Elsevier Ltd, Oxford
5. Watanabe T, Ooba N, Imamura S et al (1998) Polymeric optical waveguide circuits formed using silicone resin. J Lightwave Technol 16:1049–1055
6. Chida K, Sakaguchi S, Kimura T et al (1982) High speed coating of optical fibres with thermally curable silicone resin using a pressurized die. Electro Lett 18:713–715
7. Petrie E (2000) Handbook of Adhesives and Sealants. McGraw Hill, New York

8. Hoontrakul P, Sperling L, Pearson R (2003) Understanding the strength of epoxy-polyimide interfaces for flip-chip packages. IEEE Trans Devices and Mater Reliab 3:159–166
9. Kinloch A (1986) Structural Adhesives. Elsevier, London
10. Pizzi A, Mittal K (ed) (2003) Handbook of Adhesive technology. Marcel Dekker, New York
11. Sashida N, Hirano T, Tokoh A (1989) Photosensitive polyimides with excellent adhesive property for integrated circuit devices. IEEE Electron Compon Conf, pp 167–170
12. Itatani T, Gorwadkar S, Matsumoto S et al (2000) Positive photosensitive polyimide systhtesized by block-copolymerization for KrF lithography. Proc SPIE, pp 552–558
13. Kikuchi K, Goto M, Aoyagi M et al (2005) Efficient fabrication process for superconducting integrated circuits using photosensitive polyimide insulation layers. IEEE Trans Appl Supercond 15:94–97
14. Jain J, Samant S (2005) Novel multilayering technique using folded flexible circuits. IEEE Trans Electron Packag Manuf 28:259–264
15. Kristiansen H, Liu J (1998) Overview of conductive adhesive interconnection technologies for LCD's. IEEE Int Symp on Polym Electron Packag, pp 223–232
16. Kristiansen H, Liu J (1998) Overview of conductive adhesive interconnection technologies for LCDs. IEEE Trans Compon Packag Manuf Technol 21:208–214
17. Lau J (1995) Flip chip technologies. McGraw-Hill, New York
18. Kubo K, Touma S, Ross D (1986) Chip-on-glass LCD for automotive application. SAE Spec Publ, pp 115–119
19. Lawrence L (ed) (1985) Recent advances in liquid crystalline polymers. Elservier Applied Science Publishers, London and Newyork
20. Thompson D, Tantot O, Papapolymerou J et al (2004) Characterization of liquid crystal polymer (LCP) material and transmission lines on LCP substrates from 30–110 GHz. IEEE Trans Microw Theory Technol 52:1343–1352
21. Tentzeris M, Laskar J, Lee J-H (2004) 3D Integrated RF and millimeter-wave functions and modules using liquid crstal polymer (LCP) system-on-package technology. IEEE. Trans Adv Packag 27:332–340
22. Wang G, Thompson D, Papapolymerou J (2004) Low cost RF MEMS switches using LCP substrate. IEEE Trans Adv Packag 3:1441–1444
23. DeJean G, Bairavasubramanian R, Papapolymerou J et al (2005) Liquid Crystal Polymer (LCP): A new organic material for the development of multilayer dual-frequency/dual polarization flexible antenna arrays. IEEE Antennas and Wireless Propag Lett 4:22–26
24. Yu L, Tay F, Iliescu C et al (2006) Adhesive bonding with SU8 at wafer level for microfluidic devices. J Phys 114:189–192
25. Yu L, Iliescu C, Chen B et al (2006) SU8 adhesive bonding using contact imprinting. Int Semicond Conf, pp 189–192
26. Reuter D, Bertz A, Gessner T (2005) Selective adhesive bonding with SU-8 for zero-level-packaging. Micro- and Nanotechnology: Mater Processes, Packag, and Syst II, pp 163–171
27. Wallis G and Pomerants D (1969) Field Assisted Glass-Metal Sealing. J Appl Phys 40:3946–3949
28. Schmidt M (1998) Wafer-to-Wafer Bonding for Microstructure Formation. Proc of the IEEE 86:1575–1585
29. Wei J, Nai S, Wong C et al (2004) Glass-to-glass anodic bonding process and electrostatic force. Thin Solid Films 462–463:487–491
30. Lin C, Yang H, Wang W et al (2007) Implementation of three-dimensional SOI-MEMS wafer-level packaging using through-wafer interconnections. J Micromech Microeng 17:1200–1205

31. Guanl R, Gan Z, Fulong Z et al (2006) Anodic Bonding Study on Vacuum Micro Sealing Cavity. IEEE Elcetro Packag Technol Conf, pp 1–4
32. Jin Y, Wang Z, Lim P et al (2003) MEMS vacuum packaging technology and applications. IEEE Electron Packag Technol Conf, pp 301–306
33. Stefano L, Malecki K, Rossi A et al (2006) Integrated silicon-glass opto-chemical sensors for lab-on-chip applications. Sens and Actuators B 114: 625–630
34. Briand D, Weber P, Rooij N (2004) Silicon liquid flow sensor encapsulation using metal to glass anodic bonding. IEEE Int Micro Electro Mech Syst Conf, pp 649–652
35. Akselsen O (1992) Review Diffusion Bonding of Ceramics. J Mater Sci 27: 569–579
36. Qin C-D and Derby B (1992) Diffusion bonding of nickel and zirconia: Mechanical properties and interfacial microstructures. J Mater Res 7: 1480–1488
37. Chen S, Ke F, Zhou M, and Bai Y (2007) Atomistic investigation of the effects of temperature and surface roughness on diffusion bonding between Cu and Al. Acta Mater 55:3169–3175
38. Li H, Zheng Y, Akin D et al (2005) Characterization and modeling of microfluidic dielectrophoresis filter for biological species. J Microlelectromech Syst 14:103–112
39. Bartle P, Houldcroft P, Needham J et al (1979) Diffusion bonding as a production process. The Welding Institute, UK
40. Tummala R, Rymaszewski E, Klopfenstein A (1997) Microelectronics packaging handbook. Chapman & Hall, USA
41. Bolcar V (1968) Thermocompression bonding of external package leads on integrated circuit substrates. IEEE Trans Electron Devices 15:651–655
42. Wang Z, Tan Y, Schreiber C (2000) Development of chip-on-dot flip chip technique utilizing gold dot™ flexible circuitry. IEEE Electro Compon Technol Conf, pp 1470–1474
43. Takagi H, Kikuchi K, Maeda R et al (1996) Surface activated bonding of silicon wafers at room temperature. Appl Phys Lett 68:2222–2224
44. Suga T, Takahashi Y, Takagi H et al (1992) Structure of Al-Al and Al-Si_3Ni_4 interfaces bonded at room temperature by means of the surface activation method. Acta Metall Mater 40:S133–S137
45. Takagi H, Maeda R (2006) Direct bonding of two crystal substrates at room temperature by Ar-beam surface activation. J Cryst Growth 292:429–432
46. Takagi H, Maeda R, Hosoda N (1999) Room-twmperature bonding of lithium niobate and silicon wafers by argon-beam surface activation. Appl Phys Lett 74: 2387–2389
47. Davoine C, Fendler M, Louis C et al (2006) Impact of pitch reduction over residual strain of flip chip solder bump after refow. IEEE Int. Conf. on Therm Mech and Multiphysics Simul and Exp in Micro-Electron and Micro-Syst, pp 1–5
48. Peng C-T, Liu C-M, Lin J-C et al (2004) Reliability analysis an ddesign for the fine-pitch flip chip BGA packgeing. IEEE Trans. Compon Packag Technol 24:684–693
49. Xiao G, Chan P, Teng A et al (2001) Reliability study and failure analysis of fine pitch solder bumped flip chip on low-cost printed circuit board substrate. IEEE Electon Compon Technol Conf, pp 598–605
50. Xu Z, Suga T. (2005) Surface activated bonding –high density packaging solution for advanced microelectronic system. IEEE Int Electron Packag Technol Conf, pp 398–403
51. Shigetou A, Itoh T, Matsuo M et al (2006) Bumpless interconnect through ultrafine Cu electrodes by means of surface-activated bonding (SAB) method. IEEE Trans Adv Packag 29:218–226
52. Wang Q, Hosoda N, Itoh T et al (2003) Reliability of Au bump-Cu direct interconnections fabricated by means of surface activated bonding method. Microelectron Reliab 43:751–756
53. Wang P, Kim J, Lee C (2007) A Novel Ag-Cu Lamination Process. IEEE Adv Packag Mater Symp, pp 200–202

54. Yamada Y, Takaku Y, Yagi Y et al (2006) Pb-free high temperature solders for power device packaging. Microelectron and Reliab 46:1932–1937
55. Lee C, Wang C, Matijasevic G (1991) A new bonding technology using gold and tin multilayer composite structures. IEEE Trans Compon Hybrids and Manuf Tech 14:407–412
56. Bernier (1998) The nature of white residue on printed circuit assemblies. Kester Solder, Des Plaines, IL
57. Koopman N, Bobbio S, Nangalia S et al (1993) Fluxless soldering in air and nitrogen. Proc IEEE Electron Compon and Technol Conf, pp 595–605
58. Beranek et al (1997) Fluxless, no clean assembly of optoelectronic devices with PADS. Proc IEEE Electron Compon and Technol Conf, pp 755–762
59. Hong S, Kang C, Jung J (2004) Plama reflow bumping of Sn3.5%Ag solder for flux-free flip chip package application. IEEE Trans Adv Packag 27:90–96
60. Park C, Hong S, Jung J et al (2001) A study on the fluxless soldering of Si wafer/glass substrate using Sn3.5Ag and Sn37 Pb solder. Mater Trans 42:820–824
61. Lin W, Lee Y (1999) Study of fluxless soldering using formic acid vapor. IEEE Trans Adv Packag 22:592–601
62. Matsuki H, Matsui H, Watanabe E (2001) Fluxless bump reflow using carboxylic acid. Int Symp on Adv Packag Mater, pp 135–139
63. Dong C, Patrick R, Karwacki E (2007) Electron attachment: a new approach to H_2 fluxless solder reflow for wafer bumping. IEEE Trans Adv Packag 30:485–490
64. Matijavesic G, Lee C (1989) Void-free Au-Sn eutectic bonding of GaAs dice and its characterization using scanning acoustic microscopy. J Electron Mater 18:327–337
65. Zakel E, Gwiasda J, Kloesser J et al (1994) Fluxless flip chip assembly on rigid and flexible polymer substrates using the Au-Sn metallurgy. Proc. IEEE/CMPT Electron Manuf Tech Symp, pp 177–184
66. Kallmayer C, Opperman H, Engelmann G et al (1996) Self-aligning flip-chip assembly using eutectic gold/tin solder in different atmospheres. Proc. IEEE/CMPT Electron Manuf Tech Symp, pp 18–25
67. Okamoto H, Massalski T (eds) (1987) Phase Diagram of Binary Gold Alloys, ASM International, Metals Park, OH
68. Ishikawa M, Sasaki H, Ogawa S et al (2005) Application of gold-tin solder paste for fine parts and devices. Proc IEEE Electron Compon Technol Conf, pp 701–709
69. Matijasevic G, Lee C, Wang C (1993) Gold-tin alloy phase diagram and properties related to its use as a bond medium. Thin Solid Films 223:276–287
70. Lee C, Wang C, Matijasevic G (1993) Gold-indium alloy bonding below the eutectic temperature. IEEE Trans Compon Hybrids and Manuf Tech 16:311–316
71. Matijasevic G, Chen Y, Lee C (1994) Copper-tin multilayer composite solder for fluxless processing. Int J Microcircuits and Electron Packag 17:108–117
72. Lee C, Chen Y (1995) Indium-copper multilayer composite solder for fluxless Bonding. Mater Res Soc Spring Meet, MRS Symp Proc Electron Packag Mater Sci VIII 390:225–230
73. Chen Y, So W, Lee C (1997) A fluxless bonding technology using indium-silver multilayer composites. IEEE Trans Compon Packag and Manuf Technol 20:46–51
74. So W, Lee C (2000) A fluxless process of fabricating In-Au joints on copper substrates. IEEE Trans Compon and Packag Technol 23:377–382
75. Lee C, Chuang R (2003) Fluxless non-eutectic joints fabricated using gold-tin multilayer composite. IEEE Trans Compon and Packag Technol 26:416–422
76. Chuang R, Kim D, Park J et al (2004) A fluxless process of producing tin-rich gold-tin joints in air. IEEE Trans Compon and Packag Technol 27:177–181
77. Kim J, Lee C (2007) Fluxless Sn-Ag bonding in vacuum using electroplated layers. Mater Sci and Eng A 448:345–350

78. Kim J, Wang P, Lee C (2007) Fluxless bonding of Si Chips to Ag-copper using Electroplated Indium and Silver Structures. Proc IEEE Adv Packag Mater Symp, pp 194–199
79. Kim J, Kim D, Lee C (2006) Fluxless flip-chip solder joint fabrication using electroplated Sn-Rich Sn-Au structures. IEEE Trans Adv Packag 29:473–482

Chapter 3
Advanced Chip-to-Substrate Connections

Paul A. Kohl, Tyler Osborn, and Ate He

Abstract Transistor scaling, shrinking the critical dimensions of the transistor, has led to continuous improvements in system performance and cost. Higher density of the transistors and larger chip size has also led to new challenges for chip-to-substrate connections. The pace of change in packaging and chip-to-substrate connections has accelerated because off-chip issues are increasingly a limiting factor in product cost and performance. Chip-to-substrate connections are challenged on many fronts, including number of signal input-output (I/O) connections, I/O that operate at high-speed, power & ground I/O, and low cost.

This chapter examines various techniques and structures that have been designed to address these challenges. The mechanical compliance and electrical performance modeling of the interconnect structures is important in determining the geometry, materials, and processing necessary for an application. Various approaches have been taken to satisfy both the mechanical and electrical needs for these I/O connections. Mechanically compliant structures based on traditional solder bonded connections can drastically improve thermo-mechanical reliability but may compromise electrical performance. Additional structures improve upon the compliance of the solder ball by capping a pillar structure with solder, but still require the reliable protection of underfill. More high performance and long term improvements to satisfy both mechanical and electrical needs such as interconnects composed entirely of copper are also discussed. Finally, the future needs projected by the ITRS for ultra-high off-chip frequency and thermal management are addressed with respect to chip-to-substrate interconnects.

Keywords Input/Output · Compliant I/O · Copper interconnects · Solder-free · Electronic packaging · Flip-chip

---

P.A. Kohl (✉)
School of Chemical and Biomolecular Engineering Georgia Institute of Technology Atlanta, GA 30332-0100
e-mail: paul.kohl@chbe.gatech.edu

3.1 Introduction

Transistor scaling, shrinking the critical dimensions of the transistor, has led to continuous improvements in system performance and cost. Higher density of the transistors and larger chip size has also led to new challenges for chip-to-substrate connections. The pace of change in packaging and chip-to-substrate connections has accelerated because off-chip issues are increasingly a limiting factor in product cost and performance. Chip-to-substrate connections are challenged on many fronts, including number of signal input-output (I/O) connections, I/O that operate at high-speed, power & ground I/O, and low cost.

Chip-to-substrate interconnects provide power, electrical contacts, and a mechanical link between the chip and the substrate. The two most common chip assembly schemes today, wire bonding and flip chip bonding are illustrated in Fig. 3.1(a) and (b), respectively [1, 2, 3, 4]. The area-array flip-chip configuration provides for a higher number of I/O and the electrical environment is superior to wire bonding because it has lower inductance and capacitance. An area array of flip-chip solder balls has more I/O than peripheral wire bonds. Flip-chip connections also provide an additional thermal path for chip cooling through the I/O.

Flip chip assembly uses metal connections, often solder spheres, to connect the chip to the substrate and/or the substrate to the board as shown in Fig. 3.1(b). Chips can also be connected directly to the board in a chip-on-board (COB) configuration, Fig. 3.1(c). COB is a less costly and lower profile

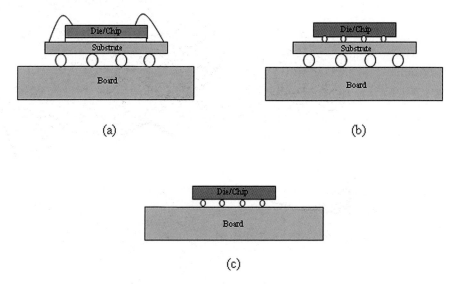

Fig. 3.1 Illustration of (**a**) chip-to-substrate wire bonding, (**b**) chip-to-substrate flip-chip interconnection, and (**c**) chip on board flip chip connection

(total height) technology, however, the substrate helps in electrical fan-out and allows for a higher number of I/O. The substrate also assists in assembly of a heat-sink which is needed on high power chips.

3.1.1 ITRS Projections for Flip-Chip Connections

Projections on the number and density of I/O can be made based on transistor scaling, and projections on chip size and performance. Table 3.1 lists some of the ITRS projections for high performance chips [5]. Although the power consumed in high performance chips has been capped at 198 W, the direct current (DC) will continue to increase because the supply voltage will drop. If the current is to be delivered at the low operating voltage, an increasing number of I/O will be involved so that the threshold for electromigration is not crossed and the power is delivered across the chip, where it is used locally.

The signal I/O will be especially challenged because the number of I/O will increase, reducing the I/O pitch, and their performance must improve so that they can support higher off-chip speed communication. The improvement in I/O design needs to occur with a decreasing cost-basis, as shown for example in the cost per pin number.

Not shown in Table 3.1 are the mechanical requirements of future I/O. The dielectric constant of the on-chip insulators is being lowered so as to decrease interconnect delay. This will lead to more fragile on-chip dielectrics which will no longer be able to support the high stress levels imposed by the I/O. Future chip-to-substrate connections will likely need to be mechanically compliant to compensate for the mechanical stresses induced by the coefficient of thermal expansion (CTE) mismatch between the semiconductor chip and the package substrate or board. Normally, one would expect the I/O induced stress to rise as the size of the solder ball is reduced and the gap between the chip and substrate is lowered. The shrinking size of the flip-chip solder balls is reflected in the pad pitch and pin-count.

The increase in chip-to-substrate speed is extremely important, particularly for processor-to-memory access. The increase in speed is however most challenging because signal degradation and distortion can occur unless the electrical

Table 3.1 Select ITRS values (2006 update) for high performance chips

Year	2005	2007	2010	2013	2016	2020
MPU Current, A	172	172	198	220	248	283
MPU Power, W	167	198	198	198	198	198
Chip-to-Board Speed, GHz	3.1	4.9	9.5	18.6	34.9	72.4
Pad Pitch, flip-chip, μm	150	130	120	110	95	85
Package Pin-count	3400	4000	4851	5616	6501	7902
Package Cost per pin, cents	1.78	1.83	1.56	1.34	1.15	0.94

characteristics (resistance, capacitance, inductance, and impedance) are acceptable. Finally, as with most electronic components, performance improvements become commercially viable only when the cost structure facilitates improved performance/cost. Thus, the higher performance I/O must be achieved at a lower per unit cost.

3.1.2 Electrical Modeling of I/O

Chip-to-substrate electrical connections have parasitic inductance, capacitance, and resistance properties that degrade their performance. The magnitude of these properties and their ultimate effect on the performance of the I/O is a function of the I/O shape, distribution, and materials. Electrically, it is most desirable to have air as the medium between the chip and substrate because of the minimal coupling induced by air. However, the local mechanical stress within the solder joints and at the joint between the solder and the planar surface is high. Filling the area around the solder joints with an epoxy helps to distribute the mechanical stresses and avoids the highest stress points and reduces solder fatigue. However, the high dielectric constant and loss of underfill degrades the electrical characteristics and increases cross-talk between I/O.

In general, the parasitic inductance is important for power integrity, the parasitic capacitance affects the signal integrity, and the resistance contributes to the signal RC delay and conductor loss. In the next sections, the electrical and mechanical attributes of the I/O will be evaluated.

3.1.2.1 Parasitic Inductance of Chip-to-Substrate I/O

The IR (current-resistance product) voltage drop and simultaneous switching noise (SSN) are two I/O problems involved in power distribution. SSN is induced by the current change that passes through the power distribution network, which is mainly due to the I/O parasitic inductance [6, 7, 8, 9]. SSN can cause problems in signal timing and integrity, resulting in false switching logic circuits [10, 11]. The voltage change due to SSN can be expressed by Eq. (3.1).

$$\Delta V = L \frac{dI}{dt} \quad (3.1)$$

Where I is the current, t is time and L is the parasitic inductance or loop inductance of the chip-to-substrate I/O. Lower power supply voltage, V_{dd}, reduces the noise margin ΔV or tolerance to SSN. Therefore, the parasitic inductance of the power/ground I/Os, L, needs to be kept as low as possible in order to maintain signal integrity. The parasitic inductance is a function of the physical geometry of the power/ground interconnect, loop distance (path

Fig. 3.2 Chip-to-substrate power and ground I/O layout

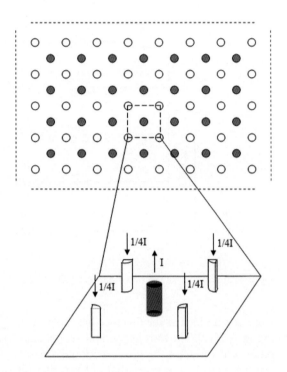

between power delivery and return path), and dielectric properties of the insulator surrounding the I/O.

The self and mutual inductance of the I/O are shown in Fig. 3.2. The DC current, I, is delivered through the center (power) I/O, and returns through the four neighboring quarter-size ground I/Os, where the current in each is I/4. Only a quarter of each of the return paths is used because they each must service four power I/O. Thus, the parasitic inductance of the I/O can be derived based on a center I/O with four surrounding cylindrical I/O. The cylindrical shape (ignoring the bulge at the center of a ball) will be used to simplify the calculations and make them more appropriate for other pillar connections.

The self inductance of the I/O pillar can be calculated from Eq. (3.2).

$$L = 0.002H \left[\ln\left(\frac{4H}{D}\right) - \frac{3}{4} \right] \times 10^{-4} \, (H) \tag{3.2}$$

Where H is height of the pillar and D is diameter [12]. The mutual inductance between two I/O can be calculated from Eq. (3.3).

$$M = 0.002H \left[\ln\left(\frac{H}{d} + \sqrt{1 + \frac{H^2}{d^2}}\right) - \sqrt{1 + \frac{d^2}{H^2}} + \frac{d}{H} \right] \times 10^{-4} \, (H) \tag{3.3}$$

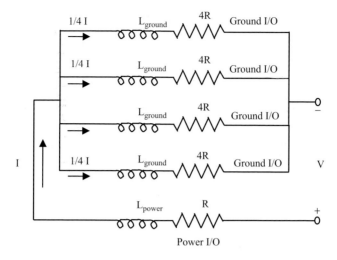

Fig. 3.3 Circuit diagram of power/ground I/O

Here, d is defined as the distance between the center points of two I/O as shown in Fig. 3.2 [12]. The four ground pillars have the same height as the center power pillar, but only 1/4 the cross-sectional area. As a result, the resistance of the ground pillar is four times that of the center pillar. The voltage drop within the full circuit (delivery and return path) can be divided into two parts: the voltage drop within the power I/O and the voltage drop within the ground I/O as shown in Eq. (3.4) and Fig. 3.3.

$$V_{circuit} = V_{power} + V_{ground} \quad (3.4)$$

For the full circuit, the voltage drop equals the product of current times the complex impedance $Z_{circuit}$, Eq. (3.5). The impedance of the circuit, Fig. 3.3, can be expressed by Eq. (3.6). If the resistance of the center power I/O is R, then the total resistance of the circuit is 2R giving Eq. (3.7).

$$V_{circuit} = IZ_{circuit} \quad (3.5)$$

$$Z_{circuit} = R_{circuit} + j\omega L_{parasitic} \quad (3.6)$$

$$V_{circuit} = 2IR + j\omega I L_{parasitic} \quad (3.7)$$

Where $V_{circuit}$ is the voltage drop in the circuit (Fig. 3.3), $Z_{circuit}$ is the overall complex impedance, $R_{circuit}$ is the resistance, ω is angular frequency, j is the imaginary unit (the square root of –1), and $L_{parasitic}$ is the parasitic inductance.

3 Advanced Chip-to-Substrate Connections

Equations (3.8), (3.9), (3.10) and (3.11) express the voltage and impedance for the center power I/O shown in Fig. 3.2.

$$V_{power} = IZ_{power} \tag{3.8}$$

$$Z_{power} = R_{power} + j\omega L_{power\_eff} \tag{3.9}$$

$$L_{power\_eff} = L_{power} - 4 \times \frac{1}{4} M_1 \tag{3.10}$$

$$V_{power} = IR + j\omega \left(IL_{power} - 4 \times \frac{1}{4} IM_1 \right) \tag{3.11}$$

Where V_{power} is the voltage drop in the power I/O, Z_{power} is the impedance, R_{power} is the resistance, L_{power} is the self inductance, M_1 is the mutual inductance between the power I/O and an adjacent ground I/O, and L_{power_eff} is the total effective inductance of the power I/O. The same procedure can be used for each ground I/O, yielding Eq. (3.12).

$$V_{ground} = \frac{1}{4} I \times 4R + j\omega \left(\frac{1}{4} IL_{ground} + \frac{1}{4} IM_3 + 2 \times \frac{1}{4} IM_2 - IM_1 \right) \tag{3.12}$$

Where V_{ground} is the voltage drop in one ground I/O, L_{ground} is the self inductance, M_2 is the mutual inductance between the two nearest ground I/Os, and M_3 is the mutual inductance between the two ground I/Os at opposite corners of the four I/O surrounding the center power I/O. Substitution of Eqs. (3.11) and (3.12) into Eq. (3.7) yields a solution for the total parasitic inductance, Eq. (3.13).

$$L_{parasitic} = L_{power} + \frac{1}{4} L_{ground} - 2M_1 + \frac{1}{2} M_2 + \frac{1}{4} M_3 \tag{3.13}$$

3.1.2.2 Parasitic I/O Capacitance

The parasitic capacitance will degrade the signal integrity by inducing crosstalk between adjacent I/O and by causing signal delay due to the RC product. Although the absolute value of the parasitic capacitance and the resistance of chip-to-substrate I/Os are small compared to the on-chip interconnect, the overall system performance will benefit from lower off-chip RC delays [13]. The capacitance is also needed to calculate the characteristic impedance (Section 3.1.2.3) of the chip-to-substrate I/O since a mismatch in the characteristic impedance between the chip and substrate could result in reflective losses for high frequency signals.

The parasitic capacitance of two cylindrical chip-to-substrate signal I/O can be calculated from Eq. (3.14) through a lumped-element circuit for high frequency signals [14, 15].

$$C = \frac{\pi \varepsilon_0 \varepsilon_r}{\ln\left[\frac{d}{D} + \sqrt{\left(\frac{d}{D}\right)^2 - 1}\right]} H \qquad (3.14)$$

Where d is the center-to-center distance between two adjacent I/O, D is the diameter of the I/O, and H is the height. For reference, the parasitic capacitance of a 125 μm diameter eutectic solder bumps was measured to be 8.8 fF [16].

3.1.2.3 Characteristic Impedance

The characteristic impedance is defined by Eq. (3.15) [17] [q].

$$Z_0 = \sqrt{\frac{R + j\omega L}{G + j\omega C}} (\Omega) \qquad (3.15)$$

Where G is the shunt conductance due to dielectric loss, Eq. (3.16), L is the self-inductance of the two copper pillars, Eq. (3.17), R is the resistance, Eq. (3.16), and R_s is the surface resistance of the pillars.

$$G = \frac{\pi \omega \varepsilon''}{\ln\left[\frac{d}{D} + \sqrt{\left(\frac{d}{D}\right)^2 - 1}\right]} H (S) \qquad (3.16)$$

$$L = \frac{\mu_0 \mu_r}{\pi} \ln\left[\frac{d}{D} + \sqrt{\left(\frac{d}{D}\right)^2 - 1}\right] H (H) \qquad (3.17)$$

$$R = \frac{2R_s}{\pi D} H (\Omega) \qquad (3.18)$$

Where μ_0 is the permeability of vacuum, and μ_r is the relative permeability, and ε'' is the imaginary part of the complex permittivity. For electrically isolated I/O, the shunt conductance, G, is essentially zero since the insulator between pillars is non-conductive. Equation (3.15) can be further simplified by comparing the magnitude of R and ωL. R can be calculated from the lumped resistance, Eq. (3.16). The surface resistance can be calculated from its definition Eq. (3.17).

$$R_s = \frac{1}{\sigma \delta_s} \qquad (3.19)$$

Where σ is the conductivity and δ_s is the skin depth is defined by Eq. (3.20) [17].

$$\delta_s = \sqrt{\frac{2}{\omega\mu\sigma}} = \sqrt{\frac{1}{\pi f \mu_0 \sigma}} \qquad (3.20)$$

The impedance of the I/O may not match the remainder of the circuit. However, that may be acceptable because the load impedance dominates the circuit impedance since the I/O length is so short [18].

3.1.3 Mechanical Modeling

The reliability of chip-to-substrate connections is a major concern in the microelectronic industry. The CTE mismatch between the silicon chip (2.5 ppm/°C) and the substrate (4–10 ppm/°C for ceramics and 15–24 ppm/°C for organic Flame Resistant 4 (FR4)/BT board) causes deformation of both the chip and substrate. This generates strains and stresses on the interconnect structures [19, 20]. Thermo-mechanical failures will occur when the shear stress exceeds the strength of the interconnect joint or when the accumulation of the inelastic strain due to cyclic loadings exceeds the material fatigue strength [21].

For solder-based chip-to-substrate solder joints, temperature fluctuations caused by either power transients or environmental changes, along with the resulting CTE mismatch between the packaging materials, results in time and temperature dependent creep deformation of solder. This deformation accumulates with repeated cycling and ultimately causes solder joint cracking and I/O failure. Finite element analysis (FEA) has been widely used to evaluate the accumulated inelastic strain of solder bumps. FEA requires empirical fatigue models to predict fatigue life of solder joints [22, 23, 24].

Finite element models (FEM) can be 1-dimensional, 2-dimensional, or 3-dimensional (lines, shapes, or surfaces). For the stress analysis of chip I/O, 3-D models are most useful since stress is a tensor that has both surface and normal (vertical to the surface) components. The size of the model is a function of the chip segment used. Figure 3.4 shows a 3-D quarter model, 3-D octant model, and 3-D slice model which have been used to investigate the thermal performance and thermo-mechanical behavior of chip packages [22, 25, 26]. The boundary conditions in 3D quarter and octant models are either structure exterior surfaces or symmetry planes. This makes the 3-D quarter and octant models most accurate, since no simplified assumptions are made to the boundary conditions [27]. However, the 3-D quarter and octant models require very large memory space and calculation time. Due to the dimension differences between the micrometer size interconnects and the millimeter size chip and board, stress modeling of individual I/O requires a large number of elements to obtain convergence. Sub-modeling has been developed to reduce the memory space needed for 3-D models [28, 29]. Sub-modeling consists of two steps. First, a 3-D quarter or octant model consisting of all the structures is calculated with a

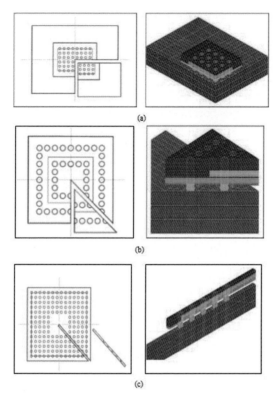

Fig. 3.4 Illustrations of **(a)** 3-D quarter model, **(b)** 3-D octant model, and **(c)** 3-D slice model

coarse mesh. Although the regions of interest on small structures have not converged, the deformations on large structures far from the regions of interest have converged. Then, the sub-model contains only the regions of interest and is analyzed with the boundary conditions transferred from the global model. A much finer mesh can be used in the sub-model, since the physical dimensions are smaller.

The 3-D slide model, Fig. 3.4(c), considers a 3-D diagonal slice of the package that passes through the full thickness of the package assembly. This captures the maximum strain and stress that will occur at the corner of the package. The correct choice of boundary conditions is important since the nodes on the two surfaces of the slice are coupled. The nodes on the same plane have identical deformation in the y-direction (normal to the surface) to meet the generalized plane deformation (GPD) constraints. The plane is neither a free surface nor a true symmetry plane. The slice plane is free to move in the y-direction, but the surface is required to remain planar [27, 30, 31, 32]. GPD modeling is a tradeoff in terms of accuracy and computational complexity. The reliability analysis of solder packages shows only a 6% difference between the 3-D Slice (GPD) model and 3-D octant model [27].

3.2 Compliant Solder-Based I/O Structures

The need for small chip footprint, especially in portable electronics, and higher I/O density has created the need for new packaging technologies. Chip-scale packages were created to address a rapidly growing segment of the market. These chip-scale packages enabled a high silicon packing density, however, CTE mismatch between the silicon and the board needed to be addressed [19, 33].

3.2.1 Peripheral-to-Flip-Chip Area Array Structures

Reducing the package footprint has been an important contributor to reducing the overall system size, weight, and cost. However, reducing the total package size lowers the area over which the mechanical strain is distributed. Numerous approaches have been taken to address the need for mechanical compliance in low-cost, small footprint packages. Some of the first compliant flip-chip structures used flip-chip attachment to convert peripheral I/O into an area array so as to shirk the footprint of the die. Pre-fabricated film or tape-based chip-to-substrate I/O has been used to connect the chip to a printed wiring boards. Amkor and Toshiba have used this style of interposer [19, 33]. The interposer (tape) is applied to the die followed by assembly. The chip can be wire bonded to the interposer followed by flip-chip attachment of the interposer-plus-chip to the printed wiring board.

A degree of compliance is given to the interposer and can be enhanced by using 3-dimensional, spring-like structures. Tessera has developed a compliant vertical link technology where the metal interconnection in the interposer lifts off the flexible foil by the injection of an encapsulant between the interposer and the IC [34]. Figure 3.5 shows the Tessera μBGA for perimeter I/O. The device is then attached to the board by use of solder balls. The Tessera μBGA provides redistribution from the perimeter I/O to area array so that the pitch on the board can be eased to coarser values. The flexible spring-like structures can bend and relieve the stress generated by the CTE mismatch between the chip and the substrate.

3.2.2 Redistribution Using Area Array Solder I/O

Redistribution of the I/O on a chip to a more convenient form can be accomplished in many ways. Redistribution can more equally distribute the I/O across a chip and can be used to build mechanical compliance into the I/O. The redistribution can be accomplished by adding an interposer layer. The interposer can be introduced through the use of a prefabricated layer or by fabricating the layer directly on the chip in wafer form. The wafer-scale interposer layer

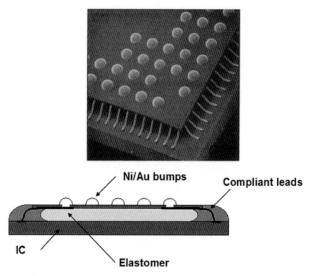

Fig. 3.5 Picture and cross section of micro BGA from Tessera

requires that one performs extra processing steps after the chip back-end processes are complete. An example of a prefabricated interposer structure is shown in Fig. 3.6 which shows the WAVE package for area array I/O devices. An interposer containing the flexible link is attached to the chip I/O. The solder ball is used to attach the chip to the substrate.

3.2.3 Wafer-Scale Compliant I/O

Compliant, solder-based I/O can be fabricated in wafer form after normal back-end chip interconnect is complete. After normal chip fabrication, a final

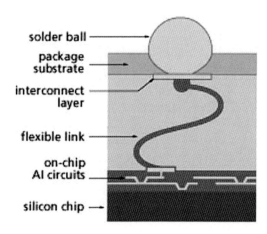

Fig. 3.6 WAVE technology developed by Tessera

polymer-metal build-up process can be used to redistribute the I/O. Once the wafer level package is complete, the chips are separated and assembled. The first uses were in lower pin-count devices and were driven by cost reductions. Since a traditional package is not required and the size of the printed wiring board can be reduced (smaller IC footprint), the packaging costs are lower. The mechanical requirements are modest since the chips are small, and the number of I/O is low.

A small amount of compliance between the chip and substrate was achieved by Fujitsu in the SuperCSP package [33]. Short copper posts were fabricated on the chip in wafer form. A polymer was then used to encapsulate the posts. The structure has a degree of compliance because the post can elastically deform and the polymer encapsulation can help distribute the stress. Most of the compliance was provided by conventional solder bumps attached to ends of the posts. The Fujitsu SuperCSP used copper-posts 350 μm diameter and 100 μm tall. Special equipment has been developed to injection mold a polymer encapsulant around the posts followed by solder ball attachment. Oki and Casio have developed a similar structure for use in low pin count devices [33]. Ibiden has developed a thin, flexible post in the second layer of polymer used in the redistribution build-up. These 'post' technologies show the importance of in-plane (x-y direction) compliance in producing a reliable, wafer-level packaged device. The solder joint on the copper post provides an inexpensive way to compensate for any z-axis (height) mismatched between the chip and substrate. The solder ball can be slightly compressed or elongated during reflow to make up for non-planarity in the parts.

Intel Corporation also uses copper posts with an attached solder ball to package microprocessors [35]. The short copper post also lifts the brittle solder joint off the chip surface to improve reliability. The highest stress point of the flip-chip solder joint is the intersection of the I/O and the chip surface [18].

Chip-to-substrate structures with higher I/O density and much greater compliance have been developed. A Sea of Leads (SoL) technology was developed as an enabling technology for future chip-to-module interconnections [36, 37]. SoL wafer level packaging technology provides an ultra-high I/O density of x-y-z compliant leads ($>10^4$ per cm^2) and can enhance the performance of a system on a chip (SoC) by routing critical on-chip global interconnects off-chip to reduce signal delay and thus increasing global clock frequency [36, 37]. The addition of an embedded air gaps into SoL adds vertical compliance (z-axis) needed for wafer level testing and mating to non-planar boards. Air-gaps also serve to lower the dielectric constant of the interconnect. Figure 3.7 shows a cross-section of process flow for an x-y-z compliant lead [38].

The exposed I/O are shown in Fig. 3.7(a). A sacrificial material shown in Fig. 3.7(b) is overcoated by a flexible material and vias are opened to the bond pad in Fig. 3.7(c). When the overcoat elastomer is curing in Fig. 3.7(d), the sacrificial material decomposes leaving a buried air-cavity. The exact compliance of the final structure depends on the size and shape of the air-cavity, and the elastic properties of the overcoat material. In-plane compliance (x-y

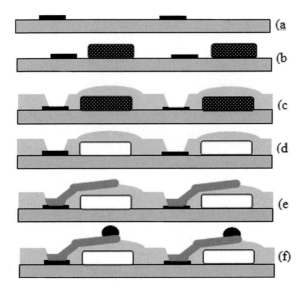

Fig. 3.7 Build-up process of embedded air-gaps

compliance) was obtained by releasing the metal lines off the polymer surface. The deflection was measured to be greater than 30 μm. While the leads are x-y-z axis compliant, they are short in length and thus exhibit minimal parasitics from DC to 45 GHz. The calculated resistance and inductance of the leads are less than 25 mΩ and 0.1 nH, respectively. Low electrical parasitics are desirable at both low and high frequencies for efficient conductive coupling of power, low power dissipation in the leads and thus low heat generation by the package. The microwave characteristics of SoL were measured at wafer-level using a two-port network analyzer with 150 μm coplanar ground-signal-ground (GSG) probes. To characterize the compliant interconnects, 15 μm thick Au leads were fabricated on a 15 μm thick polymer film. The return-loss and insertion-loss of the GSG lead interconnection were measured to be less than 20 dB and 0.2 dB, respectively, at 45 GHz [39]. In comparison, the insertion losses before and after the addition of underfill within a flip-chip package mounted on an alumina substrate with 75 μm wide × 150 μm height bumps interconnected by 600 μm long 50-Ω coplanar waveguides were found to be 0.6 dB and 1.8 dB, respectively, at 40 GHz [38, 39].

A variety of complex, three-dimensional structures have been produced by using standard lithography and metal deposition [40, 41, 42, 43, 44]. Figure 3.8 shows a one-turn helix structure produced by standard metallization and photolithographic techniques. The dimensions of the helix can be varied over wide ranges. The radius of the beam is critical to the compliance and electrical properties of the beam.

The structure has been simplified to eliminate one of the half-turn structures, shown in Fig. 3.8 [43]. The electrical properties of the compliant I/O are an

3 Advanced Chip-to-Substrate Connections

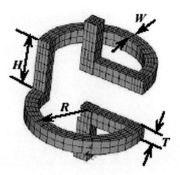

Fig. 3.8 Structure of a G-helix

essential element of the design. Of particular interest for helix structures is the self-inductance, which ranged from 0.03 to 0.15 nH for the designs studied [43].

A bimetallic beam structure has been designed and studied [44]. The bimetallic internal stress gradient in the beam creates an upward bending of the beam upon release from the surface. Solder attachment metal on the released end of the beam provides for attachment to the next layer of packaging.

Finally, more complex beam shapes have been investigated. Liao et al. have studied released beams on the chip [45]. The beams are anchored on both ends (on the chip) and released in the middle. The released portion of the beam can take a variety of forms, including 'S' shaped elements. A solder attachment bump can be placed in the middle of the release beam. This provides extensive design freedom.

3.3 Improved Mechanical Performance Solder Capped Structures

A current modification to solder type flip-chip connections that has received significant attention for research and development is the solder capped structure. Moving the more fragile solder material off the chip and/or package surface and instead placing a tougher material such as a copper bump or pillar underneath the solder connection creates a more thermo-mechanically reliable structure. Since the highest stress during thermal loading is now experienced in the higher yield stress material, the number of cycles to failure can potentially be significantly increased. Additionally, since the aspect ratio of the connection is no longer directly related to the diameter of the solder ball, finer pitch connections can be made at more practical stand-off distances. Increased allowable separation between chip and substrate is important for both increasing reliability of the connection itself and allowing for underfill to be more easily flowed in between the chip and substrate.

Several researchers have pursued structures such as the copper pillar with solder cap [46, 47, 48]. This structure, as demonstrated in Fig. 3.9, has both merits and problems. Copper has excellent properties for interconnects such as low electrical resistance, high allowable current density, and yield stress.

Fig. 3.9 High aspect ratio copper pillars capped with solder

However, typical tin based solders capping these structures lead to brittle intermetallics. Therefore, underfill is still necessary to ensure thermo-mechanical reliability. The use of copper bumps for this application has promise as an interim solution until an effective solder-free solution can be developed. The true capability of the pillar interconnect can not be realized with the solder cap since it will still be limited both electrically and mechanically by the properties of the solder portion. Other metals such as nickel have also been demonstrated for the solder-capped pillar approach [49]. The materials available to create capped solid metal pillar connections are highly variable and most likely depend on cost and application requirements. Both leaded and lead-free solders have been used, in addition to other less common solders such as tin-gold. Choosing what type of solder is again most likely dictated by needs for specific applications such as allowable reflow temperatures and the solder deposition method.

Other work has been pursued to attempt to make the mechanically rigid, high yield stress, solid metal pillar more compliant [50]. By applying a metal conductor to the exterior of a polymer pillar, the compliance of the interconnect structure could be improved. Aggarwal et al. showed that by using a compliant polyimide core material with a copper shell the compliance increased versus the use of a solid copper pillar of the same diameter (as shown in Fig. 3.10). Due to the skin effect at high operating frequency, the copper shell does not have to be very thick since the signal does not penetrate deep into the metal. For example, copper has a skin depth of approximately 1.2 μm at 3 GHz operating frequency. This portion of metal is primarily involved in the transmission of the electrical signal. As operating frequency increases the skin depth continues to decrease

3 Advanced Chip-to-Substrate Connections

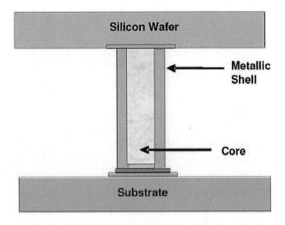

Fig. 3.10 Metal-clad polymer pillar with solder cap to improve compliance

which works in favor of these structures since the exterior copper thickness limits the mechanical compliance.

Another alternative to creating the solder bond at the end of the interconnect structure is to place the solder between two bumps or pillars. By removing the solder connection from both the chip and substrate surfaces, the highest stresses in the structure are entirely experienced in the high yield stress, higher elastic modulus material of the bump and not the more fragile solder or brittle intermetallics. It is well known that tin and copper form predominantly two intermetallics materials: Cu_3Sn and Cu_5Sn_6 that are brittle and have poor mechanical reliability.

Huffman et al. have shown that pure tin placed between two copper bumps can be used to create a solid pillar interconnect (as shown in Fig. 3.11) [51]. It was shown that 2 μm thick tin layers would form homogeneous Cu_3Sn intermetallics in the bonded region, but were very sensitive to non-planarity between

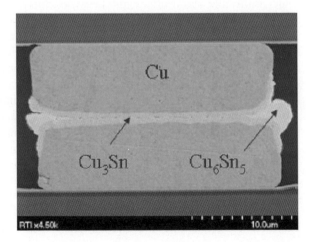

Fig. 3.11 Tin bonded between two copper bumps

chip and substrate. Small variations in the vertical separation between bumps could prevent forming of a solid bond. Thicker tin layers, 3.7–6 µm, were used to alleviate the sensitivity to non-planarity and lower bonding pressures were used to prevent squeeze out of the tin. Unfortunately, this resulted in bond intermetallics layers consisting of a tri-layer structure of $Cu_3Sn/Cu_6Sn_5/Cu_3Sn$. Higher pressures could still achieve the homogeneous Cu_3Sn intermetallic but led to significant squeeze out of tin. The results also showed that, based on die shear testing, the homogeneous bonded Cu_3Sn intermetallic led to higher shear strength. Overall the process was found to be highly sensitive to non-planarity between chip and substrate which would not be practical for production assembly since organic boards are not uniformly planar with respect to a silicon die. Iwasaki et al. have also demonstrated a technique for bonding between two bump structures that also shows need for highly planar surfaces [52]. These techniques that require such strict planarity between bonding surfaces will mostly be restricted to bonding between silicon die for applications such as 3-D integration.

3.4 Solder-Free Chip-to-Substrate Interconnects

Current industry standard flip-chip interconnects employ the use of solder balls to make electrical connection between the chip and substrate. Solder has many weaknesses for this application, and they are becoming more important as the required interconnect size continues to shrink to meet the I/O demands of modern high performance microchips. The International Technology Roadmap for Semiconductors forecasts minimum pitch for area array interconnects to shrink to 120 µm, 100 µm, and 85 µm, in 2010, 2015, and 2020 respectively [5]. Such fine pitch interconnect needs are challenging for manufacturability and reliability. Since the solder connections are formed with roughly spherical solder balls, the stand-off height between chip and substrate is limited to approximately the solder ball diameter. Therefore, as pitch decreases the gap between the two surfaces will continue to shrink. The challenges related to this loss in separation are focused into two areas: underfill and thermo-mechanical stress. During the packaging process underfilling between the chip and substrate is done with a silica particle filled epoxy material to alleviate thermal stress/strain on the weak solder connections. Flowing underfill between the surfaces is becoming more and more of a challenge as the distance between the two surfaces continues to shrink [53]. Since thermo-mechanical strain is generated between chip and substrate during operation due to CTE mismatch between Si and FR-4, chip-to-substrate interconnects must withstand this stress. As the chip stand-off distance becomes shorter, creating reliable solder connections is very challenging.

Solder has limitations in many areas including the formation of brittle intermetallics with copper that can compromise thermo-mechanical reliability.

Solder also has low electromigration resistance which is becoming more important as the diameter of interconnects continue to shrink and power requirements increase. Solder has a limited allowable range of current density due to this poor electromigration resistance. Therefore, alternative interconnect strategies are being explored that do not require solder. By removing the tin based material, the electromigration resistance and therefore the allowable current density for interconnects will increase. For example, by creating the connection between chip and substrate entirely with metallic pure copper, the allowable current density increases by approximately 10 times.

Despite the properties that make solder undesirable for future I/O needs, solder still has properties that are very useful for manufacturing. Primarily solder has exceptional capabilities for low temperature processing that is compatible with the low cost organic substrates such as FR-4. Solder also can elongate and flatten during reflow to bond misaligned and non-planar locations between chip and substrate. These properties simplify manufacturing and increase yield, and therefore any practical solution for future interconnect needs should attempt to satisfy these properties that have made solder connections tractable for so many years.

The need for solder-free solutions which satisfy the needs for higher density and improved reliability interconnects at this level will be discussed in the following sections. Several different solutions have been active topics of research. Electrically connecting the chip to the substrate without the use of low temperature melting solder materials is a very challenging problem.

3.4.1 Copper Interconnects

One of the most attractive solder-free solutions is to incorporate only copper into the interconnect structure. This would potentially allow for copper electrical connections made directly from the lowest interconnection level on-chip all the way through the package and printer wiring board. Elimination of any additional metals or other conductive materials would improve reflection loss, impedance mismatch, and generation of inter-metallic compounds. The low resistivity, high electromigration resistance, and low cost of copper make it an attractive solution for all electrical connections. The performance of such an interconnect scheme would help to minimize RC delay for the entire system. Additionally, since copper has much higher yield stress and elastic modulus versus typical tin based solders, mechanically reliable chip-to-substrate connections can be designed. Several methods have been developed and proposed to bond copper for this application. Research into copper bonding for chip-to-substrate interconnects is actively being pursued and no single solution has emerged as the definitive technique. Several of the methods for making copper connections will be discussed.

3.4.1.1 Copper Wafer Bonding

One of the first techniques developed to create a solid copper-to-copper connection between chip and external circuits was reported by Chen et al. [54]. In this process pressure and temperature are applied to encourage metallic bonding to occur between two distinct copper surfaces. Typically two wafers coated with evaporated copper and tantalum diffusion barrier are placed facing one another. Then, the stack is annealed at temperatures typically in the range from 300 to 450°C. During the anneal process pressure, typically in the range of 4000 mBar, is applied to bring the two surfaces into intimate contact across the entire wafer surface and promote bonding.

Under TEM analysis, the process exhibits excellent bond quality between the two evaporated copper films [55, 56]. As can be seen in Fig. 3.12, the two copper surfaces readily bond and form a single copper layer at a bonding temperature of 400°C followed by anneal at 400°C in a nitrogen environment. It has been shown that without sufficient post-bonding anneal in a nitrogen environment the bond is not stable and is easily broken [57]. Therefore, the optimum conditions for bonding were found to reside in the range of 400°C for 30 min followed by anneal also at 400°C for 30 min or 350°C for 30 min followed by annealing at 350°C for 60 min. Both bonding process conditions were demonstrated to generate excellent bonding between the copper films.

To analyze the strength of the bond between the copper surfaces qualitative and quantitative approaches were taken. First, the bonded wafers were subjected to dicing after bonding. Here the post copper bonded wafer pair was cut under a dicing saw which induces stress into the bond. Experiments showed that bonding above 350°C was sufficient for all diced specimens to

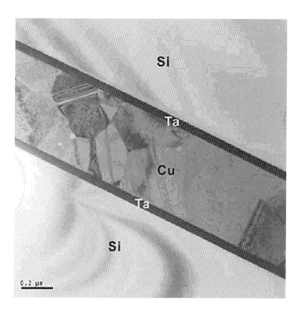

Fig. 3.12 TEM of copper wafer bonding process using 400°C bonding and annealing temperature

survive. After dicing, individual specimens were subjected to another qualitative test, the tape test. Here 3 M Scotch tape was adhered to one of the silicon pieces. By pulling on the tape either the copper to copper bond would fail and the die separate or the tape would be removed from the backside of the die. Again, high temperature annealing, >300°C, led to high number of successful tests without failure.

Something of interest discovered from these tests is that it appears annealing the structure in a nitrogen environment only enhances bonding for samples bonded at 300°C or greater. Below 300°C as the bonding temperature decreases, annealing after bonding appears to significantly degrade the bond and therefore greatly increase the number of failures in both the dicing and tape tests. This result is proposed to be the result of the thermal stress on the copper bond during the nitrogen anneal step [58].

Further quantitative testing of the process via both a normal direction pull test and shear testing confirmed that temperatures above 300°C allow for sufficient thermal activation to generate an effective bond between the copper layers. The pull test showed bond strength as high as $\sim$70 MPa for samples bonded and annealed at 400°C. Shear tests further confirmed that significantly improved bonding occurs in samples bonded and annealed at 400°C [58].

Copper wafer bonding shows excellent promise towards generating continuous copper connections between silicon and silicon [59], or possibly silicon and ceramic packages, but it will not be possible to utilize for organic substrate packaging due to the high temperature requirements. While making high quality connections for vertically integrated systems such as stacked silicon die is important for the future, organic substrate based packaging is of much more mainstream importance.

3.4.1.2 Surface Activated Bonding

While copper wafer bonding mentioned above generated excellent bonding between copper surfaces, it does have limitations. One important limitation is that of temperature tolerance. Since current flip-chip type packages utilize organic substrates, such as FR-4, temperature excursions of 400°C are not possible. Typical organic printed circuit board materials like FR-4 begin to thermally decompose and degrade when held at temperatures exceeding $\sim$250°C. This decomposition temperature varies for specific organic substrates but 250°C is a good metric. In the Surface Activated Bonding (SAB) technique, developed by Kim et al. at the University of Tokyo, bonding between copper surfaces is realized at room temperature [60].

Unlike the wafer bonding method, no elevation in temperature is required to merge the two copper surfaces and therefore create the interconnection between chip and substrate. For SAB two copper surfaces are placed into a high vacuum environment, typically in the range of 10^{-5}–10^{-7} Torr. Once under vacuum the surfaces are cleaned via an argon (Ar) ion beam with typical energy in the range of 40–100 eV. The Ar ion beam is used to remove surface oxide and any other chemical contamination on the copper. In fact the energy is more than sufficient to sputter the copper surface itself. Therefore, the cleaning in fact removes a small portion of the copper to fully decontaminate the surface. Since the sample is never removed from the UHV environment, there is little if any re-oxidation of the surface that takes place. This fact is verified by Kim et al. by Auger electron spectroscopy (AES) of the copper surfaces before and after the ion beam activation process. Furthermore, the authors comment that bonding is always performed within 60 s of the activation process to prevent any effects from residual gases oxidizing the surface.

Once the surfaces are fully cleaned and activated, they are brought into contact and external pressure is applied to force intimate contact across the surface and encourage bonding between the two copper regions. It is critical that the entire process takes place in the vacuum chamber; therefore, the authors developed a custom system that allowed for ion beam cleaning and flip-chip bonding to take place without the need to vent or open the chamber. During bonding the externally applied pressure serves as the only driving force to merge the copper surfaces. Typical applied pressures are in the range of 6–15 MPa. There is no temperature elevation during the bonding process.

The fact that temperature is not changed is important for flip-chip bonding for two reasons. First, the low temperature processing allows for no degradation of the organic substrates and continued used of these cost effective materials is possible. Second, Shigetou et al. comment in the more recent work on SAB that maintaining alignment between ultra-fine pitch copper pads would be nearly impossible with elevated temperatures due to the thermal strain. Shigetou et al. demonstrated bonding of "bumpless" copper interconnects of 10 μm pitch with ±1 μm accuracy (as shown in Fig. 3.13) [61]. This fine structure allowed for connecting ~100,000 I/O which is remarkable.

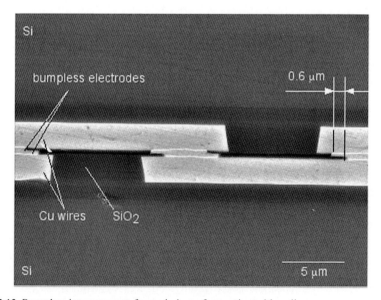

Fig. 3.13 Bumpless interconnects formed via surface activated bonding

TEM analysis confirmed that the two copper regions do in fact merge and exhibit excellent bonding despite having no elevated temperature driving force. Kim et al. proposed that the excellent bonding observed is due to the activation process. Since essentially no oxide or other chemical contamination is present on the copper surfaces they readily bond to one another. Additionally, another critical factor in this technique is that of surface roughness. In the initial work by Kim et al., surface roughness was measured by Atomic Force Microscopy (AFM) before and after the activation process [60]. It was found that the activation process had no effect on surface roughness and therefore the sputtered copper film maintained surface roughness of approximately 1.8 nm. Intimate contact between the copper regions is achieved readily owing to their smooth condition. In the later work of Shigetou et al., the copper surfaces bonded were generated via electroplating and chemical mechanical polishing (CMP). Here, the surfaces were found to have 1.2 nm rms roughness via AFM measurement [61]. Having such smooth surfaces is critical to the SAB process since the copper regions on both chip and substrate must be brought into contact during bonding. Since both silicon wafer and organic boards are not flat, they must be able to withstand sufficient external pressure to be made flat for contact during bonding.

To characterize the quality of bond made during SAB, Kim et al. performed a fully bonded wafer dicing test [60]. Similar to the test mentioned above for copper wafer bonding, two eight inch wafers were bonded via SAB. Then, the bonded pair was diced into 10 mm × 10 mm sections. Results showed that only a few edge pieces were not sufficiently bonded to survive the dicing saw stresses. Additionally, 10 mm × 10 mm chips that survived dicing were subjected to normal tensile pull testing to attempt to quantify the bond strength. However, the observations showed that samples failed in other areas and not the copper to copper bonded region. The maximum observed tensile strength was 6.47 MPa but was not indicative of the copper bond strength.

Later, Shigetou et al. demonstrated the capability of the technique for bonding "bumpless" copper interconnects. Here individual copper pads were patterned and bonded with SAB. The method demonstrated bonding and successful electrical testing. To demonstrate the capabilities of the technique to generate off-chip interconnections that satisfy the density of global wiring on-chip, 10 μm pitch was used. At such a fine pitch one of the biggest challenges for the experiments was proper alignment between 3 and 5 μm diameter copper pads. In fact, from the electrical testing experiments the authors mention that misalignment is most likely the largest contributor to contact resistance being greater than expected for a true bulk copper-like connection. With such fine dimensions, even misalignment of ~1 μm could lead to large increases in contact resistance for 3 μm connections. Yet, the demonstration of such fine pitch and high performance electrical connections is promising for future needs.

3.4.1.3 All-Copper Chip-to-Substrate Pillar Interconnects

Electroless or autocatalytic plating of copper has been used to metallize organic substrates and printed circuit boards in the electronics industry for many years [62]. Additionally due to the introduction of the dual-damascene process by IBM in 1998, copper interconnects have also been introduced into on-chip circuitry [63]. One area where copper and copper electrochemistry have not been applied until recently is chip-to-substrate interconnects. He et al. have reported a new technique that utilizes the electroless plating of copper to create chip-to-substrate interconnects [64, 65].

In the All-Copper process copper pillars are electroplated on both the chip and substrate. Then, the two pillars are flip-chip aligned and temporarily held in alignment at a fixed distance of separation. The system is then passed into an electroless copper plating bath. Copper is deposited onto both pillar surfaces until they are in intimate contact. Finally, the entire structure is annealed at temperatures ranging from 180 to 400°C under nitrogen ambient conditions. Figure 3.14 shows an image of an All-Copper interconnect after plating and annealing. The lowest temperature reported for successful bonding using this process was 180°C for 1 h and exhibited maximum allowable shear stress of approximately 165 MPa. This allowable stress before breaking is promising since the yield stress of bulk electrodeposited copper is on the order of 225 MPa. While the allowable stress before failure of the bond was found to be a function of temperature, the failures occurred at the pillar-to-substrate interface and not in the copper-to-copper bonded region. Analysis of the quality of the copper-to-copper bonded region was shown by optical microscope analysis of

Fig. 3.14 Two copper pillars bonded by the electroless plating and annealing

cross-sections. While the bonded regions showed areas that were continuous copper with no obvious interfaces or open areas, there were voids entrapped during the plating process. Optimization of the electroless plating process is necessary to create a void free all-copper structure with this method.

One of the most promising aspects of the All-Copper process is that electroless plating to join the two pillar surfaces can overcome planar and vertical misalignment. Electroless plating can fill variations of separations between pillars due to non-planarity between the chip and substrate unlike copper wafer bonding or SAB which require flat surfaces to create the copper connection. Also, since the electroless process can effectively join pillars which are misaligned even to a great extent, planar misalignment is not such an issue as in most other flip-chip type configurations. In fact, the electroless plating and annealing process was shown to create successful bonding with misalignments greater than the diameter of the structures bonded. No other solder-free bonding process has exhibited such dexterity for all aspects of misalignment between the chip and substrate.

In addition to the previously mentioned advantages of the copper pillar interconnection scheme another important aspect is that of thermo-mechanical reliability. Since high aspect ratio structures are possible with this method, stand-off between the chip and substrate can be increased giving compliance and reliability to the connection. By using high aspect ratio copper pillars, compliant chip-to-substrate connections may be realized without the need for solder materials or complex fabrication schemes to generate mechanically flexible interconnect structures.

3.4.2 Electroplated Copper Column Arrays

Another method to create metallic bonding without the use of solder is ultrasonic or thermo-sonic bonding. For this method ultrasonic energy and/or thermal energy is incorporated into the desired bond region to locally heat due to friction and generate a solid metallic bond. Gao et al. have shown this method to work for bonding electroplated copper columns with gold caps onto aluminum metallization pads [66]. The gold cap readily bonds to the aluminum pad under thermo-sonic bonding conditions giving a continuous metallic connection without the need for solder materials. While thermo-sonic bonding for gold metal has been used for wire bonding applications, it has not been adopted as a flip-chip type packaging solution. Thermo-sonic bonding of gold has been demonstrated previously using gold stud bumps by others since 1993 [67]. However, the gold stud bump approach has not been adopted.

Therefore, Gao et al. developed an alternative structure that utilizes the thermo-sonic bonding capabilities of gold without the need for such a serial and time consuming process as in gold stud bonding. By electroplating copper columns through a resist mold and then also capping the columns with gold to facilitate the thermo-sonic process, higher throughput and cost effectiveness can be realized. Electroplated copper columns are covered by an electroplated diffusion barrier of nickel and then the thermo-sonic bonding gold cap. One advantage of this process is that no chip side processing is necessary prior to bonding itself. Once the columns are gold capped, they are flip-chip aligned to the aluminum pads on the chip and then bonded under force, ultra-sonic action, and elevated temperature (as shown in Fig. 3.15). The stage temperature was chosen to be fixed at 200°C for all of the work reported. For optimization of the thermo-sonic process, ultrasonic power, bonding force, and time were used as variables and maximum shear stress of the bonded pillars as the desired output.

By using a Box-Behnken design of experiments approach, a surface response for the three variable system was performed and the bonding process optimized. The optimized process used ultra-sonic power between 8 and 16 W and time of 100–300 ms. The pressure applied was 0.012–0.013 g\$\mu$m^2 bonded. Finally the arrays were bonded to quartz test substrates and put under a few basic thermo-mechanical stability tests. While these tests served as an initial evaluation of the system under stressed conditions and without underfill, it did not reflect the more important case for flip-chip packaging with FR-4 type substrates since the CTE of quartz is much closer to that of silicon. Therefore, further testing of the system with higher CTE boards must be done in the future to accurately assess the thermo-sonic column system as a feasible solder-free solution for flip-chip packaging.

3.4.3 Compliant Gold Bump Interconnects

Copper wafer bonding and Surface Activated Bonding discussed above are solid-state bonding techniques to create connections between copper surfaces. Watanabe et al. have developed a novel process utilizing thermo-compression similar to the copper wafer bonding process that uses gold as the interconnect

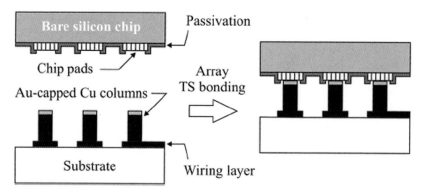

Fig. 3.15 Thermo-sonic bonding process for Cu columns

3 Advanced Chip-to-Substrate Connections

material [68, 69]. By taking advantage of the mechanically soft character of gold and also its exceptional capability to electroplate unusual geometry, the compliant gold bump interconnect was developed. By using an undercut photoresist pattern as the electroplating mold cone shaped gold bumps can be created. Due to the extremely small cross-section of the tip of the cone, the material can be made to yield and flatten during the thermo-compressive bonding cycle. Figure 3.16 shows an image of the cone shaped Au bumps and the process steps for bonding the cone shaped gold bumps to the plated gold pads. The advantage that the cone bump has versus simply bonding flat surfaces is that is it not necessary for the surfaces of chip and substrate to be perfectly parallel. Each individual bump can flatten as much as needed to compensate for the non-planarity between chip and substrate. Additionally, the sharp tipped gold cone also allows for underfill material to be dispensed directly onto the substrate prior to bonding. The sharp tip of the cone will then penetrate the underfill film and make metallic contact with the pad perfectly excluding the dielectric material. As the cone plastically deforms under pressure, the underfill continues to be forced out of the interconnect geometry. This allows for simple fabrication technique with high potential for thermo-mechanical reliability of the bonded joint.

Watanabe et al. utilized both pressure and temperature to generate a solid bond between the Au compliant structures and the underlying pads. The reported values were for chip side heating up to 300°C and substrate side heating to 100°C. Force applied to flatten the compliant bumps was reported to be 0.5 kgf per bump bonded. The structures showed confirmed electrical connection by daisy chain testing. The reported test structure consisted of 10 μm base diameter cone bumps with pitch of 20 μm. For a 2 mm × 2 mm test vehicle these dimensions lead to 10,000 I/O on a single test chip. This I/O

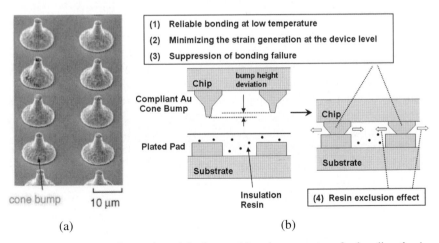

Fig. 3.16 SEM image of cone shaped Au bumps (**a**) and process steps for bonding the Au bumps to Au pads utilizing compliant bump tip

density is remarkable compared to current solder interconnect methods. However, the use of gold as the interconnect material is not the most desirable alternative material. Since gold is significantly more expensive than copper or solder, it will be less likely to be used for manufacture due to cost of materials alone. In addition electroplating of gold is much more demanding for waste disposal versus typical electroplating used for copper.

3.4.4 Electroless NiB Interconnects

Electroless copper plating for the creation of chip-to-substrate area array interconnects has been mentioned previously. Recently, a new approach to creating peripheral array interconnects utilizing an electroless plating technique has been reported by Yokoshima et al.. In this technique NiB is electrolessly deposited between copper pads to generate an electrical connection [70]. One of the most exciting aspects of this work is that the plating is done without need for any patterned material or seed layer to direct the plating process. Taking advantage of previously reported studies on "extraneous" deposition causing bridging between metal pads during electroless plating, this technique instead utilizes "extraneous" deposition to directly grow a conductive link between chip and substrate pads. Many researchers have analyzed problems dealing with plating creating bridged links between adjacent structures on surfaces [71, 72]. While most of the previous work was done to minimize or eliminate the effects of plating that generated these bridged connections, this technique attempts to harness this ability and use it to create desired connections.

Yamaji et al. have shown that by carefully choosing the correct pitch between interconnects and also the vertical spacing between the pads to be bonded, successful connections can be achieved [73]. For the demonstrated 5 μm wide copper pad test system, pitches less than 20 μm led to plating between neighboring pads and therefore undesirable electrical connections. However, when the pitch was 20 μm or greater plating did not take place between adjacent pads on the surface. Also, it was found that the pad-to-pad distance to successfully join vertically aligned pads was 5 μm or less. At these vertical separations NiB successfully deposited on the insulating material and creating a measurable electrical connection between the pads (as shown in Fig. 3.17).

While the exact mechanisms of the "extraneous" electroless deposition onto the insulating material are not known, it appears to be effective in joining pads to make electrical connections. One critical drawback of this process is it does not appear to be able to create area array connections due to the geometry for the proposed fabrication scheme. Still creating such fine pitch peripheral array connections effectively and efficiently without the need for solder is important. Future research may elucidate a path for this technique to be applied to area array type connections.

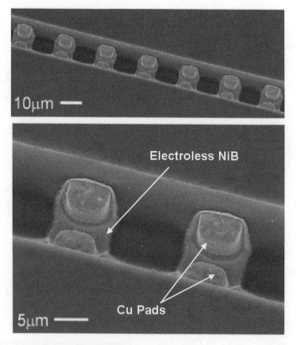

Fig. 3.17 Electroless NiB bonding between Cu pads

3.5 Distant Future Needs and Solutions for Chip-to-Substrate Connections

3.5.1 Ultra-high Off-Chip Frequency and High Bandwidth Operation

The ITRS projects that in the future off-chip operating frequency will increase dramatically. By the year 2020, the projections are for operating frequency to reach 72.4 GHz. To operate at such high frequency, the methods of interconnecting the chip to substrate must be carefully considered. Basic pin-type connections used today will not perform as well and may need to be replaced with more electrically high performance connections. There have been many methods proposed to increase performance of the chip-to-substrate connection for high frequency.

3.5.1.1 Coaxial Interconnects

Wu et al. have proposed a coaxial chip-to-substrate connection that showed promising high frequency operation up to 80 GHz based on simulations [74]. Utilizing build-up processing, a 'C' shaped ground connector was formed in conjunction with a center signal conductor on both the chip and substrate surfaces. Then, the two pieces were aligned and bonded to form the coaxial

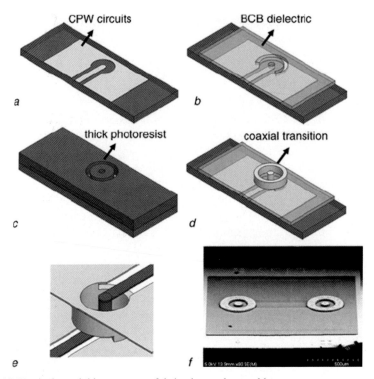

Fig. 3.18 Vertical coaxial interconnect fabrication and assembly

type connection. Figure 3.18 schematically shows the vertical coaxial interconnect fabrication and assembly processes. Experimental measurement of the high frequency performance was made by attaching the test structure to coplanar waveguides on the substrate and chip surfaces. Wu et al. have experimentally fabricated and tested the proposed vertical coaxial structures [75]. Testing the return and insertion loss S-parameters from 0 to 70 GHz was performed. Based on the results, the simulations characterized the interconnect scheme effectively. The loss caused by adding underfill to the system was greater as expected. However, the coaxial connections still showed low return and insertion loss of 13.7 dB and 0.9 dB respectively at 60 GHz. Coaxial type connections such as these could be promising for operations at such high frequencies as projected for the future by the ITRS.

3.5.1.2 Electrical and Optical Interconnects

The ITRS projects that by the 18-nm generation node high-performance chips will dissipate 200 W, require 200 A of supply current, and have clock frequencies greater that 70 GHz. These high power dissipation requirements mean that minimizing the power distribution network IR drop and noise are critical. In addition, the signaling network will be limited by increased losses from impedance mismatch, cross-talk, and organic substrates. One solution to such challenges is the incorporation of optical communications between the chip and substrate [76, 77].

By combining the high bandwidth optical I/O with typical electrical I/O connections, Bakir et al. have demonstrated a current CMOS fabrication compatible hybrid interconnect structure [78]. One of the proposed structures comprises optical connections made by polymer pins serving as vertical optical connections, with traditional solder balls providing electrical connections. By using the polymer pins for optical connections, versus the typical free-space optical I/O, the system is fully compatible with underfill dispensing for thermomechanical reliability. Bakir et al. also proposed a more integrated structure that is termed "dual-mode" pin connections [78]. For this structure the polymer pins are metallized on the exterior surfaces (as shown in Fig. 3.19). By having an electrically conductive path incorporated onto the polymer pillars, both electrical and optical connections can be made by each structure. This could potentially lead to high I/O density, but having the metal film coating the exterior of the polymer structure will limit its mechanical compliance. Therefore, there is a trade-off to be made between the electrical performance and mechanical compliance. Bakir et al. showed that compliance increases with

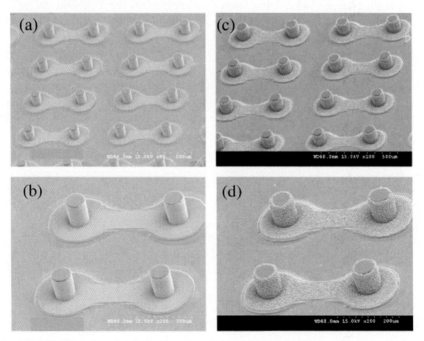

Fig. 3.19 SEM images of dual-mode metal clad polymer pins with (**a** and **b**) Au clad and (**c** and **d**) solder plated onto structures

reduced metal thickness, and that electrical resistance increases with metal thickness [78]. For practical use the dual-mode pin would have to be carefully designed based on the application specific electrical and mechanical performance needs.

3.5.2 Microfluidic Interconnects for Thermal Management

Continued power usage increases by silicon die already strain the thermal management solutions currently available. According to the ITRS, future chips will have power densities of 108 W/cm^2 by the year 2018. Such high power dissipation may require thermal management solutions beyond the conventional forced-air convection heat sinks. One alternative cooling technology that utilizes chip-to-substrate interconnects is that of on-chip microfluidic cooling. Microfluidic cooling for IC processors has been in research since the early 1980s when Tuckerman et al. suggested that it would allow heat flux as high as 790 W/cm^2 to be removed [79]. However, one of the greatest drawbacks of most directly incorporated cooling channels is that of automated bonding for the fluidic I/O connections. Since conventional microchannel fabrication uses direct wafer bonding that needs high temperature and/or high voltage to function, these methods are not compatible with standard post-BEOL CMOS chips. Dang et al. proposed that using microfluidic pipes on the front side of the die could provide an automated process to realize microfluidic cooling (Fig. 3.20) [80]. In addition to using microfluidic pipe I/O, the structure also utilized a polymer over-coating on the backside of the die to provide the water-tight seal for the microchannels without the need for direct wafer bonding (as shown in Fig. 3.21) [81]. Using this novel structure for the integrated microfluidic cooling system, low pressure drop and high heat removal capability was demonstrated. One of the most attractive reasons to move towards microfluidic cooling is that

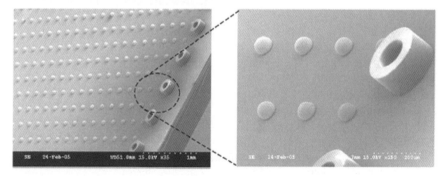

Fig. 3.20 Incorporation of microfluidic pipes with electrical I/O on chip frontside for flip-chip attachment

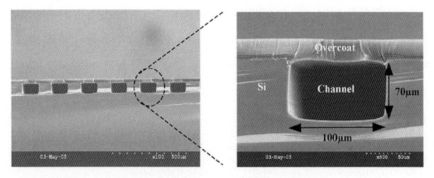

Fig. 3.21 Die backside microchannel array for microfluidic cooling

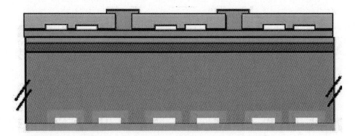

Fig. 3.22 Frontside and backside microfluidic cooling design

it could allow for increased density of 3-D stacking of silicon chips and large improvements in thermal management for these systems such as System-in-a-Package type devices.

Another microfluidic cooling study performed by Zhao et al. showed that very high heat removal rates can be realized by incorporating microchannels onto both the front and backside of the silicon die [82]. Figure 3.22 shows a schematic design of the microchannels. Additionally, they showed that depositing a small amount of copper on the interior of the channel structures could greatly enhance the heat transfer and therefore heat removal rate [82]. Using the front and backside channel architecture, 200 W/cm^2 removal rate was demonstrated.

Reference

1. S. J. Horowitz, J. J. Felten, D. J. Gerry, IEEE Transactions on Components, Hybrids, and Manufacturing Technology, vol. CHMT-2, 4, (1979) pp. 460–466
2. S. L. Khoury, D. J. Burkhard, D. P. Galloway, T. A. Scharr, Proceedings of Electronic Components and Technology Conference, vol. 1, (1990) 768–476
3. G. Pascariu, P. Cronin, D. Crowley, Proceedings of Electronics Manufacturing Technology Symposium (2003) 423–426

4. P. Wolflick, K. Feldmann, Proceedings of Electronics Manufacturing Technology Symposium (2002) 27–34
5. International Technology Roadmap for Semiconductors 2006 Update: Assembly and Packaging
6. A. Muramatsu, M. Hashimoto, H. Onodera, IEICE Transactions on Fundamentals of Electronics, Communications and Computer Science, **88**, 12 (2005) 3564–3572
7. K. Shakeri, M. Bakir, J. D. Meindl, Proceedings of the IEEE SOC Conference (2004) 78–81
8. W. D. Becker, J. Eckhardt, R. W. Frech, G. A. Katopis, E. Klink, M. F. McAllister, T. G. McNamara, P. Muench, S. R. Richter, H. H. Smith, IEEE Transactions on Components, Packaging, and Manufacturing Technology, **21**, 2 (1998) 157–163
9. O. P. Mandhana, IEEE Transactions on Advanced Packaging, **27**, 1 (2004) 107–120
10. G. A. Katopis, Proceedings of IEEE, **73**, 9 (1985) 1405–1415
11. C. T. Chen, J. Zhao, Q. Chen, Proceedings of Electronic Components and Technology Conference (2001) 1102–1106
12. F. W. Grover, Inductance Calculations, Working Formulas and Tables, New York: Dover (1962)
13. G. Troster, Proceedings of Design, Automation, and Test in Europe Conference and Exhibition (1999) 423–424
14. E. C. Jordan, K. G. Balmain, Electromagnetic Waves and Radiating Systems, Second Edition, Prentice-Hall, Upper Saddle River, NJ (2003)
15. J. D. Kraus, Electromagnetics, Fourth Edition, McGraw-Hill, Hightstown, NJ (1992)
16. G. A. Rinne, P. D. Franzon, http://www.unitive.com/casestudies/pdfs/par.pdf, (accessed January 20, 2007)
17. D. M. Pozar, Microwave Engineering, Second edition, John Wiley & Sons, New York, NY (1998)
18. A. He, T. Osborn, S. A. B. Allen, P. A. Kohl, Journal of the Electrochemical Society, *Submitted* September 2007
19. R. R. Tummala, Fundamentals of Microsystems Packaging, McGraw-Hill (2001)
20. Z. Zhang, C. P. Wong, IEEE Transactions on Advanced Packaging, **27**, 3 (2004) 515–524
21. C. Hillman, K. Rogers, A. Dasgupta, M. Pecht, R. Dusek, B. Lorence, Circuit World, **25**, 3 (1999) 28–38
22. Z. Zhang, S. K. Sitaraman, C. P. Wong, IEEE Transactions on Electronic Packaging Manufacturing, **27**, 1 (2004) 86–93
23. C. J. Zhai, Sidharth, R. Blish II, IEEE Transactions on Device and Materials Reliability, **3**, 4 (2003) 207–212
24. L. L. Mercado, V. Sarihan, R. Fiorenzo, IEEE Transactions on Advanced Packaging, **27**, 1 (2004) 151–157
25. www.me.binghamton.edu/O.M.R.L/Facilities-2D-3DANSYS.htm (accessed February 13, 2007).
26. A. Perkins, S. K. Sitaraman, Proceedings of the Electronic Components and Technology Conference (2003) 422–430
27. A. Yeo, C. Lee, J. H. L. Pang, Proceedings of Thermal and Mechanical Simulation and Experiments in Micro-Electronics and Micro-Systems Conference (2004) 549–555
28. X. Fan, M. Pei, P. K. Bhatti, Proceedings of the Electronic Components and Technology Conference (2006) 972–980
29. G. Wang, P. S. Ho, S. Groothuis, Microelectronics Reliability, 45 (2002) 1079–1093
30. K. Tunga, K. Kacker, R. V. Pucha, S. K. Sitaraman, Proceedings of the Electronic Components and Technology Conference (2004) 1579–1585
31. F. C. Classe, S. K. Sitaraman, Proceedings of the Electronics Packaging Technology Conference (2004) 82–89
32. B. A. Zahn, Proceedings of the International Electronics Manufacturing Technology Symposium (2002) 274–284

33. P. Garrou, Semi Chip Scale International '99, page D-1 (1999)
34. www.tessera.com
35. A. Longford, D. James, Presentation in Advance Packaging Conference, Semicon Europa, April (2006)
36. M. Bakir, H. Reed, H. Thacker, C. Patel, P. Kohl, K. Martin, J. Meindl, IEEE Transactions on Electron Devices, **50**, 10 (2003) 2039–2048
37. B. Dang, M. Bakir, C. Patel, H. Thacker, J. Meindl, Journal of Microelectromechanical Systems, **15**, 5 (2006) 523–530
38. D. Bhusari, H. Reed, M. Wedlake, A. Padovani, S. A. Bidstrup-Allen, P. A. Kohl, Journal of Microelectromechanical Systems, **10**, 3 (2001) 400–408
39. M. S. Bakir, H. A. Reed, A. V. Mule, P. A. Kohl, K. P. Martin, J. D. Meindl, IEEE Custom Integrated Circuits Conference (2002)
40. Q. Zhu, L. Ma, S. K. Sitaraman, Proceedings of International Conference on Thermal, Mechanics and Thermo-mechanical Phenomena in Electronic Systems (2002)
41. Q. Zhu, L. Ma, and S. K. Sitaraman, Proceedings of InterPack, The Pacific Rim International, Intersociety, Electronic Packaging Technical/Business Conference & Exhibition (2001)
42. Q. Zhu, L. Ma, and S. Sitaraman, Journal of Electronic Packaging, **126**, 2 (2004) 237–246
43. K. Kacker, T. Sokol, S. K. Sitaraman, Proceedings of the Electronic Components Technology Conference (2007) 1678–1684
44. P. Arunasalam, H. Ackler, B. Sammakia, Proceedings of the Electronics Components and Technology Conference (2006) 1147–1153
45. E.B. Liao, A.A.O. Tay, S.S.T. Ang, H.H. Fend, R. Nagarajan, V. Kripesh, R. Kumar, and M.K. Iyer, Proceedings of the Electronic Components and Technology Conference (2006) 1246–1250
46. T. Wang, F. Tung, L. Foo, V. Dutta, Proceedings of the Electronic Components and Technology Conference (2001) 945–949
47. V.S. Rao, A.A.O. Tay, V. Kripesh, C.T. Lim, S.W. Yoon, Proceedings of the Electronic Packaging Technology Conference (2004) 444–449
48. R.R. Tummala, P.M. Raj, A. Aggarwal, G. Mehrotra, S.W. Koh, S. Bansal, Proceedings of the Electronic Components and Technology Conference (2006) 102–111
49. A. Aggarwal, P.M. Raj, B.W. Lee, M.J. Yim, A. Tambawala, M. Iyer, M. Swaminathan, C.P. Wong, R. Tummala, Proceedings of the Electronic Components and Technology Conference (2007) 905–913
50. A.O. Aggarwal, P.M. Raj, R.R. Tummala, IEEE Transactions on Advanced Packaging, **30**, 3 (2007) 384–392
51. A. Huffman, M. Lueck, C. Bower, D. Temple, Proceedings of the Electronic Components Technology Conference (2007) 1589–1596
52. T. Iwasaki, M. Watanabe, S. Baba, Y. Hatanaka, S. Idaka, Y. Yokoyama, M. Kimura, Proceedings of the Electronic Components Technology Conference (2006) 1216–1222
53. W.B. Young, W.L. Yang, IEEE Transactions on Advanced Packaging, **29**, 3 (2006) 647–653
54. A. Fan, A. Rahman, R. Reif, Electrochemical and Solid-State Letters, **2**, 10, (1999) 534–536
55. K. N. Chen, A. Fan, C. S. Tan, R. Reif, Journal of Electronic Materials, **35**, 2 (2006) 230–234
56. K. N. Chen, C. S. Tan, A. Fan, R. Reif, Journal of Electronic Materials, **34**, 12 (2005) 1464–1467
57. K. N. Chen, C. S. Tan, A. Fan, R. Reif, Electrochemical and Solid-State Letters, **7**, 1 (2004) G14–G16
58. K. N. Chen, S. M. Chang, L. C. Shen, R. Reif, Journal of Electronic Materials, **35**, 5 (2006) 1082–1086
59. C. S. Tan, R. Reif, Electrochemical and Solid-State Letters, **8**, 6 (2005) G147–G149

60. T. H. Kim, M. M. R. Howlander, T. Itoh, T. Suga, Journal of Vacuum Science and Technology A, **21**, 2 (2003) 449–453
61. A. Shigetou, T. Itoh, M. Matsuo, N. Hayasaka, K. Okumura, T. Suga, IEEE Transactions on Advanced Packaging, **29**, 2 (2006) 218–226
62. M. Schlesinger, M. Paunovic, Modern Electroplating, Fourth Edition, John Wiley and Sons, New York, NY (2000)
63. P. Andricacos, C. Uzoh, J. O. Dukovic, J. Horkans, H. Deligianni, IBM Journal of Research and Development, **42**, 5 (1998) 567–574
64. A. He, M. S. Bakir, S. A. Bidstrup, P. A. Kohl, Proceedings of the Electronic Components and Technology Conference, (2006) 29–34
65. A. He, T. Osborn, S. A. B. Allen, P. A. Kohl, Electrochemical and Solid-State Letters, **9**, 12 (2006)C192–C195
66. S. Gao, A. S. Holmes, IEEE Transactions on Advanced Packaging, **29**, 4 (2006) 725–734
67. S. Y. Kang, T. H. Ju, Y. C. Lee, Proceedings of the Electronic Components Technology Conference (1993) 877–882
68. N. Watanabe, T. Asano, Proceedings of the Electronic Components and Technology Conference (2006) 125–130
69. N. Watanabe, T. Asano, Proceedings of the Electronic Components and Technology Conference (2007) 622–626
70. T. Yokoshima, Y. Yamaji, H. Oosato, Y. Tamura, K. Kikuchi, H. Nakagawa, M. Aoyagi, Electrochemical and Solid-State Letters, **10**, 9 (2007) D92–D94
71. H. Honma, H. Watanabe, and T. Kobayashi, Journal of the Electrochemical Society, **141**, 7 (1994) 1791–1795
72. T. Yokoshima, S. Nakamura, D. Kaneko, T. Osaka, S. Takefusa, A. Tanaka, Journal of the Electrochemical Society, **149**, 8 (2002) C375–C382
73. Y. Yamaji, T. Yokoshima, H. Oosato, N. Igawa, Y. Tamura, K. Kikuchi, H. Nakagawa, M. Aoyagi, Proceedings of the Electronic Components and Technology Conference (2007) 898–904
74. W. C. Wu, R. B. Huang, H. T. Hsu, E. Y. Chang, L. H. Hsu, C. H Huang, Y. C. Hu, M. I. Lai, Proceedings of the APMC (2005)
75. W. C. Wu, E. Y. Chang, C. H. Huang, L. S. Hsu, J. P. Starski, H. Zirath, Electronics Letters, **43**, 17 (2007)
76. J. D. Meindl, J. A. Davis, P. Zarkesh-Ha, C. S. Patel, K. P. Martin, P. A. Kohl, IBM Journal of Research and Development, **46**, 2/3 (2002) 245–263
77. D. A. B. Miller, Proceedings of the IEEE, **88**, 6 (2002) 728–749
78. M. S. Bakir, B. Dang, O. O. A. Ogunsola, R. Sarvari, J. D. Meindl, IEEE Transactions on Advanced Packaging, **54**, 9 (2007) 2426–2437
79. D.B. Tuckerman, R. F. W. Pease, IEEE Electron Device Letters, **2**, 5 (1981) 126–129
80. B. Dang, M. S. Bakir, J. D. Meindl, IEEE Electron Device Letters, **27**, 2 (2006) 117–119
81. B. Dang, P. Joseph, M. S. Bakir, T. Spencer, P. A. Kohl, J. D. Meindl, Proceedings of the International Interconnect Technology Conference (2005) 180–182
82. M. Zhao, Z. R. Huang, Proceedings of the Electronic Components Technology Conference (2007) 2017–2023

Chapter 4
Advanced Wire Bonding Technology: Materials, Methods, and Testing

Harry K. Charles

Abstract Wirebonding is the most dominant form of first-level chip or integration circuit interconnect method used throughout the world-wide electronics industry today. Many trillion of wirebonds are made annually using automated machines. Wirebonding is reliable, flexible, and low cost when compared to other forms of first-level microelectronic interconnection. Failures are typically at the single digit parts per million level or below. As the number of interconnections on the integrated circuit grows with increased functionality, the bonding pads are becoming much smaller and closer together. Similarly rigid inorganic substrates and package structures have given way to their more flexible organic counterparts. Everywhere in the microelectronic industry new applications, materials, and structures are appearing and challenging the performance and, hence, the dominance of wirebonding.

This chapter focuses on the basic wirebonding methods, the materials, and the testing techniques required to produce high quality wirebonds. It addresses the organic substrate problem, stacked chip bonding, and interconnection over extreme temperature ranges. Reliability of the wirebonded interconnect is explored along with testing and control methods designed to improve bond quality. High frequency bonding and the bonding to soft substrates are given special attention. Wire properties are considered along with the changing bond shapes and sizes as the number of chip's inputs and outputs increase. Methods for chip bumping using a wirebonding machine are also presented.

Keywords wirebonding · first-level interconnect · bonding wire · high temperature and high frequency bonding · interconnection of stacked and thinned Ics

---

H.K. Charles (✉)
The Johns Hopkins University, The Johns Hopkins University, Applied Physics Laboratory, 11100 Johns Hopkins Road, Laurel, MD 20723-6099
e-mail: harry.charles@jhuapl.edu

4.1 Introduction

Since the invention of the transistor in 1947 (e.g. [6]), and the birth of the integrated circuit (IC) in 1958 (e.g., [49]), semiconductor device technology has had unparalleled growth in all aspects ranging from device density and complexity to market applications. In fact, IC technology has followed a path (Moore's Law) of doubling its complexity (measured by the number of devices per single piece of silicon or chip), approximately every eighteen months to two years since its birth ([74, 61]). With today's electronic fabrication technology, over a billion transistors can be placed on a single piece of silicon (chip) less than 2 cm^2 in area. Ten to 100 billion devices per chip may be possible over the next few years. Thus, Moore's Law still holds and promises to hold for the foreseeable future, giving rise to an almost endless stream of new electronic devices and products.

With this continued rapid rise in chip density and functionality, the requirement for increased numbers of inputs/outputs (I/O) per chip has also risen dramatically. Individual transistors (the mainstay semiconductor product of the fifties) required only three to four interconnects per device. Early ICs required a dozen or so interconnect wires, but as the IC revolution continued, I/O requirements increased rapidly. Today, ICs routinely have I/O numbers in the hundreds with some chip types exceeding the 1000 mark (application specific integrated circuits, (ASICs) microprocessors, etc.) A few devices even have higher I/O numbers, typically around 1500. The complex, increased functionality of future ICs will have I/O requirements in the multiple thousands. It should still be remembered, however, that systems will still contain a wide variety of chip types – ranging from memory with I/O counts less than one hundred to special-purpose microprocessors, random logic, and ASICs with I/O numbers in the thousands. Thus, an effective interconnection system or method must be able to handle a full range of I/O number and density requirements. Maximum I/O requirement projections for individual chips in different classes of electronic products are shown in Fig. 4.1.

There are two dominate forms of first-level interconnection [21] for integrated circuits: (1) wire bonding and (2) flip-chip attachment as shown schematically in Figs. 4.2, 4.3 and 4.4, respectively. Many other interconnection methods exist to meet special needs or performance requirements. These range from tape automated bonding (TAB) which, at times, has seen significant usage within certain product lines to novel interconnection schemes involving: deposited thin films [54], G-shaped springs [57], and laser deposited (written) conductors [27, 83, 56]. Pressure contacts using deformable conducting polymers or elastomerics have been used where the need to easily remove and replace the IC are the primary concerns. Detailed description of these techniques is beyond the scope of this work.

Wire bonding is, by far, the most dominant form of first-level chip interconnection method. Many trillions of wirebonds are made annually. This staggering number of wirebonds accounts for over 90 percent of all first-level interconnects (chip to package or chip to board) produced in the world. The

4 Advanced Wire Bonding Technology: Materials, Methods, and Testing

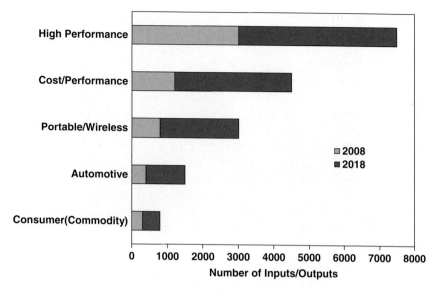

Fig. 4.1 Maximum expected I/O for different classes of electronic products both now and ten years in the future

details of the materials and methodology involved in the wire bonding of electronic and electro-optic product are the focus of this chapter. The chapter will also describe wire bonding applications and the future of wire bonding in relationship to the growing requirements for extremely high density and performance interconnects.

4.2 Interconnection Requirements

Wirebonded interconnects are usually applied to perimeter bonding pads on ICs. These perimeter bonding pads are located over non-active regions of the chip, thus preventing any damage to the IC, due to forces associated with the bonding process. Similar historical concerns gave rise to the requirement that the first bond of the wirebonded interconnect be placed on the chip (IC) and that the second bond be formed on the package or substrate. Using modern wire bonding machines, under precise computer control, researchers and some manufacturers have demonstrated bonding over active regions, as well as reverse bonding (first bond on substrate or package and second bond on the chip) without causing any chip damage or reliability concerns. This reverse bonding or reverse loop as termed by some manufacturers has been especially useful in chip stacking. Flip chip reflow soldering, on the other hand, can be used over active regions without concern of force related damage. Other comparisons of the advantages and disadvantages of wire bonding and flip chip attachment are shown in Table 4.1.

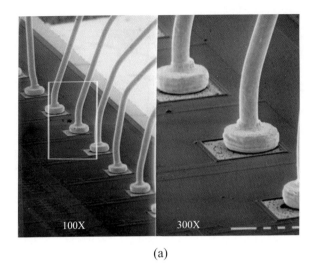

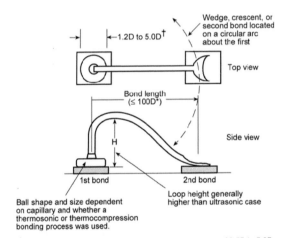

Fig. 4.2 Ball bonds (thermocompression or thermosonic). (**a**) Scanning electron microscope photo micrograph of typical ball bonds; (**b**) Schematic representation of ball bonds with important parameters indicated

Figure 4.1 illustrates current and projected I/O number requirements with time for various types (classes) of electronic products. As can be seen, I/O number requirements range from less than 100 to over 7,500 depending upon product type and the time period considered. To gain some understanding of the

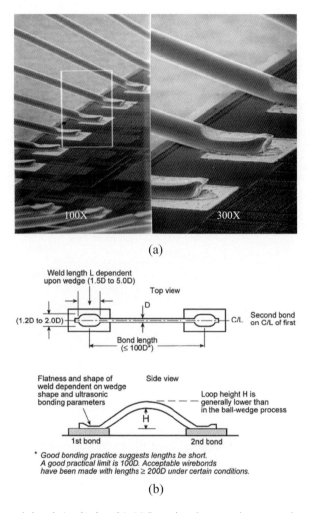

Fig. 4.3 Ultrasonic bonds (wedge bonds). (**a**) Scanning electron microscope photo micrographs of typical ultrasonic wedge bonds; (**b**) Schematic representation of ultrasonic bonds with important parameters indicated

implications of the large and increasing I/O numbers, let's consider how they might be supported from an interconnection point of view. Figure 4.5 plots the number of I/O versus chip area for the two major types of interconnect wire bonding (perimeter attachment) and flip chip (area attachment). Wire bonding, even with two rows of bonds at an extremely fine pitch (e.g., two rows of bonding pads with an effective pitch of 50 μm as shown in Fig. 4.6), requires a relatively large chip-size (225 mm^2) to reach 1000 I/O while a chip of that size could support over 18,000 I/O at the single row pitch of 100 μm using flip chipping. Area array interconnects can easily exceed 1000 I/O even on small-sized chips with a relaxed bond-to-bond spacing (pitch). It is common to normalize the I/O numbers with

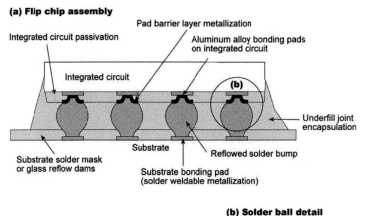

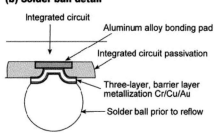

Fig. 4.4 Schematic representation of the flip chip bonding process. (**a**) Cross section of a flip chip assembly; (**b**) Detail of the solder ball and barrier layer metallization prior to reflow

respect to chip area, thus forming the I/O density (i.e., the number of I/O per unit area). Figure 4.7 plots I/O density versus chip area for various pitches of interconnect. For an area array, the I/O density is constant for a given pitch regardless of the chip size, while for a perimeter bonded chip, even with multiple rows of bonding pads, the I/O density falls off exponentially with increasing chip area.

Similarly, many other IC design, process, and material parameters affect IC bondability in addition to increased active device density and rising interconnection requirements. The aluminum-silicon alloy system (Al + 1% Si), which was standard on many early integrated circuits, has been changed by adding copper (up to 4%) to prevent electromigration as the spacing between adjacent lines has decreased. The addition of copper has produced bondability problems. Research has shown [38] that copper content above 2% prevents effective wire bonding. Another manifestation of shrinking line size is that the lines are becoming much more resistive, forcing the replacement of the aluminum-silicon alloy system with a metal having higher electrical conductivity, such as copper. Copper requires trenching encapsulation with either chromium or titanium adhesion layers [31]. The rigid organic dielectric layers on the IC are being replaced by organic materials with lower dielectric constants, such as polyimide, benzocyclobutene or Teflon®-based materials (polytetrafluoroethylene). The ultimate goal, if interconnect

4 Advanced Wire Bonding Technology: Materials, Methods, and Testing

Table 4.1 Comparison of wire bonding and flip chip interconnection factors

Factor	Wirebond	Flip Chip
Area	Requires space outside of chip perimeter for second bond	Within chip perimeter
Number I/O	Limited: one to four perimeter rows (100–1,000s possible)	Full area array. Out performs wire bonding even with a larger pitch (1,000–10,000s possible)
Flexibility	Very flexible. Ability to shift I/O. Accommodate different die orientations, die sizes, package layout, etc. (within reason, of course)	None. Substrate pattern must match I/O pattern on chip. (Some self-aligning force.)
Electrical Performance	Long round wires limits low loss frequency response to between 5-10 GHz	Short, fat solder joint pillars allow low loss frequency response above 100 GHz
Cost	Typically $0.0005–$0.001 per interconnect with full automation	Ranges from $0.01–$0.05 per interconnect[a]
Bonding Time	Sequential (10–20 bonds/second)	Gang
Bond Type	Weld: Au-Al, Au-Au, Al-Al, Au-Cu, Cu-Cu	Solder: Sn63, Sn5, Sn10, Lead free
Reliability	Monometallic systems, extremely reliable, flexible lead, eliminates or reduces any CTE issues. Bimetallic system could be susceptible to intermetallic growth and voiding.	Solder fatigue a concern due to CTE mismatches. Typically requires underfill. Intermetallic growth and voiding problems with Sn and Cu.
Environmental	Au, Al, environmentally friendly	Pb an environmental concern, hence lead free

[a] Includes extra cost for custom under bump metallurgy and a penalty for substrate re-patterning if the die shrink or pin out changes could be accommodated by the wire bond technology without changing the substrate pattern

topologies and copper passivation processes can be developed, would be to use air as the dielectric. Using copper as the IC metallization with soft organics as the intervening dielectric layers presents challenges to the first level (on-chip) interconnection processes, especially wire bonding. Copper metallization pads will necessitate copper wire bonding or a suitable barrier layer metallization cap to allow bonding with gold or aluminum wires. A gold metal flash on the copper pad to prevent oxidation is also necessary prior to forming flip chip solder balls.

4.3 Bonding Principles

4.3.1 Wire Bonding Types

Figure 4.8 illustrates an example of a modern wirebonded circuits. The wire bonding process begins by firmly attaching the backside of the integrated circuit or wirebondable component to the appropriate substrate location or package

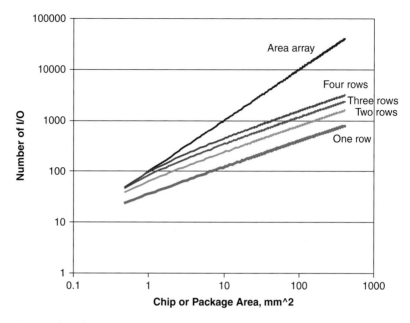

Fig. 4.5 Number of I/O's as a function of chip or package area for both perimeter (1 row to 4 rows) and area array interconnection points (bonding pads)

bottom by using an organic adhesive, a low melting point glass, the reflow of a metal alloy, or a gold-silicon eutectic alloy process [32]. Once bonded in place (this process is called die or chip attach), wires are attached to the chip bonding pads using special tools (capillaries or wedges) and various combinations of heat, pressure (force), and ultrasonic energy. Depending upon tool type and choice of welding energy (direct heat or ultrasonic heating or both), three major techniques for wire bonding have emerged over the years since microelectronic wire bonding was developed in the mid-1940s to the mid-1950s timeframe [37, 22]: thermocompression bonding, ultrasonic bonding, and thermosonic bonding.

Thermocompression bonding and thermosonic bonding methods produce a ball-wedge (first bond-second bond) type bond (Fig. 4.2(a)), where the wedge (tail, crescent, or second) bond lies on an arc about the first bond or ball bond as shown in Fig. 4.2(b). Ultrasonic bonding or wedge bonding produces a symmetric wedge-wedge (first bond-second bond) style bond as shown in Fig. 4.3(a). In ultrasonic bonding, the second bond must lie along the center line of the first (see Fig. 4.3(b)).

4.3.2 Thermocompression Bonding

A thermocompression bond (or weld) is the result of bringing two metal surfaces (bonding wire and the substrate or pad metallization, for example)

4 Advanced Wire Bonding Technology: Materials, Methods, and Testing

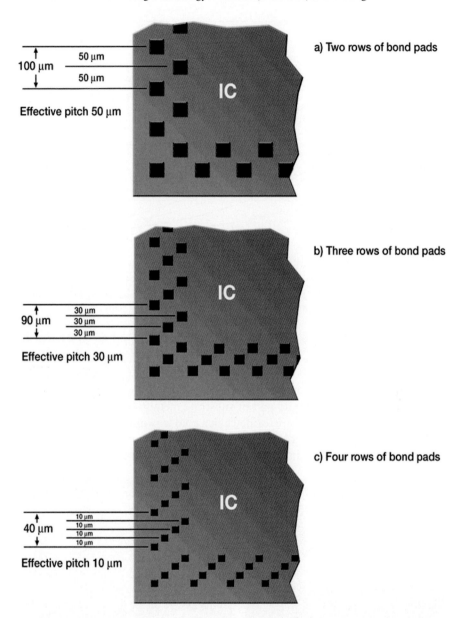

Fig. 4.6 Layouts of Multiple Rows of Bonding Pads on an Integrated Circuit. (**a**) Two rows at effective 50 μm pitch; (**b**) Three rows at an effective 30 μm pitch; (**c**) Four rows at an effective 10 μm pitch

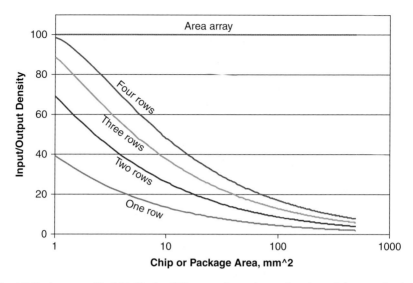

Fig. 4.7 Package or chip I/O density (I/O per unit area) as a function of the area for both perimeter (1–4 rows) and area array interconnection points (bonding pads)

together in intimate contact during a controlled time, temperature, and pressure (or force) cycle. During this "bonding cycle", the wire and, to some extent, the underlying metallization undergo plastic deformation and interdiffusion on the atomic scale. This atomic interdiffusion can result in a uniform welded interface, if both gold wire and gold pad or substrate metallization are used. Gold-aluminum intermetallics [67] are formed when gold wire and aluminum pads (or vice versa) are used. Regardless, the plastic deformation that occurs at the bonding interface ensures: intimate surface contact between the wire and the

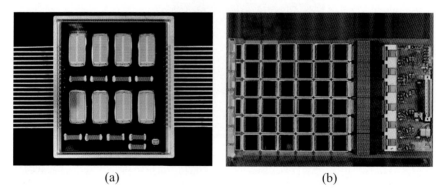

Fig. 4.8 Examples of Wirebonded Circuitry. (**a**) Static RAM Module using MCM-D technology. Unit contains 300 gold thermosonic wirebonds; (**b**) Experimental X-ray Detector for use in space. 36-detector chips with bond pads on both sides of the chip. Each chip has over 200 wirebonds per side. Total wirebonds on assembly exceed 18,000. Chips mounted on open frame to allow wirebonder access to both sides

pad, provides an increase in the interfacial bonding area, and breaks down any interfacial film layer (oxide, contamination, etc.). Surface roughness, voids, oxides, and absorbed chemical species or moisture layers can all impede the intimate metal-to-metal contact and limit the extent and strength of the interfacial weld; thus, causing a poor bond. In some cases, this interfacial contamination (usually on the pad) is so extensive that it prevents bonding altogether. The inclusion of contamination at the weld interface can lead to serious reliability problems [10].

The interfacial bonding temperatures are typically in the range of 300–400°C [45] for bonds made by thermocompression bonding. The bonding cycle, exclusive of bond positioning, takes a fraction of a second. In thermocompression bonding, the required heat for interface formation is applied by either a heated capillary (the bonding tool through which the wire feeds) or by mounting the substrate and/or package on a heated stage (column). With stage or column heat, the die and package combination must come into thermal equilibrium with the stage, which can take seconds to minutes depending upon mass. Because of the high stage or column temperatures (>300°C) involved in thermocompression bonding, IC or device die attachment is usually limited to the gold-silicon eutectic or certain metal alloy attaches. Also, long times on heated stages can cause reliability problems with previously placed wirebonds, such as uncontrolled intermetallic growth. Most modern thermocompression bonders use a combination of both capillary and column heat. The capillary is made of ceramic, ruby, tungsten carbide, or other refractory material. Special capillary shapes are needed for fine-pitch and deep access applications. Controlled capillary resistance is needed to prevent damage to electrostatically sensitive circuitry.

A typical ball bonding cycle is illustrated in Fig. 4.9. There are five major steps in the ball bonding process: (1) ball formation (Views a and b, Fig. 4.9); (2) ball attachment to IC or substrate pad (first bond) (View c, Fig. 4.9); (3) traverse to second bond location (View d, Fig. 4.9); (4) wire attachment to package or board pad (second bond) (View e, Fig. 4.9); and (5) wire separation (View f, Fig. 4.9). The initial ball formation step is accomplished by cutting the wire end as it extends through the capillary with an electronic discharge. This cutting is called flame-off due to the fact that in the early days of wire bonding an open flame hydrogen (or forming gas) torch was used to cut the wire. Once cut, the ends of the wire ball up due to surface tension and capillary action. Figure 4.10 illustrates free air balls produced with gold wire by a negative electronic flame-off system. Heat, time, and pressure or force are the major determining factors in the formation of thermocompression bonds. Typically, the forces used in thermocompression bonding are higher than in other ball bonding methods (i.e., thermosonic ball bonding), resulting in a much more flattened ball. Thus, the first bond is "nail head" shape rather than just a slightly flattened ball as obtained with standard pitch thermosonic ball bonding (e.g., see Fig. 4.2a). Fine pitch thermosonic ball bonding produces very flat, minimal diameter and height balls as described below in Sections 4.8 and 4.9.

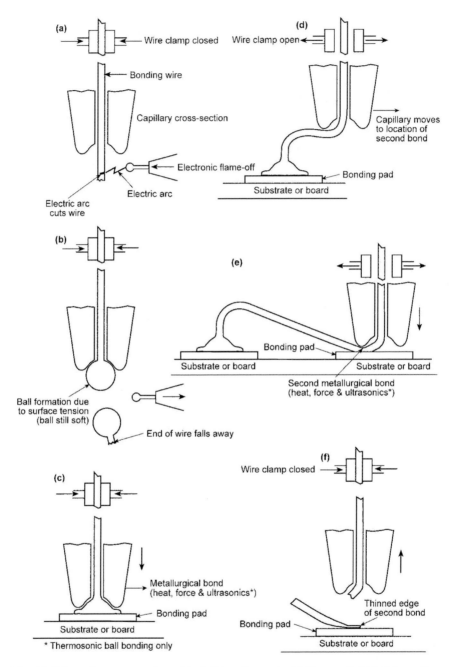

Fig. 4.9 Schematic representation of the ball bonding cycle: (**a**) Flame-off; (**b**) ball formation; (**c**) first bond; (**d**) transition to second bond; (**e**) second bond; and (**f**) separation of wire after second bond

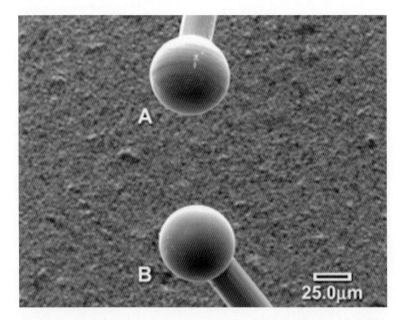

Fig. 4.10 Scanning electron photomicrograph of free air balls produced by negative electronic flame-off. The gold wire diameter is 25.4 μm. (**A**) Free Air Ball made on 100 kHz bonder (62.2 μm diameter); (**B**) Free Air Ball made on 60 kHz bonder (59.7 μm diameter). Magnification approximately 350X

Gold wire is used in most thermocompression wire bonding processes because it is easily deformed under pressure at elevated temperature and is very resistant to oxide growth that can inhibit proper ball formation. Aluminum wire, because of its rapid oxide growth, has difficulty in forming properly shaped balls on standard bonding machines. Successful aluminum wire ball bonds have been formed using an inert atmosphere around the bonding head to minimize oxide formation [30, 65]. Copper and other materials (e.g., palladium and platinum) have also been ball bonded [52] in both thermocompression and thermosonic applications. Also, wedge style thermocompression bonding with many different materials has been performed [55, 8].

4.3.3 Ultrasonic Bonding

Ultrasonic bonding (or wedge bonding) is a lower-temperature process in which the source of energy for the metal welding is ultrasonic energy produced by a transducer vibrating the bonding tool (wedge) in the frequency range of 20–300 kHz. The most common frequency is 60 kHz [38], although higher frequency ultrasonics is in use or being considered for difficult bonding situations. Thermosonic bonding at higher frequencies will be discussed in Section 4.9 below. The ultrasonic wedge bonding process is illustrated in Fig. 4.11. In

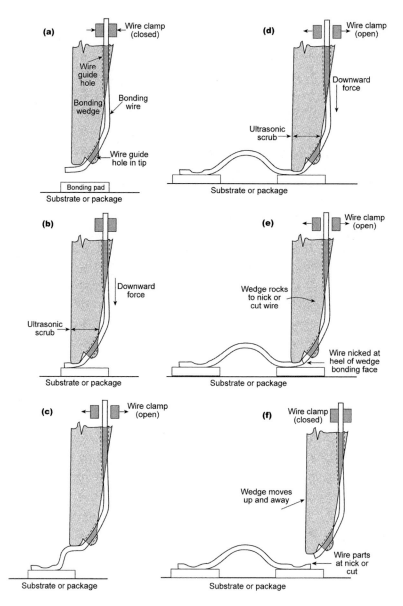

Fig. 4.11 Schematic representation of the ultrasonic (wedge) bonding cycle: (**a**) initial wire-wedge configuration; (**b**) first bond; (**c**) transition to second bond; (**d**) second bond; (**e**) wire nicking or cutting operation; and (**f**) wire separation after second bond

ultrasonic bonding, the wedge tip vibrates parallel to the bonding pad. Ultrasonic bonds are typically formed with aluminum or aluminum alloy wire on either aluminum or gold pads. Gold wire ultrasonic bonding has been performed with both round wire and flat ribbon, although it is not widely used because of cost. Gold ribbon, because of its rectangular cross section, provides a lower inductance interconnect (compared to a round wire of equivalent cross-sectional area) useful in radio frequency and microwave chip applications. In special applications, copper and palladium have been bonded by the ultrasonic process [32]). The major advantages of ultrasonic bonding include the ability to effect strong bonds with little or no applied substrate heat (implying the use of low temperature die attachment methods and/or low glass transition temperature substrates); and it typically can be performed at finer pitches (because of the elongated, narrow shape of the bond) than ball bonding methods. Automated wedge or ultrasonic bonders are typically slower than ball bonders due to the requirement that the second bond must be in-line with the first bond; i.e., follow the centerline of the wedge. Thus, either the entire package (substrate) or the bonding head must be rotated to bond in different directions. This slows down the bonding process when compared to ball bonding, which can place the second bond anywhere on a circle surrounding the first bond with only a transversal movement of the head (or stage) (See Fig. 4.2(b)).

4.3.4 Thermosonic Bonding

In thermosonic wire bonding, ultrasonic energy is combined with the ball bonding capillary technique employed in thermocompression bonding. Typically, the thermosonic bonding process is performed in a manner analogous to the thermocompression bonding process, except the capillary is not heated (or held at a lower temperature when compared to the capillary temperature in thermocompression bonding); and the stage or column temperatures are typically 150°C or less. To generate the required interfacial heat for welding at the interface of the wire and the pad, short bursts (tens of milliseconds) of ultrasonic energy are applied to the capillary when the wire and the pad are in contact. Because of the addition of ultrasonic energy (causing localized heat generation at the wire-pad interface), the requirements on stage and capillary heat (as mentioned above) and pressure (force) can be relaxed. The applied forces in thermosonic bonding are typically much less than those encountered in thermocompression bonding, thus allowing bonding over delicate or force sensitive chip or substrate regions. Since interconnections are made with the ICs (and substrates) held at temperatures of 150°C or less, they can be attached with epoxy or other organic adhesives without fear of degradation (i.e., prolonged exposures at temperatures above their glass transition temperature) due to excessive bonder stage or column temperature. Because the temperatures are lower, there is also significantly less risk of uncontrolled intermetallic growth. Thermosonic wire bonding is conducted primarily with gold wire, but aluminum [65], copper [52], and palladium [8] wires have been bonded successfully by the thermosonic process. As the metallization on high performance ICs migrates from aluminum alloys to copper [31], new pad stack configurations (e.g., copper-nickel-gold or, perhaps, just copper) will emerge. These new pad stacks will require the reevaluation of the thermosonic bonding process and, perhaps, the full consideration of the use of copper wire. Copper thermosonic wire bonding has been successfully used in the connection of the ICs to copper alloy lead frames in dual-in-line packages [41] for over two decades.

Such thermosonic bonding evaluations have already been performed for the new substrate and pad structures encountered in the development of multichip modules (MCMs) [15]. Some of the results from these structures and material evaluations are discussed in Section 4.9 below.

4.3.5 Other Techniques

It should be noted that other wire bonding or wire welding techniques have been used over the years, including DC resistance welding, AC resistance welding, and more recently, laser bonding (Mundt, et al.) or welding. While these techniques have their applications (many to terminal pins and/or circuit boards), their use and flexibility is limited when compared to the major wire bonding techniques. Laser welding at the microelectronics scale of flat ribbons is relatively new and offers some potential benefits over conventional microelectronic wire welds produced by standard wire bonding practice. Laser welds can have substantially greater penetration depths than ultrasonic welds. The laser beams can be modulated over a wide range that could produce weld penetration depths from a micrometer or two to several micrometers. Laser ribbon welding can accommodate a wide range of materials, including nickel-clad copper and gold. Reliability of laser welded nickel-clad copper is high when compared to ultrasonically bonded aluminum wire (about 3 orders of magnitude) under fatigue testing (Mundt, et al.)

4.3.6 Machine Optimization

Originally, wire bonding was done manually requiring the operator to control every step of the bonding process from flame-off to wire clamping and breaking (on large diameter wire even manual cutters were used). In manual bonding, operator skill was paramount to the fabrication of high-quality, reliable wire bonds. Even as the technology evolved and semi-automatic wirebonders appeared (flame-off and bonding cycle under machine control, but positioning or bond alignment was left to the operator), operator skill was key to producing highly successful (reliable) wirebonds [36]. Today, fully automated wirebonders dominate the scene. Both automatic thermosonic and ultrasonic wirebonders are in widespread use. Automatic wirebonders use pattern recognition to locate the bonding pads on both the chip and the package or substrate; and then, under complete computer control, the machines automatically bond all

4 Advanced Wire Bonding Technology: Materials, Methods, and Testing

connections at rates exceeding 15 wirebonds (30 welds) per second. Position accuracies at those bonding rates are typically ±2.5 to ±3 μm. Using automatic component handlers, automatic bonding machines can sustain such rates for hours. Such automation, with its concomitant accuracy and improved process control has dropped wirebond failure rates for individually packaged parts (single chip packages) into the low part per million range [36].

Failure rates associated with multichip modules, chip-on-board (or substrate) and chip-on-flex are significantly higher as a result of the complex structures and new materials present in these advanced packaging structures. Some of the bonding issues for these complex circuits and structures are described in Section 4.9 below.

Bonding machine optimization can be accomplished in several ways depending upon the availability of test samples and trained personnel. The most straight forward way is to do a fractional factorial experimental design [11] which minimizes trials and eliminates inherent operator bias. Typically, the machine set up parameters of interest include the ultrasonic energy (P), the substrate temperature (T), and the duration of the ultrasonic energy or dwell time (D). The bonding force is usually not considered (once an initial set up has been done) since it is typically held constant for a given substrate, hybrid, or module configuration. The force is usually set to a level that promotes long capillary lifetime, thus eliminating the need to change capillaries during an experimental set (which helps minimize bond variations and improves reproducibility). For the three variables mentioned above, the bonding parameter experiments would involve a simple 2^3 factorial design with each of the variables in turn being set to expected low (−1) and high (+1) range limits as shown in Table 4.2. The experimental design can be unreplicated provided sufficient number of samples (>35) exist for each treatment. Random execution order should be established for all the experimental treatments to eliminate any potential memory effects. The responses denoted as Si can be the mean shear strengths for first bond analysis (recommended) or the wirebond pull strength for each treatment. The second and third order effects are also shown in Table 4.2. The calculation of any one of these effects is simply the sum of the products for each level with the corresponding response all divided by $2(n-1)$ where $n = 3$. For example, the effect of bond power is

$$P = (-S1 - S2 - S3 - S4 + S5 + S6 + S7 + S8)/4$$

In order to determine the statistical significance of a particular effect with an unreplicated experimental design, and estimate of the sample variance is needed. A method for estimating the variance and confidence intervals at various significance levels has been described previously [19].

Using the same 2^3 factorial design concept with replicated center points, a linear model for ball bond shear strength in terms of P, T, and D can be

Table 4.2 2^3 factorial experimental design (unreplicated)

P[a]	T[b]	D[c]	PXT	PXD	TXD	PXTXD	Response[d]
−1	−1	−1	+1	+1	+1	−1	S_1
−1	−1	+1	+1	−1	−1	+1	S_2
−1	+1	−1	−1	+1	−1	+1	S_3
−1	+1	+1	−1	−1	+1	−1	S_4
+1	−1	−1	−1	−1	+1	+1	S_5
+1	−1	+1	−1	+1	−1	−1	S_6
+1	+1	−1	+1	−1	−1	−1	S_7
+1	+1	+1	+1	+1	+1	+1	S_8

[a] P = bond power (e.g., first bond power setting), where −1 represents the low power value and +1 represents the high power value
[b] T = temperature (substrate), °C. Again, −1 represents the low temperature setting and +1 represents the high temperature value
[c] D = dwell time, ms. As above, −1 represents the shortest dwell time and +1 the longest
[d] S — response function, typically the shear strength

constructed. The resultant ball shear equation simplifies the understanding of how the bonding parameters influence bond strength without the need for complex three dimensional plots, although with widespread availability of high performance computers, even on the shop floor, three dimensional contour plots may be preferred. In addition, the linear factorial design provides an efficient means for generating new models should different substrates and substrate metallization be required.

4.4 Materials

4.4.1 Bonding Wire

Microelectronic bonding wire comes in a variety of pure and alloy materials. In addition to round wire, flat-ribbon material is available for special applications such as radio frequency and microwave circuits. Round wire is by far the most common, and fine round wires with diameters as small as 5 μm are produced commercially. Large diameter round wires up to 500 μm in diameter are used for power applications. Ribbons range from 50 to 1200 μm in width and come in various thicknesses.

The major materials used for these wires (and ribbons) are gold (pure and alloys), aluminum (pure), aluminum with 1% silicon, aluminum with magnesium, and, more recently, copper. Typical properties for these wires are given in Tables 4.3 and 4.4. Other wires, such as palladium and silver have been bonded in the past as described above. Gold has been the dominant material used for the ball bonding process, while aluminum and its alloys predominate in the wedge (ultrasonic) bonding process. The gold used is extremely pure (99.99%)

Table 4.3 Mechanical properties of bonding wire

Material	Wire Diameter[a] μm	Temper[b]	Elongation %	Tensile Strength MPa	Comments
Aluminum (99.99% Pure)	18–75 (small diameter) 75–500 (large diameter)	H M S	2–6 6–12 12–18 5–10 10–20	1.9–2.5 1.7–1.9 1.5–1.9 1.4–1.5 1.0–1.4	Softer than other wire. Sags more than other wires for equivalent diameters. Difficult to handle in small diameters.
Aluminum + 1% Silicon	25–250	H M S	1–5 5–10 10–20	2.9–3.5 2.2–2.6 1.5–1.9	Standard integrated circuit bonding wire (wedge bonding). Since 1% silicon greatly exceeds the room temperature solubility of silicon in aluminum, there is a tendency for Si to precipitate at bonding temperatures – unless the alloy is homeogeneous at the nanometer level.
Aluminum + 0.5–1% Magnesium	25–250	H M S	1–5 5–15 10–20	2.9–3.5 2.2–2.6 1.5–1.9	Does not form a precipitative phase since room temperature solubility in silicon is 2%. Excellent fatigue resistance – mitigates low cycle fatigue in power devices. Sometimes small amounts of palladium (0.1–0.15%) are added.
Gold (99.99% Pure)	18–50	H SR A	1–3 3–6 4–8	3.0–4.7 3.6–4.1 3.2–3.8	Mainstay ball bonding wire. Sometimes very hard gold wire (>7 MPa tensile strength, <1% elongation) is used for wedge bonding.
Gold (98.5% Pure) + 1% Palladium	18–37		0.5–3	8.7–10.4	Formulated for stud bumping. Produces consistent uniform sized balls.

[a] Typical wire sizes available from various manufactures
[b] Temper: H = hard, M = medium, S = soft, SR = stress relieved, and A = annealed

with total impurities typically less than 10 ppm. See Section 4.9 below on wire bonding at extreme temperatures for additional discussion of wire purity. Beryllium is the key impurity used to stabilize the wire and control some of its mechanical properties. The gold wire used for stud bumping (single ended ball bonds) is not as pure, with a significant amount of palladium (~ 1%) added to

Table 4.4 Thermal and electrical properties of bonding wire materials

Material	Melting Point °C	Thermal conductivity w/m-K	Coefficient of thermal expansion $\times 10^{-6}/°C$	Electrical resistivity $\times 10^{-6}$ Ω-cm	Electrical conductivity % IACS*
Aluminum (99.99% Pure)	660	230	23–24	2.49–2.77	69–62
Aluminum + 1% Silicon	600–630	195	22–23	2.96–3.18	58–54
Aluminum + 0.5–1% Magnesium	654	180–195	22–24	3.01	57
Gold (99.99% Pure)	1063	312	14–15	2.20–2.29	78–75
Copper (99.99% pure)	1083	395	16–17	1.72–1.81	100–95
Palladium (99.99%)	1552	75	10–12	10.75–15.63	16–11

*IACS = International Annealed Copper Standard. 100% IACS = $5.81 \times 10^5/$ Ω-cm

ensure the formation of uniform balls with minimum tails (residual wire remaining on the ball after wire is broken). Aluminum with 1% silicon matches the common alloy used for semiconductor device metallization and offers improved strength and stiffness over pure aluminum in small diameter applications. Pure aluminum is used in most large-wire applications, while aluminum and magnesium is used in cases where the interconnect is subject to conditions of low-cycle fatigue or on-off power cycling [70].

Because microelectronic bond wires are drawn through a series of dies, the as-drawn wire has significant residual strain and, while strong, is often brittle (low elongation). To overcome these factors, the wire is typically strain relieved and sometimes annealed to achieve more desirable properties for the bonding process. Some of the effects of these post drawing processes can be seen in Table 4.3. Figures 4.12 and 4.13 show the effects of time after manufacture (in controlled storage) on bonding wire properties for a few wire types. It is clear that depending upon the temper of the wire, storage time can have a significant effect on wire properties and hence on the quality of the bonds themselves.

There is great interest in replacing gold bonding wire with copper wire, both for reduced cost and ease of bonding, as the IC metallizations migrate to copper. The use of copper wire precludes the need for the copper pads on the IC to have barrier layer coatings (nickel-gold or titanium-tungsten gold) to prevent intermetallic growth problems with the gold wire. Copper wire also has a high electrical conductivity and because of its strength, it resists wire sweep

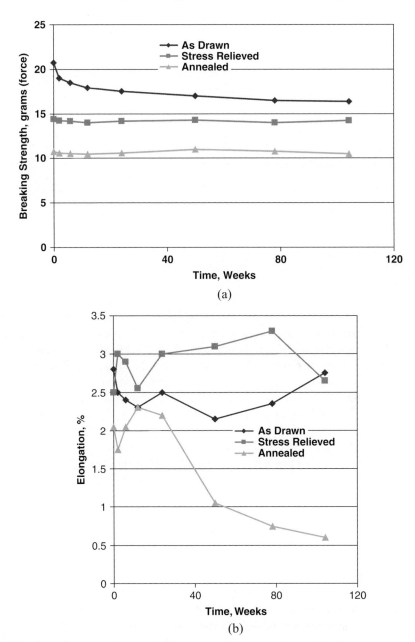

Fig. 4.12 Breaking strength and elongation of aluminum bonding wire (Al + 1% Si) as a function of storage time for various wire tempers: (**a**) breaking strength grams (force); (**b**) elongation in %

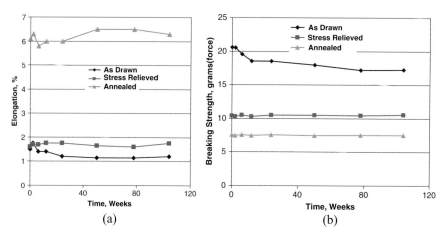

Fig. 4.13. Breaking strength and elongation of gold bonding wire (99.99% Au + Be) as a function of storage time for various wire tempers: (**a**) breaking strength in grams (force); (**b**) elongation in %

during the injection molding and/or encapsulation processes. Since copper rapidly oxidizes in air, the ball formation process must be done in an inert atmosphere requiring significant bonding machine modifications. Copper has a higher shear modulus than gold (48 GPa versus 26 GPa) and Cu balls are significantly harder than gold balls (e.g., 50 compared to 35 on the Knoop Hardness Scale), thus, creating the potential for damage to delicate chips and substrates in the bonding process. Copper ball bonding produces a significant increase in cratering [23]. Several changes to bonding machine operation have been proposed as possible solutions to the copper hardness problem including increased substrate and capillary heat, reduced ultrasonic energy and a rapid first bond touchdown (to keep the ball hot and hence softer).

Bonding to copper pads, unless barriered as described above, could require significantly more ultrasonic energy due to the formation of copper oxides. In a similar vein, copper ball bonds made to conventional aluminum alloy pads seems to be viable. Copper-aluminum intermetallics exist ($CuAl_2$ and $CuAl$) and some studies have indicated rapid increases in joint resistance during thermal aging [48]. Most studies report the reliability of the copper-aluminum system to be equal to that of the gold-aluminum system. The bondability is probably more of an issue than the reliability, even with the mitigating measures described above, because the hard copper ball is likely to "push" the soft aluminum metallization aside during the bonding process, especially with today's thin IC metallizations (~0.5 μm), resulting in a weak bond or a no stick situation. Similarly, cratering and the susceptibility of Cu to corrosion (sulfur, halogens, etc.) could inhibit the wirespread use of copper ball bonding. It is also very difficult to make small balls such as required for fine pitch wire bonding (see Section 4.9 below).

4.4.2 Bond Pads

Various metallization schemes have been used to ensure the bondability of chips, packages, and substrates. Unfortunately, most chip metallizations are selected for reasons other than the ability to form wirebonds. Historically, the typical chip metallization is aluminum containing a small percentage of silicon (typically 1%). The presence of silicon prevents the rapid diffusion of the underlying silicon (in the contact window) into the aluminum and thus reducing the pit formation in the silicon. Such pitting allows aluminum to migrate into the pits, creating aluminum conductive spikes which can damage performance or destroy device operation. Too much silicon in the aluminum can cause silicon precipitation during heat treatment and form silicon crystallites or nodules on the bonding pad surface and in contact with the underlying silicon. Such effects can cause both bonding and electrical problems.

To ensure adequate electromigration resistance as device geometries shrink [64], alloying elements are also incorporated into or sometimes placed under the standard chip metallization. For example, copper is often added to aluminum and aluminum with silicon in concentrations from 0.5 to over 5% by weight in order to prevent electromigration. Above about two weight percent copper, the wire bondability of the aluminum-copper alloy has been shown to decrease, while lower amounts have exhibited excellent bondability [81]. Aluminum with small amounts of copper, however, is subject to Al_2Cu hillock formation during thermal processing. These hillocks can cause interlayer shorts, etc. Thus, to prevent Al_2Cu hillocks, process engineers add more copper (>4 weight percent), which causes widespread hillock growth; but the hillocks are very limited in height, thus reducing the shorting potential at the expense of bondability. Higher copper levels also increase the susceptibility of aluminum to corrosion and may lead to surface oxide formation, which can further reduce bondability.

Titanium-tungsten or titanium nitride layers are sometimes added under pads to improve adhesion and to stiffen the pads on soft or flexible substrates. If process conditions are improperly controlled, these under layers can reduce bondability. Titanium also has been alloyed with aluminum metallization on chips to reduce electromigration. Again, potential titanium migration to the surface can cause bonding problems. The titanium also increases the hardness of metallizations, which in general requires more aggressive bonding parameters to effect high quality bonds. To achieve the highest bondability in the presence of titanium, bonding temperatures must be substantially increased (>180°C), which requires the use of high-temperature die attach (e.g., the gold-silicon eutectic). As recommended by [38], capping with a thin layer of pure aluminum (0.25–0.5 μm in thickness) would allow various metallizations to be used and still provide the best metallurgy for high-yield bonding. Care must be exercised to keep the pure aluminum cap metallization thin, because it has been shown that bond strength decreases with increasing aluminum layer thickness [86].

Gold metallization also can be an effective cap to ensure bondability. Gold was originally used on some semiconductor devices. The pad stack typically was titanium-palladium-gold. Such pad stacks produced excellent bondability providing the gold thickness, hardness, and morphology were carefully controlled. Today, gold is rarely used on integrated circuits, but is widely used on package bonding pads and substrates to provide a wire bondable surface. The search for bondable gold has been the subject of many articles over the last decade or two. Gold deposited by thin film deposition is inherently bondable due to its purity and fine grain structure. Most gold bonding problems have been associated with either screen printed inks used in thick film and low temperature cofired processes or with plated gold.

The bondability of thick-film metallizations, particularly gold-based films, has been of concern for many years in the microelectronics industry. Statements such as "bondable" gold still appear in various forms in the commercial advertising literature without any quantification. The implication is that if you use the particular company's bondable gold that wirebond performance should approach the ideal, i.e., wirebond pull and ball bond shear strengths close to those obtainable with thin films. Historically, authors such as [47], have shown that with clean substrates and thermocompression bonding, thick film gold substrates yielded similar ball bond shear strengths as comparable bonds made to thin-film gold. Some studies actually showed that bonding to thick film gold was less sensitive than bonding to thin films in the presence of surface contamination. The role of surface cleanliness prior to bonding on both thick and thin films cannot be over emphasized and it has been studied in great detail by several authors including [45, 46, 86].

In the past, the role of surface composition, surface morphology and actual conductor or bonding pad geometry has not been addressed in detail to the same levels as the cleanliness problem. From the studies that have been performed, [76, 33, 68], it is clear that conductor composition, morphology, and geometry are extremely important factors in thick film bondability. Certain manufacturers over the years have "flattened" or "coined" the thick film at the bonding site by using special tools placed in bonding machines. Such processes are very expensive and time consuming.

Our studies (e.g., [72]), have shown slightly different but not necessarily conflicting results. In our studies we compared different metallization ink types: pure gold, lightly alloyed gold, and heavily alloyed gold. The pure gold was an oxide bonded gold made for wire bonding using gold wire. The lightly alloyed gold was oxide bonded and especially formulated to retard strength loss (due to intermetallic diffusion/formation) that occurred when making aluminum wirebonds. The heavily alloyed gold (which contained significant amounts of platinium and palladium) was primarily made for solder reflow operations. These metallizations were screen printed and fired using a test pattern consisting of various line and bonding pad sizes, ranging from 125 to 500 μm. Thin film vacuum deposited pure gold (3 μm in thickness) was also patterned and used as a reference in these bondability studies. Pad surface and line morphology and shape were measured using a scanning electronic microscope and a stylus profiliometer, respectively. Surface impurities were analyzed by Auger electron spectroscopy and wirebond quality was assessed by both the ball bond shear test and the wirebond pull test (See Pad Cleaning below).

The metallization type had the greatest effect on both the ball bond shear strength and wirebond pull strength. Pure thin film gold demonstrated the best bondability and had the highest average shear strength. The lightly alloyed thick film gold (made for aluminum wire bonding) gave results similar to the thin film gold. The pure thick film gold and the heavily alloyed gold produced significantly poorer results (e.g., 35 g (force) shear strength compared to 48 g (force) shear strength on average) for comparably sized and placed bonds. Surface morphologies were different between all four metallizations with the thin film surface being extremely smooth, small grained with no pores. The heavily alloyed gold surface was extremely porous and very rough compared to the other metallizations. The pure thick film gold and the lightly alloyed gold had similar morphology, although the lightly alloyed gold was slightly rougher and more porous.

In a design of experiments study, parameters such as surface porosity, surface curvature and pad or line width size were determined to be secondary effects. Ball location on bonding pads or lines seemed to have little effect on the thin film and pure thick film bonding results. As surface porosity and roughness increased effects associated with ball location became slightly more dominant. Tail bonds seemed to be more affected than the ball bonds. Mechanical operations such as burnishing (scrubbing with an abrasive) or coining appeared to have little effect and in the case of the heavily alloyed gold burnishing significantly reduced the bondability.

Results of the study indicated that surface composition was the key factor in bondability. This result is consistent with findings of [38] in his Chapter 6 on plated golds. He further correlates bondability or lack there of with film hardness, i.e., soft gold is preferred. In our studies the hardness of the thick film layers increased with increasing impurity concentration, based on gold ball deformation, at given force level. No quantitative measurements of hardness were made.

4.4.3 Gold Plating

4.4.3.1 Electroplated Gold

Impurities in electroplated gold layers have long been a source of bonding problems. Impurities have caused both low bonding yields and premature failures during accelerated testing or real life operational use. Horsting (1972) [42] presented fundamental studies that related gold purity to the formation of "purple plague" and hence bond failures. Horsting believed that the accelerated diffusion of the impurities into bond intermetallic regions caused precipitates to form which acted as nucleation points for vacancies causing more rapid void formation during the normal interdiffusion of gold and aluminum. The actual impurities in the gold were not precisely determined by Horsting due to equipment limitations, qualitatively nickel, iron, cobalt, and boron were the major impurities. Later, researchers confirmed Horsting's rapid impurity diffusion theories [63].

Gold electroplating bathes typically consist of potassium-gold-cyanide solutions plus additives such as buffers, citrates, phosphates, carbonates, and lactates. Impurities such as thallium, lead, and arsenic are added to improve plating deposition rate and as modifiers to reduce grain size – hence changing surface morphology. Thallium has been the impurity most often linked to wire bonding problems [28, 29], but work by Wakabayashi (1982) [85] identified lead as another significant cause. He also indicated that under certain plating conditions, arsenic could improve bond strength. Impurities such as lead and thallium can cause the gold crystal structure to change on the bonding surface. Surface morphology can also be changed by varying the plating parameters. To date there is not conclusive proof that subtle changes in surface morphology in plated gold layers can have a correlatable effect on bondability and bond strength, unlike the experiences with thick films above.

Other plated gold phenomena such as hydrogen entrapment and film hardness can also cause bonding problems. Hydrogen entrapment can be mitigated by annealing, providing the assembly can withstand the annealing environment (at a minimum 2 days at 150°C). Annealing a gold film, while removing gases such as hydrogen, also reduces its hardness. Hardness thus becomes a key bonding indicator, if not the root cause, of bondability problems.

4.4.3.2 Electroless Autocatalytic Gold

The key to wire bonding on laminate and flexible tape substrate technologies used MCMs, chip-on-board, and other board and package implementations is the ability to do electroless gold plating on the pre-patterned copper metallization. In working with commercial plating vendors, electroless gold (autocatalytic) plating solutions can be found or developed with standard or modified chemistry that meet the deposition needs (99.99% pure gold up to 1 μm in thickness) for a variety of substrates and applications. Typical laminate processes require a nickel barrier layer over the copper. It is necessary that these autocatalytic gold processes be able to plate on nickel as well as on copper. Two major types of autocatalytic gold plating chemistries exist: (1) high deposition rate strongly basic systems containing cyanide; and (2) neutral pH systems without cyanide. The high deposition rate systems have a pH of about 12 and can erode certain circuit board materials such as polyimide during long plating runs. Several variants of these high deposition rate systems exist including ones which plate gold directly on copper and others which will plate gold onto nickel coated surfaces. Typical plating bath temperatures range from 70 to 100°C. Such systems have been used to produce bondable gold, but the high bath temperatures, the difficulty in plating on nickel (requires exacting bath chemistry at all times), and the erosion of the substrate material has made these chemistries unsuitable for most organic-based MCMs, chip-on-board, and flex circiut assemblies. Such chemistries are useful for plating circuits built on ceramic substrates.

The issues associated with the high deposition rate systems caused the development of neutral pH (nominally 7.5) autocatalytic gold processes. These baths contain no cyanide and can operate at 70°C or less and do not erode polyimide. With these systems bondable gold up to 1 μm in thickness can be deposited over nickel barrier layers. Compatible electroless nickel plating solutions exist for copper metallizations. The copper metallization must first be sensitized with a palladium-based activator. Table 4.5 presents some wirebond reliability data for gold bonds made to various thicknesses of autocatalytically plated gold (neutral pH). The data indicates that bonds remain strong even after extensive thermal aging at 150°C provided the gold is at least 0.65 μm in thickness. Other experiments have shown that a minimum of 0.5 μm is necessary to achieve uniform bonding and reliability after thermal testing.

4.4.4 Pad Cleaning

In order to make high quality, reliable wirebonds, the bonding pads must be clean. Many techniques have been tried over the years, but of all the methods, UV-ozone [86] and oxygen plasma [50] have proved to be the most effective in removing organic contamination. They are also effective against certain inorganic materials that form either a volatile oxide or, if not volatile, one that can be easily removed. While these techniques have been shown to remove a wide variety of contamination types, care must be exercised in their use. Because of the strong oxidizing environments present in O_2 plasma and UV-ozone reactors, metals such as silver, copper, and nickel may oxidize, and thus reduce their bondability. To reduce such effects in plasma reactors, argon is sometimes mixed with the oxygen. These oxygen-argon plasma cleaners are quite effective,

Table 4.5 Wirebond pull strength for various thicknesses of autocatalytic gold plating over a nickel barrier (2.5 μm thick) on a copper metallized printed wiring board

Gold Plating Thickness, μm	Number of Bonds	NDPT[a] Failures	Pull Strength[b], grams (force)	
			As Bonded	After 150°C Aging[c]
0.40	129	1	10.6	9.8
0.65	149	0	10.0	10.1
0.90	138	0	9.4	10.6

[a] NDPT = non destructive pull test (at a 2.5 g (force) limit)
[b] Sample sizes approximately 70 bonds. Standard deviations within ±10%.
[c] 160 h (polymide-glass board material)

combining reactive ion cleaning with physical sputter etching. With any kind of plasma environment, there is a possibility of active circuit radiation damage. Based on this author's experience, this probability is extremely low for oxygen-argon plasma cleaners and should not be viewed as a deterrent to their use. Similarly, because UV radiation can excite impurity states (color centers) in alumina-based ceramics, there is a tendency for white alumina ceramic substrates to appear yellow after UV-ozone treatment. The induced color change can be reversed by a subsequent thermal treatment. Table 4.6 and Fig. 4.14 show the effectiveness of UV-ozone cleaning (over solvent cleaning) in removing intentional surface contamination.

Before leaving cleaning, a few comments about ultrasonic cleaning should be made. Historically, there have been several published reports (e.g., [71]), and much anecdotal conversation describing wirebond degradation or failure due to ultrasonic cleaning. Most of the reported incidents center on wirebonds in cavity type packages, such as those used for hybrids or hermetic single chip applications.

As with all mechanical structures, a wirebond has a resonant frequency which if excited will cause the wire to vibrate and in turn may cause fatigue and ultimate failure. The resonant frequency of a given diameter bonded wire is dependent on the length and height of the loop. For reasonable geometries and relatively short lengths (<2.5 mm) the resonant frequency of a typical wirebond is quite high (>30 kHz). Historically ultrasonic cleaners operated in the 20 kHz regime, and most of the reported damage occurred with long wire bonds (>2.5 mm) placed in large industrial cleaners (high energy). Thus, the ultrasonic cleaning of cavity type devices with short wires should be safe. Today, ultrasonic cleaners span a broad frequency range from 20 kHz to over 100 kHz. According to Harman ([38], p. 230), it is unlikely that high frequency ultrasonic cleaners (>50–60 kHz) will damage wirebonds.

Table 4.6 Average ball bond shear strength (grams (force)) for various cleaning treatments and thermal aging conditions for thermosonically bonded 25.4 μm gold wire on 1 μm thickness aluminum (on silicon). Average ball diameter was 90 μm (±3 μm)

Sample Set	Cleaning Conditions	As Bonded	Thermally Aged
A	No clean[a]	50.9 (±7.1)	47.8 (±7.9)[c]
	Plasma clean[b]	52.2 (±6.5)	52.1 (±6.7)
	No Clean	50.0 (±6.2)	48.6 (±7.1)[d]
B	Contaminated[e]	38.9 (±4.1)	40.3 (±5.8)
	Solvent clean	37.3 (±6.1)	37.9 (±7.3)
	Plasma clean[f]	47.5 (±6.0)	47.9 (±6.7)
	UV-Ozone clean	53.0 (±5.1)	54.2 (±5.8)

[a] No clean as received from substrate fabrication
[b] Argon-oxygen plasma (90% Ar, 10% O2)
[c] Sample set A aged fro 96 h at 150oC
[d] Sample set B aged for 168 h at 125oC
[e] Contamination agents were photoresist and outgassing products of epoxy cure
[f] Argon-oxygen plasma (50% Ar, 50% O2)

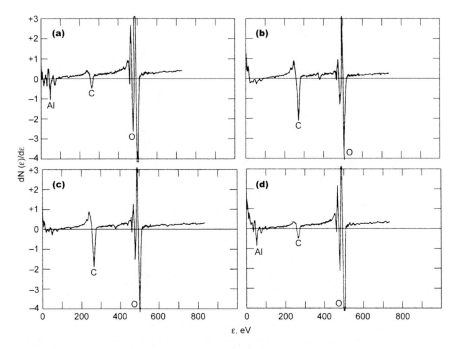

Fig. 4.14 Auger electron spectra of aluminum metallized silicon substrates both pre- and post cleaning with solvents and UV-ozone: (**a**) as processed substrate (uncontained); (**b**) substrate contaminated with photo resist; (**c**) substrate cleaned with solvent; and (**d**) substrate cleaned with UV-ozone

With pin or ribbon leaded packages in which the pin or ribbon feeds directly inside the package to form the wirebond attachment point, special care needs to be taken to ensure that the external lead structure does not resonant. Resonance in these external leads can set up vibration on the pin or ribbon end inside the package and can cause wire or wirebond failure, especially if the wire is relatively stiff. This would be especially important when parts in hermetic quad flat packages are cleaned prior to board attachment.

With today's fully encapsulated microcircuits, the cleaning of parts ultrasonically poses little risk, especially for leadless or short leaded components. The potential danger occurs when cleaning exposed wirebonds in open packages or in chip-on-board or flex applications. Another potential danger could be associated with microelectromechanical systems (MEMS) where ultrasonic resonance could cause mechanical failure of the MEMS structures themselves in addition to the potential damage to wirebonds. Again, it is a question of the resonance frequency of the structure compared to the ultrasonic agitation frequency. In all cases with exposed wires and structures, if ultrasonic cleaning methods are employed, cavitation should be avoided [38].

4.5 Testing

Since its introduction in the 1970s [39], the destructive wirebond pull test (ASTM F459-06, 2006) is the most widely accepted technique for the evaluation and control of both the wirebond's strength and the associated setup of bonding machine parameters. Despite its widespread use, due to low cost and ease of use, the destructive wirebond pull test has some significant disadvantages. First, since it is destructive, it can only provide information on a lot sample basis for production product. It can be used for pre- and post-lot qualifiers to help setup the bonding machine and, of course, as a post mortem diagnostic tool in failure analysis or as part of routine destructive physical analysis for part acceptance. Thus, it does not provide a measure of strength for most of the bonds made. Second, in fine pitch wirebonded circuitry, it is difficult to insert the hook between adjacent wires without touching bonds (wires) other than the one of interest. Third, the destructive pull test provides very little information on the strength or overall quality of the bond interfaces as long as the chief failure mode is a wire break.

Only in the case of catastrophic interface failure, such as those encountered with impurity-driven intermetallic growth [9], will the destructive wirebond pull test yield information other than the relative breaking strength of the wire assuming appropriate correction is made for both the wire and test geometries [10]. This phenomenon is especially true in standard ball bonding situations where a ball of relatively large diameter (nominally 2.5–5.0 times the wire diameter) forms an effective bonding pad attachment that is many times stronger than the breaking strength of the wire. The attachment strength, however, can vary significantly due to bonding parameters, composition of interfacial materials, and environmental stresses. The control of these variations is especially important as the ball diameters shrink. In very fine pitch ball bonding, ball diameters on the range of 1.1–1.3 times the wire diameter are quite common.

These factors have led to the development of two complementary tests: (1) the 100% nondestructive pull test (NDPT) (F458-06, 2006), and (2) the ball-shear test (F1269-06, 2006). The 100% NDPT provides a degree of confidence that each bond is strong (at least to the nondestructive preset force limit. The NDPT has been shown to have a beneficial effect in eliminating potentially ultra low strength outlyers in the pull test distribution of microelectronic wirebonds. Figure 4.15 is an illustrations of a test performed on gold metullized ceramic on two identical thermasonic wirebond sample populations. One group was NDPT applied post bonding and the other did not. Both populations were aged for 240 h at 125°C. Results showed that the NDPT, which eliminated some low strength bonds prior to burn-in, kept the resulting aged distribution from having castrophically low strength outlyers. In fact, no pull strengths below the NDPT limit were observed.

The ball-shear test can be used to investigate not only the ball-bonding pad interfaces, but also the influence of both pre- and post-bonding factors. Table 4.7 summarizes the areas of application for both the wirebond pull test and the ball-bond shear test. A careful review of Table 4.7 illustrates the complementary nature

4 Advanced Wire Bonding Technology: Materials, Methods, and Testing

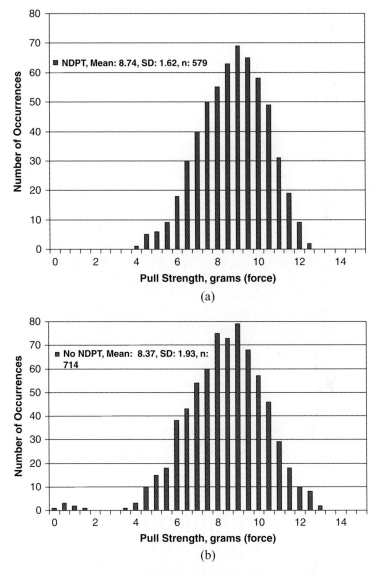

Fig. 4.15 Wire Bond pull strength histograms for 25 μm diameter thermosonically bonded gold wire on gold thin-film metallization on highly polished alumina ceramic. (**a**) after burn-in without NDPT; (**b**) after burn-in with NDPT applied post bonding. NDPT limit was 3 g (force)

of the destructive wirebond pull test and the ball-bond shear test. Figure 4.16 illustrates the improvement that can be achieved in the strength of the ball-bond pad interface by using the ball shear test (instead of the wirebond pull test) to optimize the bonding machine parameters [11]. This particular sample set was thermosonically bonded gold wire on aluminum metallized silicon.

Table 4.7 A comparison of areas of applicability between the wirebond pull test (ASTM Standard Test Method F458-84), and the ball bond shear test (ASTM Standard Test Method F1269-89)

Area of Applicability	Wirebond Pull Test	Ball Bond Shear Test
Module geometry	Yes	No
Wirebond geometry	Yes	No
Wire quality, defects, etc.	Yes	No[a]
Second Bond	Yes[b]	No
Bonding machine set-up, optimization, etc.	No[c]	Yes
Process development	No[c]	Yes
Substrate, bonding pad quality	No[c]	Yes

[a] Sensitivity to contamination, insensitive to mechanical defects
[b] Extremely dependent on geometry
[c] Insensitive unless the effect is catastrophic

As mentioned above, the most common gauge of wirebond strength, and hence quality has been mechanical testing, i.e., the wirebond pull test and the ball-bond shear test. Improvements in wirebond technology have caused both tests to have limitations. The pull test requires a hook to be placed under a wire, which is very difficult in situations where the wires are closely space without damaging adjacent wires. For successful NDPT, wires should be spaced at least two to three hook lengths apart. There is also the difficulty of applying a consistent force to the bond

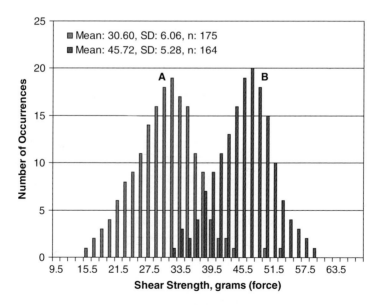

Fig. 4.16 Histograms of gold thermosonic ball bond shear strengths for bonds placed on aluminum metallization (over silicon). Histogram A are the shear test results after the bonding machine was set up using the wirebond pull test. Histogram B are the shear test results after the bonding machine was optimized using the ball shear test

interface, since the tensile and shear forces on the bond vary with the wire length and the hook position along the wire [38]. The ball shear test requires that a ram (wedge-shaped tool with a flat or slightly curved face) be placed on the major diameter of the ball. If the ball is low profile or flat such as those encountered in fine pitch wire bonding (Section 4.9), or thermocompression wire bonding, the ram can easily ride up over the ball. With close spaced bonds (40–50 µm or less separation) the ram can run into adjacent bonds causing damage.

Non-destructive ball shear (NDBS) is possible in direct analogy with NDPT. In NDBS the ram loads the ball to a preset value (nominally a fraction of the envision shearing strength of the device or system under test) and if a failure does not occur the ram is retracted and moved to the next ball bond. Experiments [12] have shown that NDBS does not affect ultimate destructive shear value, at least with balls formed from 25.4 µm diameter wire with reasonable ball diameter (greater than or equal to 2.5D) There is information in the ASTM Standard (F1269-06) that can be used to set the NDBS limit for various bonding situations. Other studies have shown that the ball can be loaded non-destructively up to 50–60% of its shearing strength without influencing the destructive shear.

Mechanical testing also tends to be time consuming and more importantly destructive. Even in non-destructive modes (see above) wires are deformed and ball edges flatten in the case of non-destructive shear testing [11], thus giving rise to concerns about future product reliability. Hence most people recommend the mechanical testing of product on a lot sample basis only and, of course for the set-up of wire bonding machines.

A new method for wirebond testing has been developed to address the mechanical test limitations [73]. The technique uses a laser to generate an ultrasonic pulse which is passed through the bond interface and detected nearby. The test is non-destructive, fast, and appears to detect bond interface anomalies. The ultrasonic wave train is thermoelastically generated by a sub-nanosecond laser pulse hitting the top of the ball or wedge bond. The ultrasonic wave travels through the ball or wedge bond and the bond interface onto the surface of the IC. The ultrasonic wave is then detected on the surface of the integrated circuit by a laser interferometer that measures changes in the surface height. This surface displacement versus time data is then numerically converted to power versus frequency data, or Power Spectral Density (PSD). The laser ultrasonic bond testing has several potential advantages over the standard mechanical tests: (1) it is non-contact and (2) it is non-destructive. All devices produced can be tested, so quality data does not have to be inferred from a lot sample. In addition, the equipment is controlled by computer so the potential exists to fully implement the test for high production rates when attached to a wirebonder for real-time bond assessment.

A schematic representation of the test configuration is shown in Fig. 4.17. Figure 4.18 presents displacement versus time curves recorded by the interferometric detection system. The numerical analysis results for a representative sample (bond aged 48 h at 250°C) are shown in Fig. 4.19. The dotted spectrum is the result of applying standard Fast Fourier Transform analysis

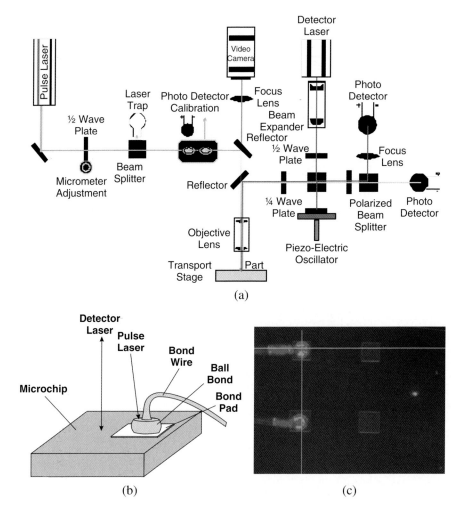

Fig. 4.17 Laser-induced ultrasonic energy wirebond evaluation system. (**a**) optical system schematic; (**b**) schematic representation of placement of the excitation and detection laser beams relative to the wirebond; (**c**) a photomicrograph showing the location of the excitation laser (cross hairs on top of ball bond) and the detection laser (white dot on right)

methods to the displacement versus time curve to extract the PSD. Further analysis using an autoregressive covariance-based technique produced the solid line shown in Fig. 4.19. The covariance method clearly shows a resonance response at 14.5 MHz. Applying this method to the other samples produced the data shown in Table 4.8. Table 4.8 presents the fundamental peak frequency and power levels for the aged samples along with shear strength data from bonds of the same population. Details of these results along with complete description of the method can be found in the paper by Romenesko, et al., [73].

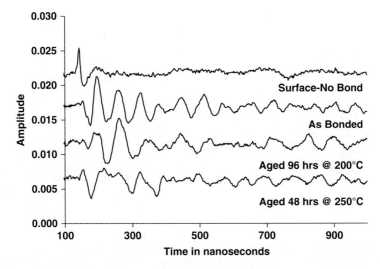

Fig. 4.18 Displacement amplitude vs. time for bonds with different aging conditions. "No bond" illustrates noise level after bond pad surface is pulsed with a laser. Traces represent averages of at least seven individual trials and have been offset in amplitude for clarity

The laser ultrasonic bond evaluation has correlated a shift in the ultrasonic frequency spectrum with both bond aging and intermetallic growth. The ultrasonic wave detected was shown to be a true surface wave and thus, non-dispersive in nature. Results proving the ultrasonic wave is a surface wave are given in

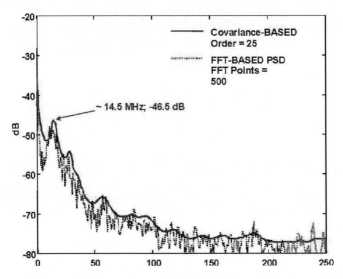

Fig. 4.19 Comparison of the power spectral density (PSD) resulting from the fast fourier transform (FFT) and the auto regressive (covariance based) numerical methods

Table 4.8 The effect of thermal aging on the power spectral density (PSD) behavior of typical thermosonic ball bonds made to various metallizations on silicon

Sample	Frequency, MHz	Power, dB	Shear strength* grams (force)
Sample 1: Al-1%Si			
Substrate	18.5	−56.0	
As bonded	16.5	−39.0	51.7 ± 1.8
Aged: 96 h @ 200°C	13.5	−44.5	60.7 ± 2.6
Sample 2: Al-1% + 0.5%Cu			
Substrate	19.5	−57.0	
As bonded	16.5	−45.5	54.3 ± 2.5
Aged: 48 h @ 250°C	14.5	−46.5	57.6 ± 2.2

*Shear strength obtained from other samples in the same sample population.

Fig. 4.20. This means that the detected frequency shifts cannot be attributable to spectral changes due to dispersion as the detection point is moved farther away from the bond pad. In addition, no significant directional dependence of the spectrum was found – again indicating that the measurements are insensitive to the detector location relative to the crystal axes of the semiconductor.

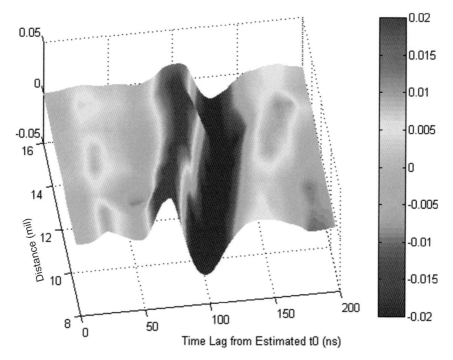

Fig. 4.20 Results of arrival time measurement with distance. Waveforms are arranged on edge and spaced by the distance to the detector, showing arrival time to be linear with distance. Vertical axis is displacement amplitude

4.6 Quality Assurance

Wire bond quality assurance (QA) is accomplished by establishing baseline standards for wirebond quality and then assuring that wirebonded product meets or exceeds these standards on a daily basis. Two principal QA tools are the wirebond pull test and the ball bond shear test described above. These test methods, coupled with visual inspection, can be used to ensure the strength of initial bonds as well as the strength of bonds receiving additional processing or thermal aging meet established standards. Section 4.5 above describes the wirebond pull and ball bond shear test in detail while Sections 4.3 and 4.5 above describe how the ball bond shear test can be used in optimizing bonding machine parameters for a given die-package/substrate configuration and metallization schemes.

Since most of the wire bond tests are practiced in a destructive mode, 100% testing is not possible, thus lot sampling and pre and post-run qualifier techniques must be used. Pre- and post- run wire bond qualifiers are used to ensure that on these standard test samples (representative of the product), the bonding machine, the bonding wire, and the operator performance have remained constant from the beginning to the end of a product run. If any change is made during a product run an additional qualifier must be bonded and tested. If no changes occurred during the day (or shift), usually beginning of the day and end of day qualifiers are sufficient. Pull test and/or shear test results are recorded for each qualifier and the data is used to produce a bond strength history over time. Run charts are usually produced and previous data samples are used to set control limits for the process. An example of a wirebond strength run chart is shown in Fig. 4.21. All deviations from the controlled process (outside the control limit) should be noted and their root causes determined. Qualifier and set-up samples should consist of at least 33 wires so that large number statistics can be employed and the strength distribution averages and standard deviations have their conventional meanings. Note: A smaller sample set can be accommodated using the students T-test as described in Section 4.9 below.

In addition to plotting the average and standard deviations on control charts, strength histograms should be produced on a regular basis to identify bimodality e.g. due to a different underlying failure mechanism. Bimodality due to geometrical differences in wire length and loop height has been discussed previously. Single mode underlying distributions are required to appropriately use the determined mean and standard deviation in a true statistical sense. Most wirebond tests put specification limits on the mean and standard deviation. For example, with 25 μm diameter gold wire these limits are: grams (force), and where equals the distribution mean and is the standard deviation. Full information on the use of the wirebond tests and how specification limits maybe set are given by Charles [12] and the references contained there in as well as the standard ASTM test references for the tests themselves (ASTM Standard:

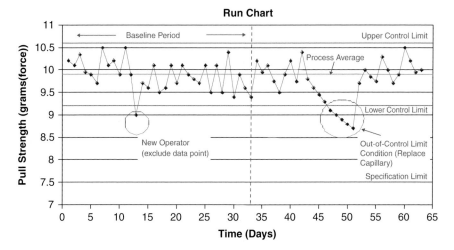

Fig. 4.21 Run chart for destructive wire bond pull test for thermosonically bonded 25.4 μm diameter gold wire on gold metallization (on ceramic). Process monitoring was used to develop control limits for the process during the baseline time limit. These limits (±2 in this case) are much tighter than the specification limit of 7.5 g (force) generated from grams (force) (Reference ASTM Standard Test Method: F459-06 (2006)). Here the run chart allowed the identification of capillary wear before the product fell below the specification limit

F459 – 06 "Standard Methods for Measuring Pull Strength of Microelectronic Wirebonds" and ASTm Standard: F1269-06 "Test Method for Destructive Shear testing of Ball Bonds").

Another quality assurance measure is working with wire vendors to ensure a continued supply of high quality wire of appropriate strength and temper with a minimum amount of drawing die lubricants. Poor quality bonding wire has been linked to several device failures over the years. Wirebond geometry has a significant influence on the apparent strength and quality of wirebonds. Careful control of contact pad geometry as well as bond length and height is necessary to ensure high quality bonds. In low profile bonds, wirebond pull testing can indicate low strength (due to resolution of force issues) while the bonds and wires are actually strong [39]. Other geometrical influences will be discussed inn greater detail in the design section below.

4.7 Reliability

Wire bonds have been shown to be a highly reliable, flexible, interconnection scheme for decades. In fact, wirebonded products have been in continuous use in space and other domains for over 25 years. Automated bonding has introduced a new level of control and precision bond placement that even further

4 Advanced Wire Bonding Technology: Materials, Methods, and Testing 151

improves the reliability and reproducibility of microelectronic products. These automated bonders coupled with improvements in bond pad metallurgy, reduction in bonding wire unwanted impurity content, more effective pad cleaning processes, high purity and stable die attach adhesives, and reduced temperature bonding processes (ultrasonic and thermosonic) have contributed to the current widespread use and reliability. In fact, wirebond defect rates, for single chip packages, are typically in the very low parts per million (ppm) range (e.g. 3 ppm or less).

Wirebonds, like all complex physical and chemical processes, can be fraught with reliability detractors if proper cautions and controls are not exercised and if phenomenological factors are not well understood. Some of the typical problems include: mechanical wire fatigue due to conditions of thermal or power cycling; interactions both chemical and mechanical with encapsulation during molding and after cure; corrosion induced by the die attach material, the atmosphere, and/or process-related contamination; and wire structural changes due to bonding parameters, such as uncontrolled grain growth associated with the heat-affected zone. An entire book by Harman [38] has been devoted to reliability issues and yield problems. Of all the issues two particular ones deserve further discussion: intermetalics and cratering.

4.7.1 Intermetallics

The most widely studied and publicized wirebond reliability probability is associated with the alloying reactions that occur at the gold wire-aluminum alloy bonding pad interface (and, to a much lesser degree, aluminum wire-gold bonding pad interface). Aluminum-gold intermetallic formation occurs naturally during the bonding process and contributes significantly to the integrity of the gold-aluminum interface. Intermetallics (in particular, $AuAl2$ or purple plague and $Au5Al2$ or white plague) are generally brittle; and, under conditions of vibration or flexing (either mechanically or thermally induced due to coefficient of thermal expansion mismatches), may break due to metal fatigue or stress cracking, resulting in bond failure [67].

At elevated temperatures, aluminum rapidly diffuses into the gold forming the $AuAl2$ phase, leaving behind Kirkendall voids [67] at the aluminum-$AuAl2$ interface. Figure 4.22 shows views of extensive intermetallic growth around and under various thermosonic wirebonds (both ball and tail bonds). Kirkendall voiding has also been observed at gold-$Au5Al2$ interfaces. Excessive intermetallic growth can lead to the coalescence of voids, which can lead to a bond crack or lift and an open circuit. Impurities in the bonding wire, on the pad metallization, or at the wirebond-pad interface have been shown to cause rapid intermetallic growth and Kirkendall voiding at temperatures below those associated with normal intermetallic formation [9]. Table 4.9 gives the formation temperature, activation energies, and some notes for the five aluminum-gold

Fig. 4.22 Scanning electron photomicrographs of advanced intermetallic growth: (**a**) underside of ball bond with regions of intermetallic voiding (Kirkendall); (**b**) residual intermetallic left on bonding pad corresponding to the voided regions of the ball in view (a) ; and (**c**) tail bond with extensive intermetallic formation under the bond edge and consuming part of the flattened bond region. Magnification approximately 75X

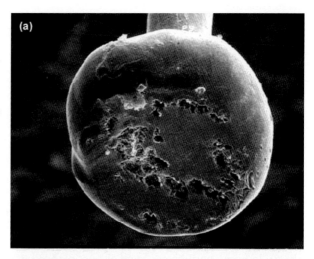

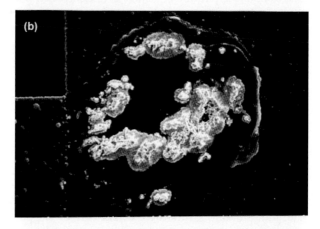

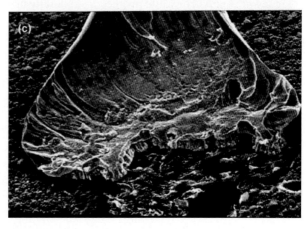

Table 4.9 Aluminum-gold intermetallic alloy properties

Alloy[a]	Formation temperature (°C)	Activation Energy[b] eV	kJ/mol^{-1}	k cal/mol^{-1}	Comments
Au_5Al_2	23–100	0.62	59.4	14.3	Tan in color
Au_2Al	50–80	1.02	98.3	23.5	Metallic gray in color (orthorhombic, randomly oriented monocrystals
$AuAl_2$	150	1.20	115.8	27.7	Deep purple in color (purple plague-resistivity 8 µ′Ω·cm)
Au_4Al	~150				Tan in color
$AuAl$	~250				White in color

[a] The intermetallic alloys typically form in the order listed (Au_5A12,... $AuAl$) consistent with their temperature of formation.
[b] A range of activation energies from 0.2 to 1.2 eV, have been observed for the aluminum-gold system depending upon growth, testing, and contamination conditions

intermetallics. The deleterious effects of intermetallics can be controlled if the time of exposure to high temperature is minimized and if proper materials and cleaning procedures are used [86]. See also Section 4.4 Pad Cleaning above. Design rules have been developed for minimizing intermetallic void failures by controlling film layer composition and thickness [24]. In addition, proper optimization of the wire bonding process has a significant influence on intermetallic growth (e.g. [25]).

4.7.2 Cratering

As mentioned previously, cratering can be a significant problem associated with the bonding and subsequent shearing of ball bonds from silicon integrated circuits. Intermetallic formation, induced stress, metallization thickness, bonding parameters, and underlying dielectric layers have all been noted to have an effect. [25] (Clatterbaugh and Charles, 1990). To help separate these phenomena, a series of cratering-related experiments (bonding, etching, metallizations, ect.) and finite elements analysis (FEM) have been performed. The results of these studies show:

1. the effects of gold-aluminum diffusion-induced strains within the weld region are negligible compared to those introduced by shear testing
2. the smaller the weld region, the more likely the underlying silicon will crater when shear tested;
3. the taller the ball bond, the more likely the underlying silicon will crater when shear tested; and
4. the stress field for an angular type weld is similar to that for a continuous circular type of the same radius.

Thus, a flatter bond with a larger weld area (or large annulus) is less prone to produce silicon cratering when shear tested.

The results of the etching experiments indicated that the occurrence of cracks in the bare silicon or in the silicon dioxide (SiO_2) on silicon due to improper bonding parameters did not occur, if the bonding machine parameters fell within the bonding window determined for the experimental configuration.. This rules out the requirement that initial substrate damage due to improper bonding parameters is necessary for ball shear-induced cratering to occur.

For the case of the borophosphosilicate glass (BPSG) on SiO_2, damage of the thin glass coating was prevalent for almost all bonding conditions. The best set of bonding conditions to resist cracking in BPSG films was found to be the combination of low power, short dwell, high stage temperature, and high force. This is consistent with the results reported by Koch and his co-investigators (Koch, et al. 1986) who studied the effects of bonding parameters for thermosonic gold ball bonding on aluminum pads over phosphosilicate glass (PSG).

The results from the metallization thickness experiment do indicate a significant reduction in the incidence of ball shear-induced cratering as the metallization thickness is increased. However, since the etching analysis of untreated wire bonded samples showed no initial substrate damage as mentioned above, a cushioning effect of the additional metallization was ruled out. A more plausible explanation would be that the additional metal would prevent the alloying of the gold ball and aluminum to the underlying silicon substrate during the thermosonic scrub. This would prevent a rigid link to the substrate available to transfer shear energy to the underlying silicon substrate.

Results from the experiment conducted to study the effect of bonding conditions on ball shear-induced cratering indicated that the greater the force and the power parameters or, equivalently, the stronger the bond, the less likely the substrate was to crater when ball shear tested. This is equivalent to stating that the larger the weld (a flatter bond), the less likely that cratering will occur. Once again, this is consistent with the results from finite element modeling and previous data indicating that the manufacture of larger, more robust bonds is less likely to cause cratering [25].

Some conclusions can be drawn from the information above concerning the effects of the gold-aluminum intermetallic on the cratering effect in silicon and thermally grown SiO_2 reported in Weiner, et al. (1983). As stated above, the effect of strains introduced due to structural misfit of intermetallic phases is of second order compared to those introduced by the ball shear ram. Also, no damage was observed in either the silicon or SiO_2 for the full range of bonding parameters used in this study. Therefore, it must be concluded that the cratering effect observed here is not a result of initial substrate damage. Several factors point to the rigid intermetallic bond between the ball and the substrate as the sufficient cause for call shear-induced catering. These factors include:

1. the absence of cratering in samples which had not been thermally annealed (i.e., there is little intermetallic formation, and, thus, the aluminum

metallization will yield before much energy can be transmitted to the underlying substrate);
2. the delay in the onset of cratering for thicker metallizations (a longer period of time would be required to penetrate the thicker metallization and alloy with the underlying silicon substrate);
3. the effect is significantly less for pads over SiO_2 (Au-Al intermetallic does not form as good a bond to SiO_2 as it will to bare silicon); and
4. as the weld area increases with additional intermetallic formation, the estimated stress concentration factors become significantly reduced, thus explaining the reduction in cratering in strong bonds after prolonged thermal aging.

In summarizing this cratering effect, a thicker bond pad metallization, a larger weld area (high ultrasonic energy and stage temperatures), and flatter bonds (high force) will produce bonds that are more resistant to any form of shear-induced cratering for bond pads on SiO_2. Weak bonds are to be avoided since they can crater at much lower shear values. Preliminary information suggests that for bonding above multilevel oxides such as PSG, BPSG, and other low-strength chemical vapor (CVD) deposited oxides, thicker pad metallization and the combination of low power, short dwell, hot stage temperature, and high force will be required to produce reasonably strong bonds and reduce damage to the underlying dielectric layers.

4.8 Design (Wire Spacing, Loop Height)

As mentioned above, wire bonding is a fully automated process with both automatic thermosic and ultrasonic bonders being used world wide. Automatic bonders use pattern recognition to locate fiducial marks on both the chip and the package or substrates. The positions to the reference points relative to the location of the bonding pads is stored in the bonding machine's computer memory, so that once aligned (registered), the machine automatically bonds all connections according to a pre-programmed sequence at rates over 15 bonds per second. Today's shrinking bond pad size and the increasing number of I/O connections on current ICs, have forced wire bonding (bonding equipment) to keep pace in order to maintain its dominant position in the chip interconnection world. Chips with I/O numbers greater than 1000 have been routinely interconnected using wire bonding techniques. High I/O chips typically use two or more rows of bonding pads along their perimeters with pad sizes as small as 25–30 μm wide and 25–40 μm in length and in-row pitches down to 35 μm. Some wirebond test chips have been seen with four rows of pads with in-row pitches of 40 μm and effective wire pitches from the four rows of 10 μm. See Fig. 4.6. These close spacings have dictated the greater use of wedge-wedge bonding or wire diameters well below 25 μm for the ball-wedge process. Wires as small as 15 μm in diameter have been routinely thermosonically ball bonded in production.

Historically, ball bonding standards required relatively large balls compared to the wire diameter (i.e. greater than or equal to 2.5 times the wire diameter) yielding effective bond areas five to six times greater than the wire cross-sectional area (see Fig. 4.2). Today ball bonds are small nail heads with sizes down to approximately 25 μm in diameter (for 15 μm diameter wire). Thus, the typical modern ball diameter standard requires the "ball" to be about 1.4–1.5 times the wire diameter (D). Such diameters would yield effective bond cross-sectional areas (wire to pad) [12] of about twice that of the wire and thus should still be robust enough to avoid ball lifts under pull testing on well made bonds. Ball (nail head) diameters down to 1.2D have been seen in production. Figure 4.23 is a scanning electron microscope photomicrograph of fine pitch thermosonic ball bonds.

In the ultrasonic bonders, the historical deformation of the bond foot was 1.5 times the wire diameter. See Fig. 4.3. Today that deformation is down to about 1.1–1.2 times the wire diameter in fine pitch applications. Thus, for the same wire diameter wedge-wedge bonds can be placed closer together provided the required bond wire geometry (height, length, first bond- second bond location, etc.) can be accomplished with the ultrasonic bonders in line step. Figure 4.24 us a scanning electron microscope photomicrograph of fine pitch ultrasonic wedge bonds.

As mentioned above, today's packaging environment is creating new geometrical challenges for wire bonding. Not only are we seeing multi-tiered pad arrangements (two, three and sometimes four tiers as discussed above, See Fig. 4.6) but also the requirement for low profile bonds necessitated by the wide spread deployment of stacked packages. Stacked packages typically have one of two staked die configurations: (1) pyramid or (2) overhanging die of the

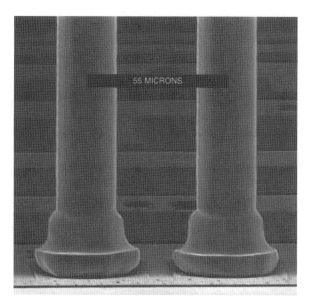

Fig. 4.23 Scanning electron photomicrograph of ultrafine pitch (55 μm) thermosonic ball bonding. The bonds were made on a K&S Model 8020 automatic ball bonder using 23 μm (0.9 mil) diameter gold alloy wire. Pad metallization was Al + 1% Si + 2% Cu on SiO2 with nominal 1μm thickness. (Photomicrograph courtesy of L. Levine, K&S)

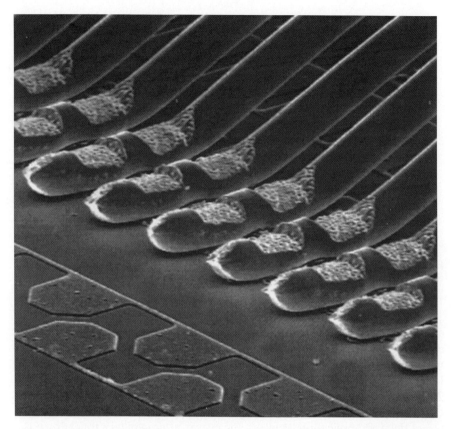

Fig. 4.24 Scanning electron photomicrograph of ultrafine pitch (40 μm) wedge bonds. The bonds were made on a K&S Model 8060 automatic wedge bonder using 20 μm (0.8 mil) diameter gold alloy wire. Pad metallization was Al + 1% Si + 2% Cu on SiO_2 with nominal 1μm thickness. (Photomicrograph courtesy of L. Levine, K&S)

same size. These two stacking configurations are shown schematically in Fig. 4.25. Stacking requires special wirebond profiles and low loop height. As die thickness decreases the spacing between the loops of the different tier wirebonds must decrease proportionally to avoid wire shorts between the different wiring layers. The top layer loop also must remain low to avoid wire exposure during molding. The maximum loop height should be no higher than the thickness of the die to maintain an optimal gap between the wire tiers. For example, if the die thickness is 100 μm, the optimal loop height would be 100 μm or less.

Some situations even require reverse bonding (i.e. the ball is on the package substrate and the tail bond is on the die). In a normal ball bonding process the ball is placed on the die contact pad and then after wire looping to the second bonding location the tail or stitch bond is formed on the substrate or package contact. In a reverse ball bonding process a stud bump is placed on the chip

Fig. 4.25 Thinned die (25.4 μm in thickness) flip chipped on since layer flexible tape (37 μm in thickness). Each chip contains over 1200 solder joints. Assembles using appropriate underfill survived 5000 cycles of temperature cycling from –40°C to + 125°C

contact pad. This bump provides elevation above the chip and acts as a force distributor for the stitch bonding process to come. Next the chip is wire bonded with the ball bond placed on the substrate and the tail bond placed on the bumped chip bonding pad. Loop height with reverse bonding can be less than 75 μm. Over hanging thin die (thickness down to 50 μm) require special bonding techniques die to die flexure (bending) upon application of bonding force. The use of delayed application of ultrasonic energy after capillary touchdown is a must.

4.9 Advanced Concepts

4.9.1 Fine Pitch

Fine pitch ball and wedge bonding is continuing to evolve rapidly as described in design Section. While most ball-bonded products are still in a pitch range of 100 μm and above, production quantities of 90 μm pitch are being manufactured. Pitches in the 60–90 μm range are in volume production, while pitches of 60 μm and below have been used on a limited scale (see Fig. 4.23). Such bonds must be made with bottleneck or stepped-neck capillaries to avoid damaging adjacent wires. Today most bonding machines are limited to minimum pitches between 35 and 70 μm. The bond is quite different from a traditional ball bond. It is quite low, almost nail head-like with a "ball" diameter in the range of 1.2–1.5 times the wire diameter. The low height of the nail head (typically 5–15 μm) makes the fine pitch ball bond difficult to shear. Most fine pitch ball bonds are still done with 25 μm diameter gold wire, although 15–20 μm wire is gaining popularity. Very fine pitches (<60 μm) and wires smaller than 25 μm in diameter are subject to greater damage in handling and molding operations than their larger more robust counterparts.

Wedge bonding leads the fine pitch parade. Wedge bonds at pitches of 40 μm have been demonstrated using 10 μm diameter gold wire. Wedge bonds at 60 μm and above are made in high volume production using 25 μm diameter gold or aluminum wires. To achieve such fine pitches, the wedge bonds typically have low deformation (= 1.2 wire diameters). Narrow, cutaway wedge tools are necessary to prevent adjacent wire damage during bonding. An example of 40 μm pitch wedge bonding is shown in Fig. 4.24.

Fine pitch is limited by the lack of chips with appropriately sized and space bonding pads that can take full advantage of the reduced size, high density wire bonding technology. Shrinking bonding pad size and pitch on chips is further hampered by limitations in test probe placement and movement. High frequency bonding (<60 kHz) has been shown to be beneficial in bonding fine pitch circuitry [34]. More details on higher frequency wirebonding will be given below.

There are many issues associated with the implementation or use of fine pitch wirebonding. Fine pitch bonding can only be accomplished successfully if the entire process (chip, package or substrate, bonding machine, and bonding practice) is designed from the beginning with fine pitch in mind. The size, placement, and shape of the bonding pads must be coordinated with the selection of the wirebonding machine, the die attach machine and process, and the package or substrate (board) layout. Square bonding pads (hexagonal or round, also) are optimal for ball bonding but pose some limitations for wedge bonding. Ideal wedge bonding pads would be long and narrow [66]; but these are seldom used because of the need to be flexible in bonding method choice and that automatic wedge bonders are, at best, a factor of two slower than automatic ball bonders (due to the need to index either the bonding head or the sample table to maintain wire alignment under the wedge). Thus, a high volume wedge bonded product will cost more than a product interconnected by ball bonding methods, even given the difference in wire cost (aluminum vs. gold, respectively).

It also should be recognized that extremely fine pitch, with any bonding technology, can result in higher costs due to added constraints, reduced throughput (generally lower bonding speeds), and typically a more fragile product. For example, the reverse bond technique described in the design section above can slow the bonding process by up to 50%. Both equipment and workers associated with the fine pitch process are typically more expensive than those associated with a conventional (low pitch) process. Automatic bonders need the latest in precision pattern recognition coupled with the most accurate placement control. Programming time is greater, and workers must be better trained to master the art of fine pitch. Die attachment machines also must have greater accuracy in the placement process than machines used in conventional pitch processes. Excess die rotation from die attach can cause shorting problems in fine pitch applications. For example, a ground wire adjacent to a power interconnection at the same loop height, loses clearance rapidly as a function of die rotation. Packages, as the wirebonding pitch declines (especially with multiple tiers of bonding pads), must be carefully designed to give maximum bonding tool access, while minimizing chances of wires touching or wire misplacement causing shorting. The ultimate design practice will force package and substrate pitches to those of the chip, thus minimizing fan out and keeping wire lengths short, which will reduce lead inductance and minimize injection molding wire sweep [79]. Copper wire could have an advantage in both electrical performance and mechanical integrity, but the ability to form minimal size balls (necessary for fine pitch) is still under development. A 60 μm pitch has been demonstrated with 25 μm diameter copper wire.

The solution to wires touching and shorting in fine pitch wirebonding could be the use of insulated wire. Insulated wire with appropriate bonding pad capping metallization could also allow chip on board assemblies to be made without insulating glob tops on overcoats. Insulated bonding wire has been around for over 20 years, but has never received widespread attention, mainly due to a host of implementation/reliability problems including wire coating contamination of the capillary, flame off inconsistencies, and low second bond strength. Recent advances (Microbonds, Inc., 2005) in wire coating technology appears to have made the spectre of coated wire viable. Coated bonding wire has obvious advantages including allowing wires to be close together, cross, and even touch. Such ability could solve wire sweep issues and die/wire shorting problems encountered in stacked die or in high density wirebonding in general. The newer coatings appear to be about 0.5 μm in the thickness on 25 μm gold wire with breakdown strengths approaching 200 V and the ability to survive baking temperatures of 300°C. Wire strength and bonding ability appear not to be reduced by the coating, but more reliability studies will need to be performed.

4.9.2 Soft Substrates

Deformable or soft substrates in modern wirebonding applications are usually associated with organic based boards or layers as follows: thin-film, multilayer structures on inorganic carriers such as encountered in multichip modules (MCM-Ds); laminate-type organic constructs such as encountered in printed wiring boards, MCM-Ls and chip-on-board structures [17]; and chips mounted to unreinforced laminates and/or flexible film layers.

MCM-D modules are made using deposited dielectric and thin-film metal layers. The carrier for these deposited films is usually silicon, although highly polished ceramics have been used in the past [13]. The dielectric materials are typically spun-on layers of polyimide. Benzocyclobutene (BCB), and several lesser-known polymers [77] also have been used. These dielectric layers usually range in thickness from 5 to 25 μm (or more), with as many as six layers being reported. Metallization schemes have been gold (with suitable adhesion layers such as chromium and tungsten), copper (again with suitable adhesion layers), and aluminum. In addition to organic dielectric layer softness, metal adhesion has been a challenge and requires careful processing to ensure metal layer integrity and inner layer adhesion.

In bonding to MCM-D structures, both thermosonic ball bonding and ultrasonic wedge bonding have been used. In bonding to MCM-Ds, two major issues arise: (1) the size of the bonding pad and (2) the number and thickness of the soft layers (polyimide, BCB, etc.) under the pad. It has been shown [16] that the pad bends or cups under the application of the bonding force. This cupping is due to the compliant nature of the organic nature. Elevated temperatures exacerbate the issue, effectively softening the polymer even more. Small bonding pads have less area over which to distribute the load and are thus more susceptible to this cupping or bending phenomenon. Pad deformations under bonding forces and the application of ultrasonic energy have been studied by Takeda, et al. [78].

Their results show that normal sized gold pads on copper traces (on polyimide flex boards) can deform as much as 20 μm under normal (but high end of the range) force and ultrasonic energy bonding conditions. They also verified that the use of a nickel underlayer (under the gold pad) can significantly reduce the deformation below 10 μm for all bonding conditions. Others have noted similar deformations but the amount of deformation was smaller. In our work, for example, we have observed that for a given bonding force, the deformation increases with organic layer thickness. Pad reinforcement structures and interlayer metallization tend to mitigate deformation. Similarly, a marked decrease in deformation was observed as the bonding force was reduced in all samples, with little or no correlation to changes in sample thickness.

In addition to unreinforced substrate materials, MCM-L and COB implementations can use fiber reinforced organic matrix material such as polyimide or epoxy. The reinforcing fibers are typically glass, although materials such as Kevlar®, quartz, and Aramid® have been used. Sometimes high-frequency circuitry is built on non-fiber reinforced substrates with very low dielectric constants such as Teflon® (polytetrafluorethylene). Most of these "laminate" technologies use copper metallization protected by thin layers of plated gold (usually with a nickel barrier layer under the gold). The thicknesses of both the metal and dielectric layers are larger than those of the MCM-D technology by factors of 5 for the metals and at least an order of magnitude or more for the dielectrics.

Other MCM-L implementations use fiber reinforced cores with non-reinforced resin layers on their surfaces [37]. Such structures can employ a variety of metallization schemes put in place and patterned by a combination of thin-film deposition (MCM-D) and printed wiring board (PWB) techniques. Via fills can be plated or actually filled with conductive organic resins [35].

Wirebonding to most MCM-L substrates including those in ball-grid arrays (BGA) and chip-scale packages (CSP) is similar to bonding to PWBs provided the substrates are made with fiber-reinforced resin laminates (e.g., polyimide-glass, epoxy-glass). Direct bonding to PWBs has been done for some time in COB applications. Many problems still exist with bonding to standard PWB fiber reinforced laminates, let alone the new problems associated with reduced pad sizes, unreinforced organic layers, and different via construction techniques found in today's MCM-Ls, BGA and CSP substrates, and integrated circuit redistribution layers. Both aluminum wedge bonds and thermosonic gold ball bonds have been used in COB applications. Wedge bonding is often preferred because it can be done without added substrate heat. Large COB assemblies will tend to warp and possibly soften if heated to or near their glass transition temperature (Tg). FR-4 (epoxy-glass) circuit boards have a Tg around 120°C, while Tg of various polymide boards exceeds 250°C. Such high-temperature resins can be thermosonically bonded provided proper substrate clamping and backside support is available for large area assemblies. Successful thermosonic bonds have been made at temperatures below 100–110°C so that even FR-4 can be bonded. Even with the thick metallizations typically encountered in the COB arena (e.g., nominally 17–35 µm), anomalies can exist in wirebonding, especially as pads shrink in size. Bonding to BGA and CSP flexible substrates is typically done with gold-ball bonding because of the need for controlled shape bonds and bonds that are very close to the chip edges to keep the package footprint as small as possible. Because of the small area and reduced thickness of the substrate, special care has to be exercised in the bonding process.

In addition to flexible and software substrates, two other difficult bonding situations occur: thinned die and stacked die (either thinned or not). Thinned die have been around for some time, especially in microwave applications where gallium arsenide (GaAs) microwave devices have been thinned to 100 µm or less to provide better thermal performance. Gallium arsenide is more susceptible to bond cratering and to mechanically induced electrical defects than silicon. For a detailed study of cratering on silicon die, see the paper by Clatterbaugh and Charles [25]. GaAs is weaker than silicon by a factor of 2. The two major material characteristics or parameters that are most relevant to cratering have been shown to be hardness and fracture toughness. Hardness is a measure of the material resistance to deformation while fracture toughness is a measure of the energy (or stress) required to propagate an existing microcrack. The Vicker's hardness for GaAs is 6.9 (± 0.6) GPa while silicon is 11.7 (± 1.5) GPa. In a similar vein, the fracture toughness of GaAs and silicon are $1.0 j/m2$ and $2.1 j/m2$, respectively. Thinned silicon die are now being mounted to flexible circuit boards. Silicon die as thin as 25 µm have demonstrated electrical integrity. An example of 25 µm thinned die mounted to a flexible tape substrate is shown in Fig. 4.25. Wirebonding, because of the thinness of the die and the softness of the flexible substrate has proven difficult and techniques are under development to allow wire bonding of these ultra thin assemblies. To date most of these assemblies have been flip chipped (i.e., attachment by solder reflow, Banda, et al., 2004).

4 Advanced Wire Bonding Technology: Materials, Methods, and Testing

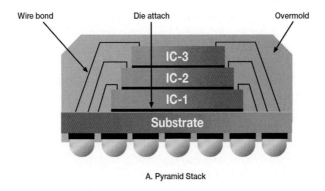

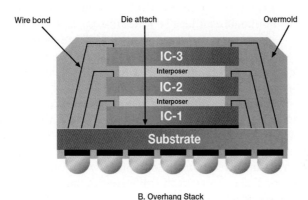

Fig. 4.26 Stacked die arrangement. View A: Pyramid Stack, View B: Overhang Stack

Stacked die (See Fig. 4.26) present their own set of issues, but in general, the problems involve multiple geometries in a given component package with closely spaced wirebonds that can overlap. In addition, sometimes the bonding must be done to chips that are cantilevered over another chip without a means of mechanical support under the bonding pad areas. Fixturing and very careful control of bonding parameters (reduced force and power, higher frequency, and temperature) has allowed successful wirebonding to stacked geometries with as many as six chips. A full discussion of the details of wirebonding to stacked chips is not possible in this work, but some insight can be gained by reading Yao, et al., [88].

4.9.3 Higher Frequency Bonding

Most of the world's current wirebonding machines have ultrasonic generators and transducers that operate at nominally 60 kHz. The choice of 60 kHz was made several decades ago based on transducer (bonding head) dimensions for microelectronic assemblies and stability during the bonding (transducer loading) operation [38]. Other frequencies from 25 to 300 kHz have been used to attach wires. Ultrasonic welding and material softening have been reported in the range between 0.1 Hz [87] and 1 MHz [53]. Today's interest in higher frequency bonding stems from reports by various authors [69, 75, 82, 40, 44, 34] that using higher ultrasonic frequencies produces better welding at lower temperatures in shorter bonding times (dwell times). It has also been indicated that higher frequency wirebonding improves bonding to pads on soft polymer layers such as Teflon®, unreinforced polymide, and flexcircuits. While all these improvements were real for the particular situations in hand, few if any controlled studies (systematic, side-by-side experiments on the same substrates with an attempt to control all variables except frequency) have been performed. The following material presents excerpts from the one such study [20, 21].

Three metallization schemes were used in this study: (1) aluminum (99.99% pure) with a titanium/titanium nitride (Ti/TiN) adhesion layer; (2) aluminum plus one percent silicon alloy (Al + 1% Si) again with a Ti/TiN adhesion layer; and (3) gold metallization with a titanium-tungsten adhesion layer (TiW). The metal bonding pad formation layers were sputter deposited to thicknesses between 1 and 2 μm on silicon base layers. The silicon wafers were p-type with a nominal resistivity of 30–50 Ω.cm. The wafers were thermally oxidized to achieve a SiO2 thickness of 1 μm prior to metal deposition or spin coating with polymide. The polymide layers were between 5 and 20 μm in thickness. The gold metallization was also deposited on highly polished ceramic (99.6% pure alumina) substrates.

Various test structures were photolitographically patterned on each of the metal layers [16, 18]. The patterns included: arrays of bonding parts of varying sizes (150–25 μm square), a daisy chain pattern consisting of almost 650 wirebonds with the resistance of the wirebonds accounting for over 60% of the total resistance of the circuit, and a radially distributed wirebond pattern for shock and vibration testing.

All wirebonding for the study was performed with two semiautomatic thermosonic ball bonders (Marpet Enterprises, Inc., Model 827) equipped with negative electronic flame off (Uthe Technology, Inc., Model 228-1) for uniform control of free air ball size. The flame offs were adjusted to produce 60 ± 2 μm diameter free air balls as shown in Fig. 4.10. Free air balls as small as 22 μm in diameter have been formed with 15 μm diameter wire. One of the MEI Model 827 wirebonding machines was equipped with a UTI Model 25ST (64.1 kHz) transducer driven by a standard UTI Model 10G ultrasonic generator. The other Model 827 wirebonding machine was equipped with a UTI Model 4ST (99.5 kHz) transducer which was driven by a UTI 10G generator tuned for 100 kHz. In order to make both transducer waveforms similar, since the Model 25ST transducer is much larger than the Model 4ST, a short 60 kHz transducer (Model 17STL (63.1 kHz) was also used. A comparison of the transducer dimensions has been given previously [18]. Squashed ball sizes were also quite uniform with diameter ranging between 76 and 82 μm depending upon substrate.

This study has yielded a large amount of data. Key observations and findings include the following. It is clear that significant differences exist between bonding at nominally 60 kHz and bonding at 100 kHz. In addition to differences in transducer electronic waveforms between the standard 60 kHz (long) and the 100 kHz transducer, there exist differences in bonding machine optimization behavior. The 60 kHz system appeared to have a larger bonding window (i.e., for a given force and substrate temperature, a wider range of ultrasonic power and dwell times produced acceptable bonds (strong, yet not over bonded or with wire damage)) when compared to the bonds produced by the 100 kHz system. The 100 kHz bonding window, in addition to being smaller than the 60 kHz window, was also sharper (i.e., a smaller change in ultrasonic power and/or dwell in relationship to the window edge was required to go from acceptable bonding to either a no-bond condition or to an over-bonded condition when compared to the 60 kHz system). Despite the smaller, sharper fall-off of the bonding window, the 100 kHz system has one obvious advantage. It formed strong bonds in times that are 30 to 60% shorter than comparable dwells for the 60 kHz system. Comparison of both bonding systems and their transducer waveforms indicate that the 100 kHz system has much faster bonding pulse rise and fall times, along with a more stable voltage (or current) amplitude envelope than that of the 60 kHz system. Switching to a short 60 kHz transducer with dimensions comparable to those of the 100 kHz transducer produced ultrasonic drive parameters (voltage and current) similar to those of the 100 kHz transducer.

Shear test data on gold substrate metallizations on high polished ceramic showed that an optimized 100 kHz system produced much stronger bonds than the 60 kHz system (See Table 4.10). As can be seen from Fig. 4.10 and Table 4.11, this difference cannot be accounted for by average ball diameters (either pre- or post-bonding), which were essentially the same for both the 60 and 100 kHz systems. When the data was analyzed for the Al + 1% Si metallization (on oxidized silicon), the 60 kHz bonds appeared stronger. Although the difference between the 60 and 100 kHz test results was relatively small (less than 7%). However, when analysis of variance techniques were applied, the difference was significant at the 99% confidence level. Similar results were observed on thermosonic ball bonds attached to an integrated circuit chip (Al + 1% Si

Table 4.10 Gold thermosonic ball bond shear strength (grams (force)) on gold and aluminum (1% S1) metallizations at both 60 and 100 kHz[a]

Metal	60-kHz	100-kHz	Δ means	Significant[b]
Au (on ceramic)	68.4±3.7	84.8±6.5	16.4	Yes (highly)
Al + 1% Si (on silicon)	54.0±3.2	50.6±2.9	3.4	Yes
Δ means	14	34.2		
Significant[b]	Yes (highly)	Yes (highly)		

[a] Nominal sample size at each frequency was 100
[b] 99% confidence that the difference in the means is significant using analysis of variance with the F-test

Table 4.11 Gold thermosonic ball bond average diameters[a] (μm) on gold and aluminum (1% Si) metallizations at both 60 and 100 kHz[b]

Metal	60-kHz	100-kHz	Δ means	Significant[c]
Gold	89.1±4.0	88.3±2.9	0.8	No
Al + 1% Si (on silicon)	91.3±2.3	92.0±2.0	0.7	No
Δ means	2.2	3.7		
Significant[c]	Yes	Yes		

[a] Average diameter $= \frac{1}{n}\sum_{i}^{n}[(X_i + Y_i)/2]$.
[b] Nominal sample size at each frequency was 100
[c] 99% confidence that the difference in the means are significant using analysis of variance with the F-test

metallization), on which both the ball shear test and the wirebond pull test gave a small edge to the 60 kHz system. Although this data set was relatively small, the student's t-test indicated that the results were significant at the 99% confidence level. Independent of frequency, the difference in ball bond shear strengths between metallization types, were relatively large and highly significant. Bonds on gold were always stronger than bonds on Al + 1% Si metallization consistent with the results shown in many previous studies [12] (Charles, 1986 and Charles, et al., 1999).

Other differences were observed such as asymmetry of ball shape with metallization type. While no differences in average ball diameters [(X-diameter + Y-diameter)/2] were observed with frequency (Table 4.11). Any variations in average ball diameters, even those between metallizations (Table 4.11) could be accounted for by variations in the free air ball size between experimental series. On the other hand the differences in the X and Y diameter measurements are highly significant and appear to depend on metallization type (Table 4.12). On gold metallization, the as bonded ball diameter in the Y-direction or the direction of the ultrasonic scrub is larger than the orthogonal nonscrub diameter (X-direction) with consistent measurements for both 60 and 100 kHz. On Al + 1% Si, the non-scrub direction (X-direction) is larger than the Y-direction by a

Table 4.12 Gold thermosonic ball bond diameters (in μm) in directions perpendicular (X-direction) and parallel (Y-direction) to the direction of the ultrasonic scrub on gold and aluminum 91% Si) metallizations at both 60 and 100 kHz[a]

Metal	Frequency	X-direction	Y-direction	Δ means	Significant[b]
Gold (on ceramic)	60 kHz	84.9±4.8	93.2±5.2	8.3	Yes (highly)
	100 kHz	82.8±3.4	93.8±4.0	11.0	Yes (highly)
Al + 1% Si (on Silicon)	60 kHz	93.9±2.9	88.7±3.4	5.2	Yes (highly)
	100 kHz	98.7±2.8	85.4±2.6	13.3	Yes (highly)

[a] Nominal sample size at each frequency was 100
[b] 99% confidence that the difference in the means are significant using analysis of variance with the F-test

significant amount for both the 60 kHz and 100 kHz bonding systems. Similar behavior was also observed for pure aluminum metallization. The cause of this phenomena is not well understood, but is believed to be associated with the dynamics of the weld formation process. On gold there is the single interdiffusion of the gold wire and gold pad materials. On aluminum and aluminum alloys the formation of gold-aluminum intermetallics is key to the bonding process. The formation of the relatively hard intermetallics may tend to lock the developing bond in the direction of the scrub on the aluminum and aluminum alloy metallizations while on the gold (being relatively ductile) the bond may be able to fully expand in the scrub direction.

Table 4.13 shows results for both 60 and 100 kHz bonded samples under conditions of thermal aging (120 h at 150°C). Aging at 150°C has been shown to be very effective [16] for assessing wirebond (ball bond) quality and reliability without introducing unwanted effects caused by substrate interactions and other heat-related phenomena. Table 4.13 again illustrates the significant improvement in shear strength using 100 kHz bonding on gold metallization, this time for gold on a silicon substrate as compared to the gold on ceramic data given in Table 4.10. The small observed differences on the Al + 1% Si metallization for 60 kHz versus 100 kHz is also consistent with the results in Table 4.10, although in this case the difference is statistically insignificant at the 99% confidence level. Again, large and significant differences were observed in the shear strengths between the two metallizations with bonds to gold being much stronger than bonds on Al + 1% Si metallization. These results are consistent regardless of the bonding frequency. After aging, the shear strength of the bonds on gold, at both frequencies, remained essentially unchanged. On the Al + 1% Si metallization the strength of the bonds increased significantly for both frequencies. Again, 100 kHz bonding produced stronger bonds on gold metallization, while 60 kHz bonding appeared to have a slight edge on Al + 1% Si. The increase strength for the aged bonds

Table 4.13 Gold thermosonic ball bond shear strength (grams (force)) on gold and aluminum (1% Si) metallizations at both 60 and 100 kHz under conditions of thermal aging[a]

Metal	Aged[b]	60 kHz	100 kHz	Δ means	Significant[c]
Gold (on silicon)	No	81.4±4.6	97.4±3.7	16.0	Yes (highly)
	Yes	82.1±3.3	96.4±4.6	14.3	Yes (highly)
Δ means		0.7	1.0		
Significant[c]		No	No		
Al + 1% Si	No	47.0±3.7	46.5±4.3	0.5	No
(on silicon)	Yes	57.8±3.3	56.1±4.1	1.7	Yes (slightly)
Δ means		10.8	9.6		
Significant[c]		Yes (highly)	Yes (highly)		

[a] Nominal sample size at each frequency was 100
[b] 120 hours at 150°C
[c] 99% confidence that the difference in the means are significant using analysis of variance with the F-test

on the Al + 1% Si metallization is consistent with similar increases reported previously under aging [14], but the timeframe for the existence of the increased strength above the as-bonded condition appears to be longer in this particular experimental series.

4.9.4 Stud Bumping

Alternative forms of flip chip technology, make use of wirebonder-produced bumps. These alternatives include "stud bump and glue" techniques using standard or anisotropically conductive adhesive as shown in Fig. 4.27.

In the stud bump and glue technique, single-ended, thermosonic wirebonds are placed on the chip bonding pads by means of an automatic wirebonder using special bonding wire (Table 4.3). The balls are then coined or tamped to a uniform height using a special tool placed in the wirebonder. The stud bumped chip is then pressed on a plate containing a thin layer of conductive adhesive (epoxy). As the chip is lifted from the plate, a small amount of conductive adhesive adheres to each bump. The chip is then placed on the corresponding substrate pads and the adhesive is cured, resulting in the geometry shown in Fig. 4.27. In an alternative method, the epoxy can be preapplied to the substrate pads by screen printing or automated dispensing.

An anisotropic adhesive is an adhesive that has small conductive particles embedded in its nonconducting organic matrix. A bumped chip is then pushed down into the adhesive, capturing a few conducting particles between the bump and the mating bonding pad on the package or substrate. When the adhesive is cured, an electrical interconnect is made. In addition, the region between the chip and the board becomes rigid, mechanically holding the chip to the board. Thus, the adhesive also serves as an underfill [43]. Figure 4.27 also illustrates this situation along with providing possible construction details for the conductive particles in the anisotropic adhesive.

4.9.5 Extreme Temperature Environments

Wire bonding has proven to be a useful interconnect for ICs and other devices operating over a wide range of temperatures. In fact, with careful selection of materials, wirebonded interconnections can be used in packaging chips and other electronic components and devices from below $-200°C$ to over $+500°C$. Such temperature extremes are found in many current and future applications including: deep space, oil and geothermal wells, rocket and jet engines, and at some locations in and on the engines of automobiles. The standard materials used in today's interconnections must be changed to address the rigors of high and low temperature environments. The commonly used aluminum-gold wirebonds present on most ICs, have been reported (Harman 2007) to be useful for

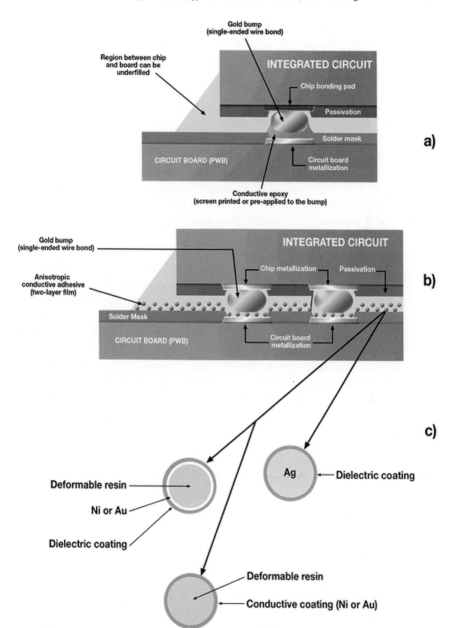

Fig. 4.27 Enlargement of tail bond region on gold pad after aging at 400°C for 168 hours. Note extensive build-up of material on the pad. SEM Photomicrograph (Magnification approximately 200X)

high temperature application, but because of the formation of brittle intermetallics and Kirkendall voids [67] the high temperature performance is limited to about 200°C. In a vacuum study for space applications of gold-aluminum bonds in plastic encapsulated microcuits, Teverovsky [80] reported a mean lifetime of 700 h at 225°C.

Wire manufacturers have begun changing gold bonding wire composition (doping) to reduce uncontrolled intermetalllic growth and void formation. Historically, impurities were controlled below the 10 ppm level, but in these new wires have stabilizing impurities greater than 100 ppm. Palladium alloying at the 1% is common practice, especially for study bumping. Recent studies (Breach et al., 2004) have shown these wires demonstrate increased resistance to intermetallic failures (longer lifetime), but again not nearly long enough for reliability application in high temperature environments. Thus, aluminum-gold interconnection systems should not be used in applications where temperatures routinely exceed 200°C.

Single metal systems, such as gold-gold or aluminum-aluminum, have been shown to be more reliable in high temperature applications and operate at higher temperatures. In fact, gold-gold interface strength has been shown to increase with both time and temperature [46].

Over the years, high temperature testing has shown that the gold-gold interface holds up well at temperatures above 200°C. Benoit, et al., reported excellent results when heating gold-gold systems to 350°C for 300 h. Recent experiments were conducted by this author to determine the high temperature limits of the gold-gold bonding system. Wirebond test patterns used for this study were reported previously [16]. Two different substrates were used: (1) high resistively (50 $\Omega \cdot$cm) bare silicon wafers and (2) silicon wafers with a SiO_2 layer. The bare silicon wafers simulated direct bonding to devices and sensors where bonding pads touch the underlying silicon. Pads on the SiO_2 were reflective of those found on ICs. In these experiments, the metallization used was sputtered gold with a titanium-tungsten (Ti-W) adhesion layer. The thickness of the gold metallization was approximately 2 μm while the thin Ti-W layer measured 0.05 μm. Patterning was done using conventional photolithography. Each test specimen consisted of 1200 bonds and several test specimens of each type were run and averaged. Both the wirebond pull test and the ball bond shear test were used to evaluate bond and wire quality. The thermal testing was conducted as follows: 100 bonds were pulled and sheared right after bonding to establish a baseline. Then all the samples were aged at 150°C for one week (168 h) in a nitrogen oven. Following aging, the samples were removed and 100 bonds out of the remaining bonds were pulled and sheared. The test samples were then returned to the oven and aged at 200°C for another week and then removed from the oven for the pulling and shearing of another 100 bonds. This process was repeated again and again raising the temperature 50°C per iteration. The process was terminated after aging at 550°C. The results of the test are summarized in Table 4.14.

4 Advanced Wire Bonding Technology: Materials, Methods, and Testing

Table 4.14 Thermal aging wirebond test results. Gold thermosonic wirebonds on gold pads with Ti-W adhesion layer (pull test and shear test results in grams (force))

Sample Aging Condition	Silicon		Silicon + SiO_2	
	Pull Test	Shear Test	Pull Test	Shear Test
As Bonded	12.3 (1.9)[a]	54.0 (2.7)	12.5 (1.0)	62.5 (3.5)
150°C b	12.5 (1.0)	60.9 (2.8)	12.3 (0.8)	58.0 (4.2)
200°C	13.9 (0.9)	60.7 (2.7)	14.0 (0.7)	64.0 (3.8)
250°C	12.5 (0.7)	57.6 (2.2)	12.5 (0.7)	62.2 (5.3)
300°C	11.3 (0.7)	56.1 (3.0)	11.2 (0.8)	57.4 (4.0)
350°C	11.1 (0.8)[c]	57.2 (3.7)[c]	11.3 (0.9)	57.1 (4.3)
400°C			10.9 (1.0)	63.3 (5.1)
450°C			10.6 (1.0)	59.7 (5.3)
500°C			10.2 (1.1)[d]	57.5 (4.9)[d]
550°C				

[a] Standard deviations are given in parenthesis.
[b] Sample aged at 150°C for 168 hours (one week). Each succeeding sample not only received the specified aging for one week, but also received all proceeding aging above it on the list.
[c] Anomalous metal migration: test terminated.
[d] Metallization lift: test terminated.

As can be seen from Table 4.14, the thermally aged samples show consistent and expected wirebond pull test and shear test results up to 350°C. At 400°C the wires on the bare silicon substrate displayed an anomalous effect (to be described below) and the test on silicon was terminated. Continued testing of the samples on SiO_2 produced consistent results to at least 500°C. At 550°C the samples began experiencing metal lifts and the tests were terminated.

A scanning electron microscope photomicrograph of the anomalous behavior of the gold wires on the gold pads on silicon is shown in Fig. 4.28. The wires became thin, embrittled, and broken with significant deposits of metallization (gold) on the bonding pad (See Fig. 4.29). This behavior was unknown to this author prior to running the test, and no obvious mechanism has come to light, but it appears to have been somewhat metallization pattern dependent. (Deposits formed on large tail bond pads rather than smaller ball bond pads and balls appeared relatively in tact.) The author re-ran the experiment with a more limited sample set concentration on the high temperature region again. Results (Table 4.15) were consistent with those shown in Table 4.14.

Again, results on Silicon were limited to 350°C while on SiO_2, wirebonds were strong at or above 500°C. Caution should be exercised in taking gold wirebonds on gold pads placed directly over silicon to temperatures in excess of 350°C until the phenomena described above is fully understood. Gold-gold bonds above 500°C on SiO_2 are believed possible with changes to adhesion layer materials and/or thicknesses.

While the author has not investigated the aluminum-aluminum bonding situation personally, there is considerable evidence in the literature (Harman, 2007), that this interface would also be robust at elevated temperatures. Given

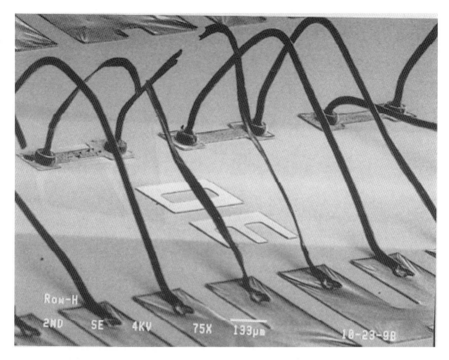

Fig. 4.28 High Temperature failures in 25.4 μm gold wire thermosonically bonded to gold pads on silicon. SEM photomicrograph (Magnification approximately 75X)

aluminum's melting point of 660°C (compared to gold at 1060°C), it should be recognized that its upper temperature use limit would be more restricted than that of gold. Aluminum-aluminum interfaces remained strong under testing for 300 h at 350°C (Harman, 2007). The above leads to the fact that for high temperature environments, one should use monometallic welds or ones which form solid solutions (such as gold with another noble metal). The aluminum-nickel system has also been shown to be robust at high temperatures and should prove useful in high temperature power device applications. Aluminum wire doped with magnesium has proven utility in resisting fatigue failures due to power cycling. Other wires such as platinum and palladium have been used at high temperatures with some success. Although, they are more difficult to bond because of their hardness (1.5–2.0 times that of Au), and may cause damage to the ICs.

The strength of the welded interface is the primary concern for wirebonds at elevated temperatures. Secondary effects (reduced strength, increased elongation) will occur in the wire due to prolonged annealing at high temperatures.

Aluminum wires may lose as much as 70% of their initial strength upon prolonged high temperature exposure, but do not pose a reliability concern unless forces within the package structure approach the breaking limits of the

4 Advanced Wire Bonding Technology: Materials, Methods, and Testing

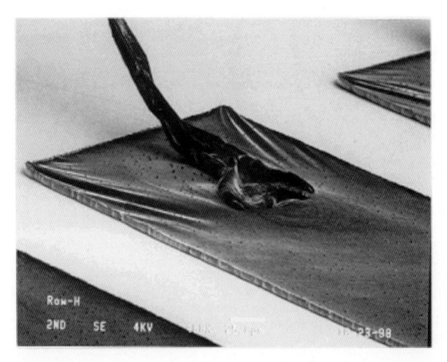

Fig. 4.29 Enlargement of tail bond region on gold pad after aging at 400°C for 168 hours. Note extensive build-up of material on the pad. SEM Photomicrograph (Magnification approximately 200X)

Table 4.15 Supplemental thermal aging wirebond test results. gold thermosonic wirebonds on gold pads with Ti-W adhesion layer (pull test and shear test results in grams (force)

Sample Aging Condition	Silicon		Silicon + SiO$_2$	
	Pull Test	Shear Test	Pull Test	Shear Test
As Bonded	11.5 (0.9)[a]	51.7 (1.8)	11.8 (1.0)	54.7 (1.9)
3000°C[b]	10.5 (0.8)	54.3 (2.1)	10.4 (0.5)	50.6 (2.3)
350°C	10.0 (0.8)[c]	56.3 (3.7)[c]	9.8 (0.7)	59.9 (5.7)
400°C			10.9 (1.0)	52.1 (5.6)
450°C			10.4 (1.0)	56.2 (4.3)
500°C			10.4 (1.2)[d]	50.8 (6.5)[d]
550°C				

[a] Standard deviations are given in parenthesis.
[b] Sample aged at 150°C for 168 hours (one week). Each succeeding sample not only received the specified aging for one week, but also received all proceeding aging above it on the list.
[c] Anomalous metal migration: test terminated.
[d] Metallization lift: test terminated.

wire. Such a condition rarely occurs in today's modern packaging configurations. Elongation in aluminum wires can increase as much as 30% under high temperature aging but again, reliability is not threatened.

Gold wires are typically stabilized and annealed at elevated temperatures so that their strength and elongation properties will be affected less by the high temperature environment. Again, as shown above under proper circumstances, the gold-gold interface will remain strong and stable to at least 500°C temperatures.

While the metallurgical interfaces are stable after long periods of annealing (storage) at high temperature, the question remains as to the fatigue resistance of these wires when subjected to temperature cycling over wide temperature ranges (ΔTs). Large ΔT environments can occur in space, oil wells, engines and under certain operational conditions (power cycling). ΔTs, in these environments can range from 200°C to over 500°C in certain applications with lower temperature in the –140°C range and high temperatures exceeding 350–400°C. Little or no data exists on fatigue of wirebonds under temperature cycling at elevated mean temperatures. Benoit, et al. [7], studied fatigue at room temperature after the wires were anneal at 300°C. His findings indicate that the annealed wires failed more rapidly than their unannealed counterparts. Much more work in this area is needed to ensure wirebond reliability in high temperature, large ΔT environments.

4.10 Summary

Wirebonding continues to be the dominant form of first-level chip connection. Over 90% of the worlds chip production is wirebonded. Because of its sheer volume, flexibility, and low cost, it will continue to dominate chip interconnect for decades to come. Wirebonding is accomplished by three basic techniques using a variety of wire and pad metallurgies. Wirebonding is robust and, on rigid substrates, has been shown to be extremely reliable (defects in the low part per million range). Bonding to softer substrates, small pads, unconventional metallurgies and stacked components has presented challenges - challenges that wirebonding has been successful in meeting. With appropriate care and understanding of the processes, wirebonding, even under these challenging conditions, can be performed reliably with high yield. Wirebonding is continually improving through advancements in automation, refinement of welding kinetics, improvements in wire and pad metallurgies, improved cleaning methods, and a better and wider spread understanding of wirebonding science. Work at extreme temperatures has begun and, at least, the gold wire-gold pad interface appears viable over a 700°C ΔT (–200°C to +500°C).

Acknowledgments The author greatly acknowledges the support of JHU/APL's Engineering and Fabrication Branch of the Technical Services Department in sample preparation and testing. Special thanks is given to Ms. Angalene Sutton for manuscript preparation.

References

1. ASTM Standard Test Method: F458-06 (2006), "Standard Non-Destructive Pull Testing of Wire Bonds," in Annual Listing of ASTM Standards, ASTM International West Conshohocken, Pennsylvania, USA
2. ASTM Standard Test method: F459-06(2006), "Standard Test Methods for Measuring Pull Strength of Microelectronic Wire Bonds" in Annual Listing of ASTM Standards ASTM International, West Conshohocken, PA, USA
3. ASTM Standard Test Method: F1269-06(2006), "Test Method for Destructive Shear Testing of Ball Bonds," in Annual Listing of ASTM Standards, ASTM International West Conshohocken, Pennsylvania, USA
4. Banda, C. V., Mountain, D. J., Charles, Jr., H. K., Lehtonen, J. S., Keeney, A. C., Johnson, R. W., Zhang, T., and Hou, Z. "Development of Ultra-thin Flip Chip Assemblies for Low Profile SiP Applications," in Proc. 37th Int. Microelectronics Symposium, Long Beach, CA, pp. 551–555 (2004)
5. Banda, C.V., Johnson, R.W., Zhang, T., Hou, Z., and Charles, Jr., H.K. "Flip Chip Assembly of Silicon Die on Flex Substrates" IEEE Trans on Electronic Packaging Manufacturing, Vol. 31, No. 1, pp 1–8, (2008)
6. Bardeen, J. and Brattain, W. H. "The Transistor, A Semiconductor Triode," Physical Review, 74, 230 (1948)
7. Benoit, J., Chen, S., Grzybowski, R., Lin, S., Jain, R., and McClusky, P., "Wire Bond Metallurgy for High Temperature Electronics" Proc. 4th International High TemperatureElectronics Conference, Albuqyerque, NM, pp 109–113 (1998)
8. Bischoff, A., Aldinger, F., and Heraeus, W. "Reliability Criteria of New Low Cost Materials for Bonding Wires and Substrates," in Proc. 34th Electronic Components Conference, New Orleans, Louisiana, USA, pp. 411–417 (1984)
9. Breach, C., Wulff, W., Ditter, K., Calpito, D. R., Garnier, M., Boillot, V., and Wei, T. C., "Reliability and Failure Analysis of Gold Ball Bonds in Fine and Ultra-fine Pitch Applications", Proceedings of Semicon Singapore, pp. 1–10 (2004)
10. Charles, Jr., H. K., Romenesko, B. M., Uy, O. M., Bush, A. G., and Von Briesen, R. "Hybrid Wirebond Testing – Variables Influencing Bond Strength and Reliability," The International Journal for Hybrid Microelectronics 5(1), 260–269 (1982a)
11. Charles, Jr., H. K., Romenesko, B. M., Wagner, G. D., Benson, R. C., and Uy, O. M. "The influence of contamination on aluminum-gold intermetallics," in Proc. Int. Reliability Physics Symposium, San Diego, California, USA pp. 128–139 (1982b)
12. Charles, Jr., H. K., Clatterbaugh, G. V., and Weiner, J. A. "The Ball Bond Shear Test: Its Methodology and Application," in Gupta D C (ed), Semiconductor Processing, ASTM STP 850, 429–457 (1984)
13. Charles, Jr., H. K. "Ball Bond Shearing: An Interlaboratory Comparison," in Proc. International Microelectronics Symposium, Atlanta, GA, pp. 265–274 (1986)
14. Charles, Jr., H. K. and Clatterbaugh, G. V. "Thin Film Hybrids.," in Minges M L. (ed), Electronic Materials Handbook, Vol. 1, Packaging, ASM International, Materials Park, Ohio, USA, 313–331 (1989)
15. Charles, Jr., H. K., Mach, K. J., and Edwards, R. L. "Multichip Module (MCM) Wirebonding," in Proc. International Symposium on Electronic Packaging Technology (ISEPT '96), Shanghai, Peoples Republic of China, pp. 336–341 (1996)
16. Charles, Jr., H. K., Mach, K. J., Edwards, R. L., Lehtonen, S. J., and Lee, D. M. "Wirebonding on Various Multichip Module Substrates and Metallurgies," in Proc. 47th Electronic Components and Technology Conference, San Jose, California, USA, pp. 670–675 (1997)
17. Charles, Jr., H. K., Mach, K. J., Edwards, R. L., Francomacaro, A. S., Lehtonen, S. J., and DeBoy, J. S. "Wirebonding: Reinventing the Process for MCMs," in Proc. International Symposium on Microelectronics, San Diego, California, USA, pp. 645–655 (1998)

18. Charles, Jr., H. K., Mach, K. J., Edwards, R. L., Francomacaro, A. S., Lehtonen, S. J., and DeBoy, J. S. "Multichip Module and Chip-On-Board Wirebonding, in Proc. 12th European Microelectronics Conf., Harrogate, Yorkshire, England, pp. 525–532 (1999)
19. Charles, Jr., H. K., Mach, K. J., Edwards, R. L., Francomacaro, A. S., DeBoy, J. S., and Lehtonen, S. J. "High Frequency Wirebonding: Its Impact on Bonding Machine Parameters and MCM Substrate Bondability," in Proc. 34th International Microelectronics Symposium, Baltimore, MD, pp. 350–360 (2001)
20. Charles, Jr., H. K., Mach, K. J., Lehtonen, S. J., Francomacaro, A. S., DeBoy, J. S., and Edwards, R. L. "High-Frequency Wirebonding: Process and Reliability Implications," in Proc. 52nd IEEE Electronic Components and Technology Conference, San Diego, CA, pp. 881–890 (2002)
21. Charles, Jr., H. K., Mach, K. J., Lehtonen, S. J., Francomacaro, A. S., DeBoy, J. S., and Edwards, R. L. "Wirebonding at High Ultrasonic Frequencies: Reliability and Process Implications," Microelectronics Reliability, Vol. 43, pp. 141–153 (2003)
22. Charles, Jr., H.K. "The Wirebonded Interconnect: Mainstay for Electronics" Chapter 3 in Micro-and Opto-Electronic Materials and Structures: Physics, Mechanics, Design, Reliability, Packaging, Vol.2. E. Suhir, Y.C.Lee, and C.P. Wong editors, Springer, pp 71–120 (2007)
23. Chen, G. K. C. "The Role of Micro-Slip in Ultrasonic Bonding of Microelectronic Dimensions," in Proc. 1972 International Microelectronic Symposium, Washington DC, October 30 – November 1, 1972, pp. 5¬A-1-1 to 5-A-1-9
24. Ching, T. B. and Schroen, W. H. "Bond Pad Structure Reliability," 24th Annual Proc. Reliability Physics Symposium, Monterey, CA, pp. 64–70 (1988)
25. Clatterbaugh, G. V., Weiner, J. A., and Charles, Jr., H. K. "Gold-Aluminum Intermetallics," Ball Bond Shear Testing and Thin Film Reaction Couples," IEEE Trans. Components, Hybrids Manufacturing Technology, CHMT-7(4), 349–356 (1984)
26. Clatterbaugh, G. V. and Charles, Jr., H. K. "The effect of high temperature intermetallic growth on ball shear induced cratering," IEEE Trans. Components, Hybrids and Manufacturing Technology, CHMT-13, No. 4, pp. 167–175 (1990)
27. Demmin, J. C. "Ultrasonic Bonding Tools for Fine Pitch, High Reliability Interconnects," in Proc. Int. Conference on Multichip Modules, Denver, Colorado, USA, pp. 397–402 (1996)
28. Ehrlich, V. J. and Tsao, J. Y. "Laster Direct Writing for VLSI," in VLSI Electronics: Microstructure Science, Vol. 7, Academic Press, pp. 129–164 (1983)
29. Endicott, H. W., James, H. K., and Nobel, F. "Effects of Gold-Plating Additives on Semiconducting Wire Bonding," Plating and Surface Finishing V, pp. 58–61 (1981)
30. Evans, K. L., Guthrie, T. T. and Hayes, R. G. "Investigations of the Effect of Thallium on Gold/Aluminum Wire Bond Reliability," in Proc ISTFA, Los Angeles, CA, pp. 1–10 (1984)
31. Gehman, B. L. "Bonding Wire for Microelectronic Interconnections," IEEE Trans. Components Hybrids and Manufacturing Technology, CHMT-3(8), 375–380 (1980)
32. Geppert, L. "Solid State," IEEE Spectrum 35(1), 23–28 (1998)
33. Glaser, A. B. and Subak-Sharpe, G. E. Integrated Engineering: Design Fabrication and Applications, Addison-Wesley, Reading, West Virginia, USA (1979)
34. Goldfarb, S., "Wire Bonds on Thick Film Conductors", proc. 21st IEEE Electronics Components Conference, Washington, DC pp 295 – (1971)
35. Gonzalez, B., Knecht, S., and Handy, H. "The Effect of Ultrasonic Frequency on Fine Pitch Al Wedge Wirebonds," in Proc. 46th Electronic Components and Technology Conference, Orlando, Florida, USA, pp. 1078–1087 (1996)
36. Gonzalez, C. G., Wessel, R. A., and Padlewski, S. A. "Epoxy-Based Aqueous-Processable Photodielectric Dry Film and Conductive Via Plug for PCB Build-Up and IC Packaging," in Proc. 48th Electronic Components and Technology Conference, Seattle, Washington, USA, pp. 138–143 (1998)

37. Harman, G. G. "Wirebonding – Towards 6σ Yield and Fine Pitch," in Proc. 42nd Electronic Components and Technology Conference, San Diego, California, USA, pp. 903–910 (1992)
38. Harman, G. G., "Metallurgical Interconnections for Extreme High and Low Temperature Environments", Chapter 4, Micro- and Opto-Electronic Materials and Structures: Physics, Mechanics, Design, Reliability, Packaging: Volume 2, Ephraim Suhir, Y. C. Lee, and C. P. Wong (Editors), Springer, 2007
39. Harman, G. G. "Wire Bonding to Multichip Modules and Other Soft Substrates," in Proc 1999 International Conference and Exhibition on Multichip Modules, Denver, Colorado, USA, pp. 292–301 (1995)
40. Harman, G. G. Wire Bonding in Microelectronics: Materials Processes, Reliability and Yield, McGraw-Hill, New York, New York, USA (1997)
41. Harman, G. G. and Canon, C. A. "The Microelectronic Wire Bond Pull Test, How to Use It, How to Abuse It," IEEE Trans. Components, Hybrids and Manufacturing Technology, CHMT-1(3), 203–210 (1978)
42. Heinen, G., Stierman, R. J., Edwards, D., and Nye, L. "Wire Bond Over Active Circuits," in Proc. 44th Electronic Components and Technology Conference (ECTC), Washington, D.C., pp. 922–928 (1994)
43. Hirota, J., Machinda, K., Okuda, T., Shimotomai, M., and Kawanaka, R. "The Development of Copper Wirebonding for Plastic Molded Semiconductor Packages," in Proc. 35th IEEE Electronics Component Conference, Washington, DC, pp. 116–121 (1985)
44. Horsting, C. "Purple Plaque and Gold Purity," 10th Annual Proc. IRPS, Las Vegas, NV, pp. 155–158. (1972)
45. Ito, S., Kuwamura, M., Akizuki, S., Ikemura, K., Fukushima, T., and Sudo, S. "Solid Type Cavity Fill and Underfill Materials for New IC Packaging Applications," in Proc. 45th IEEE Electronic Components and Technology Conference, Las Vegas, Nevada, USA (1995)
46. Jaecklin, V. P. "Room Temperature Ball Bonding Using High Ultrasonic Frequencies," in Proc. Semicon: Test, Assembly and Packaging, Singapore, pp. 208–214 (1995)
47. Jellison, J. L. "Effect of Surface Contamination on the Thermocompression Bondability of Gold," IEEE Trans. Parts, Hybrids and Packaging, Vol. PHP-11, pp. 206–211 (1975)
48. Jellison, J.L., "Kinetics of Thermocompression Bonding to Organic Contaminated Gold Surfaces" IEEE Trans. Parks, Hybrids and Packaging, PHP-13, pp 132–137 (1977)
49. Jellison, J.L., and Wagner, J. A. "role of Surface Contaminants in the Deformation Welding of Gold to Thick and Thin Films" Proc. 28th Electronic Components Conference pp 336–345, (1979)
50. Johnston, C. N., Susko, R. A., Siciliano, J. V., and Murcko, R. J. "Temperature Dependent Wear-out Mechanism for Aluminum/Copper Wire Bonds," in Proc. International Microelectronics Symposium, Orlando, FL, pp. 292–296 (1991)
51. Kilby, J. S. "Invention of the Integrated Circuit," IEEE Trans. Electronic Devices, ED-23, 648–654 (1976)
52. Klein, H. P., Durmutz, U., Pauthner, H., and Rohrich, H. "Aluminum Bond Pad Requirements for Reliable Wire Bonds," in Proc. IEEE Int. Symposium on Physics and Failure Analysis of ICs, Singapore, pp. 44–49. (1989)
53. Koch, T., Richling, W., Whitlock, J., and Hall, D., "A Bond Failure Mechanism" Proc. 24th Annual Reliability Physics Symposium, Anaheim, CA. pp 55–60 (1986)
54. Kurtz, J., Cousens, D., and Defour, M. "Copper Wire Ball Bonding," in Proc. Int. Electronic Packaging Society Conference, New Orleans, Louisiana, USA, pp. 1–5 (1984)
55. Langenecker, B. "Effects of Ultrasound on Deformation Characteristics of Metals," IEEE Transactions on Sonics and Ultrasonics, Vol. SU-13, pp. 1–8 (1966)
56. Levinson, L. M., Eichelberger, C. W., Wognarowski, and Carlson, R. O. "High-Density Interconnect Using Laser Lithography," in Proc. International Symposium on Microelectronics, Seattle, Washington, October 17–19, 1988, pp. 301–306

57. Ling, J. and Albright, C. E. "The Influence of Atmospheric Contamination in Copper to Copper Ultrasonic Welding," in Proc. 34th Electronic Components Conference, New Orleans, Louisiana, USA, pp. 209–218 (1984)
58. Liu, D., Zhang, C., Graves, J., and Kegresse, T. "Laser Direct-Write (LDW) Technology and Its Applications in Low Temperature Co-Fired Ceramic (LTTC) Electronics," in Proc. 2003 International Symposium on Microelectronics, Boston, Massachusetts, Nov. 18–20, 2003, pp. 298–303
59. Lo, George and Sitaraman "G-Helix: Lithography-Based, Wafer-Level Compliant Chip-to-Substrate Interconnect," in Proc. 54th Electronic Components and Technology Conference, Las Vegas, Nevada, June 1–4, 2004, pp. 320–325
60. Meisser, C. "Bonding Techniques for Plastic MCMs," Semiconductor International 14, 120–124 (1991)
61. Microbonds, Inc., 151 Amber Street, Unit 1 Markham, Ontario, Canada L3R3B3, www.microbonds.com
62. Miller, L. F. "Controlled Collapse Reflow Chip Joining," IBM J. Res. Dev., 13, 239–250 (1969)
63. Moore, G. E. "VLSI: Some Fundamental Challenges," IEEE Spectrum, 16(4), 30–37 (1979)
64. Mundt, R., O'Dell, G., and Ruben, D., "Laser Ribbon Bonding: A novel Interconnect Method" Proc. 37th International Microelectronics Symposium, Long Beach, CA. Session THA12-2 (2004)
65. Newsome, J.L., Oswalkm R.G., and rodrigues de Miranda, W.R., "Metallurigical Aspects of Aluminum Wire Bonds to Gold Metallization" Proc. 14th Annual Reliability Physics Symposium, Las Vegas, NV, pp 63–74 (1976)
66. Onoda, H., Itashimoto, K., and Touchi, K. "Analysis of Electromigration-Induced Failures on High Temperature Sputtered Al-Alloy Metallization," J. Vacuum Science Technology, A(13), 1546–1555 (1995)
67. Onuki, J., Suwa, M., Iizuka, T., and Okikawa, S. "Study of Aluminum Ball Bonding for Semiconductors," in Proc. 34th Electronic Components Conference, New Orleans, Louisiana, USA, pp. 7¬12 (1984)
68. Otsuka, K. and Tamutsa, T. "Ultrasonic Wire Bonding Technology for Custom LSIC with Large Number of Pins," in Proc. 31st IEEE Electronic Components Conference, Atlanta, Georgia, USA, pp. 350–355 (1981)
69. Philofsky, E. "Intermetallic Formation in Gold-Aluminum Systems," Solid State Electronics 13(10), 1391–1399 (1970)
70. Prather, J.B. Robertson, S.D., and Slemmons, J.W., "Aluminum Wire Bonding to Gold Thick-Film Conductors" Electronic Packaging and Productions, p. 68 – (1974)
71. Ramsey, T. H. and Alfaro, C. "The Effect of Ultrasonic Frequency on Intermetallic Reactivity of Au-Al Bonds," Solid State Technology, Vol. 34, pp. 37–38, (1991)
72. Ravi, K. V. and Philofsky, E. M. "Reliability Improvement of Wire Bonds Subjected to Fatigue Stresses," in Proc. 10th IEEE Reliability Physics Symposium, Las Vegas, Nevada, USA, pp. 143–149 (1972)
73. Riddle, J. "High Cycle Fatigue (Ultrasonic) Not Corrosion in Fine Microelectronic Bonding Wire," in Proc. 3rd ASM Conference on Electronics Packaging, Materials, Processes, and Corrosion in Microelectronics, Minneapolis, Minnesota, pp. 185–191 (1987)
74. Romensko, B. M., Charles, Jr., H. K., Clatterbaugh, G. V., and Weiner, J. A. "Thick-film Bondability: Geometrical and Morphological Influences," The Int. J. for Hybrid Microelectronics, Vol. 8, pp. 408–419 (1985)
75. Romenesko, B. M., Charles, Jr. H. K., Cristion, J. A., and Sui, B. K. "Gold-Aluminum Wirebond Inteface Testing Using Laser-Induced Ultrasonic Energy," in Proc. 50th Electronic Components and Technology Conference, Las Vegas, NV, pp. 706–710 (2000)

76. Schaller, R. R. "Moore's Law: Past, Present, and Future," IEEE Spectrum, 34(6), 53–59 (1997)
77. Shirai, Y., Otsuka, K., Araki, T., Seki, I., Kikuchi, K., Fujita, N., and Miwa, T. "High Reliability Wire Bonding Technology by the 120 kHz Frequency of Ultrasonic," in Proc. 1993 International Conference on Multichip Modules, Denver, Colorado, pp. 366–375, (1993)
78. Spencer, T.H. "Thermocompression Bond Kinetics – The Four Principle Variables" Int. J. hybrid Microelectronics, Vol. 5 No. 1 pp. 404–408 (1982)
79. Takahashi, T., Rutter, Jr., E. W., Moyer, E. S., Harris, R. F., Frye, D. C., St. Joor, V. L., and Oakes, F. L. "A photo-definable benzocyclobutene resin for thin-film microelectronic applications," in Proc. Int. Microelectronics Conference, Yokohama, Japan, pp. 64–70 (1992)
80. Takeda, K., Ohmasa, M., Kurosu, N., Hosaka, J. "Ultrasonic Wirebonding Using Gold Plated Wire onto Flexible Printed Circuit Board," in Proc. 1994 International Microelectronics Conference, Oamya, Japan, pp. 173–177 (1994)
81. Tay, A. A. O., Yeo, K. S., Wu, J. H. "The Effect of Wirebond Geometry and Die Setting on Wire Sweep," IEEE Trans. on Components, Packaging and Manufacturing Technology – Part B 18(1), 201¬209 (1995)
82. Teverosky, A. "Effect of Vacuum on High Temperature Degradation of Gold/Aluminum Wire Bonds in PEMS", Proc. 42nd Annual Reliability Physics Symposium, Phoenix, AZ, pp. 547–556 (2004)
83. Thomas, A. and Berg, H. M. "Micro-Corrosion of Al-Cu Bonding Pads," in Proc. 23rd IEEE Reliability Physics Symposium, Orlando, Florida, USA, pp. 153–158 (1985)
84. Tsujino, J., Mori, T., and Hasegawa, K. "Characteristics of Ultrasonic Wire Bonding Using High Frequency and Complex Vibration Systems," in Proc. 25th Annual Ultrasonic Industry Association Meeting, Columbus, Ohio, pp. 17–18, (1994)
85. Tuckerman, D. B., Ashkenas, D. J., Schmidt, E., and Smith, C. "Die Attach and Interconnection Technology for Hybrid WSI," 1986 Laser Pantography States Report UCAR-10195, Lawrence Livermore Laboratories (1986)
86. Tummula, R. R., Rymazewski, E. J., Klopfenstein, A. G. Microelectronics Packaging Handbook, Vols. I, II, & III, Chapman Hall, NY, USA (1997)
87. Wakabayashi, S., Murata, A., and Wakobauashi, N. "Effects of Grain Refinement in Gold Deposits on Aluminum Wire-Bond Reliability," Plating and Surface Finishing V, pp. 63–68 (1981)
88. Weiner, J. A., Clatterbaugh, G. V., Charles, Jr., H. K., and Romenesko, B. M. "Gold Ball Bond Shear Strengths Effects of Cleaning, Metallization and Bonding Parameters," in Proc. IEEE 33rd Electronic Components Conference. Orlando, Florida, USA, pp. 208–220 (1983)
89. The Welding Handbook, Vol. 2, eighth edition, "Ultrasonic Welding," pp. 784–812 (1991)
90. Yao, Y. F., Lin, T. Y., and Chua, K. H. "Improving the Deflection of Wirebonds in Stacked Chip Scale Packages CSP," in Proc. 53rd Electronic Components and Technology Conference, New Orleans, LA, pp. 1359–1363 (2003)

Chapter 5
Lead-Free Soldering

Ning-Cheng Lee

Abstract Due to the global trend of green manufacturing, lead-free becomes the main stream soldering choice of electronic industry. SnAgCu alloys are the prevailing choices, with SnCu(+Y), SnAg(+Y), and BiSn(+Y) families also being adopted, where Y represents minor additive elements. The soldering processing window is narrower than that of Sn63, mainly due to the elevated melting temperature of SnAgCu solder and the limited high temperature tolerance of components and board. The high surface tension of Sn aggravates the difficulty in wetting, while the high reactivity of Sn puts more constraint in contact time allowed between molten solder and base metal or solder container. The creep rate of SnAgCu is slower at low stress, but faster at high stress than Sn63. This results in a longer temperature cycling life at low joint strain applications, but a shorter cycling life at high joint strain applications. Higher Cu content stabilizes IMC structure at interface between SnAgCu solder and NiAu. The high rigidity of SnAgCu solders enhances the fragility of joints, although significant improvement has been accomplished via low Ag or high Cu content or doping approaches.

Keywords Solder · Soldering · Lead-free · Pb-free · SnAgCu · SAC · Tin-silver-copper · Surface finish · Reliability

5.1 Global Lead-Free Soldering Implementation

The electronic industry is moving toward green manufacturing as a global trend. In the area of soldering, mainly driven by European RoHS (Reduction of Hazardous Substances), lead was banned effective July 1, 2006 except in some exempt items. This European legislation is followed by China RoHS which has similar list of banned materials, and its phase 1 implementation

---

N.-C. Lee (✉)
VP of Technology, Indium Corporation of America, 1676 Lincoln Ave., Utica, NY 13502
e-mail: nclee@indium.com

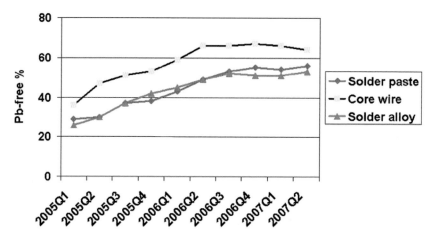

Fig. 5.1 Lead-free soldering implementation status reported by IPC [1]

was effective March 1, 2007. In Japan, the legislative activities dealt with the reclamation and recycling of electronics. The Home Electronics recycling law came into force in April 1st, 2001, and applied only to TVs, refrigerators and similar items. Although not specifically aiming at lead, this legislation effectively drove Japanese industry toward lead-free soldering process. Those legislation activities lead the trend and effectively drive the rest of the world toward lead-free soldering (see Fig. 5.1).

5.2 Prevailing Lead-Free Solder Alloys

Among the numerous lead-free solder options available, the following families are of particular interest and are the prevailing choices of industry: eutectic SnAg, eutectic SnCu, eutectic SnAgCu, eutectic SnZn, eutectic BiSn, and their modifications, as shown in Fig. 5.2. Also shown in Fig. 5.2 are the related applications including reflow soldering, wave soldering, and hand soldering. Their characteristics and potential performance in electronic applications are presented below.

5.2.1 SnCu (+ dopants, e.g. Ni, Co, Ce)

Eutectic Sn99.3Cu0.7 exhibits a melting temperature at 227°C. Sn99.3Cu0.7 is lower in tensile strength but higher in elongation than both eutectic SnAg and SnPb, reflecting the softness and ductility of SnCu [2]. The creep strength of eutectic SnCu is higher than Sn100, but lower than eutectic SnAg and SnAgCu at both 20°C and 100°C. Wetting balance test results by Hunt et al. indicated

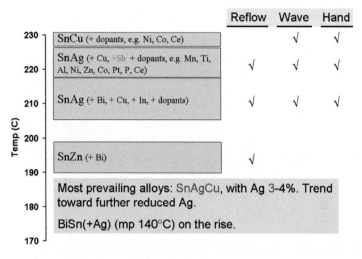

Fig. 5.2 Prevailing lead-free solder alloys and their applications

that the wetting ability decreased in the following order: eutectic SnPb > SnAgCu > SnAg > SnCu when an unactivated flux was used [3]. Eutectic SnCu is commonly used at wave soldering and hand soldering.

The mechanical and wetting properties of eutectic SnCu were enhanced by addition of small amount of dopants such as Ni, Ge, Co, and Ce. Sn99.3Cu0.7-Ni0.05+Ge (SN100C) [4] and Sn99.5Cu0.5Co <0.05 (Cobalt995) [5] were reported to exhibit a reduced wetting time, copper dissolution rate, and a more shiny smooth solder joint surface at wave soldering. Sn99.3Cu0.7Ce0.02 was reported to have enhanced elongation performance and drop test performance [6, 7].

5.2.2 SnAg (+Cu, +Sb, +dopants, e.g. Mn, Ti, Al, Ni, Zn, Co, Pt, P, Ce)

SnAgCu (SAC) is the most prevailing alloy family for electronic soldering. Sn96.5Ag3Cu0.5 (SAC305), Sn95.6Ag3.5Cu0.9 (SAC359), ternary eutectic SnAgCu), Sn95.5Ag3.8Ag0.7 (SAC387), Sn95.5Ag3.9Cu0.6 (SAC396), and Sn95.5Ag4.0Cu0.5 (SAC405) are all commonly used for reflow, wave, and hand soldering, and all exhibit a melting temperature around 217°C. Among those, SAC305 is the most popular one in Asia, and is also endorsed by IPC. Eutectic Sn96.5Ag3.5 is commonly used as well, with a melting temperature 221°C.

For SnAgCu, its hardness, tensile strength, yield strength, shear strength, impact strength, and creep resistance are all higher than eutectic SnPb (Sn63) [2]. The wetting performance is better than both eutectic SnCu and eutectic

SnAg, although poorer than that of Sn63 [8]. Addition of Sb into SAC, Sn96.2Ag2.5Cu0.8Sb0.5 (CASTIN), was reported to exhibit a slower intermetallic compound growth rate [9].

Due to its high hardness, lead-free solder joints typically suffer fragility issue upon drop test. This is a particular concern for portable electronic devices. Reducing Ag content, such as Sn98.5Ag1.0Cu0.5 (SAC105), Sn99Ag0.3Ag0.7 (SAC0307) [10], and Sn98.9Ag1.0Cu0.1 (SAC101) [11], improves the non-fragility [12]. This approach often results in an elevated liquidus temperature up to around 227°C. Further improvement has been reported by addition of small amount of Mn, Ti, Bi, Y, Ce [6, 7], Al, Ni [13], Zn [14], Co, Pt, and P [15].

5.2.3 SnAg (+ Bi, + Cu, + In, + dopants)

Lead-free alloys with Bi generally exhibit a lower melting temperature. Furthermore, they are typically better in wetting than other lead-free alloys [16], presumably due to the low surface tension of Bi (0.376 N/m for Bi versus 0.537 N/m for Sn) [17]. The low melting and good wetting features promise a user-friendly soldering process. Addition of Bi to SnAgCu system also refines the intermetallic compound (IMC) grain size and retards the excessive growth of IMC [18]. Addition of small amount of In reduces the melting temperature and increases the ductility of lead-free alloys [7].

However, Bi-containing alloys normally exhibit a high rigidity, thus may pose a concern for applications involving a high CTE mismatch or a wide range of service temperature. In addition, in the presence of lead-contamination, formation of 96°C low melting Bi52Pb30Sn18 ternary eutectic phase can result in early failure at temperature cycling test [19].

The SnAgBi-containing family is primarily used in Japanese industry, such as Panasonic (SnAgBiCu, SnAgBi, SnAgBiIn), Hitachi (SnAgBi), and Sony (SnAgBiCu) [20, 21]. Examples of those alloys supplied mainly in Japanese industry are given below [22]:

Sn97.4Ag1.3Bi0.8Cu0.5 (214–219°C, Nihon Genma)
Sn95.5Ag2.0Bi2.0Cu0.5 (211–221°C, Senju)
Sn94.25Ag2.0Bi3.0Cu0.75 (207–218°C, Senju)
Sn96.0Ag2.5Bi1.0Cu0.5 (214–221°C, Senju, Nihon Almit, Tamura Kaken, Nihon Genma)
Sn95.7Ag2.8Bi1.0Cu0.5 (214–215°C, Nihon Genma)
Sn93.6Ag2.9Bi3.0Cu0.5 (205–216°C, Tamura Kaken)
Sn92.8Ag3.0Bi1.0Cu0.7In2.5 (204–215°C, Senju)
Sn93.3Ag3.0Bi3.0Cu0.7 (206–215°C, Nihon Almit)
Sn91.5Ag3.5Bi2.5In2.5 (Matsushita)
Sn92.5Ag3.5Bi3.0Cu1.0 (208–213°C, Nihon Superior)
Sn91.7Ag3.5Bi4.8 (205–210°C, Sandia National Lab)

5.2.4 SnZn (+ Bi)

Eutectic Sn91Zn9 exhibits a melting temperature 199°C. Although attractive in its low melting temperature, its high surface tension (0.768 N/m for Zn) and high reactivity toward flux and oxygen prohibit its use for electronic soldering. Addition of Bi, such as Sn89Zn8Bi3 (189–199°C), effectively reduces the surface tension and reactivity, in addition to a further reduction of melting temperature. This enables the SnZnBi alloy to be a viable alternative for lead-free soldering in Japanese industry such as NEC and Panasonic.

However, compared with other lead-free alloys, SnZnBi is still more reactive toward flux and oxygen, therefore is limited in applications. Also, the tendency to form voids on top of CuZn IMC layer on Cu surface further confines this alloy to consumer applications [23, 24, 25]. Other alloys such as Sn86.5Zn5.5Bi3.5In4.5 (174–186°C, Indium) may also be attractive due to a greater extent of melting temperature reduction.

5.2.5 BiSn (+ Ag)

For BiSn alloys, Bi expands 3.87 vol-% during solidification. Sn contracts, but to a less amount. Thus BiSn alloys containing more than 47% Bi expand on solidification [26]. Bi58Sn42 (eutectic 138°C) has been used by IBM for wave soldering since more than 30 years ago. Unisys uses this alloy for wave soldering on 50-layer + mainframe board (1/3 inch thick), solder pot temperature about 200°C, to reduce thermal shock [27].

Bi58Sn42 has properties approaching those of Sn63 under most conditions [28]. However, Bi58Sn42 is more sensitive than Sn63 to strain rate. That is, its elongation decreases more rapidly with increasing strain rate. Glazer reported that increased elongation at low strain rates after aging resulted in ductile failure in solder, versus a quasibrittle fracture at high strain rate. The latter failure mode combines cleavage in Bi-rich phase with fracture at solder/IMC interface [26].

The ductility of Bi58Sn42 can be improved with addition of Ag, such as Bi57Sn41Ag2 [29]. On the other hand, addition of 1% Cu dramatically slowed down coarsening of Bi58Sn42 [30].

5.3 Lead-Free Solder Pastes

Solder paste is a mixture of solder powder and flux. The powder size used depends on the applications, with finer powder to be used for finer pitch of PCB assembly. The powder size is defined by IPC as shown in Table 5.1 [31].

In the advancing toward miniaturization, flux technology also has to advance in order to cope with the increasing demand on performance.

For a smaller flux/solder paste dot, oxidation of powder, pads, and parts will be more significant due to a shorter oxygen diffusion path. This situation is

Table 5.1 Particle size distributions of standard solder powders

Type	At least 80% between	At least 85% between
1	150–75 μm	–
2	75–45 μm	–
3	45–25 μm	–
4	–	38–20 μm
5	–	25–15 μm
6	–	15–5 μm
7	–	11–2 μm

further aggravated by the increasing surface area per unit volume with decreasing dot size. Figure 5.3 shows the relationship between the flux fraction burn-off and the flux volume in a thermogravimetric analysis (TGA) study for a no-clean flux [32]. The flux fraction burn-off increased rapidly with decreasing sample size after going through a heating profile, as shown in the graph. In other words, the flux fraction which remains to protect the parts from oxidation decreases with increasing miniaturization.

Therefore, either flux with a more efficient oxidation barrier capability or reflow atmosphere with a lower oxygen partial pressure is needed in order to achieve satisfactory soldering results. The relation among soldering performance, oxidation barrier capability, and oxygen partial pressure has been studied by Jaeger and Lee [33] and is shown in Fig. 5.4. Here soldering performance with a value of 1 represents a perfect fluxing, while a value of less than 1 represents less than perfect fluxing. Less than perfect fluxing will display symptoms such as poor wetting, solder balling, voiding, or poor coalescence [34].

In Fig. 5.4, K represents tendency to oxidize at reflow for solder paste, and is equal to 1 for a typical air-reflowable RMA solder paste. Apparently this type

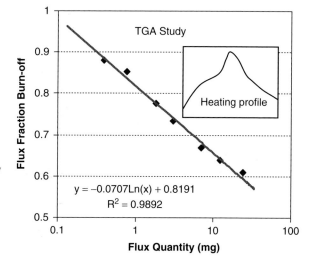

Fig. 5.3 Relation between flux fraction burn-off and flux volume in a TGA study for a no-clean flux vehicle used for solder paste. The heating profile was programmed to reflect a reflow profile with a peak temperature of 230°C [32]

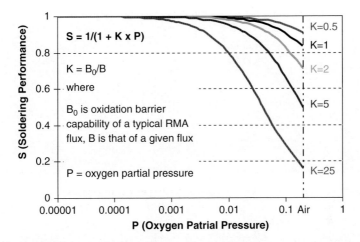

Fig. 5.4 The relation among soldering performance, oxidation barrier capability, and oxygen partial pressure for solder paste for a typical SMT print deposit

of solder paste (K = 1) already showed a compromised soldering performance when reflowed under air. This compromise will be further aggravated for smaller solder paste deposits. In order to achieve good soldering performance, either an inert reflow atmosphere or a solder paste with less tendency to oxidize (K < 1) will be required. Since inert gas is more costly than air, the only practical option remaining is a paste with less tendency to oxidize. This means a flux vehicle with an improved oxidation barrier capability.

Besides an improved oxidation barrier capability, the following flux features are also needed with further miniaturization: (1) no-clean, (2) reduced volatile, (3) halide-free, (4) greater fluxing capacity, (5) higher residue resistivity, (6) more resistant to oxidation and charring, (7) lower activation temperature, (8) slower wetting speed when solder begins to melt, (9) less spattering, (10) higher probe penetratability, (11) capability of inducing nucleation of solder upon cooling, and (12) greater slump resistance [17].

The morphology of solder powder is exemplified in Fig. 5.5 [35]. The surface of type 3 powder of Sn63 is fairly smooth, with distinct tin-rich phase (dark phase) and Pb-rich phase (light phase). Sn63 type 7 powder exhibits similar dual phases morphology, with the surface wrinkle being more noticeable under the higher magnification.

Lead-free solder powder often exhibits a rougher surface texture than Sn63 solder powder. In the case of type 3 SAC387 powder, a relatively regular, orange peel-like surface texture is easily recognizable for both alloys. This is mainly attributable to the dendrite formation of beta-tin in the high-tin alloys. The wrinkle formation is also more noticeable for type 6 powder than Sn63. In the case of Bi58Sn42, a distinct Sn-rich (dark phase) and Bi-rich (light phase) two-phase morphology is also observed, with Bi phase being the slightly

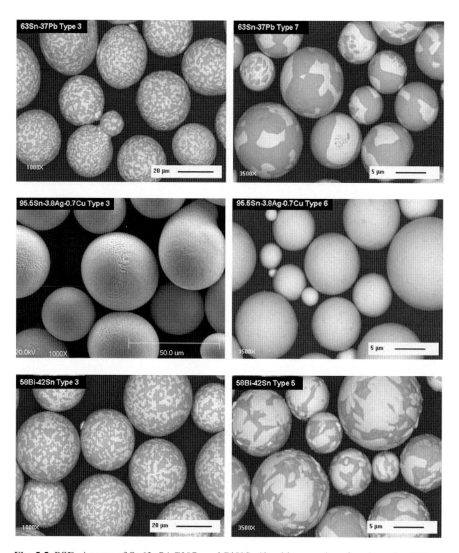

Fig. 5.5 BSE pictures of Sn63, SAC387, and Bi58Sn42 solder powder of various size [35]

dominant phase. The crystalline texture of Bi-rich phase results in a bulging formation, as indicated by the 3500X picture of type 5 powder.

The prospects of 10 major lead-free solder alloys for reflow soldering applications are shown in Fig. 5.6 [16]. Compatibility of those alloys with a variety of representative flux chemistries was considered essential, and was determined for handling-ability, including shelf life and tack time, and soldering capability, including solder balling, wetting, and solder joint appearance. Results indicated that the control Sn63 was still the most compatible alloy, rated 27.1 out of a full scale of 30 when using warm profile. The primary factor which

5 Lead-Free Soldering

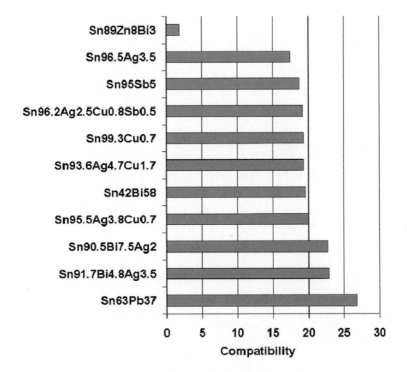

Fig. 5.6 Compatibility of alloys with reflow soldering [16]

distinguishes Sn63 from the rest alloys was the soldering performance, particularly the wetting and solder appearance. As to the solder balling, although Sn63 was also the best, it was fairly close to the best lead-free systems.

Among the lead-free options, both SnAgBi alloys studied, Sn91.7Ag3.5AgBi4.8 and Sn90.5Bi7.5Ag2, turned out to be on the top of lead-free systems, rated 22.9 and 22.8, respectively. This was mainly attributed to the better wetting and solder balling performance. Shelf life and tack time of the SnAgBi systems were also fairly good, while the solder appearance was at best considered average. The six alloys, Sn99.3Cu0.7, Sn95.5Ag3.8Cu0.7, Sn93.6Ag4.7Cu1.7, Sn96.2Ag2.5Cu0.8Sb0.5, Bi58Sn42, and Sn95Sb5, showed fairly comparable performance to each other, ranging from 19.3 to 20.3. In general, the whole group displayed a quite noticeably poorer wetting than SnAgBi systems.

Bi58Sn42 exhibited a fairly poor solder balling performance, but an outstanding solder appearance among lead-free systems. Sn96.2Ag2.5Cu0.8Sb0.5 showed a relatively poor performance in both wetting and solder appearance among these six alloys. Sn96.5Ag3.5, rated 17.1 in compatibility, was ranked below the other alloys described above, mainly due to poor performance in solder balling, and particularly the poor wetting. Sn89Zn8Bi3, rated only 2.2 in compatibility, fell far short in every category when compared with all other alloy systems. Obviously, this is attributable to the very reactive nature of zinc, which resulted in excessive oxidation of metal and excessive reaction with fluxes, and consequently an unacceptable performance for solder paste applications. High-tin-content lead-free alloys seemed to display a thicker IMC layer than Sn63 when reflowed. Overall, the reflow compatibility could be ranked in decreasing order as shown below: (1) eutectic SnPb, (2) SnAgBi, (3) SnAgCu, eutectic SnBi, SnAgCuSb, eutectic SnCu, SnSb, (4) eutectic SnAg, (5) SnZnBi.

5.4 Lead-Free Surface Finishes

5.4.1 Type of Lead-Free Surface Finishes

Table 5.2 lists the options of lead-free surface finishes for PCBs. The system is categorized per the key element used. Each category is further classified per the type of process and chemistry [35].

Table 5.2 List of lead-free surface finishes. For multi-layer finishes, the sequence of materials starts from the layer on top of base metal

Surface finish system	Finish process & chemistry
Organic Solderability Preservative (OSP)	Benzotriazole
	Imidazole
	Benzimidazole (substituted)
	Preflux (rosin/resin)
Ag	Electroless (immersion, or galvanic) Ag
Au/Ni	Electrolytic Ni/Au, or EG
	Electroless Ni/Electroless (immersion) Au, or ENIG
	Electroless Ni/Electroless (autocatalytic) Au
	Electroless Ni/Electroless (substrate-catalyzed) Au
Bi	Electroless (immersion) Bi
Pd	Electrolytic Pd or Pd-alloys
	Electroless (autocatalytic) Pd
	Electroless (autocatalytic) Pd/Electroless (immersion) Au
Pd/Ni	Electroless Ni/Electroless (immersion) Pd
	Electroless Ni/Electroless (autocatalytic) Pd
	Electroless Ni/Electroless (autocatalytic) Pd/Electroless (immersion) Au
Pd (X)/Ni	Electrolytic Ni/PdCo/Au flash
	(Electroless) Ni/(Electroless) PdNi/Electroless (immersion) Au
Sn	Electrolytic Sn
	Electroless (immersion) Sn
	Electroless (Modified immersion + autocatalytic) Sn
Sn/Ni	Electrolytic Ni/Electrolytic Sn
SnAg	Electrolytic SnAg
SnBi	Electrolytic SnBi
	Electroless (immersion) SnBi
SnCu	Electrolytic SnCu
SnNi	Electrolytic SnNi

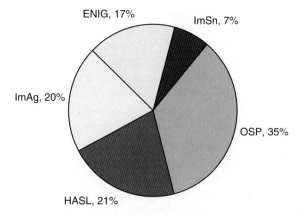

Fig. 5.7 Estimated 2007 global market share of PCB surface finishes [36]

For PCB surface finishes, OSP, hot air solder level (HASL), immersion Ag (ImAg), ENIG, and immersion Sn (ImSn) are considered the prevailing options, with estimated global market share for 2007 shown in Fig. 5.7 and the calculated annual growth rate (CAGR) shown in Fig. 5.8 [36].

5.4.2 Performance of Surface Finishes

The wetting performance of various surface finishes has been evaluated by Horaud et al. [37] via wetting balance. Results showed that the wetting time

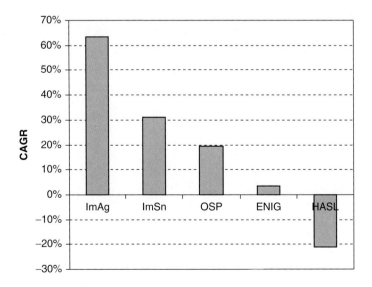

Fig. 5.8 Global calculated annual growth rate of various PCB surface finishes [36]

increased in the following sequence ImAg < ENIG, HASL < ImSn < OSP while the wetting force decreased in the following sequence ImAg > ENIG > HASL > ImSn >> OSP. On the other hand, the solder spreading displayed the following order HASL, ENIG > ImSn > ImAg > OSP. In general, for properly prepared surface finishes, the solder wettability can be generalized as "metal is better than non-metal, and noble metal is better than non-noble metal". This generalization should be taken with precaution, and HASL should be excluded from this generalization. HASL normally is very easy to wet, since the wetting process involves merely coalescence of molten solder with molten surface finish.

The generalization serves as a guideline, and can be challenged by many exceptions due to non-ideal manufacturing conditions. For instance, ENIG may show poor wetting when it suffers serious black pad symptom, while freshly manufactured ImSn can be very good in solderability.

For Pb-free soldering, type of PCB surface finish affects wetting, bond strength, voiding, aging tolerance, and may also affect reliability.

The voiding performance of solder joints is highly affected by the wettability of surface finishes, with poor wettability yielding high voiding. Therefore, tendency of forming large voids decreases in the following sequence OSP (highest in voiding) > HASL > ImAg, ENIG > ImSn [37].

OSP and ImSn are more sensitive to aging. The wettability, bond strength, and voiding of ImSn all deteriorate quickly upon aging. OSP is poorer in both wetting and voiding, but good in bond strength. The pull strength of lead-free QFP solder joints decreases in the following order: OSP > HASL > ImAg > ImSn > ENIG [38]. The sensitivity of OSP toward thermal aging can be reduced by employing new OSP chemistries with higher thermal decomposition temperature [39].

ENIG is good on wettability, aging tolerance, and voiding. However, it showed the weakest bond strength.

Figure 5.9 shows the pull strength of lead-free solder joints on various surface finishes for 50 mil pitch SOICs before and after thermal cycling [40]. The

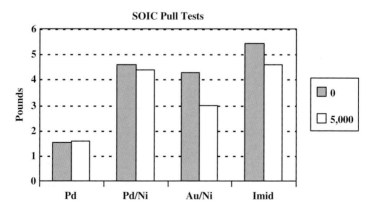

Fig. 5.9 Pull strength of lead-free solder joints on various surface finishes before and after thermal cycling test (originally published in SMI 1996) [40]

low pull strength of Pd finish is attributed to high voiding in solder joints. Thick layer of Pd or Au surface finish is detrimental to reliability and voiding, due to the formation of large quantity of intermetallics $PdSn_4$ or $AuSn_4$. In both cases, the volume of intermetallics formed is approximately 5 times of that of Pd or Au. This is considerably higher than cases of Cu_6Sn_5, Ni_3Sn_4, or Ag_3Sn, where the volume of intermetallics is typically less than 2 times of that of surface finish metals. Inclusion of large quantity of intermetallic particles in liquid solder inevitably impedes the escape of voids at soldering, thus results in high voiding and weak bond strength.

ImAg is good in overall performance. Although ImAg is neither the most robust surface finish in terms of solderability nor the highest in joint strength or reliability, it does not have obvious weakness in both features if the finish is properly prepared. As a result, it often becomes the favorable choice thus exhibits the highest CAGR. However, if not properly prepared, ImAg tends to have Cu cavity under ImAg layer, thus suffers microvoiding upon soldering [41, 42, 43]. Excessive plating time may even result in discontinuity of circuitry due to complete removal of Cu trace at perimeter of solder mask [44].

5.5 Components for Lead-Free Soldering

Besides the requirement of being lead-free, under European RoHS regulation, the polymeric materials used in electronic devices suffer dual impacts. First of all, a higher soldering temperature is needed due to the higher melting temperature of Pb-free alloys. Secondly, the ban of polybrominated biphenyls (PBB) and polybrominated diphenyl ethers (PBDE) which are commonly used as flame retardants in polymeric materials for packaging and substrates. PBB and PBDE, include penta-, octa-, and deca-brominated compounds, are banned due to formation of toxic dioxins and furans during combustion.

5.5.1 Temperature Tolerance

JEDEC/IPC J-STD-020D defines the classification temperature for SMD packages, with a higher temperature tolerance demanded for a smaller packages, as shown in Tables 5.3 and 5.4 [45]. In general, the classification temperature for Pb-free process is about 25–40°C higher than that for Sn63 process.

Table 5.3. Pb-free process – classification temperature (T_c)

Package thickness	Volume mm^3 < 350	Volume mm^3 350–2000	Volume mm^3 > 2000
< 1.6 mm	260°C	260°C	260°C
1.6 mm–2.5 mm	260°C	250°C	245°C
> 2.5 mm	250°C	245°C	245°C

Table 5.4 SnPb eutectic process – classification temperature (T_c)

Package thickness	Volume mm^3 < 350	Volume mm$^3 \geq$ 350
< 2.5 mm	235°C	220°C
> 2.5 mm	220°C	220°C

The thermal stability is comprised of both chemical structural stability and physical structural stability. The chemical structural stability reflects thermal decomposition. The physical structural stability is related to delamination and warpage. In the study of Chung et al., the evolutions of package warpage of two kinds of potential halogen-free compounds during thermal cycling test (TCT) reliability process was monitored. The compound with larger package warpage generated larger cumulate plastic work in solder joint that caused early failure during TCT process [46].

5.5.2 Moisture Sensitivity Level

Although Pb-free classification temperature raised the bar for component thermal stability, the biggest impact on electronic assembly resides in degradation in moisture sensitivity level (MSL) classification. Table 5.5 shows the classification of MSL [45].

The MSL of components was reported to degrade by one level for every 5–10°C increase in practical peak temperature (PPT), where the PPT was defined as minimum joint temperature + ΔT (board and components) + process tolerance + measurement error [47]. For some moisture sensitive components, the defect rates increased as peak reflow temperature or preheat ramp rate increased [48].

Table 5.5 Moisture sensitivity levels

Level	Floor life Time	Condition
1	Unlimited	$\leq$ 30°C/85%RH
2	1 year	$\leq$ 30°C/85%RH
2a	4 weeks	$\leq$ 30°C/85%RH
3	168 h	$\leq$ 30°C/85%RH
4	72 h	$\leq$ 30°C/85%RH
5	48 h	$\leq$ 30°C/85%RH
5a	24 h	$\leq$ 30°C/85%RH
6	Time on label (TOL)	$\leq$ 30°C/85%RH

5.6 Substrates for Lead-Free Soldering

The impact of RoHS regulation on substrates is similar to that for plastic component packages. The substrate materials need to be halogen free and capable of surviving a higher process temperature.

5.6.1 Thermal Decomposition

Khan et al. reported that there was a strong indication that high-end products built with current laminate materials would not survive lead-free processes [49]. Problems such as conductive anodic filament (CAF) were aggravated by the higher process temperature [50]. The most critical property upgrade of substrate for lead-free process is probably the resistance toward thermal decomposition. A higher glass transition temperature does not promise a higher decomposition temperature [51, 52], as shown in Table 5.6. Here glass transition temperature and decomposition temperature are two independent properties, and HGHD sample displayed the poorest thermal stability.

5.6.2 Dimensional Stability

Maintaining board dimensional stability is getting more difficult under the elevated lead-free soldering temperature. This is particularly a concern when a large board is held on rails during soldering process. For wave soldering, this board sagging issue may be rectified by employing either a high Tg resin materials, as shown in Table 5.7, or by adding a supporting cable under the board. The form stability is also important for flex print circuit (FPC). For fine pitch design, warpage of FPC can easily cause opens, and a high Tg resin is crucial for high yield performance.

Dimensional stability is also critical in preventing pad lifting at annual ring of through hole [53]. The coefficient of thermal expansion of copper is 17 ppm/°C,

Table 5.6 Materials with high and low values of Tg and Td

Material	Notation	Glass transition Temp, °C	Decomposition Temp, °C
Low Glass Transition Temp., Low Decomposition Temp.	LGLD	140	320
Low Glass Transition Temp., High Decomposition Temp.	LGHD	140	350
High Glass Transition Temp., Low Decomposition Temp.	HGLD	175	310
High Glass Transition Temp., High Decomposition Temp.	HGHD	175	350

Table 5.7 Resin type and Tg (°C)

Resin	Tg (°C)
Standard FR4 epoxies	115–125
Modified FR4 epoxies	120–130
Multi-functional epoxies	140–180
BT epoxies	160–180
Cyanate ester	230–250
Modified polyimides	220–260
Conventional polyimides	250–270

Table 5.8 Typical laminate CTEs, in ppm/°C [53]

Material	x, y axes	z axis
Polyimide E-glass	15–18	45–60
Epoxy E-glass	15–18	45–60
Modified epoxy/aramid	6.5–7.5	95–110
Modified epoxy/quartz	11.0–14.0	55–65

which is considerably lower than that of conventional laminates in the Z-axis direction, as shown in Table 5.8 [54]. Upon soldering, the laminate expands more than copper, thus causes lad lifting. When using the same soldering peak temperature, use of resin with a higher Tg reduces the adverse impact of high CTE above Tg, α_2. This effectively reduces the mismatch in dimension, and consequently alleviates the pad lifting problem.

5.7 Assembly with Lead-Free Reflow Soldering

iNEMI's study showed that the printability of lead-free solder pastes was comparable with that of Sn63 solder pastes [55]. This is expected, since the printability of a stable solder paste is governed by the rheology of solder paste, which in turn is governed by the solder powder volume fraction, solder powder shape, size, and rheology of flux vehicle, not by the alloy composition [34]. With the same stencil design, the difference in density of alloys would have impact on the weight of solder paste printed, not on the paste volume deposited.

5.7.1 Equipment

The stencil aperture design for lead-free solder paste is preferred to be slightly larger than tin-lead solder paste. This is mainly due to the poorer wetting of lead-free materials, as will be discussed later. In order to achieve comparable coverage of pads after reflow, a larger aperture design would then be desired. As to the printer, no difference in printer design is needed for transition into lead-free process.

5 Lead-Free Soldering

However, the reflow furnace would need some modification for lead-free process. Due to a higher melting temperature of lead-free alloys, a higher reflow temperature would then be required. This inevitably results in a longer excursion of heating process, and consequently more profile details to be controlled. In general, in order to maintain a comparable level of detail control as that of tin-lead process, one to two more heating zones are required for lead-free reflow furnaces. Also, due to the poorer wetting of lead-free alloys, nitrogen capability may become necessary for certain product designs, as will be discussed in the next section.2

5.7.2 Reflow Profile

Besides a higher reflow temperature for lead-free process, the shape of profile is comparable for both tin-lead and lead-free processes. In general, the shape of reflow profile of lead-free solder pastes can be categorized as soaking profile and linear-ramp profile, as exemplified in Fig. 5.10.

The advantage of linear-ramp profile has been predicted and discussed in details by Lee, with mechanisms elucidated [34]. This prediction is later supported by many studies. For instance, in the comparison of linear-ramp versus soaking profiles, the former was observed to yield a higher solder joint strength and a narrower joint strength range [56]. Furthermore, linear-ramp profile was also reported to result in doubling of the number of drops to failure when compared with soaking profile [57].

As to the reflow peak temperature, although 30°C above melting temperature has been a common practice for eutectic tin-lead reflow, the peak temperature for lead-free reflow has been pushed hard toward a lower temperature. Motorola has adopted a linear-ramp profile with peak temperature 235°C +/− 5°C and time above liquidus 70 s +/10 s for manufacturing of more than 100 millions cell phones with high yield and high quality [58]. Solectron pushed the

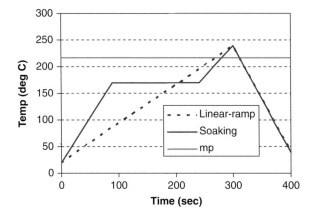

Fig. 5.10 Major categories of lead-free reflow profiles

peak temperature down to 225°C for reflowing SAC387 in air, and reported a reliability of greater than 3500 thermal cycles cycled between 0 and 100°C [59]. In the latter case, the peak temperature is less than 8°C above the liquidus temperature.

The effect of time above liquidus has also been studied. Shina et al. reported that when the time above liquidus increased from 60 to 90 then to 120 s, the joint strength increased very slightly [56].

5.7.3 Special Profiles

Although linear-ramp profile showed significant advantage over conventional soaking profile, it is not necessarily always the best choice for all manufacturing. For instance, for product designs more prone to have voiding problems, such as via-in-pad design, the best profile for minimal voiding can be situation dependent.

Soldering voiding is caused by outgassing within the solder joint when the solder is at molten stage, and can be reduced by reducing the outgassing or by improving the wetting or both [60]. The impact of reflow profile on voiding can be reflected in both directions.

5.7.3.1 Outgassing Control

Figure 5.11 shows the typical outgassing behavior of fluxes with increasing heat input. In general, with increasing heat input, the outgassing rate of virtually all fluxes increases initially, then decreases gradually after reaching the maximum point. Minimal outgassing at above melting temperature can be realized at either spot 1 or spot 2. Spot 1 represents a minimal heat input, with

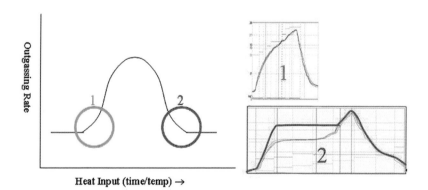

Fig. 5.11 Relation between outgassing rate and heat input of fluxes, where the heat input is a combined result of time and temperature. Spot 1 and 2 are exemplified by profile 1 and 2, respectively

short, fast ramp rate, and a low peak temperature, as exemplified by profile 1. The attempt is to complete the reflow process before major outgassing starts. Spot 2 represents a long hot soaking and a low peak temperature, as exemplified by profile 2. The attempt is to dry out volatiles before solder melts. A long hot soaking will favor the elimination of volatiles, and a low peak temperature will favor minimizing further outgassing when the solder is at molten state. The beneficial effect of long hot soaking plus a low peak temperature was confirmed by study of Solectron [61].

5.7.3.2 Wetting Control

Wetting improves with increasing fluxing reaction, which in turn increases with increasing temperature and time. Thus, a profile with a high temperature and a long time will favor better wetting [62]. Figure 5.12 shows the voiding performance of lead-free solder paste C when reflowed under various reflow profiles. Paste C is very resistant against oxidation. The soaking time increases sequentially from profile 2 m to profile 8 m. The peak temperature also varies from 230 to 255°C. Here the voiding decreases not only with increasing soaking time, but also with increasing peak temperature, thus strongly demonstrates the effect of improving wetting on reducing voiding.

5.7.3.3 Balancing Between Outgassing and Wetting

However, the wetting behavior can be complicated by flux loss and oxidation. The flux gradually dries out with increasing heat input. Furthermore, under air, oxidation also increases with increasing temperature and time. Figure 5.13 shows the voiding performance of lead-free solder paste A under various profiles. Paste A is less resistant to oxidation. As a result, increasing soaking temperature may have benefit initially for profiles with 230°C and 235°C peak temperature due to volatile elimination, it eventually results in an increasing voiding due to

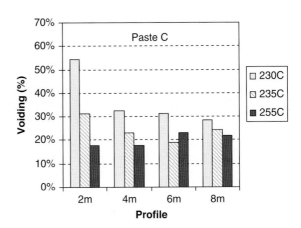

Fig. 5.12 Voiding performance of lead-free solder pastes when reflowed under various reflow profiles. Paste C is fairly resistant to oxidation

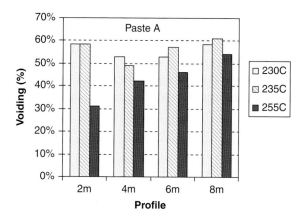

Fig. 5.13 Voiding performance of lead-free solder pastes when reflowed under various reflow profiles. Paste A is less resistant to oxidation

increasing oxidation. The increasing oxidation effect is more clearly demonstrated by profile with peak temperature 255°C [62]. The optimal profile should be a balanced one based on both wetting and outgassing considerations.

5.8 Assembly with Lead-Free Wave Soldering

5.8.1 Lead-Free Wave Soldering Process

Lead-free wave soldering process takes a longer time and a higher temperature for both preheat and wave stages than tin-lead process, as exemplified in Fig. 5.14. This is mainly due to the higher melting temperature and poorer wetting ability of lead-free alloys.

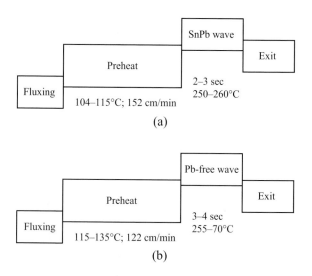

Fig. 5.14 Comparison of SnPb (a) and Pb-free (b) wave soldering process

5 Lead-Free Soldering

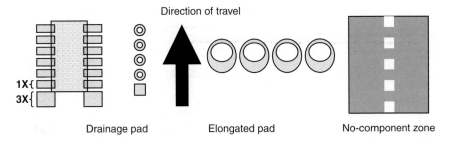

Fig. 5.15 PCB design for lead-free wave soldering

5.8.2 Design of PCB

Several modifications in PCB designs are recommended in order to cope with the transition to lead-free wave soldering process, as shown in Fig. 5.15.

(1) A center cable support under the PCB can be used to address the board sagging due to the higher temperature. To accommodate this cable support, a no-components line about 3–4 mm wide at center of the bottom side of the board is necessary.
(2) A design with hole diameter at least 10 mils (0.25 mm) larger than the pin diameter is needed in order to allow sufficient hole-fill and minimize voiding.
(3) To facilitate solder drainage, a drainage pad following the row of pads is recommended. Elongated through-hole pads along travel direction can also help reducing bridging.

For controlling fillet lifting, besides increasing the cooling rate, reduction in exposed pad size using solder mask defined pad is recommended.

Besides the design modification described above, some routine design practices used for tin-lead process should also be carried over to lead-free process. For instance, discrete components should be aligned parallel to travel direction in order to minimize bridging [63].

5.8.3 Equipment Erosion

The initial impact of lead-free conversion toward wave soldering is equipment erosion. In the wave soldering machine, tin reacts with iron favorably to form $FeSn_2$ intermetallics and results in a fast erosion of stainless steel hardware. This erosion is further aggravated by the higher solder bath temperature needed for lead-free alloys. New equipment materials have been developed to offset this tin erosion, as shown in Table 5.9.

Table 5.9 Materials developed for lead-free wave machines

Material	Advantages	Disadvantages	Longevity
304 stainless steel	Inexpensive	Minimal corrosion resistance	1 month
316 stainless steel	Inexpensive	Minimal corrosion resistance	3–6 months
Cast iron	Inexpensive	Minimal corrosion resistance	1–2 years
Surface coated stainless steel	Good resistance to corrosive effects of tin	Will degrade when coating is scratched	6–12 months
Surface coated cast iron	Good resistance to corrosive effects of tin	Will degrade when coating is scratched	3–5 years
Titanium	Excellent resistance to corrosive effects on tin	Can be expensive to fabricate	10 years

5.8.4 Through-Hole Filling of Thick PCBs

For through-hole filling, IPC-A-610D [64] specifies that 75% hole fill is required. 50% hole fill can be allowed for class 1, conditionally for class 2, never allowed for class 3. For class 2, 50% hole fill is acceptable when PTH is connected to a big heat sink, and solder wetting around barrel wall and pin is even.

While IPC spec can be met in lead-free applications for regular board thickness, industry has encountered great difficulty for thick and large boards, particularly when OSP surface finish is used [65, 66, 67]. Although reworking the out-of-spec joints is a common practice, the reliability of solder joints often is compromised due to excessive rework heating. Hewlett Packard reported an alternative criterion for hole-filling for thick boards [67]. In their approach, a minimal 5-pounds through-hole joint pull strength after various accelerated test conditioning was set as criterion. The pull strength was found to be dictated only by the pin-wetting-length within the barrel. The hole-fill requirement was then established according to board thickness and specific product applications, with a greater hole-fill required for a harsher application. Furthermore, the hole-fill requirement for the harshest condition was selected as the criterion for all applications with the same board thickness. For instance, the hole-fill requirement for 0.062 inch, 0.097 inch, and 0.130 inch board thickness was established as 75%, 47%, and 35%, respectively.

5.9 Inspection of Lead-Free Solder Joints

National Physical Laboratory (NPL) of United Kingdom evaluated ability of automated optical inspection (AOI) to inspect lead-free solder joints. Results indicated that most AOI systems could be used for the inspection of lead-free surface mount assemblies. Test data obtained were similar or better than for tin-lead assemblies. False defect rates were also comparable for both sets of assemblies [68].

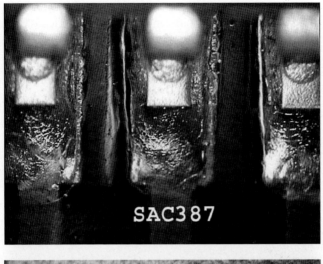

Fig. 5.16 Example of Sn95.5Ag3.8Cu0.7 and Sn99.3Cu0.7 solder joints

However, in general, lead-free solder joints are more rough and striated than tin-lead joints, as exemplified in Fig. 5.16. NEMI study indicated that distinguishing good from bad solder joints will be more difficult with lead-free, because of increasing variety of solder joints appearance. Also, more fine-tuning of optical inspection equipment will be required, especially with a mix of lead-free and tin-lead coated components. The next results will be an increased time for programming the equipment [69].

NPL studies on X-ray images indicated that automated x-ray inspection (AXI) system had no problem imaging different lead-free materials [68]. Similar observation was also reached by NEMI [69], although Cannon reported difficulty in detecting the true quality of flip chip lead-free solder joints [70]. NEMI concluded that most inspection equipment can be used for lead-free inspection, and training for operators and adjustment by suppliers are needed [69].

5.10 Rework Lead-Free Solder Joints

5.10.1 Cell Phone Rework

Goudarzi et al. studied rework of lead-free solder joints for cell phones [71]. After evaluation, the preferred equipment includes hot air unit, soldering iron, and bottom heater. Fifteen mils solder wire with 2.7% flux was selected based on reliability and visual inspection. The rework process was compared with typical tin-lead rework process, and is shown in Table 5.10.

5.10.2 BGA Rework

In general, for lead-free rework, equipment with higher power and faster response time than that for eutectic SnPb is needed. Both ramp rates up and down needed to be faster. A faster cooling rate is critical in reducing voids [72].

For reworking BGA, minimizing the temperature gradient between the top of the component and the BGA solder joints are critical, and can be achieved through nozzle design. Bottom heating is essential. On the top side, direct heating on the top of component should be avoided. The heated nitrogen is to be directed horizontally toward the side of component. In the study of Yoon et al., the reworked solder joints were examined to be acceptable by cross-sectional and X-ray analysis, and passed 3,500 thermal cycles using cycling condition from 0 to 100°C [73].

When remounting BGA, flux dip process using gel flux is recommended for lead-free. This is due to the greater resistance of gel flux versus liquid flux against evaporation, thus more adequate for the higher lead-free process temperature. The additional benefit of gel flux is better volume control of flux on pads [72]. Liquid fluxes, including flux pens, are fine in 3–5 s hand soldering, but not recommended for several minutes long lead-free reflow soldering.

Table 5.10 Rework comparison between Pb-free and SnPb systems.

		SnPb	Pb-free
Temperature profile	Top	300°C	Increased by 25°C
	Bottom	80°C	Increased by 20°C
Cycle time		Depends on amount of mass	Increased by 30%
Procedure		Same as Pb-free	Same as SnPb
Inspection		Standard inspection criteria should be used	Pb-free solder appearance may be different

5.11 Reliability of Lead-Free Solder Joints

The two most important types of reliability of lead-free solder joints for electronic devices are temperature cycling and drop test performance. The impact of materials, processes, and environment on the reliability will be discussed below. Since the reliability is governed by the microstructure, it is crucial to know the microstructure of solder joints first, particularly the intermetallic compounds (IMC) formed in the joints.

5.11.1 Microstructure

The microstructure of tin-silver-copper solders, such as SAC387, is shown in Fig. 5.17, where a number of short rod-like bright Ag_3Sn particles and some small dark grey Cu_6Sn_5 particles dispersed in the tin matrix between the grey tin dendrites. Few large Cu_6Sn_5 particles can also be seen randomly dispersed. At Ag content above approximately 2.5 wt%, large Ag_3Sn platelet formation may also be formed [74].

At reflow, upon cooling, the large Ag_3Sn platelets may form first, followed by tin dendrites formation from liquid solder. Upon further cooling, small Ag_3Sn and Cu_6Sn_5 particles further precipitate out in between tin dendrite globes [75].

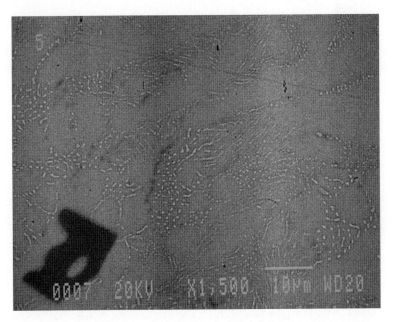

Fig. 5.17 Scanning electron micrograph of cross-sectioned SAC387 solder reflowed on Cu

For eutectic SnAg and SnCu alloys, similar small rods of Ag_3Sn particles and Cu_6Sn_5 particles are formed respectively in between tin dendrites.

5.11.2 Solder Joint Intermetallics

5.11.2.1 SAC on Cu-Based Substrate

For SAC solder, the intermetallics formed at interface on Cu are Cu_6Sn_5 near solder side and Cu_3Sn near Cu side. For wave soldering, the IMC thickness is very thin, often around or smaller than 0.1 μm and is barely discernible [76]. For reflow soldering, the IMC thickness is typically around 2 μm, with Cu_6Sn_5 as the dominant layer. Upon aging, Cu_3Sn layer grows quickly and becomes comparable to that of Cu_6Sn_5 layer, as shown in Fig. 5.18. The IMC thickness may be as high as 5 μm for some BGA joints. Kao reported that as little as 0.1% Ni addition is able to hinder Cu_3Sn formation on Cu [77].

5.11.2.2 SAC on Ni-Based Substrate

The IMC formed on NiAu is more complicated than that on Cu. Kao studied the effect of Cu content in solder on IMC structure on Ni-based substrates, as shown in Table 5.11 [77]. For joints with small solder volume, the Cu content decreased readily with growth of IMC. This resulted in shifting of equilibrium phase at interface, and eventually could cause the massive spalling of IMC layer.

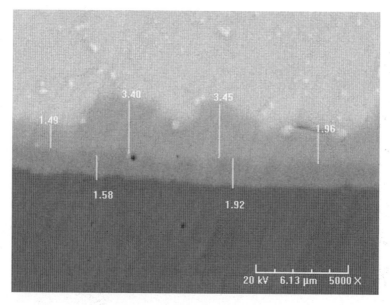

Fig. 5.18 SAC305 solder joint on Cu after aging at 150°C for 10 days. IMCs formed at interface, with Cu_6Sn_5 near solder side and Cu_3Sn near Cu side

5 Lead-Free Soldering

Table 5.11 Effect of Cu content on IMC structure formed on Ni-based substrates, if the supply of Cu is not an issue

Cu content in solder	IMC formed
≤ 0.3 wt%	Only $(Ni,Cu)_3Sn_4$ forms
0.4–0.5 wt%	Both $(Ni,Cu)_3Sn_4$ and $(Cu,Ni)_6Sn_5$ form
> 0.5 wt%	$(Cu,Ni)_6Sn_5$ forms

Lu et al. studied the effect of Cu content in solder on IMC structures formed between Cu-based substrate and Electroless Ni immersion Au (ENIG) substrate [78]. The results can be illustrated with Fig. 5.19. The IMC on Cu-substrate remained as $(Cu,Ni)_6Sn_5$. However, the IMC composition on ENIG substrate varied with increasing Cu content in solder. At 0% Cu concentration, multiple layers of IMC formed on Ni surface. With increasing Cu concentration, $(Ni,Cu)_3Sn_4$ changed to $(Cu,Ni)_6Sn_5$, and NiPSn and $Ni_3(Sn,P)$ gradually merged and eventually vanished. At 0 to 0.5 wt% Cu, no Ag_3Sn plates observed in any locations of the solder joints at time zero. On the other hand, high Cu led to flourishing growth of Cu-Sn IMCs. The latter promotes the growth of Ag_3Sn platelets.

The TEM image of the interfacial structure of SnAgCu and Au/Ni(P)/Al joint after 5 reflows was examined [79, 80]. Within the NiSnP layer with a thickness of 0.1 μm, microvoids were formed due to diffusion of Sn away from this layer.

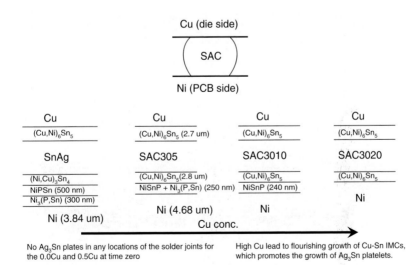

Fig. 5.19 Effect of Cu content in solder on IMC structure of solder joints formed between Cu-based substrate and ENIG substrate

5.11.2.3 IMC Growth

Song et al. studied the effect of thermal aging at 150°C on IMC formation rate of lead-free solders on OSP and ENIG, with results shown in Fig. 5.20 [81]. OSP exhibited a higher IMC growth rate than ENIG, mainly attributable to the higher diffusion rate of Cu than Ni. On OSP, SAC405 showed a lower growth rate than Sn96.5Ag3.5 (SA), presumably due to the retardation of Cu substrate diffusion in the presence of Cu in solder. On ENIG, SAC405 also appeared to be slightly lower in IMC growth rate than SA.

Xu et al. compared the effect of isothermal aging, temperature cycling, and thermal shock on IMC growth rate for SAC387/NiAu BGA specimen. The test conditions were 125°C for isothermal aging, $-40 \sim 125°C$, 15 min high temperature (T) dwell, 1 hr/cycle for temperature cycling (TC), and $-55 \sim 125°C$, 5 min high T dwell, 17 min/cycle for thermal shock (TS), respectively. For TC and TS tests, IMC thickness was measured at 500, 1000, 1500, and 2000 cycles. Results are shown by plotting IMC thickness against square root of time (hours), as shown in Fig. 5.21 [82]. As expected, IMC thickness increased with increasing testing time. It is interesting to note that IMC growth rate displayed the following order: TS > TC > isothermal aging, despite the fact that isothermal aging had the longest exposure time at 125°C during the same testing time span. Apparently, thermal stress played a more dominant role than 125°C conditioning in expediting IMC growth, and a higher thermal stress resulted in a higher IMC growth rate.

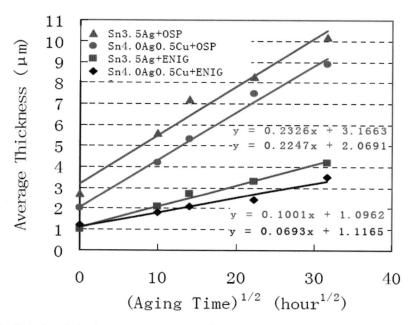

Fig. 5.20 Correlation between IMC thickness and 150°C aging time (Cu-Sn phase in both SA and SAC405 on OSP; Ni-Cu-Sn phase in SAC405 on ENIG; Ni-Sn phase in SA on ENIG) (IEEE copyright) [81]

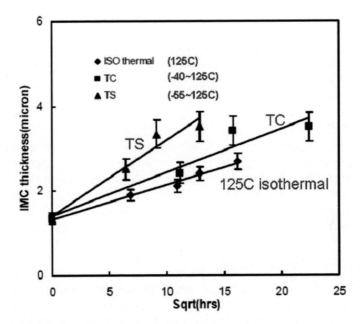

Fig. 5.21 Comparison of IMC thickness: TC, TS, Isothermal Aging, SAC387/Ni-Au couple (IEEE copyright) [82]

5.11.3 Temperature Cycling

The reliability of solder interconnects formed with SAC397 and SA was determined by Osterman et al. [83] using CLCC assemblies subjected to various thermal cycling conditions. The test results indicated:

(1) At the lower cyclic mean temperatures, Pb-free solders showed better reliability than SnPb solder.
(2) At the highest tested cyclic mean temperature which had a cyclic peak temperature of 125°C, SnPb solder outperformed the Pb-free solders.
(3) Effect of the dwell time decreased with the increasing cyclic mean temperature.
(4) At temperatures under 100°C regardless of dwell time, Pb-free solder was more reliable than SnPb solder.
(5) Reliability of the Pb-free solders showed much stronger dependence than the SnPb solder on the cyclic medium temperature.

5.11.3.1 Effect of Alloy Composition

Solder composition has a great impact on temperature cycling performance. Terashima et al. reported that in Sn-xAg-0.5Cu system, the 50% failure rate for 1, 2, 3, and 4% Ag content alloys occurred at approximately 305, 375, 545, and

605 cycles, respectively. In other words, a decrease of Ag content from 4 to 1% decreased the thermal fatigue life (50% failure) of flip chip joints on Cu pads by a factor of about 2 [84]. All alloys tested had 0.5% Cu, and the test cycle was −40/125°C, 10 min dwell. The fracture was reported to be a mixed mode, transgranular and intergranular, independent of the silver content. The positive effect of Ag content on thermal fatigue life was attributed to the reinforcement of Ag_3Sn IMC particles. A higher Ag_3Sn concentration resulted in a more rigid solder, partly through reinforcement effect, partly through refining the Sn grain size. Presence of large amount of Ag_3Sn IMC particles also suppressed microstructural coarsening more effectively, thus maintained a higher fatigue resistance. Increase in Cu content may have a similar effect through reinforcing power of Cu_6Sn_5 IMC particles. However, this approach could result a wider pasty range by raising the liquidus temperature. For instance, at 4 wt% Cu content, the liquidus should be higher than 300°C, therefore may impose challenge on soldering process.

5.11.3.2 Effect of Surface Finish

Zbrzeznya et al. studied the effect of surface finish on the accelerated temperature cycling performance of SAC387 solder joints for chip resistors [85]. The board finishes tested included Cu (immersion Ag and HASL) and Ni (ENIG), while the component finishes tested included Sn92Pb8 and 100% Sn. Results from Table 5.12 showed the joints to copper on board exhibited a significantly higher number of cycles to first failure than the joints on nickel on board. On the other hand, component finish Sn92Pb8 was moderately better than Sn-finish.

The significantly better reliability of the copper joints can be explained in terms of the copper content in the bulk. On Cu board, dissolution of Cu pad caused an enriched Cu content of joints, hence joints with refined grain size due to presence of abundant Cu_6Sn_5 IMC particles at grain boundary. This eventually resulted in a longer thermal cycling life. SAC387 joints on Ni board were depleted with Cu due to formation of $(Cu,Ni)_6Sn_5$ IMC layer at joint interface, thus suffered a coarser grain size, and consequently a shorter cycling life.

5.11.4 Fragility of Solder Joints

Although SAC solder joints have been satisfactory versus SnPb joints in thermal cycling test [83, 86, 87], the fragility of solder joints was recognized later as the unexpected weakness of lead-free alloys [87–91].

Table 5.12 Number of cycles to first failure versus board/component finish

		Board finish	
		Cu	Ni
Component finish	92Sn8Pb	5081	3250
	Sn	4000	1595

5.11.4.1 Effect of Alloy Composition

The fragility of lead-free solder joints was attributed to the higher hardness of lead-free solders. Upon impact, the shock energy could not be dissipated by the ductility of solder, thus causing the cracking of the weakest link IMC layer at interface. For SAC alloys, both Ag and Cu would increase the hardness of solder due to formation of Ag_3Sn and Cu_6Sn_5 IMC particles [89]. Since Cu content not only is relatively low, but also is desired in order to prevent IMC spalling on Ni surface upon aging, reducing Ag content turns out to be the logical shortcut to reduce the hardness of SAC solders. Therefore, in drop test, SAC105 BGA joints were reported by Liu et al. to be about 5 times better than SAC387 and SAC305 on electroplated NiAu substrate, although they were still not as good as Sn63, as shown in Fig. 5.22 [92].

The effect of Cu content on fragility is not as straight forward as Ag. Although decrease in Cu content is expected to reduce the fragility of joints due to reduction in hardness, increase the Cu content to 2 wt% was found to transform the solder into a ductile material, and consequently eliminated the fragile failure mode of Sn95Ag3Cu2 solder joints [78].

Besides varying Ag and Cu content, adding small amount of certain elements into solder was also effective in reducing the fragility of SAC solder joints. Liu et al. reported Mn, Ti, Ce, Bi, and Y were effective in reducing the fragility of SAC solder joints, with Mn and Ti being the most effective [92]. While SAC-Mn

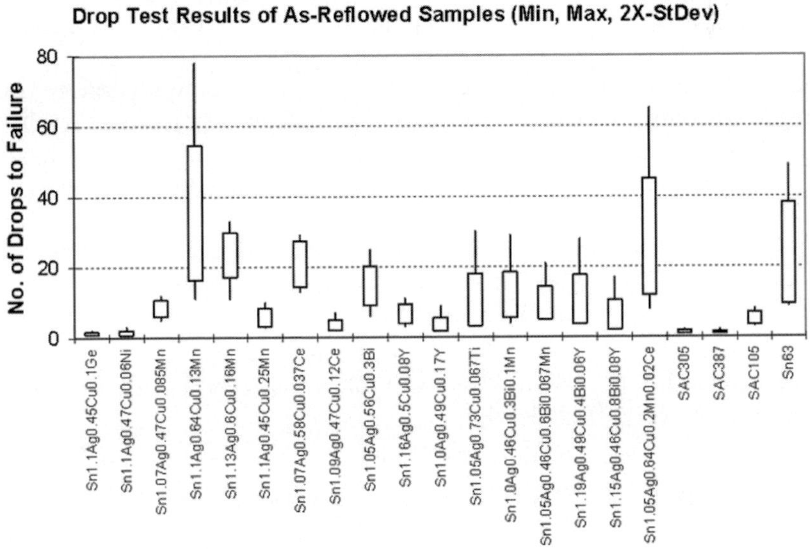

Fig. 5.22 Drop test results of as-reflowed samples. The thin line represents minimum and maximum values, the box represents two standard deviation, with mid point being the mean value [92]

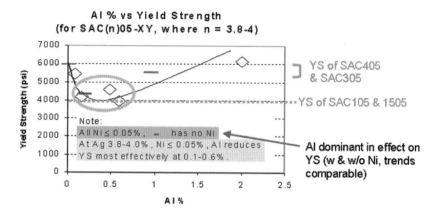

Fig. 5.23 Effect of Al addition on yield strength of SAC(n)05-XY [96]

system was found to outperform Sn63 and SAC105 in most incidences, SAC-Ti was observed to have the best performance overall for joints on ENIG/OSP, NiAu/OSP, and OSP/OSP surface finishes [93]. The superior performance of SAC-Ti was attributed to (1) the increased grain size and dendrite size, therefore reduced hardness of solder, (2) inclusion of Ti in the IMC layer, and (3) reduced IMC layer thickness. Amagai et al. found addition of 0.2 wt% In and 0.04 wt% Ni improved drop test performance by 20% [94]. Co, Ni and Pt were dissolved in IMC, which did not increase IMC grain size and thickness significantly after 4 times solder reflow processes. Upon pull test, the fracture occurred mainly in bulk solder instead of at interface [95].

The reduction of fragility of the above approaches was conducted by modifying low Ag SAC systems. As shown in Fig. 5.23, low Ag typically results in compromise in thermal cycle fatigue resistance [84]. Huang et al. investigated alloys exhibiting both high thermal fatigue and low fragility properties. Their results indicated that, for high Ag SAC alloys, adding Al 0.1–0.6% to SAC alloys was most effective in softening, and brought the yield strength down to the level of SAC105 and SAC1505, while the creep rate was still maintained at SAC305 level, as shown in Fig. 5.23 [96].

5.11.4.2 Effect of Surface Finish

The surface finish has a significant effect on fragility of lead-free solder joints. Arra et al. studied drop test performance of SAC396 solder joints, observed the drop number for NiPd, Sn85Pb15, and Sn98Bi2 to be 10, 13, and 20, respectively [97]. In another study, the drop number was found to be independent of the joint strength by variation of lead coating, as shown in Table 5.13 [98].

Darveaux et al. studied the failure mode of solder joints on various PCB surface finishes in pull test [99]. Results indicated (1) more reflow heat caused

5 Lead-Free Soldering

Table 5.13 Solder joint strength versus drop number for various lead coatings [98]

Lead coating	Average pull force (N)	Average number of drop cycles to drop off
Ni/Pd/Au	23.3	3.5
Sn98Bi2	20.0	9.5
Sn85Pb15	17.3	6.6
Sn	14.3	5.2

Fig. 5.24 Failure site of SAC solder joints on Ni and Cu in pull test

more brittle failure on Cu than on NiAu, (2) propensity toward brittle failure decreased in the following order: ImSn, ENIG > OSP > NiAu.

Song et al. also studied the failure mode of SAC BGA ball on OSP and ENIG via pull test [100]. Their results show that OSP was quite more prone to brittle failure than ENIG. The brittle failure on the ENIG was induced by weak joint between the IMC and the Ni layers and the brittleness of IMC itself, as shown in Fig. 5.24. The brittle failure on the OSP pad was induced mainly by weak joint of Cu_6Sn_5 and Cu_3Sn IMC phases.

5.12 Summary

Lead-free soldering is here to stay. SAC alloys are the prevailing choices. The soldering processing window is narrower than that of Sn63, mainly due to the elevated melting temperature of SAC solder and the limited high temperature tolerance of components and board. The reliability of lead-free solder joints is acceptable for thermal cycling performance. However, the fragility of joints remains a concern, although significant improvement has been accomplished.

Rererence

1. IPC Global Solder Statistical Program Report for 2nd Quarter 2007, August 2007
2. "Electronics Manufacturing with Lead-Free, Halogen-Free, and Conductive-Adhesive Materials" by John H. Lau, C. P. Wong, Ning-Cheng Lee, S. W. Ricky Lee. Hardcover: 700 pages, 1st edition (August 27, 2002), by McGraw-Hill.
3. C. Hunt and D. Lea, "Solderability of Lead-Free Alloys", in Proceedings of Apex 2000, Long Beach, CA, March 2000.

4. Nihon Superior data sheet, www.nihonsuperior.co.jp/english/products/leadfree/#01
5. Indium Corporation data sheet, www.indium.com/_dynamo/download.php?docid = 500
6. Weiping Liu and Ning-Cheng Lee, "Novel SACX Solders with Superior Drop Test Performance", SMTA International, Chicago, IL, September, 2006
7. Indium Corporation, patent pending.
8. Benlih Huang and Ning-Cheng Lee, "Prospect of Lead Free Alternatives for Reflow Soldering", IMAPS'99 – Chicago.
9. AIM data sheet, www.aimsolder.com/techarticles/Comparison of Lead Free.PDF
10. Alpha Metals data sheet. alpha.cooksonelectronics.com/sacxdatalibrary/pdfs/SACX%20Wetting%20versus%20SAC0307.pdf
11. www.europeanleadfree.net/pooled/articles/BF_DOCART/view.asp?Q = BF_DOCART_198565
12. M. Date, T. Shoji, M. Fujiyoshi, and K. Sato, "Pb-free Solder Ball with Higher Impact Reliability", Intel Pb-free Technology Forum, 18th – 20th July 2005, Penang, Malaysia.
13. Benlih Huang, Hong-Sik Hwang, and Ning-Cheng Lee, "A Compliant and Creep Resistant SAC-Al(Ni) Alloy", ECTC, Reno, Nevada, May 26-June 1, 2007
14. Private communication with IBM.
15. Masazumi Amagai, "A Study of Nano Particles in SnAg-Based Lead Free Solders for Intermetallic Compounds and Drop Test Performance", 56th ECTC Proceedings, pp. 1170–1190, San Diego, CA, May 30–June 2, 2006
16. Benlih Huang and Ning-Cheng Lee, "Prospect of Lead Free Alternatives for Reflow Soldering", IMAPS'99 – Chicago
17. Ning-Cheng Lee, "Future Lead-Free Solder Alloys and Fluxes – Meeting Challenges of Miniaturization", International Microsystems, Packaging, Assembly and Circuits Technology (IMPACT) conference, Taipei, Taiwan, Oct. 1–3, 2007.
18. Guo-yuan Li, and Xun-qing SHI, "Effects of bismuth on growth of intermetallic compounds in Sn-Ag-Cu Pb-free solder joints", Transactions of Nonferrous Metals Society of China, Volume 16, Supplement 2, June 2006, pp. s739–s743
19. Zequn Mei, Fay Hua, and Judy Glazer, "SN-BI-X Solders", SMTA International, San Jose, CA, Sept. 13–17, 1999.
20. K. Suganuma, "Japan Leadfree 2001".
21. T. Baggio, "The Panasonic Mini Disk Player – Turning a New Leaf in a Lead-Free Market", IPCWorks'99, Minneapolis, MN, Oct. 27, 1999.
22. "Lead-free Electronics", Editor(s): Sanka Ganesan, Michael Pecht, 2003 CALCE EPSC Press., p. 444.
23. Toshikazu Yamaguchi and Takao Enomoto, "Tin-Zinc Solder Paste", Apex, San Diego, CA, Jan. 14–18, 2001
24. Günter Grossmann, Giovanni Nicoletti, Ursin Solèr, "Results of Comparative Reliability Tests on Lead-free Solder Alloys", 52nd ECTC, S30-P1, San Diego, CA, May 28–31, 2002.
25. Hirokazu Tanaka, Yuuichi Aoki, Makoto Kitagawa and Yoshiki Saito, "Reliability Testing and Failure Analysis of Lead-Free Solder Joints under Thermo-Mechanical Stress", Apex, S28-1, Anaheim, CA, Feb. 2004
26. J. Glazer, "Metallurgy of low temperature Pb-free solders for electronic assembly", International Materials Reviews, vol. 40, No. 2, pp. 65–93 (1995).
27. "Lead-Free Solder", Manufacturing Market Insider, pp. 6 (Oct., 1993).
28. Judy Glazer, "Microstructure and mechanical properties of Pb-free solder alloys for low-cost electronic assembly: a review", J. Electronic Materials, vol. 23, no. 8, pp. 693–700 (Aug. 1994).
29. Valeska Schroeder, and Fay Hua, "Feasibility Study of 57Bi-42Sn-1Ag Solder", Apex, San Diego, CA, Jan. 14–18, 2001
30. S.G. Gonya, J.K. Lake, R.C. Long, R.N. Wild, "Lead-free tin-bismuth solder alloys", IBM, Armonk, NY, US Patent 5,368,814, Nov. 29, 1994.

31. IPC J-STD-006B Requirements for electronic grade solder alloys and fluxed and non-fluxed solid solders for electronic soldering applications, Jan. 2006.
32. Ning-Cheng Lee, "Combining Superior Anti-Oxidation and Superior Print - Is it Really Impossible?", EPP EUROPE DECEMBRE 2007, pp. 20–21.
33. Paul Jaeger and Ning-Cheng Lee, "A Model Study of Low Residue No-Clean Solder Paste", in Proceeding of Nepcon West, Anaheim, CA, 1992.
34. Ning-Cheng Lee, "Reflow Soldering Processes and Troubleshooting – SMT, BGA, CSP and Flip Chip Technologies", Newnes, pp. 269, 2001.
35. "Lead-Free Electronics: iNEMI Projects Lead to Successful Manufacturing", edited by Edwin Bradley, Carol A. Handwerker, Jasbir Bath, Richard D. Parker, and Ronald W. Gedney, 472 pages, published by Wiley-IEEE, 2007.
36. K. Wengenroth, Y. Yau, F. Fudala, R. Prendergast, and J. Abys, "Alternative final finishes for PWB", QuickStart Lead-free Workshop, Ft. Lauderdale, FL, July, 27, 2004.
37. Walter Horaud, Sylvain Leroux, Hélène Frémont, Dominique Navarro, "PCB Materials Behaviours towards Humidity ", Apex, S07-2, Anaheim, CA, Feb. 2004
38. Sammy Shina, Liz Harriman, Todd MacFadden, Donald Abbott, Richard Anderson, Helena Pasquito, George Wilkish, Marie Kistler, David Pinsky, Mark Quealy, Karen Walters, Richard McCann, and Al Grusby, "Lead Free Conversion Analysis for Multiple PWB/Component Materials and Finishes using Quality and Reliability Testing", Apex, S29-4, Anaheim, CA, Feb. 2004
39. Koji Saeki, and Michael Carano, "Next Generation Organic Solderability Preservatives (OSP) for Lead-free soldering and Mixed Metal Finish PWB's and BGA Substrates", Apex, S10-2, Anaheim, CA, Feb. 2004
40. U. Ray, I. Artaki, D.W. Finley, G.M. Wenger, T. Pan, H.D. Blair, J.M. Nicholson, and P. T. Vianco, "Assessment of Circuit Board Surface Finishes for Electronic Assembly with Lead-Free Solders", SMI 96, Sep. 10–12, 1996, San Jose, CA
41. Muffadal Mukadam, Norman Armendariz*, Raiyo Aspandiar, Mike Witkowski, Victor Alvarez, Andrew Tong, Betty Phillips, and Gary Long (Intel Corporation), "Planar microvoiding in lead-free second-level interconnect solder joints", SMTAI, Sep., 2006, Chicago, IL
42. John Swanson and Donald Cullen, "Verifying microvoid elimination and prevention via an optimized immersion silver process", APEX, S18-01, Feb. 20–22, 2007, Los Angeles, CA
43. Yung-Herng Yau, Karl Wengenroth and Joseph Abys, "A study of planar microvoiding in Pb-free solder joints", APEX, S18-02, Feb. 20–22, 2007, Los Angeles, CA
44. Jing Li Fang, Daniel K. Chan, "The Advantages of Mildly Alkaline Immersion Silver as a Final Finish for Solderability", APEX, S23-02, Feb. 20–22, 2007, Los Angeles, CA
45. IPC/JEDEC J-STD-020D June 2007 Moisture/Reflow Sensitivity Classification for Non-hermetic Solid State Surface Mount Devices.
46. Cho-Liang Chung, Liang-Tien Lu and Yao-Jung Lee, "Influence of halogen-free compound and lead-free solder paste on on-board reliability of green CSP (chip scale package)", Microelectronics and Reliability, Vol. 45, No. 12, Dec. 2005, pp. 1916–1923.
47. {Lead-Free Component Team}, "Component Implications of Lead-Free Reflow Assembly", IPC/NEMI Symposium on Lead-Free Electronics, Sep. 18–19, 2002, Montreal, Canada
48. Vijay Gopalakrishnan, Vivek Venkataraman, Robert Murcko, Krishnaswami Srihari, Scott J. Anson, "Moisture and Reflow Sensitivity Evaluations of SMT Packages as a Function of Reflow Profile at Eutectic and Lead Free Temperatures", Apex, S19-3, Anaheim, CA, Feb. 2004
49. Arshad Khan, Rex Lam, and Bruce Houghton, "Printed Circuit Board Reliability in High Temperature Lead-Free Processes", Printed Circuit Board Reliability in High Temperature Lead-Free Processes", Apex, S40-1, Anaheim, CA, Feb. 2004
50. L.J. Turbini, W.R. Bent, W.J. Ready, "Impact of higher melting lead-free solders on the reliability of printed wiring assemblies", SMTA International, Chicago, IL, Sep. 20–24, 2000.

51. Edward Kelley, "An Assessment of the Impact of Lead-Free Assembly Processes on Base Material and PCB Reliability", Apex, S16-2-1, Anaheim, CA, Feb. 2004
52. Alan Rae, "Improved Electrical Properties of Epoxy Molding Compound and Circuit Board Materials Using Halogen-Free Flame Retardant Systems", Apex, P5-05, Anaheim, CA, Feb. 2004
53. Christiane Faure, Jean-François Couderc, Gilbert Zanon, Kim Hyland, Dennis Willie, "Lead Free Assembly: Process Considerations", IPC/Soldertec Global 3rd International Conference on Lead Free Electronics", Barcelona, Spain, June 8–9, 2005.
54. C. A. Harper, Electronic Packaging and Interconnection Handbook, McGraw-Hill, New York, 1991.
55. Jasbir Bath, "Lead-free Reflow Process Experience {Lead-Free Process Team}", IPC/NEMI Symposium on Lead-Free Electronics, Sep. 18–19, 2002, Montreal, Canada
56. Sammy Shina, Hemant Belbase, Karen Walters, Tom Bresnan, Peter Biocca, Tim Skidmore, David Pinsky, Phil Provencal, Don Abbott, "Selecting Material and Process Parameters for Lead-Free SMT Soldering Using Design of Experiments Techniques", Apex, San Diego, CA, Jan. 14–18, 2001
57. Yueli Liu, Guoyun Tian, Shyam Gale, R. Wayne Johnson, and Pradeep Lall, and Larry Crane, "Lead Free Assembly of Chip Scale Packages", Apex, S34-1, Anaheim, CA, Feb. 2004
58. Vahid Goudarzi, "Pb-free Manufacturing Process Development and Implementation", Indium Corporation "Quick Start" Workshop, Libertyville, IL, Sep. 2004.
59. Sam Yoon and Roy Wu, Jasbir Bath, Chris Chou and Samson Lam, "Assembly, Rework and Reliability of Lead-free FCBGA Soldered Component", Apex, S29-2-1, Anaheim, CA, Feb. 2004
60. Ning-Cheng Lee, "Critical Parameters in Voiding Control At Reflow Soldering", Chip Scale Review, August–September 2005.
61. H. Ladhar and S. Sethuraman, "Assembly Issues with Microvia Technologies", SMTA International, Chicago, IL, Sept. 2003.
62. Yan Liu, William Manning, Benlih Huang, and Ning-Cheng Lee, "A Model Study of Profiling for Voiding Control at Lead-free Reflow Soldering", Nepcon Shanghai, China, April 11, 2005
63. Jennifer Nguyen, Robert Thalhammer, David Geiger, Harald Fockenberger and Dongkai Shangguan, "Large and Thick Board Lead-Free Wave Soldering Optimization", APEX, S34-01, Feb. 20–22, 2007, Los Angeles, CA,
64. IPC-A-610D. 7.5.5.1
65. Bala Nandagopal, Sue Teng and Doug Watson, "Effect of OSP Chemistry on the Hole Fill Performance During Pb-free Wave Soldering", APEX, S08-01, Feb. 20–22, 2007, Los Angeles, CA
66. Jennifer Nguyen, Robert Thalhammer, David Geiger, Harald Fockenberger and Dongkai Shangguan, "Large and Thick Board Lead-Free Wave Soldering Optimization", APEX, S34-01, Feb. 20–22, 2007, Los Angeles, CA,
67. Ernesto Ferrer, Elizabeth Benedetto, Gary Freedman, Francois Billaut, Helen Holder, David Gonzalez, "Reliability of Partially Filled SAC305 Through-Hole Joints", APEX, Anaheim, CA, S29-02, Feb. 5–10, 2006
68. Michael J Smith, "Test and Inspection of Lead-Free Assemblies", Apex, S27-3, Anaheim, CA, Feb., 2004
69. Jasbir Bath, "Lead-free Reflow Process Experience {Lead-Free Process Team}", IPC/NEMI Symposium on Lead-Free Electronics, Sep. 18–19, 2002, Montreal, Canada
70. Mark Cannon, "Lead Free First Article Inspection: The Key to Success", Apex, S05-5, Anaheim, CA, Feb. 2004
71. Gold Goudarzi and Olga Diaz, "Lead free repair process", QuickStart Lead Free Soldering Workshop, Oct. 07, 2003, Ft. Lauderdale, FL.

72. Paul Wood, "Lead-Free Array Rework - What's Important In Lead Free Rework", Nepcon Shenzhen, August 30th, 2005
73. Sam Yoon and Roy Wu, Jasbir Bath, Chris Chou and Samson Lam, "Assembly, Rework and Reliability of Lead-free FCBGA Soldered Component", Apex, S29-2-1, Anaheim, CA, Feb. 2004
74. Sung K. Kang, Won Kyoung Choi, Da-Yuan Shih, Donald W. Henderson, Timothy Gosselin, Amit Sarkhel, Charles Goldsmith and Karl J. Puttlitz, "Formation of Ag3Sn Plates in Sn-Ag-Cu Alloys and Optimization of their Alloy Composition", 53rd Electronic Components & Technology Conference, S02P5C, New Orleans, LA, May 27–30, 2003
75. Polina Snugovsky, Zohreh Bagheri, Matthew Kelly, Marianne Romansky, "Solder joint formation with Sn-Ag-Cu and Sn-Pb solder balls and pastes", SMTA International, September 22–26, 2002, Chicago, IL
76. Private communication with Denis Barbini, Soltec.
77. C. Robert Kao, "Cross-interaction between Cu and Ni in lead-free solder joints", TMS Lead Free Workshop, San Antonia, TX, March 12, 2006.
78. Henry Y. Lu, Haluk Balkan, Joan Vrtis, and K.Y. Simon Ng, "Impact of Cu Content on the Sn-Ag-Cu Interconnects", 55th ECTC, P.113-119, May 31–June 3, 2005
79. K. Zeng and K. N. Tu, "Reliability Issues of Pb-free Solder Joints in Electronic Packaging Technology", Dept. of Materials Science and Engineering, UCLA, Los Angeles, CA 90095-1595, USA, to be published in 2002.
80. J. Kivilahti, Helsinki University of Technology, Finland
81. Fubin Song and S. W. Ricky Lee, "Investigation of IMC Thickness Effect on the Lead-free Solder Ball Attachment Strength: Comparison between Ball Shear Test and Cold Bump Pull Test Results", 56th ECTC Proceedings, pp. 1196–1203, San Diego, CA, May 30–June 2, 2006
82. Luhua Xu and John H.L. Pang, "Effect of Intermetallic and Kirkendall Voids Growth on Board Level Drop Reliability for SnAgCu Lead-free BGA Solder Joint", 56th ECTC Proceedings, pp. 275–282, San Diego, CA, May 30–June 2, 2006
83. Michael Osterman, Abhijit Dasgupta, and Bongtae Han, "A Strain Range Based Model for Life Assessment of Pb-free SAC Solder Interconnects", 56th ECTC Proceedings, pp. 884–890, San Diego, CA, May 30–June 2, 2006
84. S. Terashima, Y. Kariya, T. Hosoi, and M. Tanaka, "Effect of silver content on thermal fatigue life of Sn-xAg-0.5 Cu flip-chip interconnects ", Journal of Electronic Materials, Vol. 32, No. 12, p.1527 (2003).
85. A.R. Zbrzeznya, P. Snugovskya, D.D. Perovicb, "Reliability of Lead-Free Chip Resistor Solder Joints Assembled on Boards with Different Finishes Using Different Reflow Cooling Rates", IPC/JEDEC 5th International Conference on Lead Free Electronic Components and Assemblies, San Jose, CA, March 18–19, 2004
86. Edwin Bradley, "Lead-Free Solder Assembly: Impact and Opportunity", 53rd Electronic Components & Technology Conference, S02P1C, New Orleans, LA, May 27–30, 2003
87. Gordon Gray, "Lead-free soldering for CSP", IPC/JEDEC 5th International Conference on Lead Free Electronic Components and Assemblies, San Jose, CA, March 18–19, 2004
88. Vijay Wakharkar and Ashay Dani, "Microelectronic Packaging Materials Microelectronic Packaging Materials Development & Integration Development & Integration Challenges for Lead Free Challenges for Lead Free", Lead-free workshop, TMS, San Antonio, TX, March 12, 2006.
89. M. Date, T. Shoji, M. Fujiyoshi, and K. Sato, "Pb-free Solder Ball with Higher Impact Reliability", Intel Pb-free Technology Forum, 18th–20th July 2005, Penang, Malaysia
90. Donald Henderson, "On the question of SAC solder alloy – Cu pad solder joint fragility", Webcast Meeting on SAC Solder Joint Fragility, Binghamton, NY, Sep. 2004.
91. P.A. Kondos & S. Mandke, "Kirkendall voiding in Cu pads and other pad issus", UIC Fragile SAC Joint Meeting. Binghamton, NY, Oct. 7, 2004.

92. Weiping Liu and Ning-Cheng Lee, "Novel SACX Solders with Superior Drop Test Performance", SMTA International, Chicago, IL, Sep. 2006
93. Weiping Liu, Paul Bachorik, and Ning-Cheng Lee, "The Superior Drop Test Performance of SACTi Solders and Its Mechanism", ECTC, Las Vegas, NV, June 2008.
94. Masazumi Amagai, Yoshitaka Toyoda, Tsukasa Ohnishi, Satoru Akita, "High Drop Test Reliability: Lead-free Solders", 54th ECTC, P.1304-1309, June 1–4, 2004, Las Vegas, Nevada.
95. Masazumi Amagai, "A Study of Nano Particles in SnAg-Based Lead Free Solders for Intermetallic Compounds and Drop Test Performance", 56th ECTC Proceedings, P. 1170–1190, San Diego, CA, May 30–June 2, 2006
96. Benlih Huang, Hong-Sik Hwang, and Ning-Cheng Lee, "A Compliant and Creep Resistant SAC-Al(Ni) Alloy", ECTC, Reno, Nevada, May 26–June 1, 2007
97. Minna Arra, DongJi Xie and Dongkai Shangguan, "Performance of Lead-Free Solder Joints Under Dynamic Mechanical Loading", 52nd ECTC, S30-P4, San Diego, CA, May 28–31, 2002.
98. Minna Arra, Todd Castello, Dongkai Shangguan, Eero Ristolainen, "Characterization of mechanical performance of Sn/Ag/Cu solder joints with different component lead coatings", SMTAI, pp.728–734, Chicago, IL, Sep. 2003.
99. Robert Darveaux, Corey Reichman, Nokibul Islam, "Interface Failure in Lead Free Solder Joints", 56th ECTC Proceedings, P. 906–917, San Diego, CA, May 30–June 2, 2006
100. Fubin Song and S. W. Ricky Lee, "Investigation of IMC Thickness Effect on the Lead-free Solder Ball Attachment Strength: Comparison between Ball Shear Test and Cold Bump Pull Test Results", 56th ECTC Proceedings, P. 1196–1203, San Diego, CA, May 30–June 2, 2006

Chapter 6
Thin Die Production

Werner Kroeninger

Abstract Thin silicon die production is becoming a challenge for all spheres of semiconductor industry. The diversity of requirements and constraints for various applications leads to a lot of different solutions for making thin dies. This chapter describes recent developments on silicon wafer thinning and singulation. Various technologies for material removal and the associated damage caused by the material removal are reviewed in details. Different surface treatment approaches and their effect on improving mechanical property of thinned silicon are also discussed in this chapter.

Keywords Thinning · Grinding · Back side treatment · Wafer-bow · Damage · Singulation · Laser · Mechanical Performance

6.1 Thin Silicon Devices

One of the major benefits of thin silicon devices is that they can enhance Integrated Circuit (IC) performance and enable innovative packages for various applications.

For an increasing number of applications, up to 95% of the original silicon wafer thickness is removed. Generally, wafers are thinned to a reduced thickness after circuit fabrication rather than fabricating circuits on a thin wafer. Intuitively, fabricating circuits on a thin wafer will save production cost. However, the real situation is different. Firstly, original wafer thicknesses are semi-standardized. These standards need to be changed if the wafer fabrication starts with thinner wafers. Secondly, since the thermal capacity of the wafers depends on wafer thickness, every process in production such as metal sputtering or insulating layer deposition needs to be modified if thin wafers are used in the production. Yet the strongest argument comes from the production of prime

W. Kroeninger (✉)
Dipl. Phys. Univ., IFAG OP FEP T UPD 5 Infineon Technologies AG, P.O. Box
10 09 44, D-93009 Regensburg, Germany
e-mail: werner.kroninger@infineon.com

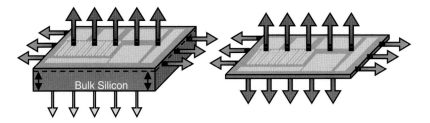

Fig. 6.1 Improved heat dissipation by reducing bulk silicon thickness of a die

wafers themselves. The majority of the cost for the prime wafers is not from the silicon crystal, but from the grinding and polishing processes that are required to achieve the surface quality. Significant process development is required if we switch from thick wafers to thin wafers.

6.1.1 Advantages of Thin Die

Wafer thinning basically removes part of the bulk silicon and leaves an active circuit layer with a thinned bulk silicon layer. An increasing number of applications can take advantage of reduced bulk silicon thickness. The following are two examples:

A personal computer (PC) processor:
After a die is assembled onto a board or substrate, the main stress which the die has to stand is the thermo-mechanical stress. The combination of die, substrate, and board has to cope with this stress. One of the important aspects of this package is how to dissipate the heat from the die effectively, and the reduced bulk silicon thickness enhances the heat dissipation (as shown in Fig. 6.1).

A smart card:
In a smart card, the die must stand consistent mechanical stress mainly from the bending of the card. To prevent the die from breaking, it needs to be flexible

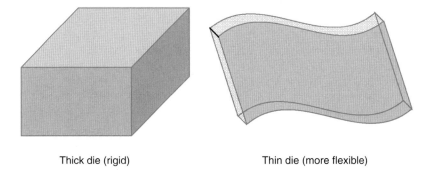

Fig. 6.2 A schematic illustration of flexibility comparison between thick and thin dies

6 Thin Die Production

enough. With thickness reduction, a die gains more flexibility (as shown in Fig. 6.2) [1, 2].

6.1.2 The Essential Considerations for Making a Thin Die

One fact of wafer thinning is that all the state-of-the-art thinning methods always leave a damage layer on the thinned surface. Thin dice consists of three main layers: active, bulk and damage layers (as shown in Fig. 6.3). The impact of the damage layer to the mechanical properties of the thin die has been the topic of continuous research efforts [3].

One way to thin bulk silicon of a device wafer is to use Silicon-on-Insulator (SOI) approach. The thinning process (for example, using etching) is selective, and thinning stops on the insulating layer. However, SOI is expensive, and still requires thin wafer handling and carrier systems.

A die breaks if the bending exceeds its limit of flexibility. This limit can be measured in a three-point bending test. The die can be bent to a certain radius, beyond which it cracks. The cracks, damage, and stress inside the die reduce its mechanical performance. The sum of all these contributions determines the exact radius. There are several methods to investigate the damage inside thinned silicon. Different techniques give information at various depths (as shown in Fig. 6.4). Because the structure of the die consists of active layers, bulk material, and damaged regions, the bending radius cannot be calculated from the elasticity modulus of silicon bulk material [4, 5].

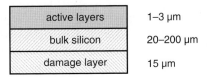

Fig. 6.3 Schematic illustration of the three main layers of a thin die

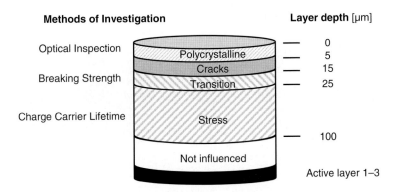

Fig. 6.4 Common methods for investigating the properties of thinned silicon

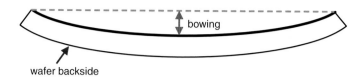

Fig. 6.5 A schematic illustration of wafer bowing after thinning

The abrasive removal of silicon during thinning leads to a damaged layer which consists of polycrystalline silicon and silicon oxides. The silicon oxide acts like wedges in the silicon leading the high compressive stress in this damage layer. On the other hand, the active layers are mainly under tensile stress. As a first approximation, wafer deformation (bowing) after thinning is circular and can be described by the amount of bowing (as shown in Fig. 6.5). As the central region of the wafer is lower than the outer region of the wafer, this type of wafer deformation is generally referred to as "negative bowing" (i.e. with a minus sign).

6.2 Wafer Thickness Reduction

6.2.1 Material Removal

The standard technique for reducing the thickness of silicon wafers is the abrasive destruction of the silicon crystal by grinding using grinding wheels. After grinding, some finishing steps need to be done to the silicon surface. These finishing steps will heal the silicon crystal and are also called damage removal or stress release.

The grinding wheels contain grinding teeth which are made of sintered composite of diamond particles, bonding material, and pores (as shown in Fig. 6.6). Synthetic diamond, rather than natural diamond, is used for forming the wheels because it shows constant form, controlled size distribution, and thus

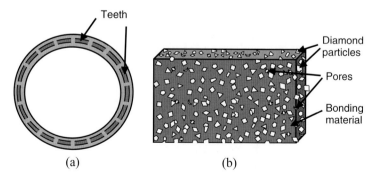

Fig. 6.6 Schematic drawing of a grinding wheel (**a**) and material composition of the teeth on the grinding wheel (**b**)

6 Thin Die Production

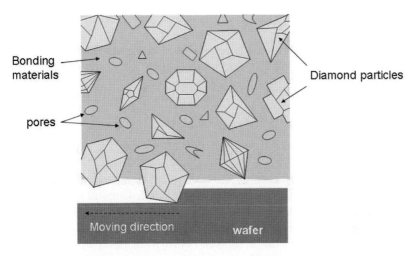

Fig. 6.7 A schematic illustration of material removal by the exposed diamond particles on the grinding teeth

improved reproducibility. The grinding of silicon wafer is done by the diamond particles which are exposed on the teeth (as shown in Fig. 6.7).

Ideally, we would like to take off the material without impact to the remaining wafer. However, in reality, because grinding was mainly done by scratching the bulk silicon material with the exposed diamonds, many defects such as cracks will be inevitably introduced by the grinding process (as shown in Fig. 6.8).

6.2.2 Grinding Process

The sintered composite which the teeth of the grinding wheel are made of is one of the factors which impact the performance of grinding wheels. There is quite a

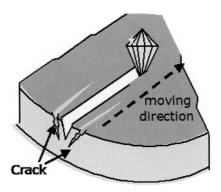

Fig. 6.8 A schematic illustration of damages introduced by grinding

variety of wheels for grinding brittle materials such as GaAs, InP, SiC and other compound semiconductors.

Especially for silicon, which takes the largest market share in brittle material grinding, lots of wheels have been developed. The desired properties for grinding wheels include:

(1) Introduce minimum damage to silicon crystal during grinding
(2) Generate minimum heat during grinding
(3) High material removal rates
(4) Low wheel consumption (long wheel life)

A standard grinding process consists of two steps: coarse grinding and fine grinding. Coarse grinding uses wheels with rough diamond particles and has higher material removal rates. Fine grinding utilizes wheels with smaller diamond particles, and has lower material removal rate and thus causes less damage to the silicon. Coarse grinding results in a much higher roughness ($R_a \sim 0.1$ μm) than fine grinding ($R_a < 0.05$ μm). The roughness difference between course and fine ground surfaces can be seen even by naked eye as shown in Fig. 6.9.

The roughness of the silicon surface is dependent on the type of grinding wheel used. The finer the diamond particles (higher mesh #), the smoother the surface produced. As can be seen from Fig. 6.10, a broad range of roughness of the ground surfaces can be generated after various grinding processes such as standard grinding (#1500), high mesh surface finish (#8000), standard grinding plus spin-etch, and standard grinding plus plasma etch. The roughness of the surface also influences mechanical integrity of the die and surface adhesion.

The number of defects (e.g. cracks and crystal disturbances), and the size, depth, and shape of the defects affect the mechanical integrity of the die. For example, cracks in sharp edges most likely will propagate. Therefore, one of the major purposes of the post grinding treatment (or stress release treatment) is to smoothen these cracks.

The stress release treatments remove the crystal layer that has been damaged by the grinding processes. Several stress release processes such as wet chemistry

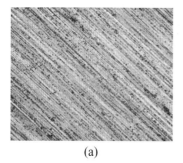
(a)

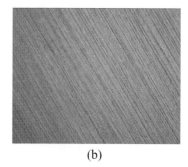
(b)

Fig. 6.9 Optical images of ground silicon surfaces by coarse (**a**) and fine (**b**) grinding

6 Thin Die Production

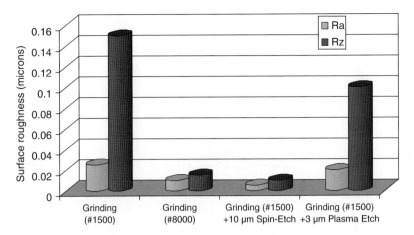

Fig. 6.10 Surface roughness comparison of surfaces ground by various wheels (with post treatment). Ra (roughness average): the average of peak and valley distances measured along the centerline of one cutoff length (usually 0.03 in.); Rz (a 10-point average): an average of the five highest peaks and the five lowest valleys measured in one cutoff length

(e.g. spin-etch and chemical mechanical polishing) [6, 7], dry-polish, and plasma-etch are in use. Even though plasma etch removes the damage, it still leaves a high surface roughness as shown in Figs. 6.10 and 6.11.

After stress release treatments, the mechanical strength of the die can be improved by a factor of four to eight (as shown in Fig. 6.12).

The parameters used for the grinding process are interdependent. Here are some examples. Using harder bond materials for the grinding wheel teeth, longer wheel life can be achieved (as shown Fig. 6.13).

To achieve high removal rates in coarse grinding, a high feed rate is generally required. However, a higher feed rate will increase the wear of the grinding

Fig. 6.11 Ground silicon surface before (*left*) and after (*right*) plasma-etch

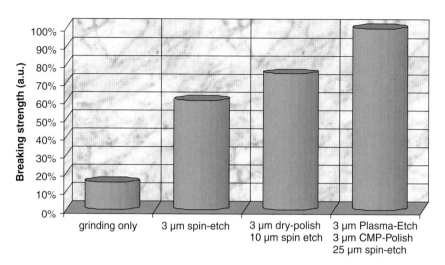

Fig. 6.12 Die breaking strength comparison after various stress release treatments
Note: a.u. – arbitrary unit

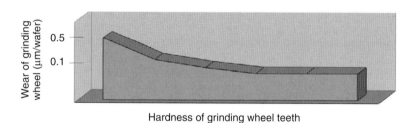

Fig. 6.13 Impact of hardness of grinding wheel teeth to the wear rate of a grinding wheel

wheel and thus reduce wheel life (as shown in Fig. 6.14 for 6-inch wafers). Therefore, there is usually a trade-off between wheel life and removal rate.

Mechanical strength of a die is affected to a large extent by the wheel roughness in the coarse grinding step. When a grinding wheel with reduced roughness (e.g. smaller diamond particles) is used, the thinned die will have higher mechanical strength (as shown in Fig. 6.15).

6.2.3 Handling of Thin Wafers

One of the most important aspects regarding thin silicon wafer is thin wafer handling. Two main parameters, bowing of the wafers and their mechanical strength, directly affect handling of the thin wafers. Bowing and mechanical strength are a function of wafer thickness as shown in Figs. 6.16 and 6.17. Wafer

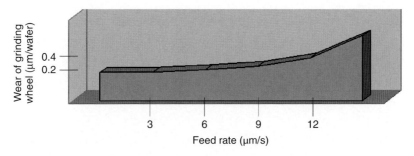

Fig. 6.14 Impact of feed rate to the wear rate of a grinding wheel

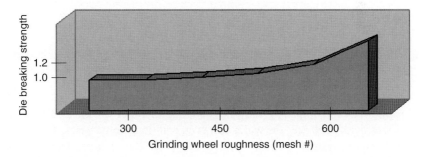

Fig. 6.15 Impact of grinding wheel roughness to the breaking strength of a thin die

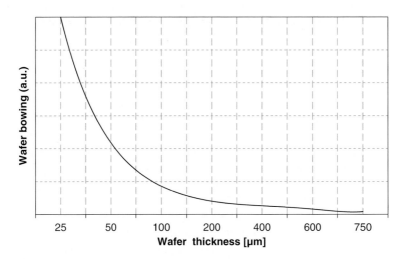

Fig. 6.16 Wafer bowing at various wafer thicknesses (8 inch wafers). Note: a.u. – arbitrary unit

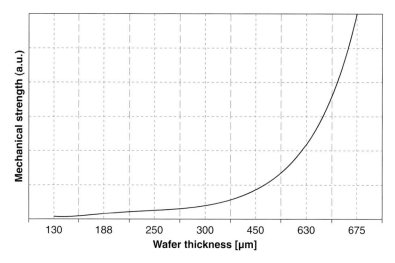

Fig. 6.17 Mechanical strength of a ground silicon wafer (8 inch) at various wafer thickness

bowing increases significantly with the reduction of wafer thickness. For wafer thicknesses below 100 μm, the wafer can even roll up in some cases. Therefore, a wafer support system is required to handle these thin wafers. The absolute value of the bowing depends on the product, its active layers, and the die size. Nevertheless, the general trend shown in Fig. 6.16 is correct in any case.

To protect the devices on a wafer, the first step for wafer thinning is to cover the active side of the wafer with a protection tape. During the process flow of thinning, a wafer reaches its lowest mechanical strength at the step of coarse grinding (as shown in Fig. 6.18). Its strength can often be improved up to an order of magnitude by stress release treatments.

For some products, especially for products with active back side, it is required to handle the ground wafers for some extra processing steps such as wafer backside metallization before wafer singulation. Therefore, the support systems need to be able to survive these processing steps.

Generally, there are two types of methods for handling the thinned wafers. The first approach is to attach another rigid and flat support plate carrier to the wafer before grinding. The carrier could be silicon, ceramic, or glass, depending on the adhering medium between the wafer and the carrier. For attaching the wafer to the carrier, several adhering media are in use such as glue, wax, electrostatic forces, or tape. All these carrier solutions are commercially available [8, 9]. Second method is to attach a support ring carrier to the wafer after thinning. Another relative new wafer thinning concept is to leave a couple of millimetres of wafer backside outer circumference unground and only thin the inner area of the wafer [10]. The unground edge ring of the thinned wafer greatly improves wafer strength and facilitates handling of the thin wafer.

6 Thin Die Production

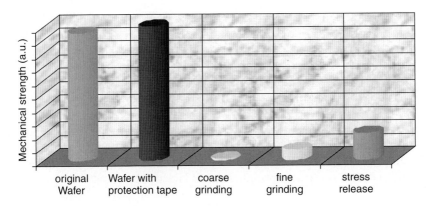

Fig. 6.18 Wafer mechanical strength at the different stages of thinning

Using a support carrier adds additional process steps, processing complexity, and costs to the process flow for die production. One of the main constraints and most important consideration for selecting the right carrier solution is that it needs to be compatible to die to substrate assembly processes.

Flip chip package, where the active side of a die is electrically connected to a substrate with bumps rather than wire bonds, is one of the recent developments in die to substrate interconnections. There exists quite a variety of bumps in terms of shape, height, and pattern layout on the die. Bump-height may range from 20 μm up to 120 μm or even more. These bumps would significantly complicate the wafer grinding process. They can potentially act as stress concentration points and induce wafer breaking. As shown in Fig. 6.19 (a), a standard back grinding tape consists of a base material and an adhesive layer. Some approaches have been developed to accommodate the bumps on the wafers. One idea is to add a layer of compliant material between the adhesive layer and the base material to cover the top portion of the bumps, as shown in Fig. 6.19 (b). The second more commonly used method is to increase the adhesive layer thickness so that the bumps are fully embedded in the adhesive layer, as shown in Fig. 6.19 (c).

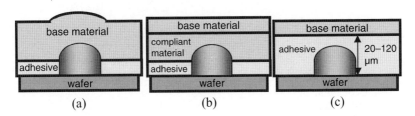

Fig. 6.19 Schematic illustrations of the adhering medium options for bumped wafer thinning

Most of these adhesive tapes are UV releasable tapes. After irradiated by UV light, the tapes will lose more than 90% of their adhesion strength. The reduced adhesion strength will facilitate the de-taping and carrier detaching processes.

6.3 Mechanical Properties of Thinned Wafers

For a thin die, mechanical performance is of most importance. Thin dies with high breaking strength and flexibility are needed for assembly and package reliability. First, the dies need to survive all the assembly processes when it is assembled onto a substrate or printed circuit board. Also, it needs to withstand the stress from the application field, such as thermo-mechanical stress for a PC processor or mechanical bending stress in a smart card.

6.3.1 Breaking Strength and Flexibility

To obtain the strength of a thin die, generally a destructive bending testing, either three-point or four-point bending, has to be conducted. A die will be bent with a force until it breaks. In principle, 3-point and 4-point bending tests provide the similar information. To obtain statistically reliable values, a lot of specimens for each split are generally tested [11].

The flexibility of a die is obtained by measuring the bending radius which the die can withstand before breaking during a 3-point bending (as shown in Fig. 6.20). The flexibility of a thin die can be greatly increased by stress release treatments on the ground surfaces (as shown in Fig. 6.21). The main mechanism for the mechanical strength improvement is to remove the damaged layer and smoothen the defects such as cracks through the stress release treatments. Increasing the removal amount of the damaged layer leads to a saturation on die flexibility as shown in Fig. 6.21.

The process of singulating a wafer into dies can potentially introduce defects on the sidewalls of the die. Assuming that the thinning process provides a high quality surface finish, the quality of the wafer singulation process can be studied

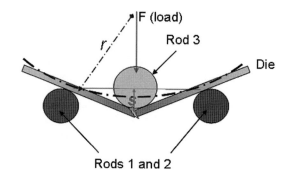

Fig. 6.20 Measuring the bending radius of a thin die using 3-point bending test (bending radius r can be calculated through s)

6 Thin Die Production

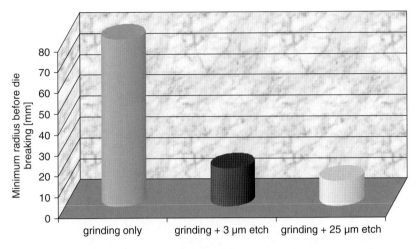

Fig. 6.21 Minimum bending radius before breaking of a thin die (120 μm) with different stress release treatments

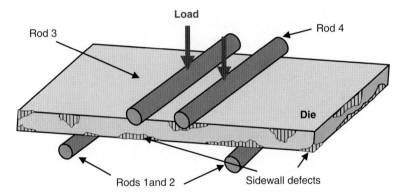

Fig. 6.22 A schematic illustration of 4-point bending test setup to study the die sidewall defects generated by wafer singulation

using a 4-point bending test. The defects on the sidewalls between the two lower rods (Fig. 6.22) will be detected by this test.

To study the quality of thinning, excluding influences from singulation, a ball ring test is generally employed. Since the ring where the die is placed is smaller than the die (as shown in Fig. 6.23), the edges of the die will not be in contact with the ring, and thus contribution of the die sidewall defects will not be captured in this test. As a result, the testing result only reflects the impact of the defects on the back side of the die. Since these defects are the residues of the wafer thinning processes, the mechanical strength value obtained from this test can be used to characterize the quality of the thinning processes [12, 13].

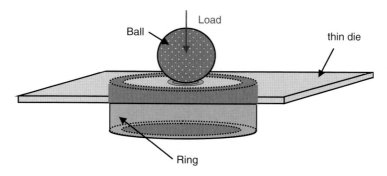

Fig. 6.23 A schematic illustration of Ball-Ring-Test on a thin die

It has been found that breaking-strength statistics of silicon dies follows Weibull distribution. For a single die, we only can predict that it will follow a certain distribution with a definite confidence level (as shown in Fig. 6.24). The Weibull distribution has two characterising main parameters. F_{median} is the breaking strength where 63.2 % of the specimens have cracked. m is the slope of an optimized line, characterising the broadness of the distribution (narrow distributions correspond to high m values) [14]. To improve the reliability of testing results, a common approach is to simply use a large number of specimens because the confidence level is increases approximately with the square root of the number of specimen used per split.

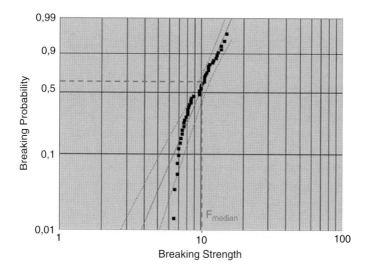

Fig. 6.24 An example of breaking strength test results which follows Weibull distribution

6 Thin Die Production

6.3.2 Characterization of the Stress and Damage Induced by Grinding

Thinning processes introduce damages to the silicon wafers. The question is if the depth of damage is a function of the final silicon wafer thickness. The question can be considered based on the following two scenarios:

Scenario I:

During thinning, the silicon is attacked by the abrasive grinding tool. The reduced thickness of the bulk silicon will become unable to withstand these mechanical attacks. The thinner the silicon the higher the probability for cracks to happen. Thus the damage zone will become deeper and deeper with silicon becoming thinner during grinding. Therefore, the damage layer depth is reciprocally proportional to the silicon thickness.

Scenario II:

Silicon is getting more flexible when it becomes thinner. It can adapt a little more to the applied pressure by the grinding wheel. As a result, the damage zone is constant and only depends on the grinding parameters (grit size, feed rate, vibration, cooling, etc.). In this case, the damage layer is nearly independent of the silicon thickness.

To investigate which of the above two scenarios is true, Raman Spectroscopy is used to study the damage depth [15]. Figure 6.25 shows a typical Raman spectrum collected at a roughly 2 μm by 2 μm area of a crystalline silicon. The frequency of the Si Raman peak depends on mechanical stress imposed on

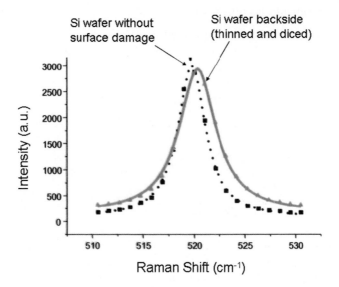

Fig. 6.25 Illustration of using Raman line shift to study the mechanical stress in the Si

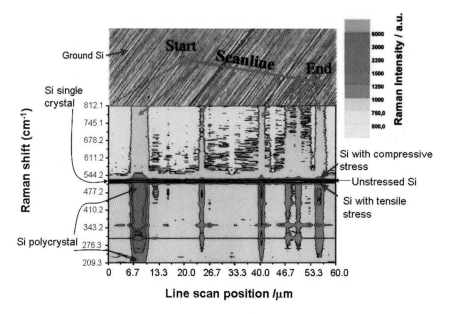

Fig. 6.26 Raman spectroscopy line scanning results on a ground Si

the Si. As can be seen from Fig. 6.25, thinned Si wafer showed Raman shift due to the damage and the compressive stress induced during wafer grinding. This technique can also be used to study the stress and damage across a line on a silicon die using Raman line scanning (as shown in Fig. 6.26).

From experimental results, it was observed

- Under different backside treatments for various silicon thicknesses ranging from 500 μm down to 50 μm, a saturation of die mechanical strength was reached by the same removal amounts.
- Raman Spectroscopy Analysis shows no significant difference between the damage layers of thin and thick dies

The results strongly indicated that for silicon wafers down to 50 μm thickness, the damage layer depth seems to be independent of silicon thickness. Therefore, the scenario II is more consistent with the experimental results.

6.3.3 Limits of Wafer Thinning

Even though there are limitations on wafer thinning process and thin wafer handling, the physical limit of die thickness is the thickness of the active layers plus a few micron thick silicon substrate. The device will still have the same functionality with 5–10 μm thick silicon substrate. Thin dies with a thickness

6 Thin Die Production

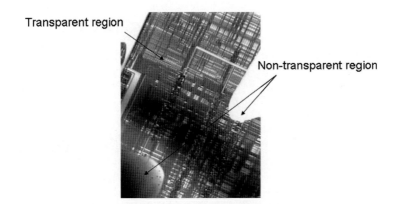

Fig. 6.27 Optical images taken from the back side of an ultra-thin wafer

of about 30 μm have been employed in stack die package for some applications such as high performance memory sticks.

The theoretical physical limit for thinning the bulk silicon is close to zero. Silicon can be thinned until it becomes transparent under visible light (as shown in Fig. 6.27). As can be seen from Fig. 6.27, the circuit structures in some regions of the front side of silicon are visible from the back side of the wafer when the bulk silicon is around 10 μm. The fact that the silicon is transparent only in some regions as shown in Fig. 6.27 was due to silicon thickness variation, which is one of major challenges for wafer thinning. For some applications (e.g. power), the precise control of die thickness is an important topic [16].

6.4 Singulation

After all the Front-End processes, electrically functional devices (dies) which are held together by the silicon wafers are obtained. For an 8-inch wafer, the typical number of dies per wafer ranges from hundreds to hundred thousands of dies. The next task is to separate these dice from each other. Today, several technologies have been used, each of which has specific advantages and constraints.

In a conventional packaging process flow, wafers are mounted on a dicing frame through a dicing tape, and singulated into dies (as shown in Fig. 6.28). The dies are then assembled into the final packages.

6.4.1 Mechanical Dicing

Mechanical dicing is well established and widely used in semiconductor industry. Its market share is above 90%. Several processes for mechanical dicing have been proposed. A standard dicing machine has two sawing spindles (blades).

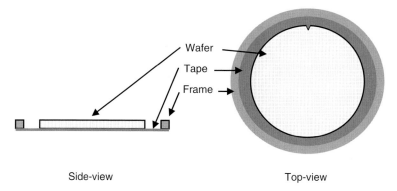

Fig. 6.28 Top and side view of a wafer mounted on dicing tape and frame

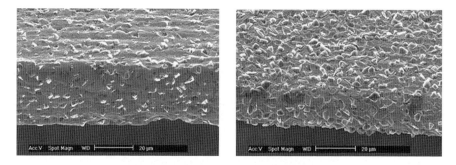

Fig. 6.29 Typical dicing blade surfaces of blades Z1 (*left*) and Z2 (*right*)

They can be used in parallel for improving throughput. It is also possible to use different blades. A wider blade (Z1, coarse) generally cuts through two thirds of the wafer thickness and then a thinner blade (Z2, fine) cuts the rest of wafer thickness. Figure 6.29 shows two examples of blades Z1 and Z2. The process of sawing is abrasive and, from mechanics point of view, similar to grinding. Blade thickness for dicing silicon typically ranges from 20 to 50 μm. This process could cause chipping and mechanical break-outs at the die edges. Figure 6.30 (a) shows a schematic illustration of dicing kerf and chipping on the die edges induced by dicing, and Fig. 6.30 (b) is an optical image of dicing streets between dies.

6.4.2 Laser Dicing

The laser has its place in industry and medicine for quite some time. Several different types of laser have been used in metrology and medical applications.

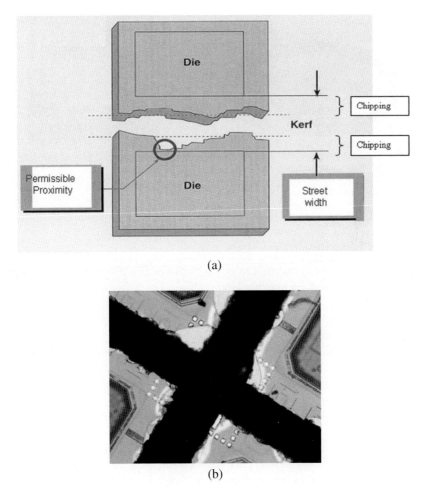

Fig. 6.30 A schematic illustration of dicing kerf and chipping on the die edges induced by dicing (**a**) and an optical image of dicing streets between dies (**b**)

Main parameters for a laser include wavelength, optical power, and pulse frequency. It is often used for joining work pieces through welding. Separation is also often done by laser.

Since the first commercial use of laser in silicon die separation in the 1980s, there have been a lot of developments and improvements. Several laser dicing techniques will be discussed in the following sections. Each of these systems is compatible with the same dicing-tapes that are used in the conventional mechanical dicing. One of the main requirements for the dicing tape is that it has to be transparent to the laser-beam.

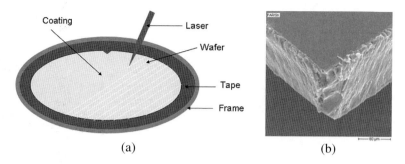

Fig. 6.31 A schematic illustration of dry laser dicing of a wafer on tape and frame (**a**) and a SEM image of the sidewall of die diced using dry laser (**b**)

6.4.2.1 Dry Laser Dicing

The main mechanism of material removal by dry laser is ablation. "Dry" means the laser is not assisted by another beam of liquid or aerosol during separation process. A lot of particles, oxides, and other residues come out of the kerfs during dry laser dicing.

As the particles could contaminate and even damage the wafer, for example, the whole wafer front side (including bumps or metal pads) needs to be protected (as shown in Fig. 6.31). During separation process, the wafer is covered by a "dicing coating" so that the residues are not able to harm the wafer surface. After the separation process, the coating is washed off.

Dicing causes damage to the silicon. Damage means deviation from an ideal crystal in the separated silicon die. The damage could be introduced by dry laser dicing, the impact of the heat, and the re-deposited material. The material removed from the dicing kerf is partly re-deposited on the die side walls. The re-deposited material can create stress in the crystal structure. The heat generated leads to disturbances in the silicon crystal and thus decreases the mechanical stability. Dry laser dicing can handle wafers with a thickness of 150 μm or less.

6.4.2.2 Water Guided Laser Dicing

For the water guided laser dicing process, a laser is coupled into a water beam which is about tens of microns in diameter, and then the laser is guided in the water beam by means of total internal reflection [17, 18].

The water serves two main purposes:

(1) Reduce the heat generated by the laser dicing
(2) Wash away any debris and prevent the re-deposition of debris on the wafer surface

Therefore, this technique can be used to dice thicker silicon and eliminate the use of dicing coatings.

6 Thin Die Production

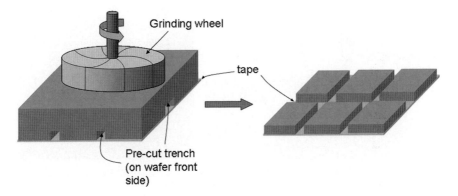

Fig. 6.32 An illustration of separation by thinning process

6.4.3 Separation by Thinning

Separation by thinning (SbT) reverses the typical sequence of wafer thinning and singulation. Wafers are pre-cut first before thinning. Figure 6.32 shows a wafer with pre-cutting trenches on its front side.

For the pre-cutting step, nearly any separation technology can be used. Classically, it is done by mechanical dicing [10]. The pre-cutting needs to be slightly deeper than the final die thickness intended. Then thinning is conducted from the backside of the wafer until die separation is achieved (as shown in Fig. 6.32).

For damage reduction, the dies can be etched by plasma after thinning. The treatment can remove damages from the backside and part of the side wall of the dies, and greatly enhance the mechanical strength of the dies. The dies are then flipped and transferred to tape and reel for assembly.

6.4.4 Separation by Creating Damage

To separate the die by creating damage is one of the oldest principles in the singulation of dice. Scribe and break for example was one of the first methods used. Scribes are made by a diamond needle on the dicing streets of a wafer. The introduced damage now is the starting point for the separation of the adjacent dies. The dies were taken apart by means of a rubber role.

6.4.4.1 Tape Expansion

A laser is also able to introduce damage into the silicon crystal [19]. The laser is focused into the material rather on the top surface to introduce a perforation line inside the silicon, as shown in Fig. 6.33 (a). For thick silicon, several passes of laser perforation may be necessary. After perforation lines are introduced

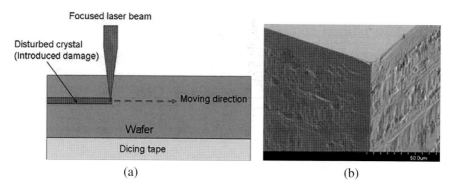

Fig. 6.33 (**a**) Creating a perforation line inside silicon, (**b**) SEM image of the sidewall of a die singulated using this technique

into a wafer, the dice are separated by tape expansion. The side wall of a singulated die is shown in Fig. 6.33 (b).

6.4.4.2 Crack Propagation

An initial crack is created on a wafer and then the separation of the dice is done by propagating the crack through induced thermal stress [20, 21].

The crack is produced by a diamond scribe. To achieve propagation, a laser is used to provide heat and an aerosol beam is utilized to cool shortly after the heat. This process has to be done in an exact sequence. The thermal tension, created by heat wave and subsequent cooling, spreads along the intended line of separation, and cracks the silicon crystal in a defined way (as shown in Fig. 6.33).

6.5 Packaging of Thin Silicon Dies

In packaging area, one of the major trends is package form factor reduction. Handling the thermo-mechanical stress within the package is becoming a challenge. Using thinner dies to mediate the stress is one of the potential solutions.

For some applications, stress on the die can be reduced by reducing the die size rather than increasing the flexibility of the die. If a die is only 250 μm by 250 μm in size, there will not be much stress applied to it in a package. Figure 6.34 shows a transponder (ID-tags) example. Both the substrate and the antenna can be flexible. The only rigid part, the die, can be kept so small that it experiences a very small amount of stresss. The only thing which needs to be worried about from mechanical stress point of view is the connections between the die and antenna.

Typically, on a power semiconductor die there is current running through the bulk silicon (as shown in Fig. 6.35). When the charge carrier moves from the front side to the back side of the die, the bulk silicon in between acts as a resistor.

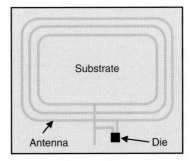

Fig. 6.34 The large part of an ID-tag is the antenna

Fig. 6.35 Schematic illustration of a power die

Thinning silicon can reduce the ohmic resistance of the bulk silicon and provide better heat dissipation [16].

References

1. W. Kröninger and E. Wittenzellner, "Thinning Silicon – Optimizing the Grinding Process Regarding Performance and Economics," Annual Fraunhofer Forum, be-flexible, IZM Munich, 2002, www.be-flexible.com
2. Workshop on Ulrathin Silicon Packaging, Sept. 2002, Fraunhofer ISIT, Itzehohe, Germany
3. "Rotationsschleifen von Si-Wafern," Promotion von Sabine Lehnicke, Institut für Fertigungstechnik Uni Hannover, Fortschrittsberichte VDI, Reihe 2, Nr. 534, 1999
4. A. Böge, "Mechanik und Festigkeitslehre," Vieweg Verlag, Braunschweig, 21. Aufl. 1990
5. H. F. Hadamovsky et al., "Werkstoffe der Halbleiterindustrie," Dt. Verlag für Grundstoffindustrie, 2. Aufl., 1990
6. J. P. John and J. McDonald, "Spray etching of Silicon in the HNO3/HF/H20 System," Journal of the Electrochemical Society, Vol. 140, No. 9, 1993
7. K. Priewasser, "Thin Dies Manufacturing Methods," Annual Fraunhofer Forum, be-flexible, IZM Munich, 2005

8. C. Landsberger, "Processing of Thin Substrates by Means of Electrostatic Carrier," Annual Fraunhofer Forum, be-flexible, IZM Munich, 2005
9. S. Pargfrieder, "Temporary bonding and de-bonding," Annual Fraunhofer Forum, be-flexible, IZM Munich, 2005
10. K. Yamagishi, "Thin wafer Handling by DBG and Taiko Process," Annual Fraunhofer Forum, be-flexible, IZM Munich, 2006
11. L. Sach, "Angewante Statistik", SpringerVerlag 1983
12. W. Kröninger and F. Mariani, "Thinning and Singulation: Root-causes of the Damage in Thin Dies," Proceedings of 56th Electronic Components and Technology Conference (ECTC), pp. 1317–1322, San Diego, CA, 2006
13. H. Blumenauer and G. Pusch, "Bruchmechanik," VEB Deutscher Verlag für Grundstoffindustrie, Leipzig, 3. Aufl. 1993
14. H. Wilker, "Weibull-Statistik in der Praxis. Leitfaden zur Zuverlässigkeitsermittlung technischer Produkte," Nordersted 2004
15. I. De Wolf, "Micro-Raman Spectroscopy to Study Local Mechanical Stress in Silicon Integrated Circuits," Semiconductor Science and Technology 11 (1996), pp. 139–154
16. T. Schmidt, "Wafer thinning – a Key Success Factor for Power Semiconductors," Annual Fraunhofer Forum, be-flexible, IZM Munich, 2005
17. T. A. Mai, "Laser-Microjet Dicing of Thin Compound Wafers and Low-k Wafers," Annual Fraunhofer Forum, be-flexible, IZM Munich, 2005
18. W. Kröninger, D. Perrottet, J.-M. Buchilly and B. Richerzhagen, "Stress Release Increases Advantages of Laser-Microjet," Semiconductor International, Packaging, Apr. 2005
19. B. Holz, "Advanced Production Technologies for Thinning and Laser Dicing of Ultra thin Wafers," Annual Fraunhofer Forum, be-flexible, IZM Munich, 2004
20. MDI Schott, "Advanced Processing," www.mdi-schott-ap.de (last visit 10.12.2007)
21. H.-U. Zühlke and P. Mende, "Thermal Laser Separation for Wafer Dicing," Annual Fraunhofer Forum, be-flexible, IZM Munich, 2006

Chapter 7
Advanced Substrates: A Materials and Processing Perspective

Bernd Appelt

Abstract This chapter reviews materials and processing for fabricating organic substrates including laminate substrates for plastic BGA (PBGA), build-up substrates for flip chip BGA (FCBGA), tape substrate for tape BGA (TBGA), coreless substrate, and some specialty substrates such as substrates for RF modules, high performance substrates with low dielectric constant, and substrate with embedded components (active dies or passives). Future trend of organic substrate development is also covered in this chapter.

Keywords Organic substrates · Copper clad laminate (CCL) · Ajinomoto Build-up Film (ABF) · Coreless, BGA, Blind via (BV) · High density interconnect (HDI) · Low dielectric constant

7.1 Introduction

Substrates have become the most expensive element of electronic packages while at the same time limiting package performance. Ceramic, multi layer substrates have always been extremely expensive but did allow for a great deal of design freedom e.g. integration of passives. The only drawbacks were a high dielectric constant and a very low coefficient of thermal expansion (CTE) as compared to printed circuit boards (PCB) but closely matched to the silicon die. Conversely, organic substrates have a CTE which is matched to PCBs but is significantly larger than that of the silicon die.

Organic substrates were originally introduced to significantly reduce the cost of packaging by taking advantage of low cost PCB manufacturing technology, materials and scale. While ceramics scaled only from a single unit to few units, organic substrates used the scale of PCBs (e.g. 410 mm × 510 mm) accommodating 100 to over 1,000 units. This approach is similar to wafer size scaling, moving from 100 to 300 mm.

B. Appelt (✉)
ASE (U.S.) Inc., 3590 Peterson Way, Santa Clara, CA 95054
e-mail: bernd.appelt@aseus.com

Today a few major categories of substrates exist:
- Ceramic substrates
    - virtually all are multi layer ceramic (MLC)
    - cavity ceramics e.g. for optical sensors
    - RF ceramics with integrated passives
- Organic substrates – which can be subdivided further
    - Laminate substrates – plastic ball grid array (PBGA)
        - PBGA with 1, 2, 4 & 6 layers of circuitry
        - high density substrates (HDI) = build-up substrates for wire bonding
    - Tape substrates – typically based on polyimide film (TBGA)
    - Build-up substrates – typically used for Flip Chip die (FCBGA)
    - Specialty substrates
        - Embedded Passives Substrates (EPS) & Embedded Die Substrates (EDS)
        - Buried Passives Substrates (BPS)
        - Cavity substrates – here the die is located in a recess

These few types of substrates are assembled into a multitude of different types of electronic packages (see Fig. 7.1, ASE Group, Inc.). It seems that almost any idea that can be expressed in a power point chart can be converted into a real component given enough conviction and commercial motivation.

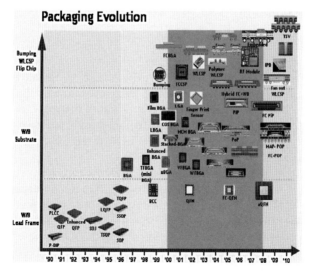

Fig. 7.1 Packages based on substrates, lead frames and wafer level packaging[1]

---

[1] Reprint Courtesy of Advanced Semiconductor Engineering Inc.

7.1.1 A Brief History: From PCBs to Substrates

The early organic substrates were indeed simple, miniature PCBs and were significantly cheaper than ceramics. They were introduced as OMPAC by Motorola [4]. Soon it became necessary to use more dedicated materials and processes to manufacture substrates to meet the rapidly increasing quality and technical requirements: smaller features (lines and spaces) and lighter weight at lower cost. Organic substrates were displacing ceramic substrates for nearly all chip applications with the exception of high reliability, high I/O or high performance applications e.g. CPU/MPUs, ASICs, RF-applications. Today however ceramic substrates are declining in usage even for those applications.

The invention of build-up PCBs by Tsukada (originally called Surface Laminar Carrier, SLC, [12]) enabled high density packaging of components on PCBs. SLC essentially employs a PCB core on top of which fine line redistribution layers are being built up (hence the name Build Up, BU, technology). This approach is similar to using redistribution layers on high end multi-layer ceramic carriers (MLCC). This SLC technology was quickly applied to flip chip ASIC die especially when the wire bond die was pad limited or when superior electrical performance due to the short interconnect was required. The real driver for this technology became the CPU for personal computers with its ever increasing hunger for I/O, performance (lower dielectric constant (D_k)) and cost reduction.

The technology trend of finer lines/spaces had been going on at a steady, evolutionary pace until the consumer and communication market (e.g. MPEG players and cell phones) came into play. The explosive growth of this market was fueled by squeezing more innovations and applications into smaller spaces. These innovations demanded a large number of new components forcing accelerated improvements and inventions in assembly and in substrates. Today, relatively conventional PBGA substrates of a total thickness of 560 μm coexist with WFBGA (very thin and fine ball grid array) substrates of 120 μm thickness.

When the organic substrates market started to boom, many PCB manufacturers tried to enter the supply chain. They were attracted by the seemingly high price and margin for these many substrates on a single manufacturing panel. Initially substrates were considered simpler versions of the much more complex PCBs which typically contained a much higher layer count with similar line and space width. But there were many challenges in making substrates (see Table 7.1).

Few succeeded, especially in the US and Europe, for a number of reasons: assembly had moved to Asia and communication among Asian substrate suppliers and assemblers was more efficient. Dedicated substrate manufacturing operations were optimized for these special needs and thus were able to meet the quality requirements at the required low cost.

Table 7.1 Comparison of manufacturing challenges of substrates and PCBs

Characteristics	Substrates on manufacturing panel	PCBs on manufacturing panel
#of units per panel	100–2,000	1–10
Circuit density	Uniform across panel as per each substrate	Many depopulated areas on a PCB per component placement
Circuit defect density impact	High	High only in substrate location
Gold fingers for wire bond chip attach	Typically 100–1,000 per substrate	None
Gold ball pads for solder ball attach	Uniform density on ball side per substrate	Only in substrate location
Defect impact on gold surface	High	Moderate
Number of circuit layers	2–4	2–10 for 'low' end 10–20$^+$ for 'high' end
Inspection content	High	Moderate

7.2 Ceramic Substrates

Ceramic substrates are relegated to special applications of either high reliability requirements e.g. multi-chip modules in large servers and military applications or RF applications where passive elements like baluns and R, L, C elements need to be incorporated directly. Ceramics also can provide hermetic packages where required. Three dimensional features are also relatively easy to mold into ceramic material. But the total volume of ceramics is rather small in comparison to organic substrates and will therefore be left for others to review.

7.3 Organic Substrates

Today two types of die assembly techniques are used. The older and mature technique is wire bonding (WB) where the die is back bonded and wires connect from the top (face) of the die to the substrate. The newer technique is flip chip (FC) bonding, where the chip is bonded face down on the substrate and the interconnection is accomplished with a small solder ball or bump. A recent variation of FC bonding is gold stud bonding.

WB substrates are arranged and processed in strip form and are singulated only after assembly. Significant cost optimization can be obtained by tuning strip size (e.g. 187 mm × 40 mm) and lay-out within the active area of the panel (e.g. 390 mm × 490 mm) as well as the substrate lay-out within the strip itself (6 = units of 27 mm × 27 mm). Such strip formats are typical for PBGA substrates. Today, assemblers are working on increasing the strip size while reducing the rails on the strips to increase panel utilization and thereby reduce cost of materials and processing.

The next step in optimizing cost of WB substrates was the development of chip scale packaging (CSP). The original definition was that the size of a chip was of the substrate and was applied mostly to smaller die with I/O below 300. In addition to shrinking the substrate, the substrates were brick walled into a matrix (MAPBGA) of three or four blocks. The block size was determined in part by overmold capabilities. The most recent trend here is also to increase and optimize strip size to maximize the number of units per panel. A few more percent of units can be added by reducing the number of mold blocks on the strip with the goal to work with a single mold block also referred to as chocolate bar.

FC substrates were shipped in unit format so all panel optimization was primarily left to the substrate supplier. Unit format was chosen for substrate yield reasons as well as assembly process requirements. The latter were based on the ceramic technology which it was displacing. Today, all high end FC substrates, FCBGA, for CPU, graphics, chipset, ASICs, etc. applications are shipped in this format. Very few FCBGA substrates are shipped in strip form to take advantage of reduced handling in assembly as is done for PBGA.

FCBGA substrates typically have high I/O and fine trace & space which are best served by blind via (BV) technology. This has lead to the use of build-up technology with a special build-up dielectric. The de facto standard is Ajinomoto Build-up Film (ABF), an unreinforced resin from Ajinomoto Fine Techno Co., Japan, which is optimized for laser drilling and fine trace/space processing. Further, FC substrates have a solderable surface finish with no or low gold content to ensure the reliability of the FC solder joint.

In the last few years, small FC die with I/O below 300 are also assembled on PBGA substrates. The lower I/O density can be accommodated on laminate technology. These FC substrates (FCCSP) are using the same strip format as MAPBGA with a FC compatible surface finish. FCCSP substrates are presently the fastest growing sector in the substrate business. FCCSP technology is still evolving at a rapid pace trying to develop the most cost effective substrate and assembly technology. One of the early introductions of FC on laminate was a SRAM application by IBM on a four-layer (4L) substrate (Laine, 2000). Today most applications use high density interconnect (HDI) substrates (see below) i.e. blind via technology although PTH technology is generally more cost effective.

HDI type designs are used because of I/O density. HDI is also used in WB substrates but only for very high end/density designs. As indicated above, WBCSP and FCCSP differ mostly in the surface finish. Therefore, this unique requirement of FCCSP will be addressed in the surface finish section.

7.3.1 2L PBGA Substrates

As mentioned before, PBGA substrates come in a few simple configurations: two layers (2L), four layers (4L) and six layers (6L) of circuitry which are interconnected by plated through holes (PTH). Recently blind holes or vias are also used as interconnects to form HDI substrates with 2L, $1+2+1$, $2+2+2$ and $1+4+1$ layers.

The basic building block is a core or copper clad laminate (CCL) which consists of glass fabric, coated with an electrically insulating organic resin sandwiched between two copper foils. Several plies of impregnated fabric, called prepreg (PP), may be used to achieve the desired thickness of the core. The fabric itself may be woven from glass fibers of varying thickness and thread count to provide more options to control the CCL thickness. After the migration to RoHS and 'green' resin systems i.e. resins that contain less than 900 ppm of Chlorine, Bromine, Antimony and no Phosphorous, there are two major resin systems in use: Mitsubishi Gas & Chemicals 'BT – NX' series and Hitachi Chemicals 'E679-FGB' series. Cu foils come in varying degrees of thickness also designated by weight (ounces of Cu per sq foot). Most common are 12 μm = 1/3 oz and 18 μm = 1/2 oz although 37 μm = 1 oz and 75 μm = 2 oz are available for thermal or power designs. The properties of the most commonly used core materials are listed in Table 7.2.

The typical, simplified process flow for a 2L substrate is shown in Fig. 7.2. Every process step listed here actually consists of many sub-processes. For example, the patterning step can be divided into pretreatment, photo resist application, expose, develop, etch, strip and inspection steps which in themselves are composed of further sub-divisions with many different chemicals and

Table 7.2 Properties of the most common dielectric CCL materials

Company	Hitachi	MGC
Type of CCL	MCL-E679FGB	HL832NX
T_g (°C) DMA	190	220
T_d (°C) TGA (5%)	–	310
CTE (ppm/°C) x/y a_1	13–15	14
CTE (ppm/°C) z a_1	23–33	30
Thermal stress T_{288} (min)	–	25
Thermal conductivity (W/mK)	0.71–0.83	0.44
D_k 1 GHz	4.6	4.7
Loss tangent 1 GHz	0.017	0.013
Volume Resistance (MOhm cm)	$1.00 E+8$	$5.00 E+8$
Surface Resistance (MOhm)	$1.00 E+7$	$5.00 E+8$
Peel Strength (KN/m) 1/3 oz	0.65	0.75
Flex Strength (MPa)	450–550	450
Flex Modulus (GPa)	37	28
Tensile Strength (MPa)	200–300	280
Young's Modulus (GPa)	20–26	29
Poisson's Ratio	0.20–0.21	–
Water Absorption (%)	0.05	0.47
Flammability UL	94-V0	94-V0
Environmental	RoHS & green	RoHS & green

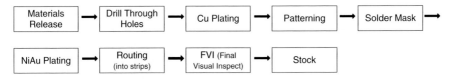

Fig. 7.2 Simplified process flow for 2L substrate

rinses. All wet chemical process steps are water based to minimize the use of organic solvents and to minimize the impact on the environment.

Common CCL thickness used to be 200 μm or greater. Now 150 μm and 100 μm have become popular in order to reduce the thickness of the finished substrate and thereby the final package thickness. Such thickness reductions require ever more careful handling of the cores during processing as well as either upgrading or new process equipment. These thin CCLs are easily torn or bent and creased manually or in the horizontal process equipment. The fragility increases especially in the pattern process when all the spaces have been etched to form the Cu traces and the reinforcement from the Cu foils no longer supports the prepreg (PP). A good reminder of the magnitude of the thickness challenge is the comparison to a typical human hair which is 100 μm thick. The thinnest core material available now is 60 μm and 50 μm thick of which 60 μm is already used in considerable volume to yield a finished substrate of a total thickness of 120 μm. The next target is to achieve a finished substrate thickness of 100 μm. Table 7.3 shows some representative substrate thickness and cross-sections.

In order to reduce the thickness further, solder resist thickness has to be reduced and controlled more tightly. Typical solder resist thickness specification averages range between 15 and 30 μm. The de facto industry standard solder masks are from Taiyo Ink Mfg. Co., Ltd., Japan, and are liquid, photo imageable inks which are applied either by screen printing or roller coating. Immediately after applying the ink, the surface will level to be relatively flat. During drying, the ink will begin to develop a conformal topography over the traces and spaces i.e. the total amount of solvent to evaporate in the space between lines is much greater than over traces. Hence, hills and valleys are formed replicating the trace/space pattern. During curing of the solder mask, the topography is typically exacerbated due to cure shrinkage. Both reactive moieties in the solder mask,

Table 7.3 Typical substrate thickness for 2L substrates

Total	100	130	160	210	260	360	560
Solder mask	15	20	30	30	30	30	30
Cu	15	15	20	25	25	25	25
Core	40	60	60	100	150	250	450
Cu	15	15	20	25	25	25	25
Solder mask	15	20	30	30	30	30	30

All dimensions in μm. Tolerance on total is +/−40 μm.

acrylate for photo reactivity and epoxy for thermal reactivity and chemical resistivity, typically exhibit large amounts of shrinkage during polymerization and cure. The topography can be minimized by several means:

a) Careful control of the drying profile prior to exposure. As the solvent evaporates, the viscosity increases and retards leveling flow. This viscosity increase can be counter acted by increasing the temperature but of course evaporation rates increase as well. High temperatures can also lead to a skinning effect where the surface evaporation is faster than the bulk diffusion rate of the solvent. The most effective temperature profiles may therefore be step profiles.
b) Adding solvents of varying boiling points can help to manage the viscosity profile very effectively but do require extensive experimentation to determine the best solvent mix and concentration. This is a common practice in impregnation of glass fabric to make prepreg.
c) Lamination of PET cover film (polyethylenetherephthalate) under temperature and pressure can provide some degree of leveling. The biggest benefit of the PET film however is an increase in image resolution. The acrylate photo reaction is usually retarded by oxygen which is present even in vacuum exposure systems and leads to a loss of resolution. The PET film minimizes rediffusion of oxygen into the solder mask during exposure and thereby yields much sharper images at higher resolution.
d) Dry film solder mask: the most effective and simplest way to minimize topography and thickness is to use a dry film solder mask. Essentially, it is a solder mask coated and supplied in the same fashion as dry film photo resist. The solder mask is coated onto a PET carrier film which protects the solder mask from contamination, handling and oxygen during exposure, and a PE (polyethylene) separator sheet to allow rolling up of the solder mask. Dry film solder mask does require vacuum lamination (vacuum, pressure, and elevated temperature) to fully encapsulate the traces without entrapping air. It does have another big manufacturing advantage. The cleanliness requirements of coating the material are handled by the material supplier instead of the substrate manufacturer. Dry film solder masks have been available for many years, the first of which where supplied for PCBs by DuPont under the trade name of Vacrel. The materials cost is significantly higher than liquid solder masks and has therefore delayed the implementation. For substrates the Dry Film solder mask standard is again from Taiyo Ink: AUS 410. Typical solder masks and properties are listed in Table 7.4.

7.3.2 4L PBGA Substrates

The simplest 4L substrate which starts with a CCL which has been patterned is laminated on both sides with PP and Cu foil to yield four layers of Cu. This raw substrate is then essentially processed like a 2L substrate (Fig. 7.3).

7 Advanced Substrates

Table 7.4 Properties of typical Taiyo Ink PSR 4000 solder mask materials

Property	AUS308	AUS310	AUS320	AUS 410	Test method
Young's modulus (GPa)	2.4	3.0	3.4	3.2	Tensile
Elongation (%)	3.0	3.5	3.5	4.9	Tensile
Tensile strength (MPa)	50	70	70	75	Tensile
T_g (°C)	100	103	114	110	TMA
CTE (ppm/°C)	60/130	60/140	60/130	50/160	TMA
Water absorption (%)	1.3	1.1	1.1	1.0	20°C/24 h
Poisson's ratio	0.29	0.28	0.29	0.32	
Dk	3.9	3.6	3.9	3.6	
Df	0.029	0.024	0.030	0.022	

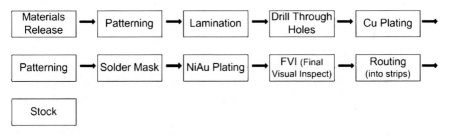

Fig. 7.3 Process flow for 4L PBGA substrates

A more complex 4L substrate can have buried PTHs (BPTH) to increase wireability. The process flow essentially follows that of a 2L substrate up to solder mask. Then the patterned core with PTHs is laminated and processed like a standard 4L substrate (Fig. 7.4). The registration requirements do increase with each level of complexity. Typically the internal lands for PTH connections are increased in size to ensure the PTHs are fully encircled by the land and avoid any hole break out. One way to ensure this improved registration is to use x-ray drills: using x-ray cameras, the internal registration fiducials are located and the new tooling holes for subsequent PTH drilling are placed accordingly.

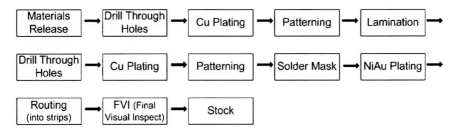

Fig. 7.4 Process flow for 4L substrate with buried via hole

7.3.3 6L PBGA Substrates

6L substrates are in moderate use only, the main reason being cost. Essentially the cost increases by approximately 50% for every layer pair added onto a 2L substrate when all other parameters remain the same. Again the simplest 6L substrate has only PTHs for interconnects and can be built either sequentially (as shown in Fig. 7.5) or in parallel (as shown in Fig. 7.6):

Sequential processing starts with a patterned 2L core on to which prepreg and Cu is laminated to form the 4L core, followed by patterning. Lamination with prepreg and Cu is repeated to form the 6L core. This structure is processed like a standard CCL. A variation is to use a 2L core with PTHs, laminate prepreg and Cu. This blank 4L core may now be drilled, plated and patterned before it is laminated into a 6L core blank. Alternatively, the 4L blank may be patterned and relaminated into a 6L core blank. The 6L core blanks are then processed like standard CCLs.

Parallel processing yields several options: with or without BPTHs. By laminating two patterned cores together with prepreg and Cu foil on the outside, a 6L substrate core blank is formed which can now be processed like a standard CCL.

The same process can be used for cores with PTHs in one or both cores to yield a 6L substrate with BPTHs which allows for complex wiring without adding much manufacturing complexity.

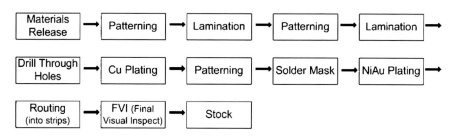

Fig. 7.5 Process flow for 6L sequential substrates

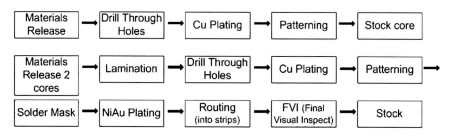

Fig. 7.6 Parallel process flow for 6L substrate

The advantages of parallel processing are cycle time reduction because the internal four layers can be built simultaneously, one lamination cycle instead of two (capacity), and yield optimization because cores can be inspected and marked. Parallel processing does require however a pinning scheme to ensure good registration during lamination so that the overlying lands for each PTH are not subject to hole break out.

It is obvious by now that BPTHs can be placed anywhere in the cross-section of 4L and 6L substrates depending on design requirements and cost optimization.

7.3.4 High Density Interconnect Substrates – HDI

With the advent of laser drilling, it became possible to drill blind vias (BV) of controlled depth. BVs can also be drilled mechanically but depth control is much more difficult. Two types of laser drills dominate in substrate or PCB manufacturing: CO_2 and UV lasers.

CO_2 lasers can drill through glass and organic matter but are stopped by Cu and have a limit of hole size currently at 65 μm or greater. Because they cannot drill through Cu, the intended hole pattern is first formed by conventional lithography i.e. the holes are etched in Cu using typical photo resists and develop-etch-strip (DES) technology. Subsequently, the CO_2 laser ablates the prepreg through this conformal Cu mask down to the capture pad. The hole shape is controlled by energy, pulse width and number of pulses.

UV lasers can ablate Cu as well as prepreg. Ablation rates are different for both and it is therefore possible to stop at the capture pad. Hole formation speed is increased when there is no top Cu to burn through. Hole sizes for UV are typically 50 μm or less to achieve the best efficiency in throughput. Overall the CO_2 lasers still have better throughput and are therefore predominant. A good overview of laser drilling has been written by John Lau, 2001.

After laser drilling, the BV need to be cleaned (desmeared) and plated. These processes actually present quite a few challenges: the fluid dynamics is rather restrictive limiting the solution flow causing problems with surface wettability, diffusion limitations, and bubble entrapment. The typical aspect ratio (depth to diameter ratio) for high volume manufacturing is still 0.7. Plating chemistry suppliers have made great advances in plating chemistries to improve on the aspect ratio and more so on enabling Cu hole fill i.e. plating the BV shut with Cu. The challenge as in PTH plating is to facilitate good throwing power into the holes. The goal is to preferentially plate the holes instead of the surface.

7.3.4.1 2L Via in Pad Substrates (2L HDI)

The process flow for 2L ViP (via in pad) substrates is the same as for standard 2L substrates with the exception of the drilling process (as shown in Fig. 7.7). As

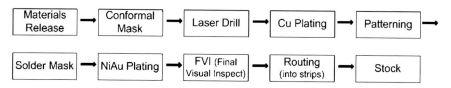

Fig. 7.7 Process flow for 2L ViP substrate

explained above, a conformal mask if formed on one side (chip side) and the BVs are drilled by CO_2 laser.

Typical core thickness is still 100 μm and therefore BV diameters are approximately 130 μm at the top. The biggest challenge today is BV reliability. Vias have to be able to survive 1,000 or more thermal cycles from −65 to 150°C without cracking. To that end a resistance shift test has been devised. Several hundred BVs in a daisy chain are thermal cycled and resistance tested periodically. Depending on the design, the resistance shift must be below a few percent to be acceptable. It has been shown that this indicator is extremely sensitive to crack formation. In turn, the crack formation is very sensitive to the cleanliness of the via hole bottom, electroless Cu quality and via shape at the bottom.

The reason for employing this design is that the BGA pad is used as the capture pad for the BV. In PTH designs, a PTH land is required next to the BGA pad. This looks like a dog bone and requires more space. The trade off for the ViP design advantage is that in laser drilling every panel is drilled one by one, thus limiting throughput. Dog bone designs on the other hand, have very high throughput because two or more cores are stacked during mechanical drilling. It is also possible to use PTHs without dog bones, but the PTH must be filled with epoxy and cap plated. The filling process does require a post-fill grinding process to remove extraneous epoxy. This is a rather stressful process requiring careful control of grinding pressure to avoid stretching and tearing of the core.

7.3.4.2 1 + 2 + 1 Substrates (4L HDI)

These four layer substrates are the first in a series of sequentially built high density substrates. In the simplest form, a 2L patterned core with PTHs is laminated with prepreg and Cu foil into a 4L blank core. Laser via formation and substrate completion follows the same process flow as for 2L ViP. Care must be taken that a good registration strategy is employed because now the laser BV must be placed inside the capture pads. The laser must therefore have access to the fiducials which were used to form the capture pads. Ideally the same fiducials are then used again to expose the pattern after plating.

The 1 + 2 + 1 substrate may also incorporate PTHs connecting top and bottom layers. In that case, PTHs are drilled after laser drilling the BVs. Modified plating parameters are required to plate BVs and PTHs at the same time.

The BPTHs described so far had a doughnut shaped land i.e. the center was not covered with Cu because the PTH was filled with resin from prepreg during lamination of the next layer. To increase wiring density, the PTH may be capped with Cu. This requires some additional process steps. After drilling and plating the core, the PTHs must be filled or plugged and plated again. The process then continues with lamination of prepreg and Cu foil, etc. The BVs may be located on top of the PTH cap, hence the name via on PTH (VoP).

Traditionally, PTH plugging has been done by screening an epoxy resin into the PTHs followed by curing and grinding off any residue of resin protruding above the Cu surface. Grinding is typically done with ceramic rollers. Care must be taken to control the pressure or the cores will be stretched in uncontrolled ways which in turn leads to high registration tolerances. Excessive grinding can also reduce the Cu thickness non-uniformly. The hole plugging epoxy is typically filled with ceramic or silica particles to reduce the thermal expansion. The concern is that high thermal expansion during later processing can put a lot of stress on the Cu cap and lead to stress cracking. Resins with very high glass transition temperature (T_g) and very high filler content are now available to minimize the CTE.

Another enhancement is available. The grinding may be followed by Cu thinning i.e. uniform etching of the Cu surface to reduce the Cu thickness. This exposes small nubs of epoxy hole plug material again which is ground off again. This cycle can be repeated until the desired Cu thinness is achieved which will allow finer traces and spaces on the core after cap plating.

The thinnest 1+2+1 substrates now in high volume production are 260 μm thick with 220 μm thick substrates emerging.

7.3.4.3 1+4+1 Substrates (6L HDI)

The most common core is a 4L core with PTHs. PTHs are capped usually and the BVs are VoP type. Consequently the process flow is essentially the same as for the corresponding 1+2+1 substrate except that a 4L core is employed. 1+4+1 has found limited application so far, mostly for designs with demanding power distribution and shielding requirements.

7.3.4.4 2+2+2 Substrates (6L HDI)

In this case, a 1+2+1 core undergoes the BV process starting with lamination a second time to build up the second build-up layer pair. The most common design for 2+2+2 is to stagger the BVs.

Recently it has become possible to plug the vias with Cu during plating. With the proper registration, this allows via stacking, and therefore, further densification of designs. The ultimate design is to stack the BVs on top of capped PTHs. For these designs usually there are no layer one to six connections with PTHs.

Aside from advanced substrate designs this structure is becoming a very common design point for modules and wireless applications.

7.4 Tape Ball Grid Array – TBGA

TBGA substrates are mostly based on polyimide (PI) although some polyester materials have been available as well. The flexibility of these thin substrates does require dedicated assembly lines. Because of the high cost of polyimide many applications have been phased over to TFBGA style substrates which can be assembled in standard assembly lines.

Film based dielectrics like polyimide do come in roll form and can be metalized in continuous form either by casting the dielectric on Cu foil or by sputtering a seed layer followed by roll to roll plating. PI based tape TBGA substrates were used for a long time already but have been declining for cost reasons. Initially the Cu foil was bonded to PI with adhesives until adhesive-less tapes became available in high volume. TBGA substrates were only manufactured as 2L type substrates. Very fine traces and spaces were produced on these dielectrics because the metallization was thin and smooth on the dielectric side.

Most TBGA substrates are either single layer metal or 2L with PTHs. 1 + 2 + 1 type structures are available now also but volumes are low.

Now other dielectric materials are available. The most recent ones are liquid crystalline polymers (LCP) from Rogers Corporation for example. LCP promises to be considerable cheaper than PI and to be much less sensitive to moisture uptake. Like PI, LCPs suffer from lack of self-adhesion and adhesives or bond plies are required to build multilayer substrates.

7.5 PBGA Substrate Trends

7.5.1 Low Cost Dielectrics

As mentioned before, Mitsubishi Gas & Chemicals BT (Bismaleimide triazine resin) and lately also Hitachi Chemicals E679-FGB (epoxy based resin) have been the de facto industry standard for PBGA substrates. The consumer and mobile industries are very cost competitive and are demanding continuously more cost reductions. The recent introduction of DDR II memory which also uses TFBGA type substrates (thin fine PBGA substrates) is amplifying the cost pressure. This trend has increased the market chances of new dielectric suppliers and their alternate dielectric materials. A number of new suppliers seem ready to fill that need. NanYa Plastics, LG Electronic Materials, Doosan Electronic Materials are some of the candidates. The properties of some of their CCL materials are listed in Table 7.5.

Table 7.5 Properties of low cost and of low CTE CCL materials

Supplier[1]	MGC	MEW	Sum	Doosan	NanYa[2]	LG[2]	Doosan[2]
CCL type	HL892	R1515B	ELC47856S	DS7409 HGS-type	NPG-200	LG-P-5006	DS4709HG G-type
T_g (°C) DMA	270	205	265	260	210	–	240
T_g (°C) TMA	250	180	220	–	165	176	180
CTE_{xy} (ppm/°C) (α1)	12–13	12	11	11–12	12	–	13–14
CTE_z (ppm/°C) (α1)	35	40	16	10–15	46	56	24
D_k @1 GHz	4.9	4.8	4.2	4.5	4.2	4.88	4.6
D_f @1Ghz	0.012	0.011	0.007	0.007	0.014	0.008	0.013
Peel strength (KN/m)	0.88	0.9	1.1	0.7	1.22	0.85	0.8
Young's modulus (GPa)	–	26.5	29	33	–	–	31
Water absorption (%)	–	0.12	0.4	0.43	0.15	0.7	0.42

[1]MGC –Mitsubishi Gas & Chemical, Japan; MEW – Matsushita Electrical Works, Japan; Sum – Sumitomo Bakelite Co., Ltd., Japan; Doosan – Doosan Electro Materials BG, Korea; NanYa–NanYa Plastics Corp., Taiwan; LG – LG Chemicals, Korea;
[2]low cost candidates

7.5.2 Low Cost Solder Masks

The same cost trend as for dielectric materials exists for all materials used in substrate manufacturing. The competition for Taiyo Ink solder masks is increasing but no clear penetration has been established as of now.

7.5.3 Thin Substrates, Thin Dielectrics

With the explosive growth of consumer and mobile applications, the demand for thin substrate has grown with equal speed. CCLs are now available down to a dielectric thickness of 50 µm. Likewise PP thickness of 40 µm for multilayer substrates is in common use. Dielectric manufacturers are developing PP of 35 µm and 30 µm thickness. This is achieved in part by changing weaving patterns and techniques. Glass bundles contain fewer fibers and are more spread out (less coiled) to make the fabric flatter. In turn, impregnation with resin is becoming more challenging because the fabric is less resistant to tension in the impregnation towers.

It would appear that the limit of reducing the thickness further is near and that new dielectric materials will need to be employed if substrates are to evolve further. Two possibilities come to mind: roll to roll processing and film based dielectrics.

Most horizontal process equipment relies on the panel being able to support itself during it's conveyance through the equipment. Roll to roll processing seems to be ideal for very thin dielectric materials which are very difficult to handle in panel format through the many process steps. Very thin glass reinforced dielectrics can be processed in roll format as has been demonstrated in smart card assembly where finished substrates in strip form were spliced together and rolled up for high speed, continuous assembly. The challenge is to supply prepreg based CCL in long rolls. Lamination typically is performed in sheet form and splicing would be necessary. Continuous lamination had been developed and was in production by Dielektra Company of Germany but has been abandoned again.

Tape or film based dielectrics may be the way to reduce the thickness of the substrate. Tape dielectrics can be processed in roll form which significantly improves handling issues of thin materials not least because of the superior mechanical properties of these dielectrics.

7.5.4 Low Expansion Dielectrics

Thermal expansion has not been an issue for WB PBGA substrates to date. FC substrates rely heavily on underfill material to provide the necessary reinforcement of the solder joint to handle the large expansion mismatch between die and substrate. For FCCSP substrates the dies have been small to date so lower CTE has not been necessary yet. But as die size grows, it will be necessary to reduce the CTE mismatch to control reliability of the package.

For all laminate dielectrics, the expansion is not uniform in all directions. Typically x and y direction (in plane) have rather similar CTEs varying only of different bundles are used in the weave for x or y direction. The expansion out of plane, z direction, is typically governed by the resin with the glass acting as filler. Therefore, CTE_z below T_g is typically three times higher than CTE_{xy} and ten or more times higher above T_g. The expansion of the resin can be reduced most effectively by adding fillers. The fillers must be chosen carefully not to degrade mechanical, chemical, and reliability properties. For example, increased moisture uptake may lead to delamination or pop corning especially at the elevated temperatures of lead-free assembly. Quartz glass and S-glass are very effective for in plane CTE reduction but are not available in yarns as fine as E-glass nor at comparable cost. So CTE_z needs to be managed with filler content and modification of resins for now. The properties of representative CCL materials (MGC, MEW, Sumitomo & Doosan) are listed in Table 7.5.

7.5.5 Surface Finishes

7.5.5.1 Electroplated Nickel Gold

The industry standard surface finish is electroplated nickel and gold (NiAu). Here the mobile industry, especially cell phone industry, developed a new reliability requirement: the drop test. The drop test became especially challenging with the introduction of lead-free solders. Lead-free solders form brittle intermetallics with nickel which fracture easily during drop testing.

7.5.5.2 OSP and AFOP

One of the early solutions was to solder directly to Cu because Cu forms different, more robust intermetallics. However the oxidation of Cu prior to solder ball placement needed to be avoided. Again the PCB industry pointed the way with an organic solder preservative (OSP). OSPs are Cu complexing compounds based on imidazole derivatives. Imidazoles of the latest generation have the proper substituents to make them high temperature stable. They can withstand multiple reflows at 260°C and are therefore lead-free process compatible. At the same time, the bond fingers do require NiAu to be free of organic residues. Here is some of the differentiation of the various suppliers of OSPs who use additives to keep the gold surface clean. For wirebond applications, the term AFOP, gold (Au) on finger and OSP on ball pad, has been coined. Because both gold and OSP are required, another photo lithography process step is needed to protect the ball pads during NiAu plating of the bond fingers. The cost of the additional process steps usually outweighs the savings in gold.

For FC applications, OSP can be used on both the ball pad and the FC pad thereby providing a low cost surface finish.

7.5.5.3 ENEPIG

The cost reduction pressure has been amplified by the new record prices for gold. This cost pressure lead to a further reduction of the gold thickness without compromising the wire bond ability and some users have reduced the minimum gold thickness from 0.5 to 0.3 µm. But this pressure has once again revived the quest for a new universal surface finish. Electroless nickel electroless palladium immersion gold (ENEPIG) chemistries have been advanced over the last ten years to make this finish attractive again.

The cost savings for ENEPIG is derived by the thinness of the immersion gold layer which is projected to be about 0.1 µm. Palladium thickness is comparable and the metal cost is much lower than gold so that the resultant cost is lower than NiAu. ENEPIG is wirebondable and solderable and therefore considered a universal finish just like NiAu. Drop test results also indicate superior performance for ENEPIG.

Because the amount of gold is much less than in NiAu, it is expected that solder embrittlement due to high gold concentration will be reduced. ENEPIG is therefore expected to be competitive with ENIG (electroless nickel immersion gold) without the concern for 'black pad' defects because of the difference in the type of chemical reactions involved in the deposition.

Another significant advantage is the electroless nature of the chemistry i.e. no buss lines are required. This will make the etching back process superfluous with a cost competitive alternative because of the simpler process flows due to fewer process steps. The same applies to buss line less processes like NPL (no plating line process) or SG (selective gold process) and mixed surface finishes like AFOP.

7.5.5.4 Tin Surface Finishes

The first use of tin finishes on substrates was on FCBGA substrates. Flip chip packages used to be based almost exclusively on ceramic substrates and later on build-up substrates. The surface finish was typically ENIG followed by printing a solder bump on to the FC pad (pre-solder). The bump was then coined to provide a top flat on top of which the bumped die could locate and be held in place until reflow by tacky flux. ENIG was plagued by a few problems like 'black pad', reduced wettability with lead-free solders, etc. The search for alternate finishes suggested tin as a solderable surface which has become popular for many FCBGA applications by using a form of electroless tin as a direct replacement of ENIG.

Immersion Tin The industry term for this electroless tin is immersion tin, iT. The chemical reaction is an exchange reaction of tin with Cu and therefore is self limiting i.e. once no further surface Cu is accessible, the tin deposition slows by orders of magnitude. The typical tin thickness is about one μm. This thickness is too low for most FCBGA packages as the die bump solder volume cannot form a high solder column. Sufficient standoff between solder mask surface and die is required for the underfill to flow in and provide void-free encapsulation of the solder bumps. Therefore, pre-solder still needs to be applied just like in the case of ENIG. For FCCSP packages iT may become acceptable for cost reasons however the stand-off will need to be managed.

iT has an acceptable shelf life and is wettable by all lead-free solders. It can be used double sided, on both the FC pads as well as the BGA pads. Reflow and IMC formation only happens above the tin melting point i.e. reflow temperatures of 240°C or above. Therefore, if eutectic solder is used for pre-solder, the bumps can be reflowed without reflowing the tin of the BGA pads.

Electroplated Tin – eT When tin is electroplated, the thickness can be controlled by the typical plating parameters. However, a plating buss is required. For buss less designs, a selective plating process can be employed: after solder mask, the panel is sputtered with a thin layer of Cu to create a buss layer. Photo resist on the buss layer then defines the tin plating area. After tin has been plated, the resist is stripped and the buss layer is flash etched. If necessary, this process can be single sided and the second side can be plated with an immersion process while the first side is protected by a peelable resist or mask. The process flow is shown in Fig. 7.8. eT can therefore be used for gold stud bumped FC assembly where thicker tin is required. It may also be usable for solder based FC if the stand-off needs to be boosted only a little over iT.

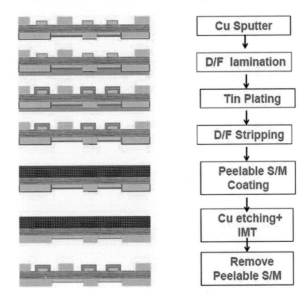

Fig. 7.8 Tin plating process for substrate with electroplated tin on bump side and immersion tin on ball side[2]

---

[2] Reprint Courtesy of Advanced Semiconductor Engineering Inc.

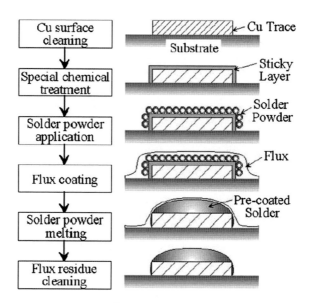

Fig. 7.9 Process flow for Super Juffit[3]

7.5.5.5 Super Juffit

Super Juffit is the process and product of Showa Denko KK, Japan. It is a buss less metallization process to provide a wide range of surface finishes. Here a tacky, organic compound is applied such that it adheres only to metal features. A fine metal powder is then applied and will selectively bond to the tacky compound. After reflow the surface finish is complete (see Fig. 7.9). The thickness of the surface finish can be controlled within a small range by the metal powder granularity to be thicker than iT but not to the full range of electroplating. The advantage of Super Juffit is that any reflowable metal or alloy can be used. Given the cost of the powders and the license, it is expected to be the more expensive option.

7.6 FCBGA Substrates

FCBGA substrates are also commonly called Build-Up (BU) substrates and have been in use for nearly ten years. The original application was actually for PCBs for the IBM laptop to accommodate the large number of components in the small available volume. Rather quickly, this BU technology was extended to FC substrates because this technology finally had the resolution required to compete with ceramic substrates: lower D_k for better electrical performance and much lower cost. The original BU was based on a standard thick core, 2L or 4L,

---

[3] Reprint Courtesy of Showa Denko K.K.

with epoxy plugged PTHs. Solder mask was applied as redistribution dielectric and BVs were photo imaged. The solder mask was then roughened for adhesion for the subsequently plated Cu. Patterning was subtractive and yielded a $1+2+1$ structure. This last cycle could be repeated to build a $2+2+2$ structure. Solder mask was again applied to protect the last layer of circuitry. The typical surface finish was ENIG followed by pre-solder and coining of the FC pads.

BU became a must have technology for many, many substrate suppliers and a large variety of different processes and materials were developed. Laser via formation was initially thought to be at a big disadvantage because drilling rates were rather slow. But laser tool technology has increased throughput by several orders of magnitude. Laserable materials can be optimized for processability and for mechanical properties of the final substrate. Photoimageable materials do have to compromise on these properties in order to have reasonable lithographic properties. So laser technology is the predominant technology today. Other technologies still in use today are Toshiba's B^2iT [8] and Matsushita's ALIVH [2] to mention only a few significant ones.

The other key event was the adoption of BU technology by Intel. Intel has very methodically driven the advancement and standardization of this technology into a mature technology. Today Intel is using BU substrates for their entire line of CPUs and North Bridge chip set and as such is by far the largest consumer of BU substrates. This market dominance set the standard for processes and materials. Even new, rapid growth applications like graphics processors, gaming processors, etc. follow this standard because initially none have the volume to drive a change in the supply chain. In some instances however more aggressive design parameters are pushed by these new players e.g. BV stacking, finer traces and spaces, finer bump pitch or finer PTH pitch.

Today, most BU substrates have cross-section ranging between $2+2+2$ and $4+4+4$. Chipset substrates tend to use the low layer counts and micro processors use the high end. The most advanced CPUs are currently at $6+10+6$. Game CPUs can be as advanced $3+2+3$ with three high stacks of BVs and 18 μm traces/spaces. Figure 7.10 outlines the process flow for a $2+2+2$ BU substrate.

The core materials and processes are essentially the same as for 2L and 4L WB substrates corresponding to $x+2+x$ or $x+4+x$ BU substrates. So it is no surprise that core design parameters are at the same level of 50 μm traces and spaces moving towards 40 μm. As semi-subtractive processing matures for laminate technology, trace/space dimensions are expected to reduce to 30 μm and 25 μm which will lead to high density wiring in the core. When this is coupled with small hole drilling, high density cores will be enabled. Small hole drilling, 100 μm, does also require thinner cores e.g. 400 μm or less. High density cores may well be able to contain the increase in BU layer pairs or even reduce the number of BU layer pairs depending on the design point.

Eight hundred microns thick cores have typically >150 μm PTHs while 400 μm thick cores can have 100 μm PTHs at a pitch of 250 μm. Ideally one would decrease the PTH pitch to match the bump pitch as was possible in ceramic substrates. This would allow for the densest designs e.g. power

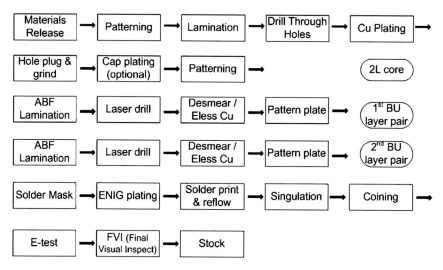

Fig. 7.10 Process flow for 2+2+2 Build-Up substrate

could drop directly from bump down to the BGA ball. Coreless substrates will be able to achieve this (see below).

The standardization of BU substrate is most pronounced in the choice of BU dielectrics. ABF is the standard today. Ajinomoto was able to develop a resin film that achieves many key process features while having the desired end properties for the final substrate in terms of thermo-mechanical properties and reliability. ABF is applied by vacuum lamination at elevated temperatures. Under these conditions it is able to completely encapsulate the circuit traces and provide a level surface. After curing, laser vias can be drilled with either CO_2 or UV lasers. The BVs are then cleaned and the surface of the ABF is roughened by permanganate desmear solution. With the appropriate process control, a rather well defined surface roughness can be created which provides good adhesion for electroless Cu plated next. The Cu is a thick electroless Cu to support semi additive processing, SAP, i.e. the circuit pattern is defined by photo resist and Cu is electro-plated into the channels and BVs. Depending on the plating chemistry (and process parameters), the BVs can be Cu filled at the same time as the traces are plated. If dimples on the plated BVs are controlled to be very shallow, then BVs can be stacked to provide more wiring capability. After plating, the photoresist is stripped and the exposed electroless Cu is flash etched and inspected by Automated Optical Inspection, AOI. This BU process cycle can be repeated as many times as BU layer pairs are needed. Yield and reliability will dictate how many layers are economical to build. Core yields and BU layer yield should be near 99% or better to achieve a raw substrate yield of 97% for a 1+2+1 without solder mask or surface finish. Such yields require excellent registration for laser drilling, patterning, and a thorough understanding of all process interactions and how to control them.

Table 7.6 Properties of ABF materials

ABF Type	SH9K	GX3	GX13
CTE (ppm/°C)	95	60	46
Tg (°C)	165	153	156
Young's modulus (GPa)	3.0	3.5	4.0
Elongation (%)	6.7	7.6	5.0
D_k @ 1 GHz	3.4	3.4	3.35
D_f @ 1 GHz	0.022	0.023	0.012
Ra (nm)	900	900	800
Filler (SiO_2)	12	18	38

The early ABF resins were only subjected to eutectic pre-solder and assembly but SH9K type did survive lead-free temperatures already. The higher temperatures result in more thermal expansion and therefore stress on the BVs. Therefore filler content was increased to reduce the CTE while at the same time the resin was converted to green i.e. reduced the chlorine and bromine content. As can be seen from Table 7.6, the other properties of ABF change only to a lesser degree.

The last critical material is the solder mask. Liquid solder masks are still in common use either Hitachi Chemicals SR7200G or the corresponding green Taiyo Ink material. Process and cleanliness challenges are the same as for laminate substrates.

The last significant process step is to apply pre-solder. This is mostly done by stencil printing. 150 μm bump pitches have been printed successfully. Below 150 μm it is anticipated that photo resist masks may be necessary, similar as in wafer bumping, to control the solder volume with sufficient accuracy. After pre-solder the substrates are singulated, coined, electrically tested and inspected. As mentioned before, pre-solder is applied over a number of different surfaces such as OSP, ENIG, iT, and ENEPIG in the future.

Another version of core may be constructed by applying ABF onto a patterned core (e.g. 2L or 4L) which is then drilled, plated and patterned. This yields a core which has low dielectric thickness in its outer layers and is capable of a finer pattern enhancing wiring density and enhanced electrical performance. ABF may also be laminated with Cu foil. This will give superior Cu adhesion but will lower the pattern density if the Cu thickness increases.

7.7 Coreless Substrates

Coreless substrates seek to emulate the via stacking capabilities of multi layer ceramic substrates at pitches controlled only by BV design constraints. Here vias can be reaching through any amount of layers, from adjacent layers to top/bottom layers. The first examples of this technology were Toshiba's B^2iT and Matsushita's ALIVH technologies. Kyocera developed a similar technology called CP core. Since then many BU substrate suppliers have developed the basic capability for coreless substrate manufacturing.

There are two common manufacturing paths for these substrates depending on whether a conductive paste is used or whether Cu plating is used to fill the vias.

The paste process technology is a parallel process. A bare, B-staged (partially cured) sheet of dielectric is laser drilled, and a conductive paste is screened into the vias to form building block A. Cu foil is laminated to both sides of A and patterned using standard subtractive processing to form block B. By stacking B and A and B and subsequent lamination a raw coreless substrate is formed. The substrate is finished as usual with solder mask and surface finish as required by assembly. This process was first devised by Matsushita for ALIVH using impregnated aramid fiber mats as the dielectric. Variations of this technology have been introduced also by other suppliers using different material sets.

The plating process technology is necessarily a sequential process. In many instances, a Cu clad carrier is used as the handling base to act as a core. This pseudo core is then subjected to BU processing as usual until the required layer count is reached. The two BU blocks are separated from the carrier e.g. by peeling or by routing. While this process is sequential, it yields essentially two substrates in a single production run. (see Fig. 7.11). The advantage of this process is that traditional BU materials and processes are used.

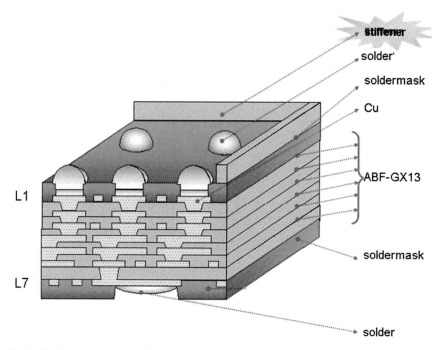

Fig. 7.11 Example of one coreless substrate cross-section (ASE Materials 2005)

---

[4] Reprint Courtesy of Advanced Semiconductor Engineering Inc.

To date, coreless substrates have found little or no implementation. It seems that warpage effects during die to substrate assembly are very large because the current materials do not provide enough stiffness. This could be minimized by attaching a stiffener prior to die attach but that is not compatible with high volume, low cost assembly yet. More innovation is required.

7.8 Specialty Substrates

Many specialty substrates have been developed to suit specific needs. A few of these specials will be addressed here briefly: RF modules, high performance, low dielectric BGAs and embedded component substrates.

7.8.1 RF Modules

Many RF modules are of the $1+2+1$ type but do typically not require high density circuitry but impedance control instead. Traces range between from 75 to 100 µm in width with impedance tolerance of 10%. Designs are now pushing to reduce the tolerance down to 5%. Pattern plating should facilitate these new requirements. Frequently, thick Cu of up to 37 µm is required in the PTHs is also required for heat dissipation.

RF modules had fostered early on a BU type of substrate using resin coated Cu foil, RCF, instead of the costly ABF as the BU dielectric. Via openings are etched into the foil which can then act as the laser drilling mask for the BVs. BVs are often Cu filled during plating. Patterning is still subtractive. For cost reasons, RCF is being replaced by PP, which makes the cross-section very similar to standard HDI substrates.

The cores have frequently cap-plated PTHs to improve wireability, enabling trace or BV on PTH. In the future, the core thickness will be reduced to 60 µm and PTH in the core will be either Cu filled or it will be Cu filled laser via. Either one will allow stacking of the BU vias. These stacked design will allow very dense wiring and thermal dissipation if the via diameter is larger e.g. 150 µm.

For electrical reasons, often special resins are used with lower D_k. These however are more expensive and sometimes more difficult to process which in turn adds process cost.

7.8.2 High Performance Substrates with Low Dielectric Constant

High I/O die for certain networking and communication servers drove the development of these high performance substrates. Two different approaches were taken by two competitors, but both used Teflon® type dielectrics achieving D_k of 2.8–3.

EI Technologies Corporation, Endicott, NY, (the former IBM Corporation site) developed the HyperBGA® technology [1] which is based on silica filled Teflon resin from Rogers Corporation. HyperBGA is the ultimate sequential substrate with nine metal layers. A sheet of Cu clad Invar, CIC, serves as the central core. Clearance holes are etched in it followed by laminating Rogers dielectric and Cu foil on both sides. This inner core is patterned and another layer pair of Rogers dielectric and Cu is added two more times. This core is then laser drilled with 50 μm through holes, plated, and patterned. Next a BU layer pair is added using an unreinforced dielectric. BVs are also 50 μm diameter. Typical surface finishes were ENIG with pre-solder (see Fig. 7.12)

This unique set of materials does require extraordinary processing of course, e.g. lamination at temperatures above 300°C, unique pre-treatment for hole cleaning and seeding for subsequent Cu plating, etc. These challenges have been mastered and a very unique substrate has been built. A stiffener is attached before die attach to ensure planarity.

The combination of CIC with Teflon dielectric manages the stress transfer from die to board by being very compliant and having an effective CTE of approx. 10 ppm/°C. Even for very large die (>18 mm × 18 mm), a very reliable substrate has thus been built with excellent electrical performance. The unique materials set and the extraordinary processing are reflected in it's price.

3 M Microelectronics, Austin, TX, (the successor of Gore Microelectronics, Eau Claire, WI) developed a rather similar substrate using Microlam® dielectric.

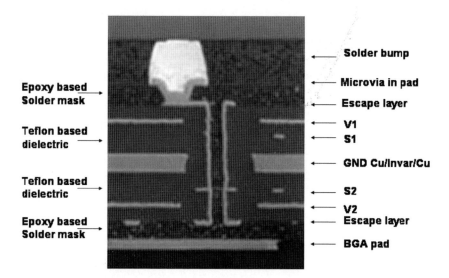

Fig. 7.12 Cross-section of a HyperBGA® substrate from IBM Corporation[5]

---

[5] Reprint Courtesy of International Business Machines Corporation, copyright 2000 ?c International Business Machines Corporation

Microlam is made from expanded Teflon, impregnated with an epoxy resin. It can therefore be laminated under standard press conditions. The plating process is also simpler because of the dominant epoxy resin. The dielectric is however not as compliant and low modulus as the Rogers dielectric and is therefore more suitable for packages of up to medium size. This substrate is called Via on Chip Pitch^{TM} substrate and has only one redistribution layer on a five layer core i.e. it is a seven layer substrate [13].

7.8.3 Substrates with Embedded Components

Embedding of components has been long been the desire of designers and developers to free up substrate or PCB surfaces. The objective is to reduce the body size of the substrate and to minimize the required solder interconnections.

There are two fundamental approaches: buried passives and embedded passives & die.

7.8.3.1 Buried Passives Substrates

PCBs with buried capacitors and resistors have been in production for more than 10 years. But still applications for substrates are limited because the electrical properties are restricted. The achievable tolerance of the electrical parameters is typically also greater then 10% which often does not meet requirements. Discussion of these applications is therefore deferred.

7.8.3.2 Embedded Die and Embedded Passives Substrates

In Fig. 7.13, a sample of embedded die technologies is from a number pioneering companies. By now, nearly all substrate suppliers and many users have developed their own flavor of technology but commercialization is very limited. In most cases, instead of a die one or more discrete passives can be embedded. Adoption is handicapped for the following reasons:

- Lack of industry standards for process and materials
- Process yield needs to be greater than 95%
- Substrate design tool availability
- Known good die availability
- Accepted business model for embedding die e.g. consigning die to embedding company
- Test solutions for embedded substrates

The different approaches all have in common that BU-like processes (laser BV and SAP) are used on top of the die to fan out from the die surface until the mounting plane is reached.

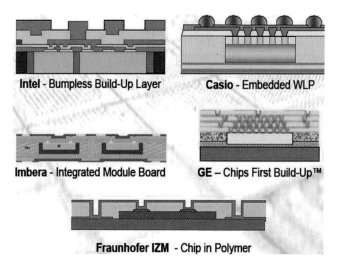

Fig. 7.13 Examples of embedded die technologies[6]

Several different approaches are used to bury the discrete component (die, capacitor or resistor):

1. A cavity is formed in a core; the die is bonded to the bottom of the cavity and the core is laminated with dielectric (RCF, ABF or PP) and Cu foil. The die pads are accessed by laser drilling [3, 5]
2. The die is bonded face-up on a carrier or core and laminated with RCF or ABF and with Cu foil. Die pads are accessed by laser vias [9]
3. The die is bonded face-down of a sacrificial foil or tape. This assembly is then either overmolded or laminated with punched prepreg. The tape is removed and RCF or ABF with Cu foil are laminated to this structure. BVs are used to access the die [10]
4. Laser vias, matching the die foot print, are drilled into a thin Cu foil mounted on a carrier. The die bumps are located on the vias and the die is bonded with adhesive to the foil. Punched PP and Cu foil are laminated over the die. The carrier is peeled or etched off and the die is accessed by laser to remove the adhesive [11]

The latter approach seems the most promising in terms of registration and achievable circuit density.

Reference

1. Alcoe D, Jimarez M, Jones G, Kindl T, Kresge J, Libous J, Stutzman R (2000) HyperB-GATM: A High Performance, Low Stress, Laminate Ball Grid Array Flip Chip Carrier, IBM Microelectronics News 2nd Q, p 36ff

---

[6] Reprint Courtesy of Advanced Semiconductor Engineering Inc.

2. Boggio B (2000) The Any Layer Interstitial Via Hole Process, Board Authority 2 (1), pp. 91–95
3. Fillion R, Bauer C (2005) High Performance, High Power, High I/O Chips First Build-Up Technology, Proc. Pan Pacific Symposium
4. Freyman B, Pennisi R (1991) OverMold Plastic Pad Array Carriers (OMPAC) a Low Cost, High Interconnect Density IC Package Solution for Consumer and Industrial Electronics, Proc. ECTC, pp. 176–180
6. Komatsu D (2005) Trend of WLP Technology and Next Generation Packaging "WLP & EWLP", Proc. MAP + RTS Conf.
7. Laine E, O'Leary P (2000) IBM Chip Packaging Roadmap, Future Fab Intl. Vol. 8
8. Lau J, Lee SWR (2001) Microvias for Low Cost, High Density Interconnects, Chapt. 4, McGraw-Hill
9. Oodaira et al. (1996) Proposed New Method (B^2iT) for Production of Printed Wiring Boards, Proc. 9th Circuit Mounting Conf., pp. 55–56
10. Ostmann A, Neumann A (2002) Chip in Polymer – Next Step in Miniaturization, Advancing Microelectronics, 29 (3)
11. Towle S, Braunisch H, Hu C, Emery R, Vandentop G (2001) Bumpless Build-Up Layer Packaging, proc. ASME Int. Mech. Eng. Congress and Exposition (IMECE) New York, Nov. 11–16, 2001, EPP24703
12. Tuominen R (2006) IMB Technology of Embedding Active Components into a Substrate, Proc. Semi Europe 2006
13. Tsukada Y (1992) Surface Laminar Circuit and Flip Chip Packaging, Proc. 42nd ECTC Conf. pp. 22–27
14. 3 M Electronics, 3 M High Performance Family of Organic Flip Chip Substrates, 3 M Electronics, Austin, TX, 80-6201-2992-6 (505.2)

Chapter 8
Advanced Print Circuit Board Materials

Gary Brist and Gary Long

Abstract Printed Circuit Board (PCB) materials refer to a set of dielectric and conductive materials used to form circuit board interconnects. The PCB industry offers a wide array of material options to meet different performance and cost requirements. Copper is the primary conductive material used in PCBs due to its cost and electrical conductivity as well as its stability and ease of processing. The most common and widely recognized class of PCB dielectric materials is FR-4 which describes a group of woven glass reinforced tetra-functional epoxy materials. The continued use and longevity of copper and FR-4 materials is due to a combination of their availability, low cost, processability, and adequate electrical, mechanical and thermal properties.

Advanced PCB materials refer to non-FR-4 dielectrics, enhanced or modified FR-4 dielectric materials as well as advanced copper foils. All advanced PCB materials offer some unique set of electrical, mechanical, thermal, or chemical properties that have been optimized for a particular application, design challenge, or manufacturing issue. As a result of the unique challenges and cost sensitivity of different markets, the number of advanced materials that have been made available has increased significantly in recent years.

The past decade has seen many new materials introduced to the market that are variations on previous materials. These new materials are tailored by adding or changing one or more components to optimize the material properties for a specific application or market. Examples include utilizing epoxy blends or adding fillers to increase a materials' glass transition temperature, change its dielectric constant or using a different glass formulation for the reinforcement fabric to reduce the materials' dissipation factor. Other examples include changing a material processing steps such as removing the twist in glass yarns to create reinforced materials that have less spatial variation to improve laser ablation processibility and spatial electrical variation.

G. Brist (✉)
Sr. PCB Technologist, Intel Corp. 5200 NE Elam Young Parkway, Hillsboro, OR 97124
e-mail: gary.a.brist@intel.com

Keywords FR-4 · Copper Foil · Rolled annealed copper · Electrodeposited copper · Conductor Loss · Moisture Absorption · Polyimide · LCP · Dielectric constant · Thermoset · Thermoplastic · Ceramic filler · Aramid paper · Reinforcement · Glass reinforcement · Glass fabric · Surface finish · Embedded resistor

8.1 Dielectric Materials

Dielectric materials are commonly classified by how they are used within the manufacturing process such as laminate cores, prepregs, bond plys, soldermasks, etc. Each dielectric material is also defined by its composition that includes a primary resin system, possible fillers mixed into the resin, and reinforcement materials.

The usage classifications for dielectrics in rigid PCBs are laminate cores, prepregs, and soldermask. In flexible PCBs the classifications are referred to dielectric films, adhesives, and coverlayers. Specialized PCB designs such as rigid flex, mixed material PCBs, PCBs with cavity or stepped structures, or HDI PCBs often use a combination of both rigid PCB materials and flexible PCB materials to solve specific design or manufacturing challenges.

Prepreg and core laminate materials refer to the reinforced epoxy dielectrics used as the primary building blocks in most multilayer PCBs. In the final PCB, these materials provide the electrical isolation between adjacent circuitry layers. Prepregs are made by impregnating a reinforcement material, usually woven glass, with a thermoset resin as seen in Fig. 8.1. The woven glass is supplied on large rolls which allow continuous processing and helps minimize manufacturing costs. The impregnated reinforced material is partially crosslinked through a heat treating process that also drives out volatiles. After heat treating, the partially crosslinked resin is solid enough to handle in shipping and PCB manufacturing and is referred to as a B-stage material. Under temperature and pressure, the resin of a B-stage material will become liquid with a viscosity

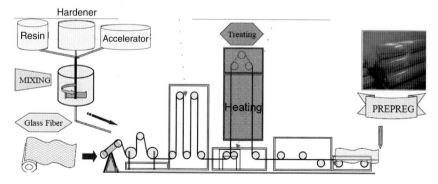

Fig. 8.1 Schematic of prepreg manufacturing process (courtesy of ITEQ)

which is dependent on its decree of crosslinking during the heat treating process. This makes B-stage materials ideal for laminating between copper foils to create core laminates and for bonding circuit layers of a PCB.

Core or laminate materials refer to the copper clad rigid reinforced epoxy dielectrics used in building a PCB. The cores are formed by laminating one or more sheets of prepreg between sheets of copper foil. See Fig. 8.2. During the laminating process of thermoset materials the resin is highly crosslinked, or cured, resulting in a rigid material. Highly crosslinked thermoset resin is known as C-stage and cannot be brought back to a liquidous state. At elevated temperatures a C-stage resin may become slightly soft depending on its degree of cure; but remains solid. This makes a core ideal for forming the individual circuit layers of a PCB as it provides a rigid platform to support the copper circuitry through the manufacturing processes and maintains spatial integrity of the circuitry during lamination at high temperature and pressure.

Copper clad cores can also be made from thermoplastic films such as polyimide, LCP (Liquid Crystal Polymer), or PEEK (Polyetheretherketone). These cores can work well for providing a platform for the copper circuitry, but special care is required during any manufacturing process that requires high temperatures close to or above the melt temperature. As a result, copper clad cores made from thermoplastic films are often bonded or laminated to a material of lower melt temperature.

Dielectric films usually refer to thin flexible dielectrics. Resins used in making dielectric films may be thermosetting or thermoplastic. The film may also be reinforced using glass, organic fibers such as aramid, or an organic matrix/membrane such as expanded poly-tetrafluoroethylene (ePTFE). The most common flexible structures use thermoplastic, non-reinforced dielectric films such as polyimide, polyester, or LCP. Thermoplastic resins are typically cast into thin sheets. Thin glass-reinforced aromatic polyethers have been used in high-speed "flex-to-install" applications.

Adhesives are typically used to bond metal layers to dielectric films to form copper clad flexible substrates. These adhesives have a lower cure or bond temperature than the thermoplastic dielectric film. Adhesives are also used as bond films between processed flexible substrates in multilayer flex applications. A variety of adhesives are available such as polyester, acrylics, (modified) epoxies, polyimide, fluorocarbon, and butyral phenolic. They differ in temperature and

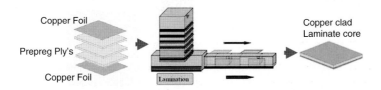

Fig. 8.2 Manufacturing of Laminate cores (courtesy of ITEQ)

chemical resistance, dielectric properties, flexibility, adhesion performance as a function of temperature, and cost. For the most part their cost and performance properties mirror those of the dielectric resins of the same chemical family.

Adhesives are widely used in conjunction with thermoplastic dielectric films. By selecting an adhesive with a cure temperature below the thermoplastic melt temperature of the dielectric film it is possible for bonding or lamination to occur without melting or distorting the thermoplastic film. This helps preserve the integrity of any circuitry formed on a metal clad dielectric film. Adhesives typically do not have the electrical or thermal properties of some dielectric films such as polyimide. As a result, copper clad "adhesiveless" polyimide or "all polyimide" materials are now available and are gaining importance. The copper clad "all polyimide" materials are not technically adhesiveless as a lower Tg polyimide layer between the polyimide core and the metal foil is used as a bonder.

Soldermask and Protective Coverlayers are applied to rigid and flexible PCBs as protection against moisture, mechanical damage, self-shorting, and contamination, or to improve flexural performance. They can be applied as "coverlays" (films) or coated as liquid "soldermask". Either type may be photo-imageable to create well defined openings in the coverlayer for component attachment. In flexible circuits, non-photoimageable coverlays (films) are punched or drilled to form the component-attach openings. Non-photoimageable liquid covercoats can also be screened on as flexible solder masks. The most common soldermask materials are photoimageable, UV curable, epoxy materials. Photo-imageable soldermasks are applied using a variety of methods such as roller coating, spray, or curtain coating.

8.1.1 Resin Systems

Resins used in PCB materials are selected based on their electrical, mechanical and thermal performance. Majority of the resins used in PCBs are thermosetting resins. And the primary thermosetting resins in use are epoxies. Other thermosetting resins such as Polyimide, Polyphenylene Ether, and Polyester are also used. These resins, once crosslinked, can not be remelted but will soften at high temperatures. Thermoplastic resins, such as PTFE, Polyamide, and LCP, are also utilized in some advanced PCB materials. Thermoplastic resins, with sufficient temperature, can be taken from solid to liquid and back to solid.

8.1.1.1 Epoxy Resins

Epoxy resin systems are the most widely used thermoset resin systems in printed circuit board applications. The epoxy resins are generally classified by the functional epoxide groups they contain. Difunctional epoxies contain two functional epoxide groups per molecule, tetrafunctional epoxies contain four, and multifunctional epoxies contain even greater numbers of functional epoxide groups per molecule. In general, epoxies with more functional epoxide groups per molecule cure with a greater degree of cross linking which typically results in higher glass transition temperatures (T_g). T_g is a common physical trait used to classify epoxy resin systems. There are three primary T_g ranges in which the epoxy resins systems are segmented, low T_g ~120–145°C, mid T_g ~150–165°C, and high T_g ~170°C and above. Higher T_g epoxies, commonly referred to as enhanced FR4 resin systems, have been developed to support thicker board constructions which require lower z-axis expansion properties for stable plated via reliability. The higher T_g materials are also used in thinner constructions which require greater thermal stability to prevent sag during assembly [18]. The primary drawback of higher T_g epoxies is an increase in flexural modulus which results in a more brittle material and degrades the mechanical performance and processibility of the material.

There are two additional elements of epoxy resin systems which play important roles, the curing agent and the flame retardant additive. The curing agent reacts with the functional epoxide groups of the epoxy molecule to form the polymer chain. Historically, dicyandiamide or "dicy" has been the most common curing agent for printed circuit board materials. Newer non-dicy curing agents have been developed to improve thermal stability for lead-free assembly, decrease cure time for faster processing, and reduce moisture sensitivity for improved electrical performance. Phenolic curing agents have been developed to increase the decomposition temperature (T_d) of the resin to aid in the survivability of material in lead-free assembly. Again, the significant drawback is the increased brittleness of the resin impacting its mechanical performance.

Epoxies are flammable materials, and pose significant risks to safety without the addition of flame retardants. Tetrabromobisphenol-A (TBBPA) has been the primary flame retardant used in printed circuit board materials for over 45 years allowing FR4 laminates to achieve a UL 94-V0 rating. Recently halogen-free flame retardants have been developed as replacements for TBBPA under the auspices of environmental friendliness. TBBPA has been grouped with other more environmentally hazardous brominated flame retardants on the basis it is a bromine containing substance. The scientific jury is still out on whether the new flame retardant materials are better for the environment than TBBPA, but market perception will continue to drive research in this area. The halogen-free flame retardants can be categorized in three main groups which demonstrate different primary fire retardant mechanisms. Phosphorous based compounds are the first group of halogen-free flame retardants, which act as char formers. Inorganic/hydrated fillers are the second group which evolve water and are endothermic. Fillers are not used as stand alone flame retardants, but combined with other flame retardants to achieve the desired result. Nitrogen based compounds comprise the third group which form an intumescent system which generates gas to extinguish the flame. There is no one dominant halogen-free flame retardant to replace TBBPA at this time. The myriad of different combinations of halogen-free laminates using phosphorous based, nitrogen based, or mixtures which can include fillers leads to concerns about the variability of the resulting material properties for functional design purposes. The "halogen-free" epoxy resin materials are typically more thermally stable with slightly higher T_g values.

Epoxies can be blended with themselves or other resin systems for the purposes of creating beneficial material properties and lowering cost of using a single alternate resin system [14]. Two of the most common epoxy/alternate resin blends for printed circuit boards are epoxy-polyphenylene oxide (PPO) and epoxy-cyanate ester. Both materials have improved electrical properties for high speed low loss product applications with dielectric constant (Dk) values ~3.5 and loss tangent (Df) values of ~0.01 for epoxy-PPO and ~0.007 for epoxy-cyanate ester.

8.1.1.2 Alternative Thermoset Resin Systems

There are several non-epoxy based thermoset resin systems used in the fabrication of printed circuit boards. The most common are polyimide, polyphenylene ether, and polyester.

Polyimide is chosen for its good flexibility, high temperature (200–240°C) compatibility (solder), low CTE, and dielectric properties. Cost and moisture absorption are drawbacks. However, advances in polyimide polymer compositions have yielded materials that have less than 1% moisture absorption, compared to standard polyimides that have 2.5–3% moisture absorption.

Polyphenylene ether (PPE) is chosen for its superior electrical properties and excellent thermal performance. PPE materials boast Dk values of ~3.6, Df values of ~.008, T_g values >220°C and T_d values >360°C. Processability of earlier formulations has been improved so the material can be processed by slightly adjusted conventional printed circuit board manufacturing practices.

Polyester is chosen for good flexibility, low CTE, good chemical resistance, and low cost, but polyester cannot be processed much above 100°C. However, processing and structural adaptations such as heat shielding, the use of heat-sink carriers, and point soldering may allow higher temperature processing.

8.1.1.3 PTFE

Fluorocarbons are chosen for good dielectric strength, low Dk, low Df, chemical resistance, low moisture absorption, but high cost, special processing requirements, and high CTE are drawbacks. PTFE (poly-tetrafluoroethylene) may be impregnated as an aqueous dispersion on woven glass and sintered at high temperature into a reinforced low loss dielectric. So called "expanded" PTFE can be impregnated with a B-stage thermoset resin to serve as a low loss prepreg layer for multilayer construction.

8.1.1.4 Thermoplastics

Several thermoplastics are available. Polyethylene Napththenate (PEN), Liquid Crystal Polymer (LCP), PEEK (Polyetheretherketone), PET (Polyethylene terephthalat) polyester, and polyamide. PEN, PEEK and LCP substrates are being offered for higher temperature resistance and enhanced dimensional stability vs. PET polyester. The PEN, PEEK and LCP are typically at a cost lower than polyimide.

LCP usage has increased in recent year due in part to its benefits of low water absorption, chemical resistance, low Df, low (or tailored) CTE, and its inherent flame retardency. Drawbacks to LCP include poor adhesion to copper, the inherent anisotropic CTEs in the x-y plane and the difficulty to form LCP structures with balanced CTEs. LCPs typically require special processing steps during via hole cleaning and electrolysis copper deposition due to its chemical resistance.

Interest in thermoplastic materials has increased in recent years for use in recyclable PCBs. The ability to melt thermoplastics during a recycling process also makes the materials difficult to process during PCB manufacturing. This property also makes it difficult to maintain or control the integrity of circuitry on thermoplastic laminate cores during multilayer or sequential lamination, as they either melt or become sufficiently softened resulting in a loss of structural/spatial integrity. In addition, drilling thermoplastic materials requires tighter process controls as the material is more sensitive to drill bit heating during processing. Thermoplastics with high melt temperatures, such as Polyamide, increase process costs due to the high process temperatures required for lamination. As a result, many high melt temperature thermoplastics utilize adhesives in forming copper clad laminate cores or in multilayer lamination. PCB assembly of thermoplastic PCBs, especially those with lower melt temperatures, may not be compatible with standard or lead-free assembly temperatures.

8.1.2 Reinforcement Materials

Reinforcement materials are used to mechanically strengthen the base resin system and provide a framework that improves the ability to handle the dielectric materials through the manufacturing processes. Different types of reinforcement materials exist and are selected based on their ability to provide the desired electrical, thermal, and mechanical properties of the final composite. Historically, the reinforcement material of choice has been woven glass fabrics as they are strong, relatively low cost and produced in continuous roll form. Other reinforcements used within the industry include chopped glass matte, Aramid® fibers, and expanded Teflon® carriers [9].

Paper based reinforcements also exist in FR-1, FR-2, and FR-3 laminates or used in conjunction with woven glass to form CEM laminates. These materials are typically used in very low cost PCBs and are not suitable for most multi-layer, plated through-hole applications.

8.1.2.1 Glass Fabrics

Glass fabrics as shown in Fig. 8.3 are formed by weaving glass yarns, consisting of glass filaments, into a sheet or fabric. These woven glass fabrics form a framework that provides strength to the resin and influences the mechanical and electrical properties of the composite material. The thickness of a glass fabric depends on the type and number of filaments in the warp and fill yarns as well as the yarns per inch in both the warp and fill of the fabric. The electrical, mechanical, and thermal performance of a glass fabric depends on the glass composition as well as the size and density of the warp and fills yarns. Typical copper traces in PCBs are similar in size to the size of glass yarns within glass fabrics. As a result, for electrical modeling of individual PCB traces, woven glass reinforced composites can not be treated as homogenous materials. This impact on electrical designs is discussed in Section 8.3.1.

There are several formulations of glass used to make reinforcement fabrics for PCB dielectrics. The most common is borosilicate electrical grade glass commonly known as E-glass. E-glass consists primarily of silica, calcium oxide, alumina, boron oxide, and alkaline oxides [1]. The industry definition of E-glass allows the percent weight of the primary ingredients to vary within a range which results in variations of electrical and mechanical properties. These variations will exist over a period of time and between suppliers of E-glass depending on the grade and source of the raw materials. As a result, the dielectric constant of E-Glass at 1 GHz can vary within a range of 5.9–6.4. Other glass formulations available include S-glass, R-Glass, T-Glass, D-glass and SI-glass (see Table 8.1). Due to a combination of market volume and physical properties that influence

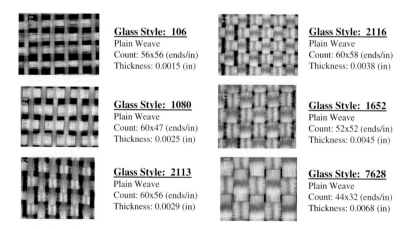

Fig. 8.3 Common glass fabric styles (Courtesy of Isola Laminates)

8 Advanced Print Circuit Board Materials

Table 8.1 Glass Formulation Comparison (Source ParkNelco)

	E-Glass	D-Glass	T-Glass	S-Glass	SI-Glass
Composition (%wt)					
SiO_2	52–56	72–76	62–65	64–66	52–56
CaO	16–25	0	0	0	0–10
Al_2O_3	12–16	0–5	20–25	24–26	10–15
B_2O_3	5–10	20–25	0	0	15–20
MgO	0–5	0	10–15	9–11	0–5
Na_2O	0–1	3–5	0–1	0	0–1
TiO_2	0	0	0	0	0.5–5
Dk / Er	5.9–6.4	~4	~6	~6	~4.4

manufacturing costs such as melt temperatures and ease of handling, these glass formulations are more expensive than common E-glass. S-glass, R-glass, and T-glass are each trade name formulations available by different companies and designed to have a higher structural strength than E-glass. D-glass and SI-glass were formulated to have lower dielectric constant and dielectric loss than E-glass for advanced electronic applications. SI-glass from Nittobo is available in some PCB laminates today, such as ParkNelco's N4000-13si, and provides a composite material with lower dielectric constant and lower dielectric loss.

The formation of the glass yarns as shown in Fig. 8.4. starts with blending the raw materials and then melting them in a high temperature furnace. The melt

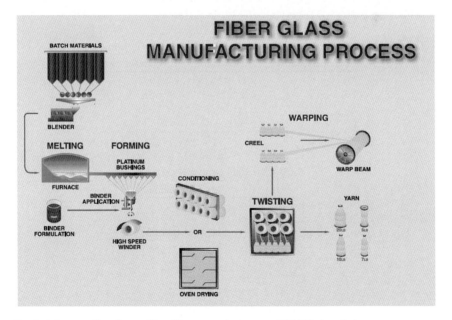

Fig. 8.4 Process flow for making Glass yarn (courtesy of PPG Industries)

temperature for E-glass is around 2600°F (1427°C). The molten glass is gravity fed through platinum bushings to form the filaments. The typical filament diameters used in common PCB reinforcements are 5–10 µm. These filaments are coated with a binder as they come out of the bushings to protect the filaments during subsequent processing. A typical binder consists of starch and oil which is not compatible with epoxy resins and must be removed by heat cleaning, high temperature ashing, after the yarn is woven into a fabric [7]. Coatings compatible with epoxy resins, such as Dielectric Solutions' DirectFinish^{TM} coating, have recently been developed that do not need removed when making laminate materials. These new coatings improve the overall strength and integrity of the glass as the high temperature ashing is not required.

After coating with a binder, the filaments are gathered into strands of yarn containing a specific number of filaments and wound onto tubes for baking. Baking drives off moisture and allows the binder to cure. The strands are then wound onto bobbins. Common filament designations and yarn definitions are given in Tables 8.2 and 8.3. The yarn is twisted as it is being wound onto the bobbin which improves weavability and helps the yarn maintain shape in the final woven fabric. Some enhanced glass fabrics use untwisted yarn which does not produce a tight yarn in the woven fabric [8]. The filaments in untwisted yarns tend to flatten and spread out within the woven fabric which creates a PCB laminate with unique functional properties (See Fig. 8.5).

The yarns are woven into fabric using air looms. Common glass fabrics used in PCB laminates are listed in Table 8.4. The typical weave pattern used in glass fabrics for PCBs is a plain weave which is an alternating one under one over pattern. This pattern is preferred due to its stability.

Table 8.2 Commonly used glass filament designations

Letter designation	Diameter (mils)	Diameter (µm)
D	0.21	5.33
DE	0.25	6.35
E	0.29	7.37
G	0.36	9.14

Table 8.3 Commonly used yarns

Yarn	Number of Filaments
D450	200
D900	100
DE150	400
DE300	200
E110	408
E225	200
G150	200
G75	400
G50	600

Table 8.4 Common glass fabric used in PCB laminates

Fabric style	Warp yarn (yarns per inch)	Fill yarn (yarns per inch)	Nominal fabric Thickness (in.)
106	D900 (56)	D900 (56)	0.0014
1067	D900 (69)	D900 (69)	0.0014
1080	D450 (60)	D450 (47)	0.0023
1500	E110 (49)	E110 (42)	0.0052
1652	G150 (52)	G150 (52)	0.0045
2113	E225 (60)	D450 (56)	0.0028
2116	E225 (60)	E225 (58)	0.0038
2165	E225 (60)	G150 (52)	0.0040
2313	E225 (60)	D450 (64)	0.0029
3313	DE300 (60)	DE300 (62)	0.0033
7628	G75 (44)	G75 (32)	0.0068
7635	G75 (44)	G50 (29)	0.0080

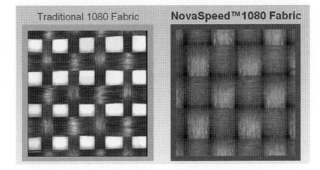

Fig. 8.5 Standard 1080 fabric compared to untwisted 1080 fabric (photo Courtesy of Dielectric Solutions)

Woven fabrics made from yarns with starch oil binders are then heat cleaned at temperatures around 1000°F (538°C) and then baked at around 700°F (371°C) for a prolonged time to remove all organic content. After glass fabric is free of organics it is then treated with a coupling agent to promote resin adhesion. Typically, the coupling agent is silane based and is tailored to the resin system with which it will be coated. Proper removal of the starch-oil and proper application of the coupling agent is essential to processability and reliability of resin coated glass woven dielectrics.

8.1.2.2 Chopped Fiber

Chopped fiber reinforcements made from either short sections of glass filaments or organic fibers are used in some PCB materials. The chopped fibers are formed in a paper-making process that parallels the production of paper from wood pulp as shown in Fig. 8.6. The chopped fiber reinforcements do not have the structural grid patterns associated with woven glass fabrics. As a result, the chopped fiber fabrics are more spatially consistent with respect to electrical

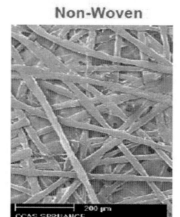

Fig. 8.6 Aramid paper (photo Courtesy of Dupont)

properties [6]. In addition, individual fiber strands are relatively short with no uniform orientation which can decrease susceptibility of fracturing in the laminate and suppress conductive anodic filament (CAF) growth.

Aramid paper reinforcement for resins is chosen for low CTE, good dielectric strength, dimensional stability and chemical resistance. High moisture absorption, high CTE in the z-axis, and medium cost are drawbacks. The ingredients are p-aramid (Kevlar®) floc, (the condensation product of p-phenylene diamine and terephthalic acid), and so called "fibrids" made of m-aramid (Nomex®), (the condensation product of m-phenylene diamine and isophthalic acid). After the paper sheet formation, the structure is densified in a calendaring operation. Thermount® is DuPont́s brand applied to laminate and prepreg, reinforced with aramid paper and manufactured by licensed laminators.

8.1.2.3 Specialty Reinforcements

Other carriers and reinforcements exist that provide a laminate with unique properties.

Expanded Teflon® is used in a family of products identified as Speedboard® and Microlam® available from W.L. Gore. The expanded Teflon® carrier is formed as a thin film with the Teflon® creating a sponge like membrane. The expanded Teflon® carrier is then impregnated with a thermoset resin which coats the membrane and fills in the air space within the membrane. The composite materials has excellent electrical properties due to the inclusion of the Teflon® carrier and can be used as a prepreg in conjunction with other PCB materials or laminated between copper to create a copper clad core. In mixed dielectric PCBs, the Speedboard® prepreg is often selected to bond low-loss PCB materials such as Roger's RO3000® or RO4000® series laminates as the expanded

Teflon prepreg has similar electrical properties. As a result, this material has found application in various RF markets.

Graphite or carbon filaments are also used as reinforcement in PCB materials. These laminates such as the ST10 and others from STABLCOR provide very high thermal conductivity along the x-y plane of the graphite reinforcement. STABLCOR material is made of a carbon fiber composite. Carbon fibers have unique thermal and mechanical property which can be very useful for the Printed Circuit Board application. Carbon fiber composites have very good thermal conductivity, have very low CTE (Co-efficient of Thermal Expansion), have very high Tensile Modulus (helps to increase stiffness) and are light weight. Carbon fibers are electrically conductive, so laminate made out of carbon fiber results in an electrically conductive laminate. Due to the electrical conductivity, these laminates are often used as a ground plane within the multilayer printed circuit boards as shown in Fig. 8.7. Carbon composite laminates are often used as alternatives to metal core boards as they provide excellent thermal properties. Isolation for non connecting vias must be obtained by drilling larger clearance areas and filling with a dielectric prior to via formation, see Fig. 8.8. Also these materials must be used in a symmetrical manner in order to maintain flatness of the PCB.

8.1.3 Fillers

Fillers are small particles added to the PCB laminate resin system to alter the performance properties of the material. Virtually all basic material performance attributes can be enhanced, including thermal, mechanical, and electrical properties. Some of the more common filler materials include glass, silica, ceramics,

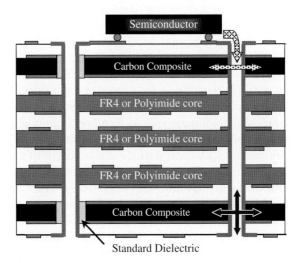

Fig. 8.7 PCB with carbon composite cores for improved thermal path (courtesy of STABLCOR Inc.)

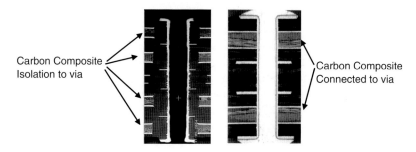

Fig. 8.8 Cross-section of isolation and connection via using carbon composite materials (courtesy of STABLCOR Inc.)

metal oxides, graphite, talc, and metal hydrates. The thermal enhancements made by fillers include increasing the thermal conductivity of the laminate. Fillers can also be used to retard the flammability of the laminate in halogen-free applications.

A common usage of fillers in PCBs is to enhance their thermo-mechanical properties for reliability. Fillers are often used to lower the CTE values of a laminate. The industry transition to lead-free processing has increased the use of fillers for this purpose to offset the increased thermal expansion demands placed on the material during the assembly process. Mechanically, some additional uses include increasing the stiffness of the material, and arresting crack propagation to aide in CAF performance.

Electrically, fillers are added to adjust the laminate properties in many different ways. Some of the common uses are to increase capacitance, lower the dielectric loss, and increase the dielectric constant of the material. Barium Titanate ($BiTiO_3$) has a dielectric constant >4400 and is added in small amounts to tailor a materials dielectric constant. Examples include its use in PTFE/ceramic composites such as the RT/duroid® 6000 and RO3000® series laminates from Rogers Corporation. For example, the RO3000® family of dielectric materials is offered in dielectric constant values of 3.0, 3.5, 6.15, and 10.2.

The addition of fillers to laminates does not come with out penalty. Cost is usually negatively impacted due to the cost of the filler itself, or from an increase in processing costs to manufacture the material and fabricate the printed circuit board. Fillers require special blending techniques to insure proper dispersion throughout the resin when fabricating the material. Manufacturing fillers with controlled or uniform particle size is very important to the manufacture and functionality of the laminate. Filled materials impact the printed circuit board fabrication process at several key process steps. Lamination process parameters need to be adjusted to allow for slower flow rates due to the increased filler content. Fillers can increase wear on drill bits and require adjustments to feed and speed rates at drilling. Surface conditioning of the filled material needs to be considered for all plating steps to insure optimum plating quality.

8.2 Conductive Materials

Conductive materials are used to form the electrically conductive elements of the PCB as well as provide a conductive finish to the exposed PCB pads and traces. Copper is the primary conductive material used to form the electrical circuits in a PCB due to its relatively low cost, high conductivity, processability and stability. Other metals such as gold (Au), silver (Ag), nickel (Ni), tin (Sn), lead (Pb), etc. are used as surface finishes to provide a solderable surface for component assembly. Metals and metal alloys such as chromates and brass are used as passivation layers to prevent corrosion or improve the adhesion between materials. Gold is also used on exposed contacts such as card edge connectors and key/button switches. Conductive materials such as graphite, Palladium and even conductive polymers are also used as seed layers to enable electroplating of copper to dielectric materials when metallizing PCB vias. Highly resistive materials are deposited as thin-films or screen printed as pastes to fabricate resistive elements that can be either embedded into the PCB or used on the surface.

8.2.1 Copper Foils

Thin copper foils are used as the base cladding for laminate cores, dielectric films, and as the external cladding during multilayer PCB lamination. The circuit features are then formed into the copper foil by removing the unwanted copper through a print-etch process. The unwanted copper can also be removed with ablation or mechanical methods; but the print-etch remains the most cost effective in volume and least damaging to the base dielectric. When forming circuitry with a semi-additive electroplating process, as done when forming the outer layers of a PTH PCB or the build-up layers of HDI PCBs, the copper foil provides the common cathode connection across the manufacturing panel.

Copper foil is made by either the electrodeposition of copper from a copper electrolytic solution or rolling copper ingots into thin sheets. Prior to being used as PCB copper cladding, the copper foils are treated to promote adhesion, improve processability through the PCB manufacturing process and prevent oxidation.

Copper foil can be obtained in a variety of thicknesses. Copper thickness is usually specified in ounces (oz.). The thickness of one ounce of copper is defined by the thickness obtained from a uniform copper sheet of one square foot weighing one ounce. The typical thickness in microns for copper foil after manufacturing and treatment is provided in Table 8.5.

Ultra thin copper foils, copper foils less than 0.25 oz (9 μm), are typically not produced as free standing foils; but, are manufactured attached to a carrier. The ultra thin copper foils, such as Gould TCU®, Oak-Mitsui MicroThin®, and OlinBrass XTF®, are plated onto a thick copper carrier foil such as shown in

Table 8.5 Common copper foil thicknesses as received (IPC 4562L)

Foil Designator	Nominal thickness (mils)	Nominal Thickness (μm)
0.25 oz (9 μm)	0.34	8.5
0.5 oz	0.68	17.1
1.0 oz	1.35	34.3
2.0 oz	2.70	68.6

Table 8.6 Design impact on material reliability

Design factor	Direction	Reliability Impact
Via Size	Smaller	Drops
Via Pitch	Smaller	Drops
Layer Count	Higher	Drops
Copper Weight	Greater	Drops
Board Thickness	Greater	Drops
Board Size	Larger	Drops

Fig. 8.9. The ultra thin foils have a release barrier between them and the copper carrier to enable separation once bonded or laminated to a dielectric.

8.2.1.1 Electrodeposited (ED) Copper

Electrodeposited (ED) copper foils are made by electroplating the copper out of a copper electrolytic solution onto a rotating drum as shown in Fig. 8.10. The thickness of the copper foil can be set by adjusting the speed of the drum and the current density of the electroplating process. The current density, as well as organic and inorganic additives for stabilizing and leveling the copper deposition, determines the grain structure and surface profile of the untreated copper foil. The grain structure of a copper foil contributes the foils' ductility, CTE/elongation, and bulk electrical resistance.

The ED process yields differences between the two foil surfaces. The drum side or shinny side surface has a profile that matches the surface topography of

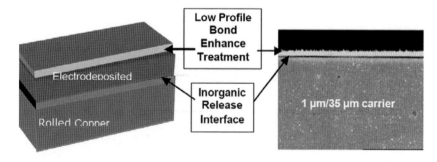

Fig. 8.9 Ultrathin copper foil with copper carrier (courtesy of Olin)

8 Advanced Print Circuit Board Materials

Fig. 8.10 Copper foil manufacturing process (courtesy of Gould)

the drum. The topography of the matte side depends on chemical makeup of the copper electrolytic solution and deposition rates.

8.2.1.2 Rolled Annealed Copper

Rolled annealed copper foil is made by roll milling copper into thin sheets. The process yields a smooth copper foil with a grain structure that is in the plane of the foil sheet. This results in a copper that has very good bending properties and therefore well suited to flexible circuit applications. Rolled annealed copper foils have also been used in RF Microwave applications due to its smoothness which provides lower conductive loss at high frequencies. It should be noted that ED copper foils are starting to gain acceptance in RF Microwave and high speed digital designs due to recent advances in producing low profile and smooth ED copper foils. Rolled annealed copper foil has the disadvantage of cost, relative to ED foils, in making thin foils. The cost of rolled annealed foils increases as the foil becomes thinner as more rolling time and energy are required. ED foils, on the other hand, require less time and energy to produce as they become thinner.

8.2.1.3 Copper Foil Treatments

Copper foils are treated in roll form, see Fig. 8.11, through several processes to improve their reliability and processability through the PCB manufacturing process. The treatment consists of several processing steps that modify the surface profile of the copper foil, see Fig. 8.12. Nodulation through either mechanical roughening or chemical roughening or leveling is done to achieve a desired surface topography. Surface roughness is important to promote adhesion to dielectric materials or to photo resists during PCB fabrication. Excessive surface roughness can be detrimental to both electrical conductor losses at high signal frequencies and to controlling the print-etch process when forming fine PCB features.

Fig. 8.11 Copper foil treatment process (Courtesy of Gould)

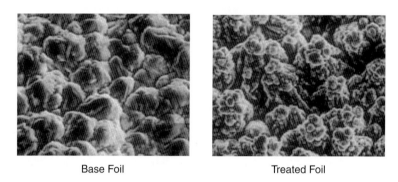

Base Foil   Treated Foil

Fig. 8.12 Copper Foil before and after treatment (courtesy of Gould)

A second treatment step involves applying a brass barrier layer to retard the thermally and chemically induced degradation of the foil during processing. The brass barrier layer is typically between 800 and 1000 Å. A third treatment step involves depositing a combination of chemical coupling and passivation agents onto the foil. These treatments ensure optimum conditions for bonding between foil and prepregs while preventing oxidation and discoloration of the foil during processing. The coupling agents include silane, Chromium, Zinc, etc. and are usually tailored to the resin system to which the foil will be laminated.

Advanced copper foils come with multiple types of treatments and most always use a different treatment on the drum side versus the matte side to provide optimized solutions. One example is Gould RTC™ foil shown in comparison to standard foil in Fig. 8.13. The reverse treated copper foil is manufactured by

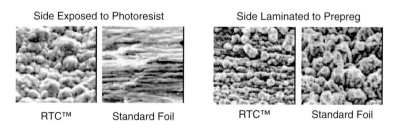

Side Exposed to Photoresist   Side Laminated to Prepreg

RTC™   Standard Foil   RTC™   Standard Foil

Fig. 8.13 Example of copper foil treatments (courtesy of Gould)

applying copper nodularization, brass thermal barrier and passivation to the shiny smooth side of the foil rather than to the roughened matte side of the foil as is done when manufacturing conventional or standard copper foil. A thin layer containing only passivation or antioxidants, which is normally applied to the shiny smooth side of standard foils, is applied to the roughened matte side of reverse treated foil. The fully treated side is laminated to a prepreg leaving the roughened matte side of reverse treated foil available for innerlayer processing. As a result, the RTC^{TM} foil has a lower profile copper surface laminated to the prepreg which improves the conductor loss of high speed and RF microwave signals and improves fine line etch control during processing. The RTC^{TM} foil also has a surface that promotes higher adhesion to photoresists which improves processability and yields when etching finer geometries during PCB fabrication.

8.2.1.4 Copper Foils for Buried Resistors

Copper foils have been developed that include a low conductivity or resistive layer on the side that is laminated to the dielectric material. This allows the creation of buried resistors to be embedded into the PCB by utilizing two print-etch processes to a PCB metal layer. There are several different products on the market. Ohmega-Ply material has been on the market for more than 20 years and is made by electro-depositing a thin-film, 0.1–0.4 µm, of Nickel-Phosphorous (NiP) alloy onto the matte, or tooth side, of ED copper foil. Gould TCR^{TM} consists of a thin-film, 0.01–0.1 µmmicron thickness, of Nickel Chromium (NiCr) or Nickel Chromium Aluminum Silicon (NiCrAlSi) deposited on the matte side. Other similar products exist. All of these products require a minimum of two print-etch processes. The first print-etch step removes both the copper and resistive layer from the combined conductor and resistor image. The second print-etch selectively removes the copper from the embedded resistor areas. The resistive thin-film remains under all conductors formed in these types of copper foils and their impact must be taken into consideration due to the skin depth of high speed, RF microwave signals. Figure 8.14 shows a top view and cross-section view of a resistor formed using Gould TCR^{TM}.

8.2.2 Surface Finishes

Surface finishes are used on the exposed copper circuit areas of the printed circuit board to prevent oxidation and corrosion of the copper prior to assembly soldering operations. They can also serve as wearable non oxidizing contact surfaces for multiple insertion or repeated electrical connections. Historically, some form of solder was used as the printed circuit board surface finish. This took one of two forms, reflowed solder plating and hot air solder leveling

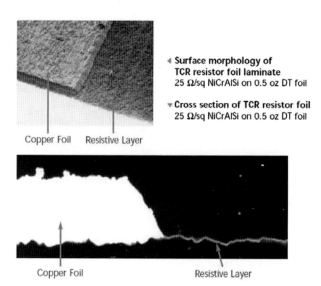

Fig. 8.14 Resistor formed using specialty copper foil (courtesy of Gould)

(HASL). Both methods provided exceptional environmental protection for the underlying copper and resulted in very receptive surfaces for subsequent soldering operations. The primary draw backs were the lack of co-planarity of the surface finish for fine pitch assembly and the stress on the laminate material from the extra thermal excursion of the surface finish process which degraded the reliability performance of the board. More recently, 2006 environmental regulations banning lead from printed circuit boards has driven the conversion of printed circuit board finishes away from the traditional lead containing HASL and solder plate to alternative non-lead containing finishes. These finishes include organic solderability preservative (OSP), immersion silver (ImAg), electroless nickel immersion gold (ENIG), immersion tin (ImSn), lead-free HASL, and many others. Many of these surface finishes were already in use in the industry as alternatives to HASL for fine pitch/small component applications requiring better surface co-planarity.

Gold surface finishes are the surface finish of choice when needing a wearable contact surface for mechanical connections to the printed circuit board. Historically, electroplated nickel and hard gold have been used for this purpose. The nickel plating provides a durable base surface and barrier to prevent the underlying copper from migrating into the surface gold layer and increasing contact resistance. The gold provides a non-oxidizing contact surface with good lubricity for the mating contact surface. Newer forms of nickel gold surface finishes have come into vogue to improve on the short comings of electroplated nickel gold. These surface finishes are electroless in nature. The shortcomings of the electrolytic nickel gold finishes include the presence of exposed copper circuitry when using the gold finish as an etch resist, the need for direct connections to all circuits requiring gold when plating after circuit etch, and the lack of uniformity of the plating which leads to excessive gold usage and higher costs. The thicker gold also reduces the reliability of any solder connections to the gold surface due to gold embrittlement of the solder joint. Electrolytic requires an applied current to drive the reduction of the plating solution ions to their metallic state on the surface of the conductors of the printed circuit board. Electroless plating is a chemical reduction reaction which reduces the plating solution ions to their metallic state at the surface of the conductors. Immersion plating is a substitution reaction trading electrons from the printed circuit board conductor surface to the plating solution ions. The immersion process oxidizes the conductor surface and replaces it by reducing the plating solution ions.

The primary surface finish types in use today are:

HASL – Legislation mandating lead-free products has drastically reduced the usage of HASL as a surface finish. There are still exemptions granted for certain products (military, aerospace, and high end computing) to use HASL and lead in their assemblies. The solderability and reliability of HASL are well understood. The primary drawback besides environmental legislation is the non co-planar surface of the finish makes it difficult to use for fine pitch and small component soldering.

Immersion Silver – The benefits of ImAg are its low cost, simple process flow, co-planar surface for fine pitch and small component soldering, easy measurement and inspection, reworkability, and wide assembly process window. The primary drawbacks for ImAg are the fact that it tarnishes and is susceptible to microvoids, corrosion, and silver migration [4].

Organic Solderability Preservative – OSP is the cheapest surface finish to apply in bare board fabrication. Other benefits include its co-planarity for fine pitch and small component soldering, and its reworkability. The primary drawbacks for OSP are its narrow assembly process window, difficulty in inspecting for defects, and non conductive nature for electrical testing.

Immersion Tin – ImSn is co-planar, has good lubricity for press fit connector pin insertion, and is reworkable. Its drawbacks include soldermask attack, susceptibility to corrosion and whiskers, reduced shelf life, and hazardous thiourea component.

Electroless Nickel Immersion Gold – The benefits of ENIG are its superior corrosion resistance, low resistance contact surface, compatibility with low cycle count mechanical contacts, surface co-planarity, via reinforcement, and no copper dissolution in lead-free assembly. The primary drawbacks are high cost, susceptibility to black-pad, more brittle solder joint interface, signal loss due to the nickel layer in high frequency RF applications, and it is non reworkable.

8.3 Electrical Considerations of PCB Materials

The electrical properties and electrical performance of PCB materials vary between resin systems, type of reinforcement used in a dielectric, the type and style of the copper foil, and even the selected surface finish. In the design of high speed digital systems and RF microwave applications the conductive circuits of the PCB are transmission lines. As a result, the choice of dielectric materials and conductive materials used in a design impacts the dielectric loss, conductor loss, propagation velocity, and dispersion of the transmission line. As most PCB dielectric materials are composites, they have both bulk and spatial electrical properties that will impact performance. The electrical properties of a PCB transmission line are also dependent on the fabrication processes used when making the PCB. A material's response to its environment, such as moisture diffusivity and moisture concentration, can also impact the performance of a design.

The electrical performance of advanced materials are often compared to a standard FR4 baseline which consists of standard copper foil laminated to a core of woven E-glass reinforcement impregnated with unmodified FR4 epoxy resin. Therefore, the electrical performance of a design using baseline of standard FR4 can be improved by changing one or more of its components. The copper foils can be changed to improve conductor loss. The reinforcement can be changed to decrease the dielectric loss or improve the spatial dielectric constant. The resin can be modified to lower the dielectric loss and reduce the dielectric constant. Figure 8.15 shows a schematic of some available options.

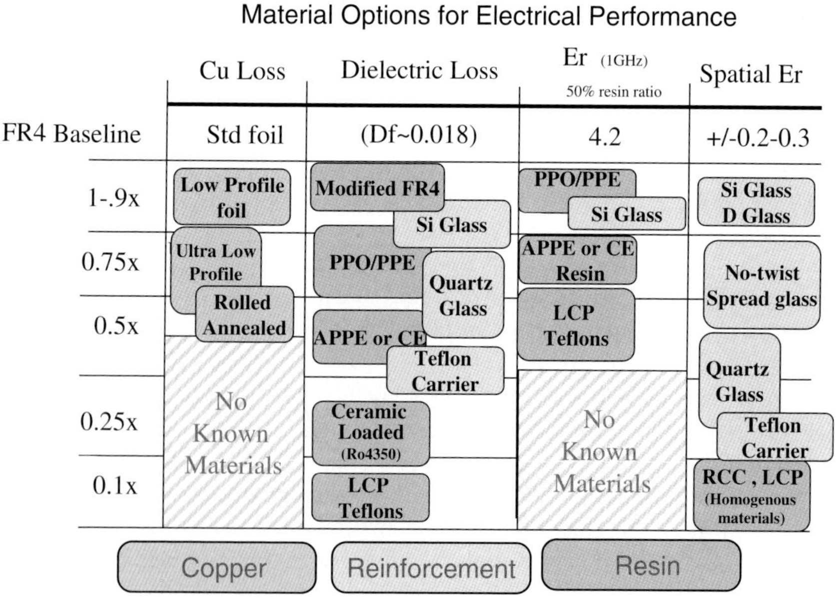

Material Options for Electrical Performance

	Cu Loss	Dielectric Loss	Er (1GHz) 50% resin ratio	Spatial Er
FR4 Baseline	Std foil	(Df~0.018)	4.2	+/-0.2-0.3
1-.9x	Low Profile foil	Modified FR4 / Si Glass	PPO/PPE / Si Glass	Si Glass / D Glass
0.75x	Ultra Low Profile	PPO/PPE / Quartz Glass	APPE or CE Resin	No-twist Spread glass
0.5x	Rolled Annealed	APPE or CE / Teflon Carrier	LCP Teflons	
0.25x	No Known Materials	Ceramic Loaded (Ro4350)	No Known Materials	Quartz Glass / Teflon Carrier
0.1x		LCP Teflons		RCC, LCP (Homogenous materials)

Copper — Reinforcement — Resin

Fig. 8.15 Sample options to improving FR4 electrical performance

8 Advanced Print Circuit Board Materials

Until recently, the primary method of choice to improve electrical performance in PCB materials was to alter the resin system. But, high frequency designs can be impacted by spatial Er which is not generally addressed when changing to a different resin. The dielectric loss benefit of changing resin can be diminished if the selected resin requires a higher copper profile to maintain adhesion.

8.3.1 Dielectric Constant

The dielectric constant of material systems reinforced with woven glass fabric is a function of the resin content. In these material systems the dielectric constant is not uniform across all laminate thickness as thickness is dependent on the selected glass fabric and the resin content. Common resins used in PCB materials have a lower dielectric constant than the glass used for reinforcement. Figure 8.16 shows the relationship between resin content and dielectric constant for Nelco's N4000-6 material using E-glass reinforcements.

The reinforcement used in thinner cores is made from smaller, finer glass yarns which results in a higher resin content and lower dielectric constant than thicker cores. The relationship of thickness and dielectric constant is not linear as there are multiple combinations of prepregs that will yield a given laminate thickness. For example, a high resin content 2113 or low resin content 2116 can be approximately the same thickness yet have a different dielectric constant due to different ratio of resin and glass. Figure 8.17 shows a typical relationship between dielectric constant and core thickness. In general, the thicker cores have higher dielectric constant. For cores above 12–13 mils in thickness, the dielectric constant is fairly constant as most of the thickness consists of multiple plys of a thick glass fabric such as 7628. The thick glass fabrics are utilized as they have a lower cost per weight and thickness as they use fewer weaving picks per inch and maximize thickness by utilizing larger yarns.

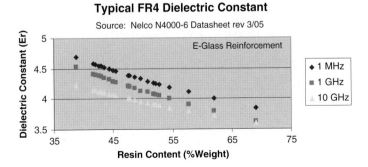

Fig. 8.16 Dielectric constant in resin, glass composite materials

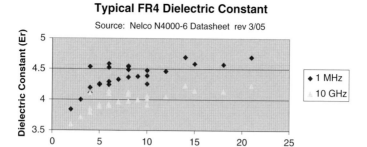

Fig. 8.17 Example of dielectric constant variation by core thickness

The variation in dielectric constant between cores of different thickness lead to high speed signal timing issues in multilayer PCBs as different fabrics can be selected for each layer in order to meet a final board thickness or other requirements. Even small differences between individual layers within a PCB design can impact signal timing. For example, in a stripline configuration the difference between a dielectric constant of 3.6 and 4.0 results in signal propagation difference of ~8.7 ps/inch. A class of advanced PCB materials, such as RO4000® series or N4380-13RF, uses only one or two types of glass fabric to ensure that the dielectric constant is same across all available thicknesses.

Homogenous materials such as pure resin or composite blends of resin and small particle fillers also yield PCB laminates with uniform dielectric constant across all available laminate thicknesses. Typical fillers include ceramic, areogels, silica, etc. Examples of near homogenous materials include TMM® which is a ceramic loaded thermoset resin and RO3000 series which is a ceramic loaded PTFE material.

The dielectric constant difference between resin and glass also lead to spatial variations across a PCB material made with woven fabrics. The fabric consists of a grid of glass yarns. As seen from the cross-section in Fig. 8.18, the region along a glass yarn results in a low resin content and the area between two glass yarns is a resin rich area. Manufacturing processes and common design practices result in a high occurrence of traces running parallel with the glass yarns. The yarn size and yarn spacing is larger than most common traces widths of 3–5 mils. As a result, the dielectric constant and propagation velocity of individual traces on a PCB will vary depending on how they are oriented relative to the glass matrix and whether the trace is directly over a glass yarn or in between two yarns. This difference can result in spatial variations in dielectric constant of 0.3–0.4 when using standard E-glass reinforcements [3]. This results in both impedance variations between individual traces and trace segments as well as differences in propagation velocity between different traces on the same dielectric layer. Figure 8.19 shows the impedance and measured effective dielectric constant for 64 equally spaced parallel traces. Homogeneous

8 Advanced Print Circuit Board Materials

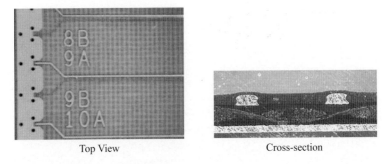

Fig. 8.18 Trace alignment to Woven glass fabric

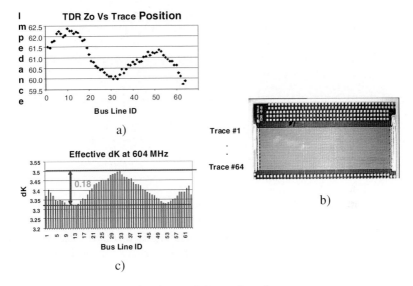

Fig. 8.19 Spatial Zo, dk variation for parallel grouping of traces

materials such as LCP and Polyimide or very fine blend composites such as Ro3003® or TMM® will not have spatial variations of the dielectric constant. Random fiber reinforcements, such as Aramid fibers will also significantly reduce spatial variations. New reinforcements such as SI glass with a lower glass dielectric constant or non-twist, spread glass fabrics have been shown to also reduce the variation.

The dielectric constant of PCB materials is not constant over frequency resulting in signal dispersion. For most all materials the dielectric constant decreases over frequency. Figure 8.20 shows the frequency dependency of random traces on a FR-4 core using 1500 fabric and a FR-4 core using 2113 fabric. As is common, the dielectric constant decreases with frequency. The difference in variation between the two fabrics is due to the spatial differences of

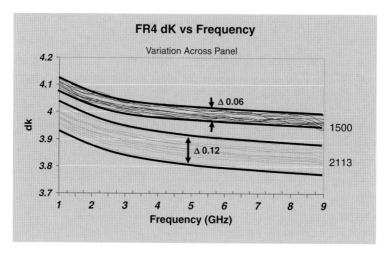

Fig. 8.20 Sample measurements of dielectric constant over frequency for 1500 and 2113 cores

the two fabrics. Some of the specialized RF materials will have a flatter response over frequency.

8.3.2 Dielectric Loss

The dissipation factor (df) of dielectric materials is a key differentiator between standard materials and advanced materials for RF applications and high speed digital designs. As electromagnetic signals propagate through a dielectric their strength is attenuated. The rate of attenuation is specified by the dissipation factor. Materials with lower df values attenuated signals less than those with higher df values. The attenuation over a fixed distance is related to the number of oscillations that occur in that distance. As a result, the dielectric loss per unit length increases approximately at a linear rate with frequency and becomes a critical element in high frequency designs. In RF designs, high signal attenuation results in need for higher power levels, added amplification, and higher thermal dissipation. Many RF designs for telecommunications and space applications rely on low loss (df < 0.005) to simplify designs and reduce power requirements. In high speed digital designs with broad frequency spectrum, attenuation also results in signal distortion as the content at higher frequencies will attenuate more per unit length than lower frequency content.

Standard FR4 has a df value in the range of 0.017–0.019 at ambient conditions. Mid loss materials, such as modified epoxies and epoxy blends such as PPO have df values in the range of 0.010–0.015. Recent development of lower loss epoxies and lower loss glass such as SIglass^{TM} have resulted in a few woven glass reinforced epoxy materials with df in the low loss range of 0.005–0.010.

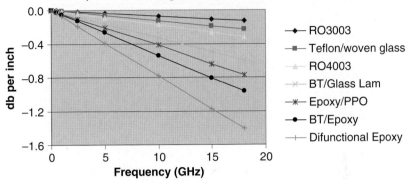

Fig. 8.21 Signal loss between materials of different dissipation factors (Courtesy of Rogers corporation)

Most all of the very low loss materials (df < 0.005) have high level of fillers, utilize PTFE, LCP or other very low loss dielectrics. Figure 8.21 shows the signal attenuation per inch comparison of a few selected materials.

For a variety of factors, the cost of PCB materials increase as the dissipation factor decreases. Mid loss, low loss, and very low loss materials do not have the volume demand of standard FR4 and most have proprietary formulations. In addition, many of low and very low loss materials use base materials or fillers that require different processing than standard FR4 or modified FR4 epoxy materials. For example, PTFE and LCP require different electrolysis and lamination cycles while addition of ceramic fillers reduce drill bit life.

8.3.3 Moisture Impact on Electrical Properties

The dielectric constant and dissipation factor of materials changes with temperature and moisture concentration. The dielectric constant and dissipation factor increases with temperature and with moisture concentration. The change in a material's electrical properties due to moisture depends on the moisture saturation level of the material and the material's moisture diffusivity. Some materials such as LCP have very low moisture diffusivity rates and are thus fairly stable across the humidity range as seen in Fig. 8.22. Other low loss materials such as Polyimide have moisture saturation which is higher than FR4. As a result, Polyimide will see a larger percentage change than FR4 in high humidity environments.

Figure 8.23 shows the percent change in dielectric constant for a 2116 FR4 structure at both the dry and saturated conditions. The PCB temperature

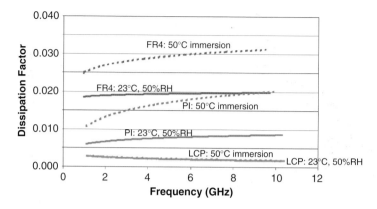

Fig. 8.22 Impact of moisture absorption in PCB dielectrics (Courtesy of Rogers corporation)

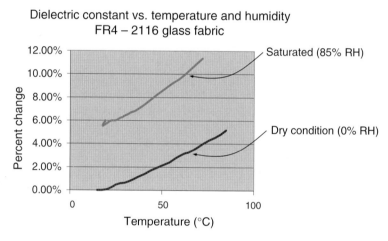

Fig. 8.23 Change in FR4 dielectric constant due to temperature

responds fairly quickly to changes in the operating environment. The time it takes for an electrical structure in a PCB to be affected by a change in relative humidity depends on the structure and moisture diffusivity. Standard FR4 and other common epoxy materials have moisture diffusivities in the range of 1.0×10^{-7} mm^2/s for heavy glass fabrics to 2.0×10^{-6} mm^2/s for finer glass fabrics [12]. As a result, ustrip and embedded ustrip transmission lines will be susceptible to changes in relative humidity over the course of minutes or hours. Stripline structures with adequate ground planes that impede the ingress of moisture can take hours or even weeks to see significant changes due to changes in relative humidity.

8.3.4 Conductor Loss

The conductor loss associated with copper traces is important to high speed designs. The conductor resistance increases from a redistribution of the currents to the outer regions of a conductor cross-section as frequency is increased. In smooth conductors, the relationship of the current density and frequency is known as the skin depth. In a smooth conductor, 67% of the current flowing in the conductor resides in the region that is one skin depth from the surface. This skin depth reduces inversely with the square root of frequency and translates into a resistance that then increases with the square root of frequency. As the signaling frequencies in today's electronic designs increase into the GHz range, the skin depth approaches the value of the copper foil roughness (see Fig. 8.24).

Copper foils are typically roughened to promote adhesion. The surface roughness of the copper foils is dependent on the surface treatments applied to the foil. Figure 8.25 shows the differences in the magnitude and spatial distribution of surface roughness that can exist between various copper foils. The peak to valley roughness of common ED copper foils are in a range of 5–10 μm with an average roughness around 0.5–1.0 μm. As a result, standard copper foils do not behave as classical smooth conductors in high frequency applications.

Figure 8.26 shows an example of the measured difference in transmission line loss between a 5inch trace design fabricated with different copper foils. Both the peak roughness and spatial density of the peaks impact the transmission line loss. For example, the RTCHP and RTC foils have approximately the same peak to valley magnitude; but, the RTCHP has a significantly higher transmission line loss due to a higher spatial density of the peaks. As a result, the selection of copper foil can have a substantial impact on high frequency designs for RF, telecom, or higher speed digital applications. The difference in transmission line loss between rough and smooth copper can be similar to the difference between standard FR4 and mid-loss (tand 0.010–0.015) materials [2].

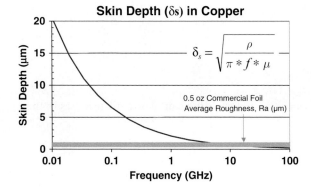

$$\delta_s = \sqrt{\frac{\rho}{\pi * f * \mu}}$$

Fig. 8.24 Skin depth in copper as a function of signaling frequency

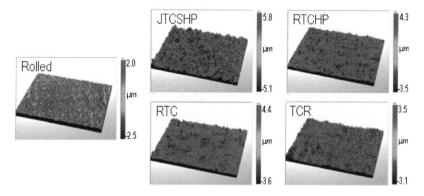

Fig. 8.25 Optical profiles of various copper foil surfaces

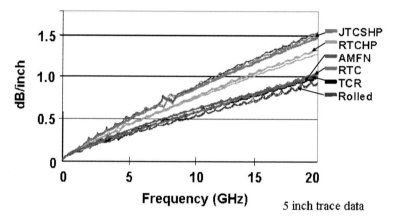

Fig. 8.26 Conductor loss comparison of different copper foils

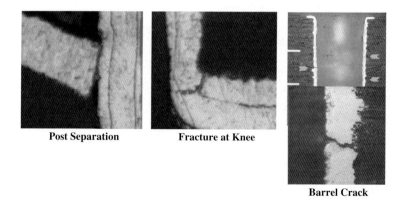

Fig. 8.27 Failure modes in PCB plated vias

 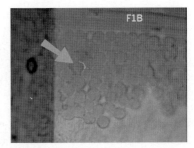

Fig. 8.28 Cross section of CAF failure in PCB dielectric

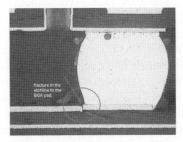 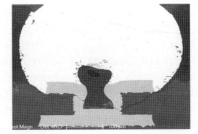

Fig. 8.29 Open failures due to pad cratering

Much work has recently gone into the proper modeling of high frequency signal propagation on rough copper foils [11].

8.4 Reliability Considerations of Printed Circuit Board Materials

Material reliability is a function of several factors: fabrication, design, assembly, and usage environment influences. Fabrication influences include proper storage and handling of the material, as well as quality fabrication operations. Proper storage prevents premature aging of and moisture adsorption in the material which can lead to fabrication defects such as delamination and blistering. Handling improperly can fracture the glass reinforcement and lead to long term reliability issues such as conductive anodic filament growth. Fabrication quality can impact many reliability concerns from via reliability to conductive anodic filament growth to pad cratering to solder joint reliability. Proper prep and curing in lamination, drill and plating of the through holes, and application of the surface finish can all impact the printed circuit board's ability to functionally perform over time. Printed circuit board design is the basis for thermo-mechanical stress influences. Minimum via size, via pitch, layer count, copper weight, board thickness, component bill of materials, and overall size all impact a material's ability to withstand the thermo-mechanical stresses imparted upon it. The influence of printed circuit design on reliability is summarized in Table 8.6

Assembly impacts involve proper storage and handling, the number of assembly operations, and the temperature at which they occur. As previously stated for fabrication, proper storage to prevent moisture adsorption prevents assembly defects which can impact the printed circuit board's long term functioning. Proper storage also prevents unwanted exposure of the printed circuit board's surface finish to oxidizing or corrosive elements which might impact the quality of the surface for soldering or electrical contact. The advent of lead-free legislation has increased assembly temperatures 30C over previous tin-lead soldering operations. The number of soldering operations (including rework) stress the material's thermo-mechanical performance to an even greater degree than previously seen with tin-lead assembly [15].

Usage environment can impact a material's reliability in many ways. Mechanical stresses such as shock and vibration can fracture the dielectric or the conductor materials leading to potential opens or shorts in the printed circuit board. Mechanical stresses include shipping, handling, and operation of the printed circuit board. Temperature extremes, either from the environment or operating conditions, can degrade the thermo-mechanical reliability of the material. Humidity and corrosive environments can adsorb moisture into the material and create corrosion and conductive filament growth leading to potential opens and shorts. We will discuss some of the more prominent reliability concerns and what material properties play an important role in mitigating risk to those concerns.

8.4.1 Via Reliability

Via reliability defects take one of 3 forms: barrel cracking, knee cracking, or post separation (Fig. 8.27). Barrel cracking and knee cracking are copper plating fractures where the repeated z-axis expansion and contraction have work hardened the copper plating and caused an open circuit. Post separation can result from one of three sources, drill smear at the copper innerlayer junction or brittle electroless copper, or contamination from the plating operation at the innerlayer junction.

Via reliability is a function of the via's plated barrel to adsorb the CTE mismatch stresses when the printed circuit board undergoes thermal cycling due to the environment or product operation. The metallic structures of the printed circuit board expand at a different rate vs the dielectric material. Tailoring the CTE properties of the dielectric to more closely match those of the copper reduces the expansion stresses on the copper [10]. Improving the tensile and elongation properties of the copper allow it to adsorb more of the expansion stresses without breaking.

8.4.2 Conductive Anodic Filament Growth (CAF)

CAF defects are filament growths which create a short between two individual circuits (Fig. 8.28). CAF requires three components to form: a pathway, moisture, and a voltage bias [17].

Material aspects which limit the pathways have the primary impact on CAF performance. CAF is reduced with improved fracture toughness of the resin to resist fabrication damage or mechanical stresses which form the cracks. Improving the wettability of the resin to the reinforcement material prevents voids in the material which form pathways [19]. The addition of fillers or a random fiber reinforcement act to arrest any cracks formed by mechanical stresses in the resin.

8.4.3 Pad Cratering

Pad cratering is a separation of the surface solder pad from the printed circuit board by way of a fracture through the resin underneath the pad (Fig. 8.29). Pad cratering, if severe enough can manifest in open circuits by breaking the trace or via attaching to the detached solder pad. It can also create pathways which can lead to CAF defects as explained above [16].

Improving the fracture toughness of the resin to prevent cracking is the primary concern for pad cratering. Improving the peel strength of the copper to the resin helps to prevent localized stress risers for crack formation. The addition of fillers or a random fiber reinforcement to act as crack arresters improves pad crater resistance.

8.4.4 Solder Joint Reliability

Solder joint reliability from a printed circuit board material perspective focuses on the quality and type of the surface finish. Poor quality of the surface finish can lead to oxidation of the soldering interface material, whether it be the underlying copper or a component of the surface finish itself, and lead to dewetting or non-wetting of the solder joint. Two additional defects related to specific surface finishes are black pad for electroless nickel immersion gold and microvoids for immersion silver. Black pad is the formation of a phosphor rich layer at the intermetallic layer between the bulk solder and the electroless nickel pad which is brittle. The solder joint fractures early at this interface, and the resulting surface is black in color, hence the name black pad. A major cause of microvoids is an overly aggressive immersion silver bath which creates caves beneath the immersion silver plate in the underlying copper. The caves oxidize and form tiny bubbles at the intermetallic surface during solder reflow operations. The tiny bubbles or microvoids create a weakened interface which fractures during post assembly stresses creating open circuits. Black pad and microvoids have resulted in many adjustments to their respective plating processes with dramatically improved results. However, like any process control related defect, they can not be eliminated.

Many thanks go to Ed Kelley of Isola Group, Doug Eng of PPG Industries, Karl Dietz of Dupont, and Doug Sober of Kaneka Texas Corp for their insight and technical input towards this chapter on Advanced PCB Materials.

References

1. E_R_D Glass data sheet, download from www.vetrotextextiles.com
2. G Brist, S Hall, S Clouser, and T Liang, (2005) Non-Classical Conductor Losses Due to Copper Foil Roughness and Treatment, ECW 10 Conference May 2005
3. G Brist, B Horine, and G Long (2004) High Speed Interconnects: The Impact of Spatial Electrical Properties of PCB due to Woven Glass reinforcement Pattern, IPC Print Circuit Expo 2004
4. D Cullen (2002) "Silver and Change: A Tale of Silver, Copper, Nickel and Gold, The Board Authority, April 2002
5. D Cullen (2002) Going Beneath the Surface of Surface Finishes, Circuitree, November 2002, pp. 50–62
6. K Dietz (2006) Tech Talk: Fine Lines in High Yield (Part CXXIV): High Performance Dielectrics, Circuitree, January 2006
7. D Eng (2001) Fiber Glass Reinforcements within Circuit Board Composites, Board Authority, September 2001
8. B Forcier (2000) Laser Drillable E-Glass Multilayer Materials An Overview of Laser Enhanced Materials, Board Authority, July 2000
9. B Forcier and F Hickman III (2000) The Design and Fabrication of HDI Interconnects Utilizing Total Integration of Fiber-Reinforced Materials, Board Authority, March 2000
10. B Forcier and B Schor (2001) High Reliability/Low CTE Epoxy Technology: An Overview of the Advantages of Low CTE Materials, Circuitree, February 2001
11. S Hall et al. (2007) Multigigahertz Causal Transmission Line Modeling Methodology Using a 3-D Hemispherical Surface Roughness Approach, IEEE Transactions on Microwave Theory and Techniques, Vol. 55 Num. 12, Dec. 2007
12. P Hamilton et al. (2007) Humidity Dependent Loss in PCB Substrates, IPC Printed Circuits Expo 2007
13. E Kelley (2004) Reengineered FR-4 Base Materials for Improved Multilayer PCB Performance, The Board Authority-Live, June 2004
14. E Kelley (2004) An assessment of the Impact of Lead-Free assembly Processes on Base Material and PCB Reliability, IPC/Soldertec Conference, Amsterdam, June 2004
15. G Long and G Brist (2005) Lead-Free Product Transition: Impact on Printed Circuit Board Design and Material Selection, ECWC 10 Conference 2005
16. G Long, T Embree, M Mukadam, S Parupalli, and V Vasudevan (2007) Lead Free Assembly Impacts on Laminate Material Properties and Pad Crater Failures, IPC APEX/EXPO Conference 2007
17. K Sauter (2002) Evaluating PCB Design, Manufacturing Process, and Laminate Material Impacts on CAF Resistance, IPC Printed Circuits Expo Technical Conference Proceedings, March 2002
18. D Sober (1997) Base Material Basics: Manufacture and Market, IPCWorks '97: Tutorial Handbook, Section 6
19. W Varnell et al. (2002) Conductive Anodic Filament Resistant Resins, IPC Printed Circuits Expo Technical Conference Proceedings, March 2002

Chapter 9
Flip-Chip Underfill: Materials, Process and Reliability

Zhuqing Zhang and C.P. Wong

Abstract In order to enhance the reliability of a flip-chip on organic board package, underfill is usually used to redistribute the thermo-mechanical stress created by the Coefficient of Thermal Expansion (CTE) mismatch between the silicon chip and organic substrate. However, the conventional underfill relies on the capillary flow of the underfill material and has many disadvantages. In order to overcome these disadvantages, many variations have been invented to improve the flip-chip underfill process. This paper reviews the recent advances in the material design, process development, and reliability issues of flip-chip underfill, especially in no-flow underfill, molded underfill, and wafer-level underfill. The relationship between the materials, process and reliability in these packages is discussed.

Keywords Flip-chip · Underfill · Interconnect · Materials · Reliability · Coefficient of thermal expansion

9.1 Introduction

The brain of the modern electronics is the integrated circuit (IC) on semiconductor chip. In order for the brain to control the system, interconnects need to be established between the IC chip and other electronic parts, power and ground, and inputs and outputs. The first-level interconnect usually connects the chip to a package made of either plastics or ceramics, which in turn is assembled onto a printed circuit board (PCB). Three main interconnect techniques are used: wire-bonding, tape automated bonding (TAB), and flip-chip. In a wire-bonded package, the chip is adhered to a carrier substrate using a die-attach adhesive with the active IC facing up. A gold or aluminum wire is then bonded between each pad on the chip and the corresponding bonding pad

Z. Zhang (✉)
Imaging and Printing Group, Hewlett-Packard Company, 1000 NE Circle Blvd, Corvallis, OR 97333, USA
e-mail: zhuqing.zhang@hp.com

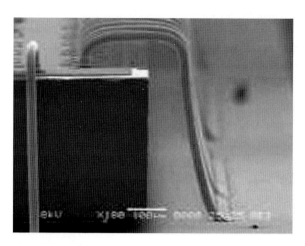

Fig. 9.1 First level interconnect using wire-bonding

on the carrier as shown in Fig. 9.1. The chip and the wire interconnections are usually protected by encapsulation. TAB, on the other hand, uses a prefabricated lead frame carrier with copper leads adapted to the IC pads. The copper is usually gold-plated to provide a finish for bonding to the IC chip pads. The chip is attached onto the carrier and either thermosonic/thermocompression bonding or Au/Sn bonding is used to establish the interconnect. Both wire-bonding and TAB interconnects are limited to peripheral arrangement and therefore low input/output (I/O) counts. Flip chip, however, can utilize the entire semiconductor area for interconnects. In a flip-chip package, the active side of an IC chip is faced down towards and mounted onto a substrate [1]. Interconnects, in the form of solder bumps, stud bumps, or adhesive bumps, are built on the active surface of the chip, and are joined to the substrate pads, in a melting operation, adhesive joining, thermosonic, or thermocompression process. Figure 9.2 shows an example of solder bumped chip surface for flip-chip interconnect. Since flip-chip was first developed about 40 years ago, many variations of the flip-chip design have been developed, among which, the Controlled Collapse Chip Connection (also known as C4) invented by IBM in 1960s is the most important form of flip-chip [2]. Compared with conventional packaging using wire-bonding technology, flip-chip offers many advantages such as high I/O density, short interconnects, self-alignment, better heat dissipation through the back of the die, smaller footprint, lower profile, and high throughput, etc. The outstanding merits of flip-chip have made it one of the most attracting techniques in modern electronic packaging, including MCM modules, high frequency communications, high performance computers, portable electronics, and fiber optical assemblies.

Until the late 1980s, flip-chips were mounted onto silicon or ceramic substrates. Low cost organic substrates could not be used due to the concern of the thermal-mechanical fatigue life of the C4 solder joints. This thermal-mechanical

Fig. 9.2 Area array solder bumps for flip-chip interconnect

issue mainly arises from the CTE mismatch between the semiconductor chip (typically Si, 2.5 ppm/°C) and the substrate (4–10 ppm/°C for ceramics and 18–24 ppm/°C for organic FR4 substrate). As the distance from the neutral point (DNP) increases, the shear stress at the solder joints increases accordingly. So with the increase in the chip size, the thermal-mechanical reliability becomes a critical issue. Organic substrates have advantages over ceramic substrates because of their low cost and low dielectric constant. But the high CTE differences between the organic substrates and the silicon chip exert great thermal stress on the solder joint during temperature cycling.

In 1987, Hitachi first demonstrated the improvement of solder fatigue life with the use of filled resin to match solder CTE [3]. This filled resin, later called "underfill", was one of the most innovative developments to enable the use of low-cost organic substrate in flip-chip packages. Underfill is a liquid encapsulant, usually epoxy resins heavily filled with fused silica (SiO_2) particles, that is applied between the chip and the substrate after flip-chip interconnection. Upon curing, the hardened underfill exhibits high modulus, low CTE matching that of the solder joint, low moisture absorption, and good adhesion towards the chip and the substrate. Thermal stresses on the solder joints are redistributed among the chip, underfill, substrate and all the solder joints, instead of concentrating on the peripheral solder joints. It has been demonstrated that the application of underfill can reduce the all-important solder strain level to 0.10–0.25 of the strain in joints which are not encapsulated [4, 5]. Therefore, underfill can increase the solder joint fatigue life by 10–100 times. In addition, it provides an environmental protection to the IC chip and solder joints. Underfill becomes the practical solution to extending the application of flip-chip technology from ceramics to organic substrates, and from high-end to cost sensitive products. Today, flip chip is being extensively studied and used by almost all major electronic companies around world including Intel, AMD, Hitachi, IBM, Delphi, Motorola, Casio, etc.

9.2 Conventional Underfill Materials and Process

The generic schematic of a flip chip package is shown in Fig. 9.3. Conventional underfill is applied after the flip chip interconnects are formed. The resin flows into the gap between the chip and the substrate by a capillary force. Therefore, it is also called "capillary underfill". A typical capillary underfill is a mixture of liquid organic resin binder and inorganic fillers. The organic binders are often epoxy resin mix, although cyanate ester or other resin has been used for underfill application as well. Fig. 9.4 shows the chemical structure of some commonly used epoxy resins. In additional to epoxy resin, a hardener is often used to form cross-linking structure upon curing. Sometimes a latent catalyst is incorporated to achieve long pot life and fast curing. Inorganic fillers typically used in underfill formulation are micron-sized silica. The silica fillers are incorporated into the resin binder to enhance the material properties of cured underfill such

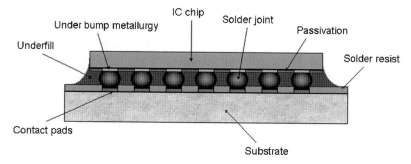

Fig. 9.3 Generic configuration of a flip chip package with underfill

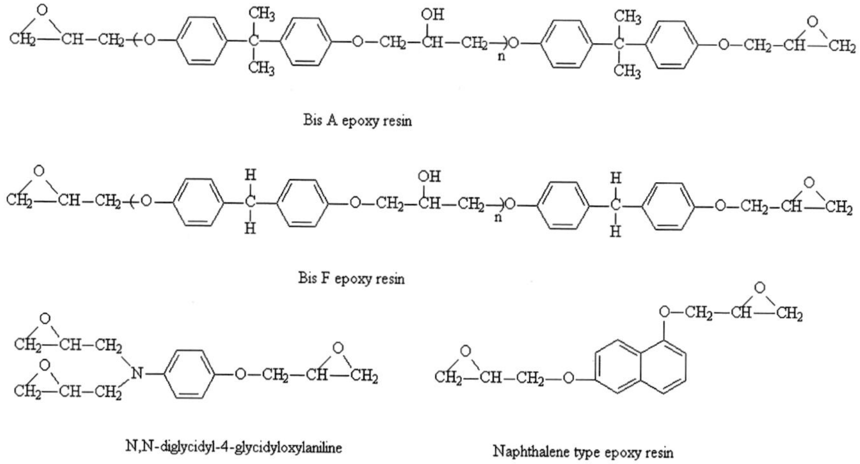

Fig. 9.4 Typical epoxy resin structures used in underfill formulations

9 Flip-Chip Underfill

as low CTE, high modulus, and low moisture uptake, etc. Other agents that can be found in an underfill formulation include adhesion promoters, toughening agents, and dispersing agents, etc. These chemicals are incorporated to help the resin mixing and enhance the cured underfill properties.

Figure 9.5 shows the process steps of flip-chip with conventional underfill. Separate flux dispensing and cleaning steps are required before and after the assembling of the chip, respectively. After the chip is assembled onto the substrate, the underfill is usually needle-dispensed and is dragged into the gap between the chip and the substrate by a capillary force. Then a heating step is needed to cure the underfill resin to form a permanent composite.

The flow of the capillary underfill has been extensively studied since it is considered to be one of the bottlenecks for the flip chip process. The capillary flow is usually slow and can be incomplete, resulting in voids in the packages and also non-homogeneity in the resin/filler system. The filling problem becomes even more serious as the chip size increases. The flow modeling of flip chip underfill is often approximated as viscous flow of the underfill adhesive between two parallel plates. One can use the Hele-Shaw model to simulate the underfill flow with the above approximation. The time required to fill a chip of length L can be calculated as [6]:

$$t_{\text{fill}} = \frac{3\eta L^2}{\sigma h \cos\theta} \qquad (9.1)$$

where η is the underfill viscosity; σ is the coefficient of the surface tension; θ is the contact angle; and h is the gap distance. It is easily seen that a larger chip with a smaller gap distance would require longer time to fill.

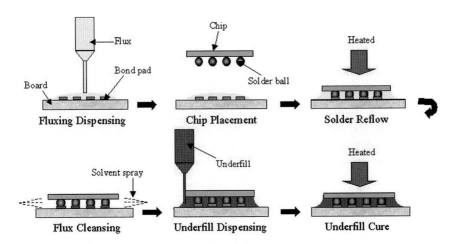

Fig. 9.5 Flip-chip process using conventional underfill

The above approximation does not take the existence of solder bumps into account. It is shown that the approximation breaks down when the spacing between bumps is comparable to the gap height [7]. Therefore, this model cannot apply to high density area array flip chip applications. Using transparent quartz dies assembled onto different substrates, Nguyen et al. observed the flow of commercial underfills and used a 3D PLICE-CAD to model the underfill flow front [8]. A comparison between the peripheral and area array chips showed that the bumps enhanced the flatness of the flow front by providing periodic wetting sites. A racing effect along the edges was observed. Voids can be formed at the merging of flow fronts. The merging of the fronts also produced streaks, which are zones of no- or slow-moving fluids, leading to higher potential for filler settling.

Recent development in underfill flow models has also considered the effect of the contact angle on solder and bump geometry. A study by Young and Yang used a modified Hele-Shaw model considering the flow resistance in both the thickness direction between the chip and substrate, and the plane direction between solder bumps [9]. It was found that the capillary force parameter would approach a constant value at very large pitch for the same gap height. As the bump pitch reduces, the capillary force will increase to a maximum as a result of underfill wetting on the solder given the contact angle on the solder is small, and then quickly drops to zero as the pitch approaches the bump diameter. Their study also showed that a hexagonal bump arrangement is more efficient to enhance the capillary force at critical bump pitch.

9.3 Reliability of Flip Chip Underfill Packages

The reliability of a flip chip package can be evaluated in a number of different methods, including thermal cycling, thermal shock, pressure-cook test, etc. The lifetime of a solder joint interconnect during temperature cycling can often be described by statistical models such as Weibull distribution. The probability density function (PDF) of the Weibull distribution is given by:

$$f(x) = \left(\frac{\beta}{x}\right)\left(\frac{x}{\theta}\right)^{\beta} \exp\left(-\left(\frac{x}{\theta}\right)^{\beta}\right) \quad (9.2)$$

where x is the thermal cycling life as a random variable; θ is the characteristic life; β is the shape parameter. The mean time to failure (MTTF), which is the expectation of the time to failure, for the Weibull distribution is:

$$\text{MTTF} = \theta \cdot \Gamma\left(1 + \frac{1}{\beta}\right) \quad (9.3)$$

where Γ is the gamma function. It is generally believed that fatigue of the solder joints is a major reason for structure and electrical failures (Tummala 2001).

The solder fatigue life can be described as a function of inelastic shear strain in Coffin-Manson equation (Manson & Coffin 1965, 1954):

$$N_f = \frac{1}{2}\left(\frac{\Delta\gamma}{2\varepsilon'_f}\right)^{1/c} \quad (9.4)$$

where N_f is the number of cycles to fatigue failure, $\Delta\gamma$ is the inelastic shear strain, ε'_f is the fatigue ductility coefficient, and c is the fatigue ductility exponent. Other strain based fatigue equations have been proposed, among which Solomon's model is often used (Soloman, 1986):

$$N_f = \left(\frac{\theta}{\Delta\gamma_p}\right)^{1/\alpha} \quad (9.5)$$

where $\Delta\gamma_p$ is the percentage inelastic shear strain, θ and α are constants.

It has been shown that the use of the underfill can increase the lifetime of the solder joints by at least an order of magnitude during thermal cycling [10]. It was found that in an underfilled flip chip package, the fatigue life is highly dependent on the material properties of the underfill. The analytic model by Nysaether et al. [11] showed that while an underfill without filler increased the lifetime by a factor of 5–10, a filled underfill with a lower CTE gave a 20–24-fold increase in lifetime. For both filled and non-filled sample, the lifetime is nearly constant regardless of distance to the neutral point (DNP), indicating that the underfill effectively couples the stress among all the solder joints.

Many numerical models have been developed to study the solder fatigue life of a flip chip package with or without underfill. The polymeric nature of the underfill material requires careful characterization for correct material property input to the numerical models. The modulus of a polymeric material is not only a function of temperature, but also a function of time, i.e., it is a viscoelastic material. Thermal mechanical analyzer (TMA) and dynamical mechanical analyzer (DMA) are typically used to characterize the viscoelastic properties of the underfill material. Dudek et al. characterized 4 commercial electronic polymers and used finite element (FE) analyses to study the effect of die size and underfill material properties on the thermo-mechanical reliability of the flip chip on board (FCOB) package [12]. They found that although the use of underfill can effectively reduce the shear strain, it can also cause bump creep strain in the transverse board direction due to stretching and compressing of the bump during thermal cycling. This load is due to the CTE mismatch between the solder and underfill/solder-mask layer. Underfill with CTE that matched the solder material (22–26 ppm/°C) gives the best thermal cycling life based on the creep strain criterion.

The function of the underfill in a flip chip package is stress redistribution, not stress reduction. A rigid underfill material mechanically couples the device and the substrate, changing partially the shear stress experienced by the solder joints into bending stress on the whole structure. Shrinkage of the underfill during cure and the CTE mismatch during cooling after cure can generate large stress on the Si chip, resulting in die crack in some cases. Palaniappan et al. performed in-situ stress measurements in the flip chip assemblies using a test chip with piezoresistive stress sensing devices [13]. The study concluded that the underfill cure process generates large compressive stress on the active die surface, indicating a complex convex bending state in the flip chip. The level of stress measured can lead to Si fracture. The residual die stress was found to be strongly dependent on underfill CTE, modulus, and Tg. A finite element analysis by Mercado et al. on the die edge cracking in flip chip PBGA packages also concluded that the energy release rate for horizontal Si facture increases with underfill modulus and CTE [14].

In addition to temperature related thermomechanical failure, moisture induced failures such as delamination and corrosion are common for a flip chip underfill package. HAST (highly accelerated stress test) is often used to determine the temperature and moisture sensitivity of the package. The test uses harsh environment conditions such as high temperature, high humidity, and high pressure. A typical test condition can be 121°C, 100% RH (relative humidity), and 2 atm pressure. It has also been known as the autoclave or pressure cooker test (PCT). The moisture absorbed by the polymeric materials can hydrolyze the interfacial bonds between underfill and the die, resulting in delamination starting from the corner of the die, which further promotes the moisture diffusion along the interface. The moisture at the interface can cause corrosion of the solder joints and metal traces on the substrate. Delamination decouples underfill with the Si die and cause stress concentration on the surrounding solder joints, leading to early fatigue failure of those joints. The absorbed moisture also causes hygroscopic swelling. Lahoti et al. studied the combined effect of moisture and temperature on the reliability of flip chip ball grid array (FCBGA) packages using FE analysis. The simulation results revealed the significance of hygroscopic induced tensile stress on the under bump metallurgy (UBM) and inter dielectric layer (ILD) [15].

Interfacial delamination of underfill to various materials such as die passivation, solder material, and solder mask on the substrate is a leading cause for failure in flip chip underfill packages. One way to improve the reliability under temperature and humid aging is to incorporate adhesion promoters, or coupling agents, into underfill to increase adhesion of underfill to the surrounding materials. Luo et al. studied 6 different coupling agents and their effect on underfill. The authors found that the incorporation of the coupling agents clearly affected the curing profile and bulk property of the underfill, such as Tg and modulus. The effect of coupling agents on adhesion and adhesion retention after temperature moisture aging was highly dependent on coupling agent type and interacting surfaces. The addition of titanate and zirconate coupling agents can improve adhesion of epoxy underfill with BCB passivated silicon. However, with the addition of same coupling agents, the adhesion

Table 9.1 Desirable underfill properties for flip-chip packages

Curing Temperature	<150°C
Curing time	<30 min
Tg	>125°C
Working life (viscosity double @ 25°C)	>16 h
CTE (α_1)	22–27 ppm/°C
Modulus	8–10 GPa
Fracture toughness	>1.3 MPa*m$^{1/2}$
Moisture absorption (8 hrs boiling water)	<0.25%
Filler contents	<70 wt%

strength of underfill with polyimide passivation decreased after aging at 85C/ 85% RH [16].

In summary, many studies have concluded that underfill material property is one of the key factors determining the reliability of the package. The general guideline on the material properties of underfill for flip chip package can be summarized in Table 9.1. However, one has to keep in mind that different failure modes coexist in a reliability test, which sometimes present conflicting requirements on underfill. For instance, to effectively couple the stress on the solder joints, high modulus underfill is desired. On the other hand, high underfill modulus can lead to high residual stress and therefore die crack. Another example is the filler loading. Low CTE requirement on underfill indicates high filler loading. However, an underfill with higher filler loading typically has a higher viscosity, causing difficulty in underfill dispensing. The result might be underfill voids and non-uniformity, which would cause reliability issues. Therefore, the choice of underfill highly depends on the application, e.g., die size, passivation material, substrate material, type of solder, and environment conditions the package will be subjected to during application, etc.

9.4 New Challenges to Underfill

As silicon technology moves to sub 0.1 μm feature size, the demands for packaging also evolves as the bump pitch gets tighter, bump size smaller, die size larger for future flip chip packages. As a result, the capillary underfill process faces tremendous challenges. As it was discussed previously, underfill flow problem aggravates as the size of the chip becomes larger and the gap between the chip and substrate gets smaller. Among the emerging development of flip chip, lead-free solder and low-K (dielectric constant) ILD (interlayer dielectric)/ Cu present new challenges to underfill [17].

High-lead and lead-tin eutectic solders have been widely used for chip-package interconnections. Recent environmental legislations towards toxic materials and consumers' demand for green electronics have pushed the drive towards lead-free solders. Alternatives have been proposed using multiple

Table 9.2 Possible lead-free alloys

Alloy	Melting point
Sn96.5/Ag3.5	221°C
Sn99.3/Cu0.7	227°C
Sn/Ag/Cu	217°C (Ternary eutectic)
Sn/Ag/Cu/X(Sb, In)	Ranging according to compositions, usually above 210°C
Sn/Ag/Bi	Ranging according to compositions, usually above 200°C
Sn95/Sb5	232–240°C
Sn91/Zn9	199°C
Bi58/Sn42	138°C

combinations of elements like tin, silver, copper, bismuth, indium and zinc, most of which require increased reflow temperature profiles during the soldering process relative to the well-known tin-lead alloys. The following Table 9.2 shows some of the common lead-free solders.

Among the several Pb-free candidate solders, the near ternary eutectic Sn–Ag–Cu (SAC) alloy compositions, with melting temperatures around 217°C, are becoming consensus candidates. The optimal composition 95.4 Sn/3.1Ag/1.5Cu has provided a combination of good strength, fatigue resistance, and plasticity [18]. In addition, the alloy has sufficient supply and adequate wetting characteristics.

The use of Sn/Ag/Cu solder presents two major challenges on the flip-chip assembly process. First, since the melting point of the alloy is more than 30°C higher than that of the eutectic Sn/Pb alloy, the process temperature is raised by 30–40°C. The high process temperature has a great impact on the substrate since the conventional FR-4 material has a Tg at around 125°C and also subjects the attached components to a higher thermal stress. Higher warpage is introduced when the board is subjected to higher temperature reflow. There have been considerable researches in high Tg substrate for lead-free process. The second challenge comes from the flux chemistry. Since the current fluxes in use are usually designed for eutectic Sn/Pb solder, they either do not have high enough activity or do not possess sufficient thermal stability at high temperature. So generally, the wetting behavior of the lead-free solders is not as good as that of the eutectic Sn/Pb solder [19, 20].

With the trends of lead-free solder interconnect, the underfill for flip-chip in package application faces new challenges of compatibility with higher reflow temperature. High temperature reflow causes component damage due to increased level of materials degradation, moisture ingress and mechanical expansion. Therefore, the thermal stability, adhesion to various interfaces, strength, and fracture toughness of the underfill need to be improved. The SAC alloy does not plastically deform as much as the eutectic PbSn solder. The creep deformation is less at lower stress level and more at higher stress level compared to the

PbSn solder, which indicates that the choice of the underfill would depend on the application needs. A temperature cycle with a large temperature difference and lower dwelling times could induce more creep and therefore requires more protection from the underfill [21]. An evaluation of underfill materials for lead-free application conducted by Intel Corporation [22] showed that majority of the failure occurred during the moisture sensitive level (MSL) 3 followed by 260°C reflow. Delamination seemed to be the common failure mechanism after the high temperature reflow. This failure was also correlated to the materials with low filler content and low coupling agent content. In general, materials with high filler content (and therefore low CTE, high modulus, and low moisture uptake) and good adhesion were compatible with the lead-free process.

As the IC fabrication moves towards small feature and high density, the interconnect delay becomes dominant. This calls out for new interconnect and interlayer dielectric (ILD) materials. The Cu metallurgy and low-K ILD has been successfully implemented to increase device speed and reduce power consumption. These low-K materials tend to be porous and brittle, having high CTE and low mechanical strength, compared to the traditional ILD materials such as SiO_2. The CTE mismatch between the low-K ILD and the silicon die creates a high thermomechanical stress at the interface. Therefore, the choice of underfill becomes critical since it not only protects the solder joints by stress redistribution, but also need to protect the low-K ILD and its interface with the silicon. Critical material properties of underfill to achieve reliability requirement for the low-K ILD package include the Tg, CTE, and the modulus. However, the optimal combination of these properties is still controversial.

Five underfills were evaluated for the low-K flip chip package by Tsao et al. [23]. Both modeling and experimental evaluation indicated that the low Tg and low stress-coupling-index underfills yielded better reliability in the low-K flip chip package. Two moderately low Tg underfills (Tg between 70 and 120°C) showed good potential in protecting both the solder joints and low-K interface. An underfill with a very low Tg (lower than 70°C), on the other hand, failed to protect the solder joints during the thermal cycling test. A study conducted by LSI Logic Corp and Henkel Loctite (now called Henkel) Corp [24], on the other hand, indicated that underfills with high Tg and low modulus is advantageous for the low-K flip chip. The low modulus of the underfill exerts lower stress on the package and therefore reducing the stress on the low-K layer, preventing underfill delamination and die cracking. The high Tg prevents solder bump fatigue by maintaining a low CTE over the temperature cycling. The high Tg, low modulus underfill developed by Henkel exhibited good manufacturability and reliability in the package qualification testing including JEDEC preconditioning, thermal cycling, biased humidity testing, and high temperature storage.

With all the new challenges to the flip-chip and underfill, capillary underfill is still the main packaging technology for flip-chip devices. However, the continuing shrinking of pitch distance and gap height will eventually post limitation on the capillary flow. The industry has started to look for alternatives to capillary underfill. The following sections describe several recent development in underfill material and process.

9.5 No-Flow Underfill

The idea of integrated flux and underfill was patented by Pennisi et al. in Motorola back in 1992 [25]. It triggered the research and development of no-flow underfill process. The first no-flow underfill process was published by Wong et al. in 1996 [26]. The schematic process steps are illustrated in Fig. 9.6. Instead of underfill dispensing after the chip assembly in the conventional process, in a no-flow underfill process, the underfill is dispensed onto the substrate before the placement of the chip. Then the chip is aligned and placed onto the substrate and the whole assembly goes through solder reflow where the chip to substrate interconnection through solder bumps is established while the underfill is cured. This novel no-flow process eliminates the separate flux dispensing and flux cleaning steps, avoids the capillary flow of underfill, and finally combines the solder bump reflow and underfill curing into a single step, hence, improving the production efficiency of underfill process. It is a step forward for the flip-chip to be compatible with surface mount technology (SMT).

The key to the success of a no-flow underfill process lies in the underfill material. The first patent on the no-flow underfill material was filed by Wong and Shi in Georgia Institute of Technology [27]. The two critical properties of the no-flow underfill to enable this new process are the latent curing ability and the build-in fluxing capability. The nature of the no-flow underfill process requires that the underfill have enough reaction latency to maintain its low viscosity until the solder joints are formed. Otherwise, gelled underfill would prevent the melting solders from collapsing onto the contact pads, resulting in a low yield of solder joint formation. On the other hand, elimination of the post cure is desired since post cure takes additional off-line process time, adding to the cost of this process. Many latent catalysts for epoxy resins have been explored for the application of no-flow underfill. In the material system that

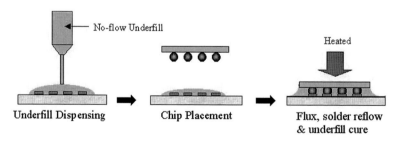

Fig. 9.6 Flip-chip process ssing no-flow underfill

9 Flip-Chip Underfill

Wong and Shi designed, Co(II) acetylacetonate was used as the latent catalyst [28, 29], which gave enough curing latency for no-flow underfill. The advantage of metal chelates lies not only in its latent acceleration, but also in the wide curing range they offer. By exploring different metal ions and chelates, the curing behavior of different epoxy resins could be tailored to the application of no-flow underfill for lead-free solder bumped flip-chip [30]. Since lead-free solders usually have a higher melting point than eutectic SnPb solder, no-flow underfill for lead-free bumped flip-chip requires higher curing latency to ensure the wetting of the lead-free solder on the contact pad. Z. Zhang et al. explored 43 different metal chelates and developed no-flow underfill compatible with lead-free solder reflow [30]. Successful lead-free bumped flip-chip on board package using no-flow underfill process has been demonstrated [31].

Despite the importance of the curing process of no-flow underfill, there is little study on the curing kinetics and its relation to the reflow profile. In an attempt to develop systematic methodology to characterize the curing process of no-flow underfill, Zhang et al. used an autocatalytic curing kinetic model with temperature dependent parameters to predict the evolution of degree of cure (DOC) during the solder reflow process [32]. Figure 9.7 shows the result of DOC calculation of a no-flow underfill in eutectic SnPb and lead-free solder reflow process. If the DOC of the underfill at the solder melting temperature is lower than the gel point, the molten solder would be allowed to wet the substrate and make the interconnection. Another approach is the in-situ measurement of viscosity of no-flow underfill using microdielectrometry by

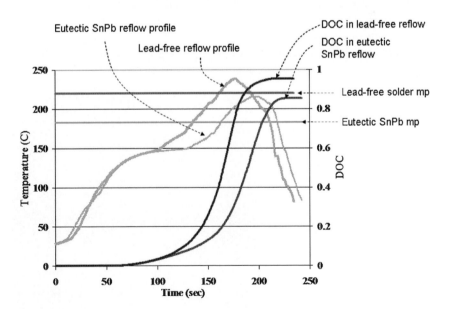

Fig. 9.7 Degree of Cure (DOC) evolution of a no-flow underfill in eutectic SnPb and lead-free reflow process

Morganelli et al. [33]. Since the viscosity is related to the ionic conductivity, the dielectric properties of the underfill can be used for the in-situ analysis of the no-flow underfill in the reflow process, which can be used to predict the solder wetting behavior.

The other key property for no-flow underfill is the fluxing capability. In a conventional flip-chip process, flux is used to reduce and eliminate the metal oxide on the solder and metal contact pads, and to prevent them from being reoxidized under high temperature. Instead of applying flux, no-flow underfill is dispensed before the chip placement. Hence, the self-fluxing capability is required to facilitate solder wetting. To achieve this goal, research has been done to develop reflow-curable polymer fluxes [34]. A comprehensive study on the fluxing agent of no-flow underfill material was carried out by Shi et al. [35, 36, 37], which included the relationship between the surface composite on Cu pad and the fluxing capability of no-flow underfill, and also the effect of the addition of the fluxing agent on the curing and material properties of no-flow underfill.

The process of no-flow underfill has always attracted much attention in the assembly industry. Voids formation is often observed in many flip-chip no-flow underfill packages. The origin of the voids could be the out-gassing of the underfill, moisture in the board, and trapped voids during assembly, etc. They are usually tacked to a solder bump or in between two bumps [38, 39]. Figure 9.8 shows an example of underfill voids in a no-flow underfill package observed with scanning acoustic microscope. Voids in the underfill, especially voids near the solder bumps lead to early failure through a number of ways including stress concentration, underfill delaminate, and solder extrusion. Study has showed that solder bridging might result from the solder bump extrusion through the micro voids trapped between adjacent bumps [40]. The material and process

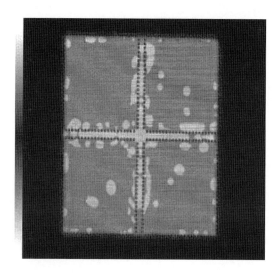

Fig. 9.8 A scanning acoustic microscopy image of a no-flow underfill package (lighter dots are underfill voids)

factors influencing the voiding behavior are complicated and interacting. It has been shown that the outgassing of anhydride could cause severe voiding, if the curing latency is high and also the reflow temperature is high; hence, the voiding becomes more prominent in a lead-free reflow process [41]. The important process parameters that affects underfill voiding in a no-flow process include the underfill dispensing pattern, the solder mask design, the placement force and speed, and the reflow profile, etc. [42, 43]. Before assembly, the substrate needs to be baked to dry out any moisture to prevent voiding from the board [39]. It has been shown that in some cases a fast gelation of underfill is desired to minimize the voiding while in other cases, extending duration at high temperature can "push" out the voids [44, 39]. In short, with the right material and process parameters, voiding in no-flow underfill can be minimized. However, the process window is usually very narrow. An important point was raised by R. Zhao [31] et al. that for a small circuit board where the temperature distribution is more homogenous, it is relatively easy to develop a "good" reflow profile while for complex SMT assemblies involving multiple components and significant thermal mass difference across the board, the optimization of the reflow process presents great challenge.

The reliability of flip-chip no-flow underfill package has been evaluated in many occasions. Discrepancies exist among these reports because the process and reliability of the no-flow underfill package depend largely on the package designs including the size of the chip, the pitch, the surface finish of the pad, etc. Among the earliest reporters on no-flow underfill, D. Gamota and C. Melton compared the reliability and typical failure mode of conventional underfill package and no-flow underfill packages [45]. They found that in a conventional underfill assembly, the failure of the assembly mainly resulted from the interfacial delamination between the underfill and the chip passivation. However, with unfilled no-flow underfill, good interfacial integrity was observed, and the assembly failed mainly due to the fracture through the solder interconnects near PCB. Since the no flow underfill was unfilled, the CTE was high. They argued that the relative localized CTE mismatch between the chip, the underfill, and the PCB resulted in a high local stress field which initiated fracture in the solder interconnects. No-flow underfill without silica fillers or very low filler loadings is not only high in CTE, but also low in fracture toughness [46]. Combined with high CTE mismatch, the low fracture toughness leads to early underfill cracking both inside the bulk and in underfill fillet. Fillet cracking causes delamination between the underfill and the die passivation and/or between the underfill and the board, while bulk cracking can initiate solder joint cracking and solder bridging [47]. These all become the common failure modes for flip-chip no-flow underfill package. Efforts have been made to enhance the toughness of the no-flow underfill materials through the incorporation of the toughening agents [48]. The effect of the glass transition temperature (Tg) of the no-flow underfill on the reliability of the package has been controversial. It is usually believed that the Tg of the underfill should exceed the upper limit of the temperature cycling (125°C, or 150°C) to ensure consistent material behavior during the reliability test. However, some tests have shown that low Tg (~70°C) underfill material performed better in liquid-to-liquid thermal shock (LLTS) [49]. The research by Zhang et al. on the development of non-anhydride based no-flow underfill [50] also showed that high Tg is not critical to reliability. Although the CTE of the underfill above Tg is much higher than that below Tg, the modulus of the underfill decreases dramatically; so the overall stress in the underfill does not necessarily increase when the environment temperature exceeds its Tg. But high Tg might result in a higher residue stress inside the underfill after the material cools down after curing, which leads to early crack in the underfill.

9.5.1 Approaches of Incorporating Silica Fillers into No-Flow Underfill

The previous research has shown that the correlation between the material properties and package reliability in the case of flip-chip underfill is very complicated. It is difficult to separate the effect of each factor since the material properties are often correlated with each other. However, it is generally agreed that low CTE and high modulus are favorable for high interconnect reliability [51]. Hence, the inclusion of silica fillers into the underfill is critical to enhance the reliability. However, since the underfill is pre-deposited on the substrate before the chip assembly in a no-flow process, the fillers are easily trapped in between the solder bump and contact pad, and thus hinder the solder joint formation [52]. Thermo-compression reflow (TCR) has been used to exclude the silica filler from solder joint [53]. The process step is illustrated in Fig. 9.9. In a TCR process, the underfill is dispensed on to a pre-heated substrate. The chip is then picked, bonded to the substrate, and held at an elevated temperature under force for a certain period of time for solder joint formation. The assembly is post-cured afterwards. It was found that the bonding force and temperature were important factors influencing yield. A detailed study was carried out by Kawamoto et al. at NAMICS Corporation to determine the effect of filler on the solder joint connection in a TCR-like no-flow underfill process [54]. The study used 2 different sizes of silica fillers at different loading levels. It was found that good solder connection can be made with underfill with up to 60 wt% filler loading without filler surface treatment. Smaller filler tended to increase the viscosity of the underfill and more fillers were trapped in the solder

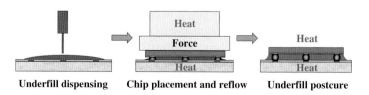

Fig. 9.9 Thermo-compression reflow for flip-chip

joints due to larger number of fillers at the same weight percentage loading. The study also found that proper surface treatment of the fillers can lower the underfill viscosity and increase yield at high filler loading.

Other approaches have been explored to incorporate silica fillers into no-flow underfill. In a novel patented process, Z. Zhang et al. used a double-layer no-flow underfill [55], in which two layers of no-flow underfill are applied. The bottom layer underfill has relatively higher viscosity and is not filled with silica fillers. It is applied onto the substrate first, and then the upper layer underfill which is filled with silica fillers is dispensed. The chip is then placed onto the substrate and reflowed, during which the solder joints are formed and the underfill is cured or partially cured. The process flow chart is illustrated in Fig. 9.10. It was demonstrated that high yield was achieved using the upper layer underfill with 65 wt% silica [56]. Further investigation on the process indicated that factors affecting the interconnection yield of the double-layer no-flow underfill are complicated and interacting with each other [57]. The process window is narrow, and the thickness and the viscosity of the bottom layer underfill are essential to the wetting of the solder bumps. And of course, it adds on another step in the flip-chip process and has a higher process cost.

The recent advances in nano-science and nano-technology have enabled innovative research in materials for electronic packaging. It was found that nano-sized silica fillers with surface modification can be mixed with thermosetting resins to provide a uniform dispersion of non-agglomerated particles. Used as no-flow underfill, the nano-composite materials allowed 50 wt% filler loading with good interconnect yield [58]. This high performance no-flow underfill developed by 3 M used 123 nm silica filler. With filler loading of 50 wt%, the

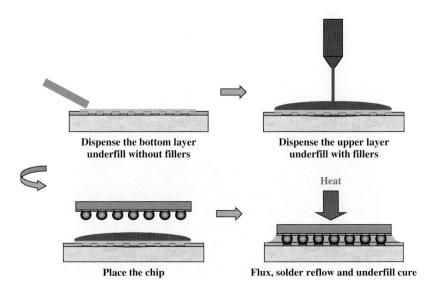

Fig. 9.10 Double-layer no-flow underfill process

CTE of the material was 42 ppm/°C and the good interconnect yield was achieved using PB10 die (5 mm × 5 mm, 64 peripheral bumps). A joint research study was conducted by 3 M and Georgia Tech on the process and reliability evaluation of the nano-silica incorporated no-flow underfill [59]. Figure 9.11 shows a SEM picture of the solder joint in presence of no-flow underfill with nano-silica fillers. A 1.5X increase of characteristics life was observed in the air-to-air thermal cycling (AATC) reliability test with the nano-silica fillers. Although the nano-composite no-flow underfill material shows good potential for a highly reliable flip chip package using a SMT-friendly no-flow underfill process, the fundamental mechanism of the nano-silica interaction with the solder joints and the underfill is still not well understood. Since nano-size particles have a large surface area and tend to form irregular agglomerations which increase the difficult to be incorporated into a binder, surface treatment of nano-silica is of great importance in formulating an underfill. A fundamental study on the surface modification of nano-size silica for underfill application was carried out by Sun et al. [60]. They found that the type of the surface treatment was the primary factor affecting the property of the formulation. Using an epoxy silane, the authors showed that the viscosity of the composite underfill was greatly reduced.

In summary, the invention of no-flow underfill greatly simplifies the flip-chip underfill process and draws flip-chip towards SMT. A successful no-flow underfill process requires careful investigation on the materials and process parameters. A lot of research efforts have been devoted to the materials, process, and reliability of flip-chip no-flow underfill assembly. Since the underfill does not contain silica filler and hence behaves differently from the conventional underfill, the failure modes and reliability concerns are sometimes also different

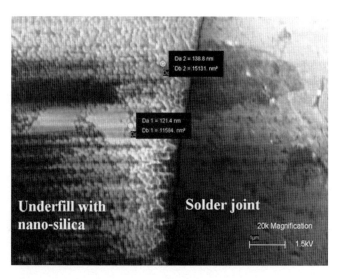

Fig. 9.11 SEM picture of a solder joint with nano-silica incorporated no-flow underfill

from the conventional flip-chip underfill assembly. There are several ways to enhance the reliability of a flip-chip no-flow underfill package. One way is to enhance the fracture toughness of the underfill without degrading other material properties to prevent underfill cracking in the thermal cycling. Or, low Tg and low modulus materials have been used to decrease the stress in the underfill. However, this approach diminishes the role of the underfill as a stress redistribution layer, and although it does decrease the stress in underfill, it can not prevent solder joint fatigue failure from the thermo-mechanical stress, especially in the case of the large chip, high I/O counts and small pitch size applications. The other way is to add silica fillers into the underfill and match the properties of a conventional underfill. In order to overcome the difficulty of filler entrapment, different approaches have been explored. However, these approaches are less SMT transparent and diminish the low cost purpose of a no-flow underfill process. Nano-silica incorporated no-flow underfill showed potential of a highly reliable flip-chip package with a SMT-friendly no-flow underfill process. However, fundamental understanding of the nano-silica and its interaction with underfill and solder is still lacking, and thus further development is needed to optimize the materials and processes.

9.6 Molded Underfill

Epoxy Molding Compounds (EMCs) have been practiced in component packaging for a long time. The novel idea of combining over-molding and the underfill together results in a molded underfill [61, 62]. Molded underfill is applied to a flip-chip in package via a transfer molding process, during which, the molding compound not only fills the gap between the chip and the substrate but also encapsulates the whole chip [63]. It offers the advantages of combining the underfilling and transfer molding into one step for reduced process time and improved mechanical stability [64]. It also utilizes EMCs which have long been proven to provide superior package reliability. Compared with the conventional underfill which is usually filled with silica at around 50–70 wt%, molded underfill can afford a much higher filler content up to 80 wt%, which offers a low CTE closely match with the solder joint and the substrate. Also, compared with the conventional molding compound, molded underfill requires fillers in smaller size, which also can contribute to lower the CTE of the material [65]. Molded underfill is especially suitable for flip-chip package to improve the production efficiency. It was reported that a four-fold production rate increase can be expected using molded underfill versus a conventional underfill process [66].

Molded underfill resembles the pressurized underfill encapsulation [67] in the mold design and process except that the materials in use are not liquid encapsulants that only fill up the gap between the chip and the substrate, but rather molding compounds that over-mold the entire components. Figure 9.12 shows

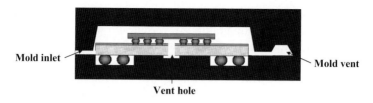

Fig. 9.12 Design of flip-chip BGA with molded underfill

a design of the mold for flip-chip ball grid array (FCBGA) components using molded underfill.

The design of the mold faces the challenge that the flip-chip geometry has a higher resistance to the mold flow so that air might be trapped under the chip. In fact, voids have been observed in the molded underfill packages using acoustic microscope [68]. Several molding processes can be used to minimize this geometry effect [69]. One way is to use mold vents as shown in Fig. 9.12 and to use also geometrical optimization to create similar flow resistance over and under the chip. One can also use vacuum assisted molding to prevent air entrapment. Another approach is to design a cavity in the substrate as shown in Fig. 9.12 . Though it requires a special design on the substrate, this method has proved to be a robust process and is commonly adopted.

Important process parameters in a molded underfill process include the molding temperature, clamp force, and injection pressure [53]. High temperature molding is favored for lower viscosity of the molding compound and hence better flow properties and less stress on the solder joint. However, the upper limit of the molding temperature is the melting point (Tm) of the solder material. Temperature near Tm combined with high injection pressure might cause the solder to melt and even the die to be "swept" away from the site. Also low Tg substrate is likely to be damaged at high molding temperature and high clamp force. The overflow of the molding compound might contaminate other contact pads or testing pads on the substrate. Bump cracking and die cracking are likely to occur as a result of high injection pressure. In short, a successful molded underfill process requires a combined effort in material selection, mold design, and process optimization. But the potential cost reduction and reliability enhancement of molded underfill is attracting great efforts in the industry.

9.7 Wafer Level Underfill

The invention of no-flow underfill eliminates the capillary flow and combines fluxing, solder reflow and underfill curing into one step, which greatly simplifies the underfill process. However, as pointed out in the previous text, no-flow underfill has some inherent disadvantages including the unavailability of a heavily filled material, which is a big concern for high reliability packages. Also, no-flow process still needs individual underfill dispensing step and therefore is not totally transparent to standard SMT facilities. An improved concept, wafer level underfill, was proposed as a SMT-compatible flip-chip process to achieve low cost and high reliability [70, 71, 72, 73]. The schematic process steps are illustrated in Fig. 9.13. In this process, the underfill is pre-applied either onto a bumped wafer or a wafer without solder bumps, using a proper method, such as printing or coating. Then the underfill is B-staged and wafer is diced into single chips. In the case of unbumped wafer, the wafer is bumped before dicing when the underfill can be used as a mask. The individual chips are then placed onto the substrate by standard SMT assembly equipment.

It is noted that in some types of WLCSP, a polymeric layer is also used on the wafer scale to redistribute the I/O and/or to enhance the reliability. However, this polymeric layer usually does not glue with the substrate and cannot be considered as underfill. The wafer level underfill discussed here is an adhesive to glue chip and substrate together and functions as a stress-redistribution layer rather than a stress-buffering layer. The attraction of the wafer level underfill lies in its low cost potential (since it does not require a significant change in the wafer back-end process) and high reliability of the assembly enhanced with the underfill. However, the wafer level underfill faces critical material and process challenges including uniform underfill film deposition on the wafer, B-stage process for the underfill, dicing and storage of B-staged underfill, fluxing capability, shelf-life, solder wetting in presence of underfill, desire for no post-cure, and reworkability, etc. Since the wafer level underfill process suggests a convergence of front-end and back-end of the line in package manufacturing, close cooperation between chip manufacturers, package companies, and material suppliers are required. Several cooperated research programs in this area been carried out [74, 75, 76]. Innovative ways of addressing the above issues and examples of wafer level processes are presented in this chapter.

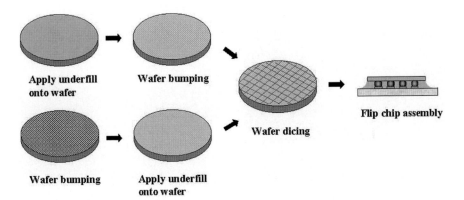

Fig. 9.13 Process steps of wafer level underfill

In most wafer level underfill process, the applied underfill must be B-staged before the singulation of the wafer. The B-stage process usually involves partial curing, solvent evaporation, or both, of the underfill. In order to facilitate dicing, storage, and handling, the B-staged underfill must appear solid-like and possess enough mechanical integrity and stability after B-stage. However, in the final assembly, the underfill is required to possess "reflowability", i.e., the ability to melt and flow to allow the solder bumps to wet the contacting pads and form solder joints. Therefore, the control of the curing process and the B-stage properties of the underfill is essential for a successful wafer level underfill process. A study conducted in Georgia Tech utilized the curing kinetics model to calculate the degree of cure (DOC) evolution of different underfills during solder reflow process [77]. Combined with the gelation behavior of the underfills, the solder wetting capability during reflow was predicted and confirmed experimentally. Based on the B-stage process window and the material properties of the B-staged underfill, a successful wafer level underfill material and process was developed. Full area array at 200 μm pitch flip-chip assembly with the developed wafer level underfill was also demonstrated [78] as shown in Fig. 9.14. (a) An optical image of wafer level underfill coated wafer with a.

The above study shows that the control of B-stage process of the wafer level underfill is critical to achieve good dicing and storage properties and the solder interconnect on board-level assembly. One way to avoid dicing in presence of non-fully cured underfill is presented in Fig. 9.15, a wafer scale applied reworkable fluxing underfill process developed by Motorola, Loctite and Auburn University [74]. Since uncured underfill materials are likely to absorb moisture that leads to potential voiding in the assembly, in this process, wafer is diced prior to underfill coating. Two dissimilar materials are applied; the flux layer coating by screen or stencil printing and the bulk underfill coating by a modified screen printing to keep the saw street clean. The separation of the flux from the bulk underfill material preserves the shelf life of the bulk underfill as well as

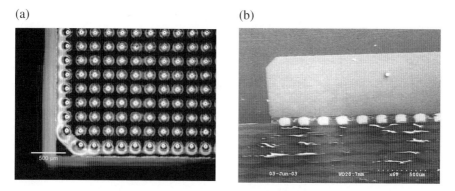

Fig. 9.14 (a) An optical image of wafer level underfill coated wafer with a 200 μm bump and (b) a cross-sectional image of solder joints after the die assembled on the subtrate

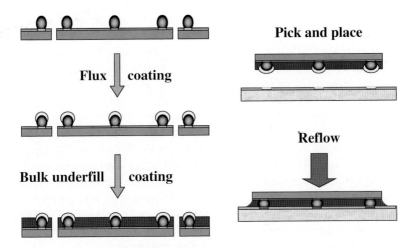

Fig. 9.15 A Wafer scale applied reworkable fluxing underfill process

prevents the deposition of fillers on top of the solder bump so as to ensure the solder joint interconnection in the flip-chip assembly. In this process, no additional flux dispensing on board is needed and hence the underfill needs to be tacky in the flip-chip bonding process to ensure the attachment of the chip to the board, as discussed in the previous text.

Underfill deposition on a wafer using liquid material via coating or printing requires subsequent B-staging, which is often tricky and problematic. The process developed by 3 M and Delphi-Delco circumvents the B-stage step using film lamination [79]. The process steps are shown in Fig. 9.16, in which the solid film

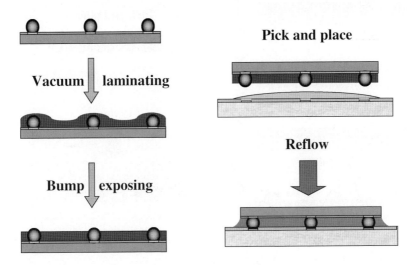

Fig. 9.16 A Wafer-applied underfill film laminating process

comprised of thermoset/thermoplastic composite is laminated onto the bumped wafer in vacuum. Heat is applied under vacuum to ensure the complete wetting of the film over the whole wafer and to exclude any voids. Then a proprietary process is carried out to expose the solder bump without altering the original solder shape. The subsequent flip-chip assembly is carried out in a no-flow underfill-like process in which a curable flux adhesive is applied on the board and then the assembly is reflowed.

Wafer level underfill can also be applied before the bumping process. Figure 9.17 shows a multi-layer wafer-scale underfill process developed by Aguila Technologies, Inc. [80]. The highly filled wafer level underfill is screen printed onto an unbumped wafer and then cured. Then this material is laser-ablated to form microvias that expose the bond pads. The vias are filled with solder paste. After reflowing, solder bumps are formed on top of the filled vias. The flip-chip assembly is similar to no-flow underfill process again with a polymer flux dispensed onto the board before chip placement.

One similarity among all these three processes is the separation of flux material from the bulk underfill. Wafer level underfill process provides the convenience of separating different functionalities by using dissimilar materials so that "the one magic material that solves everything" is not required. However, it is likely to create inhomogeneity inside the underfill layer, the impact of which on the reliability is not fully understood.

A novel photo-definable material which acts both as a photoresist and an underfill layer applied on the wafer level was reported by Georgia Tech [81]. In the proposed process as shown in Fig. 9.18, the wafer level underfill is applied on the un-bumped wafer, and then is exposed to the UV light with a mask for

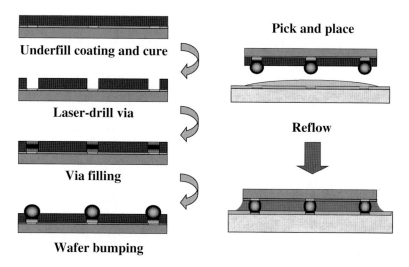

Fig. 9.17 A multi-layer wafer-scale underfill process

9 Flip-Chip Underfill

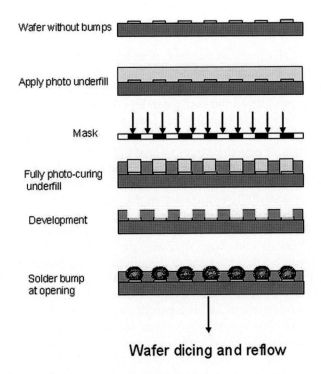

Fig. 9.18 A photo-definable wafer level underfill process

crosslinking. After development, the un-exposed material is removed and the bump pads on the wafer are exposed for solder bumping. The fully cured film is left on the wafer and acts as the underfill during the subsequent SMT assembly after device singulation. A polymeric flux is needed during the assembly for holding the device in place on the substrate and providing fluxing capability, a process similar to the dry film laminated wafer level underfill. In order to enhance the material property, the addition of silica fillers is necessary. In this case, nano-sized silica fillers were used to avoid UV light scattering which hinders the photo-crosslinking process. The nano-sized fillers also resulted in an optical transparent film on the wafer to facilitate the vision recognition during dicing and assembly process. The photo-definable nano-composite wafer level underfill presents a cost-effective way of applying wafer level underfill and has potentially fine-pitch capability.

Since wafer level underfill is a relatively new concept and most researches are still in the process and material development stage, there are few reports on the reliability of a flip-chip package using wafer level underfill. Although there is no standard process for wafer level underfill yet, the final decision might depend on the wafer and chip size, bump pitch, package type, etc. Like wafer level CSP, multiple solutions may co-exist for wafer level underfill process.

9.8 Summary

Flip-chip offers many advantages over other interconnection technologies and has been practiced in many applications. Underfill is necessary for a reliable flip-chip on organic package but is process-unfriendly and becomes the bottleneck of a high production flip-chip assembly. As silicon technology moves to nanometer nodes with feature size less than 65 nm, shrinking pitch and gap distance, as well as the introduction of lead-free solder and low-K ILD/Cu interconnect present new challenges to underfill materials and process. Many variation of the conventional underfill have been invented to address the problem, among which, the newly developed no-flow underfill, molded underfill, and wafer level underfill have attracted much attention. The no-flow underfill process simplifies the conventional flip-chip underfill process by integrating flux into the underfill, eliminating capillary flow, and combining solder reflow and underfill cure into one step. However, the pre-deposited underfill cannot contain high level of silica filler due to the interference of filler with solder joint formation. The resulting high CTE of the underfill limits the reliability of the package. Various ways have been explored to enhance the reliability through improved fracture toughness of the underfill material, low Tg and low modulus underfill, and the incorporation of fillers using other process approaches. Molded underfill combines underfill and over-mold together and is especially suitable for flip-chip in package to improve the capillary underfill flow and the production efficiency. Careful material selection, mold design, and process optimization are required to achieve a robust molded underfill process. Wafer level underfill presents a convergence of front-end and back-end in packaging manufacturing and may provide a solution for low cost and high reliable flip-chip process. Various material and process issues including underfill deposition, wafer dicing with underfill, shelf-life, vision recognition, chip placement, and solder wetting with underfill, etc., have been addressed through novel material development and different process approaches. Although the research is still in the early stage and there is no standard in the process yet, there have been considerable successes in demonstrating the process, which looks promising for the future packaging manufacturing. All of these three approaches require close cooperation between the material suppliers, package designers, assembly companies, and maybe chip manufacturers. A good understanding in both the materials and the processes and their inter-relationship is essential to achieve successful packages.

References

1. C.P. Wong, S. Lou, and Z. Zhang, "Flip the Chip", Science, Vol. 290, p. 2269, Dec, 2000.
2. E. Davis, W. Harding, R. Schwartz, and J. Coring, "Solid Logic Technology: Verdatile High Performance Microelectronics", IBM Journal of Research & Development, Vol. 8, p. 102, 1964.

3. F. Nakano, T. Soga, and S. Amagi, "Resin-Insertion Effect on Thermal Cycle Resistivity of Flip-Chip Mounted LSI Devices", The Proceedings of the International Society of Hybrid Microelectronics Conference, p. 536, 1987.
4. Y. Tsukada, "Surface Laminar Circuit and Flip-Chip Attach Packaging", Proceedings of the 42nd Electronic Components and Technology Conference, p. 22, 1992.
5. B. Han and Y.Guo, "Thermal Deformation Analysis of Various Electronic Packaging Products by Moire and Microscope Moire Interferometry", Journal Electronic Packaging, Vol. 117, p. 185, 1995.
6. S. Han and K.K. Wang, "Analysis of the Flow of Encapsulant During Underfill Encapsulation of Flip-Chips", IEEE Transactions on Components, Packaging, and Manufacturing Technology, Part B, Vol. 20, No. 4, pp. 424–433, 1997.
7. S. Han, K.K. Wang, and S.Y. Cho, "Experimental and Analytical Study on the Flow of Encapsulant During Underfill Encapsulation of Flip-Chips", Proceedings of the 46th Electronic Components and Technology Conference, pp. 327–334, 1996.
8. L. Nguyen, C. Quentin, P. Fine, B. Cobb, S. Bayyuk, H. yang, and S.A. Bidstrup-Allen, "Underfill of Flip Chip on Laminates: Simulation and Validation", IEEE Transactions on Components and Packaging Technology, Vol. 22, No. 2, pp. 168–176. 1999.
9. W.B. Young and W.L. Yang, "Underfill of Flip-Chip: The Effect of Contact Angle and Solder Bump Arrangement", IEEE Transactions on Advanced Packaging, Vol. 29, No. 3, pp. 647–653, 2006.
10. H. Bressers, P. Beris, J. Caers, and J. Wondergerm, "Influence of Chemistry and Processing of Flip Chip Underfills on Reliability", 2nd International Conference on Adhesive Joining and Coating Technology in Electronics Manufacturing, Stockholm Sweden, 1996.
11. J.B. Nysaether, P. Lundstrom, and J. Liu, "Measurements of Solder Bump Lifetime as a Function of Underfill Material Properties", IEEE Transactions on Components, Packaging and Manufacturing Technology, Part A, Vol. 21, No. 2, pp. 281–287, 1998.
12. R. Dudek, A. Schubert, and B. Michel, "Analyses of Flip Chip Attach Reliability", Proceedings of 4th International Conference on Adhesive Joining and Coating Technology in Electronics Manufacturing", pp. 77–85, 2000.
13. P. Palaniappan, P. Selman, D. Baldwin, J. Wu, and C.P. Wong, "Correlation of Flip Chip Underfill Process Parameters and Material Properties with In-Process Stress Generation", Proceedings of the 48th Electronic Components and Technology Conference, pp. 838–847, 1998.
14. L. Mercado and V. Sarihan, "Evaluation of Die Edge Cracking in Flip-Chip PBGA Packages", IEEE Transactions on Components and Packaging Technologies, Vol. 26, No. 4, pp. 719–723, 2003.
15. S.P. Lahoti, S.C. Kallolimath, and J. Zhou, "Finite Element Analysis of Thermo-hygro-mechanical Failure of a Flip Chip Package", Proceedings of IEEE 6th International Conference on Electronic Packaging Technology, 2005.
16. S. Luo and C.P. Wong, "Effect of Coupling Agents on Underfill Material in Flip Chip Packaging", Proceedings of 2000 International Symposium on Advanced Packaging Materials, pp. 183–188, 2000.
17. T. Chen, J. Wang, and D. Lu, "Emerging Challenges of Underfill for Flip Chip Application", Proceedings of the 54th Electronic Components and Technology Conference, pp. 175–179, 2004.
18. J.S. Hwang, "Lead-Free Solder: the Sn/Ag/Cu System", Surface Mount Technology, p. 18, July 2000.
19. B. Huang and N.C. Lee, "Prospect of Lead Free Alternatives for Reflow Soldering", Proceedings of SPIE – The International Society for Optical Engineering, Vol. 3906, p. 771, 1999.
20. A. Butterfield, V. Visintainer, and V. Goudarzi, "Lead-Free Solder Paste Flux Evaluation and Implementation in Personal Communication Devices", Proceedings of the 50th Electronic Components and Technology Conference, p. 1420, 2000.

21. S. Mahalingam, K. Goray, and A. Joshi, "Design of Underfill Materials for Lead Free Flip Chip Application", Proceedings of 2004 IEEE International Society Conference on Thermal Phenomena, pp. 473–479, 2004.
22. C.K. Chee, Y.T. Chin, T. Sterrett, Y. He, H.P. Sow, R. Manepali, and D. Chandran, "Lead-free Compatible Underfill Materials for Flip Chip Application", Proceedings of the 52nd Electronic Components and Technology Conference, pp. 417–424, 2002.
23. P. Tsao, C. Huang, M. Li, B. Su, and N. Tsai, "Underfill Characterization for Low-k Dielectric/Cu Interconnect IC Flip-Chip Package Reliability", Proceedings of the 54th IEEE Electronic Components and Technology Conference, pp. 767–769, 2004.
24. S. Rajagopalan, K. Desai, M. Todd, and G. Carson, "Underfill for Low-K Silicon Technology", Proceedings of 2004 IEEE/SEMI International Electronics Manufacturing Technology Symposium, 2004.
25. R. Pennisi and M. Papageorge, "Adhesive and Encapsulant Material with Fluxing Properties", U.S. Patent 5,128,746, (July 7, 1992).
26. C.P. Wong and D. Baldwin, "No-Flow Underfill for Flip-Chip Packages", U.S. Patent Disclosure, April 1996.
27. C.P. Wong and S.H. Shi, "No-Flow Underfill of Epoxy Resin, Anhydride, Fluxing Agent and Surfactant", U.S. Patent 6,180,696, (Jan. 30, 2001).
28. C.P. Wong, S.H. Shi, and G. Jefferson, "High Performance No Flow Underfills for Low-Cost Flip-Chip Applications", Proceedings of the 47th Electronic Components and Technology Conference, p. 850, 1997.
29. C.P. Wong, S.H. Shi, and G. Jefferson, "High Performance No-Flow Underfills for Flip-Chip Applications: Material Characterization", IEEE Transactions on Components, Packaging, and Manufacturing Technology, Part A: Packaging Technologies, Vol. 21, No. 3, p. 450, 1998.
30. Z. Zhang, S.H. Shi, and C.P. Wong, "Development of No-Flow Underfill Materials for Lead-Free Bumped Flip-Chip Applications", IEEE Transaction on Components, and Packaging Technologies, Vol. 24, No. 1, pp. 59–66, (2000).
31. Z. Zhang and C.P. Wong, "Development of No-Flow Underfill for Lead-Free Bumped Flip-Chip Assemblies", Proceedings of Electronics Packaging Technology Conference, pp. 234–240, Singapore, (2000).
32. Z. Zhang and C.P. Wong, "Study and Modeling of the Curing Behavior of No-Flow Underfill", Proceedings of the 8th International Symposium and Exhibition on Advanced Packaging Materials Processes, Properties and Interfaces, pp. 194–200, Stone Mountain, Georgia, (2002).
33. P. Morganelli and B. Wheelock, "Viscosity of a No-flow Underfill During Reflow and its Relationship to Solder Wetting", Proceedings of the 51st Electronic Components and Technology Conference, pp. 163–166, 2001.
34. R.W. Johnson, M.A. Capote, S. Chu, L. Zhou, and B. Gao, "Reflow-Curable Polymer Fluxes for Flip Chip Encapsulation", Proceedings of International Conference on Multichip Modules and High Density Packaging, 1998, pp. 41–46.
35. S.H. Shi and C.P. Wong, "Study of the Fluxing Agent Effects on the Properties of No-Flow Underfill Materials for Flip-Chip Applications", Proceedings of the 48th Electronic Components and Technology Conference, p. 117, 1998.
36. S.H. Shi, D. Lu, and C.P. Wong, "Study on the Relationship Between the Surface Composition of Copper Pads and No-Flow Underfill Fluxing Capability", Proceedings of the 5th International Symposium on Advanced Packaging Materials: Processes, Properties and Interfaces, p. 325, 1999.
37. S.H. Shi and C.P. Wong, "Study of the Fluxing Agent Effects on the Properties of No-Flow Underfill Materials for Flip-Chip Applications", IEEE Transactions on Components and Packaging Technologies, Part A: Packaging Technologies, Vol. 22, No. 2, p. 141, June 1999.

38. P. Palm, K. Puhakka, J. Maattanen, T. Heimonen, and A. Tuominen, "Applicability of No-Flow Fluxing Encapsulants and Flip Chip Technology in Volume Production", Proceedings of the 4th International Conference on Adhesive Joining and Coating Technology in Electronics Manufacturing, pp. 163–167, 2000.
39. K. Puhakka and J.K. Kivilahti, "High Density Flip Chip Interconnections Produced with In-situ Underfills and Compatible Solder Coatings", Proceedings of the 3rd International Conference on Adhesives Joining and Coating Technology in Electronics Manufacturing, pp. 96–100, 1998.
40. T. Wang, T.H. Chew, Y.X. Chew, and Louis Foo, "Reliability Studies of Flip Chip Package with Reflowable Underfill", Proceedings of the Pan Pacific Microelectronic Symposium, Kauai, Hawaii, February 2001, pp. 65–70.
41. Z. Zhang and C.P. Wong, "Assembly of Lead-Free Bumped Flip-Chip with No-Flow Underfills", IEEE Transactions on Electronics Packaging Manufacturing, in publication.
42. D. Miller and D.F. Baldwin, "Effects of Substrate Design on Underfill Voiding Using the Low Cost, High Throughput Flip Chip Assembly Process", Proceedings of the 7th International Symposium on Advanced Packaging Materials: Processes, Properties and Interfaces, 2001, pp. 51–56.
43. R. Zhao, R.W. Johnson, G. Jones, E. Yaeger, M. Konarski, P. Krug, and L. Crane, "Processing of Fluxing Underfills for Flip Chip-on-Laminate Assembly", Presented at IPC SMEMA Council APEX 2002, Proceeding of APEX, San Diego, CA, pp. S18-1-1 – S18-1-7, 2002.
44. T. Wang, C. Lum, J. Kee, T.H. Chew, P. Miao, L. Foo, and C. Lin, "Studies on a Reflowable Underfill for Flip Chip Application", Proceedings of the 50th Electronic Components and Technology Conference, pp. 323–329, 2000.
45. D. Gamota and C.M. Melton, "The Development of Reflowable Materials Systems to Integrate the Reflow and Underfill Dispensing Processes for DCA/FCOB Assembly", IEEE Transactions on Components and Packaging Technologies, Part C, Vol. 20, No. 3, p. 183, July 1997.
46. X. Dai, M.V. Brillhart, M. Roesch, and P.S. Ho, "Adhesion and Toughening Mechanisms at Underfill Interfaces for Flip-Chip-on-Organic-Substrate Packaging", IEEE Transactions on Components and Packaging Technologies, Vol. 23, No. 1, March 2000, pp. 117–127.
47. B.S. Smith, R. Thorpe, and D.F. Baldwin, "A Reliability and Failure Mode Analysis of No Flow Underfill Materials for Low Cost Flip Chip Assembly", Proceedings of 50th Electronic Components & Technology Conference, 2000, pp. 1719–1730.
48. K.S. Moon, L. Fan, and C.P. Wong, "Study on the Effect of Toughening of No-Flow Underfill on Fillet Cracking", Proceedings of the 51st Electronic Components and Technology Conference, 2001, pp. 167–173.
49. H. Wang and T. Tomaso, "Novel Single Pass Reflow Encapsulant for Flip Chip Application", Proceedings of the 6th International Symposium on Advanced Packaging Materials: Process, Properties, and Interfaces, 2000, pp. 97–101.
50. Z. Zhang, L. Fan, and C.P. Wong, "Development of Environmental Friendly Non-Anhydride No-Flow Underfills", IEEE Transactions on Components and Packaging Technologies, Vol. 25, No. 1, March 2002, pp. 140–147.
51. S.H. Shi, Q. Yao, J. Qu, and C.P. Wong, "Study on the Correlation of Flip-Chip Reliability with Mechanical Properties of No-Flow Underfill Materials", Proceedings of the 6th International Symposium on Advanced Packaging Materials: Processes, Properties and Interfaces, 2000, pp. 271–277.
52. S.H. Shi, and C.P. Wong, "Recent Advances in the Development of No-Flow Underfill Encapsulants – a Practical Approach towards the Actual Manufacturing Application", Proceedings of the 49th Electronic Components and Technology Conference, p. 770, 1999.

53. P. Miao, Y. Chew, T. Wang, and L. Foo, "Flip-Chip Assembly Development Via Modified Reflowable Underfill Process", Proceedings of the 51st Electronic Components and Technology Conference, 2001, pp. 174–180.
54. S. Kawamoto, O. Suzuki, and Y. Abe, "The Effect of Filler on the Solder Connection for No-Flow Underfill", Proceedings of the 56th Electronic Components and Technology Conference, 2006, pp. 479–484.
55. Z. Zhang, J. Lu, and C.P. Wong, Provisional Patent 60/288,246: "A Novel Process Approach to Incorporate Silica Filler into No-Flow Underfill", 5-2-2001.
56. Z. Zhang, J. Lu, and C.P. Wong, "A Novel Approach for Incorporating Silica Fillers into No-Flow Underfill", Proceedings of the 51st Electronic Components and Technology Conference, 2001, pp. 310–316.
57. Z. Zhang and C.P. Wong, "Novel Filled No-Flow Underfill Materials and Process", Proceedings of the 8th International Symposium and Exhibition on Advanced Packaging Materials Processes, Properties and Interfaces, 2002, pp. 201–209.
58. K.M. Gross, S. Hackett, D.G. Larkey, M.J. Scheultz, and W. Thompson, "New Materials for High Performance No-Flow Underfill", Symposium Proceedings of IMAPS 2002, Denvor, September, 2002.
59. K. Gross, S. Hackett, W. Schultz, W. Thompson, Z. Zhang, L. Fan, and C.P. Wong, "Nanocomposite Underfills for Flip Chip Application", Proceedings of the 53rd Electronic Components and Technology Conference, 2003, pp. 951–956.
60. Y. Sun, Z. Zhang, and C.P. Wong, "Fundamental Research on Surface Modification of Nano-Size Silica for Underfill Applications", Proceedings of the 54th Electronic Components and Technology Conference, 2004, pp. 754–760.
61. P.O. Weber, "Chip Package with Molded Underfill", U.S. Patent 6,038,136, (March 14, 2000).
62. P.O. Weber, "Chip Package with Transfer Mold Underfill", U.S. Patent 6, 157,086, (December 5, 2000).
63. K. Gilleo, B. Cotterman, and T. Chen, "Molded Underfill for Flip Chip in Package", High Density Interconnection, p. 28, June 2000.
64. T. Braun, K.F. Becker, M. Koch, V. Bader, R. Aschenbrenner, and H. Reichl, "Flip Chip Molding – Recent Progress in Flip Chip Encapsulation", Proceedings of 8th International Advanced Packaging Materials Symposium, March, 2002, pp. 151–159.
65. F. Liu, Y.P. Wang, K. Chai, and T.D. Her, "Characterization of Molded Underfill Material for Flip Chip Ball Grid Array Packages", Proceedings of the 51st Electronic Components and Technology Conference, 2001, pp. 288–292.
66. L.P. Rector, S. Gong, T.R. Miles, and K. Gaffney, "Transfer Molding Encapsulation of Flip Chip Array Packages", IMAPS Proceedings, 2000, pp. 760–766.
67. S. Han and K.K. Wang, "Study on the Pressurized Underfill Encapsulation of Flip Chips", IEEE Transactions on Components, Packaging, and Manufacturing Technology, Part B: Advanced Packaging, Vol. 20, N0. 4, pp. 434–442, 1999.
68. L.P. Rector, S. Gong, K. Gaffney, "On the Performance of Epoxy Molding Compounds for Flip Chip Transfer Molding Encapsulation", Proceedings of the 51st Electronic Components and Technology Conference, 2001, pp. 293–297.
69. K.F. Becker, T. Braun, M. Koch, F. Ansorge, R. Aschenbrenner, and H. Reichl, "Advanced Flip Chip Encapsulation: Transfer Molding Process for Simultaneous Underfilling and Postencapsulation", Proceedings of the 1st International IEEE Conference on Polymers and Adhesives in Microelectronics and Photonics, 2001, pp. 130–139.
70. S.H. Shi, T. Yamashita, and C.P. Wong, "Development of the Wafer-Level Compressive-Flow Underfill Process and Its Required Materials", Proceedings of the 49th Electronic Components and Technology Conference, p. 961, 1999.
71. S.H. Shi, T. Yamashita, and C.P. Wong, "Development of the Wafer-Level Compressive-Flow Underfill Encapsulant", Proceedings of the 5th International Symposium on Advanced Packaging Materials: Processes, Properties and Interfaces, p. 337, 1999.

72. K. Gilleo and D. Blumel, "Transforming Flip Chip into CSP with Reworkable Wafer-Level Underfill", Proceedings of the Pan Pacific Microelectronics Symposium, p. 159, 1999.
73. K. Gilleo, "Flip Chip with Integrated Flux, Mask and Underfill", W.O. Patent 99/56312, (November 4, 1999).
74. J. Qi, P. Kulkarni, N. Yala, J. Danvir, M. Chason, R.W. Johnson, R. Zhao, L. Crane, M. Konarski, E. Yaeger, A. Torres, R. Tishkoff, and P. Krug, "Assembly of Flip Chips Utilizing Wafer Applied Underfill", Presented at IPC SMEMA Council APEX 2002, Proceedings of APEX, San Diego, CA, pp. S18-3-1 – S18-3-7, 2002.
75. Q. Tong, B. Ma, E. Zhang, A. Savoca, L. Nguyen, C. Quentin, S. Lou, H, Li, L. Fan, and C.P. Wong, "Recent Advances on a Wafer-Level Flip Chip Packaging Process", Proceedings of the 50th Electronic Components and Technology Conference, pp. 101–106, 2000.
76. S. Charles, M. Kropp, R. Kinney, S. Hackett, R. Zenner, F.B. Li, R. Mader, P. Hogerton, A. Chaudhuri, F. Stepniak, and M. Walsh, "Pre-Applied Underfill Adhesives for Flip Chip Attachment", IMAPS Proceedings, International Symposium on Microelectronics, Baltimore, MD, 2001, pp. 178–183.
77. Z. Zhang, Y. Sun, L. Fan, and C.P. Wong, "Study on B-Stage Properties of Wafer Level Underfill", Journal of Adhesion Science and Technology, Vol. 18, No. 3, pp. 361–380 (2004).
78. Z. Zhang, Y. Sun, L. Fan, R. Doraiswami, and C.P. Wong, "Development of Wafer Level Underfill Material and Process", Proceedings of 5th Electronic Packaging Technology Conference, Singapore, pp. 194–198, December 2003.
79. R.L.D. Zenner and B.S. Carpenter, "Wafer-Applied Underfill Film Laminating", Proceedings of the 8th International Symposium on Advanced Packaging Materials, pp. 317–325, 2002.
80. R.V. Burress, M.A. Capote, Y.-J. Lee, H.A. Lenos, and J.F. Zamora, "A Practical, Flip-Chip Multi-Layer Pre-Encapsulation Technology for Wafer-Scale Underfill", Proceedings of the 51st Electronic Components and Technology Conference, pp. 777–781, 2001.
81. Y. Sun, Z. Zhang, and C.P. Wong, "Photo-Definable Nanocomposite for Wafer Level Packaging", Proceedings of the 55th Electronic Components and Technology Conference, p. 179, 2005.

Chapter 10
Development Trend of Epoxy Molding Compound for Encapsulating Semiconductor Chips

Shinji Komori and Yushi Sakamoto

Abstract Epoxy molding compounds (EMCs) have been used extensively as an encapsulation and protection material for semiconductor packages and must meet ever-evolving packaging requirements including moisture sensitivity level (MSL), moldability, and environmental and other reliability 1000 requirements. This chapter provides an overview of most recent development on various aspects of EMCs including advanced material development, molding process, and approaches to improve moldability, moisturized reflow resistance, warpage control of molded area array packages, and stress management for molding die with low-k interlayer dielectrics (ILD).

Keywords Epoxy molding compounds (EMCs) · Flame retardant · Moldability · Moisturized reflow resistance · Warpage · Low-k ILD

10.1 Introduction

Recently, semiconductor packages have been used in almost all electronic equipment and products including information terminals (such as cell phones and personal computers), digital cameras, video game consoles, and household electrical appliances (for example, refrigerators, laundry machines), and the range of applications is expanding year by year.

Epoxy molding compounds (EMCs) for semiconductor are used for protecting semiconductor chips from external environment, specifically from external physical forces such as impact and pressure, and external chemical forces such as moisture, heat, and ultraviolet ray, maintaining the electric insulating property, and providing the semiconductor package with a form allowing easier mounting on a print circuit board.

S. Komori (✉)
Information and Telecommunication Materials Laboratories (ITML), Electronic Device Materials Research Laboratory 1, Sumitomo Bakelite Co., Ltd., 20-7, Kiyohara Industrial Park Utsunomiya-city Tochigi prefecture, 321-3231 Japan

Along with the progress in high-density surface mounting technology for the purpose of more mounting of electronic components and also for the reduction in size and weight, increase in performance, and reduction in cost, semiconductor packages are made thinner, miniaturized, and denser, in accordance with the high-density mounting requirements. Specifically, QFP (Quad Flat Package), SOP (Small Outline Package) and the like are replaced with thinner TQFP (Thin Quad Flat Package) and TSOP (Thin Small Outline Package), which are in turn replaced with area-mounted packages, including bump-connected BGA (Ball Grid Array), CSP (Chip Scale Package) and the like which have a smaller mounting area and offer higher speed than the lead frame based packages. Production volumes of these newer packages are expanding consistently year after year. Some examples of the semiconductor packages are shown in Fig. 10.1, and a cross-sectional view of a couple of typical semiconductor packages is shown schematically in Fig. 10.2. In addition, the recent trend of the packaging technology is shown in Fig. 10.3.

On the other hand, with the increasing attention onto measures to cope with environmental issues such as environment protection, recycle, and reuse, there is a strong need for environmentally friendly products. The major approaches, which have been pursued for achieving environmentally sound semiconductor packages, include the removal of lead from the solder used in connecting

Fig. 10.1 Examples of some semiconductor packages

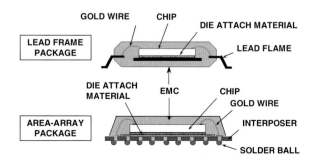

Fig. 10.2 Cross-sectional view of semiconductor packages

10 Development Trend of Epoxy Molding Compound

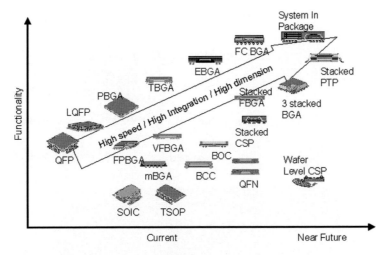

Fig. 10.3 Trend of the packaging technology

external terminals of the semiconductor packages, and the removal of halogen- and antimony-based flame retardants from the EMCs.

The trend in research and development of environmentally friendly EMCs that can withstand lead-free soldering and do not contain the halogen or antimony flame retardants will be reviewed in this chapter.

10.2 Introduction to Epoxy Molding Compounds

Various raw material ingredients are added to an epoxy molding compound (EMC) to meet the requirements on reliability, physical properties and moldability. Examples of some typical ingredients are epoxy resins, phenolic resins, fused silica as filler, coupling agents, curing promoter, and a release agent, all of which are important raw materials that influence the adhesion strength and moldability of the resulting product.

A typical composition of an EMC for semiconductor is shown in Fig. 10.4. These raw materials are mixed and kneaded under heat into homogeneous mixture in a kneader or a roll mixer. Generally, the materials are kneaded while cooled into a sheet shape, which is then pulverized. The powdery material is palletized into pellets, which are used in the transfer molding step (Figs. 10.5 and 10.6).

10.2.1 Epoxy Resin

Epoxy resins are advantageous because their physical properties are well balanced. For example, they have high adhesion strength, small shrinkage, superior chemical resistance and moisture resistance, relatively high heat

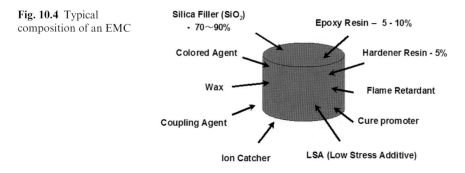

Fig. 10.4 Typical composition of an EMC

Fig. 10.5 An example of an epoxy molding compound in powder and pellet forms

resistance, and superior electrical properties. Also, use of epoxy is advantageous from the point of processing efficiency. For example, they offer relatively low temperature curing, short curing time and a low melt viscosity before curing.

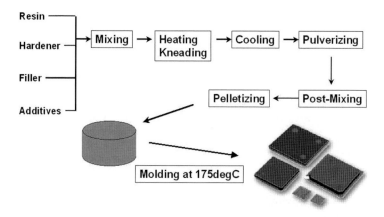

Fig. 10.6 Transferring molding process flow for an EMC

10 Development Trend of Epoxy Molding Compound   343

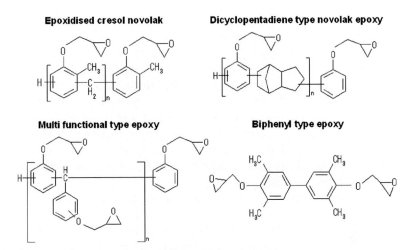

Fig. 10.7 Schematic of chemical structures of some common epoxy resins

Novolac epoxy resins that are solid epoxy resins have been commonly used for increasing cross linking density of epoxy network after curing. Fillers are used increasingly at a higher loading for reducing water absorption by semiconductor packages and reducing the dimensional change. Crystalline low-viscosity epoxy resins with bi-phenol structures have become widely utilized recently because high filler loading can potentially be incorporated in these resin systems (Fig. 10.7).

10.2.2 Hardener

Phenol novolac resins have been commonly employed as the hardeners for EMCs because of their excellent performance in terms of heat resistance, moisture resistance, electrical properties, curing property, and storage stability. It is easy to tailor the melt viscosity of an EMC that contains phenol novolak hardener because the molecular weight distribution of the material can be controlled by adjusting the number of repeating units in the novolak structure.

10.2.3 Inorganic Filler

Silica fillers are commonly used in EMCs for reducing the coefficient of thermal expansion (CTE) and reducing the moisture absorption of the EMCs. Crystalline silica powders are made of natural silica stone, while amorphous silica obtained after melting natural silica stone is called fused silica.

Fused silica is commonly manufactured from high purity natural silica stone. It is widely used for EMC because of its high purity, high chemical resistance, low thermal expansion, high electric insulating property, and low price.

The fused silica can be roughly categorized into flake silica and spherical silica. The flake silica is obtained by pulverizing melted silica stone, while the spherical silica is obtained by spontaneous sphericalization by surface tension when pulverized natural silica particles are melted and liquefied by spraying them into a flame of liquid petroleum gas and oxygen at high temperature. In contrast to the natural silica, silica prepared from an alkali silicate, silicon tetrachloride, or an alkoxysilane is called synthetic silica, which can also be either flake or spherical silica.

Because alpha (α) particles cause malfunctions (soft error) of a memory device represented by DRAM (Dynamic Random Access Memory), it is necessary to reduce the amount of α particles presented in an EMC as low as possible. As a result, the natural low alpha-particle fused silica with a radioactive element content less than 1.0 parts per billion (ppb) obtained from a natural silica stone containing a smaller amount of radioactive elements such as uranium (U) and thorium (Th), and synthetic silica with a radioactive element content less than 0.1 ppb are used favorably as the fillers for this application.

The flowability and the flash property of an EMC are influenced significantly by the particle shape, particle size distribution, maximum diameter, average diameter, and specific surface area of the fused silica. In particular, spherical fused silica is preferably used in a high-filler-loading EMC, because it is effective in reducing the viscosity of the EMC material during molding.

On the other hand, in the case of semiconductor package with larger internal heat generation, it is necessary to increase the thermal conductivity and the heat dissipation property of the package. High-thermal-conductivity filler such as crystalline silica, alumina, or silicon nitride is used as the filler for EMCs for this application.

10.2.4 Curing Promoter

The curing promoter is a catalyst promoting the reaction between the epoxy resin and the hardener. A promoter having a rapid curing property is desirable from the viewpoint of productivity. However, its impact to the electrical properties of an EMC after curing and storage stability at low temperature is an important consideration as well.

The transfer molding temperature is normally 170–180°C, considering the balance among the heat-resistant temperature of the semiconductor chip, curing chemistry, and the melt viscosity of the EMC. Organic phosphine compounds such as triphenylphosphine, organic phosphonium compounds, and amine compounds including amidine compounds such as DBU (1,8-diaza-bicyclo (5,4,0) undecene-7) are frequently used as curing promoters in this temperature range.

10.2.5 Silane-Coupling Agent

The silane-coupling agents that have long been utilized for strengthening the interface between an inorganic filler and an organic resin matrix, and are also used in semiconductor EMCs. Silane-coupling agents which have organic functional groups reacting with the epoxy resin and the phenol resin hardening agent include epoxysilane, aminosilane, and the like, and are effective in increasing the adhesion strength of the cured product [2].

In addition, surface treatment of the filler with a silane-coupling agent is known to be effective in improving the dispersion of the filler in the EMC and reducing the resin's viscosity during molding [3].

On the other hand, silane-coupling agents have been added for improving the adhesion strength between an EMC and the semiconductor chip surface and the relevant materials (silver, copper, gold, or 42 alloy as material of lead frame, polyimide as passivation material or the like). For example, mercaptosilane is effective in increasing the adhesion strength of an EMC to a lead frame with silver plating. It was reported that the mechanism of the adhesion strength improvement was due to the chemical bonding between lead frame and the silane [4].

10.2.6 Flame Retardant

The product of the EMC for semiconductor should satisfy the flammability rating of UL94 V-0, and, for that purpose, a flame retardant is usually added into EMC formulations.

In particular, a combined use of a brominated epoxy resin (monobromophenol novolak epoxy resin or tetrabromobisphenol A epoxy resin) and an antimony-based flame retardant such as antimony trioxide is effective synergically in increasing the flame-retarding efficiency, and has been widely used.

Recently however, due to environmental concerns, there is a strong need for a flame retardant without the halogen and antimony compounds because the halogen compounds may generate halogen gases during combustion and the antimony compounds may have chronic toxicity. As a result, application of various environmentally sound flame retardants has been studied in the field of EMC for semiconductor, and some of them were already commercialized.

In addition, flame resistant systems containing no flame retardant that meet the requirements of UL94 V-0 flame resistance specification were developed by making the resin structure self-extinguishing [5-6]. EMCs based on this technology were also commercialized [7].

It is possible to achieve the desired flame resistance by using a multi-aromatic ring-containing resin without increasing the filler loading. A resin having a high aromatic carbon ring content has a higher oxygen index, and thus is resistant to combustion. Multi Aromatic Resins (MARs), which contain many aromatic

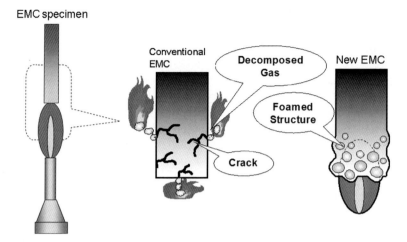

Fig. 10.8 A schematic illustration on flame retarding mechanism of new EMC with long cross-linking chains with comparison to conventional EMC

rings, are resistant to combustion and get carbonized easily instead. In addition, resins having a longer cross linking chains are more flexible at high temperature and generate foams with the volatile components formed, forming a surface protection film which blocks oxygen and heat (Fig. 10.8). Therefore, this type of resin itself is flame retardant and it is possible to prepare an environmentally friendly EMC without using such flame retardant as halogen or antimony compounds.

10.2.7 Other Additives

A colorant such as carbon black and a releasing agent such as natural or synthetic wax are also formulated in EMCs as almost essential components. An ion getter that captures ionic impurities such as Na^+ and Cl^-, a silicone- or synthetic rubber-based low-stress agent and the likes may also be added as needed.

10.3 Molding Process of EMC

A transfer molding method is normally used in encapsulating or packaging a semiconductor chip with an EMC.

As shown in Fig. 10.9, the transfer molding process include the following steps: (a) place a substrate, lead frame, and interposer with semiconductor chips in the cavities of a heated mold; (b) close the mold die tightly and feed EMC pellets through the mold pot under pressure with a plunger into each cavity;

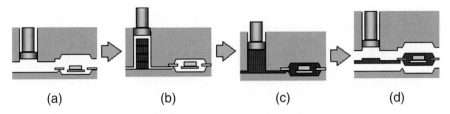

Fig. 10.9 Illustration of transfer molding process steps

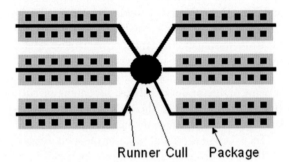

Fig. 10.10 Schematic illustration of EMC feeding method in a conventional transfer molding process

(c) keep the EMC compressed in the cavity under pressure until it is cured; and (d) open up the mold and release molded package.

There are two types of transfer-molding methods: (a) a conventional method where large-sized pellets are fed from one pot into multiple cavities on a large mold die (as shown in Fig. 10.10); and (b) a multi-plunger method where small tablets are fed from several pots into one or a few cavities present in a mold (Fig. 10.11).

The mold temperature is generally set between 170 and 180°C. The molding time is generally 120 s in the conventional method and 60–90 s in the multi-plunger method. The multi-plunger method is more advantageous and used more widely because of the uniformity of the cured EMC in different cavities, small amount of undesirable wastes such as cull and runner, and ease of automation.

Fig. 10.11 Schematic illustration of EMC feeding method in a multi-plunger molding process

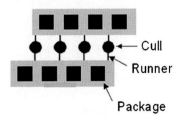

The semiconductor package after molding is normally post-cured to fully cure the EMC and achieve optimal properties. The post-curing condition is generally at 170–180 °C for 2–8 h.

10.4 Moldability

Improvement of moldability is important not only for increasing productivity but also for enhancing the reliability of semiconductor packages.

In transfer molding process, the uncured EMC pellets are melted in the mold and fed as a low-viscosity fluid into cavities. The heated epoxy resin hardens as it cures and finally forms a crosslinked structure. The hardening (or curing) process of the EMC during molding must be controlled carefully. Too much or too fast curing may lead to an excessive increase in viscosity and impact the flow of the molten EMC in the mold, possibly causing molding defect and damage of the constituents of semiconductor devices. If the hardening or curing is too slow, the EMC is not cured sufficiently when the mold is open, causing poor mold releasing and also leading to breakdown of the molded product, stain on mold die, and the like.

Therefore, it is critical to tailor the melt viscosity, flowability, curing property of the EMC properly according to the mold die and the semiconductor devices.

The releasing property of EMC need to be consistently good even after repeated molding processes for a long period of time for production. Also, easy releasing is essential for reducing the stress imposed to devices due to the sticking between EMC and mold die.

It is also important to reduce the melt viscosity and to enhance the curing property at the same time. High melt viscosity may break gold wires and/or sweep the gold wires that might cause opens and shorts.

Figure 10.12 shows the impact of the pressure applied to the mold die during molding, as an indicator of the viscosity of the molten EMC, to the gold wire deformation rate. It can be seen that higher molding pressure (due to higher melt viscosity of the EMC) causes more gold wire deformation.

For very thin semiconductor packages, the thickness of the molded product could be as thin as about dozens of microns. In such a package, it is essential

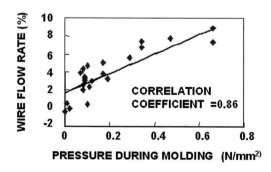

Fig. 10.12 Relationship between wire deforming rate and the viscosity of the molten EMC

that the molding compound be void-free after cure to provide the required reliability to the package. During molding, the viscosity of EMC needs to be kept low so that it can flow smoothly through the narrow gaps inside the mold die. One way to achieve this low viscosity is to slow down the curing reaction by reducing the amount of hardening accelerator in the EMC formulation. However, an extreme reduction of the viscosity of the resin or the amount of the accelerator may result in deterioration in productivity and releasing property.

10.5 Moisturized Reflow Resistance

10.5.1 Moisturized Reflow Resistance

When surface mounted packages such as QFP and SOP are employed for the first time in late 1980s, defects generated by infrared reflow process in the steps of mounting and soldering a semiconductor package on a circuit board attracted attention as a significant technical problem [8]. The defects include delamination at the package/substrate interface and package crack that are caused by drastic heating of the moisturized semiconductor package at high temperature in the reflow step.

Area-array packages such as BGA and CSP, which were commercialized in the later half of 1990's, tend to have the defects during the reflow step more frequently than the conventional lead frame packages such as QFP, demanding EMCs with even better moisturized reflow resistance. In area array packages, additional water absorption can occur due to the interposer substrate, also there are additional interfaces which have weaker adhesive to the EMC such as solder resist and gold plating. Moisture can easily penetrate through these interfaces. In addition, due to the non-symmetric structure (e.g. EMC only on one side of the substrate), the package tends to have higher warpage. Therefore, the desirable EMC need simultaneously to provide improvement in reflow resistance and reduction of package warpage.

In addition, lead-free solders such SnAg based solders are replacing conventional SnPb eutectic solders. Because the lead-free solders (such as SnAg) have a melting point 30–40°C higher than the conventional SnPb eutectic solders. The reflow temperature, which was traditionally around 230°C for eutectic SnPb solder, has to be raised to around 260°C. As a result, the dimensional change of the package during reflow and the stress imposed by moisture vapor pressure are increased, possibly generating more package defects.

10.5.2 Mechanism

The possible defect modes of molded packages during reflow are shown in Fig. 10.13.

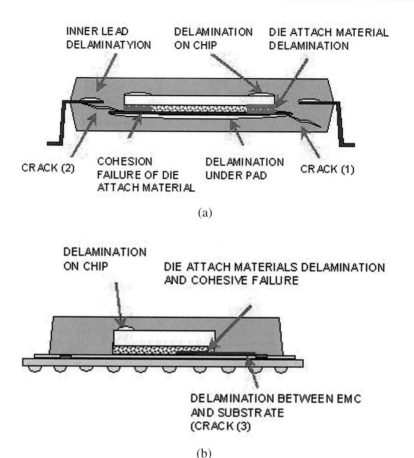

Fig. 10.13 Illustration of possible defect modes of leaded packages (**a**) and area-array packages (**b**) during reflow

Due to the mismatch of the coefficient of thermal expansion (CTE) of the various components in a semiconductor package, these components will thermally expand at differently rates during reflow, and thus internal stress is generated in the package. There is a clear correlation between the package defect rate and the amount of moisture absorbed by the package. As shown in Fig. 10.14, packages with higher moisture content before reflow tend to have higher defect rate after reflow.

One of the reasons for the delamination induced by water absorption is interfacial adhesion strength degradation by swelling of the EMC [9], other possible factors including the increase of thermal expansion coefficient, and decrease of Tg and adhesion strength seem to accelerate delamination and cracking during moisturized reflow.

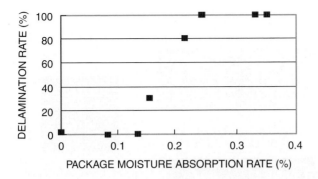

Fig. 10.14 Correlation between package delamination rate after reflow and the package moisture content before reflow in a BGA package (package: 35 mm ×35 mm PBGA; stress condition: 85°C/85%RH; IR reflow: peak temp = 260°C; delamination rate = # of packages with delamination / total number of package evaluated)

There are roughly two types of package cracking. The first type of cracking, which is labeled as crack (1) in Fig. 10.13 and is a delamination of the entire backside of the die, is caused by the moisture vapor pressure applied on the delaminating interface. The second type of cracking, which is shown as crack (2) and crack (3) in Fig. 10.13, is caused by propagation of the delamination which is generated by moisture vapor pressure and happens at the interfaces between the die attach material and the lead frame or between die attach adhesive and the backside of silicon chip.

The delamination and cracking in the package can be examined by microscopic inspection of a cross-section of the package (Fig. 10.15) or by nondestructive inspection such as a scanning acoustic microscope (SAM) (Fig. 10.16).

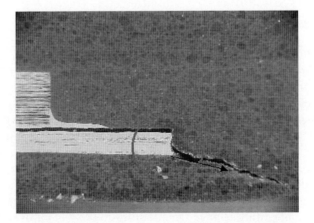

Fig. 10.15 A cross-sectional microscopic image of a crack in a molded package

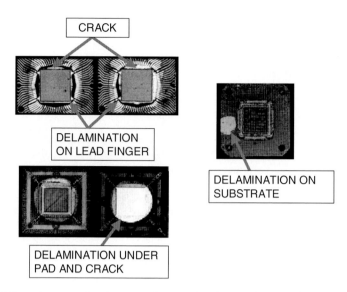

Fig. 10.16 Scanning acoustic microscopy images of various package crack and delamination

Because the reflow defect is highly related to the moisture content of a molded package, the moisture content of the package is preferably low before reflow. Even though the molded package might have low moisture content right after molding process, it will absorb moisture from the air after molding and before it is mounted onto a board through solder reflow. Therefore, semiconductor manufacturers generally will specify a guaranteed usable period of time for a molded package before it is processed through the reflow step.

10.5.3 Improvement on Moisturized Reflow Resistance

Moisturized reflow resistance of molded packages has been improved to certain extent by various approaches such as modifications of the die attach material, lead frame, and interposer. However, tailoring the properties of the EMC seems to be more effective.

The higher reflow temperature required for lead-free solder (for example 260°C) has a significantly negative effect on the properties of an EMC. Table 10.1 listed the properties of EMC at 260°C versus them at 240°C. The negative effect can be quantified by "stress/resistance ratio" which is expressed as the product of vapor pressure, thermal expansion and modulus over the product of adhesion strength and flexural strength. As can be seen from Table 10.1, temperature change from 240 to 260°C has an overall negative effect to the EMC properties by a factor of 2.

10 Development Trend of Epoxy Molding Compound

Table 10.1 Influence on the properties of EMC due to the higher reflow temperature

Parameters	Unit	240°C	260°C	Effect
Generate vapor pressure	MPa (RV)	3.2 (100)	4.5 (139)	–
Thermal expansion	RV	100	137	–
Adhesion strength	RV	100	90	–
Flexural modulus	RV	100	90	+
Flexural strength	RV	100	90	–

Stress/Resistance ratio change due to reflow temperature change from 240°C to 260°C
= stress increase/resistance change
= vapor pressure × thermal expansion × flexural modulus/(adhesion strength × flexural strength)
= $1.39 \times 1.37 \times 0.9/(0.9 \times 0.9) = 2.12$

10.5.3.1 Resin System with Low Water Absorption and Low Modulus

The resin system in an EMC has significantly influence on the overall properties of the EMC. The correlations among several key properties (flexural strength, elastic modulus, and water absorption) of a resin system are shown in Fig. 10.17, and will be discussed in details here.

Generally, the flexural strength of a resin system has a linearly proportional correlation with the elastic modulus and the water absorption (as shown in Fig. 10.17a & b). For example, a resin system consisting of a common cresol novolak epoxy resin and a phenol novolak resin has a high strength at high temperature, a high elastic modulus, and a high water absorption. As a result, solder reflow resistance of an EMC based on this resin system is low. To achieve good reflow resistance, an ideal resin system should have the property in the dotted circle regions shown in Fig. 10.17.

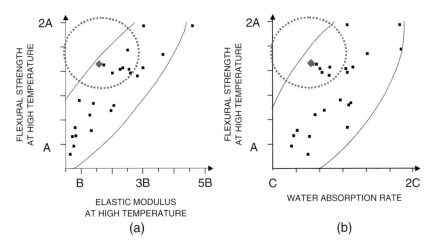

Fig. 10.17 Correlation diagram between flexural strength vs. elastic modulus (**a**) and flexural strength vs. water absorption (**b**)

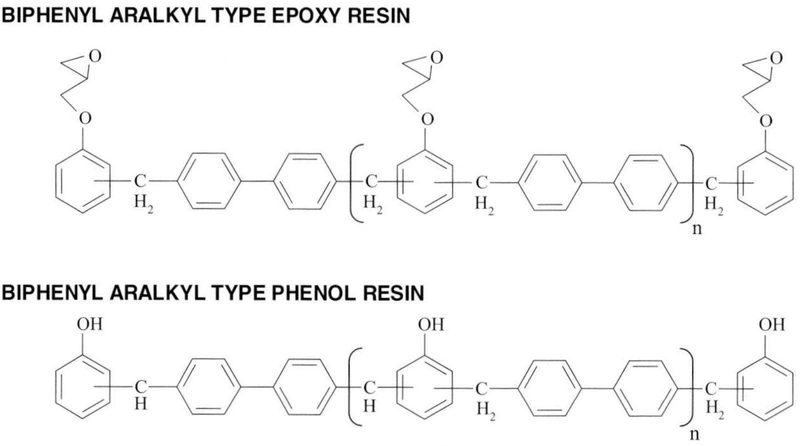

BIPHENYL ARALKYL TYPE EPOXY RESIN

BIPHENYL ARALKYL TYPE PHENOL RESIN

Fig. 10.18 Chemical structures of biphenyl aralkyl resin systems

According to the above concept, it is effective for improving EMC solder reflow resistance by introduce a multi-aromatic ring resin (MAR resin) and biphenyl aralkyl structure-containing resin into the base epoxy resin-phenol resin system. The chemical structures of these two resin systems are shown in Fig. 10.18. These new resin systems have low elastic modulus at high temperature due to its long chain between crosslinking-points and low water absorption due to the presence of hydrophobic structures (e.g. the aromatic rings). Using these resin systems, it is possible to formulate EMCs with higher solder reflow resistance because these resin systems have different correlation between the strength and the elastic modulus or water absorption.

Table 10.2 listed the property comparison of these resin systems (MAR and biphenyl) with conventional resin systems (ECN: Epoxidized cresol novolak, and DCP: Dicyclo pentadien type novolak epoxy). These new resin systems show higher resistance/stress ratios, and thus are more resistant to lead-free solder reflow.

Table 10.2 Resistance ability/stress comparison of resin systems (each value is relative to the value of Biphenyl system)

Resin system	ECN system	DCP system	Biphenyl system	MAR system
Epoxy	ECN	DCP	Biphenyl	MAR
Hardener	PN	PN	Xylylene Novolak	MAR
Water absorption	1.15	0.95	1.00	0.85
Thermal expansion	0.80	1.00	1.00	1.00
Flexural modulus	1.85	1.30	1.00	0.85
Adhesion strength	0.80	1.10	1.00	1.20
Flexural strength	1.40	1.05	1.00	0.95
Resistance/Stress ratio	0.66	0.94	1.00	1.56

Resistance/stress ratio = [adhesion strength × flexural strength]/[water absorption × flexural modulus × thermal expansion]

10 Development Trend of Epoxy Molding Compound

10.5.3.2 High Filler Loading Technique

In addition to the modification of resin structure, it is also effective to increase the filler loading for improving moisturized reflow resistance of an EMC.

As shown in Fig. 10.19, increasing filler loading leads to decreasing the moisture absorption, and thus decreasing the vapor pressure generated during reflow.

However, increase in filler loading raises the melt viscosity of the EMC and also decrease the adhesion strength because of the deterioration of moldability and wettability of EMC to the die, lead flame, substrate and the like.

Some work has been done to modify fused silica filler so that it will not increase melt viscosity much even at a high filler loading. For example, by utilizing a mixture of filler with optimized size distribution, a higher filler packing density (lower porosity), shown in Fig. 10.20(a), and higher filler loading can be achieved without increasing the melt viscosity of the EMC. Figure 10.20(b) shows a cross-sectional image of a cured EMC with fillers with optimized size distribution.

10.5.3.3 Improvement of Adhesion Strength

An epoxy resin with a small epoxy equivalence, which generates a higher OH group density after curing, is generally considered to have a better adhesion strength. However, from stress reduction point of view, high OH density is rather disadvantageous because it increases water absorption and elastic modulus at high temperature. The increase in filler loading, which is advantageous for reducing moisture absorption and coefficient of thermal expansion, may deteriorate the wettability and adhesion of EMC to the die, lead frame, and substrate because of the high melt viscosity of the EMC during molding.

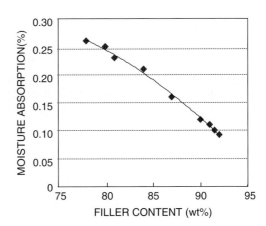

Fig. 10.19 Relationship between filler loading and water absorption of an EMC

Diameter of the filler	Filler Content	
R	100%	91.5%
0.414R	0	6.5%
0.225R	0	1.85%
Schematic illustration of filler packing	Porosity: 29.5%	Porosity: 19.0%

(a)

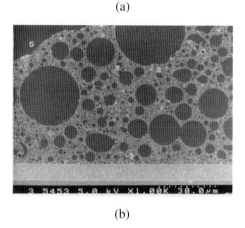

(b)

Fig. 10.20 Schematic illustration of the impact of filler size distribution to filler packing density (**a**), and a cross-sectional image of an EMC with fillers with optimized size distribution and high filler loading

A releasing agent wax, which is added for improving EMC releasability from the mold, has a possible effect of lowering the adhesion strength and causing delamination because the stress generated during mold release may tear off the bonded interfaces. Therefore, it is important to select the right types of releasing agent.

As in the example of the silane-coupling agent described above, it is also effective to add an additive to improve the adhesion strength of EMC to the die, lead frame, and substrate by forming chemical bonds at interfaces.

10.6 Warpage Improvement of Area-Array Packages

An area-array package which is molded only on one side, differently from lead frame packages, often show warpage when it is cooled down to room temperature after molding because of the CTE mismatch between the

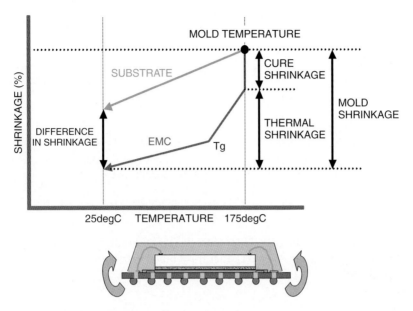

Fig. 10.21 Mechanism of warpage of area array packages

substrate and the EMC (Fig. 10.21). The package warpage leads to deterioration in the reliability of the connection between the circuit board and the solder bumps.

It is necessary to reduce both cure shrinkage and residual thermal shrinkage in order to decrease the warpage in an area array package. An EMC system (such as triphenolmethane-type epoxy resin and a hardener) with a high-Tg and low curing shrinkage is desired [10]. Another technique for reducing cure shrinkage is to increase the filler loading in the EMC.

Area-array packages have lower moisturized reflow resistance than lead frame packages and thus there is a need for improving the moisturized reflow resistance, in addition to warpage reduction. Therefore, EMCs with low curing shrinkage and excellent moisturized reflow resistance is highly desired.

In addition to reduce molding shrinkage, decreasing the elastic modulus of an EMC at a high temperature above the Tg is also effective in reducing the area array package's warpage (Fig. 10.22). A resin with a low water-absorption and low elastic modulus can be used to improve both warpage and moisturized reflow resistance [7].

There has been ongoing effort in reducing area array package warpage by modifying the interposer and the die attach material, in addition to the EMC.

As described above, it is a challenging task to improve both moldability and reliability simultaneously. There has been ongoing research work to develop

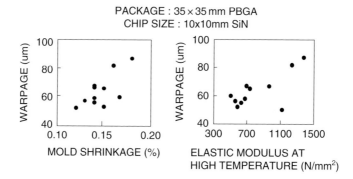

Fig. 10.22 Effect of mold shrinkage and elastic modulus to area array package warpage

new raw materials for the resin, filler, cure promoter, additives, and new mixing methods for formulating the next generation high performance EMCs for area array packages.

10.7 Challenges on Molding Low-K Dies

Recent high performance demands on ICs have led to the use of a low-k dielectric layers. The low-k layers, a fragile dielectric layer when compared with the conventional SiO_2 dielectric layer, requires low stress EMC properties over a wide range of temperatures.

Beyond the 0.13 μm generation, circuitry in the die becomes easy to damage as the low-k layers have more porosity than previous generations. This trend will continue in future generations with the low-k dielectric become even weaker mechanically (Fig. 10.23).

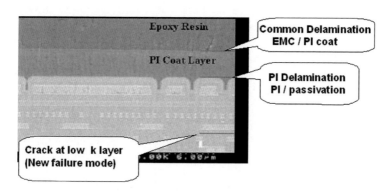

Fig. 10.23 Failure modes in low-k die

10.7.1 Stress Management

IC packages are composed of various materials with different CTEs:

Material	CTE (ppm/°C)
Molding compound	8–17
Lead frame	Copper – 17; Alloy42 – 7
BT substrate	13–17
Si die	3

The CTE mismatch between these components causes internal stress in the package. The stress generated is proportional to CTE, temperature and modulus (as shown in the following equation and also in Fig. 10.24):

$$S = \int E(T) \times \alpha(T) dT$$
$$S = (\alpha_1 - \alpha_i) \times E_1 \times (T - Tg) + (\alpha_1 - \alpha_i) \times E_2 \times (Tg - T_1)$$

S: Internal stress
α_1, α_2: CTE of EMC at below and above Tg
α_i: CTE of material (e.g. die)

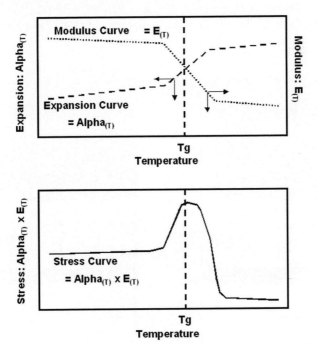

Fig. 10.24 Impact of EMC thermal expansion and modulus to the thermal stress

E_1 an E_2: modulus of EMC below and above Tg
Tg: glass transition of EMC
T_1: Molding temperature
T: Stress calculation temperature

Also, to reduce thermal stress, low stress additives and low stress resins can be employed. The low stress additives are used to decrease flexural modulus.

During cross-linking, low stress additives absorb package stress. However, the addition of low stress additive often causes higher water absorption. MAR resins have several unique properties such as low water absorption, low modulus, and self-extinguishing in flame tests, and thus are most suitable for molding low-k die applications.

10.7.2 FEM Study

FEM mechanical simulation has been done study the stress distribution in a package (Fig. 10.25).

As can be seen from Fig. 10.25, the stress generated is mainly concentrated on the chip surface. Therefore, stress reduction to protect low-k layer is of great importance.

Table 10.3 shows the FEM simulation results of stress under various combinations of CTE, modulus and Tg. It can clearly be seen that low CTE, low modulus and high Tg are the desirable properties for EMC in order to reduce the package stress.

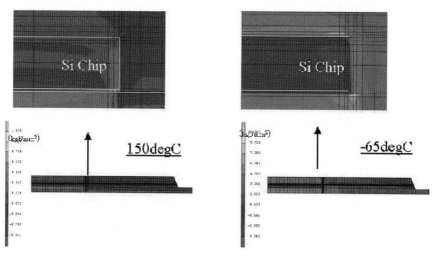

Fig. 10.25 FEM stress simulation results of a PBGA package

Table 10.3 FEM simulation results of stress for various sets of material properties of EMCs data (at −65°C)

EMCs		CTE (ppm/°C)		Modulus (kg/mm²)		Tg (°C)	Max. main stress (kg/mm²)	Shear stress (kg/mm²)
		α_1	α_2	< Tg	>Tg			
G770	Reference	9	43	2600	60	140	6.9	11.7
Model 1	CTE down	7	34	2600	60	140	5.0	8.6
Model 2	CTE up	11	52	2600	60	140	8.9	14.7
Model 3	E down	9	43	2100	50	140	6.2	10.1
Model 4	E up	9	43	3100	70	140	7.6	13.1
Model 5	Tg down	9	43	2600	60	110	9.6	15.9

10.7.3 EMC Evaluation

The relationship between filler content and internal stress was also studied (Fig. 10.26). The internal stress increases at lower temperatures and decreases with higher filler loading. Therefore, higher filler loading and lower modulus are necessary to reduce the stress. The MAR system shows lower modulus than the bi-phenyl system in case of same filler loading (as shown in Fig. 10.27). Therefore, MAR based EMCs will impose lower stress to the die than bi-phenyl based EMCs.

The following are the results of an experiment with a PBGA package (Fig. 10.28). The dielectric layer of a die is an organic porous low-k material and its modulus is 4 GPa. This die was intended to mimic the structure of a low-k die. The die was molded with a low stress additives (LSA) based EMC.

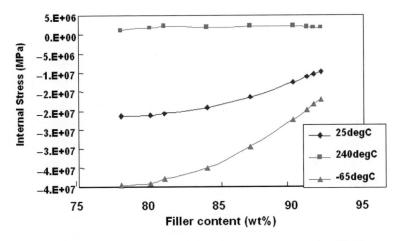

Fig. 10.26 Relationship between internal stress and the filler content of an EMC

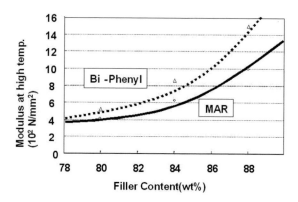

Fig. 10.27 Relationship between modulus and filler content of MAR and biphenyl based EMCs

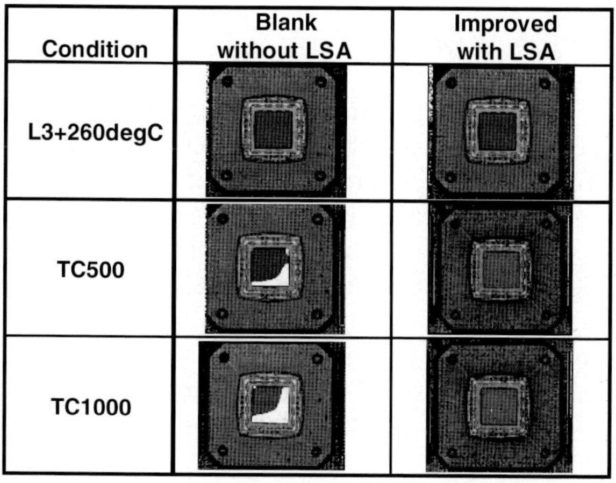

Condition	Blank without LSA	Improved with LSA
L3+260degC		
TC500		
TC1000		

Fig. 10.28 Scanning acoustic microscope (SAM) images of PBGA packages with a regular EMC and an EMC with LSA

After 500 thermal cycles, the package without LSA shows delamination in the low-k layer on the chip (white regions in the SAM images), which further propagated after 1000 cycles. However, the package with LSA based EMC showed no delamination, because the addition of LSA improved the temperature cycle resistance and reduced stress.

10.8 Future Trend

In the future, semiconductor packages will be becoming thinner, more miniaturized in size, and more sensitive to stresses. Also, advanced packages such as stack die package, SiP (System in Package), or the like makes the narrow gap filling during EMC molding more challenging, and thus the flowability of the EMC in the narrow gaps becomes more critical.

In addition, more strict requirements in reliability, productivity, and environmental factors are being imposed for the EMC for semiconductor. More research effort need to be conducted to develop more advanced EMCs to meet all these requirements.

References

1. Kazuo TAKAHASHI, Kogyo Zairyo, 42, 112 (1994)
2. Hiroyuki HOZOJI, Osamu HORIE, Shoji OGATA, Shunichi NUMATA, and Tokuyuki KINJO, Japanese Journal of Polymer Science and Technology, 47, 483 (1990)
3. The Society of Powder Technology, Japan Ed., "Comminution, Classification and Surface Modification", p. 566, N.G.T. (2001)
4. Akinobu KUSUHARA, Masumi SAKA, and Toshitsune ISHIGURO, Journal of the Adhesion Society of Japan, 35, 153 (1999)
5. M. IJI and Y. KIUCHI, Polym. Adv. Tchnol., 12, 393 (2001)
6. Masatoshi IJI, Yukihiro KIUCHI, Isao KATAYAMA, and Takayuki UNO, Electronic Materials, 2000, April, 86 (2000)
7. Yushi SAKAMOTO and Hiroki OSUGA, Electronics Mounting Technology, 18, 44 (2002)
8. Semiconductor and Integrated Circuits Division, Hitachi, Ltd. Ed., "Mounting Technology of Surface Mounted LSI Packages and Improvement of the Reliability Thereof" p. 502, Oyo Gijutsu Shuppan (1989)
9. Naotaka NAKA, Makoto KITANO, Tetsuo KUMAZAWA, and Asao NISHIMURA, Journal of Japan Society Mechanical Engineers, 63, 614 (1997)
10. Ken OOTA, Masumi SAKA, J. Polym. Eng. Sci., 41, 1373 (2001)

Chapter 11
Electrically Conductive Adhesives (ECAs)

Daoqiang Daniel Lu and C.P. Wong

Abstract Recently, significant advances have been made to improve electrically conductive adhesive (ECA) technology. Recent material development of various anisotropic conductive adhesives/films (ACAs/ACFs) and their applications are reviewed first and then recent research achievements in material development and in both electrical and mechanical aspects of isotropic conductive adhesives (ICAs) including electrical conductivity improvement, contact resistance mechanism elucidation and approaches to stabilize the contact resistance, and mechanical impact performance enhancement are reviewed in details.

Keywords Electrically conductive adhesives (ECAs) · ICAs · ACAs/ACFs · Conductivity · Conductive particles · Contact resistance · Corrosion inhibitor · Impact performance · Lead-free · Flip chip · Chip scale package (CSP) · Ball grid array (BGA) · Surface mount technology (SMT)

11.1 Introduction

Electrically conductive adhesives (ECAs) are composites of polymeric matrices and electrically conductive fillers. Polymeric matrices have excellent dielectric properties and thus are electrical insulators. The conductive fillers provide the electrical properties and the polymeric matrix provides mechanical properties. Therefore, electrical and mechanical properties are provided by different components, which is different from metallic solders that provide both the electrical and mechanical properties. ECAs have been with us for some time. Metal-filled thermoset polymers were first patented as electrically conductive adhesives in the 1950s [1–3]. Recently, ECA materials have been identified as one of the alternatives for lead-bearing solders for

---

D.D. Lu (✉)
Intel Corporation, Henkel Corporation, Henkel Loctite (China) Co., Ltd., No.90 Zhujiang Road, Yantai ETDZ, Shandong, China 264006,Chandler, AZ
e-mail: daniel.lu@cn.henkel.com

microelectronics packaging applications. There are two types of conductive adhesives: anisotropically conductive adhesives (ACAs) and isotropically conductive adhesives (ICAs).

11.2 Description of Anisotropically Conductive Adhesives

11.2.1 Overview

Anisotropically conductive adhesives (ACAs) represent the first major division of polymer bonding agents. The anisotropic class of adhesives provides unidirectional electrical conductivity in the vertical or Z-axis. This directional conductivity is achieved by using a relatively low volume loading of conductive filler (5–20 vol.%) [4–6]. The low volume loading is insufficient for interparticle contact and prevents conductivity in the X-Y plane of the adhesive. The Z-axis adhesive, in film or paste form, is interposed between the two surfaces to be connected. Application of heat and pressure to this stack-up causes conductive particles to be trapped between opposing conductor surfaces on the two components. Once electrical continuity is achieved, the dielectric polymer matrix is hardened by chemical reaction (thermosets) or by cooling (thermoplastics). The hardened dielectric polymer matrix holds the two components together and helps maintain the contact pressure between component surfaces and conductive particles. A series of sketches in Fig. 11.1 illustrates the attachment steps in achieving ACA joints. Anisotropically conductive adhesives have been developed for use in electrical interconnection, and various designs, formulations and processes have been patented in Europe, Japan, and the USA [6].

11.2.2 Categories

Broadly, ACAs fall into two categories: those that are anisotropically conductive before processing and those whose anisotropy arises as a result of processing. Their characteristics can be summarized as follows: (a) Pre-processing anisotropy results from materials characterized by an ordered system of conductor elements interspersed in an adhesive matrix film. They are always in the form of tape or sheet and are complicated to manufacture, requiring an adhesive film to be laser-drilled or etched then filled with conducting materials. They provide predictable contacts and are typically applied to a substrate as preforms. (b) Post-processing anisotropy results from materials that are a homogeneous mix of conductive fillers and adhesive matrix and which have no internal structure or order prior to processing. All adhesive pastes and some tapes fall into this category.

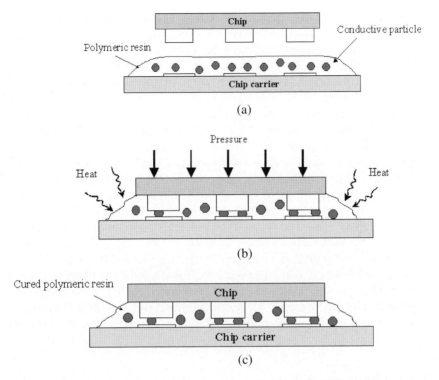

Fig. 11.1 A series of schematics illustrating the steps in forming an ACA joint. (**a**) Component parts: a bumped die and mating carrier with ACA spread over the surface. (**b**) Die is mounted with the carrier and held in place when cured. (**c**) Side view of the completed assembly

11.2.3 Adhesive Matrix

The adhesive matrix is used to form a mechanical bond at an interconnection. Both thermosetting and thermoplastic materials are used. Thermoplastic adhesives are rigid materials at temperatures below the glass transition temperature (Tg) of a polymer. Above the Tg, polymers exhibit flow characteristics. Thus, the Tg must be sufficiently high to avoid polymer flow during the application conditions, but the Tg must be low enough to prevent thermal damage to the associated chip carrier and devices during assembly. The principal advantage of thermoplastic adhesives is the relative ease with which interconnections can be disassembled for repair operations [7, 8]. However, thermoplastic ACAs suffer from many disadvantages. One of the most serious issues is that adhesion is not sufficient to hold the conductive particles in position, causing the contact resistance to increase after thermal shocks [7, 8]. Moreover, a phenomenon called "spring back" increases the contact resistance while the adhesive layer recovers from the stress caused by pressing of an ACA onto the components during bonding. This phenomenon, a creep characteristic exhibited by thermoplastic elastomers, occurs much after an ACA film has been heated to create the electrical joints. The contact resistance sometimes increases to more than three times the initial resistance during spring back (i.e. unloading) [7].

Thermosetting adhesives, such epoxies and silicones, form a three-dimensional cross-linked structure when cured under specific conditions. Cure techniques include: heat, UV-light, and added catalysts. As a result of the cure reaction that is irreversible, the initial uncross-linked material is transformed into a rigid solid. The thermosetting ACAs are stable at high temperatures and, more importantly, provide a low contact resistance. This results from a compressive force that maintains the conductive particles in intimate contact after the cure. That is, the shrinking caused by the cure reaction achieves a low contact resistance with long time stability. The ability to maintain strength at high temperature and robust adhesive bonds are the principal advantages of these materials. However, because the cure reaction is not reversible, rework or repair of interconnections is not an option [7, 8]. The choice of adhesive matrix and its formulation is critical to the long-term life properties of a composite. In practice, many options exist for the adhesive matrix. Acrylics can be used in low-temperature applications (under 100°C), while epoxies are more robust and can be used at higher temperatures (up to 200°C). Polyimide is used in the harshest environments where the temperature approaches 300°C [6].

11.2.4 Conductive Fillers

11.2.4.1 Solid Metal Particles

Conductive fillers are used to provide the adhesive with electrical conductivity. The simplest fillers are metal particles such as gold, silver, nickel, indium, copper, chromium and lead-free solders (SnBi) [6, 7, 9–11]. The particles are usually spherical and range 3–15 μm in size for ACA applications [12]. Needles or whiskers are also quoted in some patents [6].

11.2.4.2 Non-metal Particles with Metal Coating

Some ACA systems employ non-conductive particles with a thin metal coat. The core material is either plastic or glass with a metal coating consisting of gold, silver, nickel, aluminum or chromium. The basic particle shape of these systems is also spherical. Plastic-cored particles deform when compressed between opposing contact surfaces, thus provide a large contact area. Polystyrene (PS) is often selected as the core material because the coefficient of thermal expansion of metal-coated PS beads is very close thermoset adhesives. The combination of epoxy resin and metal-plated PS beads results in a large improvement in thermal stability [7]. In addition, glass can also be selected as the core material. Glass-cored particles coated with metal lead to a controlled

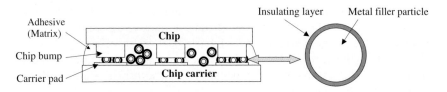

Fig. 11.2 Schematic depicting the cross section of an interconnection using a MCF filled ACA

bond-line thickness because the glass core is not deformable. Since the conductive particle size is known, the conductivity of the joint can be predicted.

11.2.4.3 Metal Particles with Insulating Coating

To achieve fine pitch connections, metal spheres or metal-coated plastic spheres coated with insulating resin fillers were developed. The insulating resin layer is only broken under pressure to expose the underlying conductive surfaces, referred to as a microcapsule filler (MCF). Higher filler loading can be achieved with MCFs for fine pitch applications to avoid creating electrical short circuit conditions between printed circuit features [7, 12]. A typical cross section of an ACA interconnection with microcapsule filler material is illustrated in Fig. 11.2.

11.3 Flip Chip Applications Using Anisotropically Conductive Adhesives

In traditional flip chip packages, solder bumps provide electrical connections between a chip and chip carrier. To achieve high reliability, organic underfill materials are usually required to fill the gap between the chip and chip carrier. The cured underfill creates a monolithic structure that evenly distributes the stress over all the material in the gap, not just on the solder connections. In the past several years, much research has been conducted to develop flip chip packages using ACAs in place of solder bumps. The primary advantages of ACA over lead-bearing solder for flip chips include ACAs' fine pitch capability, lead–free, low processing temperature, absence of flux residue, and generally lower cost. Also, ACA flip chip technology does not require an additional underfilling process because the ACA resin acts as an underfill.

ACA flip-chip technology has been employed in many applications where flip chips are bonded to rigid chip carriers [13]. This includes bare chip assembly of ASICs in transistor radios, personal digital assistants (PDAs), sensor chips in digital cameras, and memory chips in laptop computers. In all the applications, the common feature is that ACA flip-chip technology is used to assemble bare chips where the pitch is extremely fine, normally less than 120 μm. For those fine applications, it is apparent that the use of ACA flip-chips instead of soldering is more cost effective.

ACA flip-chip bonding exhibits better reliability on flexible chip carriers because the ability of flex provides compliance to relieve stresses. For example, the internal stress generated during resin curing can be absorbed by the deformation of the chip carrier. ACA joint stress analysis conducted by Wu et al. indicated that the residual stress is larger on rigid substrates than on flexible substrates after bonding [14].

11.3.1 ACA Flip Chip for Bumped Dies

11.3.1.1 Two Filler Systems

Y. Kishimoto et al. reported [15] anisotropic conductive adhesive pastes using two different fillers: Au-coated rubber particles (soft) and nickel particles (hard). The ACAs were used to bond a flip chip with Au plated bumps to a board with copper metallization. With the application of pressure, the soft particles were brought into contact with surface pads and were deformed which lowered this contact resistance. The hard particles, however, deformed the bumps and pads, thus were also in intimate contact with the surfaces to help reduce this contact resistance. The study showed that their choice of both hard and soft fillers in ACA materials had similar voltage-current behavior, and both exhibited stable contact resistance values after 1000 cycles of thermal cycling and 1200 h of 85°C/85%RH aging conditions [15].

11.3.1.2 Coated Plastic Filler

Casio developed an advanced anisotropic conductive adhesive film called the Microconnector (Fig. 11.3) [16–18]. This adhesive contains conductive particles made by coating plastic spheres with a thin layer of metal, followed by an additional 10 nm-thick layer of insulating polymeric material. The insulating layer consists of a large number of insulating micro powder particles that electrically insulate the outer surface of the spheres. The thin insulation layer is formed by causing insulating micro powder particles to adhere to the surface of the metal layer via electrostatic attraction. The base adhesive resin is thermoplastic or thermosetting, producing compressive force when cured. When heat and pressure are applied during bonding, the insulating layer, which is in contact with the bump surface of an IC, is broken. However, the insulating layer remains intact on conductive particles not crushed by the bonding pads, thereby producing only Z-axis electrical interconnections and preventing lateral short circuit conditions. With an additional insulating layer, a fine pitch and low contact resistance can be achieved without the risk of lateral short-circuiting by increasing the filler percentage (i.e. amount of particles per unit volume base adhesive resin or film). Casio is manufacturing pocket TV's with a liquid crystal using this material [18].

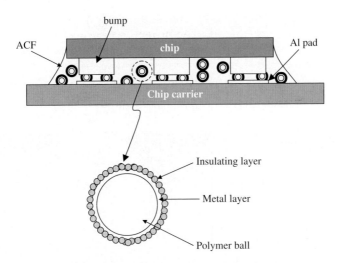

Fig. 11.3 Schematic depicting Casio's ACF technology – Microconnector

11.3.1.3 Solder Filler Systems

Unlike most commercial ACAs, where the electrical conductivity is based on the degree of mechanical contact achieved by pressing conductive particles to contact pads on board and chip bumps, solder-filled ACAs establish microscopic metallurgical interconnections. The advantage of these joints is that the metallurgical bonds that are established prevent electrical discontinuities from occurring should the adhesive polymeric matrix undergo relaxation during the operational lifetime. Therefore, solder-filled ACAs combine the benefits of both soldering and adhesive joining resulting in more reliable ACA joints. Furthermore, better electrical performance is achieved due to lower contact resistance established through the metallurgical bonds [19].

Joints made with SnBi-filled and Bi-filled ACAs experience brittle intermetallic compound formation and have problems with typical conductor and coating materials such as copper, nickel, gold, and palladium [20]. Bi and SnBi are, however, compatible with tin, lead, zinc and aluminum. Because Zn and Al are easily oxidized, only Sn and Pb are suitable surface finish materials for SnBi- and Bi-filled ACA applications. High quality interconnections were formed by metallurgically bonding SnPb-bumped chips on SnPb-coated substrates utilizing a Bi particle-filled ACA [21]. The joints once formed at a relatively low temperature could withstand a high temperature. The joint formation process is illustrated in Fig. 11.4. At the bonding temperature, 160°C, as the Bi particles have locally penetrated thin oxide layers on both SnPb surfaces, the liquid lentil formation occurs immediately. After the Bi particles have dissolved completely into the liquid lentils between the solid SnPb bumps and coating, more Sn and Pb will dissolve into the liquid lentils until the liquid has reached its equilibrium composition at the bonding temperature. After

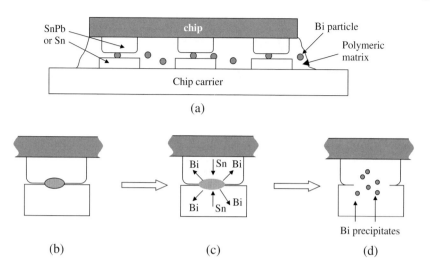

Fig. 11.4 Schematic illustration of the formation of electrical interconnects between a bumped chip and a mating carrier using a Bi-filled ACA. (**a**) The chip is aligned and placed on a chip carrier. (**b**) The Bi-particle is deformed between a chip bump and a carrier pad when a bonding pressure is applied. (**c**) The Bi-particle dissolves into the liquid lentils upon exposure of heat. (**d**) Bi diffuses into the Sn-Pb matrix and form fine solid precipitates

solidification, the dissolved Bi will precipitate out as very fine particles from the saturated solution. Since the melting is transient, the remelting of the solid lentils will happen at higher temperatures than the first melting. The remelting point of the solid lentils can be controlled by the concentration of Bi present in the joint. The formed ACA joints exhibited a stable resistance after 2000-h 85°C/85%RH aging or after 1000-h temperature cycle testing (−40–125°C). Even though this work is still preliminary, it demonstrates an interesting idea and concept. For lead-free applications, different materials such as pure Sn can be used for chip bumps and surface finish on the substrate [21].

11.3.1.4 Ni Filler

Toshiba Hino Works developed a flip chip bonding technology using an anisotropically conductive film (ACF) filled with nickel spheres and LSI chips with gold ball bumps for mobile communications terminals. A resin sealing process at the sides of the LSI chip was added to improve mechanical strength. An FR-5 glass epoxy chip carrier was utilized to improve heat resistance. The assembled pager sets passed qualification consisting of drop, vibration, bending, torsion, and high temperature testing. The process has been demonstrated capable of mass production utilizing full automation of the flip-chip bonding method capable of producing 30,000 pager modules per month [22].

11.3.2 ACA Bumped Flip Chips on Glass Chip Carriers

ACAs are probably the most common approach for flip chip on glass applications. The ACA flip chip on glass technology not only provides assemblies with a higher interconnection density and a thinner and smaller size but also has fewer processes and lower costs as compared with TAB (tape automated bonding) technology. Also, bonding IC chips directly to the glass of the LCD panel using ACAs is a better choice when the pitch becomes less than 70–100 μm. Small size and high resolution LCDs such as viewfinders, video-game equipment displays, or light valves for liquid-crystal projectors use flip-chip on glass technology for the IC connections.

11.3.2.1 Selective Tacky Adhesive Method

Sharp developed a flip-chip bonding approach that utilizes ACA technology depicted in Fig. 11.5 [23, 24]. The novel feature of the Sharp technology is the method of attaching electrical conductive particles onto IC termination pads. This "bumping" procedure consists of coating the wafer with a 1–3 μm thick UV curable adhesive. Coated wafers are irradiated with UV light in a standard photolithographic process while the Al pads on the IC are optically masked. As a result of this process, the thin adhesive film above the Al pads remains uncured and tacky, whereas the adhesive on other chip areas is cured. Due to the tackiness of the adhesive on the Al pads, conductive particles only easily adhere to these sites. The conductive particles utilized by Sharp are gold-coated polymer spheres. UV-curable adhesive is dispensed on LSI chips before being aligned with a glass carrier. While still applying pressure to maintain contact between the LSI chip and glass carrier, a light-setting adhesive is irradiated

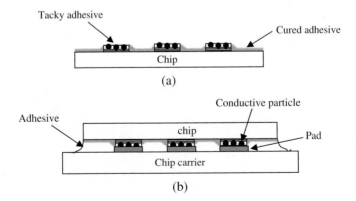

Fig. 11.5 Schematic depicting an ACA flip chip technology scheme utilized by Sharp. (**a**) Conductive particles adhere to uncured tacky adhesive on pad areas of a chip. (**b**) Chip remains in contact with the glass chip carrier since the adhesive exerts a compressive force after it is cured with light (UV)

with UV light. Even upon releasing the pressure, the chip terminations remain electrically connected to their mating carrier pads. This is due to the deformed conductive particles which remain in contact with these termination pads as a result of the compressive force exerted by the cured adhesive. This process has several advantages, among them is that no bump plating is required; and the bonding process can be done by irradiating with UV light at room temperature, thus other materials are not damaged due to the effects of heat. This packaging concept can potentially achieve a high throughput.

11.3.2.2 The MAPLE Method

Seiko Epson Corporation developed a new flip-chip on glass technology called the "MAPLE" (Metal-Insulator-Metal Active Panel LSI Mount Engineering) method. The MAPLE method is to bond ICs directly to a glass panel substrate using a thermosetting anisotropic conductive film containing uniformly distributed conductive Au particles. While typical flip-chip on glass technologies require several alignment steps, this bonding process is very simple. First an ACA sheet is placed on a glass panel. After aligning the IC bumps with mating glass panel pads and temporarily bonding, proper IC interconnections are established by permanently bonding at high temperature and pressure. It is necessary for the bonding press-tool surface to be flat and parallel to the IC [25]. Comparing Metal-Insulator-Metal (MIM) panel modules made with TAB, MIM panel modules utilizing the MAPLE approach had smaller panel fringe size, thinner panel thickness, fewer assembled sides, fewer processes, and simpler module structure. The panel modules utilizing MAPLE passed all the required reliability tests. The MAPLE approach is being used in mass production of MIM panel modules.

11.3.3 ACA Bumped Flip Chips for High Frequency Applications

In many low frequency applications, conductive adhesive joining has proved to be a cost effective and reliable solution. The high frequency behavior of ACA interconnections has attracted much attention in the past several years. The high frequency behavior of ACAs in flip chip packages has been reported by several investigators. Rolf Sihlbom et al. demonstrated that ACA-bonded flip chips can provide performance equivalent to solder flip chips in the frequency range of 45 MHz–2 GHz on FR4 chip carriers and 1–21 GHz on a high-frequency Telfon-based chip carrier. The different particle sizes and materials in the conductive adhesives gave little difference in high frequency behavior of ACA joints [26, 27].

Myung-Jin Yim et al. developed a microwave frequency model for ACF-based flip chip joints based on microwave network analysis and S-parameter measurements. By using this model, high frequency behavior of ACF flip chip interconnections with two filler particles, Ni and Au-coated polymer particles, was simulated. It was predicted that Au-coated polymer-particle-filled ACF flip-chip interconnections exhibited comparable transfer and loss characteristics to solder bumped flip chips up to about 13 GHz and thus they can be used for up to 13 GHz, but Ni-filled ACF joints can only be used for up to 8 GHz because the Ni particle has a higher inductance compared the Au-coated particle. Polymeric resins with a low dielectric constant and conductive particles with low inductance are desirable for high-resonance frequency applications [28].

11.3.4 ACA for Unbumped Flip Chips

Although ACAs are typically utilized with flip-chip bumped die, they are also used for unbumped flip chips in some cases. For unbumped flip chips, a pressure-engaged contact must be established by bringing the particles to the aluminum chip pads rather than a bump. The pressure must be sufficient to break the oxide on the aluminum pads. A sufficient quantity of particles must be trapped in the contact pad area and remain in place during bonding and curing to achieve a reliable interconnection. In addition to maximizing the number of particles in the contact area, the number of particles located between adjacent pads must be minimized to prevent electrical shorts. An additional factor that must be considered in the case of unbumped flip chips is adhesive flow during bonding and curing. It is essential to control the temperature heating rate to be sufficiently slow when the polymeric resin is curing so the conductive filler particles can migrate from the chip carrier side to the chip side pad [29].

11.3.4.1 Gold-Coated Nickel Filler

An application utilizing gold-coated nickel particles has been reported to provide reliable connection to unbumped flip chips [30]. Another study showed ACAs containing larger particles could accommodate planarity issues due to surface roughness, non-flat or non-parallel pads, compared to ACAs containing smaller particles. It was very difficult to obtain 100% consistency in conduction with unbumped flip-chip dice using ACAs with small diameter balls [31].

11.3.4.2 Ni/Au Coated Silver Filler

A flip-chip technology developed by Toshiba Corporation utilized an ACF to attach bare umbumped chips (with Al pads) onto a PCB with bumps formed from a silver paste screen printed on the PCB [32]. After curing, Ag bumps were formed (70 μm diameter, 20 μm height) which were subsequently over plated with Ni/Au. It was determined that an ACF with a low CTE (28 ppm/°C), low water absorption rate (1.3%), and utilizing a Au-plated plastic ball worked best. It was also found that Ni/Au plated Ag paste-formed bumps exhibited a lower initial connection resistance and a lower connection resistance increase as compared to Ag paste-formed bumps which were not overplated with Ni/Au.

11.3.4.3 Conductive Columns

Nitto Denko Corporation developed an anisotropic conductive film for fine-pitch flip chip applications [33]. The features of this ACF were: (1) connectable between bumpless chips and a fine-pitch printed circuit board; (2) high electrical conductivity; (3) repairable (easy to peal off chips from a printed circuit board at elevated temperatures); (4) high reliability; and (5) can be stored at room temperature. The other notable features are: usable at pitches down to 25 μm; the conductive elements are micro-metallic columns as opposed to random-shaped particles; and that this adhesive matrix consists of a thermoplastic polymer resin; the conductive columns are coated with an insulator; and a high Tg polymer which completely separated the columns from the adhesive.

It is easy to change the diameter of the conductive columns in order to make the film compatible with a variety of pitches. Sn/Pb or other solder materials are plated on both the top and bottom of the conductive columns (usually copper). The plated solder on the both ends of the conductive columns melts and forms metallurgical connections between the conductive columns and metal pads on a chip and the mating chip carrier, which ensures a good connection. Figure 11.6 (a) illustrates the cross section of the film structure. A rough surface, a result of plating, has the advantage of providing a good connection with the mating terminal pads. A typical terminal pad structure of a chip without bumps is shown in Fig. 11.6 (b). To achieve a good connection, the height of the conductive columns must be larger than the thickness of the passivation layer (t_p). Since t_b, the distance from a Cu pad surface of the chip carrier to the passivation layer surface of the chip is usually smaller than t_a (the distance from solder mask surface of the chip carrier to the passivation layer surface of the chip), the conductive columns will assume an inclined position during bonding if the thickness of the conductive columns is larger than the ACF thickness (t_{ACF}). It is important to adjust the thickness of board or chip carrier pads and ACF thickness to achieve good connection and adhesion. Reliability results indicate that the ACF possessing an adhesive matrix with a high Tg (282°C) exhibits high reliability; the contact resistance remained unchanged after 1000 cycles of accelerated thermal cycle testing (−25–125°C).

11.3.5 ACA Flip Chip for CSP and BGA Applications

Aiming at the CSP application market, Merix Corporation and Auburn University collaborated to develop anisotropic conductive adhesives called Area Bonding Conductive (ABC) adhesives. ABC adhesives are two-region

11 Electrically Conductive Adhesives (ECAs)

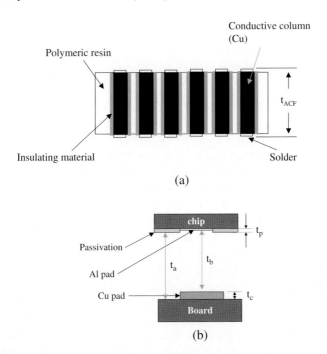

Fig. 11.6 Illustration of a scheme for fine-pitch, flip chip interconnection, (**a**) An anisotropic conductive film filled with micro metallic columns. (**b**) A typical cross sectional structure of a chip without bumps and the mating chip carrier

thermoset adhesives with electrically conductive adhesive pads surrounded by a continuous oxide filled dielectric adhesive to form a total area bond. Both regions are solvent free, B-staged, non-tacky epoxies supplied on a Mylar carrier release film. In contrast to conventional ACAs, conductive areas of ABC adhesive are only at bond pad locations. The ABC adhesives potentially can provide a reliable, low cost, low temperature, low pressure process for flip chip and CSP applications [34].

11.3.5.1 Double-Layered ACF Film

Motorola developed a low-cost and low-profile flip chip on flex CSP package using ACFs [35]. The package has the flexibility to utilize the existing wire-bonding pad configuration without adding prohibitive redistribution and wafer solder bumping costs, and eliminated the need for under-chip encapsulation. Two types of ACF film were studied: double layer films with the second layer loaded with Ni-Au plated polystyrene-divinylbenzene (PS-DVB) spheres, and solid Ni particles. The film structure, consisting of a nonfilled and a conducting particles-filled adhesive layers, is illustrated in Fig. 11.7. The

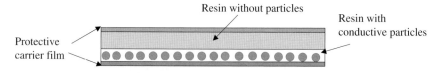

Fig. 11.7 A schematic of a double layer ACF

double-layer design reduces the particle density in x-y spacing of interconnection pads, which serve to enhance the x-y plane insulation characteristics. At the same time, the double-layer film provides more adhesive volume that helps to entrap more particles on the bonding interconnection pads. Both calculated and observed values show that the number of conducting particles trapped is much higher than those in single-layer ACF. This indicate that even though particle density in a double-layer ACF is low, conducting particles to effect electrical contact with both interconnection pads are trapped more effectively between interconnection pads in the double-layer ACF. The die has Ni/Au plated bonding pads and the chip carrier is flexible polyimide, which can provide adequate compensation for the planarity differences. Its compliant nature under compressive bonding operation allows the copper traces at the bonding area to deform and compensate for non-planarity or irregularity that exists. The ACF adhesive system provided the system with stable contact resistance after 500 cycles of the liquid to liquid temperature shock (LLTS) aging ($-55-125°C$).

11.3.5.2 Ceramic Chip Carriers vs. Organic Chip Carriers

A ceramic chip carrier and an organic chip carrier, whose configuration is equivalent to that of micro-ball grid array (μBGA) style chip scale package (CSP) and broadly representative of BGA and flip-chip devices, were evaluated using ACAs with conducting particles of various sizes [36]. The ceramic chip carrier has AgPd thick film bonding pads and the organic chip carrier is a conventional PCB (1 oz Cu clad FR5 laminate) with sub-micron Au-coated Cu pads. It was determined that uniform conductivity and high yield were more readily achieved with organic chip carrier rather than ceramic chip carriers. This is because bonding pads on the FR5 chip carrier have better coplanarity compared to the thick-film ceramic chip carrier. It was demonstrated that the optimum process conditions and adhesive material choice were very different for organic and ceramic chip carriers. ACAs with finer particles exhibited worse overall performance on both chip carriers, while ACAs with larger and polymer-cored particles exhibited better performance because the deformable polymer-cored particles compensated the gap variations between the chip bumps and the chip carriers.

11.3.6 SMT Applications

ACAs have been investigated as replacement for SnPb solder in surface mount attachment for fine pitch applications. In addition to providing a lead-free attachment solution, cost effectiveness is an additional key benefit. The key attractive advantage is the cost effectiveness of using ACA to bond fine pitch surface mount components. A limitation of ACA adhesives is the need to cure under contact pressure. The concept of using an ACA as a solder replacement on rigid chip carriers utilizing conventional surface mount technology has been demonstrated by J. Liu et al. [37]. Fine pitch SM components were bonded to FR4 boards with ACAs using a fine pitch bonder and then components with larger pitches were bonded with ICAs using standard SM equipment. The study demonstrated that standard surface-mounting tools could be used to assemble conductive adhesives. The connection resistance of solder-plated, plastic components (0.65 mm pitch) with ACAs bonded did not change after an accelerated temperature cycling test (ATC) conducted at −40–85°C. However, similar parts failed under conditions of −55–125°C for 1000 cycles [38]. The mechanical stability problem may have been the result of an improper joint geometry, i.e. not optimized for ACA bonding.

11.3.7 Failure Mechanism

Since the adhesive matrix is a non-conductive material, interconnection joints rely to some extent on pressure to assure contact for conventional ACAs. Adhesive interconnections therefore exhibit different failure mechanisms compared to soldered connections, where the formation of intermetallic compounds and coarsening of grains are associated with the main mechanisms. Basically there are two main failure mechanisms that can affect the contacts. The first is the formation of an insulating film on either the contact areas or conductive particle surfaces. The second is the loss of mechanical contact between the conductive elements due to either a loss of adherence, or relaxation of the compressive force.

11.3.7.1 Oxidation of Non-noble Metals

Electrochemical corrosion of non-noble metal bumps, pads, and conductive particles results in the formation of insulating metal oxides and significant increase in contact resistance. Electrochemical corrosion only occurs in the presence of moisture and metals that possess different electrochemical potentials. Humidity generally accelerates oxide formation and so too the increase in contact resistance. Reliability test results for flip chip on flex (FCOF) using gold bumps and ACFs filled with Ni particles indicated that the connection resistance increased with time under elevated temperature and humidity storage conditions [39]. In this case, the gold bump acts as cathode and the Ni particle as an anode. A nickel oxide, which is electrically insulating, eventually forms on the surface of the Ni particles.

11.3.7.2 Loss of Compressive Force

The compressive forces acting to maintain contact among the conductive components are partly due to curing shrinkage achieved when curing the polymeric matrix of ACAs. Both the cohesive strength of the adhesive matrix and the interfacial adhesion strength between the adhesive matrix and the chip and chip carrier must be sufficient to maintain the compressive force. However, the thermal expansion of adhesives, their swelling due to moisture adsorption, and mechanical stresses due to applied loads tend to diminish this compressive force created as a result of curing. Moreover, water not only diffuses into the adhesive layer but also penetrates to the interface between adhesive and chip/chip carrier causing a reduction in adhesion strength. As a result, the contact resistance increases and can even result in a complete loss of electrical contact [40].

11.4 Description of Isotropic Conductive Adhesives (ICAS)

11.4.1 Percolation Theory of Conduction

Isotropic conductive adhesives (ICAs) are composites of polymer resin and conductive fillers. The conductive fillers provide the composite with electrical conductivity through contact between the conductive particles. With increasing filler concentrations, the electrical properties of an ICA transform it from an insulator to a conductor. Percolation theory has been used to explain the electrical properties of ICA composites. At low filler concentrations, the resistivities of ICAs decrease gradually with increasing filler concentration. However, the resistivity drops dramatically above a critical filler concentration, Vc, called the percolation threshold. It is believed that at this concentration, all the conductive particles contact each other and form a three-dimensional network. The resistivity decreases only slightly with further increases in the filler concentrations [41–43]. A schematic explanation of resistivity change of ICAs based on percolation theory is shown in Fig. 11.8. In order to achieve conductivity, the volume fraction of conductive filler in an ICA must be equal to or slightly higher than the critical volume fraction. Similar to solders, ICAs provide the dual functions of electrical connection and mechanical bond in an interconnection joint. In an ICA joint, the polymer resin provides mechanical stability and the conductive filler provides electrical conductivity. Filler loading levels that are too high cause the mechanical integrity of adhesive joints to deteriorate. Therefore, the challenge in formulating an ICA is to maximize conductive filler content to achieve a high electrical conductivity without

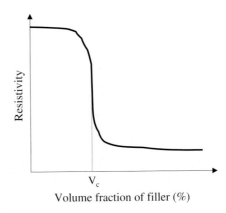

Fig. 11.8 Effect of filler volume fraction on the resistivity of ICA systems

adversely affecting the mechanical properties. In a typical ICA formulation, the volume fraction of the conductive filler is about 25–30% [44, 45].

11.4.2 Adhesive Matrix

Polymer matrices of isotropic conductive adhesives are similar to anisotropically conductive adhesives. An ideal matrix for ICAs should exhibit a long shelf life (good room temperature latency), fast cure, relatively high glass transition temperature (Tg), low moisture pickup, and good adhesion [46].

11.4.2.1 Matrix Materials

Both thermoplastic and thermoset resins can be used for ICA formulations. The main thermoplastic resin used for ICA formulations is polyimide resin. An attractive advantage of thermoplastic ICAs is that they are reworkable, e.g., can easily be repaired. A major drawback of thermoplastic ICAs, however, is the degradation of adhesion at high temperature. Another drawback of polyimide-based ICAs is that they generally contain solvents. During heating, voids are formed when the solvent evaporates. Most of commercial ICAs are based on thermosetting resins. Epoxy resins are most commonly used in thermoset ICA formulations because they possess superior balanced properties. Silicones, cyanate esters, and cyanoacrylates are also employed in ICA formulations [47–51].

11.4.2.2 Achieving Latency and Rapid Curing

Most commercial ICAs must be kept and shipped at a very low temperature, usually –40°C, to prevent the ICAs from curing. Pot life is a very important factor for users of the ICAs. In order to achieve desirable latency at room temperature, epoxy hardeners must be carefully selected. In some commercial ICAs, solid curing agents are used, which do not dissolve in the epoxy resin at room temperature. However, these curing agents can dissolve in the epoxy at a higher temperature (curing temperature) and react with the epoxy resin. Another approach to achieve latency is to employ an encapsulated imidazole as a curing agent or catalyst. An imidazole is encapsulated inside a very fine polymer sphere. At room temperature, the polymer sphere does not dissolve or react with the epoxy resin. But at a higher temperature, after the polymer shell is broken, the imidazole is released from the sphere to cure the epoxy or catalyze the cure reaction. Fast cure is another attractive property of a desirable ICA. Shorter cure times increase throughput resulting in lower processing cost. In epoxy-based ICA formulations, proper hardeners and catalysts such as imidazoles and tertiary amines can be used to achieve rapid cure.

11.4.2.3 Effect of Low Tg Materials

Conductive adhesives with low Tgs can lose electrical conductivity during thermal cycling aging [52, 53]. Electrical conductivity in metal-powder, filled conductive adhesives is achieved through the contact of adjacent metal particles with each other, thus producing a continuous electrical path between a component lead and metallized pad. When a joint is subjected to thermal cycling conditions, it experiences repeated cyclic shear motion of the lead relative to the chip carrier pad. The amount of shear strain is primarily dependent on the thermal cycling conditions and thermal expansion mismatch between the component and chip carrier. Neglecting lead-deformation and substrate compliance, the majority of the shear strain produced is accommodated by visco-elastic or visco-plastic deformation of the conductive adhesive. When a conductive adhesive deforms to accommodate the shear strain produced, the metal particles move, thus changing the position of contact point(s) between adjacent metal particles. If the organic matrix is too compliant, it will flow to fill the area left behind the moving metal particles. When the direction of the shear strain is reversed during thermal cycling, adjacent metal particles move back to their original contact locations, which now are partially covered with the compliant, dielectric organic-matrix material. As the number of thermal cycles increases, the contact resistance between adjacent particles increases, thus increasing the interconnection joint resistance [52].

11.4.2.4 Effect of Moisture Absorption

Moisture absorption can influence the reliability of conductive adhesive interconnection joints. Moisture in polymer composites is known to have an adverse effect on both mechanical and electrical properties of epoxy laminates [53, 54]. Studies relating to the reliability and moisture sensitivity of electronic packages indicate similar degrading effects. It was determined that moisture absorption can cause an increase in contact resistance, especially if the metallization on the

Table 11.1 Effects of Moisture on ICA Joints

Major effects
Degrade bulk mechanical strength
Decrease interfacial adhesion strength and cause delamination
Promote the growth of voids present in joints
Give rise to swelling stress in joints
Induce the formation of metal oxide layers resulting from corrosion

bond pads and components are not noble metals [55]. Effects of moisture absorption on conductive adhesive joints are summarized in Table 11.1. In order to achieve high reliability, conductive adhesives with low moisture absorption are required. High adhesion strength to pad and component metallization is a necessary property for conductive adhesives used for interconnections in electronic assemblies. Epoxy-based ICAs tend to have better adhesion strength than polyimide and silicone-based ICAs. However, a silicone matrix tends to have lower moisture absorption than epoxy resins [48].

11.4.3 Conductive Fillers

Because polymer matrices are dielectric materials, conductive fillers in ICA formulations provide the material with electrical conductivity. In order to achieve high conductivity, the filler concentration must be at least equal or higher than the critical concentration predicted by percolation theory.

11.4.3.1 Pure Silver vs. Ag-Coated Fillers

Silver (Ag) is by far the most popular conductive filler, although gold (Au), nickel (Ni), copper (Cu), and carbon are also used in ICA formulations. Silver is unique among all of the cost-effective metals by nature of its conductive oxide (Ag_2O). Oxides of most common metals are good electrical insulators and copper powder, for example, becomes a poor conductor after aging. Nickel and copper-based conductive adhesives generally do not have good conductivity stability because they are easily oxidized. Even with antioxidants, copper-based conductive adhesives show an increase in volume resistivity on aging, especially under high-temperature and humidity conditions. Silver-plated copper has been utilized commercially in conductive inks, and should also be appreciable as a filler in adhesives. While composites filled with pure silver particles often show improved electrical conductivity when exposed to elevated temperature and humidity or thermal cycling, this is not always the case with silver-plated metals, such as copper flake. Presumably, the application of heat and mechanical energy allows the particles to make more intimate contact in the case of pure silver, but silver-plated copper may have coating discontinuities that allow oxidation/corrosion of the underlying copper and thus reduce electrical paths [44].

11.4.3.2 Particle Shape and Size

The most common morphology of conductive fillers used for ICAs is flake because flakes tend to have a large surface area, and more contact spots and thus more electrical paths than spherical fillers. The particle size of ICA fillers generally ranges from 1 to 20 μm. Larger particles tend to provide the material with a higher electrical conductivity and lower viscosity [56]. A new class of silver particles, porous nano-sized silver particles, has been introduced in ICA formulations [57, 58]. ICAs made with this type of particles exhibited improved mechanical properties, but the electrical conductivity is less than ICAs filled with silver flakes. In addition, short carbon fibers have been used as conductive fillers in conductive adhesive formulations [59, 60]. However, carbon-based conductive adhesives show much lower electrical conductivity than silver-filled ones.

11.4.3.3 Silver-Copper Fillers

A powder with a specific structure was introduced as a filler for conductive adhesives in 1992 [61]. The powder particle consists of two metallic components, copper and silver. Silver is highly concentrated on the particle surface and the concentration gradually decreases from the surface to the inner of the particle, but always contains a small amount of silver. Conductive adhesive paste filled with this powder exhibits excellent oxidation resistance, i.e. can be exposed to oxygen content about 100 ppm in a nitrogen atmosphere without oxidizing. It also exhibits higher solderability than commercially-available copper pastes, sufficient adhesion strength even after heating and/or cooling test, and the least migration, almost the same degree as pure copper paste [61].

11.4.3.4 Low-Melt Fillers

In order to improve electrical and mechanical properties, low-melting-point alloy fillers have been used in ICA formulations. A conductive filler powder is coated with a low-melting-point metal. The conductive powder is selected from the group consisting of Au, Cu, Ag, Al, Pd, and Pt. The low-melting-point metal is selected from the group of fusible metals, such as Bi, In, Sn, Sb, and Zn. The filler particles are coated with the low-melting-point metal, which can be fused to achieve metallurgical bonding between adjacent particles and between the particles and the bond pads that are joined using the adhesive material [62, 63].

11.5 Flip Chip Applications Using Isotropic Conductive Adhesives

A key factor in achieving a low-cost, flip chip technology is the use of isotropic conductive adhesives. In comparison to the classical flip chip (FC) technologies, the use of ICAs for the bumping and joining provide numerous advantages (Table 11.2).

11 Electrically Conductive Adhesives (ECAs)

Table 11.2 Advantages of flip chip technologies utilizing ICAs

Advantages
Process simplification and reduction of indexing steps by eliminating activation and purification processes
A smaller temperature load on elements and wiring carriers
The availability of a large spectrum of material combinations
A broad range of applicable adhesive systems allows a the selection of processing parameters and joining characteristics
Few requirements for under bump metallization (UBM) since alloy phase formation does not have to be considered

Motorola successfully demonstrated an ICA flip chip bumping process using stencil printing technology both through mathematical modeling and experimentation [64]. Both GaAs and Si flip chip devices with Au thin-film metallization, and alumina and FR4 chip carriers also with Au metallization were used in this study. The electrical performance of chip and chip carrier combinations (i.e. GaAs/Al2O3, GaAs/FR4, and Si/FR4) utilizing conductive adhesive polymer bumps, showed no difference from Au and AuSn bumps (all of the flip-chip dies are mounted onto the chip carriers using an ICA). However, premature failure was observed in HAST and thermal shock tests.

The polymer bumping method is a low-cost and efficient process conducted at the wafer-level and suitable for large-scale production. Data of joint resistance stability under accelerated aging conditions such as 85°C/85% relative humidity and temperature cycling demonstrates polymer flip chip interconnections are capable of long-term stability. The polymer flip chip assembly is compatible to a large range of rigid carriers, and heat-sensitive, flexible chip carriers.

11.5.1 Process

Several flip-chip bumping and joining techniques have been reported in the literature. Flip chips using ICAs are often called polymer flip chips (PFC). The PFC process is a stencil printing technology in which an ICA is printed through a metal stencil to form polymer bumps on bond pads of IC devices subsequent to the under bump metallization deposition on aluminum termination pads. The sequential processes to achieve PFC interconnects are UBM deposition, stencil printing an ICA, bump formation (ICA solidification), flip chip attach to achieve electrical connections, and underfill for enhanced mechanical and environmental integrity [64–66].

11.5.1.1 Formation of Protective Chip Pad Layer

As with virtually all flip chip processes, the Al bond pads must be protected to eliminate the formation of non-conductive aluminum oxide. This insures a low and stable resistance at bond-bond pad interface. The polymer flip chip process utilizes an electroless plating technique, Ni/Au or Pd, to cover the Al bond pads prior to polymer bumping. The typical metal thickness is 0.5–1.0 μm for Pd and 3.0–5.0 μm for Ni/Au.

11.5.1.2 Print ICA

The PFC process combines high precision stencil printing techniques with highly conductive ICAs. These polymers can be thermosetting or thermoplastic. First, the polymer bumps are formed by deposition of an ICA through the metal mask directly onto the metallized bond pads on a wafer. Printed conductive adhesive bumps can offer an attractive alternative to the other bumping technologies in terms of cost and manufacturability. The printing process typically involves a screen or stencil with openings through which bumps are deposited. A screen consists of an interwoven wire mesh with an emulsion that covers the wire mesh. The emulsion is photolithographically patterned to match the bump sites. Stencils are made of metal foil. Holes for bump deposition are made by etching, electroforming (plating), or laser drilling.

During the printing process, the paste is typically dispensed some distance away from the stencil apertures. Typically, the stencil is separated from a substrate by the snap-off distance. The squeegee is lowered, resulting in contact of the stencil to the substrate or wafer surface. As the squeegee moves across the stencil surface, a stable flow pattern develops in the form of a paste roll. The consequent hydrodynamic pressure developed by the squeegee pushes the paste into the patterned stencil openings. The stencil lifts away from the substrate surface with the paste remaining on the substrate.

11.5.1.3 Curing

The polymer bumps are then either fully cured or partially cured to the so-called B-stage for thermosetting polymer bumps. For thermoplastic polymer bumps, after stencil printing the solvent is removed to form solid bumps. Bump heights are typically 50–75 μm and process can accommodate pitches down to 5 mils. Bump densities of up to 80,000 bumps/wafer have been formed with excellent co-planarity.

Once the bumped wafers are diced, chips are picked from the wafers, flipped over, and then placed on and bonded to chip carriers. Different process procedures are utilized to bond thermosetting polymer bumps to similar thermoplastic bumps as noted in Fig. 11.9. Final processing involves a heat cure for thermosetting bumps, while thermoplastic bump connections only require a few seconds under heat and pressure to melt the thermoplastic.

11 Electrically Conductive Adhesives (ECAs)

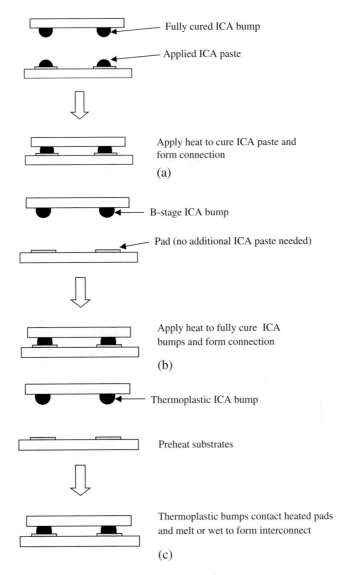

Fig. 11.9 Schematic illustrating various die attachment assembly processes utilizing ICAs. (**a**) chip with cured ICA bumps mated with uncured ICA on carrier pads. (**b**) chip with partially cured (B-staged) ICA bumps mated with bare carrier pads. (**c**) chip with thermoplastic ICA bumps mated with bare but preheated carrier pads

11.5.1.4 Underfill

An underfill is then injected into the gap between the chip and chip carrier and then cured to complete the flip chip process. The function of the underfill is to provide mechanical integrity and environmental protection to a flip chip assembly. Studies have demonstrated that both thermoset and thermoplastic ICAs can offer low initial joint resistances of less than 5 milliohms and stable joint resistances (Au-to-Au flip chip bonding) during all the accelerated reliability testing. The reliability results have indicated that there is no substantial difference in the performance of thermoset ad thermoplastic bumps and both types of polymers offer reliable flip chip electrical interconnections [66].

11.5.2 Metal-Bumped Flip Chip Joints

ICAs can also be used to form electrical interconnections with chips that have metal bumps. Isotropic conductive adhesive materials utilize much high filler loading than ACAs to provide electrical conduction isotropically (i.e. in all directions) throughout the material. In order for these materials to be used for flip-chip applications, they must be selectively applied to only those areas that are to be electrically interconnected. Also, the materials are not to spread during placement or curing to avoid creating electrical shorts between circuit features. Screen or stencil printing is most commonly used to precisely deposit the ICA pastes. However to satisfy the scale and accuracy required for flip-chip bonding requires very accurate pattern alignment. To overcome this difficult requirement, Matsushita developed a transfer method [67].

Raised studs or pillars are required on either the die or chip carrier. Matsushita uses a conventional ball bonder to form Au-stud bumps. Bumping is significantly faster than creating complete wire bonds. A ball bumping process eliminates the need for traditional sputtering and plating processes used for standard bump formation. To prevent the bond area from becoming too large, the bumps are formed in a conical shape. The bumps are pressed level by a flat surface, which adjusts both height and planarity. The ICA is selectively transferred on the bump tips by contacting the face of the die to a flat thin film of the ICA which is produced by screen printing and whose transfer thickness is controlled by changing the printed film thickness. Then the die is picked, aligned, and placed on a chip carrier. The whole assembly is exposed to heat to cure the ICA and form connections between the die and chip carrier. Finally, an underfill (an insulating adhesive) is dispensed between the die and the chip carrier and cured. This method offers the options of oven curing an assembly since bonding pressure is not required. A specially formulated ICA is used to avoid silver migration, containing 20% palladium in a silver palladium alloy. A schematic of the process flow of forming joints with stud-bumped flip chips using ICAs is shown in Fig. 11.10.

11.5.2.1 Comparison with Soldered Joints

Another process for bonding a flip chip with metal bumps consists of screen-printing an ICA on a chip carrier, aligning and placing the chip, curing the ICA

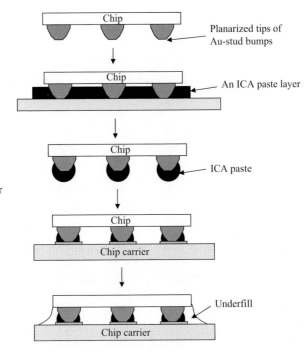

Fig. 11.10 A schematic of the process flow of joints formed with stud-bumped flip-chips using ICAs. (**a**) Tips of the gold stud bumps formed with a wire bond tool are planarized. (**b**) Planarized bumps are dipped into a thin layer of ICA. (**c**) The chip is withdrawn, leaving the bumps coated with ICA. (**d**) The chip is placed on mating pads of a chip carrier with on pressure required during curing. (**e**) An underfill (an insulating adhesive) is dispensed and cured

to form bonds, and underfilling. By using this approach, SINTEF Electronics conducted a comparison study between an ICA-bonded and solder-bonded flip chips on FR4 chip carrier with Ni/Au metallization. The number of thermal cycles (−55–125°C) to failure for both solder and ICA flip chip circuits was compared. The study showed that stable contacts could be maintained for at least 1000–2000 cycles for ICA flip chip joints. This is comparable to the lifetime for solder flip chip joints. However, the variation among ICA samples was very high and optimization of assembly processes is needed in order to achieve more reproducible joint resistance [68].

11.5.3 ICA Process for Unbumped Chips

Another polymer flip chip bumping process is known as micro machined bumping [69, 70]. Initially Cr/Au contact metal pads for conductive-polymer bumps are deposited on Si wafers, followed by patterning a thick photoresist to create bump holes. A high-aspect ratio and straight sidewall patterns are very important in shaping the conductive-polymer bumps. After the lithography, thermoplastic conductive polymer materials, usually thermoplastic paste filled with Ag flake, is applied by either dispensing or screen printing the paste into the bump-hole patterns. The wafer is heated in a convection oven to remove the solvent. Due to the difference in curing conditions between the thick photoresist and conductive-polymer, the photoresist can be carefully stripped to expose the dried polymer bumps. Finally, the wafer is diced into individual chips.

Chips with thermoplastic bumps are placed on chip carriers and preheated to approximately 20°C above the melting point of the polymer causing the bumps to reflow onto the matching chip carrier pads. Mechanical and electrical bonds are established as the chip carrier cools below the polymer melting temperature. To enhance the mechanical bonding strength, a small amount of pressure can be applied by placing a weight on the chip.

This flip-chip bonding technique has high potential to replace conventional solder flip-chip techniques for sensor and actuator systems, optical micro electromechanical systems (MEMS), optoelectonic multichip modules (OE-MCMs), and electronic system applications [70].

11.6 Applications of ICAS in Microelectronic Packaging

11.6.1 Surface Mount Applications

Tin-lead solders (Sn-Pb) are the standard materials used to interconnect electronic components on printed circuit boards (PCBs). The most common reflow soldering process is surface mount technology (SMT), which uses tin/lead solder pastes. The pressure to reduce the industrial use of lead is growing, particularly in Europe since it poses as a hazard to human health [71]. Thus, the use of tin-lead solder paste in SMT processes must be reduced or eliminated to both satisfy legislative actions and market-driven pressures as well.

11.6.1.1 Advantages

Lead-free and environmentally sound interconnect bonding processes are urgently needed. Among the possibilities are electrically conductive adhesives (ECAs) and lead-free solders [72–74]. Compared to soldering technology, ECA technology can offer numerous advantages such as fewer processing steps which reduces processing cost, lower processing temperature, which makes the use of heat-sensitive and low-cost chip carriers possible, and fine-pitch capability [73].

11.6.1.2 Disadvantages

However, conductive adhesive technology is still in its infancy, and concerns and limitations do exist. The main limitations of commercial ICAs include lower conductivity than solder materials, an unstable contact resistance with non-noble metal finished components, and poor impact performance. The electrical conductivity ($\sim 10^{-4}$ ohm·Cm) of an ICA is lower than Sn-Pb solders ($\sim 10^{-5}$ ohm·Cm). Although generally adequate for most electronics applications, the electrical conductivity of ICAs must be improved. Contact resistance between an ICA and non-noble metal (such as Sn/Pb, Sn, and Ni) finished components in noted to dramatically increase with time especially under elevated temperature and humidity aging conditions [75–78]. In addition, printed circuit board assemblies are often subject to significant mechanical shock during assembly, handling, and throughout their product life. Packages cannot survive without adequate impact resistance. However, most microelectronic commercial ICAs exhibit poor impact performance. Components assembled using ICAs tend to separate from the substrate when the package experiences a sudden shock [77, 79]. For conductive adhesive technology to provide an acceptable solution as a solder replacement, new conductive adhesives with the desired overall properties must be developed [77]. There has been considerable effort to improve the properties of ICAs, and to make them more reliable materials. These improvements are described in the following sections.

11.6.1.3 CSP Applications

Matsushita Electric Industrial Co., Ltd. developed solderless joining technologies using nickel-filled isotropic conductive adhesives to mount a ceramic chip scale package (CSP-C) onto a FR4 board [80]. Nickel was selected instead of Ag because, unlike Ag, nickel does not migrate. A significant coefficient of thermal expansion (CTE) mismatch existed between the CSP-C ceramic chip carrier (CTS = 7 ppm) and the FR4 organic chip carrier (CTE = 16 ppm). This CTE mismatch resulted in large stress to be generated within the solder joints during accelerated thermal cycle (ATC) testing which resulted in early failure due to solder fatigue. ICAs usually exhibit better thermo-mechanical properties than solders. In addition, metal-migration between joints is a great concern because the joints in a CSP area array package are arranged with a close pitch (i.e. in close proximity).

The packaging procedure was as follows: (a) the ICA was screen-printed on the area array lands of the FR4 motherboard; (b) the CSP-C was mounted; (c) and the ICA was cured to form bonds. The Ni-filled conductive adhesive demonstrated a much higher resistance to metal migration compared to Ag-filled ICAs, and equivalent to solder joints. Also, the thermal fatigue life of the Ni-filled ICA joints was 5 times greater than comparable solder joints.

11.6.2 High Frequency Performance of ICA Joints

Only very limited work has been conducted to investigate the high frequency behaviors of ICA joints. J. Felba et al. [81] investigated a formulation of isotropically conductive adhesive that performed well as a solder replacement in microwave applications. The study involved in various different adhesive base materials and several types of main (silver flakes, nickel and graphite) and additional (soot and silver semiflake powder) filler materials. In order to assess the usefulness of a given adhesive formulation, an additional gap in the gold strip of a standard microstrip bandpass filter was made and bridged by an adhesive bonded silver jumper. Both the quality factor (Q-factor) and loss factor (L) of the filter with the bonded jumper were measured at a frequency of 3.5 GHz in a preliminary experiment and at 3.5 GHz and 14 GHz in a final experiment. It was determined that silver flake powders are the best filler materials for ICA for microwave applications because ICAs filled with the silver flake powders exhibit the highest Q-factor and lowest loss factor. Also, addition of soot should be avoided since it decreases the quality factor [82].

A study at Georgia Tech of a flip-chip test vehicle mounted on a FR4 chip carrier with a gold-plated copper transmission lines [81]. The performance of eutectic Sn-Pb and ICAs were evaluated and compared using this test device. Both ICAs and eutectic Sn-Pb solder were determined to exhibit almost the same behavior at a frequency range of 45 MHz–2 GHz and the measured transmission losses for both materials were minimal. It was also found that the S11 characteristics of both Sn-Pb and ICAs after exposure to 85°C/85% relative humidity aging for 150 h did not vary from the previous signals prior to aging, but S12 value of the Sn-Pb joints deviated more than that of ICA joints after the aging.

11.6.3 Fatigue Life of ICA Joints

There have been several studies investigating the fatigue life of ICA joints. Aiming to understand the performance of ICA interconnects under fracture and fatigue loading, J. Constable et al. [83] investigated performance of ICA interconnects under fracture and fatigue loading by monitoring resistance changes (micro-ohm sensitivity) of ICA joints during pull and fatigue testing (cyclic loading up to 1000 cycles). Observation of the fracture surface suggested that the ICA joint life depended upon the adhesive failure of the bond to the metal surface. It was observed that fracture strains for the ICAs were in the range of 20–38%, and resistance remained approximately constant in the elastic region, but the resistance started to increase rapidly as soon as the pull-force departed from linear elastic behavior. For fatigue tests, linear displacement was ramped up the pre-programmed maximum displacement and ramped back to the starting position. It was observed that the shear strain for ICA joints surviving 1000 cyclic loading was typically 10%, which is about an order of magnitude greater than solders. This suggests that using conductive adhesives may be advantageous for some flip chip applications. It is believed that since silver filler particles of ICAs cannot accommodate this large strain, the silver filler particles must move relative to one another as the epoxy matrix is strained. The most common pattern of resistance change was only increased to a point corresponding to about a 70% loss in interface contact resistance before sudden failure. This was an indication that the interface crack slightly propagated into the adhesive [83].

In an effort to gain a fundamental understanding of the fatigue degradation of ICAs, R. Gomatam et al. [84] studied the behavior of ICA joints under temperature and humidity conditions. The fatigue life decreased at elevated temperature and high humidity conditions. It was also observed that the fatigue life of the ICA joints decreased considerably as the temperature cycle frequency was decreased. This effect was attributed to the fact that as the frequency was decreased, the propagating crack was exposed to higher loads for longer periods of time, effectively resulting in high creep loading [84].

11.7 Improvement of Electrical Conductivity of ICAS

Electrical conductivity of ICAs is inferior to solders [85]. Even though the conductivity of ICAs is adequate for most applications, a higher electrical conductivity of ICAs is still needed. To develop a novel ICA for modern electronic interconnect applications, a thorough understanding of the materials is required.

11.7.1 Eliminate Lubrication Layer

An ICA is generally composed of a polymer binder and Ag-flake filler material. A thin layer of organic lubricant is present on the surface of the Ag flakes. This lubricant layer plays an important role in the performance of ICAs, including the dispersion of Ag flakes in adhesives, and the rheology of the adhesive formulations [86, 85, 87,88]. The organic layer consist of a Ag salt formed between the Ag surface and the lubricant, which typically is a fatty acid such as stearic acid [88, 89]. This lubricant layer affects the conductivity of an ICA because it is electrically insulating [88, 89]. To improve conductivity, the organic lubricant layer must be partially or fully removed through the use of chemical substances that can dissolve the organic lubricant layer [88–90]. However, the viscosity of an ICA paste may increase if the lubricant layer is removed. An ideal chemical substance (or lubricant remover) should be latent (does not remove the lubricant layer) at room temperature, but be active (capable of removing the lubricant layer) at a temperature slightly below the cure temperature of the polymer binder. The lubricant remover can be a solid short-chain acid, a high-boiling-point ether such as diethylene glycol monobutyl ether or diethylene glycol monoethyl ether acetate, and a polyethylene glycol with a low molecular weight [88–90]. These chemical substances can improve electrical conductivity of ICAs by removing the lubricant layer on the Ag-flake surfaces and providing an intimate flake-flake contact [88, 90].

11.7.2 Increase Shrinkage

In general, ICA pastes exhibit low electrical conductivity before cure, but the conductivity increases dramatically after they are cured. ICAs achieve electrical conductivity during the cure process, mainly through a more intimate contact between Ag flakes caused by the shrinkage of polymer binder [91]. Accordingly, ICAs with high cure shrinkage generally exhibit the best conductivity. Therefore, increasing the cure shrinkage of a polymer binder is another method for improving electrical conductivity. For ICAs based on epoxy resins, a small amount of a multifunctional epoxy resin can be added into the formulation to increase cross-linking density, shrinkage, and thus increase conductivity [91].

11.7.3 Transient Liquid Phase Fillers

Another approach for improving electrical conductivity is to incorporate transient liquid-phase sintering metallic fillers into ICA formulations. The filler used is a mixture of a high melting-point metal powder (such as Cu) and a low-melting-point alloy powder (such as Sn-Pb). Upon reaching its melting point, the low melt-point powder liquefies dissolving the high melting-point particles. The liquid exists only for a short period of time and then forms an alloy and solidifies. The electrical conductivity is established through a plurality of metallurgical connections formed in-situ from these two powders in a polymer binder. The polymer binder fluxes both the metal powders and the metals to be joined and facilitates the transient liquid bonding of the powders to form a stable metallurgical network for electrical conduction, and also forms an interpenetrating polymer network providing adhesion. High electrical conductivity can be achieved using this method [91–94]. The ICA joints formed include metallurgical alloying to the junctions as well as within the adhesive itself. This provides a stable electrical connection during elevated temperature and humidity aging. In addition, the ICA joints showed good impact strength due to the metallurgical interconnection between the conductive adhesive and the components. One critical limitation of this technology is that the numbers of combinations of low melt and high melt fillers are limited. Only certain combinations of metallic fillers that are mutually soluble exist to form this type of metallurgical interconnections.

11.8 Improvement of Contact Resistance Stability

Contact resistance between an ICA (generally a Ag-flake-filled epoxy) and non-noble metal finished components increases dramatically during elevated temperature and humidity aging, especially at 85°C/85% relative humidity. The National Center of Manufacturing and Science (NCMS) defined the stability criterion for solder replacement conductive adhesives as a contact resistance shift of less than 20% after aging at 85°C/85%RH conditions for 500 h [76].

11.8.1 Causes for Resistance Increase

Two main mechanisms, simple oxidation and corrosion of the non-noble metal surfaces, have been proposed in the literature as the possible causes for the increase in contact resistance of ICA joints during elevated temperature and humidity aging. Simple oxidation of the non-noble metal surface is claimed as the main reasons for the observed increased resistance. Corrosion is claimed as the possible mechanism for resistance increase only by several investigators [74, 75, 95–97]. One study strongly indicates that galvanic corrosion rather than simple oxidation of the non-noble metal at the interface between an ICA and non-noble metal is the main reason for the shift in contact resistance of ICAs (Fig. 11.11) [98, 99]. The non-noble acts as the anode, and is reduced to a metal ion ($M - ne = M^{n+}$) due to the loss of electrons. The noble metal acts as a cathode, and its reaction generally is $2H_2O + O_2 + 4e = 4OH^-$. Then M^{n+} combines with OH^- to form a metal hydroxide or metal oxide. As a result of this electrochemical (corrosion) process, a layer of metal hydroxide or metal oxide is formed at the interface that is electrically insulating, causing the contact resistance to increase dramatically [98, 99].

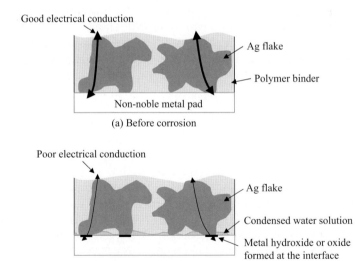

Fig. 11.11 Schematic depicting the effect of galvanic corrosion of a non-noble metal pad on electrical conduction of a silver-filled ICA. (**a**) Good electrical conduction before corrosion. (**b**) Poor electrical conduction due to the formation of a metal hydroxide or oxide formation as a result of galvanic corrosion

11.8.2 Approaches to Stabilize Contact Resistance

11.8.2.1 Reduce Moisture Absorption

Galvanic corrosion requires the presence of moisture. An electrolyte solution must be formed at the interface before galvanic corrosion can occur. Therefore, one way to prevent galvanic corrosion at the interface between an ICA and the non-noble metal surface is to lower the moisture absorption of the ICA. ICAs that have low moisture absorption generally exhibit more stable contact resistance on non-noble surfaces compared with those with high moisture absorption [100, 101]. Without an electrolyte, galvanic corrosion rate is very low. The electrolyte in this case is mainly from the impurity of the polymer binder (generally epoxy resins). Therefore, ICAs formulated with high purity resins should perform better.

11.8.2.2 Use of Corrosion Inhibitors

Another method of preventing galvanic corrosion is to introduce organic corrosion inhibitors into ICA formulations [99–102]. In general, organic corrosion inhibitors act as a barrier layer between the metal and environment forming a film over the metal surfaces [103–106]. Some chelating compounds are especially effective in preventing metal corrosion [105]. Most organic corrosion inhibitors react with the epoxy resin at a specific temperature. Therefore, if an ICA is epoxy- based, the corrosion inhibitors must not react with the epoxy resin during curing which would cause them to be consumed and lose their effect. Organic corrosion inhibitors are thoroughly discussed in the literature [104, 106]. Figure 11.12 shows the effect of a chelating corrosion inhibitor on the contact resistance between an ICA and a Sn/Pb surface. It can be seen that this corrosion inhibitor is very effective in stabilizing the contact resistance.

11.8.2.3 Use of Oxygen Scavengers

Since oxygen accelerates galvanic corrosion, oxygen scavengers can be added into ICA formulations to slow down the corrosion rate [103]. When oxygen molecules diffuse through the polymer binder, they react with the oxygen scavenger and are consumed. However, when the oxygen scavenger is completely depleted, then oxygen can again diffuse into the interface and accelerate the corrosion process. Therefore, oxygen scavengers only delay the galvanic corrosion process. Similar to corrosion inhibitors, the oxygen scavengers used must not react with the epoxy resin at its cure temperature [103, 107–110].

11.8.2.4 Sharp-Edge Filler Particles

Another approach of improving contact resistance stability during aging is to incorporate some electrically-conductive particles, which have sharp edges and

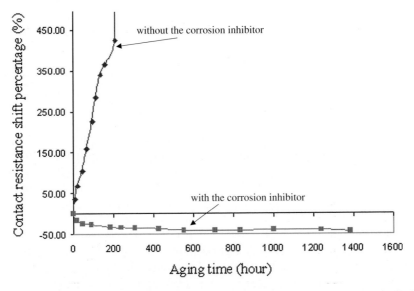

Fig. 11.12 Effect of a corrosion inhibitor on contact resistance between an ICA and a Sn-Pb surface with time, aging condition: 85°C/85% relative humidity

referred as oxide-penetrating fillers, into the ICA formulations. Force must be provided to drive the oxide-penetrating particles through the oxide layer of adjoining particles and metal pads, and keep them in position. This can be accomplished by employing polymer binders that show high shrinkage when cured as discussed in Section 11.7.2 [111]. This concept is used in Poly-Solder (a silver-loaded ICA material patented by Poly-Flex Circuits), which has good contact resistance stability with standard surface-mounted devices (SMDs) on both solder-coated and bare circuit boards [111].

11.9 Improvement of Impact Performance

The ability to resist performance degradation when subjected to mechanical shock is a critical property that solder replacement ICAs must possess. There are ongoing efforts to develop ICAs which exhibit acceptable impact strength; are capable of passing the standard drop test used to evaluate the impact strength of components attached to a printed circuit board (PCB). Among the methods are decreasing the filler loading to improve the impact strength [112], but then reduces the electrical conductivity of the conductive adhesives. A development study was reported where conductive adhesives were formulated using low modulus resins that absorb the impact energy developed during a drop [113]. Also, conformal coating of the surface-mounted devices has been used to improve mechanical strength. A study demonstrated that conformal coating improved the impact strength of conductive adhesives joints [114].

11.9.1 Epoxide-Terminated Polyurethane Systems

A class of conductive adhesives based on an epoxide-terminated polyurethane (ETPU) was developed [115, 116]. This class of conductive adhesives exhibits properties typical of polyurethane materials, such as high toughness and good adhesion. The modulus and glass transition temperature of an ICA can be adjusted by incorporating epoxy resins such as bisphenol-F based epoxies. Conductive adhesives based on ETPU exhibit a broad loss factor (tan δ) peak with temperature and a high tanδ value at room temperature. The tanδ value of a material is a good indication of the damping property and impact performance of a material. In general, the higher the tanδ value, the better the damping property (impact strength) of the material. ICAs based on ETPU resins also exhibit a much higher loss factor over a wide frequency range compared to ICAs based on bisphenol-F epoxy resins (shown in Fig. 11.13). This indicates that ICAs based on ETPU resins should exhibit good damping property and improved impact performance in a variety of electronic packages. This class of conductive adhesives has demonstrated superior impact performance and a substantial improvement in contact resistance stability with non-noble metal surfaces, such as Sn/Pb, Sn, and Cu [115, 116].

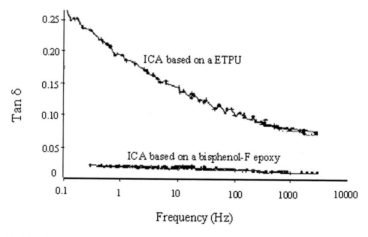

Fig. 11.13 The effect of frequency on the loss factor for two ICA materials

References

1. H. Wolfson and G. Elliot, "Electrically Conducting Cements Containing Epoxy Resins and Silver," U.S. Patent, 2,774,747, 1956
2. K.R. Matz, "Electrically Conductive Cement and Brush Shunt Containing the Same," U.S. Patent, 2,849,631, 1958
3. D.P. Beck, "Printed Electrical Resistors," U.S. Patent, 2,866,057, 1958
4. K. Gilleo, "Assembly with Conductive Adhesives," *Soldering and Surface Mount Technology*, No. 19, pp. 12–17, February 1995
5. P.G. Hariss, "Conductive Adhesives: A Critical Review of Progress to Date," *Soldering and Surface Mount Technology*, No. 20, pp. 19–21, May 1995
6. A.O. Ogunjimi, O. Boyle, D.C Whalley, and D.J. Williams, "A Review of the Impact of Conductive Adhesive Technology on Interconnection," *Journal of Electronics Manufacturing*, 2: pp. 109–118, 1992
7. S. Asai, U. Saruta, M. Tobita, M. Takano, and Y. Miyashita, "Development of an Anisotropic Conductive Adhesive Film (ACAF) from Epoxy Resins," *Journal of Applied Polymer Science*, 56: pp. 769–777, 1995
8. D.D. Chang, P.A. Crawford, J.A. Fulton, R. McBride, M.B. Schmidt, R.E. Sinitski, and C.P. Wong, "An Overview and Evaluation of Anisotropically Conductive Adhesive Films for Fine Pitch Electronic Assembly," *IEEE Transactions on Components, Hybrids and Manufacturing Technology*, 16(8): pp. 320–326, December 1993
9. H. Ando, N. Kobayashi, H. Numao, Y. Matsubara, and K. Suzuki, "Electrically conductive adhesive sheet," European Patent, 0, 147, 856, 1985
10. K. Gilleo, "An Isotropic Adhesive for Bonding Electrical Components," European Patent 0, 265, 077, 1987
11. R. Pennisi, M. Papageorge, and G. Urbisch, "Anisotropic Conductive Adhesive and Encapsulant Materials," US Patent 5,136,365, 1992
12. H. Date, Y. Hozumi, H. Tokuhira, M. Usui, E. Horikoshi, and T. Sato, "Anisotropic Conductive Adhesives for Fine Pitch Interconnections," *Proceedings of ISHM'94*, (Bologna, Italy), pp. 570–575, September 1994
13. J. Liu, "ACA Bonding Technology for Low Cost Electronics Packaging Applications-Current Status and Remaining Challenges," *Proceedings of 4th International Conference on Adhesive Joining and Coating Technology in Electronics manufacturing*, (Helsinki, Finland), pp. 1–15, June 2000
14. C.M.L. Wu, J. Liu, and N.H. Yeung, "Reliability of ACF in Flip Chip with Various Bump Height," *Proceedings of 4th International Conference on Adhesive Joining and Coating Technology in Electronics Manufacturing*, (Helsinki, Finland), pp. 101–106, June 2000
15. Y. Kishimoto and K. Hanamura, "Anisotropic Conductive Paste Available for Flip Chip," *Proceedings of 3rd international Conference on Adhesive Joining and Coating Technology in Electronics Manufacturing*, (Binghamton, New York), pp. 137–143, September 1998
16. K. Sugiyama and Y. Atsumi, "Conductive Connecting Structure," US patent 4999460, March 12, 1991
17. K. Sugiyama and Y. Atsumi, "Conductive Connecting Method," US patent, 5123986, June 23, 1992
18. K. Sugiyama and Y. Atsumi, "Conductive Bonding Agent and a Conductive Connecting Method," US patent 5180888, January 19, 1993
19. R. Nagle, "Evaluation of Adhesive Based Flip-chip Interconnect Techniques," *International Journal of Microelectronics Packaging*, 1: pp. 187–196, 1998
20. J.K. Kivilahti, "Design and Modeling of Solder-filled ACAs for Flip-Chip and Flexible Circuit Applications," in Conductive Adhesives for Electronics Packaging, J. Liu ed., Port Erin, British Isles, Electrochemical Publications Ltd., 1999, pp. 153–183

21. M. Vuorela, M. Holloway, S. Fuchs, F. Stam, and J. Kivilahti, "Bismuth-Filled Anisotropically Conductive Adhesive for Flip-Chip Bonding," *Proceedings of 4th International Conference on Adhesive Joining and Coating Technology in Electronics manufacturing*, (Helsinki, Finland), pp. 147–152, June 2000
22. A. Torii, M. Takizawa, and M. Sawano, "The Application of Flip Chip Bonding Technology Using Anisotropic Conductive Film to the Mobile Communication Terminals," *Proceedings of Int'l Electronics Manufacturing Technology/Int'l Microelectronics Conference*, (Tokyo, Japan), pp. 94–99, April 1998
23. H. Atarashi, "Chip-on-Glass Technology Using Conductive Particles and Light-Setting Adhesives," *Proceedings 1990 Japan Int. Electron. Manufact. Technol. Symp.*, (Tokyo, Japan), pp. 190–195, June 1990
24. H. Matsubara, "Bare-Chip Face-Down Bonding Technology Using Conductive Particles and Light-Setting Adhesives," *Proceedings of Int'l Microelectronics Conference*, (Yokohama, Japan), pp. 81–87, 1992
25. K. Endoh, K. Nozawa, and N. Hashimoto, "Development of 'The Maple Method," *Proceedings of Japan Int'l Electronics Manufacturing Technology Sypmposium*, (Kanazawa, Japan), pp. 187–191, 1993
26. R. Sihlbom, M. Dernevik, Z. Lai, J.P. Starski, and J. Liu, "Conductive Adhesives for High-Frequency Applications," *IEEE Transactions on Components, Packaging, and Manufacturing Technology*, Part A, 20(3): pp. 469–477, September 1998
27. M. Dernevik, R. Sihlbom, K. Axelsson, Z. Lai, J. Liu, and P. Starski, "Electrically Conductive Adhesives at Microwave Frequencies," *Proceedings of 48th IEEE Electronic Components & Technology Conference*, (Seattle, Washington), pp. 1026–1030, May 1998
28. M.J. Yim, W. Ryu, Y.D. Jeon, J. Lee, J. Kim, and K. Paik, "Microwave Model of Anisotropic Conductive Adhesive Flip-Chip Interconnections for High Frequency Applications," *Proceedings of 49th Electronic Components and Technology Conference*, (San Diego, CA), pp. 488–492, May 1999
29. K. Gustafsson, S. Mannan, J. Liu, Z. Lai, D. Whalley, and D. Williams, "The Effect on Ramping Rate on the Flip Chip Joint Quality and Reliability Using Anisotropically Conductive Adhesive Film on FR4 Substrate," *Proceedings of 47th Electronic Components and Technology Conference*, (San Jose, CA), pp. 561–566, May 1997
30. G. Connell, "Condutive Adhesive Flip Chip Bonding for Bumped and Unbumped Die," *Proceedings of 47th Electronic Components and Technology Conference*, (San Jose, CA), pp. 274–278, May 1997
31. C.N. Oguibe, S.H. Mannan, D.C. Whalley, and D.J. Williams, "Flip-chip Assembly Using Anisotropic Conducting Adhesives: Experimental and Modelling Results," *Proceedings of 3rd international Conference on Adhesive Joining and Coating Technology in Electronics Manufacturing*, (Binghamton, New York), pp. 27–33, September 1998
32. H. Hirai, T. Motomura, O. Shimada, and Y. Fukuoka, "Development of Flip Chip Attach Technology Using Ag Paste Bump Which Formed on Printed Wiring Board Electrodes," *Proceedings of Int'l Symp on Electronic Materials & Packaging*, (Hong Kong, China), pp. 1–6, November–December 2000
33. Y. Hotta, M. Maeda, F. Asai, and F. Eriguchi, "Development of 0.025 mm Pitch Anisotropic Conductive Film," *Proceedings of 48th IEEE Electronic Components & Technology Conference*, (Seattle, Washington), pp. 1042–1046, May 1998
34. G. Connell, R.L.D. Zenner, and J.A. Gerber, "Conductive Adhesive Flip-Chip Bonding for Bumped and Unbumped Die," *Proceedings of 47th Electronic Components and Technology Conference*, (San Jose, CA), pp. 274–278, May 1997
35. L. Li and T. Fang, "Anisotropic Conductive Adhesive Films for Flip Chip on Flex Packages," *Proceedings of 4th International Conference on Adhesive Joining and Coating Technology in Electronics Manufacturing*, (Helsinki, Finland), pp. 129–135, June 2000
36. A. Ogunjimi, S. Mannan, D. Whalley, and D. Williams, "Assembly of Planar Array Components Using Anisotropic Conductive Adhesvies – A Benchmark Study Part I:

Experiment," *IEEE Transactions on Components, Packaging and Manufacturing Technology*, Part C, 19(4): pp. 257–263, October 1996
37. J. Liu, L. Ljungkrona, and Z. Lai, "Development of Conductive Adhesive Joining for Surface-Mount Electronics Manufacturing," *IEEE Transactions on Components, Packaging, and Manufacturing Technology*, part B, 18(2): pp. 313–319, May 1995
38. J. Liu, "Reliability of Surface-mounted Anisotropically Conductive Adhesive Joints," *Circuit World*, 19(4), pp. 4–15, 1993
39. Y.C. Chan, K.C. Hung, C.W. Tang, and C.M.L. Wu, "Degradation Mechanisms of Anisotropic Conductive Adhesive Joints for Flip Chip on Flex Applications," *Proceedings of 4th International Conference on Adhesive Joining and Coating Technology in Electronics manufacturing*, (Helsinki, Finland), pp. 141–146, June 2000
40. H. Kristiansen and J. Liu, "Overview of Conductive Adhesive Interconnection Technologies for LCDs," *IEEE Transactions on Components, Packaging, and Manufacturing Technology*, Part A, 21 (2): pp. 208–214, June 1998
41. P.B. Jana, S. Chaudhuri, A.K. Pal, and S.K. DE, "Electrical Conductivity of Short Carbon Fiber-Reinforced Carbon Polychloroprene Rubber and Mechanism of Conduction," *Polymer Engineering and Science*, 32: pp. 448–456, March 1992
42. A. Malliaris and D.T. Tumer, "Influence of Particle Size on the Electrical Resistivity of Compacted Mixtures of Polymers and Metallic Powders," *Journal of Applied Physics*, 42: pp. 614–618, 1971
43. G.R. Ruschau, S. Yoshikawa, and R.E. Newnham, "Resistivities of Conductive Composites," *Journal of Applied Physics*, 73(3): pp. 953–959, 1992
44. K. Gilleo, "Assembly with Conductive Adhesives," *Soldering and Surface Mount Technology*, No. 19, pp. 12–17, February 1995
45. P.G. Hariss, "Conductive Adhesives: A Critical Review of Progress to Date," *Soldering and Surface Mount Technology*, No. 20, pp. 19–21, May 1995
46. J.C. Jagt, "Reliability of Electrically Conductive Adhesive Joints for Surface Mount Applications: A Summary of the State of the Art," *IEEE Transactions on Components, Packaging, and Manufacturing Technology*, Part A, 21(2): pp. 215–225, 1998
47. M.A. Lutz and R.L. Cole, "High Performance Electrically Conductive Adhesives," *Hybrid Circuits*, No. 23, pp. 27–30, September 1990
48. J.M. Pujol, C. Prudhomme, M.E. Quenneson, and R. Cassat, "Electroconductive Adhesives: Comparison of Three Different Polymer Matrices. Epoxy, Polyimide, and Silicone," *Journal of Adhesion*, 27: pp. 213–229, 1989
49. J. Ivan, J. Gonzales, and M.G. Mena, "Moisture and Thermal Degradation of Cyanate-ester-based Die Attach Material," *Proceedings of 47th Electronic Components and Technology Conference*, (San Jose, CA), pp. 525–535, May 1997
50. I.Y. Chien and M.N. Nguyen, "Low Stress Polymer Die Attach Adhesive for Plastic Packages," *Proceedings of 1994 Electronic Components and Technology Conference*, (San Diego), pp. 580–584, May 1994
51. D.P. Galloway, M. Grosse, M.N. Nguyen, and A. Burkhart, "Reliability of Novel Die Attach Adhesive for Snap Curing," *Proceedings of the IEEE/CPMT International Electronic Manufacturing Technology (IEMT) Symposium*, (Austin, TX), pp. 141–147, October 1995
52. R.L. Keusseyan, J.L. Diiday, and B.S. Speck, "Electric Contact Phenomena in Conductive Adhesive Interconnections," *International Journal of Microcircuits and Electronic Packaging*, 17(3): pp. 236–242, 1994
53. M.K. Antoon, J.L. Koenig, and T. Serafini, "Fourier-Transform Infrared Study Of The Reversible Interaction Of Water And A Crosslinked Epoxy Matrix," *Journal of Polymer Science (Physics)*, 19: pp. 1567–1575, 1981
54. M.K. Antoon and J.L. Koenig, "Irreversible Effects Of Moisture On The Epoxy Matrix In Glass-Reinforced Composites," *Journal of Polymer Science (Physics)*, 19: pp. 197–212, 1981

55. C.G.L. Khoo and J. Liu, "Moisture Sorption in Some Popular Conductive Adhesives," *Circuit World*, 22(4), pp. 9–15, 1996
56. S.M. Pandiri, "The Behavior of Silver Flakes in Conductive Epoxy Adhesives," *Adhesives Age*, pp. 31–35, 1987
57. B. Gunther and H. Schafer, "Porous Metal Powders for Conductive Adhesives," *Proceedings of the 2nd International Conference on Adhesive Joining & Coating Technology in Electronics Manufacturing*, (Stockholm, Sweden), pp. 55–59, June 1996
58. S. Kotthaus, R. Haug, H. Schafer, and B. Gunther, "Investigation of Isotropically Conductive Adhesives Filled with Aggregates of Nano-sized Ag-Particles," *Proceedings of the 2nd International Conference on Adhesive Joining & Coating Technology in Electronics Manufacturing*, (Stockholm, Sweden), pp. 14–17, June 1996
59. P.K. Pramanik, D. Khastgir, S.K. De, and T.N. Saha, "Pressure-sensitive Electrically Conductive Nitrile Rubber Composites Filled with Particulate Carbon Black and Short Carbon Fibre," *Journal of Materials Science*, 25: pp. 3848–3853, 1990
60. P.B. Jana, S. Chaudhuri, A.K. Pal, and S.K. De, "Electrical Conductivity of Short Carbon Fiber-Reinforced Polychloroprene Rubber and Mechanism of Conduction," *Polymer Engineering and Science*, 32(6): pp. 448–456, 1992
61. A. Yokoyama, T. Katsumata, A. Fujii, and T. Yoneyama, "New Copper Paste for CTF Applications," *IMC 1992 Proceedings*, pp. 376–38, 1992
62. S.K. Kang, R. Rai, and S. Purushothaman, "Development of High Conductivity Lead (Pb)-Free Conducting Adhesives," *Proceedings of 47th Electronic Components and Technology Conference*, (San Jose, CA), pp. 565–570, May 1997
63. S.K. Kang, R. Rai, and S. Purushothaman, "Development of High Conductivity Lead (Pb)-Free Conducting Adhesives," *IEEE Transactions on Components, Packaging and Manufacturing Technology*, Part A, 21(1): pp. 18–22, March 1998
64. J. Lin, J. Drye, W. Lytle, T. Scharr, R. Subrahmanyan, and R. Sharma, "Conductive Polymer Bump Interconnects," *Proceedings of 46th Electronic Components and Technology Conference*, (Orlando, FL), pp. 1059–1068, May 1996
65. T. Seidowski, F. Kriebel, and N. Neumann, "Polymer Flip Chip Technology on Flexible Substrates-Development and Applications", *Proceedings of 3rd international Conference on Adhesive Joining and Coating Technology in Electronics Manufacturing*, (Binghamton, New York), pp. 240–243, September 1998
66. R.H. Estes, "Process And Reliability Characteristics of Polymer Flip Chip Assemblies Utilizing Stencil Printed Thermosets And Thermoplastics," *Proceedings of 3rd International Conference on Adhesive Joining and Coating Technology in Electronics Manufacturing*, (Binghamton, New York), pp. 229–239, September 1998
67. Y. Bessho, "Chip on Glass Mounting Technology of Lsis for LCD Module," *Proceedings of Int'l Microelectronics Conference*, pp. 183–189, May 1990
68. J.B. Nysaether, Z. Lai, and J. Liu, "Isotropically Conductive Adhesives and Solder Bumps for Flip Chip on Board Circuits – A Comparison of Lifetime Under Thermal Cycling," *Proceedings of 3rd International Conference on Adhesive Joining and Coating Technology in Electronics Manufacturing*, (Binghamton, New York), pp. 125–131, September 1998
69. K.E. Oh, "Flip Chip Packaging with Micromachined Conductive Polymer Bumps," *IEEE Journal on Selected Topics in Quantum Electronics*, 5(1): pp. 119–126, January–February 1999
70. M. Gaynes, R. Kodnani, M. Pierson, P. Hoontrakul, and M. Paquette, "Flip Chip Attach with Thermoplastic Electrically Conductive Adhesive," *Proceedings of 3rd International Conference on Adhesive Joining and Coating Technology in Electronics Manufacturing*, (Binghamton, New York), pp. 244–251, September 1998
71. B. Trumble, "Get the Lead Out!," IEEE Spectrum, pp. 55–60, Vol. 35, May 1998
72. B.T. Alpert and A.J. Schoenberg, "Conductive Adhesives as a Soldering Alternative," *Electronic Packaging & Production*, pp. 130–132, Vol. 31, November 1991

73. R. Cdenhead and D. DeCoursey, "History of Microelectronics – Part One," *International Journal of Microelectronics*, 8(3): p. 14, 1985
74. G. Nguyen, J. Williams, F. Gibson, and T. Winster, "Electrical Reliability of Conductive Adhesives for Surface Mount Applications," *Proceedings of International Electronic Packaging Conference*, (San Diego, CA), pp. 479–486, September 1993
75. J.C. Jagt, P.J.M. Beric, and G.F.C.M. Lijten, "Electrically Conductive Adhesives: A Prospective Alternative for SMD Soldering?," *IEEE Transactions on Components, Packaging, and Manufacturing Technology*, Part B, 18(2): pp. 292–298, 1995
76. M. Zwolinski, J. Hickman, H. Rubon, and Y. Zaks, "Electrically Conductive Adhesives for Surface Mount Solder Replacement," *Proceedings of the 2nd International Conference on Adhesive Joining & Coating Technology in Electronics Manufacturing*, (Stockholm, Sweden), pp. 333–340, June 1996
77. H. Botter, "Factors That Influence the Electrical Contact Resistance of Isotropic Conductive Adhesive Joints During Climate Chamer Testing," *Proceedings of the 2nd International Conference on Adhesive Joining & Coating Technology in Electronics Manufacturing*, (Stockholm, Sweden), pp. 30–37, June 1996
78. C.P. Wong, D. Lu, S. Vona, and Q.K. Tong, "A Fundamental Study of Electrically Conductive Adhesives," *Proceedings of the 1st IEEE International Symposium on Polymeric Electronics Packaging*, (Norrkoping, Sweden), pp. 80–85, 1997
79. J. Bolger and S. Morano, "Conductive Adhesives: How and Where They Work," *Adhesives Age*, pp. 17–20, June 1984
80. H. Takezawa, M. Itagaki, T. Mitani, Y. Bessho, and K. Eda, "Development of Solderless Joining Technologies Using Conductive Adhesives," *Proceedings of 4th International Symposium and Exhibition on Advanced Packaging Materials, Processes, Properties and Interfaces*, (Braselton, GA), pp. 11–15, March 1999
82. J. Felba, K.P. Friedel, and A. Moscicki, "Characterization and Performance of Electrically Conductive Adhesives for Microwave Applications," *Proceedings of 4th International Conference on Adhesive Joining and Coating Technology in Electronics manufacturing*, (Helsinki, Finland), pp. 232–239, June 2000
81. S. Liong, Z. Zhang, and C.P. Wong, "High Performance Measurement for Isotropically Conductive Adhesives," *Proceedings of 51th Electronic Components and Technology Conference*, (Orlando, FL), pp. 1236–1240, May 2001
83. J.H. Constable, T. Kache, H. Teichmann, S. Muhle, and M.A. Gaynes, "Continuous Electrical Resistance Monitoring, Pull Strength, and Fatigue Life of Isotropically Conductive Adhesive Joints," *IEEE Transactions on Components and Packaging Technology*, 22(2): pp. 191–199, June 1999
84. R. Gomatam, E. Sancaktar, D. Boismier, D. Schue, and I. Malik, "Behavior of Electrically Conductive Adhesive Filled Adhesive Joints Under Cyclic Loading, Part I: Experimental Approach," *Proceedings of 4th International Symposium and Exhibition on Advanced Packaging Materials, Processes, Properties and Interfaces*, (Braselton, GA), pp. 6–12, March 2001
86. L. Smith-Vargo, "Adhesives That Posses a Science All Their Own," *Electronic Packaging & Production*, pp. 48–49, August 1986
85. E.M. Jost and K. McNeilly, "Silver Flake Production and Optimization for Use in Conductive Polymers," *Proceedings of ISHM*, (Bournemouth, England), pp. 548–553, June 1987
87. S.M. Pandiri, "The Behavior of Silver Flakes in Conductive Epoxy Adhesives," *Adhesives Age*, pp. 31–35, Vol. 30, October 1987
88. D. Lu, Q.K. Tong, and C.P. Wong, "A Study of Lubricants on Silver Flakes for Microelectronics Conductive Adhesives," *IEEE Transactions on Components, Packaging and Manufacturing Technology*, Part A, 22(3): pp. 365–371, 1999
89. D. Lu, Q. Tong, and C.P. Wong, "A Fundamental Study on Silver Flakes for Conductive Adhesives," *Proceedings of 4th International Symposium and Exhibition on Advanced Packaging Materials, Processes, Properties and Interfaces*, (Braselton, GA), pp. 256–260, March 1998

90. A.J. Lovinger, "Development of Electrical Conduction in Silver-filled Epoxy Adhesives," *Journal of Adhesion*, 10: pp. 1–15, 1979
91. D. Lu, Q.K. Tong, and C.P. Wong, "Conductivity Mechanisms of Isotropic Conductive Adhesives (ICAs)," *IEEE Transactions on Components, Packaging, and Manufacturing Technology*, Part C, 22(3): pp. 22(3)–227, 1999
92. C. Gallagher, G. Matijasevic, and J.F. Maguire, "Transient Liquid Phase Sintering Conductive Adhesives as Solder Replacement," *Proceedings of 47th Electronic Components and Technology Conference*, (San Jose, CA), pp. 554–560, May 1997
93. J.W. Roman and T.W. Eagar, "Low Stress Die Attach by Low Temperature Transient Liquid Phase Bonding," *Proceedings of ISHM*, (San Francisco, CA), pp. 52–57, October 1992
94. C. Gallagher, G. Matijasevic, and A. Capote, "Transient Liquid Phase Sintering Conductive Adhesives," US Patent 5863622, August 1998
95. H. Botter, "Factors That Influence the Electrical Contact Resistance of Isotropic Conductive Adhesive Joints During Climate Chamer Testing," *Proceedings of the 2nd International Conference on Adhesive Joining & Coating Technology in Electronics Manufacturing*, (Stockholm, Sweden), pp. 30–37, June 3–5, 1996
96. K. Gilleo, "Evaluating Polymer Solders for Lead Free Assembly, Part I," *Circuits Assembly*, pp. 50–51, February 1994.
97. K. Gilleo, "Evaluating Polymer Solders for Lead Free Assembly, Part II," *Circuits Assembly*, pp. 51–53, January 1994
98. D. Lu, Q.K. Tong, and C.P. Wong, "Mechanisms Underlying the Unstable Contact Resistance of Conductive Adhesives," *IEEE Transactions on Components, Packaging, and Manufacturing Technology*, Part C, 22(3): pp. 228–232, 1999
99. Q.K. Tong, G. Fredrickson, R. Kuder, and D. Lu, "Conductive Adhesives with Superior Impact Resistance and Stable Contact Resistance," *Proceedings of the 49th Electronic Components and Technology Conference*, (San Diego, CA), pp. 347–352, May 1999
100. D. Lu and C.P. Wong, "Novel Conductive Adhesives for Surface Mount Applications," *Journal of Applied Polymer Science*, 74: pp. 399–406, 1999
101. D. Lu and C.P. Wong, "Novel Conductive Adhesives with Stable Contact Resistance," *Proceedings of 4th International Symposium and Exhibition on Advanced Packaging Materials, Processes, Properties and Interfaces*, (Braselton, GA), pp. 288–294, March 1999
102. C. Cheng, G. Fredrickson, Y. Xiao, K. Tong, and D. Lu, US patent 6,344,157, 2002
103. H. Leidheiser, Jr., "Mechanism of Corrosion Inhibition with Special Attention to Inhibitors in Organic Coatings," *Journal of Coatings Technology*, 53(678): pp. 29–39, 1981
104. G. Trabanelli and V. Carassiti, "Mechanism and Phenomenology of Organic Inhibitors," in Advanced Corrosion Science and Technology, M.G. Fontana and R.W. Staehle ed., Vol. 1, Plenum Press, New York, NY, 1970, pp. 147–229
105. G. Trabanelli, "Corrosion Inhibitors," in Corrosion Mechanisms, F. Mansfeld ed., Marcel Dekker, Inc., New York, NY, 1987, pp. 119–164
106. O.L. Riggs, Jr., "Theoretical Aspects of Corrosion Inhibitors and Inhibition," C.C. Nathan ed., NACE, pp. 2–27, 1973
107. P.A. Reardon, "New Oxygen Scavengers and Their Chemistry under Hydrothermal Conditions," Corrosion'86, Paper no. 175, NACE, (Houston, TX), 1986
108. M.G. Noack, "Oxygen Scavengers," Corrosion'89, Paper no. 436, NACE, (Houston, TX), 1989
109. P.A. Reardon and W.E. Bernahl, "New Insight into Oxygen Corrosion Control," Corrosion'87, Paper no. 438, NACE, (Houston, TX), 1987
110. S. Romaine, "Effectiveness of a New Volatile Oxygen Scavenger," *Proceedings of the American Power Conference*, (Chicago, IL), pp. 1066–1073, April 1986

111. D. Durand, D. Vieau, A.L. Chu, and T.S. Weiu, "Electrically Conductive Cement Containing Agglomerates, Flake and Powder Metal Fillers," US Patent 5180523, November 1989
112. S. Macathy, *Proceedings of Surface Mount International*, (San Jose, CA), pp. 562–567, August 1995
113. S.A. Vona and Q.K. Tong, "Surface Mount Conductive Adhesives with Superior Impact Resistance," *Proceedings of 4th International Symposium and Exhibition on Advanced Packaging Materials, Processes, Properties and Interfaces*, (Braselton, GA), pp. 261–267, March 1998
114. J. Liu and B. Weman, "Modification of Processes and Design Rules to Achieve High Reliable Conductive Adhesive Joints for Surface Mount Technology," *Proceedings of the 2nd International Symposium on Electronics Packaging Technology*, (Shanghai, China), pp. 313–319, December 1996
115. D. Lu and C.P. Wong, "High Performance Electrically Conductive Adhesives," *IEEE Transactions on Components, Packaging, and Manufacturing*, Part C, 22(4): pp. 324–330, 1999
116. D. Lu and C.P. Wong, US patent 6,740,192, 2004

Chapter 12
Die Attach Adhesives and Films

Shinji Takeda and Takashi Masuko

Abstract This chapter outlines the strong correlation between developments in electronic packaging technologies and required properties of die attach materials. An overview of die attach materials is summarized with the trends in the market. Die attach paste, adhesive tape for a lead on chip (LOC), die attach film, and the prospects of advanced die attach film are described in each section. The technical requirements of the die attach materials, which include high purity, fast curing, low stress and high package crack resistance are discussed.

Die attach films have become the main stream of die attach materials owing to their excellent properties and reliability. The future of advanced die attach films is explained with the introduction of adhesive film with dicing / die attach dual functionality.

The effects of adhesive properties such as peel strength and water absorption to improve package crack resistance are reported in detail. The development of die attach films with excellent reliability for advanced packages such as BGA/CSP is also reviewed.

Keywords die attach · adhesives · films · package crack resistance · advanced BGA/CSP · low stress · peel strength · water absorption · chip warpage · polyimide · epoxy resin

12.1 Die Attach Materials

12.1.1 Trends in Electronic Packaging

The section outlines the developments in electronic packaging technology which has been the forefront of research in the semiconductor industry. The characteristics and targets of die attach materials are strongly influenced by the packaging

---

S. Takeda (✉)
Research and Development Division, Hitachi Chemical Co., Ltd., 48 Wadai, Tsukuba, Ibaraki, 300-4247, Japan
e-mail: shin-takeda@hitachi-chem.co.jp

technology. Therefore, it is important to fully understand the packaging technology for the development of die attach materials, since the materials are intimately associated with the packaging schemes.

The growing demand for personal portable devices has lead to an unprecedented increase in the competitiveness of the market and has consequently caused huge demands for further downsizing in terms of compactness. Packages have been changing due to demands of rapid miniaturization of electronic devices resulting in high packaging density and high operating speeds (Fig. 12.1) [1–4].

The introduction of advanced materials in combination with devising packaging schemes provides a solution for the electronic packaging problem. In the 1980s, the mainstream electronic packaging was pin-inserting-type packages and Dual Inline Packages (DIP). The synthesis of high performance Epoxy Molding Compounds (EMC) has lead to surface- mounting-type packages including Quad Flat Packages (QFP) (Fig. 12.2) and Thin Small Outline Packages (TSOP). In early 1990s, Ball Grid Array (BGA) was developed because narrow-pitched-pin-type had practical mounting process issues [2–4]. The BGA package was miniaturized to the size of a silicon chip, which is Chip

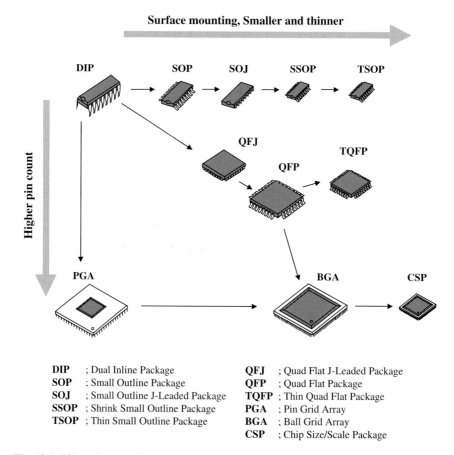

DIP	; Dual Inline Package	QFJ	; Quad Flat J-Leaded Package
SOP	; Small Outline Package	QFP	; Quad Flat Package
SOJ	; Small Outline J-Leaded Package	TQFP	; Thin Quad Flat Package
SSOP	; Shrink Small Outline Package	PGA	; Pin Grid Array
TSOP	; Thin Small Outline Package	BGA	; Ball Grid Array
		CSP	; Chip Size/Scale Package

Fig. 12.1 LSI package trends

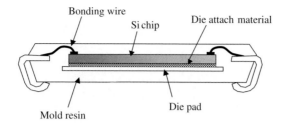

Fig. 12.2 Internal structure of QFP

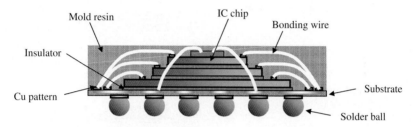

Fig. 12.3 Internal structure of Stacked-CSP

Scale Package (CSP) [2–5]. Moreover, the packaging schemes progressed with further miniaturization, high speed and function systems as well as Stacked-CSP, which has a three-dimensional stacking structure (Fig. 12.3) [4, 6, 7].

In the case of a mobile phone structure, the mounting area and height is particularly limited. Recently, a System in Package (SiP) was developed for a mobile phone, which has different function chips, logic and memory which are stacked in a package. Stacked package types are increasingly being used in mobile phone technology [4].

12.1.2 Trends in Die Attach Materials

Die attach materials are used as adhesives between a silicon chip (die) and a substrate. The packaging process in Fig. 12.4 illustrates a conventional process flow. The Die attaching process is an important part of manufacturing of Integrated Chip (IC) devices because die attach materials plays a key role in the reliability and performance of the semiconductor packaging. Due to an overwhelming development in the packaging process, the subsequent enhancement and modification of the die attach properties have ensued [1].

Die attach materials such as gold, solder and polymeric adhesives such as pastes or films have been extensively used. The requirements of the die attach are dependant on the configuration of the packaging. The die attach material most widely used in the 1980s was the Gold-Silicon eutectic. A silicon chip is attached on a gold-coated lead frame at 400°C and the eutectic is produced. The Au-Si proved problematic with its high die stress due to the mismatch of

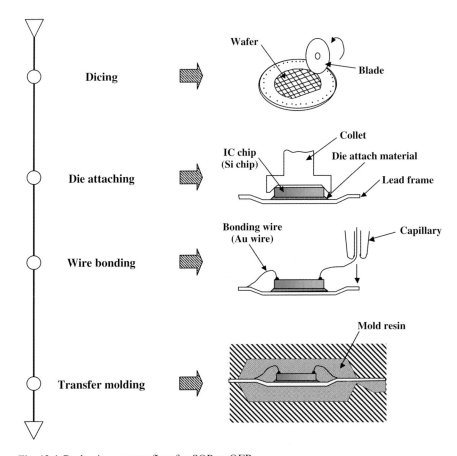

Fig. 12.4 Packaging process flow for SOP or QFP

thermal expansion between the die attach material and a substrate. Economic factors also contributed to the decline in Au-Si since the price of gold significantly increased resulting in the demand for cheaper materials [8–10].

A solder is advantageous because of the high thermal conductivity and non-water-absorptivity. The reliability has been proven to be enhanced with the properties. The formation of a homogeneous microstructure has also improved the resistance to thermal fatigue failure [11].

However, there are a number of disadvantages with a solder. The application of the solder involves the use of a flux to ensure good connectivity, which is subsequently removed chemically. The defluxing process is costly and causes environmental concerns due to the emission of chlorinated substances. Furthermore, the high attaching temperature caused the oxidation of the copper lead frame which increased the stress between the chip and the frame.

A polymeric adhesive such as a silver paste was developed to overcome the problems of the Au-Si eutectic and a solder.

The newly synthesized paste composed of a silver filler and a resin became an attractive alternative for the existing materials. The silver paste (die attach paste) has lower associated costs and ideal properties including lower levels of stress as well as a reduced attaching temperature. Further details are discussed in a later part of the chapter.

12.1.3 Demands on Die Attach Materials

The technical requirements on a die attach material include high purity (low contamination), quick curing, low stress and package crack resistance during reflow soldering [9, 10]. High purity material is responsible for the reliability performance of devices, it is not exclusive to die attach materials but also applies to most materials of electronic packaging. It was reported that a trace amount of ionic contamination would lead to the corrosion of aluminum in a package [12]. In the early stages of a silver paste development, the susceptibility to contamination was a major technical issue. This contamination problem was solved by the purification of raw materials and basic resins in conjunction with the introduction of special compounding techniques.

Short curing time is extremely important in terms of processability and it contributes in-line-curing process. With regards to the silver paste, both the stability at room temperature and quick curing system (within 1–2 minutes) are required. A hardener and an accelerator are selected and optimized for the short time curing process. However, an adhesive in the form of a silver paste suffers from voids due to the raising temperature in a curing system. The voids reduce the reliability of a packaging and remain a pressing problem for the application of silver pastes [9].

Low stress properties are another fundamental feature for an advanced packaging system. In the case of the advanced packaging, a large IC chip is attached on a copper lead frame or polymer-based substrate such as glass-epoxy and polyimide having circuits on the surface. As the briefly mentioned, CTE (Coefficient of Thermal Expansion) mismatch between the silicon chip and a substrate is a serious problem (Fig. 12.5). The stress can be alleviated by using a low stress die attach material [13, 14].

Package crack resistance has probably the greatest importance as a property for the current die attach materials, as the trend in semiconductor packaging suggests a movement towards higher integration and higher pin counts. This requires a larger chip in a smaller and thinner package thus making crack resistance imperative in the efficient operation of the system. A major reliability problem associated with surface mount devices is package cracks produced by soldering stress, typically exposing the entire package to temperature as high as 240–260°C. The mechanism of package crack resistance during reflow soldering is described in the section. Nevertheless, the die attach film serves to overcome these problems [15–17].

Recently, lead-free soldering has set a global trend in the electronic industry due to the growing concerns of lead poisoning and contamination in the

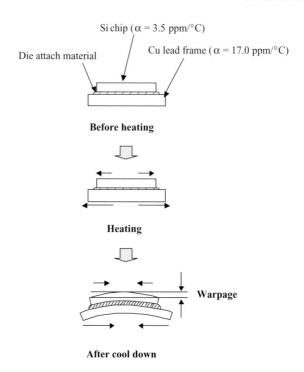

Fig. 12.5 Silicon chip warpage

environment. The reflow soldering temperature has risen due to higher melting point of lead-free solder substitutes. The trend poses very serious problems on the current die attach materials especially in terms of package crack resistance.

12.1.4 Die Attach Paste

A die attach paste consists of a filler such as silver, alumina and silica, and a basic resin such as epoxy resin, polyimide, polyacrylate and silicone resin. A filler is dispersed in a basic resin which has adequate flow properties for workability at room temperature. Heating cures the thermoset resin and a silicon chip is fixed on a lead frame after the curing process. The majority of a die attach pastes are silver pastes composed of a silver filler and an epoxy resin. Silver filler has a unique shape flake-type shape and an average diameter of filler is 2–10 μm. A silver paste has demonstrated good spreadability in a basic resin as well as good handling properties (thixotropy) due to this special flake-like shape with it corresponding surface coating reagents [11].

An insulating filler paste such as a silica paste is used for advanced packages including BGA, stacked-CSP, in order to keep insulation between a silicon chip and electrical pattern on a polymer substrate [18].

12 Die Attach Adhesives and Films

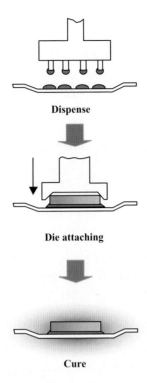

Fig. 12.6 Die attaching process using a die attach paste

As shown in Fig. 12.6, the illustration depicts the process of a die attach paste involving three main stages, namely, dispensing, attaching and curing [9, 11]. Similarly, there are practical difficulties in the die attach paste process, for example, bleeding out of a low viscosity paste, spreadability of a high viscosity paste, dryness of a paste including a low boiling point solvent, and outgassing of a paste including a reactive dilute reagent.

The most popular base resin is an epoxy resin system which shows good adhesion strength and exhibits ideal properties as a die attach material. Contaminants such as chlorine and sodium were included in an epoxy resin in the early years because the resin was synthesized from epichlorohydrine and sodium phenolate. Since highly purified materials were developed for this semiconductor field, the contamination problem was solved [18].

A die attach paste using polyacrylate as a basic resin was developed in 1990s. The benefits of a polyacrylate paste include a rapid curing system (1–2 minutes curing) due to radical polymerization, a longer shelf life due to a stable peroxide initiator as well as good handling properties owing to a low viscosity material (acrylate monomer) [18].

More recently, various kinds of pastes are being developed by die attach manufacturers to meet specific customer requirements. Customers may demand a reflow resistant paste with a stronger adhesion strength and low moisture absorption, electrical and thermal high conductivity paste having high content of filler, and non-solvent type paste having void-free and low stress for each requirement suited for their respective application [18].

12.1.5 Adhesive Tape for LOC Package

An adhesive tape is used for a LOC (Lead on Chip) package as shown in Fig. 12.7 [10, 19, 20]. A LOC package is smaller than a standard package such as a QFP (Quad Flat Package) because of the structure of a LOC package which lead pins are located above a silicon chip. Inner leads are attached on a surface of a chip by an adhesive tape. The structure of a LOC package is suitable for high density mounting and it is applied for TSOP (Thin Small Outline Package) in the field of DRAM (Dynamic Random Access Memory) [19, 20].

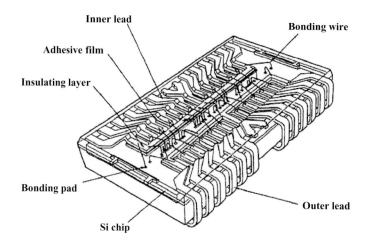

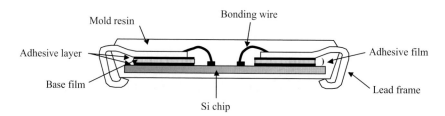

Fig. 12.7 Internal structure of LOC package

12 Die Attach Adhesives and Films

A basic resin of the adhesive tape for LOC is a polyimide and consists of three principle layers [10, 19, 20]. The adhesive is coated on both sides of a polyimide film in a similar nature to the material Kapton®. The adhesive tape is a hot melt type adhesive and it is attached at high temperatures, 200–400°C, and high pressures, 10–30 N/chip. Since the adhesive tape is directly attached onto a silicon chip surface, high purity, low outgassing and heat resistance properties are important.

Manufacturers have developed variations of the polymer structure including a flexible segment has been introduced into a hard polymer chain backbone in order to decrease adhesive temperature [19, 20].

12.1.6 Die Attach Film

Die attach films have become the key technology to realize excellent reliability, high performance, high speed, high density of devices, as well as smaller and thinner packages.

A die attach film which composed of a thermoplastic resin and a silver filler was announced by Du Pont® in 1988 [21]. Nitto Denko® proposed new concept which is a die attach – dicing film which reduced number of die attach process and dicing process in 1991 [22].

In 1994, Hitachi Chemical® developed a novel die attach film 'HIATTACH' which became possible to attach at low temperature, low pressure and short time (within a second), and also shows excellent package crack resistance during reflow soldering [23, 24]. Reliability performance of devices has been considerably improved by the die attach film. Furthermore, new die attach process including a new die attach machine for the die attach film, which is completely different from the process of a die attach paste, was developed (Fig. 12.8).

Low stress type of die attach films were developed by several manufacturers in 1997 in order to meet requirements for an advanced packaging.

There have been significant developments in the production of die attach films and the ongoing research in this field is paramount to the electronic packaging industry.

The development of die attach films is introduced in details in the next section.

12.1.7 The Future of Advanced Die Attach Film

12.1.7.1 Die Attach Film for Advanced BGA/CSP

Recently, Mold Array Package (MAP) which is able to reduce number of processes, resulting in lower cost [25]. The size of advanced BGA/CSP is increasing and is causing various technical issues. The main issue is the large warpage of a substrate and poor connectivity inside/outside of a package due to CTE mismatch of the materials in the structure which are composed of a substrate, solder balls, and a printed circuit board [25]. In this area, the most

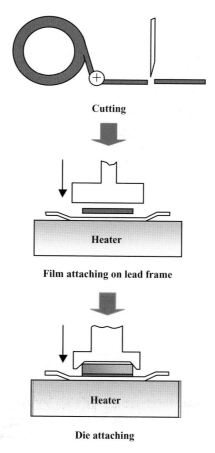

Fig. 12.8 Die attaching process using die attach film

important requirement for a die attach film is to reduce the stress of a package. The die attach film for the purpose of low attaching temperature and low modulus is under development to meet the above requirements for the advanced BGA/CSP [26, 27].

12.1.7.2 Dicing / Die Attach Dual Functioning Film

Stacked multi-chip package (Stacked-MCP; Fig. 12.3), in which plural chips are stacked up, has been widely noticed and is under further development by the demands on advanced packages being smaller, thinner, and higher performance [28, 29]. The requirements of a die attach film for Stacked-MCP are to reduce the number of manufacturing steps and to handle thinner wafers easily. To meet the requirements, a dicing / die attach double functioning film has been developed and launched in 2005 [30, 31]. The film was composed of two layers, a UV reaction type dicing film and a thermosetting type die attach film.

Figure 12.9 shows conventional process of manufacturing Stacked-MCPs. Two lamination steps are necessary for a dicing film and a die attach film.

12 Die Attach Adhesives and Films

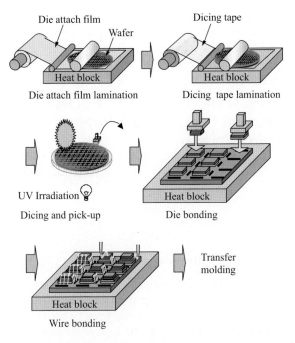

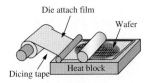

Fig. 12.9 Conventional process for manufacturing Stacked-MCPs two lamination steps are required for die attach film and dicing tape

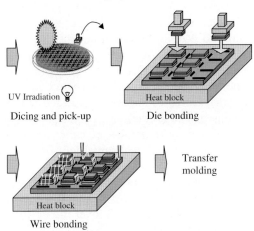

Fig. 12.10 Novel process for manufacturing Stacked-MCPs using double functioning tape

Figure 12.10 shows new process for the stacked package. The dicing/die attach film is laminated by one step. The double functioning film realized to reduce the number of manufacturing steps and to handle thinner wafers easily in the process [30].

12.2 Development of Die Attach Film: High Reliability Performance for Package Crack Resistance and Advanced Package

In this section, development of die attach films including design concepts and properties is described in details.

12.2.1 Introduction

12.2.1.1 Technical Issues of Silver Paste

Silver paste has been used commonly as die attach materials in a plastic package. As the trend in semiconductor packaging is shifting towards higher integration and higher pin counts which requires larger chips in smaller and thinner packages. Major technical issues of a silver paste are:

1) Package crack/delamination during reflow soldering
2) Wettability/spreadability with attachment large die size
3) Voids
4) Productivity/processability with dispensing system

The die attach film was developed to overcome these issues.

12.2.1.2 Package Crack

A major reliability problem associated with surface mount devices is package cracks generated by soldering stress, typically exposing the entire package to high temperature around 245°C. Absorbed moisture in a package vaporizes, expands, and causes delamination/fracture/package cracks, commonly termed as the 'popcorn' phenomenon (Fig. 12.11) [32–34]. Though various formulation changes in epoxy molding compounds to reduce amount of moisture absorption has achieved some degree of success, die attach materials still has this problem totally. Because package cracks are produced by moisture absorption and delamination of die attach materials, moisture absorption of die attach materials should be minimized, and peel strength of those should be increased to prevent package cracks.

Lead (Pb)-free soldering for electronic industry is a segment of global trend toward Pb-free environment [35]. The environmental law regulations against Pb require the makers to either eliminate or recover the Pb from waste of electrical and electronic equipments including imported products. Projects towards

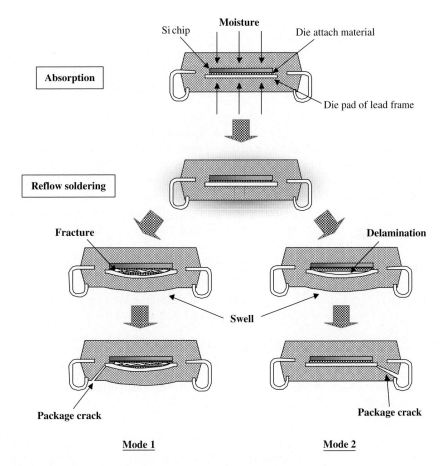

Fig. 12.11 Generation of package crack during reflow soldering

Pb-free at particularly major Japanese companies are becoming more active to eliminate Pb. Melting points of Pb-free solder alternatives, such as Sn/Ag and Sn/Cu alloys are 30–40°C higher than that of a traditional solder Sn/Pb. Therefore, maximum temperature of IR reflow tends to increase from 245°C to 275°C. Since this higher reflow temperature might cause serious damage to die attach materials, a high performance die attach film is demanded to overcome the Pb-free issues.

12.2.1.3 Advanced Packages

New advanced packages became popular to satisfy customer's demands for smaller, thinner, and lower-cost packages for laptop PC, mobile phones, and other consumer electronic systems. One of features of the new advanced package is that a polymer-based substrate such as glass-epoxy and polyimide, which have circuit on the surface, is used instead of a metal lead frame. In this field, lowering adhesive temperature, reducing stress, and avoiding contamination to the circuits on the surface are demanded. The requirements are leading to a number of problems on a current silver paste. Therefore, a high performance die attach film for the advanced packages are necessary to meet the requirements.

A die attach area is generally very close to a wire boding pad in the case of BGA/CSP compared with a conventional package such as LQFP. Therefore, contamination of a wire bonding pad would be easily caused by bleeding of a die attach paste used. In particular, it should be more serious for a stacked CSP because a die attach area of the package is much closer to a wire bonding pad. For these reasons, a die attach film is more suitable for BGA/CSP than a paste type due to no bleeding and no contamination. In addition, a die attach film can be expected to show advantages of insulating reliability, voids-free and no slanting of die attach layer.

Some key technical problems of new advanced packages BGA/CSP are listed here:

1) Chip warpage caused by thermal expansion mismatch between a silicon chip and a polymer-based substrate having large coefficient of thermal expansion (CTE)
2) Low heat resistance of a polymer-based substrate
3) Moisture absorption of a die attach layer caused by a thin package

Based on the above technical problems, a list of corresponding requirements for die attach materials is summarized here:

a) Decreasing chip warpage (low stress material)
b) Lowering die attach temperature, <200°C (low Tg material)
c) Improving package crack resistance (low water absorption and high peel strength material)

12.2.2 Design of Base Resin for Die Attach Film

New polyimides having various chemical structures were synthesized as the base resin of a die attach film.

Traditionally, polyimides have been used as materials for application in future aircraft and spacecraft because of their superior thermal, mechanical and electrical properties [36]. They have been applied in microelectronic devices as film or varnish for the purpose of miniaturization and improvement in performance and reliability [37]. Most polyimides must be processed in the stage of their soluble poly (amic acid) precursors, which are subsequently imidized by heating at high temperatures over 350°C. Because they are insoluble and infusible in their imidization form, polyimides are difficult to process. Furthermore, they must be processed or prepared at high temperatures over

300°C due to their high Tg or high softening temperature, resulting in the thermal damage of the peripheral materials. Their application fields have been limited for this reason [38]. In order to obtain soluble polyimides and/or to prepare polyimides at lower temperatures, extensive study has been carried out on the synthesis of polyimides having lower Tg or lower softening temperatures by introducing flexible structures such as alkylene, ether, and siloxane connecting groups into the polymer backbone [39–41]. These polyimides melt down above their Tg or softening temperatures because of their inherent thermoplasticity. It was difficult to develop polyimide-based materials which can be prepared at low temperatures and have high mechanical strength at high temperatures.

To solve the problem, material design as blending or mixing different polymers and fillers has been studied to develop new materials having various superior properties which cannot be achieved by one component materials [42–46]. According to such material design, there has been much effort to develop new composite materials, as high performance die attach adhesives for microelectronics applications, having lower fluidity and higher mechanical strength at high temperatures above Tg, by blending an epoxy resin as a crosslinking agent and a filler as a reinforcement. An extensive study of the mechanical strength of the composite materials has proven that this approach was very successful.

A number of researchers have reported the studies on polyimide–epoxy resin composites [44–46]. However, the main purpose of these studies was to improve the inherent brittleness of the epoxy matrices. A few studies have been reported on the restriction of the flow of polyimide-based composite materials at high temperatures above Tg by blending epoxy resins and fillers.

In this study, new composite films composed of a polyimide, an epoxy resin and a silver flake which has been used as a filler in silver paste die attach materials, were examined to develop new die attach adhesives which can be attached at lower temperatures and have both good stress relaxation property and high mechanical strength at high temperatures above Tg. In the previous work [47, 48], the effect of contents of the epoxy resin and the silver flake on adhesion properties, morphology and rheological properties of the die attach films has been demonstrated in details. In this section, the die attach films using various polyimides were reviewed and the relationships between the chemical structures of the polyimides and the various properties of the composite films were described.

12.2.3 Die Attach Film for Package Crack Resistance

12.2.3.1 Water Absorption of Base Resin

Water absorption properties of three kinds of polyimides are shown in (Table 12.1). Polyimides generally have high water absorption, e.g. 1.3% of type A (EBTA-BAPP). To reduce water absorption, hydrophobic chemical structure was introduced into polymer backbone, type B (EBTA/DBTA-BAPP)

Table 12.1 Water absorption of polyimides

Polyimide	Monomers [mol%] Dianhydride	Diamine	Water absorption[a] [wt%]
A	EBTA (100)	BAPP (100)	1.3
B	EBTA (50) / DBTA (50)	BAPP (100)	0.7
C	DBTA (100)	BAPP (100)	0.2

[a] Immersed for 24 h at RT.

n=2 : EBTA
n=10 : DBTA

BAPP

and type C (DBTA-BAPP). A new type C polyimide showed significantly low water absorption, 0.2%, due to the hydrophobic structure. Type C was used for a base resin of a die attach film.

12.2.3.2 Peel Strength

Peel strength is the most important property to prevent delamination of die attach materials in terms of package crack resistance. Relationship between die attach film composition and peel strength was studied [48].

Die attach films composed of a polyimide, an epoxy resin and a silver filler were prepared. The relationship between the adhesion behaviors and the bulk and surface properties of the die attach film will be discussed.

The die attach film was cut into 5 mm square pieces. Each square piece was inserted between a 5 mm square silicon chip of 400 μm thick and a copper bare lead frame of a larger size. The silicon chip and the lead frame were compression-bonded under 0.4 MPa pressure at 300°C for 5 sec, followed by heating in a 180°C oven for 1 h to cure the die attach film completely. Prior to the measurement, half of the samples prepared were treated in a thermo-hygrostat at 85°C and 85% RH for 168 h. Using a push-pull gauge, the peel strength was measured by peeling the silicon chip from the lead frame at the pulling speed of 0.5 mm/s as shown in (Fig. 12.12). The measurement was carried out on the 250°C specimens (the pulling started 20 s after the specimen was set on the 250°C heater plate).

Figure 12.12 shows the relationships between the epoxy resin content of the die attach film, the silver filler content fixed at 40 wt%, and the peel strength to the copper lead frames. Table 12.2 summarizes the failure mode of the adhesive specimens in measuring the peel strength.

The peel strength to the copper lead frame also remarkably increased with increasing epoxy resin content up to 10 phr (Fig. 12.13). However, both the peel strength at dry and wet conditions clearly decreased with increasing epoxy resin

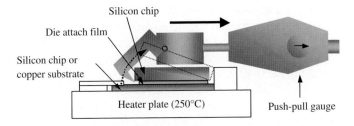

Fig. 12.12 Measurement of peel strength

Table 12.2 Failure modes of the adhesive specimens (Fig. 12.13)

Film No.	Epoxy resin content [phr]	Failure mode* Dry	Wet
1	0	Co	Co
2	10	Co	Co
3	20	Ad/L	Ad/L
4	30	Ad/L	Ad/L
5	40	Ad/L	Ad/L

*Dry: Before moisture exposure
Wet: After moisture exposure
Co: Cohesive failure of film
Ad/L: Interfacial failure between film and lead frame

content at 20 phr and over. The failure mode changed from the cohesive failure of the film (up to 10 phr) to the interfacial failure between the film and the copper lead frame (20 phr and over) regardless of the conditions (Table 12.2).

The elastic modulus of the film without epoxy resin component (No.1) was lower than that of the other films containing 10 to 40 phr of the epoxy resin

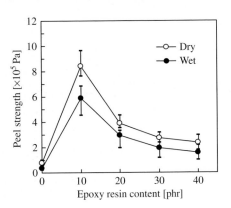

Fig. 12.13 Relationship between epoxy resin content and peel strength to copper lead frame

component (Nos. 2–5) at both 20°C and 250°C. No. 1 film melted down at 250°C because of its thermoplasticity and having no network structure (Table 12.4).

The remarkable increase in peel strength with increasing epoxy resin content up to 10 phr (Fig. 12.13) must be resulted from the increased elastic modulus of the film at high temperature due to the formation of the network structure (Table 12.4).

While the fracture strength of the film increased with increasing epoxy resin content, the interfacial adhesion strength between the film and the copper lead frame decreased inversely (Fig. 12.13). This can be explained by the decreased thermodynamic work of adhesion W_A^{Cu}, and the decreased surface energy γ_s of the film (Table 12.5) [49–54].

Figure 12.14 shows the relationships between the silver filler content of the die attach film, the epoxy resin content fixed at 10 phr, and the peel strength to the copper lead frames. Table 12.3 summarizes the failure modes.

The peel strength to the copper lead frame behaved somewhat differently at dry and wet conditions (Fig. 12.14). The peel strength at dry condition was almost independent of the silver filler content up to 40 wt%. Then it significantly decreased at 60 wt% and over. On the other hand, the wet peel strength

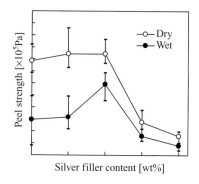

Fig. 12.14 Relationship between silver filler content and peel strength to copper lead frame

Table 12.3 Failure modes of the adhesive specimens (Fig. 12.14)

Film No.	Silver filler content [wt%]	Failure mode*	
		Dry	Wet
6	0	Co	Si/Ad
7	20	Co	Si/Ad
8	40	Ad/L	Ad/L
9	60	Ad/L	Ad/L
10	80	Ad/L	Ad/L

*Dry: Before moisture exposure
Wet: After moisture exposure
Co: Cohesive failure of film
Si/Ad: Interfacial failure between silicon chip and film
Ad/L: Interfacial failure between film and lead frame

showed a remarkable peak at 40 wt%. For the copper lead frame, the failure mode was complicated (Table 12.3). While the failure mode at dry condition was the cohesive failure of the film up to 20 wt%, that at wet condition was the interfacial failure between the silicon chip and the film. The failure mode was the interfacial failure between the film and the copper lead frame at 40 wt% and over, regardless of the conditions.

The increase in peel strength with increasing silver filler content up to 40 wt% might be resulted from the increased elastic modulus at high temperature and the increased Tg of the film due to the restriction of microbrownian motion of the base resin by the silver filler. Although the fracture strength of the film increased with increasing silver filler content, the interfacial adhesion strength between the film and the silicon chip or the copper lead frames decreased with increasing filler content at 60 wt% and over. This might be due to the decreased thermodynamic work of adhesion W_A, induced by the decreased base resin content and the increased thermal stress σ_{max} (Tables 12.4 and 12.5).

The drastic decrease in wet peel strength at 20 wt% and less silver filler content and also the change in failure mode from the cohesive to interfacial failure for the silicon chip can be well explained by the increased contact angle for distilled water θ^{H_2O}. That is, the interface between the silicon chip having a highly hydrophilic surface and the film is flooded more easily by the absorbed moisture, because of the increased surface hydrophilicity of the film with decreasing silver filler content (Table 12.5).

12.2.3.3 Package Crack Resistance

A die attach film DF-A was developed through an investigation of the relationship above and then the composition was optimized. Table 12.6 shows properties

Table 12.4 Bulk properties of the films

Film No.	Epoxy resin content [phr]	Silver filler content [wt%]	Elastic modulus [Mpa] 20°C	Elastic modulus [Mpa] 250°C	Tg [°C]	$\sigma_{max}(\times K)$ [Mpa]*	Water absorption [vol%] 24 h	Water absorption [vol%] 168 h
1	0		2910	Melt down	119	351	0.06	0.17
2	10		4010	6	119	412	0.23	0.35
3	20	40	3930	4	113	383	0.25	0.40
4	30		3580	4	109	350	0.32	0.48
5	40		3960	5	109	368	0.44	0.58
6		0	2310	2	117	307	0.32	0.49
7		20	2480	4	117	318	0.28	0.43
8	10	40	4010	6	119	412	0.23	0.35
9		60	4280	16	126	456	0.17	0.26
10		80	10200	150	131	737	0.10	0.15

* Maximum stress at the die corners (K: geometric constant)

Table 12.5 Surface properties of the films

Film No.	Epoxy resin content [phr]	Silver filler content [wt%]	Contact angle [degree] θ^{H_2O}	$\theta^{CH_2I_2}$	Surface energy [mN/m] γ_s^d	γ_s^p	γ_s	Work of adhesion [mN/m] W_A^{Si}	W_A^{Cu}
1	0		70.7	23.0	41.6	6.8	48.4	94.3	88.7
2	10		73.7	34.7	37.1	6.5	43.6	89.9	84.1
3	20	40	75.8	35.0	37.5	5.5	43.0	88.1	83.7
4	30		78.5	36.8	37.0	4.5	41.5	85.1	82.3
5	40		81.4	42.3	34.6	4.0	38.6	81.7	79.3
6		0	73.7	26.4	41.0	5.5	46.5	90.9	87.1
7		20	73.8	32.1	38.4	6.1	44.5	90.2	85.1
8	10	40	73.7	34.7	37.1	6.5	43.6	89.9	84.1
9		60	76.1	35.4	37.3	5.4	42.7	87.6	83.4
10		80	77.7	37.2	36.8	4.9	41.7	86.0	82.5

Table 12.6 Characteristics of DF-A and silver paste

Item		Unit	DF-A	Silver paste	Test condition
Composition	Base resin		Polyimide and epoxy resin	Epoxy resin	
	Silver filler content	wt%	40	70	
Die attach condition	Temp.	°C	230	–	
	Pressure	N/chip	0.5	1.0	
	Time	sec	1	<1	
	Cure	°C-min	180-30	180-60	
Package crack resistance (85°C85%RH)	245 °C	–	504 h OK	24 h OK	QFP 14×20×1.4^t mm
	265 °C		168 h OK	24 h NG	Chip size:8×10 mm
	275 °C		168 h OK	24 h NG	EMC:CEL-9200
Water absorption		vol%	0.2	1.2	RT for 24 h
Peel strength	245 °C	×10^5 Pa	8.1	1.0	Chip size:5×5 mm
	265 °C		8.1	1.0	Lead frame:Cu
	275 °C		8.3	0.8	

of DF-A and a current silver paste. DF-A is composed of a modified polyamide base resin having hydrophobic structure, a thermosetting resin of optimum content, and a silver filler of 40 wt%. Water absorption of DF-A is 1/6 times as small as that of a current silver paste. Peel strength of DF-A is 8 times greater than that of the paste.

To evaluate the package crack resistance of DF-A, a 14 × 20 × 1.4 mm Low profile Quad Flat Package (LQFP) was used. The LQFP consisted of a silicon chip (8 × 10 × 0.3 mm), a die attach film DF-A (thickness 30 μm), a copper lead frame with a flat die stage and an epoxy molding compound (Hitachi Chemical Co., Ltd. CEL-9200). The package was exposed to moisture conditioning at 85°C and 85% RH for 24–504 hours and then was tested at high temperature 265–275°C during reflow soldering. As a result, no package crack was observed in packages using DF-A after reliability test mentioned above.

Consequently, DF-A showed excellent package crack resistance at high reflow temperature (265–275°C) and anti-popcorn property due to low water absorption and high peel strength [55].

12.2.4 Die Attach Film for Advanced Package

12.2.4.1 Low Tg and Low Water Absorption of Polyimide Base Resin

A variety of new polyimides were prepared as the base resin of a die attach film and their properties were investigated [56].

All of the polyimides as-synthesized had Mn of 23000–36000 and Mw of 68000–121000, and only polyimide PI-6 with polysiloxane diamine PSX had slightly lower Mn and Mw values (Table 12.7). PSX having longer siloxane units, which will lower its miscibility with NMP during polyimide synthesis due to the low polarity of the monomer, may result in the decrease in polymerizability in the solvent.

The Tg of PI-1 measured by DSC was 120°C, while those of general polyimides such as ULTEM^{TM} are over 200°C [57]. The lower Tg is due to the introduction of the decamethylene connecting group, a long and flexible molecular chain, into the polyimide backbone. Furthermore, the introduction of a dodecamethylene connecting group (PI-2) or several aliphatic ether connecting groups (PI-3 and 4) or siloxane linkages (PI-5 and 6) lowered the Tg to less than 120°C. In particular, the introduction of the polysiloxane linkage lowered the Tg to 30°C (PI-6).

The water absorption of the polyimides was dependent on their chemical structures (Table 12.8). Among the polyimides examined, PI-3 containing many hydrophilic ether linkages had the highest water absorption, and PI-6 containing hydrophobic polysiloxane linkage had the lowest one. As a result, a close correlation between the solubility parameter (SP) and the water absorption of the polyimides was found. The increased SP of a polyimide towards the SP of water (23.4), results in the increased water absorption of the polyimide.

These differences in water absorption can be explained by comparing the solubility parameters (SPs) of water with those of the polyimides. According to the solubility theory, a polymer and a solvent (water) having near SP values are miscible with each other [58]. The SP of the solvents and polyimides were calculated from their chemical structures according to the Okitsu method [59].

Table 12.7 Preparation and characterization of the polyimides

Polyimide	Monomers [mol%] Dianhydride	Diamine	Yield [%]	Molecular weight distribution Mn	Mw	Mw/Mn	Tg [°C]
PI-1	DBTA (100)	BAPP (100)	95.0	32500	121000	3.73	120
PI-2	DDBTA (100)	BAPP (100)	94.8	33200	102600	3.09	107
PI-3	DBTA (100)	BAPP (50) / TODE (50)	95.7	36700	115500	3.14	64
PI-4	DBTA (100)	BAPP (50) / DODE (50)	96.3	28900	88600	3.07	57
PI-5	DBTA (100)	BAPP (50) / TSX (50)	95.0	26900	80800	3.01	64
PI-6	DBTA (100)	BAPP (50) / PSX (50)	92.5	23800	68600	2.89	30

n=10 : DBTA
n=12 : DDBTA

BAPP

TODE

DODE

n=1 : TSX
n=10 : PSX

12.2.4.2 Low Stress (Silicon Chip Warpage)

Two materials having different coefficient of thermal expansion (CTE) are attached with an adhesive material by heat–pressing to yield thermal stress at the interface. If the adhesive material cannot relax the thermal stress, an internal stress will remain at the interface to yield a residual strain. When using a silicon chip (Si) and a copper substrate (Cu) as the adherents which have different CTE (3.5 and 17.0 ppm/°C, respectively), a warpage on the silicon chip is generated as the result of the residual stress or the resultant strain (Fig. 12.15). The maximum thermal stress σ_{max} at the die corners was determined by the following equation [60]:

$$\sigma_{max} = K \cdot \Delta \alpha \cdot \Delta T \cdot (Ea \cdot Es \cdot L/d)^{1/2} \quad (12.1)$$

12 Die Attach Adhesives and Films

Table 12.8 Solubility[a] and water absorption of the polyimides

Polyimide	Solvent[b] / SP[c] [(MPa)$^{1/2}$]							SP[c] [(MPa)$^{1/2}$]	Water asorption [wt%]
	MIBK /8.4	THF /9.1	CHN /9.9	DMAc /10.8	NMP /11.3	DMSO /12.0	DMF /12.1		
PI-1	±	++	++	++	++	±	+	10.8	0.12
PI-2	±	++	++	++	++	±	+	10.6	0.10
PI-3	±	++	++	++	++	±	+	11.3	0.33
PI-4	±	++	++	++	++	±	+	11.2	0.28
PI-5	±	++	++	++	++	±	+	10.9	0.10
PI-6	++	++	++	++	++	±	+	9.9	0.01

[a] ++: soluble at room temperature, +: soluble on heating at 60°C, ±: soluble on heating, −: only swelling on heating
[b] MIBK: methyl isobuthyl ketone, THF: tetrahydrofuran, CHN: cyclohexanone, DMAc: N,N-dimethylacetamide, NMP: N-methyl-2-pyrrolidinone, DMSO: dimethyl sulfoxide, DMF: N,N-dimethylformamide
[c] solubility parameter

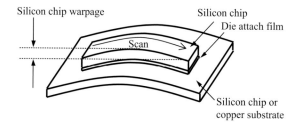

Fig. 12.15 Schematic illustration of silicon chip warpage

where K is the geometric constant, $\Delta\alpha$ the difference in the CTE, ΔT the difference between the Tg of the adhesive material and room temperature, Ea and Es the elastic moduli of respective adhesive material and copper substrate at room temperature, L the side length of the silicon chip, and d the thickness of the adhesive material. Tg and Ea are the two main factors to affect the thermal stress for the adhesive material.

With reference to the above theoretical interpretation, the stress relaxation properties of the composite films used as the adhesive materials were investigated. Figures 12.16 and 12.17 show the silicon chip warpage as a function of the storage modulus E' at 20°C and the tan δ peak temperature (taken as the Tg) of the die attach film. The warpage decreased (the stress relaxation property of the die attach film improved) with decreasing E' and tanδ peak temperature. In particular, the film based on PI-6 was the most effective in relaxing the stress among the films examined. The correlation coefficient value between the chip warpage and the E' was 0.466 (Fig. 12.16), whereas that between the warpage and the tan δ peak temperature was 0.969 (Fig. 12.17). Therefore, the influence of the Tg on the warpage is more remarkable than that of the E'. The difference in the thermal strain between the silicon chip and copper substrate during cool

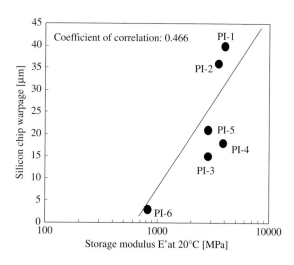

Fig. 12.16 Relationship between the storage modulus E' of the die attach film and the silicon chip warpage. Die attaching temperature: 250°C

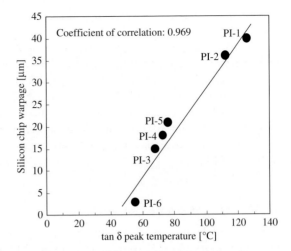

Fig. 12.17 Relationship between the tan δ peak temperature of the die attach film and the silicon chip warpage. Die attaching temperature: 250°C

down after the die attachment is partially released by the movement of the film. This result indicates that lowering the Tg, rather than the E' of the adhesive film, is more effective in stress relaxation and in reducing residual strain.

The peel strengths of the composite films at 250°C are summarized in Table 12.9. The failure modes were all cohesive failure of the films themselves. The peel strength showed different behavior between the two groups of the specimens. Good correlation between the adhesion strength and the storage modulus E' at 250°C was observed for Si/Si specimens: the peel strength tended to increase with increasing E'. However, such correlation was not observed for the Si/Cu specimens. The difference in adhesion behavior could be explained by the difference in the warpage. While the E' of the film mainly affected the peel strength of the Si/Si specimens because there was little or no warpage caused by the thermal stress, the stress relaxation property of the film affected the peel strength of the Si/Cu

Table 12.9 Adhesion properties of the die attach films[a]

Polyimide used in the film	Peel strength at 250°C [MPa]		E' at 250°C [MPa]	Silicon chip warpage [μm]	
	Si/Si[b]	Si/Cu[c]		Si/Si[b]	Si/Cu[c]
PI-1	0.84	0.53	6.1	1	40
PI-2	0.43	0.26	2.2	1	36
PI-3	0.51	0.47	3.0	0	15
PI-4	0.58	0.51	3.3	0	18
PI-5	0.59	0.46	3.3	0	21
PI-6	0.50	0.49	2.4	0	3

[a] Die attaching temperature: 250°C
[b] Between silicon chips
[c] Between silicon chip and copper substrate

specimens because there was a thermal stress and the resultant warpage. For the films based on PI-1 and 2, the peel strength of the Si/Cu specimen was remarkably low compared with that of the Si/Si specimen, because the silicon chip warpage was much larger for the former than the latter. On the other hand, for the film based on PI-6, the peel strength of the Si/Cu specimen was almost similar to that of the Si/Si specimen, because the silicon chip warpage for both specimens was much smaller. Thus, the E' and the stress relaxation property of the die attach film would be the two main factors to affect the adhesion strength in the case of specimens attaching two adherents having different CTEs.

12.2.4.3 Low Attaching Temperature

Figure 12.18 shows the relationship between the attaching temperature and the peel strength for the Si/Si specimens using the die attach film based on PI-1 or 4. For the film based on PI-1, the peel strength straightly decreased with decreasing attaching temperature. On the other hand, that for the film based on PI-4 almost did not decrease with decreasing attaching temperature between 250°C and 140°C. These results could be explained by the failure modes of the specimens shown in Table 12.10. For the film based on PI-1, the failure mode changed from cohesive failure to interfacial failure with decreasing attaching temperature because the adhesion at the interface decreased remarkably. On the other hand, for the film based on PI-4, the failure modes were cohesive failure between 250°C and 160°C because the adhesion at the interface was better. The film based on PI-4 can keep good adhesion between 160°C and 250°C, resulting

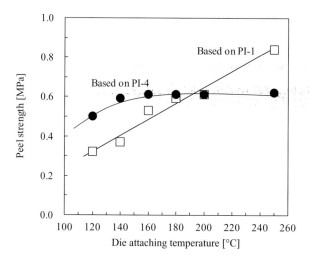

Fig. 12.18 Relationship between die attaching temperature and the peel strength of the die attach films

12 Die Attach Adhesives and Films

Table 12.10 Failure modes of the specimens

Film	Die attaching temperature [°C]					
	120	140	160	180	200	250
Based on PI-1	Ad[a]	Ad	Ad	Ad	Ad/Co	Co[b]
Based on PI-4	Ad/Co[c]	Ad/Co	Co	Co	Co	Co

[a] Ad: Interfacial failure between film and silicon chip
[b] Co: Cohesive failure of film
[c] Ad/Co: Combined

in the stable peel strength. The difference in the adhesion between the films based on PI-1 and 4 is related to the difference in the Tg of the polyimide used. Because the Tg of PI-4 (57°C) is lower than that of PI-1 (120°C), the film based on PI-4 can exert sufficient flow to wet the surface of the silicon chip even at lower attaching temperatures, resulting in improving the adhesion at lower temperatures.

12.2.4.4 Properties of Die Attach Film

A die attach film DF-B was developed based on the study in the above sections and its composition was optimized. Table 12.11 shows properties of DF-B and a current die attach paste.

Table 12.11 Characteristics of DF-B and insulating paste

Item		Unit	DF-B	Insulating paste	Test condition
Composition	Base resin		Polyimide and epoxy resin	Epoxy resin	
Warpage of silicon chip		μm	20	40	Chip size: 5×13 mm Substrate: Cu
Die attach condition	Temp.	°C	180	–	
	Pressure	N/chip	1.0	1.0	
	Time	sec	1	<1	
	Cure	°C-min	180-30	180-60	
Peel strength (120 °C)	Polyimide Substrate	×10^5 Pa	2.4	0.5	Chip size 5×5 mm without solder resist
	Glass epoxy substrate		7.2	2.0	
Water absorption		vol%	0.2	0.9	RT for 24 h
Package crack resistance (85°C85% RH168h)	F-BGA	–	OK	NG	PKG size 18×18×0.8^t mm without solder resist
	Stacked CSP		OK	–	PKG size 8×11×1.4^t mm without solder resist

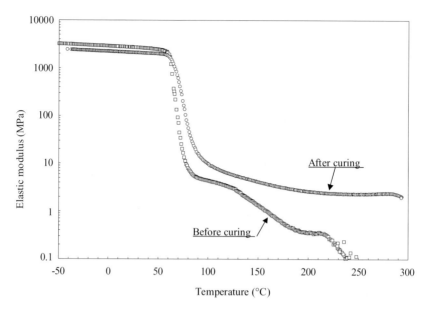

Fig. 12.19 Dependence of dynamic modulus on temperature

DF-B was composed of a modified polyimide base resin with a low Tg, a thermosetting resin, and an insulating filler. DF-B exhibited relatively low attaching temperature, low stress, and low chip warpage. Figure 12.19 shows a dynamic mechanical analysis of DF-B. As the Tg of base resin is relatively low (57°C), DF-B melts and flows at high temperature before curing. This was the reason for attaching ability of DF-B at 180°C. Modulus of DF-B in rubber region at high temperature (above 100°C) increased after curing because curing formed polymer network. It should result in its heat resistance and its good reliability.

Peel strength of DF-B was 4–5 times greater than that of a current die attach paste in both cases of a polyimide substrate and a glass-epoxy substrate, as shown in Table 12.11. Water absorption of DF-B was 1/4 times as small as that of a paste. Thus, DF-B displayed high peel strength.

To evaluate package crack resistance of DF-B, F-BGA and stacked-CSP was used. F-BGA (size: 18 × 18 × 0.8 mm) consisted of a silicon chip, a DF-B film (thickness 40 μm), a polyimide substrate having one electrical layer without a solder resist, and an epoxy molding compound. Stacked-CSP (size: 8 × 11 × 1.4 mm) consisted of two silicon chips, two DF-B films (thickness 25 μm), a polyimide substrate having one electrical layer without a solder resist, and an epoxy molding compound. Those packages were exposed to moisture conditioning at 85°C and 60%RH for 168 h and then were tested at 245°C during reflow soldering. No package crack was observed in both packages using DF-B after reliability test described above.

Consequently, DF-B has excellent reliability for new advanced packages such as BGA/CSP [55].

References

1. S. Kayama, M. Tanimoto, S. Uchida, H. Tsukada, T. Suto, "ASIC Packaging Technology Handbook", Science Forum (1992)
2. E. Hagimoto, "CSP Technology", Kogyochosakai (1997)
3. T. Kasuga, "CSP/BGA Technology", Nikkankogyoshinbunsha (1998)
4. A. Dotani, "Semicon Japan 2003 Navigator", Nikkei Microdevices, 49 (2003)
5. H. Asakura, Nikkei Microdevices, (4) 74 (1999)
6. K. Fujita, "Current Die Bonding Technology", Proceedings of VLSI Assembly Technology Forum Part II, ISS Industrial Systems, 37 (1998)
7. K. Takahashi, SEMI Technology Symposium, 539 (2000)
8. D. Makino, N. Ichimura, K, Suzuki, Electronic Parts and Materials, 20(11), 69 (1981)
9. M. Yamazaki, "Die Bonding Technology for high performance of LSI Package", Proceedings of LSI Assembly Technology Forum, ISS Industrial Systems, 37 (1996)
10. Y. Kanno, "Current Die Bonding Technology for Novel Package", Proceedings of VLSI Assembly Technology Forum Part II, ISS Industrial Systems, 1 (1998)
11. I. Maekawa, Triceps, (12), 21 (1988)
12. G. Ito, Keikinzoku, Japan Institute of Light Metals, 18(3), 177 (1969)
13. J. C. Bolger, 14th National SAMP Technical Conference, October, 12 (1982)
14. J. C. Bolger, S. L. Morano, Adhesive Age, June, 17 (1984)
15. M. Harada, Gekkan Semiconductor World, (9), 119 (1992)
16. O. Kobayashi, Gekkan Semiconductor World, (5), 53 (1994)
17. S. Ishio, T. Maruyama, K. Miyata, Y. Soda, A. Namii, K. Toyozawa, K. Fujita, M. Kada, Technical Report of the Institute of Electronics, Information and Communication Engineers, ICD94-155 (11), 65 (1994)
18. K. Yamada, T. Dohdoh, Electronic Parts and Materials, (4), 93 (2007)
19. T. Uno, Gekkan Semiconductor World, (9), 114 (1992)
20. T. Kawamura, T. Suzuki, H. Sugimoto, N. Imai, M. Kzuya, Hitachi Cable, (12), 37 (1993)
21. W. M. Wasulko, A. G. Stauffer, Microelectric Manufacturing and Testing, 9 (1988)
22. Y. Akada, K. Nakamoto, K. Akazawa, Nitto Denko Technical Report, 29(2), 69 (1991)
23. S. Takeda, T. Masuko, M. Yusa, Y. Miyadera, "Die Bonding Adhesive Film", Hitachi Chemical Technical Report, No. 24, 25 (Jan. 1995)
24. S. Takeda, T. Masuko, Y. Miyadera, M. Yamazaki, I. Maekawa, "A Novel Die Bonding Adhesive-Silver Filled Film", Proceedings of 47th Electronic Components & Technology Conference (ECTC), May 18–21, 1997 San Jose, California, USA, 518 (1997)
25. M. Yasuda, Hitachi Chemical Technical Report, (40), 7 (2003)
26. T. Kato, M. Uruno, Seikei-Kakou, 12(5), 246 (2000)
27. T. Kato, O. Suwa, S. Fujii, M. Yamazaki, T. Masuko, Hitachi Chemical Technical Report, (43), 25 (2004)
28. R. Haruta, Journal of Japan Institute of Electronics Packaging, 10(5), 353 (2007)
29. S. Akejima, Journal of Japan Institute of Electronics Packaging, 10(5), 375 (2007)
30. T. Matsuzaki, T. Inada, K. Hatakeyama, Hitachi Chemical Technical Report, (46), 39 (2006)
31. K. Ebe, H. Senoo, O. Yamazaki, Journal of the Adhesion Society of Japan, 42(7), 280 (2006)
32. M. Harada, Gekkan Semiconductor World, (9), 119 (1992)
33. T. Yoshida, Gekkan Semiconductor World, (5), 72 (1994)

34. S. Ishio, T. Maruyama, K. Miyata, Y. Soda, A. Namii, K. Toyozawa, K. Fujita, M. Kada, Technical Report of the Institute of Electronics, Information and Communication Engineers, ICD94-155 (11), 65 (1994)
35. H. Li, A. Johnson, C. P. Wong, IEEE Transactions on Components and Packaging Technologies, 26(2), 466 (2003)
36. A. K. S. Clair, T. L. S. Clair, Polymer Engineering and Science, 22(1), 9 (1982)
37. D. Makino, "Recent Progress of the Application of Polyimides to Microelectronics", pp. 380–402, Polymers for Microelectronics, Kodansha (1994)
38. D. Wilson, "Recent Advances in Polyimide Composites", High Performance Polymers, 5, 77 (1993)
39. F. W. Harris, M. W. Beltz, SAMPE Journal, 23, 6 (1987)
40. N. Furukawa, Y. Yamada, Y. Kimura, High Perfomance Polymers, 8, 617 (1996)
41. J. L. Hedrick, H. R. Brown, W. Volksen, M. Sanchez, Polymer, 38(3), 605 (1997)
42. L. Li, D. D. L. Chung, Composites, 22(3), 211 (1991)
43. Y. Nakamura, Journal of the Adhesion Society of Japan, 38(11), 442 (2002)
44. K. Gaw, M. Kikei, M. Kakimoto, Y. Imai, Reactive and Functional Polymers, 30, 85 (1996)
45. C. C. Su, E. M. Woo, Polymer, 36(15), 2883 (1995)
46. M. Kimoto, Journal of the Adhesion Society of Japan 36(11), 456 (2000)
47. T. Masuko, S. Takeda, Journal of the Adhesion Society of Japan, 40(4), 136 (2004)
48. T. Masuko, S. Takeda, Journal of the Network Polymers of Japan, 25(4), 181 (2004)
49. A. Kawai, H. Nagata, M. Takata, Japan Journal of Applied Physics, 31, 1993 (1992)
50. F. M. Fowkes, Industrial and Engineering Chemistry, 56, 40 (1964)
51. M. Imoto, Journal of the Adhesion Society Of Japan, 26(1), 39 (1990)
52. T. Hata, T. Kitazaki, T. Saito, The Journal of Adhesion, 21, 177 (1987)
53. R. A. Gledhill, A. J. Kinloch, The Journal of Adhesion, 6, 315 (1974)
54. H. Yamabe, Journal of the Adhesion Society of Japan, 29(1), 12 (1993)
55. S. Takeda, T. Masuko, "Novel Die Attach Films Having High Reliability Performance for Lead-Free Solder and CSP" Proceedings of 50th Electronic Components and Technology Conference (ECTC), May 21–24, 2000, Las Vegas, Nevada, USA, p 1616 (2000)
56. T. Masuko, S. Takeda, Y. Hasegawa, Journal of Japan Institute of Electronics Packaging, 8(2), 116 (2005)
57. S. H. Hsiao, P. C. Huang, Journal of Polymer Research, 4(3), 183 (1997)
58. R. F. Fedors, Polymer Engineering and Science, 14(2), 147 (1974)
59. T. Okitsu, Secchaku, 40(8), 342 (1996)
60. J. C. Bolger, "Polyimide Adhesives to Reduce Thermal Stress in LSI Ceramic Packages", 14th National SAMPE Technical Conference, October, pp. 257–266, (1982)

Chapter 13
Thermal Interface Materials

Ravi Prasher and Chia-Pin Chiu

Abstract Increasing electronic device performance has historically been accompanied by increasing power and increasing on-chip power density both of which present a cooling challenge. Thermal Interface Material (TIM) plays a key role in reducing the package thermal resistance and the thermal resistance between the electronic device and the external cooling components. This chapter reviews the progress made in the TIM development in the past five years. Rheology based modeling and design is discussed for the widely used polymeric TIMs. The recently emerging technology of nanoparticles and nanotubes is also discussed for TIM applications. This chapter also includes TIM testing methodology and concludes with suggestion for the future TIM development directions.

Keywords Thermal resistance · density factor · thermal interface materials (TIMs) · rheology

Nomenclature

R_{cs}	Contact resistance between two bare solids
Rc_{TIM}	Contact resistance of an ideal TIM
H	Hardness
P	Pressure
m	Mean asperity slope
k_c	Thermal conductivity for composite
k_{TIM}	Thermal conductivity of the TIM
k_p	Thermal conductivity of particles (fillers)
k_m	Thermal conductivity of the polymer matrix
R_{TIM}	Thermal resistance of TIM (same as impedance)
BLT	Bond line thickness

R. Prasher (✉)
Intel Corporation, 5000 W. Chandler Blvd., AZ 85226, USA
e-mail: ravi.s.prasher@intel.com

R_b	Thermal boundary resistance
R_c	Contact resistance of TIM
DF	Density factor
R_{jc}	Junction to case thermal resistance
K	Consistency index in Eq. (13.5)
r	Radius of the substrate
C	Empirical constant in Eq. (13.7)
R_{bulk}	Bulk thermal resistance
E_a	Activation Energy
A	Acceleration factor
A_c	Actual contact area
A_{nc}	Non-contact area occupied by air gaps
G	Shear modulus
G'	Storage shear modulus
G''	Loss shear modulus

Greek

σ	Surface roughness
$\Psi_{J\text{-}a}$	Junction to ambient thermal resistance
Ψ_{cs}	Case to sink thermal resistance
Ψ_{sa}	Sink to ambient thermal resistance
ϕ	Volume fraction of particles in TIMs
α	Biot number
τ_y	Yield stress of the TIM

13.1 What is Thermal Interface Resistance?

When two solid surfaces are in contact, as shown in Fig. 13.1, roughness on each of the surfaces typically limits the actual contact area between the two solids to a very small fraction of the apparent area especially for lightly loaded interfaces [1, 2]. Consequently, the heat flow across such an interface involves solid-to-solid conduction in the area of actual contact, A_c, and conduction through the air gap occupying the non-contact area, A_{nc}, of the interface. This constriction of heat flow is manifested as thermal contact resistance (R_c) at the interface. The solid-solid contact resistance (R_{cs}) between two nominally flat surfaces 1 and 2 assuming plastic deformation of the asperities is given by [1]

$$R_{cs} = \frac{0.8\sigma}{mk_h}\left(\frac{H}{P}\right)^{0.95} \tag{13.1}$$

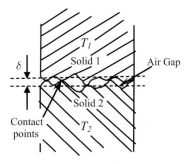

Fig. 13.1 Schematic showing that real area of contact is less than apparent area of contact

where $\sigma = (\sigma_1^2 + \sigma_2^2)^{0.5}$, σ is the root mean square roughness, $m = (m_1^2 + m_2^2)^{0.5}$, m = mean asperity slope, H the microhardness of the softer material, P the applied pressure, and $k_h = 2k_1k_2/(k_1 + k_2)$ is the harmonic mean thermal conductivity of the interface. m is a measure of the slope of the rough surface and m is given by $m = \tan(\theta)$, where θ is the slope of the rough surface. R_{cs} is 6.2°C cm^2/W for $P = 10$ psi (68 kPa) and 0.7°C cm^2/W for $P = 100$ psi (680 kPa) for Cu/Si interface. In this calculation H of copper which is the softer material is taken as 1280 MPa [3]. σ of the Cu is assumed 1 μm and σ of the Si is assumed 0.1 μm for a polished Si. m is calculated by the relation $m = 0.076(\sigma \times 10^6)^{0.5}$ [4].

The low pressure range is applicable in non-CPU products or for heat sink on top of a large heat spreader. This calculation shows that R_{cs} is very large for bare contact between two bulk solids.

Equation (13.1) is the best-case result because it assumes plastic deformation of interfaces at all pressures and nominally flat surfaces. Figure 13.2 shows two typical thermal architectures used in electronics packaging. Although Fig. 13.2

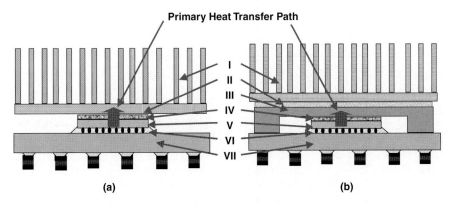

Fig. 13.2 Schematic Illustration of the Two Thermal Architectures (**a**) Architecture I typically used in Laptop Applications and (**b**) Architecture II typically used in Desktop and Server Applications. I – Heat Sink, II – TIM, III – IHS, IV – TIM, V – Die, VI – Underfill, and VII – Package Substrate

shows the Si die or the chip to be flat in reality, the Si die or the chip surface is typically warped due to the coefficient of thermal expansion (CTE) mismatch between the die and the package substrate which will lead to further increase in the interface resistance.

The most common way to reduce R_c is to fill the interfacial gap between the asperities in Fig. 13.1 with some high-thermal-conductivity soft materials that are typically referred as thermal interface materials (TIM) as shown in Fig. 13.3.

From Fig. 13.3 it can be inferred that the total thermal resistance (R_{TIM}) of real TIMs can be written as [5]

$$R_{TIM} = \frac{BLT}{k_{TIM}} + R_{c1} + R_{c2} \qquad (13.2)$$

where R_c represents the contact resistances of the TIM with the two bounding surfaces and BLT is the bondline thickness of the TIM. In recent years there has been a great drive in the industry in reducing R_{TIM} The, heat flux from the chip is non-uniform [6, 7] because both the core and cache are on the same die. Majority of power is dissipated from the core i.e. from a much smaller area of the chip. Even within the core, heat flux (q) is non-uniform. Mahajan et al. [6] discussed the issue of the non-uniform heat fluxes from the die and indicated that cooling solutions not only need to maintain the average chip temperature below a design point, it is also important to maintain the temperature of the hottest spot below a certain design point. Therefore, the thermal problem near the chip is very severe. The total thermal resistance for non-uniform heating can be written as [8]

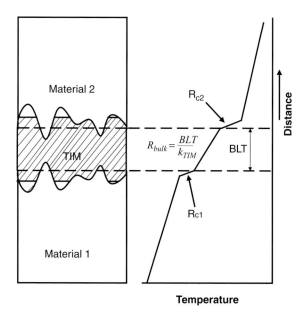

Fig. 13.3 Schematic representing a real TIM

$$\Psi_{j-a} = DF \times R_{jc} + \Psi_{cs} + \Psi_{sa} \qquad (13.3)$$

where Ψ_{j-a} is the junction to ambient thermal resistance, R_{jc} the junction to case thermal impedance for an uniformly heated die, Ψ_{cs} is the case to sink resistance and Ψ_{sa} is the sink to ambient thermal resistance. DF in Eq. (13.3) is called "Density Factor" that accounts for the non-uniformity of q and the die size [8]. The unit of DF is cm^{-2}. For a 1 cm^2 uniformly heated chip DF is 1. DF is typically larger than 1 for most microprocessor due to high non-uniformity of q and small die sizes, but theoretically DF can approach zero for very large die. According to Equation (13.3), reduction in both R_{jc} and DF leads to reduction in Ψ_{j-a}. Note that the reduction of R_{jc} leads to greater reduction in Ψ_{j-a} if DF is larger than 1. Since R_{jc} is primarily due to TIM thermal resistance, this has lead to a great drive in the electronics cooling industry to develop better TIMs.

This chapter focuses mainly on polymeric TIMs due to their widespread use [9]. These polymeric TIMs are typically filled with highly-thermal-conductive particles to enhance their apparent thermal conductivity. This chapter is divided into 7 sections including this introduction section. Section 13.2 is devoted to the recent development in thermal interface modeling and their related physics. Section 13.3 emphasizes reliability performance of polymeric TIMs. Solder alloy based TIMs and implications of nanotechnology for TIM technologies are briefly discussed in sections 13.4 and 13.5, respectively. Section 13.6 reviews some fundamental metrology for TIM performance measurement. The chapter concludes with Section 13.7 highlighting some key unresolved fundamental issues and possible directions in the future.

In this chapter, p denotes the particle, m the matrix, R_b the interface resistance between the particle and the matrix, d the diameter, k the thermal conductivity and ϕ denotes the volume fraction of the particles.

13.2 Recent Development in Thermal Interface Modeling

In order to accurately model the physics of TIM performance, we need to understand: (1) k_{TIM}, (2) BLT, and (3) R_c according to Equation (13.2). Equation (13.2) shows that R_{TIM} can be reduced by reducing the BLT, increasing the thermal conductivity k_{TIM}, and reducing the contact resistances R_{c1} and R_{c2}. Table 13.1 summarizes the characteristics of various TIMs [9, 10] and their advantages and disadvantages.

Since most TIMs are loaded with solid particle fillers, the physics to describe TIM thermal performance becomes complicated. Prasher [5] first attempted to separate the bulk resistance (R_{bulk}) and R_c by proposing a physical model, as shown in Fig. 13.3.

Prasher and co-workers have introduced various models for BLT, k_{TIM} and R_c in a series of papers [5, 11–14]. They mainly focus on grease, gel and phase change materials (PCM) since these TIMs are most widely used as compared to elastomers [10]. In the following sections, modeling of k_{TIM}, BLT and R_c for different types of TIMs is illustrated sequentially.

Table 13.1 Summary of Characteristics of some typical thermal interface materials (PCM: phase change material)

TIM Type	RTIM of fresh samples (°C cm² W⁻¹)	General Characteristics	Advantages	Disadvantages
Greases	0.1	Typically silicone based matrix loaded with particles to enhance thermal conductivity	• High bulk thermal conductivity • Thin BLT with minimal attach pressure • Low viscosity enables matrix material to easily fill surface crevices • No curing required • TIM delamination is not a concern	• Susceptible to grease pump-out and phase separation • Considered messy in a manufacturing environment due to a tendency to migrate
PCM	0.1	Polyolefin, epoxy, low molecular weight poly esters, acrylics typically with BN or Al₂O₃ fillers	• Higher viscosity leads to increased stability and hence less susceptible to pump-out • Easier application and handling than greases • No cure required • Delamination is not a concern	• Lower thermal conductivity than greases • Surface resistance can be greater than greases. Can be reduced by thermal pre-treatment • Requires attach pressure to increase thermal effectiveness and thus could lead to increased mechanical stresses
Gels	0.08	Al, Al₂O₃, Ag particles in silicone, olefin matrices that require curing	• Conforms to surface irregularity before cure • No pump out or migration concerns	• Cure process needed • Lower thermal conductivity than grease • Lower adhesion than adhesives; delamination can be a concern
Adhesives	Data not available	Typically Ag particles in a cured epoxy matrix	• Conform to surface irregularity before cure • No pump out • No migration	• Cure process needed • Delamination post reliability testing is a concern • Since cured epoxies have modulus, CTE mismatch induced stress is a concern

13.2.1 Model to Predict Thermal Conductivity (k_{TIM})

Since most polymeric TIMs are typically filled with highly-thermal-conductive particles to increase k_{TIM}, these TIMs can be treated as composites. In general thermal conductivity of any composite can be written as [10].

$$k_c = f(k_m, k_p, R_b, \phi) \tag{13.4}$$

where k_m is the thermal conductivity of the matrix, k_p is the thermal conductivity of the particles, R_b the interface resistance between the particle and the matrix, and ϕ is the volume fraction of the particles. Many literatures can be found for modeling thermal conductivity of composites (k_c). Prasher [10] has extensively discussed the merits and demerits of various models. Table 13.2 lists various models to predict k_c.

Prasher [10, 12] found that Bruggeman asymmetric model (BAM) matches the experimental data of various polymeric TIM. Figure 13.4 shows the comparison between the experimental k_{TIM} data from various sources and the BAM model prediction. It can be seen that BAM is very successful in modeling k_{TIM}. BAM matches the data by assuming α (Biot number) of 0.1. Assuming k_m of 0.2 W m^{-1} K^{-1} and particle diameter (d) of 10 μm (typical in commercial TIMs), $\alpha = 0.1$ gives $R_b = 5 \times 10^-$ K m^2 W^{-1}. R_b at the interface between the particle and the matrix could arise due to phonon acoustic mismatch or incomplete wetting of the

Table 13.2 Models to predict the thermal conductivity of particle laden TIMs [10]

Name of the Model	Formula	Remarks
Maxwell-Garnett with R_b	$\dfrac{k_c}{k_m} = \dfrac{[k_p(1+2\alpha)+2k_m] + 2\phi[k_p(1-\alpha)-k_m]}{[k_p(1+2\alpha)+2k_m] - \phi[k_p(1-\alpha)-k_m]}$ $\alpha = \dfrac{2R_b k_m}{d}$ $\dfrac{k_c}{k_m} = \dfrac{(1+2\alpha)+2\phi(1-\alpha)}{(1+2\alpha)-\phi(1-\alpha)}$ for $k_p \gg k_m$	Spherical particles Typically valid for $\phi < 0.4$
Bruggeman symmetric model	$(1-\phi)\dfrac{k_m - k_c}{k_m + 2k_c} + \phi\dfrac{k_p - k_c}{k_p + 2k_c} = 0$ (R_b not included)	Spherical particles Typically good at higher ϕ
Bruggeman asymmetric model	$(1-\phi)^3 = \left(\dfrac{k_m}{k_c}\right)^{(1+2\alpha)/(1-\alpha)} \times \left\{\dfrac{k_c - k_p(1-\alpha)}{k_m - k_p(1-\alpha)}\right\}^{3/(1-\alpha)}$ $\dfrac{k_c}{k_m} = \dfrac{1}{(1-\phi)^{3(1-\alpha)/(1+2\alpha)}}$ for $k_p \gg k_m$	Spherical particle

k_m = thermal conductivity of matrix, k_p = thermal conductivity of particles, R_b = interface resistance between particles and the matrix, ϕ = volume fraction of the particles.

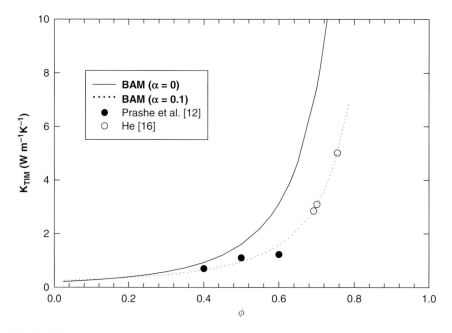

Fig. 13.4 Comparison between experimental data and Bruggeman Asymmetric model

interface by the polymer. R_b due to phonon acoustic mismatch is of the order of 10^{-8} K m^2 W^{-1} [15] at room temperatures, resulting in α of 0.0002 for the case with d of 10 μm and k_m of 0.2 W m^{-1} K^{-1}. Prasher et al. [15] also showed that phonon acoustic mismatch at room temperature is negligible when compared to incomplete particle wetting; however, phonon acoustic mismatch could possibly become important especially for nano-size particles. R_b due to incomplete particle wetting by polymer could become a function of the volume fraction. This is because completely wetting the surfaces with increasing volume fraction becomes more difficult due to manufacturing and processing limitations.

13.2.2 Rheological Model to Predict TIM Bondline Thickness (BLT)

Prasher et al. [12] measured the viscosity of various silicone based TIMs and indicated that these TIMs behave like Herschel-Bulkley (H-B) fluid. The viscosity (η) for H-B fluid is given by

$$\eta = \frac{\tau_y}{\dot{\gamma}} + K(\dot{\gamma})^{n-1} \quad (13.5)$$

where τ_y is the yield stress of the polymer, $\dot{\gamma}$ is the strain rate, K the consistency index and n is an empirical constant. Prasher et al. [12] further showed that

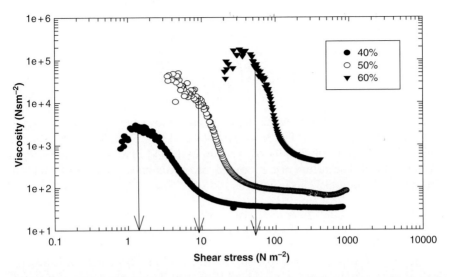

Fig. 13.5 Viscosity vs. shear stress to obtain yield stress for different volume fraction of particles in silicone based greases [25]

steady state BLT depends only on τ_y. Figure 13.5 shows τ_y for three different PLP TIMs loaded with different volume fractions of particles. In this figure, τ_y increases with volume fraction of the particle. By applying law of conservation of momentum and mass, the BLT by using Eq. (13.5) can be expressed by

$$\text{BLT} = \frac{2}{3}r\left(\frac{\tau_y}{P}\right) \qquad (13.6)$$

where r is the radius of the substrate and P the applied pressure. However, Prasher et al. [12] found that Eq. (13.6) under predicted the actual TIM BLT by a huge margin. Thus, they decided to introduce an empirical model

$$h_L = C\left(\frac{\tau_y}{P}\right)^m \qquad (13.7)$$

where C and m are empirical constants. They found m is 0.166 and C is 0.31×10^{-4}. Subsequently, Prasher [13] offered an explanation to Eq. (13.7) by applying finite size scaling argument to a percolating system of particles.

A heterogeneous system can be macroscopically treated as homogenous only if the thickness (BLT in this case) is much larger than the diameter of the particles. At high pressures the BLT of TIMs typically ranges from 20 μm to 50 μm. If the particle diameter is of the order of 10 μm then the TIM can not be treated a macroscopically homogeneous system. Prasher [13] used the finite size scaling of elasticity modulus [17] for a thin percolating systems as a clue to scale τ_y of the TIM. Prasher [13] also considered the fact that if BLT >> d (at low pressures) then any BLT model should reduce to Eq. (13.6). Based on these arguments Prasher's model (Called scaling-bulk (S-B) model) is given as

$$\text{BLT} = \frac{2}{3} r \left[c \left(\frac{d}{\text{BLT}} \right)^{4.3} + 1 \right] \left(\frac{\tau_y}{P} \right) \quad (13.8)$$

where $c = 13708$. This equation at high pressures shows that $m = 0.188$ which is very close to the m obtained in the empirical Eq. (13.7). Equation (13.8) reduces to Eq. (13.6) for very small value of P/τ_y and to Eq. (13.7) for large value of P/τ_y with $m = 0.188$. The author also proposed an approximate version of Eq. (13.8) for quick calculations. This is given as

$$h_L = \frac{2r}{3} \left(\frac{\tau_y}{P} \right) + \left(\frac{cr}{1.5} \right)^{0.188} d^{0.811} \times \left(\frac{\tau_y}{P} \right)^{0.188} \quad (13.9)$$

Figure 13.6 shows the comparison of Eq. (13.8) (S-B model) with experimental data obtained from various TIMs [10]. The author also compared Eq. (13.8) with a variety of other suspensions for d as large as 80 μm and as small as 2 μm and showed that Eq. (13.8) matches very well with the data [13]. Equation (13.8) can be applied to phase change material, greases and pre-cure gels as these TIMs are well described by the H-B model.

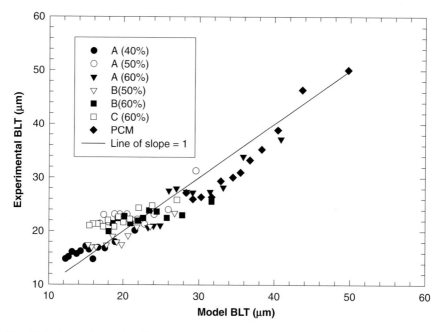

Fig. 13.6 Comparison of scaling bulk model with experimental data for the phase change material [12, 13]

13.2.3 Effect of Particle Volume Fraction on Bulk TIM Thermal Resistance

The bulk thermal resistance of the TIM is given by,

$$R_{bulk} = \frac{BLT}{k_{TIM}} \qquad (13.10)$$

By combining Eqs. (13.7) and (13.10), R_{bulk} can be arranged as:

$$R_{bulk} = \frac{1}{k_{TIM}} C \left(\frac{\tau_y}{p}\right)^m \qquad (13.11)$$

Figure 13.5 shows that τ_y depends on particle volume fraction (ϕ). If electrostatic interaction is assumed to be negligible compared to the Van der walls interaction in the particle laden polymer, then τ_y can be expressed as [18]

$$\tau_y = A \left[\frac{1}{(\phi_m/\phi)^{1/3} - 1}\right]^2 \qquad (13.12)$$

where A is constant, and ϕ_m is the maximum particle volume fraction. Equation (13.12) can also be rearranged as

$$\tau' = \frac{\tau_y}{A} = \left[\frac{1}{(\phi_m/\phi)^{1/3} - 1}\right]^2 \qquad (13.13)$$

where τ' is the dimensionless yield stress. Using τ', Eq. (13.11) can be written as

$$\frac{R_{bulk} p^m}{C A^m} = \frac{\tau'^m}{k_{TIM}} \qquad (13.14)$$

Using the BAM and Eq. (13.14), Prasher et al. [12] showed that R_{bulk} reaches a minima with respect to the volume fraction of the fillers. This was experimentally verified by them, as shown in Fig. 13.7.

Figure 13.7 shows that there is an optimal volume fraction for the minimization of the TIM thermal resistance. Prasher [13] recently performed parametric studies on the thermal resistance for various factors such as ϕ, diameter of the filler and the applied pressure. The key conclusion from Prasher's parametric study was that there is an optimal volume fraction for a given pressure and filler shape, above which thermal resistance of the TIM increases.

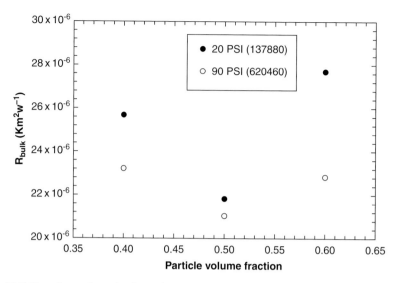

Fig. 13.7 Experimental results for resistance vs. particle volume fraction for silicone based thermal greases [12]

13.2.4 Model to Predict Thermal Contact Resistance (R_c)

Prasher [5] proposed an incomplete wetting by applying surface chemistry and assuming pure liquid like behavior for TIMs. This model assumed that the TIM is unable to fill all the valleys due to trapped gases in the valleys of the rough surface, as shown in Fig. 13.8. By applying force balance among the externally applied pressure, capillary force due to the surface tension of the TIM and back pressure due to the trapped air, it was possible to calculate the penetration length of the TIM in the interface. A constriction resistance parameter was defined based on A_{real} and $A_{nominal}$ as shown in Fig. 13.8. For $k_{TIM} \ll k_{substrate}$ the surface chemistry model is given by

$$R_{c_{1+2}} = \left(\frac{\sigma_1 + \sigma_2}{2 k_{TIM}}\right)\left(\frac{A_{nominal}}{A_{real}}\right) \qquad (13.15)$$

where σ_1 and σ_2 are the surface roughness of the two substrates sandwiching the TIM. A_{real} can be calculated from penetration length of the TIM. The surface chemistry model was in good agreement with PCM and greases as shown in Fig. 13.9. However, considering the later finding by Prasher [12, 13] that these TIMs posses yield stress and viscosity, which means that they are semi-solid and semi-liquid, the pure liquid-based surface chemistry model is not good enough for the modeling of the contact resistance of TIMs.

Intuitively speaking, the area covered by the TIM in the valleys of the interface as shown Fig. 13.1 will eventually depend on the pressure and yield stress. This

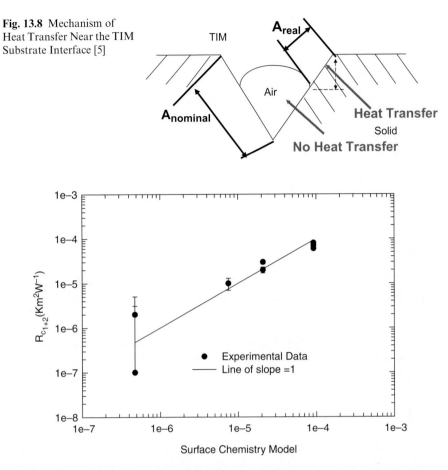

Fig. 13.8 Mechanism of Heat Transfer Near the TIM Substrate Interface [5]

Fig. 13.9 Comparison of the surface chemistry model with experimental results for phase change materials [5]

relation could be somewhat similar to that obtained for bare metallic contacts where the contact resistance depends on the pressure and the hardness of the softer material. For TIMs most likely hardness will have to be replaced by the yield stress. Internal studies at Intel on various state-of-the-art TIMs have suggested, however, that bulk resistance of the TIM is more dominant than R_c. For cured gels, Prasher and Matyabus [14] proposed a semi-empirical model for R_c, which has similar form as Eq. (13.1). This model is given as

$$\frac{R_c k_{TIM}}{\sigma} = c\left(\frac{G}{P}\right)^n \tag{13.16}$$

where $G = \sqrt{G'^2 + G''^2}$. G' is the storage shear modulus and G'' is the loss shear modulus of the TIM. $G' > G''$ is for cured gels, while $G' < G''$ is for uncured gels

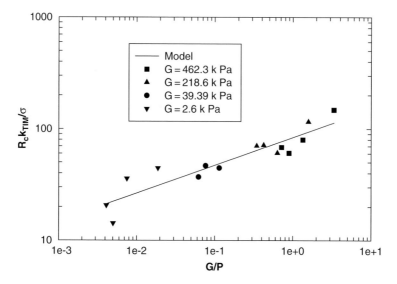

Fig. 13.10 $R_c k_{TIM}/\sigma$ vs. G/P for Gel TIM [14]

that are nothing by greases. Figure 13.10 shows the comparison of this model with experimental data from four gels with different formulations.

13.3 Reliability Consideration for Polymeric TIMs

Most research on polymeric TIMs until now has focused on the behavior of freshly made TIMs. In reality, these TIMs are likely exposed to high temperatures and harsh conditions during the product life time. Assuming that a product life is 7 years, this translates into approximately 61000 hours under continuous operation or 35000 hours for 14 hours per day. If the product operation temperature is 100°C, then the polymers in the TIM are being exposed to relatively high temperatures for the product life time. Polymers degrade under such high temperatures [19]. However, it is unlikely to test these TIMs for such long times to understand their behavior for exposure to high temperatures before launching the product. Therefore, to understand the degradation behavior, accelerated life time testing is performed. Under accelerated testing the TIM is exposed to much higher temperatures than the "use condition" (or operational) temperature. For example, if the product operation temperature is 100°C, TIM could be tested at 125°C and 150°C for a much shorter period of time than the product life. The thought behind this is that higher temperature will accelerate the degradation and engineers would be able to generate TIM degradation models in a limit time frame. Figure 13.11 shows the thermal resistance (R_{jc}) of a PCM TIM as function of time and temperature [10].

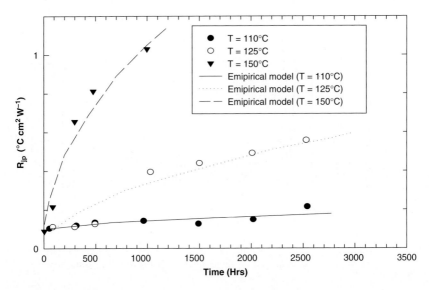

Fig. 13.11 Degradation of thermal resistance with time [10]. Lines are the empirical curve fit of the form given by Eq. (13.17)

The lines represent an Arrhenius type model obtained by curve fitting empirical data to the following equation:

$$R_{jc}(t) = R_{jc}(t=0) + A\sqrt{t}\exp\left(\frac{-E_a}{k_b T}\right) \quad (13.17)$$

where E_a is the activation energy, A is the acceleration factor and k_b is the Boltzmann constant, t represents the time, and the first term on the right hand side is the R_{jc} value at $t=0$ (i.e. fresh TIM without exposing to high temperature). The research discussed in previous sections have focus on R_{jc} at $t=0$. Equation (13.17) shows some type of a diffusion process with a square root dependence on time [19]. Once A and E_a are obtained from matching the data at different (or higher) temperatures the use temperature can be put into Eq. (13.17) to obtain the value of R_{jc} at the use temperature and end-of-life of the product. In the industry, TIMs are typically designed for end-of-life performance so that it should be very careful to choose the appropriate TIM based on their reliability performance. This is because some TIM gives the best R_{jc} at $t=0$, but that TIM might degrade so that at end-of-life it gives worse performance than the other TIMs.

There is no mechanistic understanding of the degradation of the thermal performance of the TIMs due to large exposures to high temperatures. Form of Equation (13.17) suggests some type of diffusion process, however it is not clear what is really diffusing. Even if it is assumed that the oxidation of the TIMs

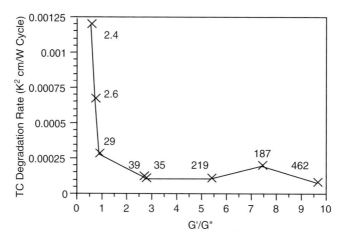

Fig. 13.12 Effect of G'/G'' on the degradation rate, as measured by thermal performance of gel TIMs subjected to temperature cycling. The labels present the value of G for each sample [14]

follows a diffusion process, no attempt has been made to relate this to thermal performance. This area to research TIM reliability behavior is wide open.

In addition to high temperature, thermal greases suffer from another type of degradation that is commonly known as pump-out [9]. Thermal grease pump-out typically occurs after temperature cycling or power cycling. Recently Prasher and Matyabus [14] related the pump out problem to the ratio of G' and G''. They found that G' of grease should be greater than G'' to avoid pump out. This is exactly what a gel does. Gel is nothing but a cured grease. Figure 13.12 shows the rate of degradation of thermal performance vs. G'/G''. It was observed that the degradation rate is approaching to a much lower constant after $G' > G''$.

Other types of reliability tests include testing under humidity, temperature cycling, and mechanical shock and vibration. For most of these reliability tests, testing mechanistic is not fully understood and most reliability analysis is empirical in nature.

13.4 Solder Alloy Based TIMs

Solders are also actively pursued as TIMs due to their higher thermal conductivity [20–22]. Chiu et al. [20] characterized various solder TIMs and demonstrated some solders have very small thermal resistance. However, voiding is the biggest concern for solder because voids formed by air are poor in thermal conduction. Pritchard et al. [21] performed numerical studies to understand the voiding effects on the over all thermal performance of solders. Hu et al. [22] also performed experimental characterization of voids in solder type TIMs. There are not many literatures in microscopic modeling for thermal performance of solders primarily because of the lack of a driving force. This is probably because (1) most of the solders thermal performance is inherently very good, and (2) solders are not preferred over polymer TIMs because of cost and the complicated assembly processes involved in implementing solder TIMs as compared to polymer TIMs.

13.5 Nanotechnology Based TIMs

Ever since carbon nanotube (CNT) [23] was demonstrated to have very high thermal conductivity, various CNT based composites have been proposed and evaluated [24–27]. Although CNTs are promising candidates as TIM fillers, CNTs have big limitation that interface resistance plays a significant role due to large inherent conductivity of the CNT [28]. Huxtable et al. [29] have experimentally measured R_b between CNTs and various liquids, showing that R_b is substantial (8.33×10^{-8} K m^2 W^{-1}). Prasher et al. [30] recently calculated the interface resistance of multiwalled CNT (MWCNT) for both horizontal and vertical contacts and showed the MWCNTs behave as graphite. Prasher's calculation shows the contact resistance of the vertical contact is smaller than the contact resistance of horizontal contact.

Nan et al. [28] recently proposed simplified effective medium model to compute k of CNT based composites. Hu et al. [31] have performed a feasibility study of CNT based TIM, showing that the potential of achieving percolation threshold at very small volume fraction. Another concern of CNT based TIM is that the BLT of the resultant TIM will be high due to the high yield stress of fiber based composites. Therefore, the overall thermal resistance is the more appropriate metric as compared to k_{TIM} for a fair comparison between CNT based TIMs and conventional TIMs.

Xu and Fisher [32] have grown CNTs directly on the back of Si, and then pressed a heat spreader against the as-grown CNTs. Figure 13.13 shows the schematics of their concept. They have also combined PCM TIM and as-grown CNT to reduce the thermal resistance. Thermal resistance for this concept was

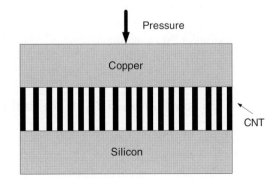

Fig. 13.13 Schematic showing the use of carbon nanotube (CNT) grown in the back of Si (Xu and Fisher [32])

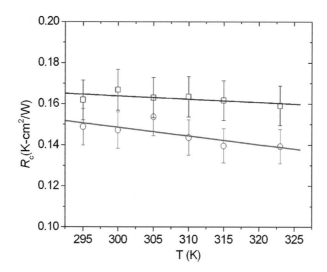

Fig. 13.14 Thermal resistance of aligned carbon nanotube on Si. The other surface is copper (Hu et al. [33])

experimentally measured by Hu et al. [33], as shown in Fig. 13.14. At this time there is no physics based model for this concept. However vertically grown CNT-based TIM looks promising. It seems that this concept will not suffer from the reliability problems that occurs on polymer based TIMs if this concept does not combine polymers. This concept has good potential and could lead to various creative ideas, such as those shown by Xu and Fisher [34], where they used PCM TIMs in combination with vertically grown CNT or that by Tong et al. [35] where they used thin layer of indium in combination with vertically grown CNT.

In addition to CNT, researchers also proposed nano particles as TIM fillers [36, 37]. However, nano particles suffer from the same problem as CNTs because R_b plays a dominant role in nano particles based composites. Putnam et al. [38] measured R_b between polymer and alumina in the range of 2.5×10^{-8} to 5×10^{-8} K m^2 W^{-1}. This means that the critical radius ($\alpha = 1$) below which the thermal conductivity of the nanocomposite is less than the conductivity of the matrix varies between 5 nm and 10 nm. The yield stress of particle laden polymers increases with decreasing particle diameter [18], leading to higher BLT for nano particle based TIM than conventional TIMs. Therefore, it is not clear if nano-composites can really be of great use as TIM fillers at this moment.

13.6 Characterization of TIM Thermal Performance

Characterization of thermal interface materials in electronic applications is necessary to ensure timely product launches. This section will briefly review the methodology to test TIM. Many TIM testing apparatus [39, 40] have been developed based on ASTM D5470-93 [41, 42]. This testing apparatus can be used as a simple incoming TIM monitor or a quick benchmarking tool for new TIM without having to go through the time and expense of completing package level measurements. The tester was typically designed to test materials at controlled bond line thickness and controlled pressures, while being able to directly measure the bond line thickness. Chiu et al. [42] has demonstrated that the tester is capable of evaluating TIM thermal impedance with a reproducibility of 0.03°C-cm^2/W at a 95% confidence level. The testing apparatus was further modified for experimental validation of TIM characteristics between non-flat surfaces by Chiu et al. [42].

In addition to the steady-state measurement by ASTM D5470-93, several transient thermal analysis techniques [43, 44] were also used for TIM characterization. However, all these methodologies can not capture the interaction of TIM with the actual packages and heat sinks in the application environment. Several papers have described how to characterize TIM behavior by using different thermal test vehicles [45–47]. In order to understand reliability performance of TIM, Chiu et al. [48] used an accelerated reliability test method to predict thermal grease pump-out in Flip-Chip applications. Bharatham et al. [49] studied the impact of different application pressure on a phase-change TIM on a bare-die FCBGA (Flip Chip Ball Grid Array) package with a heat sink solution. All these testing methods were used to capture the possible reliability issues which are hard to predict by theory or numerical modeling. It is very important for a packaging engineer to validate TIM performance on the actual product with various reliability tests before launching the products.

13.7 Future Directions

The future research should focus more on understanding the reliability and performance degradation of TIMs. Current commercial TIMs are capable of providing a thermal resistance between 0.03°C cm^2 W^{-1} and 0.1°C cm^2 W^{-1} for fresh samples [50]. However, due to degradation at large exposures to high temperatures as discussed earlier, the thermal performance can degrade severely depending on the operating temperature and time of exposure. There is no mechanistic understanding of these degradations and there is strong demand for fundamental physics based modeling that can relate the degradation of the polymer properties to thermal properties of the polymer composites.

Use of nano particles and nanotubes is almost inevitable. However, any researcher in this area should benchmark thermal performance of their new concepts with the current commercially available TIMs [50]. Research should also focus on minimizing the overall thermal resistance instead of just focusing on increasing thermal conductivity. This is because although k_{TIM} increases with increasing volume fraction, bulk thermal resistance reaches a minima due to competing effects between the BLT and k_{TIM} that is clearly shown in Fig. 13.7.

A good physics based model for the contact resistance between the particle laden TIMs and the substrate is still lacking from the literature. Contact resistance will become important for thin highly conducting TIMs. Modeling of the thermal resistance of vertically-grown CNT array will be also needed in the future due to their promise as TIM.

References

1. M.M. Yovanovich, and E.E. Marotta, "Thermal Spreading and Contact Resistances," in Heat Transfer Handbook, A. Bejan and A.D. Kraus eds., John Wiley & Sons, Hoboken, New Jersey, 261–395, 2003
2. C.V. Madhusudana, Thermal Contact Conductance, Springer-Veralag, New York, 1996
3. A. Iwabuchi, T. Shimizu, Y. Yoshino, T. Abe, K. Katagiri, I. Nitta, and K. Sadamori, "The Development of a Vikers-Type Hardness Tester for Cryogenic Temperatures down to 4.2 K," *Cryogenics*, 36(2), 75–81, 1996
4. M.A. Lambert, and L.S. Fletcher, "Thermal Contact Conductance of Non-flat, Rough, Metallic Coated Metals," *Journal of Heat Transfer*, 124, 405–412, 2002
5. R. Prasher, "Surface Chemistry and Characteristic Based Model for the Thermal Contact Resistance of Fluidic Interstitial Thermal Interface Materials," *Journal of Heat Transfer*, 123, 969–975, 2001
6. R. Mahajan, C-P. Chiu, and G. Chrysler, "Cooling a Chip," *Proceedings of IEEE*, 94(8), 1476–1486, 2006
7. A. Watwe, and R. Prasher, "Spreadsheet Tool for Quick-turn 3D Numerical Modeling of Package Thermal Performance with Non-Uniform Die Heating," Proceedings of 2001 *ASME International Mechanical Engineering Congress and Exposition*, Paper No. 2-16-7-5, New York, November 11–16, 2001
8. J. Torresola, G. Chrysler, C. Chiu, R. Mahajan, D. Grannes, R. Prasher, and A. Watwe, "Density Factor Approach to Representing Die Power Map on Thermal Management," *IEEE Transactions on Advanced Packaging*, 28(4), 659–664, 2005
9. R. Mahajan, C-P. Chiu, and R. Prasher, "Thermal Interface Materials: A Brief Review of Design Characteristics and Materials," *Electronics Cooling*, 10(1), 2004
10. R.S. Prasher, "Thermal Interface Materials: Historical Perspective, Status and Future Directions," *Proceedings of IEEE*, 98(8), 1571–1586, 2006
11. R.S. Prasher, P. Koning, J. Shipley, and A. Devpura, "Dependence of Thermal Conductivity and Mechanical Rigidity of Particle Laden Polymeric Thermal Interface Materials on Particle Volume Fraction," *Journal of Electronics Packaging*, 125(3), 386–391, 2003
12. R.S. Prasher, J. Shipley, S. Prstic, P. Koning, and J-L. Wang, "Thermal Resistance of Particle Laden Polymeric Thermal Interface Materials," *Journal of Heat Transfer*, 125(6), 1170–1177, 2003
13. R.S. Prasher, "Rheology Based Modeling and Design of Particle Laden Polymeric Thermal Interface Material," *IEEE Transactions on Component and Packaging Technologies*, 28(2), 230–237, 2005
14. R.S. Prasher, and J.C. Matayabus, "Thermal Contact Resistance of Cured Gel Polymeric Thermal Interface Materials," *IEEE Transactions on Components and Packaging Technology*, 27(4), 702–709, 2004
15. R. Prasher, and P. Phelan, "Microscopic and Macroscopic Thermal Contact Resistances of Pressed Mechanical Contacts," *Journal of Applied Physics*, 100, 063538, 2006
16. Y. He, "Rapid Thermal Conductivity Measurement with a Hot Disk Sensor: Part 1. Theoretical Considerations," Proceedings of the 30th North American Thermal Analysis Society Conference, Sept. 23–25, 2002, Pittsburgh, PA, USA, 499–504, 2002

17. A. Sepehr, and M. Sahimi, "Elastic Properties of Three-Dimensional Percolation Networks with Stretching and Bond-Bending Forces," *Physical Review B*, 38(10), 7173–7176, 1988
18. A.V. Shenoy, "Rheology of Filled Polymer System," Kluwer Academic Publishers, MA, USA, pp. 1–390, 1999
19. T.L. Tansley, and D.S. Maddison, "Conductivity Degradation in Oxygen Polypyrrole," *Journal of Applied Physics*, 69(11), 7711–7713, 1991
20. C-P. Chiu, J.G. Maveety, and Q.A. Tran, "Characterization of Solder Interfaces Using Laser Flash Metrology," *Microelectronics Reliability*, 42, 93–100, 2002
21. L.S. Pritchard, P.P. Acarnley, and C.M. Johnson, "Effective Thermal Conductivity of Porous Solder Layers," *IEEE Transactions on Components and Packaging Technologies*, 27(2), 259–267, 2004
22. X. Hu, L. Jiang, and K. E. Goodson, "Thermal Characterization of Eutectic Alloy Thermal Interface Materials with Void-like Inclusions", Proceedings of Annual IEEE Semiconductor Thermal Measurement and Management Symposium, pp. 98–103, March 9–11, 2004, San Jose, CA, USA
23. P. Kim, L. Shi, A. Majumdar, and P.L. McEuen, "Thermal Transport Measurements of Individual Multiwalled Nanotubes," *Physical Review Letters*, 87(21), 215502-1215502-4, 2001
24. J. Hone, M.C. Llaguno, M.J. Biercuk, A.T. Johnson, B. Batlogg, Z. Benes, and J.E. Fisher, "Thermal Properties of Carbon Nanotubes and Nantube-based Materials," *Applied Physics A: Materials Science and Processing*, 74, 339–343, 2002
25. M.J. Biercuk, M.C. Llaguno, M. Radosavljevic, J.K. Hyun, A.T. Johnson, and J.E. Fischer, "Carbon Nantube Composites for Thermal Management," *Applied Physics Letters*, 80(2), 2767–2769, 2002
26. E.T. Thostenson, Z. Ren, and T.-W. Chou, "Advances in the Science and Technology," *Composite Science and Technology*, 61, 1899–1912, 2001
27. C.H. Liu, H. Huang, Y. Wu, and S.S. Fan, "Thermal Conductivity Improvement of Silicone Elastomer with Carbon Nanotube Loading," *Applied Physics Letters*, 84(21), 4248–4250, 2004
28. C.-W. Nan, G. Liu, Y. Lin, and M. Li, "Interface Effect on Thermal Conductivity of Carbon Nanotube Composites," *Applied Physics Letters*, 85(16), 3549–3551, 2004
29. S. Huxtable, D.G. Cahill, S. Shenogin, L. Xue, R. OZisik, P. Barone, M. Usrey, M.S. Strano, G. Siddons, M. Shim, and P. Keblinski , "Interfacial Heat Flow in Carbon Nanotube Suspensions," *Nature Materials*, 2, 731–734, 2003
30. R.S. Prasher, "Thermal Boundary Resistance and Thermal Conductivity of Multiwalled Carbon Nanotubes," *Physical Review B*, 77, 075424, 2008
31. X. Hu, L. Jiang, and K.E. Goodson, "Thermal Conductance Enhancement of Particle-Filled Thermal Interface Materials Using Carbon Nanotube Inclusions", 9th Intersociety Conference on Thermal and Thermomechanical Phenomena in Electronic System, June 1–4, 2004, Las Vegas, NV, USA
32. J. Xu, and T.S. Fisher, "Enhanced Thermal Contact Conductance Using Carbon Nanotube Arrays," 2004 Inter Society Conference on Thermal Phenomena, Las Vegas, 549–555, 2004
33. X. Hu, A. Padilla, J. Xu, T.S. Fisher, and K.E. Goodson, "3-Omega Measurements Vertically Oriented Carbon Nanotubes on Silicon," *Journal of Heat Transfer*, 128, 1109–1113, 2006
34. J. Xu, and T.S. Fisher, "Thermal Contact Conductance Enhancement with Carbon Nanotube Arrays," 2004 International Mechanical Engineering Congress and Exposition, Anaheim, CA, Nov. 13–20, Paper number IMECE2004-60185, 2004
35. T. Tong, Y. Zhao, L. Delzeit, Al. Kashani, M. Meyyappan, and A. Majumdar, Dense Vertically Multiwalled Carbon Nanotube Arrays as Thermal Interface Materials, *IEEE Transactions on Components and Packaging Technologies*, 30(1), 92–100

36. P.C. Irwin, Y. Cao, A. Bansal, and L.S. Schadler, "Thermal and Mechanical Properties of Polyimide Nanocomposites," 2003 Annual Report Conference on Electrical Insulation and Dielectric Phenomena, 120–123, 2003
37. L. Fan, B. Su, J. Qu, and C.P. Wong, "Effects of Nano-sized Particles on Electrical and Thermal Conductivities of Polymer Composites," 9th International Symposium on Advanced Packaging Materials, 193–199, 2004
38. S.A. Putnam, D.G. Cahill, B.J. Ash, and L.S. Schadler, "High-precision Thermal Conductivity Measurements as a Probe of Polymer/nanoparticle Interfaces," *Journal of Applied Physics*, 94(10), 6785–6788, 2003
39. R. Aoki, and C.-P. Chiu, "Testing apparatus for thermal interface materials," *Proceedings of the SPIE – The International Society for Optical Engineering*, 3582, 1036–1041, 1999
40. G.L. Solbrekken, C.-P. Chiu, B. Byers, and D. Reichebbacher, "The Development of a Tool to Predict Package Level Thermal Interface Material Performance," 7th Intersociety Conference on Thermal and Thermomechanical Phenomena in Electronic Systems, 2000. ITHERM 2000, Vol. 1, 23–26 May, 48–54, 2000
41. "Standard Test Method for Thermal Transmission Properties of Thin Thermally Conductive Solid Electrical Insulation Materials," ASTM D5470-93
42. C.-P. Chiu, G.L. Solbrekken, and T.M. Young, "Thermal Modeling and Experimental Validation of Thermal Interface Performance Between Non-Flat Surfaces," 7th Intersociety Conference on Thermal and Thermomechanical Phenomena in Electronic Systems, 2000. ITHERM 2000, Vol. 1, 23–26 May, 52–62, 2000
43. C.-P. Chiu, and G. Solbrekken, "Characterization of Thermal Interface Performance Using Transient Thermal Analysis Technique," 1999 ISPS Conference
44. C.-P. Chiu, J.G. Maveety, and Q.A. Tran, "Characterization of Solder Interfaces Using Laser Flash Metrology," *Microelectronics Reliability*, 42(1), 93–100, 2002
45. C.-P. Chiu, G.L. Solbrekken, V. LeBonheur, Y.E. Xu, "Application of Phase-Change Materials in Pentium® III and Pentium® III XeonTM Processor Cartridges," Proceedings International Symposium on Advanced Packaging Materials Processes, Properties and Interfaces (Cat. No.00TH8507). Reston, VA, USA: IMAPS – Int. Microelectron. & Packaging Soc, 265–270, 2000
46. T.J. Goh, A.N. Amir, C.-P. Chiu, and J. Torresola, "Cartridge Thermal Design of Pentium(R) III Processor for Workstation: Giga Hertz Technology Envelope Extension Challenges," Proceedings of 3rd Electronics Packaging Technology Conference (EPTC 2000) (Cat. No.00EX456). Piscataway, NJ, USA: IEEE, 65-71, 2000
47. T.J. Goh, A.N. Amir, C.-P. Chiu, and J. Torresola, "Novel Thermal Validation Metrology Based on Non-Uniform Power Distribution for Pentium® III XeonTM Cartridge Processor Design with Integrated Level Two Cache," Proceedings of 51st Electronic Components and Technology Conference, 29 May–1 June, 1181–1186, 2001
48. C.-P. Chiu, B. Chandran, K. Mello, and K. Kelley, "An Accelerated Reliability Test Method to Predict Thermal Grease Pump-Out in Flip-Chip Applications," Proceedings of 51st Electronic Components and Technology Conference, 29 May–1 June, 91–97, 2001
49. L. Bharatham, W.S. Fong, C.J. Leong, and C.-P. Chiu, "A Study of Application Pressure on Thermal Interface Material Performance and Reliability on FCBGA Package, 2006 EMAP
50. E. Samson, S. Machiroutu, J.-Y. Chang, I. Santos, J. Hermarding, A. Dani, R. Prasher, D. Song, and D. Puffo, "Some Thermal Technology and Thermal Management Considerations in the Design of Next Generation Intel^R Centrino^{TM} Mobile Technology Platforms," *Intel Technology Journal*, 9(1), 2005

Chapter 14
Embedded Passives

Dok Won Lee, Liangliang Li, Shan X. Wang, Jiongxin Lu, C. P. Wong, Swapan K. Bhattacharya, and John Papapolymerou

Abstract Driven by ever growing demands of miniaturization, increased functionality, high performance, and low cost for microelectronic products and packaging, new and unique solutions in IC and system integration, such as system-on-chip (SOC), system-in-package (SiP), system-on-package (SOP), have been hot topics recently. Despite the high level of integration, the number of discrete passive components (resistors, capacitors, or inductors) remains very high. In a typical microelectronic product, about 80% of the electronic components are passive components, which are unable to add gain or perform switching functions in circuit performance, but these surface-mounted discrete components occupy over 40% of the printed circuit/wiring board (PCB/PWB) surface area and account for up to 30 percent of solder joints and up to 90 percent of the component placements required in the manufacturing process. Embedded passives, an alternative to discrete passives, can address these issues associated with discrete counterparts, including substrate board space, cost, handling, assembly time, and yield [1, 2]. Figure 14.1 schematically shows an example of realization of embedded passive technology by integrating resistor and capacitor films into the laminate substrates.

By removing these discrete passive components from the substrate surface and embedding them into the inner layers of substrate board, embedded passives can provide many advantages such as reduction in size and weight, increased reliability, improved performance and reduced cost, which have driven a significant amount of effort during the past decade for this technology. This chapter provides a review on most recent development in embedded inductors, capacitors, and resistors.

Keywords Passives · magnetic inductor · quality factor · embedded capacitor · composites · thin film resistor

---

D.W. Lee (✉)
Stanford University, Stanford, CA, USA

Fig. 14.1 Schematic representation of the size advantages of the embedded passives as compared to discrete passives

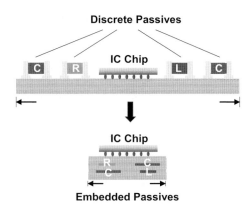

14.1 Embedded Inductors

14.1.1 Introduction

14.1.1.1 Limitations of Discrete Inductors and Air Core Spiral Inductors

Discrete wire-wound inductors are commercially available for a wide range of inductance from 1 nH to 100 μH as shown in Fig. 14.2(a) [3]. DC resistance is relatively small due to a large cross-section area of the wound wire. Hence a large inductance of ~100 nH can be obtained for a DC resistance of $\leq 1\ \Omega$ (Fig. 14.2(b)). This allows a high quality factor of $Q > 30$ at a frequency of 100 MHz or beyond. While tightly wound wire generates a large inductance, it also causes a large parasitic capacitance which limits the useful operating frequency of the discrete inductor due to its self-resonant frequency. Figure 14.2(a) illustrates that the self-resonant frequency of the ferrite inductor

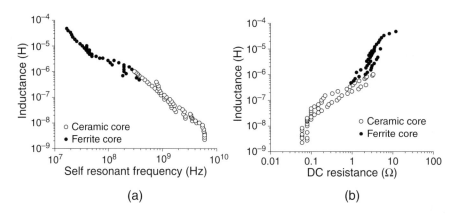

Fig. 14.2 Characteristics of commercially available wire-wound inductors (based on Ref. [3])

is limited to ~300 MHz. Ceramic core inductors can operate at higher frequencies because their cores are made of a ceramic dielectric material with high linearity and low hysteresis, but their inductance are limited to a range from ~10 nH to ~1 μH. Moreover, a discrete inductor is known to be bulky and occupies a large volume with the area of a few mm^2 and a thickness more than 1 mm. The bulkiness is especially troublesome in portable electronics applications where the number of passive elements is more than that of active devices.

Embedded inductors are "integrated" inductors designed and fabricated based on integrated circuit (IC) or printed circuit board (PCB) technologies, and they consume significantly less device area. Because embedded inductors are placed directly on Si wafers or in packages for integrated circuits, they truly enable the realization of the integrated electronics such as system-on-a-chip (SoC) and system-in-package (SiP). Spiral inductors with air core, e.g., planar metallic coils embedded in a nonmagnetic insulator such as silicon dioxide, have been the mainstream design of the embedded inductors to date. However, their device performance are limited and often of inferior quality than those of the discrete inductors. Calculated device properties of spiral inductors based on the results reported in the literature are shown in Fig. 14.3 [4–6]. The spiral inductors occupy smaller area than discrete inductors; however, their practical inductance goes only up to ~200 nH (Fig. 14.3(a)). Moreover, due to the greatly reduced cross-section area of the conductor, the coil resistance increases significantly. The inductance is limited to a few nH for the DC resistance below 1 Ω (Fig. 14.3(b)). This also limits the quality factor to be about 15 or below at the useful frequency range [7]. The resistance can be reduced with the use of wider coil width w, but utilization of the wider coil leads to the significant increase in the device area as illustrated in Fig. 14.3. While the performances of the spiral inductor can be acceptable for low-inductance and high-frequency

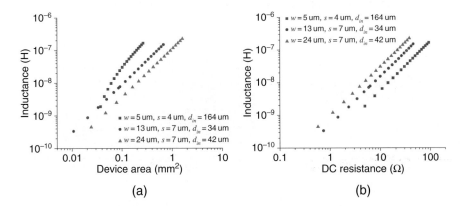

Fig. 14.3 Calculated device properties of the integrated spiral inductors with the square layout and various values of the turn coil width w, turn spacing s, and the inner diameter d_{in} (based on Refs. [4–6])

applications, embedded inductors with the properties comparable to those of the discrete inductors are still needed for numerous integrated electronics applications.

14.1.1.2 Advantages of Magnetic Inductors as Embedded Inductors

Use of magnetic cores with high permeability in thin film inductors has been proposed to significantly increase the inductance while maintaining a low resistance and small inductor area. The classical model predicts that, with the inclusion of the magnetic core, the inductance of the solenoid inductor is enhanced by the relative permeability (μ_r) of the magnetic core material [8]:

$$L_{\text{Solenoid}} = \frac{\mu_0 \mu_r N^2 w_M t_M}{l_M} \qquad (14.1)$$

where N is the number of coil turns, and w_M, t_M and l_M are the width, the thickness, and the length of the magnetic core, respectively. If the inductance enhancement is large enough, a smaller number of turns would be needed to meet a given inductance requirement, leading to a significant decrease in both the coil resistance and the device area consumed. Unfortunately, the classical model grossly overestimates the inductance enhancement attainable in practical embedded inductors, as we will see amply in the rest of this chapter.

14.1.1.3 Inductor Designs

Various designs for the magnetic inductors have been investigated and reported in the literature. The most widely considered designs are the transmission line, spiral inductor, and solenoid inductor as shown in Fig. 14.4.

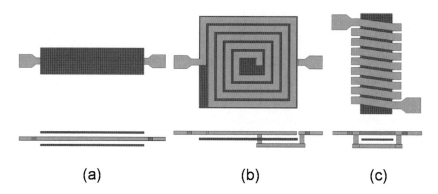

Fig. 14.4 Schematic top and cross-section views of magnetic inductor designs: (**a**) transmission line; (**b**) spiral inductor; and (**c**) solenoid inductor

A transmission line has a simple structure, in which a metallic strip is sandwiched between two magnetic layers as shown in Fig. 14.4(a) [9, 10]. The magnetic layers can enclose the line completely using magnetic flanges or vias in order to enhance the magnetic flux closure between two layers. It is relatively easy to fabricate and can be ideal for applications where a low resistance is crucial. However, the inductance generated is limited to a few nH due to the simple geometry.

A spiral inductor with magnetic planes is a simple modification from the mainstream spiral inductor with air core [11, 12]. A spiral coil is placed on top of a single magnetic layer or sandwiched between two magnetic layers (Fig. 14.4(b)). Two magnetic layers are often considered in order to obtain significant enhancement in inductance, and they can be patterned to reduce the eddy current losses. Coil designs and the fabrication processes are relatively well developed due to the similarity with the mainstream spiral inductor. However, the fabrication involves the processing of four metallization layers. Moreover, because the magnetic flux generated by the spiral coil is not well directed, the use of the magnetic material is not efficient in this case. Hence the inductance enhancement due to the magnetic contribution is limited.

A solenoid inductor is realized by forming the coil using two conductor layers to wrap around the magnetic core (Fig. 14.4(c)) [13, 14]. This design resembles the discrete wire-wound inductor, and it is very efficient magnetically. The magnetic contribution to the device properties is also well understood compared to other designs. The design, however, involves relatively complicated processing, including the accurate alignment between layers and the formation of multiple via contacts between two conductor layers. The solenoid inductor design is the main focus throughout the chapter because of its ability to make full use of the magnetic core.

14.1.1.4 Requirements and Survey for the Magnetic Core Materials

The inductance of the magnetic inductor is in principle proportional to the relative permeability of the magnetic core according to Eq. (14.1). Hence a high permeability is clearly one of the desired properties for the magnetic core material. The permeability reduces to unity when the operating frequency is above the ferromagnetic resonance frequency (FMR) of the magnetic material [15]. Therefore a high FMR is also desirable to be useful at high frequencies.

Use of magnetic core comes with the cost of introducing magnetic power losses, which result in a significant increase in the inductor resistance at high frequencies. Main magnetic loss mechanisms that need to be considered include: (a) hysteresis loss [16]; (b) eddy current loss [17]; and (c) ferromagnetic resonance loss [17]. The area inside the magnetic hysteresis loop corresponds to the energy loss per cycle, and hysteresis power loss is proportional to the loop area. Hence a small coercivity H_C is required to minimize the loop area and in turn the hysteresis loss. The eddy current loss is due to the eddy current formed inside of the magnetic core and is in principle inversely proportional to the electrical resistivity of the magnetic material. The ferromagnetic resonance loss appears as the operating frequency approaches the FMR. Hence this loss can be avoided if the FMR is large enough.

The above considerations suggest that the requirements for the magnetic core materials include high permeability, large FMR frequency, small coercivity, and large resistivity. Several types of magnetic core materials have been investigated in the literature. One is the amorphous alloys such as CoZrTa [11, 18] and CoZrNb [19]. They have soft magnetic properties with high permeability. Due to their amorphous structure, their resistivity is larger than that of polycrystalline metals. However, it is not sufficiently large enough for high frequency applications where the eddy current loss is still severe. The eddy current loss can be reduced by using the lamination structures such as CoZrNb/AlN [20] and CoFeSiB/SiO$_2$ [21]. The eddy current loss will be reduced due to the insulating layers of AlN or SiO$_2$, but the multilayer structures involve more complex fabrication processes. For high frequency applications, nanogranular magnetic materials such as CoFeHfO [22] and FeAlO [23] can be considered. Nanogranular magnetic particles are dispersed in an insulating matrix, so the resistivity tends to be very high. However, since the magnetic materials are separated into individual grains, which tend to have significant magnetocrystalline anisotropy, the permeability of nanogranular magnetic materials is relatively small. The magnetic and electric properties of possible magnetic core materials are summarized in Table 14.1. A suitable magnetic core material can be chosen depending on the application and the frequency range of interest.

14.1.2 Modeling and Design Considerations of Magnetic Inductors

Key device properties of the embedded inductors are the inductance, resistance, and quality factor. Analytical models for these properties are discussed below.

Table 14.1 Summary of magnetic and electrical properties of magnetic core materials (based on Ref. [24])

Materials	μ_r	FMR (GMz)	H$_{chard}$(Oe)	Resistivity ($\mu\Omega$–cm)	Reference
CoZrTa	600~780	~1.5	<1	~100	[18]
CoZrNb	~850	~0.7	<1	~120	[19]
FeCoN	1200	1.5	<1	50	[25]
CoFeHfO	140~170	~2.4	<1	~1600	[22]
FeAlO	500~700	~1.5		50~2000	[23]
CoFeSiO	~200	~2.9	~6	~2200	[26]
CoFeAlO	~300	~2.0	1~5	200~300	[27]

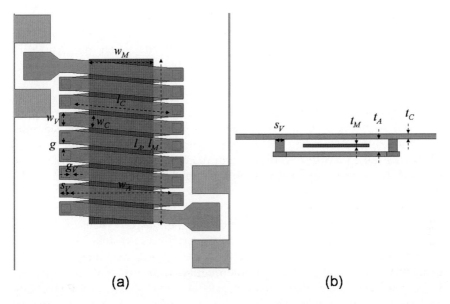

Fig. 14.5 Schematic design of an integrated solenoid inductor: (**a**) top view and (**b**) cross-section view

A schematic design of an embedded inductor with magnetic core is shown in Fig. 14.5. Design parameters used in the models are as follows: the number of turns N, width of air core w_A, length of air core l_A or magnetic core l_M, thickness of air core t_A, width of magnetic core w_A, thickness of magnetic core t_A, widths of coil w_C or w_V, length of coil l_C, thickness of coil w_C, gap between turns g, size of via s_V, and gap surrounding the via g_V. Analytical models were put forward based on the established analytical models for discrete solenoid inductors [7, 28], the careful analyses of experimental results [14, 29], and the comparison with the results from the finite element electromagnetic field simulation tools [30, 31].

14.1.2.1 Inductance

Inductance of discrete solenoid inductors can be described by the classical Wheeler formula [7]:

$$L = \frac{10\pi\mu_0 N^2 r^2}{9r + 10l_A}, \quad r \equiv \sqrt{\frac{(w_A + s_V)(t_A + t_C)}{\pi}} \quad (14.2)$$

where πr^2 represents the effective cross-section area of the air core. The expression for the inductance of the embedded solenoid inductors with air core, L_{AC}, is modified from the Wheeler formula based on the comparison with the experimental and simulation results:

$$L_{AC} = L_{\text{Winding}} + L_{\text{Parasitic}} \text{ where } L_{\text{Winding}} = \frac{10\pi\mu_0 N^2 a^2}{9a + 10l_A},$$

$$a \equiv \sqrt{\frac{(w_A + 2s_V)(t_A + 2t_C)}{\pi}}$$

(14.3)

The winding inductance L_{Winding} depends on the modified cross-section area of the air core $A_{AC} = (w_A + 2s_V)(t_A + 2t_C)$ which completely includes the vias and coils surrounding the air core. The parasitic inductance $L_{\text{Parasitic}}$ represents the effects of deviations from the classical winding, including the probe pads at the ends of winding and the ground ring surrounding the inductor.

The classical expression given by Soohoo (Eq. (14.1)) has been widely used to estimate the inductance of magnetic inductors [8]; however, it usually greatly overestimates the actual inductance values. One effect overlooked in the classical expression is the demagnetization field. For a finite-sized magnetic core, a demagnetization field is formed inside the magnetic core, and, as a result, a higher magnetic field is needed to overcome the demagnetization field to obtain the same magnetization, which effectively reduces the relative permeability μ_r of the magnetic core. In addition, for embedded inductors, the contribution from winding to the inductance can be comparable to the magnetic contribution, and hence the winding contribution and magnetic contribution should be carefully separated. Taking all of these factors into account, we can describe the inductance of the embedded solenoid inductor with magnetic core, L_{MI}, by the following expression:

$$L_{MI} = L_{AC} + \Delta L, \text{ where } \Delta L \equiv \frac{\mu_0 \mu_r N^2 w_M t_M}{l_M[1 + N_d(\mu_r - 1)]}$$

$$= \frac{\mu_0 \mu_{\text{eff}} N^2 w_M t_M}{l_M}, \mu_{\text{eff}} \equiv \frac{\mu_r}{[1 + N_d(\mu_r - 1)]}$$

(14.4)

Note that ΔL is the net increase in the inductance due to the magnetic contribution and N_d is the demagnetization factor. The demagnetization field is not uniform inside the magnetic core having the orthorhombic shape. It is not trivial to estimate N_d analytically for $\mu_r > 1$, but numerical solutions are available in the literature [32, 33]. The effective permeability μ_{eff} for various geometries of magnetic core with $\mu_r = 1000$ is plotted in Fig. 14.6(a). The plot shows that $l_M \gg (w_M t_M)^{1/2}$ is preferred to maintain a high effective permeability. Figure 14.6(b) illustrates that the demagnetization effect is more severe for a higher permeability. While the magnetic domain structure also needs to be considered in order to fully account for the effective permeability of the magnetic cores under various geometries [31], it has been confirmed experimentally that the above expression can adequately estimate the inductance of the embedded magnetic inductors.

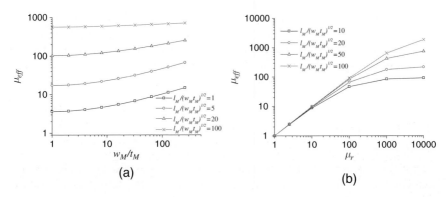

Fig. 14.6 Effective permeability for (**a**) $\mu_r = 1000$ and (**b**) $w_M/t_M = 64$ (based on Ref. [33])

The increase in inductance due to the magnetic contribution can be estimated using Eqs. (14.3) and (14.4). Assuming $l_A \gg a$ and $L_{\text{Winding}} \gg L_{\text{Parasitic}}$, the inductance enhancement is approximately:

$$\frac{L_{MI} - L_{AC}}{L_{AC}} = \frac{\Delta L}{L_{AC}} \approx \mu_{\text{eff}} \frac{A_{MI}}{A_{AC}}, \text{ where } A_{MI} \equiv w_M t_M \qquad (14.5)$$

where A_{MI} is the cross-section area of the magnetic core. Since $\mu_{\text{eff}} < \mu_r$ and $A_{MI} < A_{AC}$, the actual enhancement can be greatly reduced from μ_r. Even so, the inductance enhancement by a factor more than 30 has been reported [14].

Other properties of the relative permeability need to be considered carefully. One is the uniaxial anisotropy of the permeability [34]. The magnetic core materials of interest have high permeability values along their hard axis direction and unity along other orthogonal directions at high frequencies. Hence assuming the isotropic permeability would be unrealistic and result in the overestimation of the inductance. It has been shown experimentally that the measured permeability is proportional to $\cos 2\alpha$ where α is the angle between the external excitation magnetic field and hard axis of magnetic material [29]. One can be misled to believe that the permeability can be significant at an angle away from the hard axis. However, the permeability has the tensor nature, and the observed $\cos 2\alpha$-dependence corresponds to only one component in the tensor matrix. In order to fully describe the properties of a magnetic material, all the components of the permeability tensor must be assigned accurately. Due to the tensor properties of the permeability, a toroidal-like embedded inductors with closed uniaxial magnetic core will have an inductance much smaller than that is expected from closed isotropic magnetic core.

Substrate effect on the magnetic properties also needs to be taken into account. The soft magnetic properties and uniaxial anisotropy of the magnetic core material can deteriorate with increasing surface roughness [24, 35]. This deterioration can be due to the fact that the domain structure and domain wall energy of the film are more variable for rougher surfaces [36]. The soft magnetic films deposited on rougher surfaces tend to have larger anisotropy values and hence smaller permeability values. Hence, to fabricate the embedded inductors on package substrates, we may need to apply surface planarization to smoothen the substrate surface before the magnetic film deposition.

14.1.2.2 Resistance

The expression for the resistance of the embedded solenoid inductors with air core, R_{AC}, was obtained by representing it as the series resistance of three line segments: the coil crossing over the air core, the connection to the via, and the via:

$$R_{AC} = 2N\rho \left[\frac{l_C}{w_C t_C} + \frac{(s_V + 2g_V)}{w_V t_C} + \frac{t_C + t_A}{s_V^2} \right] \quad (14.6)$$

where ρ is the electrical resistivity of the coil material. The contact resistance between two conductor layers is not included in Eq. (14.6).

The resistance of the embedded inductor with magnetic core, R_{MI}, should include the magnetic power losses. According to the classical electromagnetism, the hysteresis power loss is related to the ratio of the imaginary and real parts of the permeability of the magnetic material [37]. At high frequencies, other magnetic losses including the eddy current loss and ferromagnetic resonance loss need to be considered, and their effects are included in the measured permeability spectra. Hence, by modifying the classical expression for the hysteresis power loss, the magnetic power loss P_{Magnetic} can be described as follows:

$$P_{\text{Magnetic}} = \omega \left(\frac{\mu''}{\mu'} \right) E_{\text{Mag.cont.}} \quad (14.7)$$

where μ''/μ' is the permeability ratio of the magnetic core and $E_{\text{Mag.cont.}}$ is the contribution in the energy stored due to the inclusion of the magnetic core. Equation (14.7) can be expressed in terms of device properties using the following relations:

$$P_{\text{Magnetic}} = \frac{1}{2} R_{\text{Magnetic}} I^2 \quad (14.8)$$

$$E_{\text{Mag.cont.}} = E_{MI} - E_{\text{Coil}} = \frac{1}{2} L_{MI} I^2 - \frac{1}{2} L_{AC} I^2 = \frac{1}{2} \Delta L I^2 \quad (14.9)$$

where R_{Magnetic} is the contribution to the resistance due to the magnetic power losses, E_{MI} is the total energy stored in the magnetic inductor, and E_{Coil} is the energy stored due to the coils of the inductor. Substituting Eqs. (14.8) and (14.9) into Eq. (14.7), we obtain the resistance of the embedded solenoid inductor with magnetic core, R_{MI}, as follows:

$$R_{MI} = R_{AC} + R_{\text{Magnetic}} = R_{AC} + \omega \left(\frac{\mu''}{\mu'}\right) \Delta L \tag{14.10}$$

As the frequency increases, R_{Magnetic} becomes more significant than R_{AC}, because it is proportional to both the frequency and μ''/μ'. The latter ratio also increases rapidly with frequency. The magnetic core contribution to the resistance is proportional to the net inductance enhancement over the air core as shown in Eq. (14.10), which imposes a fundamental engineering trade-off in magnetic inductor design: The more inductance enhancement we obtain by using a magnetic core at low frequencies, the more resistive losses we introduce at high frequencies.

14.1.2.3 Quality Factor

Quality factor Q is the figure of merit that determines the efficiency of the inductor device. Q is proportional to the ratio of energy stored to the energy lost, per unit time. Assuming that the contributions from the coils and the magnetic core can be separated, the quality factor can be expressed as follows:

$$Q = 2\pi \frac{\text{Peak} - \text{to} - \text{peak energy}}{\text{Power dissipation} \cdot T} = 2\pi f \frac{E_{\text{Coil}} + E_{\text{Mag.cont.}}}{P_{\text{Coil}} + P_{\text{Magnetic}}} \tag{14.11}$$

where $P_{\text{Coil}} = R_{AC} I^2/2$ is the ohmic power loss from the coils. Using Eqs. (14.7)~(14.10), we obtain the quality factor of the magnetic inductor, Q_{MI}, as the following:

$$Q_{MI} = \frac{\omega L_{MI}}{R_{MI}} = \omega \frac{L_{AC} + \Delta L}{R_{AC} + \omega \left(\frac{\mu''}{\mu'}\right) \Delta L} \tag{14.12}$$

It should be emphasized that μ''/μ' is the permeability ratio of the magnetic core used in the embedded inductor. The permeability of the patterned and processed magnetic core is usually not the same as that of the bulk magnetic film.

At low frequencies Q_{MI} can be significantly larger than the quality factor of the air core inductor, Q_{AC}, because of the large inductance enhancement ΔL, but it starts to decrease sooner and faster than Q_{AC} due to the larger resistive losses at higher frequencies.

If ΔL is very small, Q_{MI} becomes close to Q_{AC} according to Eq. (14.12), which is intuitively correct, since a magnetic inductor would behave like an air core inductor for a very small magnetic contribution. On the other hand, if ΔL is very large compared to L_{AC}, Q_{MI} approaches the permeability ratio μ''/μ' of the magnetic core. Equation (14.12) suggests that Q_{MI} is higher than Q_{AC} below the frequency at which Q_{AC} and μ''/μ' cross each other, and Q_{MI} becomes less than Q_{AC} beyond this cross-over frequency. Hence the cross-over point can be considered as the useful bandwidth of the magnetic inductor. This important design criterion will be further examined in the following sections.

14.1.3 Embedded On-package and On-chip Inductors: Experiments and Analyses

14.1.3.1 Survey of Inductor Works in Literature

There have been many researches on the embedded inductors with magnetic core for radio frequency (RF) circuits and power delivery applications. Table 14.2 summarizes the recently reported embedded inductors with magnetic core. While many of them were built on the silicon substrate, there are continuing efforts to fabricate the embedded inductors on the organic package substrates as well. The reported inductance ranges from 3 nH to 5 μH; however, the inductance includes the contribution from the coil. The actual magnetic contribution to the inductance can be represented by the relative inductance enhancement $\Delta L/L_{AC}$, and it has been limited to about $\sim$100% or below. Recently, large enhancements by a factor more than 10 were reported [9, 14, 41–43], showing that the use of the magnetic core is indeed effective in increasing the inductance. The inductance density can go as high as 532 nH/mm^2, but other properties such as the DC resistance and the device area for some inductors may have been excessive. However, even with the practical constraints on the resistance ($R_{DC} < 1~\Omega$) and the device area (< 1 mm^2), the inductance density can reach above 200 nH/mm^2 using magnetic core [43].

14.1.3.2 On-chip Inductor Results

Embedded inductors fabricated on Si substrate are shown in Fig. 14.7 [14, 43]. Copper conductors were formed by electroplating through the photoresist masks. CoTaZr magnetic cores were deposited using RF-sputtering and then patterned by wet etching. Polyimide was used as the interlayer insulating material. Fabricated magnetic inductors have $N = 8.5$ turns with the probe pads and ground ring, and the inductor in Fig. 14.7(b) has the lateral dimensions reduced from that shown in Fig. 14.7(a).

Figure 14.8(a) shows the measured inductance data for different numbers of coil turns. The inductance increases with the number of turns and has a value of 70.2 nH at 10 MHz with $N = 17.5$ turns, which is an enhancement by a factor of 34 from 2.0 nH of the air core inductor with the identical geometry. The device area for $N = 17.5$ is 0.88 mm^2, corresponding to the inductance density of 80 nH/mm^2, with the DC resistance of 0.67 Ω. By shrinking the lateral dimensions while maintaining the vertical dimensions unchanged, the inductance density further increases to 219 nH/mm^2 without affecting the coil resistance

Table 14.2 Summary of embedded inductors with magnetic core

Inductor design	Core material	Substrate	L_{MI} (nH)	$\Delta L\ IL_{AC}$	R_{DC} (Ω)	Q_{Max}	L/Area (nH/mm^2)	Reference
Transmission line	CoZrO$_2$	Si	~3		0.014		5.6	[10]
Transmission line	CoTaZr	Si	~17	19		~3.8 @ 170 MHz		[9]
Spiral	CoTaZr	Si	47.9	0.65	59	~2.7 @ 1 GHz	532	[11]
Spiral	Fermite	Si	1500		0.67	70 @ 5 MHz	42	[38]
Spiral	NiFe	Si	3200	1.3	5.9	1.3 @ 1 MHz	246	[39]
Spiral	CoNbZr	Si	8.5~13.7	0.07~0.71	~5	3.03~11.8 @ 1 GHz	59~95	[40]
Spiral	FeHfN	Si	~4.8	0.30	~0.9	~10.2 @ 900 MHz		[12]
Solenoid	FeCoBSi	Si	45	10	~4			[41]
Solenoid	NiFe	Si	~500	≥8.1	0.095	~20 @ 2 MHz	16	[42]
Solenoid	CoTaZr	Si	70.2	34	0.67	6.3 @ 26 MHz	80	[14]
Solenoid	CoTaZr	Si	48.4	32	0.67	6.5 @ 30 MHz	219	[43]
Spiral	Fermite-polymer	Polymide	1330		2.6	18.5 @ 10 MHz	53.2	[44]
Spiral	NiFe-based	Polymide	5060		1.76	10.1 @ 1.4 MHz	213	[45]
Solenoid	CoFeSiB/SiO$_2$	MPS	5000		1.4		421	[46]
Solenoid	CoFeHfO	PCB	3.25	0.13	0.012	22 @ 250 MHz	0.23	[47]

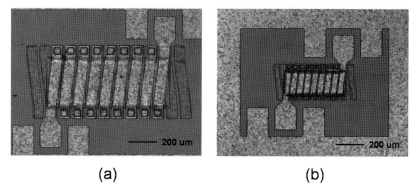

Fig. 14.7 Optical microscope images of on-chip magnetic inductors: (**a**) before and (**b**) after shrinking the lateral dimensions by a factor of two [14, 43]

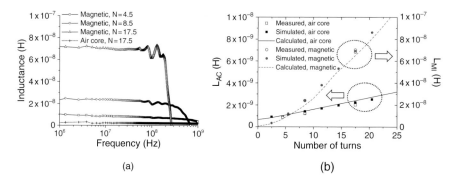

Fig. 14.8 (**a**) Measured inductance data of on-chip inductors and (**b**) comparison of measured inductance data with the simulation and calculation results for the on-chip inductors with or without magnetic core [14]

significantly [43]. These properties are significant improvement over those of the mainstream air core spiral inductors and are comparable to those of the discrete wire-wound inductors.

The measured inductance values for the embedded inductors with or without magnetic core are compared with the simulation and calculation results in Fig. 14.8(b). The simulation data are obtained using Ansoft HFSS^{TM} [30], and it has been shown that the software includes the demagnetization effect [31]. The calculation data are obtained using Eqs. (14.3) and (14.4). Good agreements for both cases confirm that the analytic models can accurately describe the inductances of air core and magnetic inductors and that the demagnetization effect plays a major role in determining the effective permeability for the magnetic inductors. The expected inductance enhancement using Eq. (14.5) was about 37×, which is very close to the observed enhancement of 34×.

The measured resistance and quality factor data of the magnetic inductors are shown in Fig. 14.9. Low-frequency resistances are similar for the air core and magnetic inductors; however, the resistance for the magnetic inductor increases significantly with the frequency due to the magnetic power losses. The resistance increase is greater for higher number of coil turns, which is due to the trade-off described in Section 1.2. A higher number of turns generate a larger inductance gain, which in turn results in a higher magnetic core contribution to the resistance at high frequencies as shown in Eq. (14.10). Consequently, at low frequencies, the quality factor is higher for the higher number of turns because of the larger inductance enhancement, but it starts to fall sooner due to the larger resistive losses at high frequencies.

Figure 14.10(a) shows the permeability spectra taken from a 2 μm thick blanket CoTaZr film and CoTaZr magnetic core structure processed in parallel

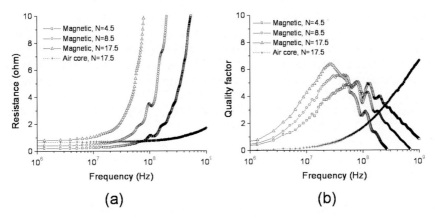

Fig. 14.9 Measurement data of on-chip inductors: (**a**) resistance and (**b**) quality factor [14]

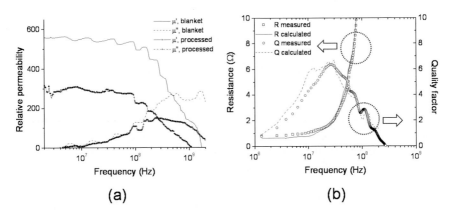

Fig. 14.10 (**a**) Permeability spectra of 2 um thick CoTaZr blanket film and processed magnetic core and (**b**) comparison of the calculated resistance and quality factor with the measurement results for the on-chip magnetic inductor with $N = 17.5$ [14]

with the inductor fabrication. The permeability spectra are not identical to each other mainly due to the demagnetization effect and the dependence on the substrate surface as discussed in Section 14.1.2. Resistance and quality factor were calculated using Eqs. (14.10) and (14.12), respectively, with the permeability spectra of the processed magnetic core, and they were compared with the measurement data for $N = 17.5$ as shown in Fig. 14.10(b). The excellent agreement between the calculation and measurement results confirms that the analytical models discussed in Section 14.1.2 can accurately describe the frequency-dependent device behaviors.

14.1.3.3 Other On-chip Inductors in Literature

There have been numerous studies on the embedded magnetic inductors fabricated on silicon substrate as shown in Table 14.2. However, the analytical models discussed in Section 14.1.2 cannot be applied to most of the reported experimental results, because not all the information is available for the analysis. Nonetheless, a number of the reports from other groups also demonstrate the trade-off between the inductance gain and the resistance increase. Figure 14.11(a) shows the device properties of the embedded inductors with different magnetic core thicknesses [42]. One solenoid-like embedded inductor has 5 μm thick Cu conductors and 1.4 μm thick laminated NiFe core ("thinner"), whereas the other one has 20 μm thick conductors and 16 μm thick core ("thicker"). The thicker inductor has much larger inductance than the thinner inductor due to the thicker magnetic core. At low frequencies near 1 MHz, the resistance of the thicker inductor is smaller than that of the thinner inductor due to the thicker conductors; however, it starts to increase faster with frequency because of the magnetic power losses. Hence, while the quality factor for the thicker inductor is

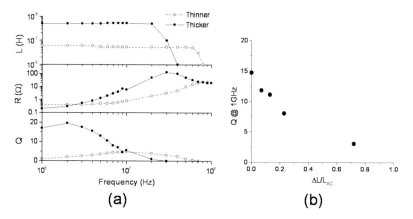

Fig. 14.11 (a) Device properties of on-chip inductors with different thicknesses of magnetic core and conductor (based on Ref. [40]) and (b) comparison between the inductance enhancement and the quality factor at 1 GHz (based on Ref. [40])

significantly greater at 1 MHz, it soon decreases as the frequency increases and becomes smaller than that for the thinner inductor at 10 MHz and above.

Figure 14.11(b) shows the device properties of the spiral inductors with different magnetic core structures [40]. CoNbZr magnetic cores with the same thickness are put on top and bottom of the planar spiral conductor, and their shapes were varied to study the effect on the magnetic flux closure. Depending on the core structure, the relative inductance enhancement, $\Delta L/L_{AC}$, increases from 7% to 71%. A larger inductance enhancement leads to a higher resistance increase, which in turn results in a smaller quality factor at the high frequency of 1 GHz as shown in Fig. 14.11(b).

The above examples, along with the on-chip inductor data presented in the previous subsection, indicate that the trade-off between the inductance gain and the resistance increase applies regardless of the inductor design and the method to enhance the inductance.

14.1.3.4 On-Package Inductor Results

Embedded inductors built on the package substrates have relatively loose constraint on the area consumption and hence can afford to have larger device areas [44–47]. However, a rougher surface of the package substrate can deteriorate the magnetic properties of the magnetic core material [33], and there are limitations on the fabrication processes and the design rules compared to the state-of-the-art silicon processing [24].

One example of the embedded inductors fabricated on the package substrate is shown in Fig. 14.12 [24, 47]. In this solenoid design using CoFeHfO magnetic core, two sets of copper coils are connected symmetrically to enhance the closure of the magnetic flux, which in turn increases the inductance and also to reduce the resistance by a factor of two. However, the inductance gain is limited to ~12% as shown in Fig. 14.13(a). This is mainly due to the reduction of the permeability because of the substrate effect, the small ratio of the magnetic core area to the air core area, A_{MI}/A_{AC}, and the large contribution of the parasitic inductance $L_{Parasitic}$ to the air core inductance L_{AC}. Nonetheless, the observed inductance gains are in reasonable agreement with the HFSS simulation data and calculation results using Eq. (14.4) (Fig. 14.13(b)). The inductance gain will be larger if thicker magnetic core or more coil turns are used.

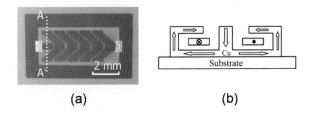

Fig. 14.12 (a) Optical microscope image and (b) schematic cross-section (A-A') view of on-package magnetic inductor [47]

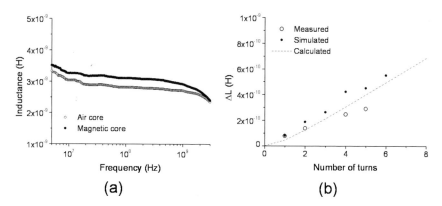

Fig. 14.13 (a) Inductance data of on-package inductors and (b) comparison of measured inductance data with the simulation and calculation results for the on-package magnetic inductors [48]

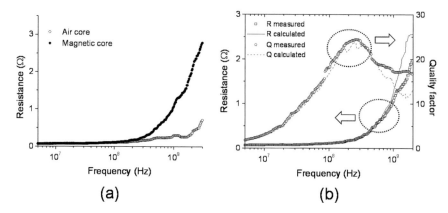

Fig. 14.14 (a) Measured resistance data of on-package inductors and (b) comparison of the calculated resistance and quality factor with the measurement results for the on-package magnetic inductor with $N = 5$ [47]

Even though the inductance gain above is very small, the resistance of the magnetic inductor starts to increase significantly as the frequency approaches 300 MHz as shown in Fig. 14.14(a). The resistance increase in turn causes the decline of the quality factor at 300 MHz (Fig. 14.14(b)). The analytical models are applied to the measurement results, and very good agreement is shown in Fig. 14.14(b), which indicates that even a very small inductance gain can lead to a significant resistance increase when the frequency is high enough. It also confirms that the intrinsic trade-off between the inductance gain and the resistance increase can be accurately described by the analytical models whether the magnetic contribution to the inductance is large or small. It should be noted that even using CoFeHfO with low magnetic power losses [46], the magnetic contribution to the resistance becomes significant below 1 GHz despite very small inductance gains. It is also evident in Table 14.2 that the peak quality factors of the embedded magnetic inductors occur at frequencies of 1 GHz or below.

14.1.4 Future Directions of the Embedded Magnetic Inductors

14.1.4.1 Fundamental Trade-offs of Magnetic Inductors

Examples shown in the previous section clearly demonstrate that the inductance gain due to the use of magnetic core comes with the cost of introducing the magnetic power losses at high frequencies. The magnetic power losses increase the device resistance with the frequency and result in the drop of the quality factor at sufficiently high frequencies. For a given ratio of permeability, μ''/μ', a larger inductance gain at low frequencies leads to a lower peak frequency for the quality factor as illustrated in Fig. 14.15(a). This can be understood as the trade-off between the inductance gain and the bandwidth of the magnetic inductor. When the frequency is high enough, the quality factor of the magnetic inductor becomes smaller than that of the air core inductor. The analytical models suggest that, at the frequency f_{MI}, the quality factor of the air core inductor is equal to the ratio of permeability, μ''/μ', if the magnetic power losses is the main loss mechanism at high frequencies (Fig. 14.15(a)). Hence f_{MI} can be considered as the useful bandwidth of the magnetic inductor.

Due to these trade-offs, the reported device properties tend to have either large inductance gain ΔL with large μ''/μ' or small ΔL with small μ''/μ'. The frequency-dependent quality factors of the two cases are illustrated in Fig. 14.15(b). While most of the reported device properties belong to these two cases, the case of large ΔL with small μ''/μ' leads to maximum peak quality

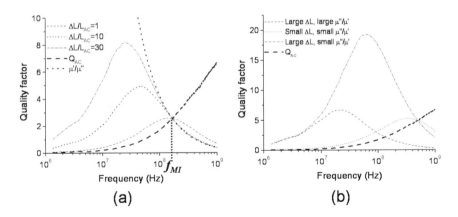

Fig. 14.15 Schematic plots of the quality factors for the embedded magnetic inductors

factors and can be realized based on the recent studies and understanding of the fundamental trade-offs for the magnetic inductors. Some of the ideas will be given below.

14.1.4.2 Direction for the Magnetic Inductor Designs

A design for the magnetic inductor can be chosen based on the applications and their requirements. A transmission line is good for the low resistance applications, but the achievable inductance values are limited. A spiral inductor has a limited inductance enhancement due to inefficient guidance of the magnetic flux, but it can be useful up to relatively high frequencies. A solenoid inductor can generate a large inductance gain and inductance density, but the magnetic power losses are introduced earlier in frequency. The solenoid design is relatively well understood magnetically as discussed in Section 14.1.2, and the analytical models can be used to optimize the design parameters.

Desirable device properties include high inductance, low coil resistance, and small device area consumption. However, there are also trade-offs between these parameters as illustrated in Fig. 14.16. The plots are obtained using the analytical models, and the Cu conductors with $N = 20$ and $t_C = 10\ \mu m$ and the

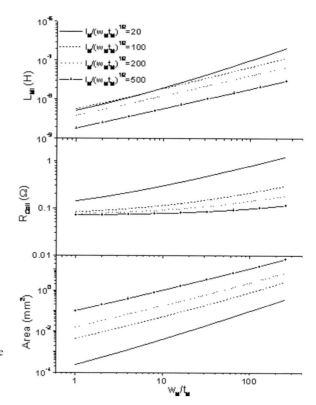

Fig. 14.16 Calculated device properties of the embedded solenoid inductors with magnetic core

magnetic core with $\mu_r = 1000$ and $t_M = 2$ μm are considered. While the effective permeability prefers to have a large $l_M/(w_M t_M)^{1/2}$ as shown in Fig. 14.7(a), it tends to decrease the inductance. A large $l_M/(w_M t_M)^{1/2}$ decreases the coil resistance but increases the area consumption. A large w_M/t_M can increase the inductance significantly, but increases the coil resistance and the area consumption as well. While it is not trivial to find an ideal solution producing all the desired properties, the analytical model can be used to find an optimal design for given requirements. In addition, a merit of the magnetic inductor is that the inductance gain can be traded for reduced coil resistance and device area consumption.

14.1.4.3 Desired Magnetic Core Properties

Low-frequency device properties and the physical properties of the device can be optimized and chosen by the design process shown above. However, in order to predict and control the frequency responses of the device properties, the magnetic properties of the core material need to be considered, since the magnetic power losses are the main loss mechanism for the magnetic inductors. In addition, the trade-offs discussed above are closely related to the magnetic core properties. The inductance gain is strongly dependent on the relative permeability, whereas the resistance increase and the bandwidth are closely related to the permeability ratio.

There are several parameters that can affect the frequency responses and illustrate the trade-offs. One is the anisotropic field H_K. The relative permeability is inversely proportional to H_K, while the ferromagnetic resonance (FMR) frequency is proportional to $H_K^{1/2}$ according to the Kittel equation [13]. By increasing H_K, the relative permeability and hence the inductance gain would be reduced. But the FMR loss is introduced at higher frequencies, hence reducing the ratio μ''/μ' at a given operation frequency. This effectively reduces the magnetic power losses and extends the bandwidth. The thickness of the magnetic core can also be a critical parameter to consider. A thicker magnetic core produces a larger inductance enhancement; however, it also generates more eddy current loss and increases μ''/μ' at a given frequency [17], thus increasing the device resistance. Another parameter is the structure of the magnetic core material. In comparison to the amorphous magnetic alloys, the granular magnetic materials have relatively high FMR frequencies and high electrical resistivities, thus reducing the magnetic losses significantly (Table 14.1). However, the nanogranular magnetic particles are not connected to one another, and hence their permeability values are relatively small.

A high inductance gain can be obtained by optimizing the design parameters based on the analytical models. Thick granular magnetic alloy or laminated structure using amorphous or granular magnetic films can be considered as the magnetic core material to reduce the magnetic power losses at high frequencies [48]. In addition, since the magnetic properties of the processed and patterned magnetic core are of importance, the fabrication processes need to be considered carefully to maintain the desired magnetic properties after the fabrication.

14.1.4.4 Potential Applications for the Integrated Magnetic Inductors

Embedded inductor has become one of the key passive elements in radio-frequency (RF) ICs, microwave applications, and mobile power delivery applications. The embedded inductor with magnetic core shows a great potential for applications in the integrated filters, power conversion, and EMI noise reduction, enabling the realization of the RF system-on-chip circuits. The device properties of the embedded magnetic inductor can be optimized for various applications and frequency ranges of interest. The embedded magnetic inductor can be especially ideal for power delivery applications whose switching frequency is about 10 MHz [49]. However, due to the intrinsic magnetic losses and the fundamental trade-offs of the magnetic inductors, the usefulness of the magnetic inductor is severely constrained at frequencies beyond 1 GHz. Novel magnetic materials and inductor designs with much reduced magnetic losses at GHz frequency range would be needed to extend the useful bandwidth of the magnetic inductor further into the GHz frequencies.

14.2 Embedded Capacitors

14.2.1 Embedded Capacitor Dielectrics Material Options

To meet the stringent materials requirements for dielectric materials to realize the embedded capacitor applications, considerable attention has been devoted to the research and development of the candidate high-k materials. To date, no one perfect dielectric materials has yet been identified for embedded capacitor applications because they all compromise on certain issues including electrical and mechanical performance, or processing constraints. However, a very wide range of material candidates are potentially useful.

14.2.1.1 Ferroelectric Ceramic Materials

Ferroelectric ceramic materials such as barium titanate ($BaTiO_3$), $BaSrTiO_3$ (barium strontium titanate), $PbZrTiO_3$ (lead zirconate titanate) etc., have been used as dielectric materials for decoupling capacitors because this type of materials possess high-k up to thousands [1, 50]. By far the highest specific capacitances exceeding 1800 nF/cm^2 are achievable with these materials. However, very high processing temperature in excess of 600°C required for sintering makes it unsuitable to process them directly into low cost organic boards. And the dielectric properties of ferroelectrics are typically a stronger function of temperature, frequency, film thickness and bias, which results in significant nonlinearities in their performance.

14.2.1.2 Ferroelectric Ceramic/Polymer Composites

Study on ferroelectric ceramic/polymer composites with high-k has also been actively explored as a major material candidate. The methodology of this approach is to combine the advantages from the polymers that meet the requirements for the low cost organic substrate process, i.e. low temperature processibility, mechanical flexibility and low cost, with the advantages from the ferroelectric ceramic fillers, such as desirable dielectric properties [51, 52, 53–59]. The major advantage of this type of materials lies in the processing since the high temperature steps required to reach high k from the ferroelectric phase can be done in advance of application to the organic substrate. However, some challenging issues in these polymer composites for high k applications have been addressed, such as limited dielectric constants and capacitance density, low adhesion strength resulting in air gaps and lowered capacitance. Most of the k values of polymer-ceramic composites developed to date are below 100 at room temperature due to the low k polymer matrix (usually in the range of 2–6). By employing polymer matrix with relatively high k, the k values of polymer-ceramic composites can be effectively enhanced, because the k of polymer matrix shows very strong influence on the k of the final composites [52, 57]. For instance, poly(vinylidene fluoride-trifluoroethylene) (P(VDF-TrFE)) copolymer, a class of relaxor ferroelectric, can have a relatively high room temperature k ($\sim$40) after irradiation treatment [60]. Bai et al. prepared Pb(Mg1/3Nb2/3)O3-PbTiO3/P(VDF-TrFE) composites with k values above 200 [5]. Rao et al. reported a lead magnesium niobate-lead titanate (PMN-PT) + BaTiO3/high-k epoxy system (effective k: 6.4) composite with k value about 150, in which ceramic filler loading as high as 85% by volume [61]. Studies have shown that the dielectric constant of 0–3 composites composite is dominated by the matrix, therefore, a relatively large volume fraction of high dielectric constant of the ferroelectric inorganic phase is needed. The high filler loading of ceramic powders is still the technical barriers for real application of polymer-ceramic composite in the organic substrate, because it results in poor dispersion of the filler within the organic matrix and almost no adhesion towards other layers in PCB as well due to the low polymer content. Table 14.3 summarizes the type, composition, and dielectric properties (room temperature values if not otherwise specified) of the ceramic/polymer composite material candidates for embedded capacitors.

14.2.1.3 Conductive Filler/Polymer Composites

Conductive filler/polymer composite is another approach towards ultra-high k materials for embedded capacitor application of next-generation microelectronic packaging. Ultra-high k values have been observed with conductive

Table 14.3 Summary of ceramic/polymer composite candidates

Materials	Dielectric Constant	Dissipation Factor	Filler Size	Filler Loading	Ref.
BaTiO$_3$/epoxy	40 (1 Hz)	0.035	100–200 nm	60 vol%	55
PZT/PVDF	50		20 μm	50 vol%	54
Pb(Mg$_{1/3}$Nb$_{2/3}$)O$_3$-PbTiO$_3$/ P(VDF-TrFE)	~200 (10 kHz)	0.1 (10 kHz)	0.5 μm	50 vol%	52
bimodal BaTiO$_3$/ epoxy	90 (100 kHz)	0.03 (100 kHz)	916 nm + 60 nm	75 vol%	60
PMN-PT + BaTiO$_3$/ high-k epoxy	~150 (10 kHz)		900 nm/50 nm	85 vol%	61
CaCu$_3$Ti$_4$O$_{12}$/ P(VDF–TrFE)	243 (1 kHz)	0.26 (1 kHz)		50 vol%	62
BaTiO$_3$/ P(VDF-HFP)	37 (1 kHz)	< 0.07 (1 MHz)	30–50 nm	50 vol%	63

filler/polymer composites when the concentration of the conductive filler is approaching the percolation threshold, which can be explained by the percolation theory for conductor-insulator percolation system [64]. The effective dielectric constant of a percolation system near critical filer loading can be described as [63]:

$$\varepsilon = \varepsilon_D/|f - f_c|^q \tag{14.13}$$

where f and f_c are the concentration and the percolation threshold concentration of the conductive filler within the polymer matrix, respectively; ε_D is the dielectric constant of the polymer matrix; and q is a scaling constant related to the material properties, microstructure, and connectivity of the phases in the conductive filler/polymer system [65]. Sometimes the effective dielectric constant of the metal-insulator composite could be three or four orders higher than the dielectric constant of the insulating polymer matrix. This phenomenon can be interpreted in terms of a "supercapacitor network" with very large area and small thickness: when the concentration of the metal is close to the percolation threshold, large amount of conducting clusters are in proximity to each other but they are insulated by thin layers of dielectric material. This percolative approach requires much lower volume concentration of the filler compared to traditional approach of high dielectric constant particles in a polymer matrix. Therefore, this material option represents advantageous characteristics over the conventional ceramic/polymer composites, specifically, ultra-high k with balanced mechanical properties including the adhesion strength. Various metal particles or other conductive fillers, such as silver, aluminum, nickel, carbon black, have been used to prepare the polymer-conductive filler composites or three-phase percolative composite systems [64, 66–73]. High dielectric loss, low dielectric strength, and narrow processing

Table 14.4 Summary of the conductive filler/polymer composite candidates

Materials	Dielectric Constant	Dissipation Factor	Filler Size	Filler Loading	Ref.
Ag flake/epoxy	~1000 (10 kHz)	0.02 (10 kHz)	1.5 μm	11.23 vol%	64
Al/epoxy	109 (10 kHz)	0.02 (10 kHz)	3 μm	80 wt%	66
Ni-BaTiO$_3$/PVDF	300 (10 kHz)	0.5 (10 kHz)	Ni:0.2 μm, BT: 1 μm	Ni: 23 vol%, BT: 20 vol%	67
Ni-BaTiO$_3$/PMMA	150 (1 MHz)		Ni: 4 μm, BT: 1 μm	Ni: 12 vol%, BT: 20 vol%	68
Carbon black/epoxy	13000 (10 kHz)	3.5 (10 kHz)	~30 nm	15 vol%	69
Ag/carbon black/epoxy	2260 (10 kHz)	0.45 (10 kHz)	Ag: 13 nm	Ag: 3.7 wt%, carbon black: 30 wt%	70
Al/Ag-epoxy	160 (10 kHz)	0.045 (10 kHz)	Al: 3 μm, Ag: <20 nm	Al: 80 wt%	71
Ag/epoxy	~300 (1 kHz)	0.05 (1 kHz)	40 nm	22 vol%	72
Ag@C/epoxy	>300 (1 kHz)	< 0.05 (1 kHz)	80–90 nm core	25–30 vol%	73

window are technical barriers for this category of materials. Because the highly conductive particles are easy to form, a conductive path in the composite as the filler concentration approaches the percolation threshold. Currently, much work has been focused to solve these problems of the polymer-conductive filler composites and much progress has been made. Details of some conductive filler/polymer composite materials reported in recent years are summarized in Table 14.4.

14.2.2 New Concepts and Current Trend

14.2.2.1 Nanodielectrics

With the increased enthusiasm and activity toward the research on the nanotechnology, a new class of dielectric materials, nanodielectrics, is emerging. It is anticipated that nanocomposites are highly promising nanodielectrics [74]. Polymer composite materials based on nanoparticles, one category of nanocomposites, provide a potential solution to meet the present and future technological demand in terms of the good processibility and mechanical properties of polymers combined with the unique electrical, magnetic or dielectric properties of nanoparticles [75]. Additionally, nano-sized particles are preferred for high-k dielectric composite materials because they could help achieve thinner dielectric films leading to a higher capacitance density. Therefore, more nanoparticles of ceramic, metallic, or even organic semiconductor have been introduced to prepare high-k dielectric materials recently.

14.2.2.2 Approaches to Dielectric Performance Enhancement

Filler Size Effect

It is worth to mention there are several issues of nanoparticle-based dielectric composite materials that need to be addressed. Although finer particle size is required to obtain a thin dielectric film and increase the capacitance density, for ferroelectric ceramics, extremely fine particles may lead to the change of crystal structure from tetragonal, which results in the high permittivity, to cubic or pseudocubic. Generally, the tetragonality and hence the permittivity of ceramic particles decreases with the particle size. Uchino et al. [76] and Leonard et al. [77] found that the tetragonality of BaTiO3 powders disappears finally when the particle size decreases to approximately 100 nm and 60–70 nm, respectively. Cho et al. prepared BaTiO3/epoxy composite embedded capacitor films (ECFs) with average particle size of 916 nm (P1) and 60 nm (P2), the k values of ECFs made of P1 were higher than those made of P2. So, the coarser particle is more useful than the finer particle to obtain high k of ECFs using unimodal powder. But by adopting bimodal fillers, fine nanoparticle can effectively enhance the k values by maximizing packing density and removing the voids and pores formed in the dielectric films. A dielectric constant of about 90 was obtained at a frequency of 100 kHz using these two different size BaTiO$_3$ powders [60].

Controlled Dispersion

On the other hand, uniform dispersion of nanoparticles in nanocomposite materials is required because nanoparticle agglomerates will lead to undesirable electrical or materials properties. Therefore, dispersion of nanoparticles is an extremely important contributor for achieving improved dielectric properties and reproducibility. Addition of surfactant or dispersant such as phosphate esters can improve the dispersion of nanoparticles in polymer matrix and thereby the overall film quality and dielectric performance of the nanocomposites [78].

Chemical modification of nanoparticles is a useful approach to facilitate the dispersion of nanoparticles as well. For instance, Kim et al. reported that surface modification of BaTiO$_3$ and related perovskite-type metal oxide nanoparticles with phosphonic acid ligands leads to well-dispersed BaTiO$_3$/polymer nanocomposite films with high dielectric strength [63]. This methodology is straightforward and easily adapted to a wide range of systems by choosing appropriate ligand functionality. Another example is related with CuPc oligomer, a class of organic semiconductor materials with k as high as 105.

Zhang et al. prepared CuPc/P(VDF-TrFE) composites, which showed a k of 225 and a loss factor of 0.4 at 1 Hz at low applied-field [79]. The high dielectric loss is due to the long-range intermolecular hopping of electron. Wang et al. further chemically modified CuPc and bonded to P(VDF-TrFE) backbone to improve the dispersion of CuPc in polymer matrix. Compared to the simple blending method, the CuPc oligomer particulates in grafted sample are of relatively uniformly size in the range of 60–120 nm, which is about 5 times smaller than that of blended composite. Furthermore, dielectric loss was reduced and dielectric dispersion over frequency was weakened [80].

Control of Dielectric Loss for Conductive Filler/Polymer Nanocomposites

Conductive filler/polymer nanocomposites have been identified as a promising method to fulfill the material requirements for embedded capacitors. However, the dielectric loss of this type of materials is very difficult to control, because the highly conductive particles are easy to form a conductive path in the composite as the filler concentration approaches the percolation threshold. To solve this drawback, currently much work has been directed to the control of the dielectric loss of this system.

Core-shell structured filler was proposed to be utilized as fillers instead of using conductive filler directly because the non-conductive shell could serve as electrical barriers between the conductive cores to form a continuous interparticle-barrier-layer network and thus achieve high-k and low loss. The core-shell structure can be formed either pristine or by synthesis. Xu et al. developed high-k polymer composite materials using self-passivation Al as the filler. The self-passivated insulating aluminum oxide layer on the Al metallic core showed significant effects on dielectric properties of the corresponding composites. For composite containing 80 wt.% Al, a k of 109 and a low dielectric loss tangent of about 0.02 (@10 kHz were achieved [66]. Shen et al. reported new polymer composite containing core/shell hybrid particles, more specifically metal Ag cores coated by organic dielectric shells. The organic dielectric shells act as interparticle barriers to prevent the direct connection of Ag particles and facilitate the dispersion of fillers in the polymer matrix as well, leading to stable high-k (>300) and rather low dielectric loss tangent (<0.05) for the polymer dielectric [73].

The surfactant layer coated on the nanoparticle surfaces during nanoparticle synthesis could also serve as a barrier layer to prevent the formation of conduction path to control the dielectric loss. Qi et al. reported a Ag/epoxy nanocomposite with 22 vol.% of Ag possessing a high-k of 308 and a relatively low dielectric loss of 0.05 at a frequency of 1 kHz [72]. The 40 nm Ag nanoparticles coated with a thin layer of mercaptosuccinic acid were randomly distributed in the polymer matrix. As displayed in Fig. 14.17, the k and dielectric loss increase with the filler concentration up to 22 vol.%. The decrease of k after that point is not due to conduction, but attributed to the porosity caused by the absorbed surfactant layer and solvent residue, especially at a higher Ag

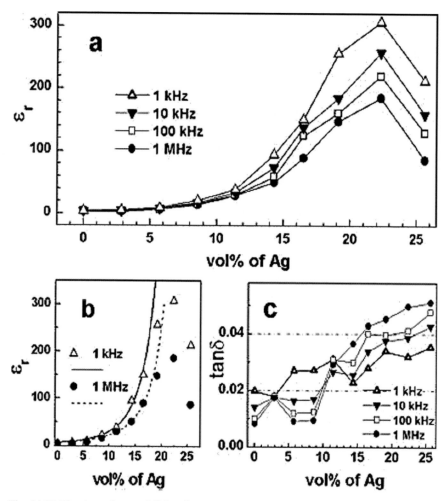

Fig. 14.17 The dependence of dielectric constant and dielectric loss tangent values on silver volume fraction and frequency [72]

content. In addition, no rapid increase of the dielectric loss tangent values was observed. Therefore, the observed highest k value was not considered as a real percolation threshold and the formation of a conducting filler network was prevented by the surfactant coating layer.

Surface modification of nanoparticles with organic molecules was employed to change the surface chemistry of nanoparticles and thus interaction between nanoparticles and polymer matrix. The surface coating layer on the nanoparticles via surface modification was demonstrated to decrease the dielectric loss and enhance the dielectric breakdown strength, which can be attributed to the interparticle electrical barrier layer formed via surface modification of nanoparticles preventing the metal cores from direct contact. Therefore, surface modification of nanoparticles is believed to be an effective approach to adjust the electrical features at the nanoparticle surface and the interface between the nanoparticle and the polymer matrix, and thus tailor the corresponding property of interest of nanocomposites [81, 82].

Another novel approach is to take advantage of the unique properties of the metal nanoparticles to control the dielectric loss of the conductive filler/polymer composite. Lu et al. reported incorporation of ultra-fine sized Ag nanoparticles in the Ag/carbon black/epoxy nanocomposites increased the k value and decreased the dielectric loss tangent as shown in Fig. 14.18. The remarkably increased k of the nanocomposites was due to the piling of charges at the extended interface of the interfacial polarization-based composites. The reduced dielectric loss might be due to the Coulomb blockade effect of the containing Ag nanoparticles, a well-known quantum effect of metal nanoparticles. In addition, the presence of the capping agent and its ratio with respect to the metal precursor were found to have great effect on the size and size distribution of the synthesized Ag nanoparticles in the nanocomposites (see Fig. 14.19). Smaller size and narrower size distribution of Ag nanoparticles resulted in more evident Coulomb blockade effect and thereby reduced dielectric loss [70].

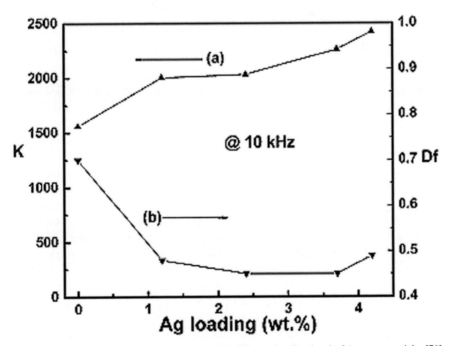

Fig. 14.18 Variation of k and Df at 10 kHz with different loading level of Ag nanoparticles [70]

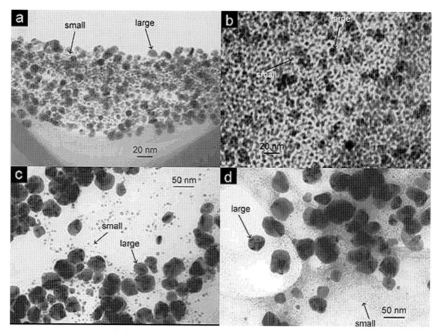

Fig. 14.19 Transmission electron microscopy (TEM) micrographs of Ag nanoparticles in the epoxy matrix in the presence of a capping agent with [capping agent]/[Ag precursor] ratio (**a**) $R = 1$, (**b**) $R = 0.6$, (**c**) $R = 0.4$ and (**d**) $R = 0.2$ [70]

14.2.3 Conclusions

Generally, high-k dielectric materials which can be realized for embedded capacitor applications are required to have high dielectric constant, low dissipation factor, high thermal stability, simple processibility, and good dielectric properties over broad frequency range. However, no such ideal materials that simultaneously satisfy the above-mentioned prerequisites have been realized at present. Nanocomposite materials have the potential to meet both present and future technological demands, and these materials have been studied extensively. A wide variety of research efforts have been directed to achieve high dielectric performance for the materials candidates, that is, to maximize the capacitance density, obtain high dielectric constant, and suppress the dielectric loss. Several techniques to improve the overall dielectric properties of these materials include: (i) optimized formulation of dielectric materials with high filler loading of high dielectric constant ceramics for ceramic-polymer nanocomposites and appropriate loading level of conductive fillers in the neighborhood of percolation threshold for conductive filler-polymer nanocomposites; (ii) improvement in morphology of dielectric materials, such as filler size and distribution, packing, and dispersion in the polymer matrix; (iii) appropriate processing methods to obtain thinner dielectric films; (iv) modification of the filler interface to facilitate dispersion in the polymer matrix and suppress the dielectric loss of the composite materials.

14.3 Embedded Resistors

14.3.1 Introduction

Resistors have several applications in high frequency circuits including use as attenuators, terminations, power dividers, and oscillators. For server applications, the resistors are used primarily for pull up and pull down applications and require resistance values in the range of 1 K-ohms to 40 K-ohms. The widely implemented termination resistors require resistance values 50 ohms and below. However, there are issues with implementing embedded resistors such as standardization, process yield and tolerance, competition with SMT, and cost. Discretes are not at a stalemate. Surface mount components are also moving toward better specifications and smaller footprints. Sizes as small as 0201 are in use, 01005 is on the horizon.

Embedded resistors are implemented either by thick film or thin film processes. Thick film resistors are typically metal oxides applied using a screen printing or an equivalent process [83, 84]. Such deposition processes tend to leave defects in the film that can lead to instability over time. Thin film processes use a variety of metal alloys deposited by sputtering process to produce a very uniform film on the order of a few thousand Angstroms. The later method is preferred for microwave resistors, however, the vacuum requirement of the sputtering process limits the sputtered area, thus effectively increase per unit cost of thin film resistors. An alternative to this approach involves the use of commercially available resistive films on a carrier substrate [85–88]. In contrast to surface mount components, integration of resistive components within the substrate can reduce the system volume and mass, eliminate the need for some discrete components and assembly, enhance electrical performance and reliability, and reduce the overall implementation cost [89, 90].

The strategic parameters for embedded resistors are listed below:

a) Low Temperature Coefficient of Resistance (TCR) (<50 ppm/C)
b) High frequency stability up to 40 GHz
c) Lower cost
d) PCB (printed circuit board) -compatible materials and processes
e) Higher power handling (>10 Watts/cm^2)
f) Higher reliability (ΔR <5%) under JEDEC standard
g) Wide spectrum of sheet resistance (5 ohms/sq to >1 kohms/sq)
h) Co-integration with L and C in one layer
i) Improve process tolerance to 10–15% without trim and 1–5% with trim
j) Higher process yield >99%

However, there are several barriers toward achieving the target. Some of them are discussed below.

14.3.2 Technology Barriers

14.3.2.1 Design

Resistors are implemented by designing individual components, followed by designing components into a system [90]. For the layout and artwork generation, most resistors can be shaped and located according to simple design rules, however, automated software oftentimes is limited to checking minimum size and clearances around the buried object. The lack of proper CAD/CAM systems with electrical modeling and simulation capability could be a major obstacle to using embedded resistors for frequency sensitive applications as cost and time-to-market considerations restrict "trial and error" designs iterations.

The ability to achieve better than 1% precision values is critical in new high-end systems (i.e. super servers and workstations) that are pushing data bus performance >2 GHz clock rates [90]. In these systems the resistor must be nearly "ideal" and must accurately match the impedance of the trace on the board to prevent signal reflection and ensure clean signal integrity. Modeling is becoming more important as speeds increase and board designs need to become more sophisticated.

14.3.2.2 Materials Specification

According to existing roadmaps [90], the application range for resistors is extremely wide (~1 ohm to >200 M-ohms). No single material is available today that can cover this entire resistance range with required properties. Thin film materials have good stability but are limited to low resistivity and moderate tolerance. Therefore, material development is critically needed for applications requiring higher R values (which most mobile products require) and tighter tolerances better than 10%. Tolerances lower than 5% will require resistor trimming, thereby adding to the overall cost that can be significant. Temperature coefficient of resistance (TCR) is also an important parameter that needs to be lower than 50 ppm/C which is not currently available for higher resistance values. The power density of the available materials is rated at 10 W/in^2 or more, which is adequate in the near term, but as densities increase, this power density may become inadequate.

14.3.2.3 Process Optimization

Resistive materials can be broadly classified as thick film and thin film. Thin films are usually deposited on a substrate to build up to a desired thickness for a specific sheet resistivity. These resistors are patterned by subtractive technology, a process that includes a sequence of print and etch operations. Thin film resistors offer superior performance and reliability of a metallic conductor, however, the span of resistor values on a single layer is limited by its having a single sheet resistivity. Serpentine patterns are used to create high value resistors [90]. Partial square designs are used for terminating resistors within the footprint of an array package. Occasionally, more than one layer may be required to accommodate both low and high value resistors. Thick film resistors are usually formed by additive processes and have the advantage of high sheet resistivities greater than 1 k-ohms per square. Additive processes offer multiple sheet resistivities on the same layer, however, the thick film polymers lack the stability exhibited by thin film metallic alloys.

Several materials in the research and development stage require processes new to circuit board fabricators. Yield loss per device must be extremely small because embedded resistors cannot be repaired [90]. Multi-layer structure with embedded resistors requires different materials, and, therefore, it's necessary to have them processed without significant deviation in tooling. As the number of layer increases, each layer becomes progressively more expensive than the preceding layer, since significant investment has already been made to complete fabrication of the preceding layers. Thus, it is imperative to come up with materials and processes particularly for R, L, and C within the same layer, thus reducing the layer count and decreasing the cost.

Rapid Prototyping is perhaps the most serious bottleneck. The turnaround time for a revision of resistor values and placement can be as short as 72 hours [90]. Test methodology to sort out any failure effectively during and after manufacturing is another barrier since the components are hidden within the core.

The assembly temperatures are much higher in lead free assembly. The resistors need not be affected by the short time exposure at temperatures >250°C. It has not been a problem for thin film resistors, but, for polymer thick film (PTF), the higher assembly temperature may be an issue.

14.3.3 Fundamental of Resistors

A resistor controls electric current by resisting the flow of charge through itself [91]. Usually, it contains a strip of the resistive material with two conducting pads at the ends as shown in Fig. 14.20.

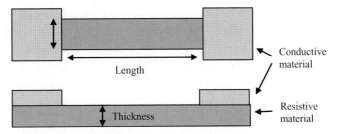

Fig. 14.20 Typical resistor geometry

The unit of resistance is ohm (Ω), and it measures how well it resists or opposes the flow of current. It is calculated by using Eq. (14.14).

$$R = \frac{\rho L}{Wd} \quad (14.14)$$

R is the resistance (Ω), ρ is the resistivity of material (Ω-cm), L is the length of the strip (cm), W is the width of the strip (cm), and d is the thickness of the strip (cm). As shown in Eq. (14.1), resistance is dependent on the resistivity of the material and the dimensions of the strip. Higher resistance can be achieved by using higher resistivity materials, increasing the length, and using smaller cross-sections. The resistivity is an intrinsic material property that is dependent on the composition and microstructure of the material.

Sheet resistance is the resistance of a square strip. It is another common way of expressing the resistance as shown in Eq. (14.15).

$$R = \left(\frac{\rho}{d}\right)\left(\frac{L}{W}\right) = R_s N_s \quad (14.15)$$

R_s is the sheet resistance (Ω/sq) and N_s is the number of squares. As long as the ratio of L and W or the number of squares remains same, the resistance value does not change. The decision of how big each square would be will depend on several parameters, such as available real estate, heat dissipation, tolerance, parasitic capacitance, standing waves and internal reflections, and reflections at the resistor and interconnect interface [90, 91].

Usually, heat dissipation and tolerance are the most significant parameters. As power is applied and current flows through, most of the energy is dissipated as heat. If too much heat is generated, then resistors can be permanently damaged. Therefore, large area should be utilized to improve heat dissipation and also tolerance. Other parameters, such as parasitic capacitance, can be a problem for high frequency applications. Long serpentine resistors should be avoided in order to minimize the parasitics. In most cases, resistor footprint should utilize large area with minimum number of squares to achieve the optimal performance and reliability.

The Temperature Coefficient of Resistance (TCR) and Voltage Coefficient of Resistance (VCR) measure change in resistance with respect to change in temperature and voltage, respectively, as shown in Eqs. (14.16) and (14.17).

$$\text{TCR} = \frac{1}{R_{T1}} \frac{R_{T2} - R_{T1}}{T_2 - T_1} \quad (14.16)$$

$$\text{VCR} = \frac{1}{R_{v1}} \frac{R_{v2} - R_{v1}}{V_2 - V_1} \quad (14.17)$$

For most applications, it is desirable to have near zero TCR and VCR, and thus achieving stable resistance throughout the operating conditions. Some applications such as thermistors, however, exploit the changes in resistance with temperature and require high TCR. Also, TCR is sometimes purposely made to a certain value to offset an existing material or component that is temperature dependent. For instance, capacitance in RC network usually increase with temperature, so negative TCR can be implemented to maintain a steady RC time constant. TCR is usually measured between −55 and 125°C, and VCR is usually measured between 5 V and 50 V.

14.3.4 Materials and Processing Technologies

The materials systems used for embedded resistors are generally categorized as metals, alloys, semiconductors, cermets, and polymer thick films [91]. From processing standpoint, these material systems are classified as thin film, printable, and plated. Resistive alloys, ceramic-metal nanocomposites, and carbon filled polymers [92] are some examples of classic resistive materials. Table 14.5 shows some examples of the current state-of-the-art on embedded resistors materials [83, 90].

Most of the resistive alloys are useful for low end resistance values. They are usually sputtered, but they can be also electrolytically or electrolessly plated. Among resistive alloys, NiCr, NiCrAlSi, CrSi, TiN_xO_y, and TaN_x are potential candidates [89]. NiCr and NiCrAlSi foils are commercially available from Gould (Ticer) Electronics Inc. They can provide sheet resistance between 25 and 250 Ω/sq with relatively low TCR. TaN_x is another attractive resistive alloy formed by reactive sputtering of Ta in a nitrogen atmosphere and can achieve stable resistivities up to 250 $\mu\Omega$-cm with TCR of around −75 ppm/°C [93]. Sputtered TiN_xO_y offers relatively higher resistivities up to 5 $k\Omega$/sq with TCR of +/−100 ppm/°C [94]. Ohmega-Ply® and MacDermid's M-Pass™ are some of the mature commercialized Ni alloy based plated resistor that can provide up to 250 Ω/sq and 100 Ω/sq respectively. Insite™ (Rohm and Haas) is processed by doping titanium on to Cu using combustion chemical vapor deposition. The resulting foil can provide sheet resistances of 500 and 1,000 Ω/sq with material tolerance of 10% and TCR of 200 ppm/°C. Patterning of resistors involves simple print and etch steps.

High end resistances above 100 $k\Omega$ can be achieved by ceramic-metal nanocomposites also known as a cermet [83]. These resistors are commonly used in ceramic packages, but they can also be sputtered at a relatively low temperature for organic packages. The most commonly used is Cr-SiO. Depending on the ratio, Cr-SiO can achieve up to 10 $m\Omega$-cm with near zero TCR and good stability [83]. DuPont Interra™ ceramic thick film resistor is based on lanthanum boride (LaB_6) material that can achieve up to 10 $k\Omega$/sq with TCR of +/−200 ppm/°C. This material has been used for many years in ceramic packages, and it is known to be highly stable and reliable. The fabrication of the resist foil begins with conditioning the copper foil with thin layer of copper/glass paste to increase the adhesion between the Cu and LaB_6. Then, LaB_6 paste

Table 14.5 Examples of Embedded Resistors [83, 89]

Industry/Institution	Materials	Approach	Range of values ohms/sq	TCR (ppm/°C)
Intarsia			10–100	
Boeing	Ta_2N		20	
NTT		Sputter		(+/−100)
GE			25–125	(−75 to −100)
Osaka University				
Metech		Polymer thick film process	Insulating to conducting	
Acheson Colloids	Conductive Polymer composites			
Electra				
Ashai Chemical				
W R Grace				
DOW Corning				
Raychem Corporation				
Ormet Corporation				
Ohmega Ply	NiP alloy	Electroplate	25–500	
Singapore Inst. Microelectronics	TaSi	DC Sputter	10–40	
LSI Logics			8–20	
University of Arkansas/Sheldahl	CrSi	Sputter		−40
W. L. Gore and Associates	TiW	Sputter	2.4–3.2	
Shipley	Doped Pt on Cu foil	PECVD	Up to 1000	100
Deutsche Aerospace	NiCr	Sputter	35–100	
GOULD Electronics	NiCr NiCrAlSi		25–100	
Georgia Institute of Technology	Ni-W-P		10–500	
MacDermid	NiP	Electroless plate	25–100	
DuPont	LaB_6	Screen print and foil transfer	Up to 10 K	+/−200

is screen printed and fired onto the Cu foil at 900°C to activate the resist material. The resulting fired film thickness is 14–18 μm, and it could exhibit material tolerance of 15% before trimming.

Polymer thick films (PTF) are metal or carbon filled particles in liquid resin targeted to a specific sheet resistance. There are many suppliers including W R Grace and Ashai Chemicals. They provide wide range of resistances from 1 Ω/sq to 10^7 Ω/sq at a relatively low cost. They are commonly available in viscous liquid form that can be easily screen printed or stenciled, and they have relatively low curing temperature [95]. Some of drawbacks are, however, tolerance, stability, and reliability. The oxidation between polymer and copper interface can cause drift in resistance values, and they are vulnerable to delamination or cracking due to CTE mismatch [83]. An example of the lift off process for deposition of carbon filled resistive ink on is shown in Fig. 14.21.

Thin film resistors can also be realized by direct electroless plating that can be adopted in the PCB manufacturing industries with no additional investment. Electroless plating onto a nonconductive substrate requires the surface to be sensitized and then activated. Conventionally, tin chloride ($SnCl_2$) and palladium chloride ($PdCl_2$) dissolved in diluted hydrochloric acid (HCl) are used as sensitizers and activators, respectively [96–99]. Acid hypophosphite-based baths are more commonly used due to low pH that most polymers can withstand. The electroless process offers uniform resistor thickness in the sub-micron range, low profile, and excellent adhesion. The measured sheet resistance values are in the range of 10–1000 ohms/sq and the thickness of the Ni-alloy deposit is of the order of 2000–5000 A (angstrom). Plating can also be done at room temperature without any appreciable change in bath pH [100].

Electroless plated resistors on epoxy surface has been optimized and commercial products are available from MacDermid [1]. Chahal et al. [98] characterized electroless plated NiP and NiWP on epoxy surface at frequencies up to 15 GHz as shown in Fig. 14.22. The high frequency measurements were performed with an HP 8510C vector network analyzer and ground-signal-ground (G-S-G) coplanar waveguide 200 μm pitch probes. The measured results of structures with different sheet resistivities and resistance values are presented in Fig. 14.22b. The temperature coefficient of resistance in NiWP was near zero thus providing a great benefit to the circuit designers. Electroless plated NiP and NiWP resistors have also been realized on liquid crystal polymer (LCP) and Benzocyclobutene (BCB) substrates [99].

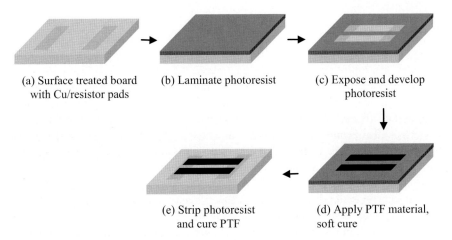

(a) Surface treated board with Cu/resistor pads
(b) Laminate photoresist
(c) Expose and develop photoresist
(e) Strip photoresist and cure PTF
(d) Apply PTF material, soft cure

Fig. 14.21 Illustration of a lift off process to realize PTF resistors on PWB [95]

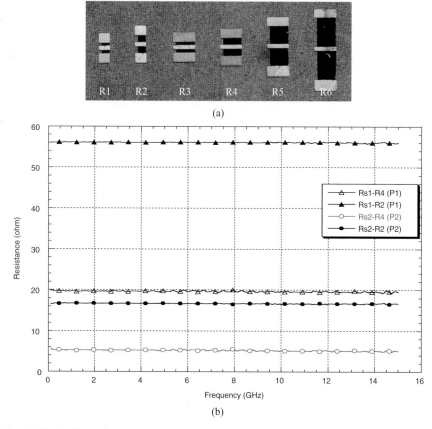

Fig. 14.22 (**a**) Photomicrograph of G-S-G resistor structures, NiP/NiWP resistor film in dark, (**b**) measurement results [98]

14.3.5 Thin Film Resistors on LCP for RF Applications

The key advantages of liquid crystal polymer (LCP) in comparison to other microwave organic dielectrics are low loss (tan δ = 0.002–0.005) up to mm-wave frequency range, near hermetic nature (water absorption <0.04%), relatively lower cost, and coefficient of thermal expansion that can be matched close to Silicon or GaAs as well as printed wiring board. Also, LCP is a flexible material, which allows for the realization of conformal RF modules in non-orthogonal and non-planar surfaces. In addition, multilayer circuits are possible with LCP due to two types of LCP material with different melting temperatures. Thus, vertically integrated designs can be realized increasing the functionality and space savings. LCP can provide organic solutions for applications in the range of 2 to 75 GHz [87].

Laminated thin film resistors on LCP were designed, fabricated, and characterized by Horst et al. [86–88]. Several GSG structures were fabricated using the

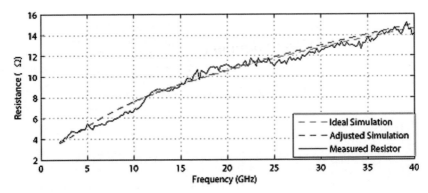

Fig. 14.23 Resistance vs. frequency plot [86]

process provided by the manufacturer. Authors used HFSS (ANSOFT) field solver model used to provide a model for the measurement values. Figure 14.23 shows both an ideal and an adjusted simulated value across frequency from 2 to 40 GHz. The ideal values use resistances calculated from the nominal dimensions. The adjusted values eliminate this variable by substituting the measured dimensions of the fabricated resistor to calculate the resistance value. The smaller resistors whose dimensions are more difficult to etch accurately have a higher deviation from their nominal value, but, once this variable is removed from the measurements, the resistors behave close to the predicted model.

Authors [86] simulated different termination topologies to provide a 50 Ω load with the smallest parasitic response across the broadest range of frequencies. The coplanar waveguide (CPW) topology was determined to provide the best response. Attenuators were designed and fabricated using the resistor foils. Wilkinson power dividers were designed across a wide range of frequencies from X band to W band. At V-band, the circuit yielded 0.3-dB excess insertion loss, 19-dB isolation, and 50% bandwidth. At the W-band, the circuit measured 0.75-dB excess insertion loss, 24-dB isolation, and 39% bandwidth [88]. Grzyb et al reported Wilkinson power divider, balun, hand-pass filter and branch-line coupler with thin film NiCr resistor and Ta_2O_5 capacitors deposited on low loss LCP and BCB substrates. Their designed elements show very good performance and agreement with the full-wave simulations [101].

14.3.6 Conclusions

In this section, design, packaging and materials aspect of embedded resistors have been addressed. Thin film resistors play an important role toward miniaturization of electronic systems. Currently, the embedded resistor technology is sufficiently matured compared to embedded capacitor and inductor. Nevertheless, new materials and process optimization would be required to meet the future technology needs.

Acknowledgments The authors wish to thank National Science Foundation, Intel Corporation, National Semiconductor, and Stanford Graduate Fellowship (L. Li) for the financial support which made this work possible. Helpful collaborations with and previous work by A. M. Crawford, D. Gardner, G. Vandentop, H. Braunisch, R. Nair, K-P Hwang, Y. Min, M. Mao, T. Schneider, and R. Bubber, and a software license from Ansoft are also gratefully acknowledged.

References

1. Ulrich RK, Schaper LW (2003) Integrated passive component technology. IEEE Press, Wiley-Interscience, Hoboken, NJ, USA
2. Prymark J, Bhattacharya S, Paik K, Tummala RR (2001) Fundamentals of microsystems packaging. McGraw-Hill, New York
3. www.johansontechnology.com; www.vishay.com.
4. Yue CP, Ryu C, Lau J, Lee TH, Wong SS (1996) A physical model for planar spiral inductors on silicon. IEEE Int Electron Devices Meeting, San Francisco, pp 155–158
5. Mohan SS, Yue CP, Hershenson M, Lee TH, Wong SS (1998) Modeling and characterization of on-chip transformers. IEEE Int Electron Devices Meeting, San Francisco, pp 531–534
6. Mohan SS, Hershenson M, Boyd SP, Lee TH (1999) Simple accurate expressions for planar spiral inductances. IEEE J Solid-state Circuits 34:1419–1424
7. Lee TH (2004) The design of CMOS radio-frequency integrated circuits, 2nd edn. Cambridge University Press, New York
8. Soohoo RF (1979) Magnetic thin film inductors for integrated circuit applications. IEEE Trans Magn 15:1803–1805
9. Gardner DS, Schrom G, Hazucha P, Paillet F, Karnik T, Borkar S (2007) Integrated on-chip inductors with magnetic films. IEEE Trans Magn 43:2615–2617
10. Prabhakaran S, Sullivan CR, Venkatachalam K (2003) Measured electrical performance of V-groove inductors for microprocessor power delivery. IEEE Trans Magn 39:3190–3192
11. Crawford AM, Gardner DS, Wang SX (2002) High-frequency microinductors with amorphous magnetic ground planes. IEEE Trans Magn 38:3168–3170
12. Viala B, Couderc S, Royet AS, Ancey P, Bouche G (2005) Bidirectional ferromagnetic spiral inductors using single deposition. IEEE Trans Magn 41:3544–3549
13. Zhuang Y, Rejaei B, Boellaard E, Vroubel M, Burghartz JN (2003) Integrated solenoid inductors with patterned sputter-deposited Cr/Fe10Co90/Cr ferromagnetic cores. IEEE Electron Device Lett 24:224–226
14. Lee DW, Hwang KP, Wang SX (2008) Fabrication and analysis of high-performance integrated inductor with magnetic core. IEEE Trans Magn 44
15. Kittel C (1996) Introduction to solid state physics, 7th ed. Wiley, New York
16. O'Handley RC (1999) Modern magnetic materials: principles and applications. Wiley, New York
17. Riet EV, Roozeboom F (1997) Ferromagnetic resonance and eddy currents in high-permeable thin films. J Appl Phys 81:350–354
18. Lee DW, Wang SX (2006) Multiple magnetic resonances in permeability spectra of thick CoTaZr films. J Appl Phys 99:08F109-1-3
19. Yamaguchi M, Baba M, Arai KI (2001) Sandwich-type ferromagnetic RF integrated inductor. IEEE Trans Microwave Theory and Tech 2331–2335
20. Shirakawa K, Kurata H, Kasuya M, Ohnuma S, Toryu J, Murakami K (1993) Thin film inductor with multilayer magnetic core. IEEE Transl J Magn Jpn 169–176
21. Kurata H, Shirakawa K, Nakazima O, Murakami K (1994) Solenoid-type thin-film micro-transformer. IEEE Transl J Magn Jpn 9:90–94

22. Li L, Crawford AM, Wang SX, Marshall AF, Mao M, Thomas S, Bubber R (2005) Soft magnetic granular material Co-Fe-Hf-O for micromagnetic device applications. J Appl Phys 97:10F907-1-3
23. Shimada Y, Yamaguchi M, Ohnuma S, Itoh T, Li WD, Ikeda S, Kim KH, Nagura H (2003) Granular thin films with high RF permeability. IEEE Trans Magn 39:3052–3056
24. Li L (2007) Nanogranular soft magnetic materials and on-package integrated inductors. Ph.D thesis, Stanford University, Stanford, CA, USA
25. Sun NX, Wang SX, Silva TJ, Kos AB (2002) High-frequency behavior and damping of Fe-Co-N-based high-saturation soft magnetic films. IEEE Trans Magn 38:146–150
26. Ikeda K, Kobayashi K, Fujimoto M (2002) Multilayer nanogranular magnetic thin films for GHz applications. J Appl Phys 92:5395–5400
27. Ohnuma S, Kobayashi N, Masumoto T, Mitani S, Fujimori H (1999) Magnetostriction and soft magnetic properties of $(Co_{1-x}Fe_x)$-Al-O granular films with high electrical resistivity. J Appl Phys 85:4574–4576
28. Thompson MT (1999) Inductance calculation techniques – Part II: Approximations and handbook methods. Power Control and Intelligent Motion 25:40–45
29. Li L, Lee DW, Wang SX, Hwang KP, Min Y, Mao M, Schneider T, Bubber R (2007) Tensor nature of permeability and its effects in inductive magnetic devices. IEEE Trans Magn 43:3168–3170
30. Ansoft Corporation (2007) Ansoft student licensing program, Pittsburg
31. Lee DW, Wang SX (2008) Effects of geometries on permeability spectra of CoTaZr magnetic cores for high frequency applications. J Appl Phys 103:07E907-1-3
32. Chen DX, Pardo E, Sanchez A (2002) Demagnetizing factors of rectangular prisms and ellipsoids. IEEE Trans Magn 38:1742–1752
33. Chen DX, Pardo E, Sanchez A (2005) Demagnetizing factors for rectangular prisms. IEEE Trans Magn 41:2077–2088
34. Riet EV, Klaassens W, Roozeboom F (1997) On the origin of uniaxial anisotropy in nanocrystalline soft-magnetic materials. J Appl Phys 81:806–814
35. Li L, Wang SX, Hwang KP, Min Y, Mao M, Schneider T, Bubber R (2006) Package compatibility and substrate dependence of granular soft magnetic material CoFeHfO developed by reactive sputtering. J Appl Phys 99:08M301-1-3
36. Li M, Wang GC, Min HG (1998) Effect of surface roughness on magnetic properties of Co films on plasma-etched Si(100) substrates. J Appl Phys 83:5313–5320
37. Harrington RF (1961) Time-harmonic electromagnetic fields. McGraw-Hill, New York
38. Fukuda Y, Inoue T, Mizoguchi T, Yatabe S, Tachi Y (2003) Planar inductor with ferrite layers for DC-DC converter. IEEE Trans Magn 39:2057–2061
39. Brandon EJ, Wesseling E, White V, Ramsey C, Del Castillo L, Lieneweg U (2003) Fabrication and characterization of microinductors for distributed power converters. IEEE Trans Magn 39:2049–2056
40. Yamaguchi M, Bae S, Kim KH, Tan K, Kusumi T, Yamakawa K (2005) Ferromagnetic RF integrated inductor with closed magnetic circuit structure. IEEE MTT-S Int Microwave Symp Digest, Long Beach, pp 351–354
41. Frommberger M, Schmutz C, Tewes M, McCord J, Hartung W, Losehand R, Quandt E (2005) Integration of crossed anisotropy magnetic core into toroidal thin-film inductors. IEEE Trans Microw Theory Tech 53:2096–2100
42. Orlando B, Hida R, Cuchet R, Audoin M, Viala B, Pellissier-Tanon D, Gagnard X, Ancey P (2006) Low-resistance integrated toroidal inductor for power management. IEEE Trans Magn 42:3374–3376
43. Lee DW, Hwang KP, Wang SX (2008) Design and fabrication of integrated solenoid inductors with magnetic cores. 58th Electronic Components and Technology Conference, Lake Buena Vista, pp. 701–705
44. Brandon J, Wesseling E, Chang V, Kuhn W (2003) Printed microinductors and flexible substrates for power applications. IEEE Trans Comp Package Technol 26: 517–523

45. Waffenschmidt E, Ackermann B, Wille M (2005) Integrated ultra thin flexible inductors for low power converters. IEEE 36th Power Electronics Specialists Conf. (PESC '05), Recife, pp 1528–1534
46. Sato F, Ono T, Wako N, Arai S, Ichinose T, Oba Y, Kanno S, Sugawara E, Yamaguchi M, Matsuki H (2004) All-in-one package ultracompact micropower module using thin-film inductor. IEEE Trans Magn 40:2029–2031
47. Li L, Lee DW, Hwang KP, Min Y, Hizume T, Tanaka M, Mao M, Schneider T, Bubber R, Wang SX. Small Resistance and High Q Magnetic Integrated Inductors on PCB. Submitted to IEEE Trans Adv Pack
48. Li L, Lee DW, Mao M, Schneider T, Bubber R, Hwang KP, Min Y, Wang SX (2007) High-frequency responses of granular CoFeHfO and amorphous CoZrTa magnetic materials. J Appl Phys 101:123912-1-4
49. Ghahary A (2004) Fully integrated DC-DC converters. Power Electronics Technology:24–27
50. Tohge N, Takahashi S, Minami T (1991) Preparation of $PbZrO_3$-$PbTiO_3$ ferroelectric thin films by the sol-gel process. J Am Ceramic Soc 74(1):67–71
51. Gregorio R, Cestari M, Bernardino FE (1996) Dielectric behavior of thin films of beta-PVDF/PZT and beta-PVDF/$BaTiO_3$ composites. J Mater Sci 31:2925–2930
52. Bai Y, Cheng ZY, Bharti V, Xu HS, Zhang QM (2000) High-dielectric-constant ceramic-powder polymer composites. Appl Phys Lett 76:3804–3806
53. Mazur K (1995) Polymer-ferroelectric ceramic composites in ferroelectric polymers: chemistry, physics, and applications. In: Nalwa HS (ed) Marcel Dekker Inc., New York
54. Dasgupta DK, Doughty K (1988) Polymer-ceramic composite materials with high dielectric constants. Thin Solid Films 158:93–105
55. Liang S, Chong S, Giannelis E (1998) Barium titanate/epoxy composite dielectric materials for integrated thin film capacitors. Proceedings of 48th Electronic Components and Technology Conference, pp 171–175
56. Windlass H, Raj PM, Balaraman D, Bhattacharya SK, Tummala RR (2001) Processing of polymer-ceramic nanocomposites for system-on-package applications. Proceedings of the 51st Electronic Components and Technology Conference, pp 1201–1206
57. Rao Y, Ogitani S, Kohl P, Wong CP (2002) Novel polymer-ceramic nanocomposite based on high dielectric constant epoxy formula for embedded capacitor application. J Appl Polymer Sci 83:1084–1090
58. Dang ZM, Lin YH, Nan CW (2003) Novel ferroelectric polymer composites with high dielectric constants. Adv Mater 15:1625–1629
59. Cho SD, Lee JY, Hyun JG, Paik KW (2004) Study on epoxy/$BaTiO_3$ composite embedded capacitor films (ECFs) for organic substrate applications. Mater Sci Eng B 110(3):233–239
60. Zhang QM, Bharti V, Zhao X (1998) Giant electrostriction and relaxor ferroelectric behavior in electron-irradiated poly(vinylidene fluoride-trifluoethylene) copolymer. Science 280:2101–2104
61. Rao Y, Wong CP (2004) Material characterization of a high-dielectric-constant polymer-ceramic composite for embedded capacitor for RF applications. J Appl Polymer Sci 92:2228–2231
62. Arbatti M, Shan XB, Cheng ZY (2007) Ceramic-polymer composites with high dielectric constant. Adv Mater 19:1369–1372
63. Kim P, Jones SC, Hotchkiss PJ, Haddock JN, Kippelen B, Marder SR, Perry JW (2007) Phosphonic acid-modified barium titanate polymer nanocomposites with high permittivity and dielectric strength. Adv Mater 19:1001–1005
64. Rao Y, Wong CP, Xu J (2005) Ultra high k polymer metal composite for embedded capacitor application. US Patent 6864306
65. Pecharroman C, Moya JS (2000) Experimental evidence of a giant capacitance in insulator-conductor composites at the percolation threshold. Adv Mater 12:294–297

66. Xu J, Wong CP (2005) Low loss percolative dielectric composite. Appl Phys Lett 87:082907
67. Dang ZM, Shen Y, Nan CW (2002) Dielectric behavior of three-phase percolative Ni–BaTiO$_3$/Polyvinylidene fluoride composites. Appl Phys Lett 81:4814–4816
68. Choi HW, Heo YW, Lee JH, Kim JJ, Lee HY, Park ET, Chung YK (2006) Effects of BaTiO$_3$ on dielectric behavior of BaTiO$_3$-Ni-polymethylmethacrylate composites. Appl Phys Lett 89:132910
69. Xu J, Wong CP (2004) Super high dielectric constant carbon black-filled polymer composites as integral capacitor dielectrics. Proceedings of the 54th IEEE Electronic Components and Technology Conference, Las Vegas, NV, USA, pp 536–541
70. Lu J, Moon KS, Xu J, Wong CP (2006) Synthesis and dielectric properties of novel high-K polymer composites containing in-situ formed silver nanoparticles for embedded capacitor applications. J Mater Chem 16(16):1543–1548
71. Lu J, Moon KS, Wong CP (2006) Development of novel silver nanoparticles/polymer composites as high k polymer matrix by in-situ photochemical method. IEEE Proceedings of the 56th Electronic Components and Technology Conference, San Diego, CA, pp 1841–1846
72. Qi L, Lee BI, Chen S, Samuels WD, Exarhos GJ (2005) High-dielectric-constant silver-epoxy composites as embedded dielectrics. Adv Mater 17:1777–1781
73. Shen Y, Lin Y, Li M, Nan C-W (2007) High Dielectric performance of polymer composite films induced by a percolating interparticle barrier layer. Adv Mater 19:1418–1422
74. Frechette MF, Trudeau ML, Alamdari HD, Boily S (2004) Introductory remarks on nanodielectrics. IEEE Transactions on Dielectrics and Electrical Insulation 11:808–818
75. Nicolais L, Carotenuto G (2005) Metal-polymer nanocomposites. John Wiley & Sons, Inc., Hoboken, New Jersey, USA
76. Uchino K, Sadanaga E, Hirose T (1989) Dependence of the crystal-structure on particle-size in BaTiO$_3$. J Am Ceramic Soc 72:1555–1558
77. Leonard MR, Safari A (1996) Crystallite and grain size effects in BaTiO$_3$. Proceedings of the IEEE 10th International Symposium on Ferroelectric Applications 2:1003–1005
78. Bhattacharya SK, Tummala RR (2000) Next generation integral passives: materials, processes, and integration of resistors and capacitors on PWB substrates. J Mater Sci: Mater Electron 11:253–268
79. Zhang QM, Li HF, Poh M, Xia F, Cheng ZY, Xu HS, Huang C (2002) An all-organic composite actuator material with a high dielectric constant. Nature 419:284–287
80. Wang J, Shen Q, Yang C, Zhang Q (2004) High dielectric constant composite of P(VDF-TrFE) with grafted copper phthalocyanine oligmer. Macromolecules 37: 2294–2298
81. Lu J, Wong CP (2007) Tailored dielectric properties of high-k polymer composites via nanoparticle surface modification for embedded passives applications. IEEE Proceedings of the 57th Electronic Components and Technology Conference, Reno, NV, USA, pp 1033–1039
82. Lu J, Wong CP Manuscript in preparation
83. Ulrich R, Schaper L (eds) (2003) Integrated passive component technology. IEEE Press, New York
84. Wasserman Y (1995) Integrated single-wafer RP solutions for 0.25-micron technologies. IEEE Trans-CPMT-A 17(3):346–351
85. Wang J, Davis MK, Hilburn R, Clouser S (2003) Power dissipation of embedded resistors. 2003 IPC Printed Circuits Expo, Long Beach, CA, USA, March 23–27
86. Horst S, Bhattacharya SK, Johnston S, Papapolymerou J, Tentzeris M (2006) Modeling and characterization of thin film broadband resistors on LCP for RF applications. 56th Electronic Components and Technology Conference, San Diego, CA, USA, pp 1751–1755
87. Horst S, Anagnostou D, Ponchak G, Tentzeris E, Papapolymerou J (2007) Beam-shaping of planar array antennas using integrated attenuators. 57th Electronic Components and Technology Conference, Reno, NV, USA, pp 165–168

88. Horst S, Bairavasubramanian R, Papapolymerou J, Tentzeris M (2007) Modified Wilkinson power divider for millimeter-wave integrated circuits. IEEE MTT 55(11): 2439–2446
89. Bhattacharya S, Tummala R (2000) Next generation integral passives: materials, processes, and integration of resistors and capacitors on PWB substrates. J Mater Sci: Mater Electron 11(3): 253–268
90. iNEMI 2004 Roadmap [www.iNEMI.org]
91. Halliday D, Resnick R, Walker J (1997) Fundamentals of physics. John Wiley & Sons, New York
92. Bhattacharya S (ed) (1986) Metal-filled polymers: properties and applications. Marcel Dekker, Inc., New York
93. Coates K, Chien CP, Hsiao YYR, Kovach DJ, Tang CH, Tanielian MH (1998) Development of thin film resistors for use in multichip modules. 1998 International Conference on Multichip Modules and High Density Packaging, IEEE, pp 490–495
94. Shibuya A, Matsui K, Takahashi K, Kawatani A (2001) Embedded TiNxOy thin-film resistors in a build-up CSP for 10 Gbps optical transmitter and receiver modules. Proceedings of the 51st Electronic Components and Technology Conference, pp 847–851
95. Lee KJ, Damani M, Pucha R, Bhattacharya SK, Sitaraman S, Tummala R (2007) Reliability modeling and assessment of embedded capacitors on organic substrates. IEEE Transactions on Component and Packaging Technology. 30(1):152–162
96. Koiwa I, Usada M, Osaka T (1990) Effect of heat-treatment on the structure and resistivity of electroless Ni-W-P alloy films. J Electrochem Soc 137(11):1222–1228
97. Aoki H (1991) Study of mass production of low Ohm metal film resistors prepared by electroless plating. IEICE Transactions E. 74(7):2049–2054
98. Chahal P, Tummala R, Allen M, White G (1998) Electroless Ni-P and Ni-W-P thin film resistors for MCM-L based technologies. ECTC 232–239
99. Bhattacharya SK, Varadarajan M, Chahal P, Jha G, Tummala R (2007) A novel electroless plating for embedding thin film resistors on BCB. J Electron Mater 36(3):242–244
100. Dhar S, Chakrabarti S (1996) Electroless Ni plating on n- and p-type porous Si for ohmic and rectifying contacts. Semicond Sci Technol 11:1231–1234
101. Grzyb J, Klemm M, Troster G (2003) MCM-D/L Technology for Realization of Low Cost System-on-Package Concept at 60–80 GHz. 33rd IEEE European Microwave Conference, Munich, Germany, pp 963–966

Chapter 15
Nanomaterials and Nanopackaging

X.D. Wang, Z.L. Wang, H.J. Jiang, L. Zhu, C.P. Wong, and J.E. Morris

Abstract This chapter provides a brief overview on nanomaterials and nanopackaging, and then a review on recent advances on nanoparticles and their applications, lead-free nanosolder, carbon nanotubes (CNT) and their applications for interconnect, thermal management, and integration into microsystems. Also, operation principle, fabrication techniques, and packaging of piezoelectric nanogenerators based on vertically aligned Zinc oxide (ZnO) nanowires (NWs) are reviewed in great details and possible solutions for improving the nanogenerator's performance by improving the packaging technique are discussed.

Keywords Nanotechnology · nanopackaging · nanoparticle · nanosolder · carbon nanotube · zinc oxide · nanogenerator · nanowire · piezotronics · semiconductor nanomaterial

15.1 Nanopackaging: Nanotechnologies in Microelectronics Packaging

15.1.1 Introduction

The future importance of nanoelectronics and "electro-nanotechnologies" is sufficiently well recognized to have become the subject of industrial and government policy roadmaps. Nanotechnology is conventionally defined by the crucial functional element being less than 100 nm in dimension. On that basis, we are well into the Nanoelectronics Era, having passed by the 90 nm and 65 nm CMOS nodes, with 45 nm systems in commercial production, and with 32 nm devices functioning in the R&D laboratories. Of course, according to this definition, solder has always qualified as a nanotechnology, since grain sizes are typically 10's of nm, along with many thin film applications of thicknesses lees than 100 nm. To date, there has been very little published on the packaging

X. Wang (✉)
Georgia Institute of Technology, Atlanta, GA, USA

of these nano-scale CMOS systems, other than by Mallik and Mahajan in [1]. However, the other side of Nanopackaging, i.e. the application of nanotechnologies to electronics packaging, is alive and well. Nanotechnology drivers are the varied ways in which materials properties change at small dimensions, and these properties can be put to work to solve past packaging problems, and to develop new approaches to future Nanoelectronics packaging issues. Electron transport mechanisms at small dimensions include ballistic transport, severe mean free path restrictions in very small nanoparticles, various forms of electron tunneling, electron hopping mechanisms, and more.

In addition, candidate next-generation nanoelectronics technologies, (e.g. single-electron transistors, quantum automata, molecular electronics) are generally hyper-sensitive to dimensional change, if based on quantum-mechanical electron tunneling, and appropriate packaging will be essential to the success or failure of these technologies. Packaging strategies must therefore be developed in parallel with the basic nanoelectronics device technologies in order to make informed decisions as to their commercial viabilities [1].

15.1.2 Nanoparticles

15.1.2.1 Nanoparticle Fabrication and Properties

The nanoparticle fabrication technique to be selected will depend primarily of the intended function. Noble metal nanoparticles, for example, have been fabricated by an "eco-friendly" ultrasonic processing technique [1, 2], and Ag/Cu with "polyol" [3]. Alternatively, a pre-cursor may be used, e.g. $AgNO_3$ for Ag nanoparticles, and there are techniques to control the particle shapes, e.g. spherical, cubic, or wires [1, 4]. Nanoparticles tend to cluster into aggregates, and so the crucial step is often the use of a dispersant to counter this tendency [1, 10, 14]. The thermal or sputter evaporation of metals and condensation on an insulating substrate will also yield a surface distribution of nanoparticles [1, 5, 6–8].

The enhanced chemical activities of nanoparticles, which make them effective as catalysts, are due to the high surface-to-volume ratio, and hence to the high proportion of unsatisfied chemical bonds. In addition, other physical property changes [1] include:

- Melting point depression: The melting points of small metal nanoparticles drop significantly with decreasing size [1, 9] at dimensions under 5 nm [10]
- Sintering: The thermally activated surface self-diffusion process drives net diffusion away from convex surfaces of high curvature [5], and into concave surfaces, yielding low temperature bonding between nanoparticles in contact
- Coulomb block, or blockade: An external field or thermal source of electrostatic energy is required to charge an individual nanoparticle; this effect is the basis of single-electron transistor operation [11]

- Single grain structures, such as nanoparticles, may achieve theoretical maximum mechanical strengths [12]
- Nanoparticles one to two orders smaller than the wavelength of visible light provide unique optical scattering properties [13], and absorption peaks which "color" thin films or suspensions of such nanoparticles.

15.1.2.2 Nanoparticle Applications

Embedded passive components are seen to be the solution to the problem of high proportions of PWB surface space being occupied by discrete passives. The "cermet" (ceramic-metal) resistors used in specialized on-chip applications are adaptable to the embedded-PWB role. The structure consists of metallic nanoparticles embedded in a dielectric (or polymer) with electron tunneling as the transport mechanism between particles. At low fields, the coulomb block array is randomly charged by thermal energy, giving a high negative temperature coefficient of resistance (TCR), offset by the inclusion of positive TCR metallic paths. Examples of structure-related properties are provided for the $Cr_x(SiO)_{1-x}$ and $(Cr_xSi_{1-x})_{1-y}N_y$ systems in [1].

High dielectric constant, k, and minimal thickness are required for embedded capacitors. The former requirement is met by the inclusion of high dielectric constant particulates, and the latter requirement suggests nanoparticles, e.g. barium titanate, or metals [1]. Nanoparticle surface energies must be reduced to avoid aggregation [14]. The target k is 50–200; $k\sim150$ has been achieved with metal nanoparticles at the expense of high leakage (dielectric loss), since this is a similar structure to the cermet resistor, albeit at lower metallic load. An alternative approach to leakage is to use aluminum particles, to take advantage of the native oxide coating [15], with $k\sim160$ achieved [16]. Ag/Al mixtures have also been studied [17].

Note that thermally conductive materials have very similar structural requirements to the passive components', with metallic or SiC nanoparticles as fillers [18].

Inductive components are also required, especially for RF applications. Classical magnetism theory turns out to be inapplicable for nanograin dimensions less than the ferromagnetic exchange length (10's of nm) which can sustain high permeability and low coercivity [1].

The simple addition of nanoparticles to traditional isotropic electrically conductive adhesives (ICA/ECAs) filled with micron-sized Ag fillers in an epoxy matrix might be expected to lower conductivity by providing bridges between particles, but does not in fact improve conductance, due to mean free path restrictions and added interface resistances [1]. The same principles limit the performance of alumina loaded thermal composites [19]. The addition of silver nanoparticles does achieve dramatic reductions, however, by sintering wide area contacts between flakes [1, 20], a principle also applicable to microvia fill in PWBs [21], which can also profitably use ICA materials [1]. Nanoparticle filler sintering is the key step in any effective use of nanoparticles in these technologies, and can also improve anisotropic conductive adhesive performance [22, 34], aided by contact conductance enhancement by the addition of self-assembly molecular surface treatments [1, 20, 23, 24].

PWB surface electrical interconnect is achievable by screen or "ink-jet" printing of nanoscale metal colloids in suspension [1, 25–28]. As above, electrical continuity is established by sintering Ag nanoparticles [29–33], which can also be used for die-attach [34].

Silica fillers are added to flip-chip underfills to reduce the coefficient of thermal expansion, and nanoparticles resist settling better [35] and scatter light less than larger fillers, permitting UV optical curing [36] and other advantages of optical transparency [37]. The higher viscosity of the nano-filled material can be reduced by silane surface treatments [38]. Physical properties have been successfully modeled in terms of structural parameters [1]. Nanoparticles with functionalized surfaces may be employed to increase the modulus, glass transition temperature (Tg), and dielectric property such as voltage endurance of polymer composites due to the strong interaction between the nanoparticles and the polymeric matrix and larger interaction zone [39, 40].

The addition of Pt, Ni, or Co nanoparticles to lead-free SnAg-based solder [41, 42] eliminates Kirkendall voids, reduces intermetallic compound (IMC) growth, and reduces IMC grain sizes, significantly improving drop-test performance [1, 43], promoting finer grain growth, increased creep resistance, and better contact wetting [44]. Nanoparticles in solder grain boundaries also inhibit grain boundary sliding and thermomechanical fatigue.

15.1.3 Other Nanoscale Topics

Micro-spring contact technology is novel in itself, but its down-sizing to 10 nm wide cantilevers, still 10 μm long, is truly remarkable [1, 45]. Nano-imprinting technology is also being used to fabricate optical interconnect waveguides in organic PCBs [46]. There are many nanowire applications being pursued. The 10–50 μm long Ag/Co nanowires of 200 nm diameter can be maintained in a parallel vertical orientation by a magnetic field while polymer resin flows around them [47], to form an anisotropic conductive film for z-axis contacts [48–50]. Nanowire principles and applications have been reviewed in [1/Ch. 20].

The atomic force microscope (AFM) correlates adhesion to surface feature measurements [51], and confocal microscopy has been applied to packaging research [52]. The new atomic force acoustic microscope [53] adapts the AFM to the well-known acoustic technique for package failure detection in another example of nanoscale instrumentation.

The computer modeling of microelectronics or nanoelectronics packages including nano-scale elements must be based on nanoscale elements. Nano-filler composite models must include two-phase models of the composite structure, with nano-scale materials properties if each element [54, 55]. Molecular

Dynamics modeling software has been particularly useful in the prediction of macroscale effects from the understanding of nanoscale interactions [56], but effective use requires the development of software interfacing from the nanoscale modeling results to the microscale of the whole package [1].

15.2 Nano Solder Particles

A variety of lead-free solder alloys have been investigated as potential replacements for tin/lead solders. Some lead-free candidates and their respective melting points are listed in Table 15.1. Two alloy families, tin/silver/copper (SnAgCu) and tin/copper (SnCu) seem to be generating the most interest. SnAgCu alloy composition (with or without the addition of a fourth element) appears to be the most popular replacement and has been chosen to be the benchmark, with SnPb being the baseline, that all other potential alloys for the industry have been tested against. Concerns with this alloy family include higher processing temperatures, poorer wettability due to their higher surface tension, and their compatibility with lead bearing finishes. The SnCu alloy composition is a low cost alternative for wave soldering, and compatible with most lead bearing finishes. Process considerations must be addressed with this alloy due to its higher melting temperature than most SnAgCu alloys.

As shown in Table 15.1, the higher melting points of the SnAgCu and SnCu alloys require electronic components on assembly boards to be reflowed at a temperature which is 30–40°C higher than that of eutectic SnPb solder. The higher reflow temperature leads to a number of undesirable consequences such as higher residual stress in the packages, which adversely affects their reliability. Also, the higher reflowing temperature also increases the tendency of the "popcorning" for plastic encapsulated ICs during the reflow process and potentially induces more serious warpage in organic substrates. Furthermore, heat-sensitive components might not survive the high process temperatures of

Table 15.1 Lead-free alloys

Alloy	Melting Point
Sn96.5Ag3.5	221°C
Sn96Ag3.5Cu0.5	217°C
Sn20Au80	280°C (mainly used in interconnects for optoelectronic packaging)
Sn99.3/Cu0.7	227°C
SnAgCuX(Sb, In)	Ranging according to compositions, usually above 210°C
SnAgBi	Ranging according to compositions, usually above 200°C
Sn95Sb5	232–240°C
Sn91Zn9	199°C
SnZnAgAlGa	189°C
Sn42Bi58	138°C

lead-free assembly. Therefore, lowering the processing temperature of the lead-free metals has attracted much attention recently.

The melting point of many materials can be dramatically reduced by decreasing the size of the materials. The melting and freezing behavior of finite systems have been of considerable theoretical and experimental interests for many years. As early as 1888, J.J. Thomson suggested that the freezing temperature of a finite particle depends on the physical and chemical properties of the surface. It was not until 1909, however, that an explicit expression for a size-dependent solid-liquid coexistence temperature first appeared. By considering a system consisting of small solid and liquid spheres of equal mass in equilibrium with their common vapor, it was shown that the temperature of the triple point was inversely proportional to the particle size. A similar conclusion was later reached based on the conditions for equilibrium between a solid spherical core and a thin surrounding liquid shell. Systematic experimental studies of the melting and freezing behavior of small particles began to appear in the late 1940s and early 1950s: first in a series of experiments on the freezing behavior of isolated micrometer sized metallic droplets, and later in an electron diffraction study of the melting and freezing temperatures of vapor-deposited discontinuous films consisting of nano-sized islands of Pb, Sn, and Bi. These studies demonstrated that small molten particles could often be dramatically undercooled, and that solid particles melted significantly below their bulk melting temperatures. The surface pre-melting process has been suggested as one of sources of the melting point depression of the nanoparticles [57].

Tin (Sn) and its alloys are easily oxidized due to their low chemical potential. For nano-sized tin and its alloys, oxidation happens more easily due to the higher surface area to volume ratio of nanoparticles. The presence of oxides of the nanoparticles causes poor wetting and interconnection formation. Therefore, capping each nanoparticle to prevent oxidation is critical and capping agents can cover the particle surfaces to serve as an effective barrier against the penetration of oxygen.

Various approaches to synthesize the single element nanoparticles have been reported, which can be largely categorized into "bottom-up" (chemical reduction) and "top-down" method (physical method), respectively. The chemical reduction methods include the inert gas condensation, sol-gel, aerosol, micelle/reverse micelle, and irradiation by UV, γ-ray, and microwave, and so on [58–64]. For bimetallic or multicomponents nanoparticles, chemical "bottom-up" and physical "top-down" methods have also been used as well. The chemical methods use the co-reduction of dissimilar metal precursors or successive reduction of two metal salts, which is usually carried out to prepare a core-shell structure of bimetallic nanoparticles. The binary alloys reported in the form of nano alloys or core-shell structures such as Ag-Au are an alloy system that forms solid solutions and their structures can be controlled by the reduction order. On the other hand, in the case of the alloys that do not favor the formation of solid solutions such as eutectic alloys (Sn based alloys), little research has been reported because more sophisticated synthetic methods are required due to their oxidative nature. The physical method can be used to synthesize monometallic [65] and bimetallic nanoparticles [66]. Using this technique, nanoparticles can be synthesized in gram quantities directly from the bulk materials without complex reaction procedures. This is suitable for the low melting point metal precursors and their alloys.

Effective capping capability can reduce or eliminate agglomeration of nanoparticles as well as protecting them from oxidation. It was found that 1,10-phenanthroline was an effective capping agent in forming crystalline SnAg alloy nanoparticles as shown Fig. 15.1(a), (b) and (c) [67]. When the SnAg alloy nanoparticles were formed, they were instantly coordinated to 1,10-phenanthroline through the pair of chelating nitrogen donor sites adjoining the two heterocyclic aromatic rings. The HRTEM characterizations (Fig. 15.1 (c)) showed that the particles were covered by capping agents which could provide an effective barrier against the penetration of atmospheric oxygen to the nanoparticles. At the same time, when using $NaBH_4$ as a reducing agent, hydrogen generated during a reduction reaction was found to be helpful in the creation inert environments.

The thermal properties of the synthesized SnAg alloy nanoparticles were studied by differential scanning calorimeter (DSC). Both the particle size dependent melting point depression and latent heat of fusion have been observed (Fig. 15.2). As can be seen from Fig. 15.2, as much as 25°C decreasing of melting point as the SnAg particle size was reduced to about 5 nm. It has already been found that surface melting of small particles occurs in a continuous manner over a broad temperature range, whereas the homogeneous melting of the solid core occurs abruptly at the critical temperature T_m [68, 69]. For smaller size metal nanoparticles, the surface melting is strongly enhanced by curvature effects. Therefore both the melting point and latent heat of fusion will decrease with the particle size.

Although the nano alloys melt at a lower temperature, their wettability would be poorer than that of eutectic SnPb solders due to their intrinsically higher surface tension. Besides the intrinsic behavior of the nano alloys, the melting or sintering behavior of the nano alloy particles surrounded by liquids such as flux vehicles has not been investigated. Since all the theoretical approaches for the melting behavior of fine particles assume the particles are placed in a free space, its behavior surrounded by the flux vehicle will be very interesting from a practical point of view.

A nano solder paste was formulated by dispersing the SnAg alloy nanoparticles into a low viscosity acidic type flux. Their wetting properties on the cleaned copper surface were studied. The synthesized particles were surface coated by capping agents to prevent oxidation. During the reflow process, the capping agents need to be debonded from the particle surfaces. Otherwise, they will hinder the wetting of particles on substrates. The desorption of these capping agents depends upon their affinity to the nanoparticle surfaces and their intrinsic thermal stability.

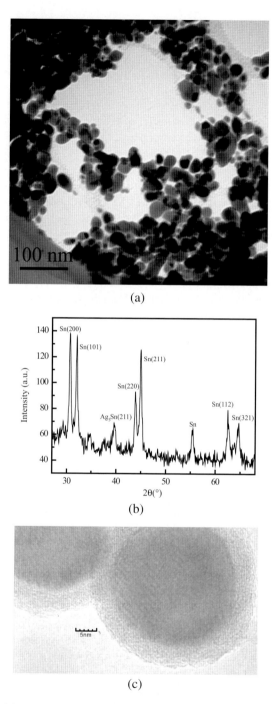

Fig. 15.1 TEM (**a**), XRD (**b**), and HRTEM (**c**) images of synthesized SnAg alloy nanoparticles

15 Nanomaterials and Nanopackaging

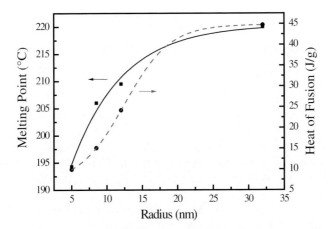

Fig. 15.2 Size dependent melting of SnAg alloy nanoparticles by DSC

To study their wetting behavior, the nano solder pastes were placed on top of the cleaned copper foil surface and then reflowed at 230°C in an oven with an air atmosphere for 5 minutes. A cross-sectional image of the sample after the reflow process was shown in Fig. 15.3. It was observed that the SnAg alloy nanoparticles with an average particle size of 64 nm completely melted and wetted on the cleaned copper foil surface. The energy dispersive spectroscopy (EDS) results revealed the formation of the intermetallic compounds (IMC) (Cu_6Sn_5), which showed scallop-like morphologies in Fig. 15.3. The thickness of the IMC was approximately 4.0 μm.

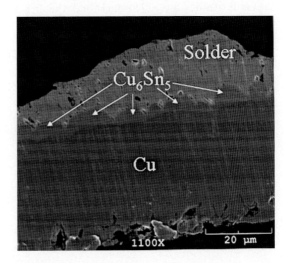

Fig. 15.3 A cross-sectional SEM image of a solder interface formed using the SnAg alloy nanoparticles (with an average particle size of 64 nm) on a cleaned copper foil surface after the reflow process

15.3 Carbon Nanotubes

15.3.1 Introduction

As originally proposed, Moore's law states that the number of transistors in semiconductor devices or integrated circuits (ICs) doubles approximately every two years [70]. One of the historical consequences of increasing the number of devices on a chip and thus microprocessor performance is an associated increase in power consumption. Heat dissipation challenges create opportunities for fundamental research in materials and thermal management strategies. Specifically, it has been suggested that future cooling approaches may be based on micro- and nanotechnologies [71]. For thermal management applications, the distinctive properties of one-dimensional structures and materials have gained much attention. Among such materials, carbon nanotubes (CNTs), due to their unique thermal properties, give rise to new opportunities in thermal management of microelectronic devices and ICs. Also, the extraordinary electrical and mechanical properties of CNTs make them a promising candidate for electrical interconnects [72, 73].

In general, CNTs can be grown by arc-discharge, laser ablation, and chemical vapor deposition (CVD) methods. However, for device applications, growth of CNTs by CVD methods is particularly attractive, due to features such as selective spatial growth, large area deposition capabilities, and aligned CNT growth.

15.3.2 Carbon Nanotubes for Interconnect Applications

15.3.2.1 Electrical Properties of Carbon Nanotubes

Previous studies have demonstrated that a carbon nanotube behaves like a quantum wire due to geometrical confinement of the tube circumference [74]. The conductance of a multi-wall nanotube (MWNT) or a single wall nanotube (SWNT) is determined by two factors: the conducting channels per shell and the number of shells. A SWNT consists of one shell. A SWNT rope or MWNT can be viewed as a parallel assembly of single SWNTs. Due to the structural imperfection of grown CNTs, the conductance for a SWNT, a SWNT rope, or MWNT can written as

$$G = G_0 M = (2e^2/h)MT \tag{15.1}$$

where M is an apparent number of conducting channels, and T is the transmission probability for an electron through the contacts and the tube. Ideally, T is unity and $M = 2$ for a perfect ballistic SWNT less than 1 μm long. In actual operation, T may be significantly lower than 1 due to electron-electron coupling, intertube coupling effects, scattering from defects and impurities, structural distortions, and coupling with substrates or contact pads. Therefore, the experimentally measured conductance is much lower than the quantized value. Therefore, the high electrical resistance of a single nanotube necessitates the use of nanotube bundles aligned in parallel.

15.3.2.2 Carbon Nanotubes as Interconnects

There are two types of interconnects employed in microelectronic devices: horizontal and vertical. Horizontal interconnects link transistors in different locations on an integrated circuit; many layers of these horizontal interconnects (up to 12) can exist on a state-of-the-art circuit [75]. Each layer is then separated by an inter-layer dielectric, generally porous SiO_2 or SiO_2 doped with C or F to lower its dielectric constant [76]. These materials are rather weak mechanically and are thermally unstable above $\sim 450°C$. As dimensions decrease for on-chip interconnects, the current density carried by each interconnect increases. The International Technology Roadmap for Semiconductors (ITRS) predicts that in 2010 the current density will reach 5×10^6 A/cm^2, a value which can only be supported by CNTs, since they are capable of a current density of $\sim 10^9$ A/cm^2 [77].

Vertical interconnects pass through holes (vias) in the dielectrics to connect horizontal interconnects to the source, drain, or gate metallization of transistors. In existing microelectronic technology, the vias are fabricated from copper. Via regions are the most common source of failures in interconnect structures due to the high current densities and heterogeneous current distributions that cause electron-induced material transport (electromigration) [78]. Carbon nanotubes are expected to offer a substantially higher resistance to electromigration than do copper lines. Thus, CNT connections between metallization layers may solve the problems of electromigration and heat removal. Researchers from Fujitsu and Infineon have investigated this area extensively [79–81]. In one approach, a hole is etched in the interlayer dielectric, and catalyst is deposited into the bottom of the hole; excess catalyst is removed from the top of the hole. Alternatively, a catalyst layer is deposited under the interlayer dielectric, and is exposed by etching a hole in the dielectric. In both approaches, CNTs are then grown within the hole by CVD or by plasma-enhanced CVD (PECVD).

When interconnects and vias are further reduced in size to meet requirements for future ICs, CNT vias will offer still more advantages. Vias consisting of only one MWNT are conceivable, since multi-walled nanotubes can be produced with diameters from 5 to 100 nm; indeed, Infineon has demonstrated such a process [78].

To take full advantage of CNT ballistic conductivity, one must open the CNT ends after growth [82] to permit better wetting and contact by Sn/Pb, etc. CNT flip-chip electrical interconnection is also under study [83–86], with μm-scale CNT clusters successfully developed as flip-chip "nano-bumps" [1, 87]. Au and Ag incorporation into CNTs has also been studied for electrical contacts with minimal galvanic corrosion [88]. Metal and carbon loaded polymers have long been used for high-frequency con-ductors in electromagnetic shielding, and both carbon fibers and multi-walled CNTs have been studied in polymer matrices for the purpose [1, 89, 90]. CNT replacement of ICA metal filler, however, [92–94] does not even match the electrical conductivity of standard materials [91, 95].

15.3.3 Carbon Nanotubes for Thermal Management

Several investigations have indicated that CNTs have unusually high thermal conductivity in the axial direction. For example, molecular dynamics simulations of a SWNT by Berber et al. indicated that the thermal conductivity of a SWCNT can be as high as 6600 W/mK at room temperature [96]. Dai et al. presented a method for extracting the thermal conductivity of an individual SWNT from high-bias electrical measurements in the temperature range from 300 to 800 K by reverse fitting the data to an existing electrothermal transport model [97]. The thermal conductivity measured was nearly 3500 W/mK at room temperature for a SWNT of length 2.6 μm and diameter 1.7 nm. Kim et al. developed a microfabricated suspended device hybridized with MWNTs (~1 μm) to allow the study of thermal transport where no substrate contact was involved [98]. The thermal conductivity and thermoelectric power of a single carbon nanotube were measured, and the observed thermal conductivity is >3000 W/mK at room temperature.

Hone et al. measured the thermal conductivity of aligned and unaligned SWNTs from 10 to 400 K [99]. Thermal conductivity increased smoothly with increasing temperatures for both aligned and unaligned SWNTs. At room temperature, the thermal conductivity of aligned SWNTs was greater than 200 W/mK, compared to ~30 W/mK of unaligned ones; above 300 K, the thermal conductivity increased and then leveled off near 400 K. Yi et al. measured the thermal conductivity of millimeter-long aligned MWNTs [100]. The thermal conductivity was low, only ~25 W/mK, at room temperature, due to a large number of CNT defects. However, thermal conductivity could reach ~2000 W/mK if the aligned MWNTs were annealed at 3000°C to remove the defects. Yang et al. investigated the thermal conductivity of MWNT films prepared by microwave CVD using a pulsed photothermal reflectance technique [101]. The average thermal conductivity of carbon nanotube films, with the film thickness from 10 to 50 μm, was ~15 W/mK at room temperature and independent of tube length. However, by taking into account a small volume filling fraction of CNTs, the effective nanotube thermal conductivity can reach 200 W/mK.

Aligned CNTs have been grown directly on silicon surfaces for thermal management. Xu et al. grew aligned CNTs on silicon wafers using plasma-enhanced CVD [102]. The thermal testing performed was based on a one-dimensional reference bar method in high-vacuum with radiation shielding, and temperature measurements were carried out with an infrared camera. Dry CNT arrays have a minimum thermal interface resistance of 19.8 mm^{2}K/W, while CNT arrays with a phase change material (PCM) produced a minimum resistance of 5.2 mm^2K/W. Xu et al. used a photothermal metrology to evaluate the thermal conductivity of aligned CNT arrays grown on silicon substrates by plasma-enhanced CVD [103]. The effective thermal resistance was 12~16 mm^2K/W, which is comparable to the resistance of commercially available thermal grease.

The high CNT thermal conductivity is used directly for conductive cool-ing of chips, and indirectly in convective cooling [104, 105]. For conductive systems, CNT alignment is the problem [104], since the thermal conductivities of random arrays show no advantages over conventional materials [107]. Composites filled with CNTs have also been studied for thermal interface materials, e.g. CNT/carbon-black mixtures in epoxy resin [108]. The use of a liquid crystal resin matrix can impose structural order on the CNT alignment to yield a seven-fold improvement in thermal conductivity [109]. Recently, electrospun polymer fibers filled with CNTs, or with SiC or metallic nanoparticles, have shown advances in both mechanical and thermal properties [110].

Micron-scale clusters of vertically grown nanotubes [111, 112] define microchannels for convective cooling coolant flow, similarly to the metal or silicon structures they aim to replace, with similar thermal performances. The problem is that the flowing coolant is only in contact with the outer-most CNTs of the clusters, and the internal CNTs are not even in good contact with each other. The system has been modeled [104], and the solution is clearly to spread the CNTs apart by an optimal separation to permit coolant contact with each one [111]. The problem then is whether individual CNTs can withstand the coolant flow pressure without detaching from the substrate [1].

15.3.4 Integration of Carbon Nanotubes into Microsystems

For electronic device applications, chemical vapor deposition (CVD) methods are particularly attractive. However, the CVD technique suffers from several drawbacks. One of the main challenges for applying CNTs to circuitry is the high growth temperature (>600°C). Such temperatures are incompatible with microelectronic processes, which are typically performed below 400–500°C in backend-of-line sequences. Another issue is the poor adhesion between CNTs and the substrates, which will result in long term reliability issues and high contact resistance. At the device level, CNTs must be integrated and interconnected with metal electrodes to allow signal input and output. Typical approaches for CNT growth on such substrates involve the deposition of catalysts such as Fe or Ni on metal layers such Ti or Ti/Au. Unfortunately, results indicate that electrical contact is not necessarily improved, suggesting that attachment of CNTs onto the electrodes produces poor mechanical and electrical properties yielding high contact resistance. On the other hand, to meet manufacturing requirements and throughput for IC applications, a large number of CNTs must be positioned simultaneously rather than aligning CNTs one by one.

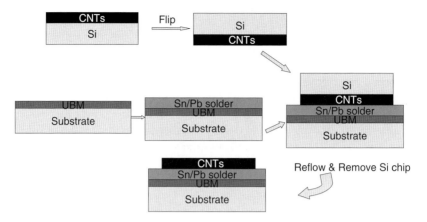

Fig. 15.4 Schematic diagram of "CNT transfer technology"

To overcome the above disadvantages, Zhu et al. proposed a methodology termed "CNT transfer technology", which is enabled by open-ended CNT structures [106, 113]. This technique is similar to flip-chip technology as illustrated schematically in Fig. 15.4.

The eutectic tin-lead paste is stencil-printed on a copper substrate. After reflow, the tin-lead solder is polished to 30 µm thick. The silicon substrates with CNTs are then flipped and aligned to the corresponding copper substrates, and reflowed in a reflow oven to simultaneously form electrical and mechanical connections. This process is straightforward to implement and offers a strategy for both assembling CNT devices and scaling up a variety of devices fabricated using nanotubes (e.g., flat panel displays). This process offers an approach to overcome the serious obstacles of integration of CNTs into integrated circuits and microelectronic device packages by offering low processing temperatures and improved adhesion of CNTs to substrates. Figure 15.5 shows the demarcation between the broken

Fig. 15.5 SEM of the copper substrates on which the CNTs were assembled after some CNTs were pulled from the surface by tweezers; this figure demonstrates the excellent mechanical bond strength of CNTs transferred to the copper substrate by the solder reflow process

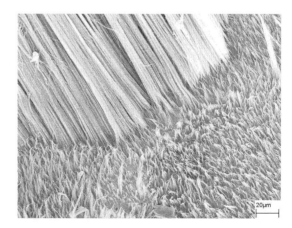

CNTs and the intact and connected ones. When pulled from the substrate, the CNTs break along the axis rather than at the CNT-solder interface. The excellent mechanical bonding strength of CNTs on the substrate anchors the CNTs and thereby improves the CNT/substrate interfacial properties.

15.3.5 Summary and Future Needs

Scaling of microelectronic devices has led to an interest in utilizing carbon nanotubes for electrical interconnects and thermal management approaches. Carbon nanotubes are also promising for vertical interconnects (for on-chip or packaging levels) and heat removal for microelectronics packaging. CNTs may be able to meet some of the ITRS projections for device interconnects and thermal requirements. However, a number of materials and CNT process integration issues need to be addressed before a CNT technology platform can be developed, including growth of structurally perfect carbon nanotubes, chirality control of carbon nanotubes, and positioning of carbon nanotubes in predefined locations simultaneously. The barriers to CNT implementation in the packaging of microelectronic devices and ICs offer numerous opportunities for new developments and approaches. Clearly, more effort is required in order to take CNT technologies from the research laboratory to high volume production.

15.4 Nanogenerators: Principle, Fabrication and Packaging

15.4.1 Introduction

Developing novel technologies for wireless nanodevices and nanosystems are of critical importance for in-situ, real-time and implantable biosensing, remote and wireless sensing, defense technology and commercial applications. The power sources required by such devices are highly desired to be life-time self-charging and in comparable small sizes. Harvesting energy from environment, including solar, thermal, and mechanical energies provides a perfect solution for these applications. There are huge emergent needs for nanoscale sensing devices for biological sensing and defense applications. Among those possible energy sources, mechanical wave and vibration energies more ubiquitously exist under various circumstances around us [114]. Relying on the piezoelectric effect of a ceramic beam when driven to vibrate by a small mass via gravitation, mechanical energy can be converted into electricity. Many types of MEMS microgenerator have been developed using piezoelectric thin film cantilevers [115]. However, their relatively large size, bio-incompatible nature, and low sensitivity to small vibrations seriously restrict the application in nanotechnologies and the advantages provided by nanomaterials.

Nanowires (NWs) and nanobelts (NBs) of inorganic materials are the forefront in today's nanotechnology research [116]. Among the known one-dimensional nanomaterials, zinc oxide (ZnO) has three key advantages [117]. First, it exhibits both semiconducting and piezoelectric properties, providing a unique material for building electro-mechanical coupled sensors and transducers [118, 119]. Secondly, ZnO is relatively bio-safe and biocompatible, and it can be used for biomedical applications with little toxicity. Finally, ZnO exhibits the most diverse and abundant configurations of nanostructures know up to today, such as nanowires, nanobelts, nanosprings, nanorings, nanobows and nanohelices [120]. Recently, ZL Wang et al. developed for the first time a novel approach of converting mechanical energy into electric power using aligned ZnO nanowires. This discovery sets the foundation for nanoscale power conversion, which will lead to a new adaptable, mobile, and cost-effective energy harvesting technology [121]. In the following sections, the operation principle, fabrication techniques, and packaging of nanogenerators are reviewed in details and possible solutions for improving the nanogenerator's performance by improving the packaging technique are discussed.

15.4.2 Nanogenerator from ZnO Nanowires

15.4.2.1 Piezoelectric Property of ZnO Nanowires

Zinc Oxide has a Wurtzite hexagonal structure (space group P63mc) with lattice parameters $a = 0.3296$, and $c = 0.52065$ nm. The structure of ZnO can be simply described as a number of alternating planes composed of tetrahedrally coordinated O^{2-} and Zn^{2+} ions, stacked alternatively along the c-axis (Fig. 15.6). The tetrahedral coordination in ZnO results in non-central symmetric structure. The lack of a centre of symmetry, combined with large electromechanical coupling, results in strong piezoelectric and pyroelectric properties and the consequent use of ZnO in actuators [122], piezoelectric sensors [123, 124], piezoelectric diodes [125], and nanogenerators [126]. To illustrate the piezoelectric effect, one considers an atom with a positive charge that is surrounded tetrahedrally by anions (Fig. 15.6a) Strong piezoelectric effect can be observed along ZnO [0001] crystal direction. Once the {0001} is the biggest surface of a ZnO nanobelt, effective piezoelectric coefficient d_{33} has been measured from 14.3 pm/V to 26.7 pm/V (Fig. 15.6b), which is much larger than that of the bulk (0001) ZnO of 9.93 pm/V [127].

A ZnO nanowire (NW) is a beam-like structure that always grows along its [0001] direction and exhibits a hexagonal cross-section. When such a NW is laterally bent, a positive voltage can be created on the tensile side surface; while a negative voltage shows on the compressive side. In order to identify the magnitude of this voltage drop across the cross-section of the NW, the perturbation theory was applied for calculating the piezoelectric potential distribution in a nanowire as pushed by a lateral force at the tip [128]. The

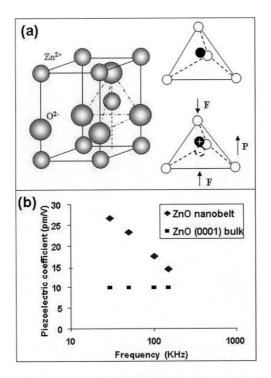

Fig. 15.6 (a) Wurtzite structure model of ZnO. The tetrahedral coordination of Zn-O and the corresponding piezoelectric effect are shown. (b) Comparison of piezoelectric coefficient d33 of ZnO nanobelt and bulk

analytical solution given under the first order approximation produces a result that is within 6% from the full numerically calculated result using finite element method. Under a simplified condition, where we assume that the nanowire has a cylindrical shape with a uniform cross-section of diameter $2a$ and length l, the maximum potential at the surface of the NW is given by equation:

$$\varphi_{max}^{(T,C)} = \pm \frac{3}{4(\kappa_0 + \kappa_\perp)} [e_{33} - 2(1+\nu)e_{15} - 2\nu e_{31}] \frac{a^3}{l^3} \nu_{max} \qquad (15.2)$$

where φ is the electric potential; κ_0 and $\kappa_\perp$ are the dielectric constants of vacuum and ZnO crystal along its c-plane, respectively; e_{33}, e_{15} and e_{31} are the linear piezoelectric coefficients; ν is Poisson ratio; and ν_{max} is the maximum deflection at the NW's tip. This equation clearly shows that the electrostatic potential is directly related to the aspect ratio of the NW instead of its dimensionality.

For a typical NW that was grown through the vapor-liquid-solid (VLS) process [129], the diameter d is 50 nm and length l is 600 nm. When it is bent 145 nm to the right by an 80 nN lateral force, which is a common scanning situation in AFM, ± 0.3 V piezoelectric potential can be induced on the two side surfaces, as shown in

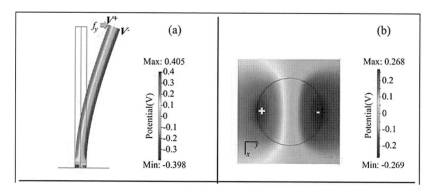

Fig. 15.7 Potential distribution for a ZnO nanowire with $d = 50$ nm, $l = 600$ nm at a lateral bending force of 80 nN. (**a**) and (**b**) are side and cross-sectional (at $z = 300$ nm) output of the piezoelectric potential in the NW given by finite element calculation

Fig. 15.7b, the calculated potential distribution across the NW cross section. Figure 15.7a is the potential distribution along the bent NW generated by finite element calculation using above equation. Calculation also shows that the piezoelectric potential in the NW almost does not depend on the z-coordination along the NW. Therefore the potential is uniform along z direction except for regions very close to the ends of the NW. This means that the NW is approximately like a "parallel plate capacitor". The maximum potential at the surface of the NW is directly proportional to the lateral displacement of the NW and inversely proportional to the length-to-diameter aspect ratio of the NW. For a larger size NW with $d = 300$ nm and length $l = 2$ μm [130], the surface piezoelectric potential can reach ± 0.6 V, when it's deflected by a 1000 nN force.

15.4.2.2 Nanogenerator from a Single ZnO Nanowire

The existence of piezoelectric potential in a bent ZnO NW was first demonstrated by atomic force microscopy (AFM) [131]. These charges can be accumulated and then released when a Schottky contact is introduced between the charged ZnO surfaces and the contacting electrode. As we discussed in the first section, across the width of the NW at the top end, the piezoelectric potential between V_s^- and V_s^+ is distribution from compressed surface to the stretched surface. In experimental design, a Pt coated Si AFM tip was used to deflect the ZnO NW and connect it to the external circuit. The contact between Pt and ZnO was Schottky, which dominates the entire transport process. In the first step, the AFM conductive tip that induces the deformation is in contact with the stretched surface of positive potential V_s^+ (Fig. 15.8a). Since the Pt metal tip has a potential of nearly zero, $V_m = 0$, the metal tip – ZnO interface is negatively biased for $\Delta V = V_m - V_s^+ < 0$. With consideration the n-type semiconductor characteristic of the as-synthesized ZnO NWs, the Pt metal-ZnO

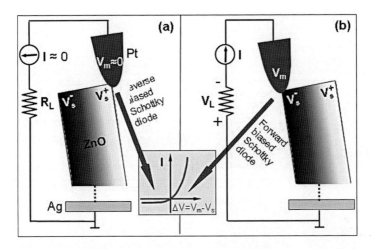

Fig. 15.8 Schematic diagram showing the metal-semiconductor contacts between the AFM tip and the ZnO N with reverse (**a**) and forward (**b**) biased Schottky rectifying behavior

semiconductor (M-S) interface in this case is a reversely biased Schottky diode (Fig. 15.8a), resulting in little current flowing across the interface. In the second step, when the AFM tip is in contact with the compressed side of the NW (Fig. 15.8b), the metal-ZnO interface is positively biased for $\Delta V = V_L = V_m - V_s^- > 0$. The metal-ZnO interface in this case is a positively biased Schottky diode, resulting in a sudden increase in the output electric current, e.g., a sharp increase in output voltage V_L (positive). The current is the result of ΔV driven flow of electrons from the semiconductor ZnO NW to the metal tip. The flow of the free electrons from the loop through the NW to the tip will neutralize the ionic charges distributed in the volume of the NW and thus reduce the magnitude of the potential V_s^- and V_s^+. Therefore, the output voltage V_L starts to drop and reaches zero until all of the ionic charges in the NW are fully neutralized.

This proposed mechanism has been clearly observed when the Pt coated AFM tip scans over a long ZnO wire that was large enough to be seen under optical microscope. As shown in Fig. 15.9a, one end of the ZnO wire was affixed on an intrinsic silicon substrate by silver paste, while the other end was left free. The wire was laid on the substrate but kept a small distance from the substrate to eliminate the friction with the substrate except at the affixed side. This wire was deflected and measured simultaneously in AFM contact mode under a constant normal force of 5 nN between the tip and sample surface. The output voltage across an outside load of resistance $R_L = 500$ MΩ was continuously monitored during the scan. The topography image directly captured if the tip passed over the belt or not because it was a representation of the normal height received by the cantilever. When the tip pushed the wire but did not go over and across it, as judged by the flat output signal in the topography image (Fig. 15.9b), no voltage output was produced, indicating the stretched side

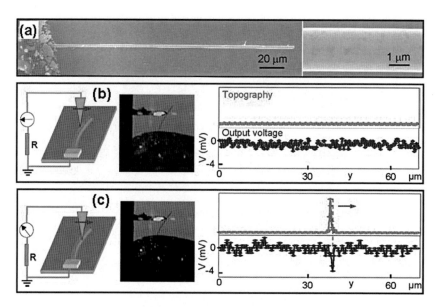

Fig. 15.9 In-situ observation of the process for converting mechanical energy into electric energy by a piezoelectric ZnO wire. (**a**) SEM images of a ZnO wire with one end affixed on a silicon substrate. (**b, c**) Two characteristic snapshots and the corresponding topography (*upper curve*) and output voltage (*lower curve*) images when the tip scanned across the middle section of the wire. The schematic illustration of the experimental condition is shown at the left hand side, with the scanning direction of the tip indicated by an arrowhead

produced no piezoelectric discharge event. Once the tip went over the wire and in touch with the compressed side, as indicated by a peak in the topography image, a sharp voltage output peak is observed (Fig. 15.9c). By analyzing the positions of the peaks observed in the topography image and the output voltage image, we noticed that the discharge occurred after the tip nearly finishing acrossing the wire, which clearly indicates that the compressed side was responsible for producing the negative piezoelectric discharge voltage.

Similar scanning process has also been applied on vertically aligned ZnO NW arrays and many sharp output voltage peaks have been observed [128]. The experimental design is schematically shown in Fig. 15.10a. The aligned ZnO NWs with an average height of ~1 μm were grown on GaN substrate, which is connected to an external load through silver paste. In the AFM contact mode, a constant normal force was maintained between the tip and sample surface. The tip scanned over the top of vertically aligned ZnO NWs, which were thus bent and then released. Meanwhile, the corresponding output voltages across the load were recorded. In the output voltage image shown in Fig. 15.10b, many sharp output voltage peaks (like discharge peaks) have been observed, which are typically about 4–50 times higher than the noise level and most of the voltage peaks were ~ 6–9 mV in magnitude.

15 Nanomaterials and Nanopackaging

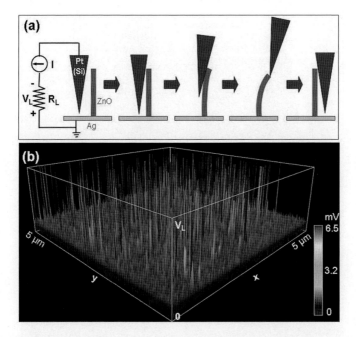

Fig. 15.10 (**a**) Experimental set up and procedures for generating electricity by deforming a piezoelectric NW using a conductive AFM tip. (**b**) Output voltage image map of ZnO NW arrays

15.4.2.3 Direct Current Nanogenerator

Although the AFM based approach has explored the principle and potential of the nanogenerator, for technological applications, an innovative design has to be made to drastically improve the performance of the nanogenerator in following aspects. First, the use of AFM for making the mechanical deformation of the NWs must be eliminated so that the power generation can be achieved by an adaptable, mobile and cost-effective approach over a larger scale. Secondly, all of the NWs are required to generate electricity simultaneously and continuously, and all the electricity can be effectively collected and output. Finally, the energy to be converted into electricity has to be provided in a form of wave/vibration from the environment, so the nanogenerator can operate "independently" and wirelessly.

In addressing these challenges, X.D. Wang et al. have developed an innovative approach by using ultrasonic waves to drive the motion of the NWs, leading to the production of a continuous current [132]. The prototype of such a nanogenerator is schematically shown in Fig. 15.11a. An array of aligned ZnO NWs was covered by a zigzag Si electrode coated with Pt. The Pt coating not only enhanced the conductivity of the electrode, but also created a Schottky contact at the interface with ZnO. The NWs were grown on either GaN

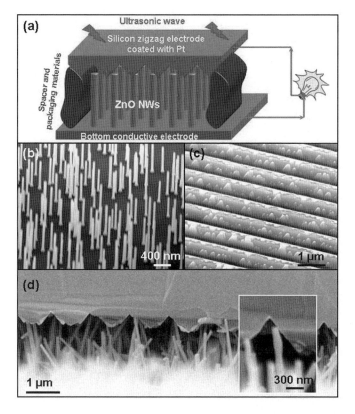

Fig. 15.11 Nanogenerators driven by an ultrasonic wave. (a) Schematic diagram showing the design and structure of the nanogenerator. (b) Low-density aligned ZnO NWs grown on a GaN substrate. (c) Zigzag trenched electrode coated with Pt. (d) Cross-sectional SEM image of the nanogenerator; Inset: A typical NW that is forced by the electrode to bend

substrates (Fig. 15.11b) or sapphire substrates that were covered by a thin layer of ZnO film [133], which served as a common electrode for directly connecting the NWs with an external circuit. The density of the NWs was $\sim 10/mm^2$, and the height and diameter were ~ 1.0 mm and ~ 40 nm, respectively. The top electrode was composed of parallel zigzag trenches fabricated on a (001) orientated Si wafer [134] and coated with a thin layer of Pt (200 nm in thickness) (Fig. 15.11c). The electrode was placed above the NW arrays and manipulated by a probe station under an optical microscope to achieve precise positioning; the spacing was controlled by soft-polymer stripes between the electrode and the NWs at the four sides. The resistance of the nanogenerator was monitored during the assembly process to ensure a reasonable contact between the NWs and the electrode by tuning the thickness of the polymer film. Then the assembled device was sealed at the edges to prevent the penetration of liquid. The cross-sectional image of the packaged NW arrays in Fig. 15.11d

15 Nanomaterials and Nanopackaging

shows a "lip/teeth" relationship between the NWs and the electrode. Some NWs are in direct contact with the top electrode, but some are located between the teeth of the electrode. The inclined NWs in the scanning electron microscopy (SEM) image were primarily caused by the cross sectioning of the packaged device.

In this design, the zigzag trenches on the top electrode act as an array of aligned AFM tips. Figure. 15.12a to c, shows four possible configurations of contact between a NW and the zigzag electrode. When subject to the excitation of an ultrasonic wave, the zigzag electrode can move down and push the NW,

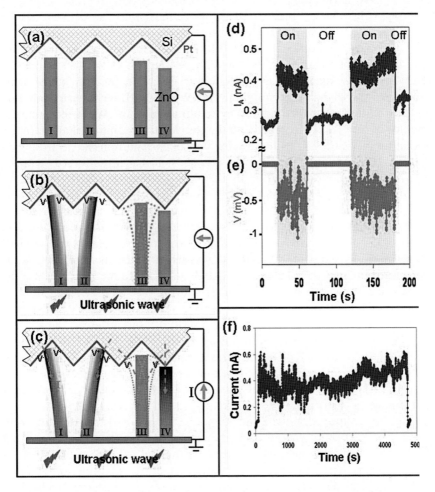

Fig. 15.12 (**a** to **c**) The mechanism of the nanogenerator driven by an ultrasonic wave. (**d, e**) Current and voltage measured on the nanogenerator, respectively, when the ultrasonic wave was turned on and off. (**f**) Continuous current output of the nanogenerator for an extended period of time

which leads to lateral deflection of NW I and creates a strain field across the width of NW I, with the NW's outer surface being in tensile strain and its inner surface in compressive strain. The inversion of strain across the NW results in an inversion of piezoelectric field E_z along the NW, which produces a piezoelectric-potential inversion from V^- (negative) to V^+ (positive) across the NW (Fig. 15.12b). When the electrode makes contact with the stretched surface of the NW, which has a positive piezoelectric potential, the Pt metal–ZnO semiconductor interface is a reversely biased Schottky barrier, resulting in little current flowing across the interface. This is the process of creating, separating, preserving, and accumulating charges. With further pushing by the electrode, the bent NW I will reach the other side of the adjacent tooth of the zigzag electrode (Fig. 15.12c). In such a case, the electrode is also in contact with the compressed side of the NW, where the metal/semiconductor interface is a forward-biased Schottky barrier, resulting in a sudden increase in the output electric current flowing from the top electrode into the NW. This is the discharge process. Analogous to the situation described for NW I, the same processes apply to the charge output from NW II. NW III is chosen to elaborate on the vibration/resonance induced by an ultrasonic wave. When the compressive side of NW III is in contact with the electrode, the same discharge process as that for NW I occurs, resulting in the flow of current from the electrode into the NW (Fig. 15.12c). NW IV, which is short in height, is forced (without bending) into compressive strain by the electrode. In such a case, the piezoelectric voltage can also be created at the top of the NW, thus contributes to the electricity output.

The current and voltage outputs of the nanogenerator are shown in Fig. 15.12d and e, respectively, with the ultrasonic wave being turned on and off regularly. A jump of $\sim$0.15 nA was observed when the ultrasonic wave was turned on, and the current immediately fell back to the baseline once the ultrasonic wave was turned off. Correspondingly, the voltage signal exhibited a similar on and off trend but with a negative output of $\sim$–0.5 mV. The size of the nanogenerator is $\sim$2 mm^2 in effective substrate surface area. The number of NWs that were actively contributing to the observed output current is estimated to be 250 to 1000 in the current experimental design. The nanogenerator worked continuously for an extended period of time of beyond 1 hour (Fig. 15.12f). Our current progress has increased the output current to 800 nA and voltage to 10 mV.

The above approach presents an adaptable, mobile, and cost-effective technology for harvesting energy from the environment, and offers a potential solution for exploring new self-powered technology for life-time unattended sensor systems, battery-free electronics and even in-situ, real-time and implantable biological devices. In the following section, the assembly techniques of nanogenerators will be described in details, including the fabrication of the core part of the nanogenerator – aligned ZnO NWs and the packaging of the nanogenerator devices.

15.4.3 Aligned Growth of ZnO Nanowire Arrays

Aligned ZnO NW arrays are the fundamental building blocks of the nanogenerators. Vapor-Liquid-Solid (VLS) process has been recognized as a simple but very efficient self-assembly technique for growing aligned NWs with controlled size, orientation and distribution. The quality of aligned NWs directly determines the energy conversion efficiency of the nanogenerator. Therefore, growing the aligned ZnO NWs with uniform length, size and ordered distribution is essential to achieve high output nanogenerator. In general, the crystal structure of the substrate controls the NW's growth direction; the catalyst controls the NW's distribution; and the vapor concentration controls the NW's quality, which includes the height, size uniformity, surface smoothness, and even defects density. In this section, the growth technique and mechanisms of aligned ZnO NWs will be reviewed.

15.4.3.1 Alignment of ZnO Nanowires

The large-scale perfect vertical alignment of ZnO nanowires has been firstly demonstrated on *a*-plane ((11 $\bar{2}$ 0) crystal surface) orientated single-crystal aluminum oxide (sapphire) substrates in 2001 [135]. The general idea of this technique is to use gold nanoparticles as catalysts, in which the growth is initiated and guided by the Au particle and the epitaxial relationship between ZnO and Al_2O_3 leads to the alignment.

Unlike the normal vapor-liquid-solid (VLS) process, a moderate growth rate is required for the alignment since the catalyst needs to be molten, form alloy, and precipitate step by step to achieve the epitaxial growth of ZnO on sapphire surface. Therefore, a relatively low growth temperature was always applied to reduce the vapor concentration. Mixing ZnO with carbon powder, which is so-called carbon-thermal evaporation, can reduce the vaporization temperature from 1300°C to 900°C,

$$ZnO(s) + C(s) \xrightleftharpoons{900°C} Zn(v) + CO(v) \tag{15.3}$$

The above reaction is reversible in a relatively lower temperature. So when the Zn vapor and CO were transferred to the substrate region, they could react and turn back to ZnO, which could be absorbed by gold catalyst and eventually formed ZnO NWs through VLS process.

Wang et al. developed another process for making aligned ZnO NWs [136]. The source materials contained equal amounts (by weight) of ZnO powder and graphite powder (0.3 gram each). The source materials were grounded and well mixed together. The mixture was loaded into an alumina boat that was placed at the center of an alumina tube with the substrate being positioned 10 cm downstream from the tube's center. Both ends of the tube were water cooled to achieve a reasonable temperature gradient. A horizontal tube furnace was used to heat the tube to 950°C at a rate of 50°C/min and the temperature was held for 20–30 minutes under a pressure of 300–400 mbar at a constant argon flow at 25 sccm. Then the furnace was shut down and cooled to room temperature under a flow of argon.

The typical morphology of the aligned ZnO NWs on sapphire substrate is shown in Fig. 15.13a, a SEM image recorded at a 30° tilted view. All of the NWs are perpendicular to the substrate surface and the darker dot on the top of each nanowire is the gold catalyst. In this process, the growth spots were dictated by the existence of catalyst. If the applied catalyst was just a thin layer of gold, the distribution of the NWs would be random (Fig. 15.13a). This is because the thin layer of gold would melt into randomly distributed nanoparticles before it catalyze the growth of aligned ZnO NWs. Once the catalyst was pre-patterned in particular shape, such as a hexagonal network (Fig. 15.13b), the as-grown aligned NWs would exhibit the same honeycomb-like distribution (Fig. 15.13c) [136]. All of the ZnO nanowires have about the same height, about 1.5 μm, and their diameters range between 50 and 150 nm. By changing the growth time the height of the ZnO nanowires could be varied from a few hundred nanometers to a few micrometers.

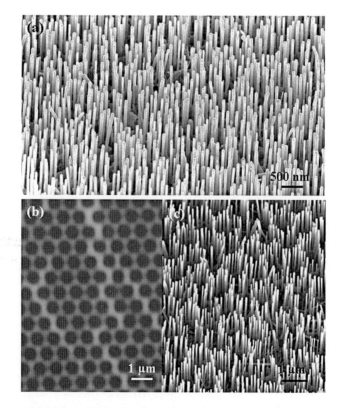

Fig. 15.13 (a) SEM image of aligned ZnO nanowires grown on a sapphire substrate using a thin layer of gold as catalyst; (b) SEM image of gold catalyst patterns using a PS sphere monolayer as a mask; (c) SEM image of aligned ZnO nanorods grown with a honeycomb pattern

The orientation of ZnO NWs can be well aligned on sapphire substrates. However, two intrinsic problems associated with this technique limit its application to nanogenerator. Since Al_2O_3 is a non-conductive material, it is difficult to electrically connecting all the aligned ZnO NWs together. Besides, a lateral growth of side branches close to the substrate surface is almost inevitable during the early stages of growth. For nanogenerator application, it is highly desirable to grow aligned ZnO NWs on a conductive or semiconductive substrate in order to fabricate the assembly of nano-electronic devices. Therefore, semiconducting nitrides, such as GaN, AlGaN, and AlN have been chosen as the supporting substrates [129, 137]. The ZnO NWs were also grown through a VLS process using a mixture of equal amounts (by weight) of ZnO and graphite powders that were loaded in an alumina boat located at the center of an alumina tube. To facilitate the reaction, 2% (1 sccm) oxygen was mixed with argon carrier gas at a flow rate of 49 sccm and the substrates were placed down stream in a temperature zone at ~850°C. A horizontal tube furnace was used to heat the source materials to 950°C at a rate of 50°C/min and the temperature was held at the peak temperature for 30 minutes under a pressure of 30 mbar. Then the system was slowly cooled to room temperature under an argon flow.

A typical low-magnification SEM image of the as-synthesized ZnO nanowires grown on GaN is shown in Fig. 15.14a. All of the ZnO nanowires are straight and perpendicular to the substrate with a high uniformity across the

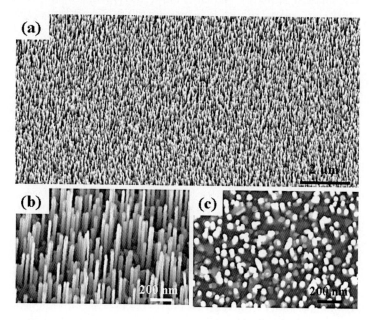

Fig. 15.14 SEM images of aligned ZnO nanorods grown on a GaN substrate. (**a**) Low-magnification 30° side view image. (**b**) High-magnification 30° side view image. (**c**) High-magnification top view image

entire substrate, indicating that this technique can be scaled up for large-area production. As shown from a higher magnification SEM image in Fig. 15.14b, the ZnO nanowires exhibit uniform diameter. Figure 15.14c shows a top view of the aligned ZnO nanowires, where only the very bright gold catalyst tips can be observed. It also confirms that almost every single nanowire is perpendicular to the substrate and that there are no side branches, which is generally unavoidable when sapphire is used as substrate.

15.4.3.2 Epitaxial Relationships

The aligning results on sapphire and nitrides substrates clear show that the crystal structure of substrate is crucial for the orientation of NWs grown by VLS process. Epitaxial relationship between the substrate surface and ZnO nanowires determines whether there will be an aligned growth and how well the alignment can be. NWs grown on silicon substrates are always randomly orientated because gold catalyst tends to form alloy with silicon at relatively low temperature and destroys the single-crystalline silicon surface [138]. The successful alignment of ZnO NWs on sapphire and nitride substrates is attributed to the very small lattice mismatches between the substrates and ZnO.

In the case of sapphire, (11 $\bar{2}$ 0) plane orientated substrate is always used because the smallest lattice mismatch is along the c-axis of Al_2O_3 and a-axis of ZnO. The epitaxial relationship between ZnO NW and a-plane sapphire substrate is schematically shown in Fig. 15.15a, where $(0001)_{ZnO}$ || $(11\bar{2}0)_{Al_2O_3}$, $[11\bar{2}0]_{ZnO}$ || $[0001]_{Al_2O_3}$. The lattice mismatch between $4[01\bar{1}0]_{ZnO}$ (4 × 3.249 = 12.996 Å) and $[0001]_{Al_2O_3}$ (12.99 Å) is almost zero, which confined the growth orientation of ZnO NWs. Nevertheless, since the (11 $\bar{2}$ 0) plane of Al_2O_3 is a rectangular lattice but the (0001) plane of ZnO is a hexagonal lattice, this epitaxial relationship can only hold in one direction. As illustrated in Fig. 15.15a, along the [1 $\bar{1}$ 00] direction, the lattice mismatch is fairly large, which will introduce some distortion on their lattice and cause stress around their interface. As a result, the lateral growth of ZnO side branches was always observed for a-plane sapphire substrates, especially on the edge of the growth area, where is no space restriction for lateral growth (Fig. 15.15b).

The nitride substrates such as GaN, AlN and AlGaN all have the same wurtzite structure as ZnO. Therefore, the deposited ZnO NWs are confined to their six equivalent <01 $\bar{1}$ 0> directions and only grow along the [0001] direction, exactly following the substrate's crystal orientation, as shown in Fig. 15.16a. In this case, the epitaxial confinement is evenly distributed along the entire 2D atomic plane. As a result, even though the lattice mismatch becomes larger and larger from GaN (1.8%) to AlN (4.3%), the aligned growth was still kept very well and the possibility for ZnO nanorods to undergo lateral growth is rare (Fig. 15.16b). Therefore, c-plane oriented $Al_xGa_{1-x}N$ substrates are ideal for the growth of aligned ZnO NWs.

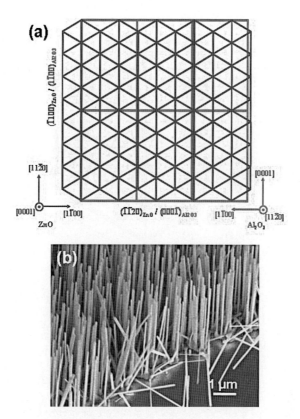

Fig. 15.15 (a) Schematic of the epitaxial relationship between c-plane of ZnO and a-plane of Al_2O_3. (b) Edge region image of the aligned ZnO nanowires grown on sapphire substrate

15.4.3.3 Control of Nanowire Growth

The quality of aligned NWs is generally controlled by the vapor concentration around the deposition region. In the VLS process, there are many variables can be controlled to affect the ZnO vapor concentration, including furnace temperature, temperature gradient inside furnace, chamber pressure, pre-pumping pressure, oxygen concentration, flow rate of carrier gas, and location of substrate. Among those effects, some are just the equipment physical parameters, such as the temperature gradient, pre-pumping pressure, and location of substrate, which can be easily kept as constant during the experiments. However, other effects showed a cross interaction between each other.

Song et al. performed a systematic investigation to identify the optimal growth conditions, and from the experiment results, it was found that the oxygen partial pressure and the system total pressure played key roles in the growth of ZnO NWs [139]. With different oxygen volume percentage and different chamber pressure, the quality and growth behavior of the ZnO nanowires were strongly affected. Over 100 growth experiments under different growth conditions were

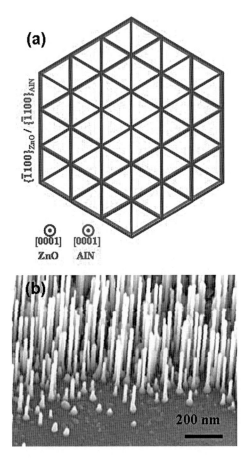

Fig. 15.16 (a) Schematic of the epitaxial relationship between c-plane of ZnO and c-plane of AlN. (b) Edge region image of the aligned ZnO nanowires grown on AlN substrate

conducted to quantitatively define the best combination of the O_2 partial pressure and the chamber pressure for growth of aligned ZnO nanowires. For consistency, all of the samples were collected at the 880°C temperature zone, which is 10 cm away from the source materials. The O_2 volume percentage in the chamber varied from 1 to 4 vol %, and the system pressure varied from 1.5 to 300 mbar. Since the growth system was first pumped down to 2×10^{-2} mbar and then the system pressure was brought back to a growth pressure at a value between 1.5 and 300 mbar, the oxygen coming from the residue air only contributed 0.28–0.0014% toward the entire oxygen content, which was much less than the percentage of O_2 in the flow gas. Therefore, the partial pressure of O_2 was considered to be the volume percentage of the O_2 introduced in the flow gas.

The experimental results are summarized in Fig. 15.17, which is a "phase diagram" for the O_2 volume percentage in the chamber and system pressure,

15 Nanomaterials and Nanopackaging

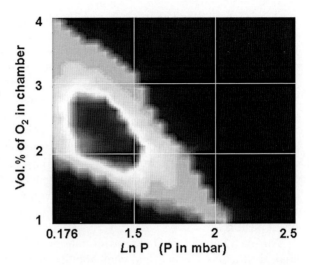

Fig. 15.17 Phase diagram" that correlates oxygen volume percent in the growth chamber (e.g. partial pressure) and the growth chamber pressure (plotted in logarithm and P is in unit of mbar) for growing aligned ZnO nanowires. The point matrix was broadened and smoothed by MatLab to form a quasi-continuous phase diagram

under which the optimum conditions for growing aligned ZnO nanowires are presented. The term of "phase diagram" used here is given a new meaning of representing the "contour map" for controlled synthesis of nanowires. This phase diagram was determined for the VLS process. As shown in Fig. 15.17, the horizontal axis is the logarithm of the total chamber pressure; the vertical axis is the oxygen volume percentage in the chamber, and the quality of the grown ZnO nanowires is represented by different contrast. The quality of the nanowires is characterized by their uniformity, density, length, and alignment. In the phase diagram, the dark region at the center of the bright triangle represents the best growth condition, where a perfect alignment of ZnO nanowires with a high density and uniform length and thickness were achieved. The growth is good in the bright area around the central dark region, where the density is lower and the nanowires are shorter. In the gray area around the edge of the bright triangle, the growth is poor, where only a little amount of short nanowires was found. No growth was found in the dark region outside of the bright triangle. This phase diagram provides the road map for growing high quality aligned ZnO nanowires.

15.4.4 Assembly and Packaging of Nanogenerator

15.4.4.1 Surface Protection of ZnO Nanowire-Based Devices

The operation principle of nanogenerator is to convert mechanical vibration energy into electricity, which allows it to work under a large variety of conditions, such as in liquid or even inside human body. However, it is also required that ZnO NWs can remain stable and functioning under harsh conditions. In order to investigate the stability of ZnO NWs in liquid and bio-fluid, Zhou et al. conducted a systematic study on the dissolving behavior of ZnO NWs in various solutions with moderate pH values, including deionized (DI) water (pH ≈ 4.5–5.0), ammonia (pH ≈ 7.0–7.1, 8.7–9.0), NaOH solution (pH ≈ 7.0–7.1, 8.7–9.0), and horse blood serum [140].

The first study was conducted to investigate how an individual ZnO wire interacted with DI water with pH ≈ 4.5–5.0. Figure 15.18a shows a SEM image of a big ZnO wire (ca. 1 μm in thickness). It has a hexagonal cross-section with very smooth side surfaces. A droplet of DI water was then applied on top of the entire ZnO wire. After covering by the DI water for about 30 min, the ZnO wire was intensively etched showing very rough and irregular surfaces (Fig. 15.18b). In addition, the thickness of the wire was reduced to only 200 nm and the wire shape was not hexagonal anymore, indicating anisotropic etching around the wire. Ammonia and NaOH solution also showed a very similar dissolving phenomenon.

The interaction of ZnO wires with pure horse blood serum (pH ≈ 8.5) was studied to assess the bio-fluid compatibility of the ZnO NWs. Fig. 15.18c–d shows the SEM images of a ZnO wire dipped in pure horse blood serum for 1 and 6 h, respectively. The dissolving rate was much slower than those of the

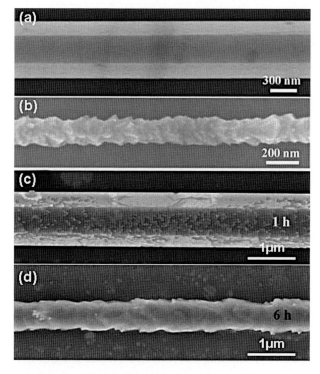

Fig. 15.18 (**a, b**) SEM images of an individual ZnO wire before and after interacting with deionized water, respectively. (**c, d**) SEM images of the ZnO wire after 1, and 6 hour(s) in pure horse blood serum (pH ≈ 8.5), respectively

aqueous solutions. The surface of ZnO wire only turned to be slightly rough after one hour immersing in horse blood serum, but the hexagonal shape could still be clearly distinguished (Fig. 15.18c). After 6 hours in the horse blood serum, 2/3 of the ZnO wire was dissolved and the shape became irregular (Fig. 15.18d).

This study showed that the ZnO wires can survive in the fluid for a few hours before they eventually degrade into mineral ions. It indicates that the NWs need to be protected if it will be directly interfacing with liquid. For nanogenerator application, the ZnO NWs are required to be stable and robust, and are expected to last for a long time. The quick degradation of ZnO NWs will seriously damage the performance and lifetime of the nanogenerator when it is required to work under water or in bio-fluids. Therefore, packaging the nanogenerator to achieve a good protection of the ZnO NWs without sacrificing their moving freedom is essential for the nanogenerator's versatility in various working conditions.

15.4.4.2 Nanogenerators in Biofluid

By improving the packaging technique, Wang et al. designed a nanogenerator that was able to generate electricity in biocompatible fluid as driven by ultrasonic wave [141]. The nanogenerator was modified from the original model composed by vertically aligned ZnO NWs and a Pt coated zigzag top electrode, as described previously. The nanogenerator core was completed packaged by a polymer to prevent the infiltration of liquid into the nanogenerator. The polymer also has certain flexibility to maintain the freedom of relative movement between ZnO NWs and the top electrode. As shown in Fig. 15.19a, the nanogenerator was placed inside a container filled with 0.9% NaCl solution, which is a typical bio-compatible solution. The substrate and the top electrode were connected by waterproof extension cord to the outside of the container and marked as the positive and negative electrode, respectively.

In the experiment, the bio-fluid was filled into a glass container with a diameter of 11 cm and a height of 9.5 cm. The ultrasonic wave source was placed at the center beneath the container. Once the ultrasonic wave was transporting inside the container, it was reflected by the container's wall and the water surface. As a result, the wave intensity was enhanced in certain region inside the fluid. The liquid-proof nanogenerator was placed inside the fluid and held by a clap that can be freely moved in any directions, while the output current was continuously monitored. First, the nanogenerator was placed at the center region above the water surface and the ultrasonic wave was kept on. The corresponding current signal is shown in Fig. 15.19b. Then the nanogenerator was slowly moved into the fluid along the z direction (depth direction) and a jump in current for $\sim$1 nA was immediately detected once it touched water surface. When the nanogenerator reached $\sim$3.3 cm below the water surface, the current quickly jumped to $\sim$20 nA. The output could be kept at such a high level as long as the nanogenerator stayed at this depth. After 15-second steady high output, the nanogenerator was moved further down and the current dropped back to 1-2 nA level again. When reached the bottom of the container, the nanogenerator was pulled upwards to the water surface. The same

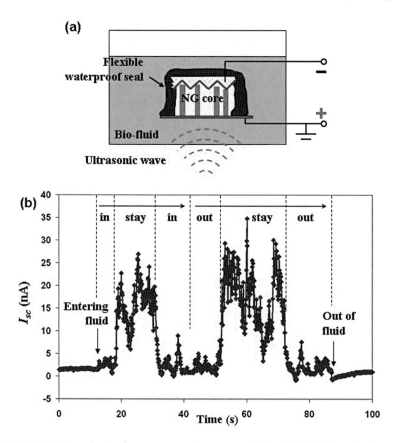

Fig. 15.19 (**a**) Schematic of a nanogenerator that operates in biofluid. (**b**) Short circuit current signal measured during the movement of nanogenerator along the horizontal direction (from water surface to the bottom and then back to surface)

20–25 nA high current output was observed again once the nanogenerator reached 3.3 cm depth. The current signal dropped back to its baseline after the nanogenerator was pulled out of water.

Corresponding lifetime testing showed a more than two hour steady output when such a nanogenerator was agitated by the ultrasonic waves in the fluid. This model successfully showed the feasibility of nanogenerator operation inside a liquid media and set a solid foundation for self-powering implantable and wireless devices and systems.

15.4.4.3 Packaging of Nanogenerators

Nanogenerator under liquid presented how packaging can improve the nanogenerator's adaptability to various working circumstances. In addition, packaging is also essential for achieving high output.

Investigation of the nanogenerator model revealed that the low output was mainly resulted from the very low percentage of the NWs that were actually generating power. Therefore, in order to increase the output power, an essential task is to enable a large number of NWs discharge continuously and simultaneously. From the principle of nanogenerator, the perfect position for a NW is at the center between two teeth of the top zigzag electrode. Although perfect aligning ZnO NWs with precisely controlled locations has been demonstrated, how to keep the NWs at the desired position is still challenging for packaging the nanogenerators. This is one of the key factors for governing the output power of the nanogenerators.

First, a precise spacing control is required between the top zigzag electrode and the bottom substrate. In Wang et al.'s nanogenerator model, this spacing was controlled by a layer of soft polymer, which not only keeps the top and bottom electrode separated but also provides certain level of flexibility allowing the top electrode moving up-and-down [134]. However, in this set-up, most NWs were stuck between the teeth and only a very small percentage (~1%) NWs were actually at the right position and active for electricity generation. This was one of the main reasons for the low output power. The precise spacing control can be realized by introducing a spacer between the top and bottom electrodes. Considering the length of NWs is ~2 μm and the depth of the trench on the top electrode is ~800 nm, the thickness of the spacer should be ~400 nm smaller than the length of NW so that the NWs can be located at the center of the trenches. Meanwhile, such a spacer also needs to be water proof and flexible for proper functioning of the nanogenerator. Polymers like SU8 will be a good choice for the spacer fabrication.

On the other hand, besides the spacing control, the ZnO NWs need to be grown selectively on the substrate according to the pattern of the top electrode. Due to the uneven top electrode, although the space is well controlled everywhere, the distance between the top electrode surface and bottom substrate still varies. Therefore, with a random distribution, most ZnO NWs are very likely to be directly compressed or bent during assembling, which lowers the number of active ZnO NW and increases the device capacitance. Both effects result in a lower output power. Controlling growth pattern of ZnO NWs can be realized by pre-patterning the catalyst using the same pattern as the top electrode. In the final packaging, the ZnO NW arrays need to be well aligned with the top electrode to match their patterns. Considering the width of a typical trench is ~2 μm, micromanipulation is required to achieve this alignment.

When the above two packaging strategies are fulfilled, it is expected >80% NWs will be involved in electricity generating, resulting in an output power of roughly 10 μW/cm^2.

15.4.5 Summary

The piezoelectric nanogenerators are built on vertically aligned ZnO NWs. The operation principle relies on the piezoelectric potential generated on a bent ZnO NW. The rectifying effect of the Schottky junction between the metal electrode and ZnO crystal can selectively accumulate and release charging, resulting a continuous mechanical-to-electric energy conversion. The nanogenerator can convert small mechanical vibration energy and hydraulic energy into electricity under various circumstances, such as under ground, in water, or even inside human body. The application and the energy conversion efficiency largely depend on the fabrication and packaging techniques. In general, majority of the NWs (ideally one hundred percent) need to be involved in power generation and all the NWs have to be well isolated from the corrosive surroundings but still remain good moving freedom. This requires a high dimensional uniformity of the NWs as well as a perfect positioning and alignment of the top electrode on the NW arrays. Upon addressing these crucial challenges, the nanogenerators can potentially become a new self-powered technology for lifetime unattended sensor systems, battery-free electronics, and even in-situ, real-time, and implantable biological devices.

References

1. J.E. Morris (editor), "Nanopackaging: Nanotechnologies and Electronics Packaging," Springer, 2008
2. Y. Hayashi, H. Takizawa, M. Inoue, K. Niihara, and K. Suganuma, "Ecodesigns and Applications for Noble Metal Nanoparticles by Ultrasound Process", IEEE Transactions on Electronic Packaging Manufacturing, 28(4), 338–343, 2005
3. H. Jiang, K. Moon, and C.-P. Wong, "Synthesis of Ag-Cu Alloy Nanoparticles for Lead-Free Interconnect Materials", Proc. 10th IEEE/CPMT International Symposium on Advanced Packaging Materials, Irvine, CA, USA, 173–177, 2005
4. S. Pothukuchi, Y. Li, and C.-P. Wong, "Shape Controlled Synthesis of Nanoparticles and Their Incorporation into Polymers", Proc. 54th IEEE Electronic Component & Technol. Conf., Las Vegas, NV, USA, 1965–1967, 2004
5. M. Ohring, "Materials Science of Thin Films: Deposition & Structure (Second edition)," Academic Press, 395–397, 2002
6. F. Wu and J.E. Morris, "Characterizations of $(SiOxCr1-x)N1-y$ Thin Film Resistors for Integrated Passive Applications", 53rd Electronic Components & Technol. Conf., New Orleans, LA, USA, 161–166, 2003
7. J.E. Morris, "Recent Progress in Discontinuous Thin Metal Film Devices," Vacuum, 50(1–2), 107–113, 1998
8. J.E. Morris, F. Wu, C. Radehaus, M. Hietschold, A. Henning, K. Hofmann, and A. Kiesow, "Single Electron Transistors: Modeling and Fabrication" Proc. 7th Int. Conf. Solid State & Integrated Circuit Technology (ICSICT), Beijing, China, 634–639, 2004
9. H. Jiang, K. Moon, H. Dong, and F. Hua, "Thermal Properties of Oxide Free Nano Non Noble Metal for Low Temperature Interconnect Technology", Proc. 56th IEEE Electronic Component & Technol. Conf., San Diego, CA, USA, 1969–1973, 2006
10. J.R. Sambles, "An Electron Microscope Study of Evaporating Gold Particles: The Kelvin Equation for Liquid Gold and the Lowering of the Melting Point of Solid Gold Particles," Proc. Roy. Soc. Lond. A, 324, 339–351, 1971
11. J.E. Morris, "Single-Electron Transistors," in "The Electrical Engineering Handbook Third edition): Electronics, Power Electronics, Optoelectronics, Microwaves, Electromagnetics, and Radar," Richard C. Dorf (editor), CRC/Taylor & Francis, 3.53–3.64, 2006

12. R.A. Flinn and P.K. Trojan, "Engineering Materials & their Applications (Second edition)," Houghton-Mifflin, 75–77, 1981
13. T. Yamaguchi, M. Sakai, and N. Saito, "Optical Properties of Well-Defined Granular Metal Systems," Phys. Rev. B, 32(4), 2126–2130, 1985
14. R. Das, M. Poliks, J. Lauffer, and V. Markovich; "High Capacitance, Large Area, Thin Film, Nanocomposite Based Embedded Capacitors", Proc. 56th IEEE Electronic Component & Technol. Conf., San Diego, CA, USA, 1510–1515, 2006
15. J. Xu and C.-P. Wong, "High-K Nanocomposites with Core-Shell Structured Nanoparticles for Decoupling Applications", Proc. 55th IEEE Electronic Component & Technol. Conf., Orlando, FL, USA, 1234–1240, 2005
16. J. Xu and C.-P. Wong, "Effects of the Low Loss Polymers on the Dielectric Behavior of Novel Aluminum-filled High-k Nano-composites," Proc. 54th IEEE Electronic Component & Technol. Conf., Las Vegas, NV, USA, 496–506, 2004
17. J. Lu, K. Moon, and C.-P. Wong, "Development of Novel Silver Nanoparticles/Polymer Composites as High K Polymer Matrix by In-situ Photochemical Method", Proc. 56th IEEE Electronic Component & Technol. Conf., San Diego, CA, USA, 1841–1846, 2006
18. L. Ekstrand, H. Kristiansen, and J. Liu, "Characterization of Thermally Conductive Epoxy Nano Composites," Proc 28th Int. Spring Seminar on Electronics Technology (ISSE'05), Vienna, Austria, 19–23, 2005
19. L. Fan, B. Su, J. Qu, and C.-P. Wong; "Electrical and Thermal Conductivities of Polymer Composites Containing Nano-Sized Particles", Proc. 54th IEEE Electronic Component & Technol. Confer., Las Vegas, NV, USA, 148–154, 2004
20. H. Jiang, K. Moon, L. Zhu, J. Lu, and C.P. Wong, "The Role of Self-Assembled Monolayer (SAM) on Ag Nanoparticles for Conductive Nanocomposite", Proc. 10th IEEE/CPMT International Symposium on Advanced Packaging Materials, Irvine, CA, USA, 266–271, 2005
21. R. Das, J. Lauffer, and F. Egitto, "Electrical Conductivity and Reliability of Nano- and Micro-Filled Conducting Adhesives for Z-axis interconnections," Proc. 56th IEEE Electronic Component & Technol. Conf., San Diego, CA, USA, 112–118, 2006
22. K. Moon, S. Pothukuchi, Y. Li, and C.-P. Wong, "Nano Metal Particles for Low Temperature Interconnect Technology", Proc. 54th IEEE Electronic Component & Technol. Conf., Las Vegas, NV, USA, 1983–1988, 2004
23. Y. Li, K. Moon, and C.-P. Wong, "Improvement of Electrical Performance of Anisotropically Conductive Adhesives", Proc. 10th IEEE/CPMT International Symposium on Advanced Packaging Materials, Irvine, CA, USA, 221–226, 2005
24. Y. Li, K. Moon, and C.-P. Wong, "Electrical Property of Anisotropically Conductive Adhesive Joints Modified by Self-Assembled Monolayer (SAM)", Proc. 54th IEEE Electronic Component & Technol. Conf., Las Vegas, NV, USA, 1968–1974, 2004
25. S. Joo and D.F. Baldwin, "Demonstration for Rapid Prototyping of Micro-Systems Packaging by Data-Driven Chip-First Process Using Nano-Particles Metal Colloids," Proc. 55th IEEE Electronic Component & Technol. Confer., Orlando, FL, USA, 1859–1863, 2005
26. A. Moscicki, J. Felba, T. Sobierajski, J. Kudzia, A. Arp, and W. Meyer; "Electrically Conductive Formulations Filled Nano Size Silver Filler for Ink-Jet Technology", Proc. 5th International Conference on Polymers and Adhesives in Microelectronics and Photonics, Wroclaw, Poland, 40–44, 2005
27. J. Kolbe, A. Arp, F. Calderone, E.M. Meyer, W. Meyer, H. Schaefer, and M. Stuve, "Inkjettable Conductive Adhesive for Use in Microelectronics and Microsystems Technology," Proc. 5th International Conference on Polymers and Adhesives in Microelectronics and Photonics, Wroclaw, Poland, 160–163, 2005
28. J.G. Bai, K.D. Creehan, and H.A. Kuhn, "Inkjet Printable Nanosilver Suspensions for Enhanced Sintering Quality in Rapid Manufacturing," Nanotechnology, 18 1–5, 2005

29. W. Peng, V. Hurskainen, K. Hashizume, S. Dunford, S. Quander, and R. Vatanparast, "Flexible Circuit Creation with Nano Metal Particles", Proc. 55th IEEE Electronic Component & Technol. Conf., Orlando, FL, USA, 77–82, 2005
30. J.G. Bai, Z.Z. Zhang, J.N. Calata, and G.-Q. Lu, "Low-Temperature Sintered Nanoscale Silver as a Novel Semiconductor Device-Metallized Substrate Interconnect Material," IEEE Trans. Components & Packaging Technol., 29(3), 589–593, 2006
31. M. Nakamoto, M. Yamamoto, Y. Kashiwagi, H. Kakiuchi, T. Tsujimoto, and Y. Yoshida, "A Variety of Silver Nanoparticle Pastes for Fine Electronic Circuit Patter Formation," Proc. 6th International Conference on Polymers and Adhesives in Microelectronics and Photonics, Tokyo, 105–109, 2007
32. D. Wakuda, M. Hatamura, and K. Suganuma, "Novel Room Temperature Wiring Process of Ag Nanoparticle Paste," Proc. 6th International Conference on Polymers and Adhesives in Microelectronics and Photonics, Tokyo, 110–113, 2007
33. A. Moscicki, J. Felba, P. Gwiazdzinski, and M. Puchalski, "Conductivity Improvement of Microstructures made by Nano-Size-Silver Filled Formulations," Proc. 6th International Conference on Polymers and Adhesives in Microelectronics and Photonics, Tokyo, 305–310, 2007
34. J.G. Bai, Z.Z. Zhang, J.N. Calata, and G.-Q. Lu, "Characterization of Low-Temperature Sintered Nanoscale Silver Paste for Attaching Semiconductor Devices," Proc. 7th IEEE CPMT Conference on High Density Microsystem Design and Packaging and Component Failure Analysis (HDP'05), Shanghai, China, 272–276, 2005
35. P. Lall, S. Islam, J. Suhling, and G. Tian, "Nano-Underfills for High-Reliability Applications in Extreme Environments", Proc. 55th IEEE Electronic Component & Technol. Conf., Orlando, FL, USA, 212–222, 2005
36. Y. Sun, Z. Zhang, and C.-P. Wong, "Photo-Definable Nanocomposite for Wafer Level Packaging", Proc. 55th IEEE Electronic Component & Technol. Conf., Orlando, FL, USA, 179–184, 2005
37. Y. Sun and C.-P. Wong, "Study and Characterization on the Nanocomposite Underfill for Flip Chip Applications", Proc. 54th IEEE Electronic Component & Technol. Confer., Las Vegas, NV, USA, 477–483, 2004
38. Y. Sun, Z. Zhang, and C.-P. Wong, "Fundamental Research on Surface Modification of Nano-size Silica for Underfill Applications", Proc. 54th IEEE Electronic Component & Technol. Conf., Las Vegas, NV, USA, 754–760, 2004
39. T. Ramanathan et al., "Functionalized SWNT/Polymer Nanocomposites for Dramatic Property Improvement," www.interscience.wiley.com
40. M. Roy et al., "Polymer Nanocomposite Dielectrics – The Role of the Interface," IEEE Transactions on Dielectrics and Electrical Insulation, 12(6), 1273–1273, 2005
41. W. Guan, S.C. Verma, Y. Gao, C. Andersson, Q. Zhai, and J. Liu, "Characterization of Nanoparticles of Lead Free Solder Alloys," Proc. 1st IEEE Electronics Systemintegration Technol. Conf., Dresden, Germany, 7–12, 2006
42. K.M. Kumar, V. Kripesh, and A.A.O. Tay, "Sn-Ag-Cu Lead-free Composite Solders for Ultra-Fine-Pitch Wafer-Level Packaging," Proc. 56th IEEE Electronic Component & Technol. Conf., San Diego, CA, USA, 237–243, 2006
43. M. Amagai, "A Study of Nano Particles in SnAg-Based Lead Free Solders for Intermetallic Compounds and Drop Test Performance", Proc. 56th IEEE Electronic Component & Technol. Confer., San Diego, CA, USA, 1170–1190, 2006
44. V. Kripesh, K. Mohankumar, and A. Tay, "Properties of Solders Reinforced with Nanotubes and Nanoparticles", Proc. 56th IEEE Electronic Component & Technol. Conf., San Diego, CA, USA, 2006
45. K.M. Klein, J. Zheng, A. Gewirtz, D.S. Sarma, S. Rajalakshmi, and S.K. Sitaraman, "Array of Nano-Cantilevers as a Bio-Assay for Cancer Diagnosis", Proc. 55th IEEE Electronic Component & Technol. Conf., Orlando, FL, USA, 583–587, 2005

46. B. Lee, R. Pamidigantham, and C.S. Premachandran, "Development of Polymer Waveguide using Nano-Imprint Method for Chip to Chip Optical Communication and Study the Suitability on Organic Substrates", Proc. 56th IEEE Electronic Component & Technol. Conf., San Diego, CA, USA, 2006
47. R.-J. Lin, Y.-Y. Hsu, Y.-C. Chen, S.-Y. Cheng, and R.-H. Uang, "Fabrication of Nanowire Anisotropic Conductive Film for Ultra-fine Pitch Flip Chip Interconnection", Proc. 55th IEEE Electronic Component & Technol. Confer., Orlando, FL, USA, 66–70, 2005
48. S. Fiedler, M. Zwanzig, R. Schmidt, E. Auerswald, M. Klein, W. Scheel, and H. Reichl, "Evaluation of Metallic Nano-Lawn Structures for Application in Microelectronics Packaging," Proc. 1st IEEE Electronics Systemintegration Technol. Conf., Dresden, Germany, 886–891, 2006
49. H.P. Wu, J.F. Liu, X.J. Wu, M.Y. Ge, Y.W. Wang, G.Q. Zhang, and J.Z. Jiang, "High Conductivity of Isotropic Conductive Adhesives Filled with Silver Nanowires," Int. J. Adhesion & Adhesives, 26, 617–621, 2006
50. H. Wu, X. Wu, J. Liu, G. Zhang, Y. Wang, Y. Zeng, and J. Jing, "Development of a Novel Isotropic Conductive Adhesive Filled with Silver Nanowires," J. Composite Mater., 40(21), 1961–1969, 2006
51. C.K.Y. Wong, H. Gu, B. Xu, and M.M. Fyuen, "A New Approach in Measuring Cu-EMC Adhesion Strength by AFM", Proc. 54th IEEE Electronic Component & Technol. Confer., Las Vegas NV, USA, 491–495, 2004
52. M. Luniak, H. Hoeltge, R. Brodmann, and K.-J. Wolter, "Optical Characterization of Electronic Packages with Confocal Microscopy," Proc. 1st IEEE Electronics Systemintegration Technol. Conf., Dresden, Germany, 1318–1322, 2006
53. B. Koehler, B. Bendjus, and A. Striegler, "Determination of Deformation Fields and Visualization of Buried Structures by Atomic Force Acoustic Microscopy," Proc. 1st IEEE Electronics Systemintegration Technol. Confer., Dresden, Germany, 1330–1335, 2006
54. B. Michel, R. Dudek, and H. Walter, "Reliability testing of Polytronics Components in the Micro-Nano Region", Proc. 5th International Conference on Polymers and Adhesives in Microelectronics and Photonics, Wroclaw, Poland, 13–15, 2005
55. S. Koh, R. Rajoo, R. Tummala, A. Saxena, and K.T. Tsai, "Material Characterization for Nano Wafer Level Packaging Application", Proc. 55th IEEE Electronic Component & Technol. Confer., Orlando, FL, USA, 1670–1676, 2005
56. E.D. Dermitzaki, J. Bauer, B. Wunderle, and B. Michel, "Diffusion of Water in Amorphous Polymers at Different Temperatures Using Molecular Dynamics Simulation," Proc. 1st IEEE Electronics Systemintegration Technol. Confer., Dresden, Germany, 762–772, 2006
57. L. Allen, R.A. Bayles, W.W. Gile, and W.A. Jesser, "Small Particle Melting of Pure Metals", Thin Solid Film, 144, 297–308, 1986
58. R. Birringer, H. Gleiter, H. P. Klein, and P. Marquart, "Nanocrystalline Materials an Approach to a Novel Solid Structure with Gas-like Disorder?", Phys. Lett. 102A, 365–369, 1984
59. B.I. Lee and E.J.A. Pope, "Chemical Processing of Ceramics", Marcel Dekker, 1994
60. O.G. Raabe, in "Fine Particles", ed. by B.Y.H. Liu, Academic Press, Inc., 60, 1975
61. J. Thomas, "Preparation and Magnetic Properties of Colloidal Cobalt Particles", J. Appl. Phys. 37, 2914–2915, 1966
62. G.L. Rochfort and R.D. Rieke, "Preparation, Characterization, and Chemistry of Activated Cobalt", Inorg. Chem., 25, 348–355, 1986
63. K. Klabunde, Y. Li, and B. Tan, "Solvated Metal Atom Dispersed Catalysts", Chem. Mater., 3, 30–39, 1991
64. F. Mafune, J.Y. Kohno, Y. Takeda, and T. Kondow, "Dissociation and Aggregation of Gold Nanoparticles under Laser Irradiation", J. Phys. Chem. B, 105, 9050–9056, 2001

65. Y.B. Zhao, Z.J. Zhang, and H.X. Dang, "Preparation of Tin Nanoparticles by Solution Dispersion", Mater. Sci. Eng. A359, 405–407, 2003
66. Y.B. Zhao, Z.J. Zhang, and H.X. Dang, "Synthesis of In-Sn Alloy Nanoparticles by a Solution Dispersion Method", J. Mater. Chem. 14, 299–302, 2004
67. H.J. Jiang, K. Moon, F. Hua, and C.P. Wong, "Synthesis and Thermal and Wetting Properties of Tin/Silver Alloy Nanoparticles for Low-Melting Point Lead-Free Solders", Chem. Mater. 19, 4482–4485, 2007
68. R. Garrigos, P. Cheyssac, and R. Kofman, "Melting for Lead Particles of Very Small Sizes-Influence of Surface Phenomena", Z. Phys. D, 12, 497–500, 1989
69. W.Y. Hu, S.G. Xiao, J.Y. Yang, and Z. Zhang, "Melting Evolution and Diffusion Behavior of Vanadium Nanoparticles", Eur. Phys. J. B, 45, 547–554, 2005
70. G.E. Moore, "Progress in Digital Integrated Electronics," International Electron Devices Meetings, Washington, DC, USA, 11–13, 1975
71. R.S. Prasher et al., "Nano and Micro Technology Based Next Generation Package-Level Cooling Solutions," Intel. Technol. J., 9(4), 285–292, 2005
72. F. Kreupl et al., "Carbon Nanotubes in Interconnect Applications," Microelectron. Eng., 64(1–4), 399–408, 2002
73. A.P. Graham et al., "How do Carbon Nanotubes Fit into the Semiconductor Roadmap?" Appl. Phys. A-Mater. Sci. & Process., 80(6), 1141–1151, 2005
74. S. Frank et al., "Carbon Nanotube Quantum Resistors," Science, 280(5370), 1744–1746, 1998
75. T.W. Wu and E.C. Chen, "Crystallization Behavior of Poly(Epsilon-Caprolactone)/Multiwalled Carbon Nanotube Composites," J. Polymer Sci. Part B-Polymer Phys., 44(3), 598–606, 2006
76. S. Mizuno et al., "Dielectric Constant and Stability of Fluorine Doped PECVD Silicon Oxide Thin Films," Thin Solid Films, 283(1), 30–36, 2006
77. B.Q. Wei, R. Vajtai, and P.M. Ajayan, "Reliability and Current Carrying Capacity of Carbon Nanotubes," Appl. Phys. Lett., 79(8), 1172–1174, 2001
78. A.P. Graham et al., "Carbon Nanotubes for Microelectronics?" Small, 1(4), 382–390, 2005
79. M. Nihei, "Electrical Properties of Carbon Nanotube Bundles for Future Via Interconnects," Japanese J. Appl. Phys. Part 1-Regular Papers Short Notes & Review Papers, 44(4A), 1626–1628, 2005
80. W. Hoenlein et al., "Carbon Nanotubes for Microelectronics: Status and Future Prospects," Mater. Sci. Eng. C-Biomimetic and Supramolecular Syst, 23(6), 663–669, 2003
81. Y. Awano, "Carbon Nanotube Technologies for LSI via Interconnects," IEICE Transactions on Electronics, E89-C(11), 1499–1503, 2006
82. L. Zhu, Y. Xiu, D. Hess, and C.-P. Wong, "In-situ Opening Aligned Carbon Nanotube Films/Arrays for Multichannel Ballistic Transport in Electrical Interconnect", Proc. 56th IEEE Electronic Component & Technol. Confer., San Diego, CA, USA, 171–176, 2006
83. A. Naeemi, G. Huang, and J. Meindl, "Performance Modeling for Carbon Nanotube Interconnects in On-chip Power Distribution," Proc. 57th IEEE Electronic Component & Technol. Conf., Reno, NV, USA, 420–428, 2007
84. Y. Chai, J. Gong, K. Zhang, P.C.H. Chan, and M.M.F. Yuen, "Low Temperature Transfer of Aligned Carbon Nanotube Films Using Liftoff Technique," Proc. 57th IEEE Electronic Component & Technol. Conf., Reno, NV, USA, 429–434, 2007
85. C.-J. Wu, C.-Y. Chou, C.-N. Han, and K.-N. Chiang, "Simulation and Validation of CNT Mechanical Properties – The Future Interconnection Method," Proc. 57th IEEE Electronic Component & Technol. Confer., Reno, NV, USA, 447–452, 2007
86. A. Ruiz, E. Vega, R. Katiyar, and R. Valentin, "Novel enabling wire bonding technology," Proc. 57th IEEE Electronic Component & Technol. Confer., Reno, NV, USA, 458–462, 2007
87. G.A. Riley, "Nanobump Flip Chips," Adv. Packaging, 18–20, 2007

88. R.T. Pike, R. Dellmo, J. Wade, S. Newland, G. Hyland, and C.M. Newton, "Metallic Fullerene and MWCNT Composite Solutions for Microelectronics Subsystem Electrical Interconnection Enhancement", Proc. 54th IEEE Electronic Component & Technol. Conf., Las Vegas, NV, USA, 461–465, 2004
89. J. Ding, S. Rea, D. Linton, E. Orr, and J. MacConnell, "Mixture Properties of Carbon Fibre Composite Materials for Electronics Shielding in Systems Packaging," Proc. 1st IEEE Electronics Systemintegration Technol. Conf., Dresden, Germany, 19–25, 2006
90. J.-C. Chiu, C.-M. Chang, W.-H. Cheng, and W.-S. Jou, "High-Performance Electromagnetic Susceptibility for a 2.5 Gb/s Plastic Transceiver Module Using Mutli-Wall Carbon Nanotubes", Proc. 56th IEEE Electronic Component & Technol. Confer., San Diego, CA, USA, 183–186, 2006
91. C.-M. Chang, J.-C. Chiu, C.-Y. Yeh, W.-S. Jou, Y.-F. Lan, Y.-W. Fang, J.-J. Lin, and W.-H. Cheng, "Electromagnetic Shielding Performance for a 2.5 Gb/s Plastic Transceiver Module Using Dispersive Multiwall Carbon Nanotubes," Proc.57th IEEE Electronic Component & Technol. Conf., Reno, NV, USA, 442–446, 2007
92. J. Li and J.K. Lumpp, "Electrical and Mechanical Characterization of Carbon Nanotube Filled Conductive Adhesive," Proc. IEEEAC, 2006, paper #1519
93. L. Xuechun and L. Feng, "The Improvement on the Properties of Silver-Containing Conductive Adhesives by the Addition of Carbon Nanotube," Proc. 6th IEEE CPMT Conference on High Density Microsystem Design and Packaging and Component Failure Analysis (HDP'04), Shanghai, China, 382–384, 2004
94. A.M. Bondar, A. Bara, D. Patroi, and P.M. Svasta, "Carbon Mesophase/Carbon Nanotubes Nanocomposite – Functional Filler for Conductive Pastes," Proc. 5th International Conference on Polymers and Adhesives in Microelectronics and Photonics, Wroclaw, Poland, 215–218, 2005
95. A. Bara, A.M. Bondar, and P.M. Svasta, "Polymer/CNTs Composites for Electronics Packaging," Proc. 1st IEEE Electronics Systemintegration Technol. Conf., Dresden, Germany, 334–336, 2006
96. S. Berber, Y.K. Kwon, and D. Tomanek, "Unusually High Thermal Conductivity of Carbon Nanotubes," Phys. Rev. Lett., 84(20), 4613–4616, 2000
97. E. Pop et al., "Thermal Conductance of an Individual Single-Wall Carbon Nanotube above Room Temperature," Nano Lett., 6(1), 96–100, 2006
98. P. Kim et al., "Thermal Transport Measurements of Individual Multiwalled Nanotubes," Phys. Rev. Lett., 87(21), 215502-1–215502-4, 2001
99. J. Hone et al., "Electrical and Thermal Transport Properties of Magnetically Aligned Single wall Carbon Nanotube Films," Appl. Phys. Lett., 77(5), 666–668, 2000
100. W. Yi et al., "Linear Specific Heat of Carbon Nanotubes," Phys. Rev. B, 59(14), R9015–R9018, 1999
101. D.J. Yang et al., "Thermal Conductivity of Multiwalled Carbon Nanotubes," Phys. Rev. B, 66(16), 165440.1–165440.6, 2000
102. J. Xu and T.S. Fisher, "Enhancement of Thermal Interface Materials with Carbon Nanotube Arrays," Int. J. Heat and Mass Transfer, 49, 1658–1666, 2006
103. Y. Xu et al., "Thermal Properties of Carbon Nanotube Array Used for Integrated Circuit Cooling," J. Appl. Phys., 100(7), 074302, 2006
104. T. Wang, M. Jonsson, E. Nystrom, Z. Mo, E.E.B. Campbell, and J. Liu, "Development and Characterization of Microcoolers using Carbon Nanotubes," Proc. 1st IEEE Electronics Systemintegration Technol. Conf., Dresden, Germany, 881–885, 2006
105. J. Xu and T.S. Fisher, "Enhanced Thermal Contact Conductance Using Carbo Nanotube Array Interfaces," IEEE Trans. Components & Packaging Technol., 29(2), 261–267, 2006
106. L. Zhu, Y. Sun, J. Xu, Z. Zhang, D.W. Hess, and C.-P. Wong, "Aligned Carbon Nanotubes for Electrical Interconnect and Thermal Management", Proc. 55th IEEE Electronic Component & Technol. Conf., Orlando, FL, USA, 44–50, 2005

107. H.A. Zhong, S. Rubinsztajn, A. Gowda, D. Esler, D. Gibson, D. Bucklet, J. Osaheni, and S. Tonapi, "Utilization of Carbon Fibers in Thermal Management of Microelectronics", Proc. 10th IEEE/CPMT International Symposium on Advanced Packaging Materials, Irvine, CA, USA, 259–265, 2005
108. K. Zhang, G.-W. Xiao, C.K.Y. Wong, H.-W. Gu, M.M.F. Yuen, P.C.H. Chan, and B. Xu, "Study on Thermal Interface Material With Carbon Nanotubes and Carbon Black in High-Brightness LED Packaging with Flip-Chip Technology", Proc. 55th IEEE Electronic Component & Technol. Conf., Orlando, FL, USA, 60–65, 2005
109. T.-M. Lee, K.-C. Chiou, F.-P. Tseng, and C.-C. Huang, "High Thermal Efficiency Carbon Nanotube-Resin Matrix for Thermal Interface Materials", Proc. 55th IEEE Electronic Component & Technol. Conf., Orlando, FL, USA, 55–59, 2005
110. J. Liu, M.O. Olorunyomi, X. Lu, W.X. Wang, T. Aronsson, and D. Shangguan, "New Nano-Thermal Interface Material for Heat Removal in Electronics Packaging," Proc. 1st IEEE Electronics Systemintegration Technol. Conf., Dresden, Germany, 1–6, 2006
111. Z. Mo, R. Morjan, J. Anderson, E.E.B. Campbell, and J. Liu, "Integrated Nanotube Microcooler for Microelectronics Applications", Proc. 55th IEEE Electronic Component & Technol. Confer., Orlando, FL, USA, 51–54, 2005
112. L. Ekstrand, Z. Mo, Y. Zhang, and J. Liu, "Modelling of Carbon Nanotubes as Heat Sink Fins in Microchannels for Microelectronics Cooling", Proc. 5th International Conference on Polymers and Adhesives in Microelectronics and Photonics, Wroclaw, Poland, 185–187, 2005
113. L. Zhu et al., "Well-Aligned Open-Ended Carbon Nanotube Architectures: An Approach for Device Assembly," Nano Lett., 6(2), 243–247, 2006
114. J.A. Paradiso and T. Starner, "Energy Scavenging for Mobile and Wireless Electronics", Pervasive Computing 05, 18–27, 2005
115. E.K. Reilly, E. Carleton, and P.K. Wright, "Thin Film Piezoelectric Energy Scavenging Systems for Long Term Medical Monitoring", Proceedings of the International Workshop on Wearable and Implantable Body Sensor Networks, 38–41, 2006
116. X.D. Wang, J.H. Song, and Z.L. Wang, "Nanowire and Nanobelt Arrays of Zinc Oxide from Synthesis to Properties and to Novel Devices", J. Mater. Chem. 17, 711–720, 2007
117. Z.L. Wang, "Piezoelectric Nanostructures: From Growth Phenomena to Electric Nanogenerators", MRS Bulletin, 32, 109–116, 2007
118. Z.L. Wang, "The New Field of Nanopiezotronics", Materials Today, 10, 20–28, 2007
119. Z.L. Wang, Nanopiezotronics. Adv. Mater., 19, 889–892, 2007
120. Z.L. Wang, "Zinc Oxide Nanostructures: Growth, Properties and Applications", J. Phys. Condens. Matter., 16, R829–R858, 2004
121. Z.L. Wang, X.D. Wang, and J.H. Song, "Piezoelectric Nanogenerators for Self-Powered Nanodevices", IEEE Perv Comp, 7, 49–55, 2008
122. B. Buchine, W.L. Hughes, and F.L. Degertekin "Bulk Acoustic Resonator Based on Piezoelectric ZnO Belts", Nano Lett., 6, 1155–1159, 2006
123. X.D. Wang, J. Zhou, and J.H. Song, "Piezoelectric Field Effect Transistor and Nanoforce Sensor Based on a Single ZnO Nanowire", Nano Lett., 6, 2768–2772, 2006
124. C.S. Lao, Q. Kuang, and Z.L. Wang "Polymer Functionalized Piezoelectric-FET as Humidity/Chemical Nanosensors", Appl. Phys. Lett. 90, 262107 (2007)
125. J.H. He, C.L. Hsin, and J. Liu "Piezoelectric Gated Diode of a Single ZnO Nanowire", Adv. Mater., 19, 781–784, 2007
126. Z.L. Wang and J.H. Song, "Piezoelectric Nanogenerators Based on Zinc Oxide Nanowire Arrays", Science, 312, 242–246, 2006
127. M.H. Zhao, Z.L. Wang, and S.X. Mao, "Piezoelectric Characterization on Individual Zinc Oxide Nanobelt under Piezoresponse Force Microscope", Nano Lett., 4, 587–590, 2004

128. Y.F. Gao and Z.L. Wang, "Electrostatic Potential in a Bent Piezoelectric Nanowire. The Fundamental Theory of Nanogenerator and Nanopiezotronics", Nano Lett., 7, 2499–2505, 2007
129. X.D. Wang, J.H. Song, and P. Li, "Growth of Uniformly Aligned ZnO Nanowire Heterojunction Arrays on GaN, AlN, and Al0.5Ga0.5 N Substrates", J. Am. Chem. Soc., 127, 7920–7923, 2005
130. P.X. Gao, J.H. Song, and J. Liu, "Nanowire Nanogenerators on Plastic Substrates as Flexible Power Source", Adv. Mater., 19, 67–72, 2007
131. J.H. Song, J. Zhou, and Z.L. Wang, "Piezoelectric and Semiconducting Coupled Power Generating Process of a Single ZnO Belt/Wire. A Technology for Harvesting Electricity from the Environment", Nano Lett., 6, 1656–1662, 2006
132. X.D. Wang, J.H. Song, and J. Liu, "Direct Current Nanogenerator Driven by Ultrasonic Wave", Science, 316, 102–105, 2007
133. X.D. Wang, J.H. Song, and C.J. Summers, "Density-Controlled Growth of Aligned ZnO Nanowires Sharing a Common Contact: A Simple, Low-Cost, and Mask-Free Technique for Large-Scale Applications", J. Phys. Chem. B, 110, 7720–7724, 2006
134. J. Frühauf and S. Krönert, "Wet Etching of Silicon Gratings with Triangular Profiles", Microsyst. Technol., 11, 1287–1291, 2005
135. M.H. Huang, S. Mao, and Y. Feick, "Room-Temperature Ultraviolet Nanowire Nanolasers", Science, 292, 1897–1820, 2001
136. X.D. Wang, C.J. Summers, and Z.L. Wang, "Large-Scale Hexagonal-Patterned Growth of Aligned ZnO Nanorods for Nano-Optoelectronics and Nanosensor Arrays", Nano Lett., 4, 423–426, 2004
137. H.J. Fan, F. Fleischer, and W. Lee, "Patterned Growth of Aligned ZnO Nanowire Arrays on Sapphire and GaN Layers", Superlattices Microstruct, 36, 95–105, 2004
138. M.H. Huang, Y. Wu, and H. Feick, "Catalytic Growth of Zinc Oxide Nanowires by Vapor Transport", Adv. Mater., 13, 113–116, 2001
139. J.H. Song, X.D. Wang, and E. Riedo, "Systematic Study on Experimental Conditions for Large-Scale Growth of Aligned ZnO Nanowires on Nitrides", J. Phys. Chem. B, 109, 9869–9872, 2005
140. J. Zhou, N.X. Xu, and Z.L. Wang, "Dissolving Behavior and Stability of ZnOWires in Biofluids: A Study on Biodegradability and Biocompatibility of ZnO Nanostructures", Adv. Mater., 18, 2432–2435, 2006
141. X.D. Wang, J. Liu, and J.H. Song, "Integrated Nanogenerators in Biofluid", Nano Lett., 7, 2475–2479, 2007

Chapter 16
Wafer Level Chip Scale Packaging

Michael Töpper

Abstract Wafer Level Packaging (WLP) based on redistribution is the key technology which is evolving to System in Package (SiP) and Heterogeneous Integration (HI) by 3-D packaging using Through Silicon Vias (TSV). Materials and process technologies are key for a reliable WLP. It is not only the choice for the right polymer or metal but the interfaces could be even more critical like under bump metallurgy or the adhesion of polymers. This chapter focuses on the materials and processes for WLP which are the basic for all new 3-D integration technologies.

Keywords Wafer level packaging · WL-CSP · Redistribution · Bumping · Thin film · Polymers · Photo-resists · Adhesion · BCB · PI · UBM

16.1 Introduction

Electronic packaging and assembly is the basic technology to link the small dimensions of the IC to an interconnecting substrate – usually the Printed Circuit Board (PCB) or a Multilayer Ceramic (MLC) like Low Temperature Co-Fired Ceramics (LTCC) [1, 2]. These substrates combine a number of ICs and passive components to build the final microelectronic system for the users [3]. New applications with their expanding performance and functionality in conjunction with new device technologies are pushing the requirements and innovation for electronic packaging. Milestones for this progress have been Surface Mount Technology (SMT), Flip Chip in Package (FCIP), Flip Chip on Board (FCOB) and Wafer Level Packaging (WLP) which is evolving to System in Package (SiP) and Heterogeneous Integration (HI) by 3-D packaging using Through Silicon Vias (TSV) [4]. Therefore the technology boundaries between front-end semiconductor technology, packaging and system engineering are becoming seamless. Heterogeneous integration bridges the gap between nanoelectronics and its derived applications bringing together nanoelectronics,

---

M. Töpper (✉)
Fraunhofer IZM, Gustav-Meyer-Allee 25, D-13355 Berlin, Germany
e-mail: michael.toepper@izm.fraunhofer.de

microsystem technologies, bio-electronic and photonic component technologies. The focus of this chapter will be on the materials for WLP which is the basic for all new 3-D integration technologies coming up in the near future.

16.2 Definition of Wafer Level Chip Size Packaging

The evolution of Single Chip Packages (SCPs) has started from small metal boxes and developed over the Dual Inline Package (DIP) for through hole assembly and SMT packages such as the Plastic Quad and Flat Package (PQFP) to the Ball Grid Array (BGA) [5]. BGA packages use rigid or flexible interposer for the redistribution from the peripheral pads to the area array. A further miniaturization to a maximum of 1.2 × the chip size brought up the concept of the Chip Size Package (CSP) [6]. Typical area array pitch for CSP is currently 0.5 mm moving now to 0.4 mm which results in additional technology pressure on the PCB. DIP and PQFP represent packages with peripheral I/Os while BGAs and CSPs are area array packages which are assembled in Flip Chip (FC) fashion. FC is a face-down assembly technique originally developed by IBM as C4 (Controlled Collapse Chip Connection) which provides excellent performance and represents a must for dice with high I/O counts like microprocessors.

A major requirement for the flip chip interconnections are modified pads on the IC. The so called UBM (Under Bump Metallization) or BLM (Ball Limiting Metallurgy) is the basis for a low-ohmic electrical and mechanical contact between chip and substrate. The self alignment function of the assembly is one of the major advantages of flip chip assembly using solder. Chips can be misregistered as much as close to 50% off the pad center and the surface tension of the molten solder will align the pads of the chip to the substrate metallization. The disadvantage of flip chip assembled IC is that the bumps are the only mechanical links between chip and substrate. As a consequence the stress caused by the CTE (coefficient of thermal expansion) mismatch of the semiconductor die and the substrate act only upon the bump interconnects. Therefore, underfill (epoxy resins with filler particles) have to be filled in between the flip chip and the substrate which translates into extra costs in the assembly process.

The pin count per Si area of the electronic component is a crucial number for the package selection. It can range from less than 100 for RF MEMS, Discretes, Power IC, Analog IC, Passives, MEMS (inertial MEMS, pressure sensors), image sensors to more than 1000 for ASIC and even over 4000 for microprocessors. Peripheral packages can accommodate smaller pin count devices but area array packages are necessary for large I/O numbers if the package size needs to be small. Using area array interconnection enables a much larger interconnection pitch for the same amount of I/Os. If x and y are the die lengths and p_p and p_a are the peripheral and area array pad pitch, the maximum number of I/Os is $n = 2(x/p_p - 1) + 2(y/p_p - 1)$ for a peripheral layout and $n = (x/p_a - 1)(y/p_a - 1)$ for an area array layout. A 5×5 mm large die, for example, could have a maximum of $9 \times 9 = 81$ I/Os in an area array design with a pitch of 0.5 mm. This can be assembled with standard SMT equipment. In a peripheral package the pitch would have to be 0.23 mm which would be difficult to assemble with standard SMT and board technology and would lead to a cost increase.

A large variety of CSP types has been developed since this idea was brought up to the market [7, 8]. Most of these packages using standard die level packaging technology based on leadframes, flexible or rigid interposers with the standard interconnection technologies (wire bonding, TAB and flip chip).

The biggest disadvantage of these highly miniaturized CSPs is the CTE mismatch between the Si and the PCB. Larger packages can withstand the CTE mismatch to a larger number of temperature cycles due to the solder balls with diameter of over 350 μm. The displacement of the solder balls during the temperature cycling is still in the elastic region of the solders. The redistribution capability offers a doable link between the dimension of the FE (Front End) to the PCB.

Low cost and miniaturization with increased functionality is driving the packaging industry towards adopting the concepts of Wafer Level Packaging (WLP) [9]. The idea of WLP is to finalize as much of the packaging sequence on wafer level as possible. WLPs are by definition true chip size packages using only a fan-in routing which is also a limitation. There is only the chip area available for the redistribution. The steady die shrink together with the reduction of the peripheral pad pitch is reaching a limit for the 0.5 mm area array pitch. Wafer level processes are independent of the number of dice and the number of bond pads per die and wafer. With WLP the back end will benefit from productivity advances in the front end such as larger wafer diameters and die shrink [10]. This is not the case with all the other types of CSPs where each die has to be individually mounted on a carrier or an interposer [11].

Typically dice have a peripheral pad layout, therefore a redistribution process becomes necessary to reroute the peripheral pads to the solderable area array pads. An example for a microprocessor is given in Fig. 16.1.

A redistribution layer is a combination of polymer and metal layers. Polyimide (PI), Polybenzoxazole (PBO) or Benzocyclobutene (BCB) is used for the dielectric isolation. If the process is done at the back end of a front end line also inorganic interlayers like silicon nitride may be used. Aluminum or copper are the metals of choice for the rerouting metallization.

The first redistribution technology was published in 1994 by Sandia. In 1995 Fraunhofer IZM and Technical University of Berlin (TUB) initiated several German and European projects to explore this technology for different

Fig. 16.1 Wafer level chip size package: The peripheral pads are redistributed by Cu/BCB technology into an area array. Solder balls are attached to the pads of the redistribution

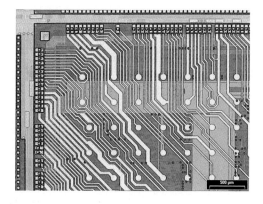

applications. By the years more and more companies started to offer WLP services in large volume. Flip Chip International (former Flip Chip Technology) in Phoenix, Arizona, and Amkor (starting out of MCNC as Unitive) at the Research Triangle Park, North Carolina (and since 1999 also in Taiwan) created standards in this technology under the trade names UltraCSP (FCI) and Xtreme (Unitive) and have shipped WLPs in million pieces per week.

The redistribution technology for WL-CSP technology can be expanded to a higher integration level for 3-D integration. A base chip on wafer-level can be used as an active substrate for FC-bonding of a second die. The electrical and mechanical interconnection is done using eutectic solder balls which are deposited by electroplating. The base chip is redistributed to an area array of UBMs. A low electrical resistivity of the redistribution is achieved by electroplating copper. The dielectric isolation is achieved using the low-κ Photo-BCB. An example for this 3-D WL-CSP is given in Fig. 16.2. It shows such a stacked FC-BGA with a flip chip mounted microcontroller on a Silicon chip with redistributed IC pads. The interconnection from the interposer to the board is done using wire bonding.

Redistribution technology provides the possibility to integrate passive components like resistors, capacitors and inductors into the thin film wiring on wafer level. The potential of integrated passives is obvious if one compares it with the revolutionary change in electronics going from single transistors to integrated circuits. Moore's law was the result of the constant development in on-wafer technologies. The main difference is that passives cannot be scaled down to sub-microns due to physical limits. In addition, the possible reduction in footprint is limited for integrated passives. For example, discrete ceramic SMT capacitors are using multilayer structures (10–20 layers) which are unacceptable for integrated passives. Therefore, SMT capacitors can achieve higher capacitance values at a given board space.

Many of the wafer level processes appear to be similar to front end processes. But process requirements are very different. Therefore, standard front end equipment is often not a good choice for wafer bumping or wafer level

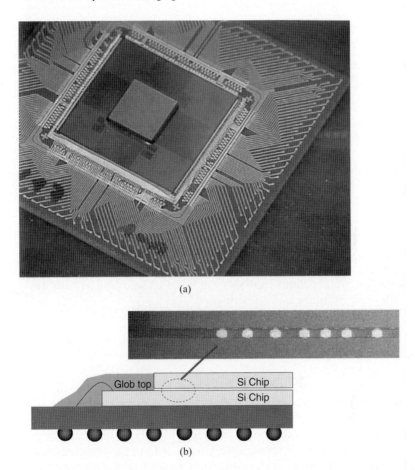

Fig. 16.2 Chip-on-chip integration using FC-bonding (courtesy of Fraunhofer IZM and Infineon): optical image (**a**) and cross-sectional image (**b**) of chip-on-chip package

packaging because it leads to over-investment and over-engineering. This has to be kept in mind when setting up a wafer level packaging line.

16.3 Materials and Processes for Bumping and Redistribution Technology

There are different possibilities to classify materials for WLP. One principal difference is whether they are used as permanent or auxiliary materials. The interaction between the materials and the processing equipment is essential from the reliability point of view (Fig. 16.3).

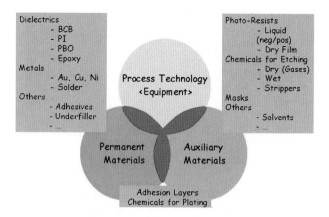

Fig. 16.3 Interaction of materials and equipment for wafer level packaging [12]

In the following sections the focus will be on metals, photo-resists and polymers for wafer bumping and redistribution.

16.3.1 Metals for Wafer Bumping

A bump formed in most cases on wafer level is defined as a usually conducting 3-dimensional interconnect element between die and substrate [13]. The interconnect process between chip and substrate is based on soldering, thermo-compression bonding and adhesive bonding [14]. Many different bump metallurgies are used ranging from pure Au, Cu, Sn or In to alloys such as eutectic or high-melting PbSn, AuSn, AgSn, SnCu and AgSnCu depending on the application. Legislations like the one by the European Union and other countries demand the ban of lead from electronic products by 2006 with the exception of high-lead solders for microprocessor applications. This ban of lead for electronic products by RoHS (*Restriction of the use of certain hazardous substances in electrical and electronic equipment*) has changed the materials for choice dramatically.

16.3.1.1 Under Bump Metallization

The UBM has to provide a low contact resistance to the chip pad and the solder, good adhesion to the chip metallization and the chip passivation and a hermetic seal between UBM and IC pad [15]. It has to be a reliable diffusion barrier between the IC pad and bump with low film stress and it needs to be sufficiently resistant to stress caused by thermal mismatch during die assembly. In the case of PbSn bumping, common UBM stacks are Cr–Cr:Cu–Cu–Au (original C4 from IBM); Ti–Cu; Ti:W–Cu; Ti–Ni:V; Cr–Cr:Cu–Cu; Al–Ni:V–Cu;

Ti:W(N)–Au. Usually, these UBM stacks are sequentially deposited by sputtering or evaporation. The advantage of sputtering over evaporation is the higher kinetic energy of the deposited atoms (0.1–0.5 eV for evaporation and 1–100 eV for sputtering), which guarantees a much higher adhesion. For 200 and 300 mm wafers the evaporation distance has increased to a nearly unacceptable level which decreases further the deposition efficiency which is proportional to the square of distance.

The UBM etch process removes the UBM metallization between the bumps. For cost and technology reasons wet chemical etching is common. For a UBM stack consisting of different metal layers different etch chemistries are required for each layer. Among the requirements for the etching step are to achieve a uniform etching result and a minimum bump undercut, and to monitor the remaining metallization thickness in order to stop the etching process or to switch the etch chemistry in case a layer of a UBM stack is fully removed. It is important to design the etching process in such a way that the bump surface is not oxidized or modified in any other way. In addition, the design of the UBM stack has to take the UBM etching process into account in order to achieve reliable and good process results.

A schematic drawing of Ti:W-Cu for PbSn is given in Fig. 16.4. In the case of Sn-based bumps deposited onto a copper based UBM, intermetallics compounds (IMCs) between Sn and Cu are formed by the reflow process providing the required adhesion of the bump to the chip pad. IMCs are brittle in nature due to the ordered crystal structure which is in contrast to the solid solutions like the PbSn. The metals which are mostly used in packaging – Cu, Ni, Au and Pd – form binary intermetallics with Sn-based solders of the Hume Rothery type [17]. These compounds are based on electron valence bonding. The crystal structure is controlled by the number of electrons in the bonding. The composition of each phase can be calculated based on the concentration of the valence electrons. For example the Cu_3Sn and the Cu_6Sn_5 phases are found for intermetallics of Cu and Sn and Ni_3Sn_4 and Ni_3Sn phases are formed between Ni and Sn. The growth rate depends on temperature, the different activation energies

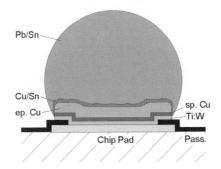

Fig. 16.4 Example of a UBM stack (Ti:W-Cu) for PbSn bumps [16]

of compound forming and diffusion processes. In general, the intermetallic growth rate is much higher for Cu compared to Ni. This is becoming more important for the lead-free solder due to their higher Sn content.

16.3.1.2 Bumping Technologies

The main bumping processes are electroplating, stencil printing, evaporation, placing preformed solder spheres and C4NP. The selection of UBM and bump metal mainly depends on the melting point of the solder, the thermal and mechanical reliability of the interfaces between UBM and bump, the integrity of adjoining pad metallization, the bumping process capability, the operating conditions of the assembly, and the reliability demands on the whole package. The main process steps are summarized in Fig. 16.5.

A major requirement for all bumping deposition technologies is the control of the bump volume across the wafer as well as the solder composition in order to achieve a uniform bump height distribution and to avoid an incomplete reflow process. The maximum reflow temperature is typically a few 10°C above the melting point of the solder and it is important that an optimized temperature profile is maintained during the reflow process. All interconnects of a chip are linked at the same time for the FC-assembly process. There is the risk that the chip will not work if one joint (in the case of a signal line) is failed.

Evaporation is the deposition method which was originally developed by IBM for FC bonding. It became popular under the acronym C4 (controlled collapse chip connection) summarizing the main steps of soldering (collapse) in combination with a necessary solder mask (controlled). The technology allows

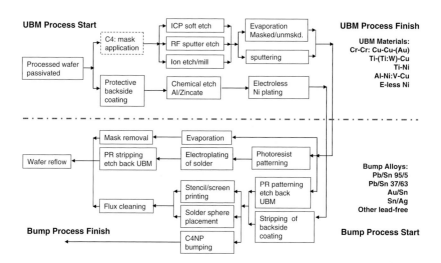

Fig. 16.5 Overview of bumping technologies [18]

a wide range of solder materials with excellent quality. The transition to 200 mm wafer size was the end for evaporation due to yield and cost.

Electrodeposition of metal through a lithographically defined plating mask introduced by Hitachi in 1981 [19] is the bumping method of choice to meet the requirements of defined bump shape and size. Mostly ECD (Electro-Chemical Deposition) is used as an acronym for this technology. In general electroplating is a relative slow deposition technique with typical plating speeds ranging from 0.2 µm to only a few µm per minute depending on the deposited material [20]. Plating techniques can use constant voltage (potentiostatic), constant current (galvanostatic) or pulse plating. Pulse plating is the method of choice for fine-pitch application providing a more uniform, smooth deposit with less porosity. Among the most important parameters that influence the uniformity of plating height and solder composition as well as bump morphology is the electrical field distribution across the wafer as it defines the plating current. Therefore the voltage has to be applied over many points along the wafer parameter. In this case the photo resist is completely removed along the perimeter of the wafer (edge bead removal) and the electrode in form of a ring is attached to the wafer. A sealing ring is put on top of the resist surface to prevent the electrode from being contaminated by the electrolyte. The current distribution has approximately rotational symmetry but can show a radial variation. In this case the anode design in fountain platers offers another means to control the plating current uniformity by compensating for radial field variations. In addition, the ratio of open area (i.e. total plated area) versus full wafer area influences the plating current uniformity. It is important to have a uniform distribution of bumps across the wafer surface which may require to deposit bumps in some areas on the wafer without any die underneath (dummy bumps). A proper tool design ensuring uniform current density and equalized bath agitation across the wafer allows a bump height homogeneity of less than ± 5% at 300 mm in diameter.

In the case of PbSn plating, tin and lead salts are dissolved in the electrolyte and dissociate into their anions and cations. Sulfuric acid is added to increase the conductivity of the electrolyte. Additives are added to refine the deposited solder. The positively charged Sn^{2+} and Pb^{2+} cations migrate to the cathode (wafer) due to the applied voltage in the plating cell and are deposited by a discharge reaction on its surface. For this to occur the cations need to be reduced to the metallic state by accepting electrons from the cathode. The electroplating process is a complex process. The mechanism of the metal deposition consists of several steps. The hydrated metal ions have to diffuse to the wafer surface covered by a Helmholtz double layer. In addition the metal ions can be chemically attached to complexing molecules. Additives are controlling the metal growth to achieve fine-grained solders. Obviously, the plating time can be shortened by increasing the plating current density. However, the maximum current density is limited because high plating speeds translate into higher challenges in plating bath maintenance.

Though structure dimensions are not as critical as in advanced front-end processing like 65 nm node, they are tiny enough to be defined best by photo-imaging. In contrast to subtractive etching or lift-off technique, ECD bump formation precisely replicates the photoresist pattern in lateral dimensions. The ECD bumping technique can accommodate a wide range of wafer types, passivation materials and pattern configurations. All kinds of semiconductor such as silicon, SiGe, GaAs, and InP as well as ceramic and quartz substrates can be processed. Besides, there is no restriction in passivation types such as silicon oxide, oxy-nitride, silicon nitride, and polymers like PI or BCB in the main. For standard I/C wafers, the bumping sequence will be applied directly onto the I/O pads. If the original I/O layout has been redistributed, the bumping sequence will take place onto the routed metal layer covered by a dielectric solder resist mask. The placement of solder bumps on top of a polymer layer offers reduced self-capacitance, which is desirable for RF applications. As the ECD bump fabrication can be incorporated into the thin-film technology, the sequence can be divided into fundamental process steps which are sputtering of the UBM, lithographical printing, electroplating of the bump, removal of the photoresist, and differential etching of the plating base schematically shown in Fig. 16.6.

In this process, the UBM that provides the adhesion layer in combination with a diffusion barrier and a wettable surface is sputtered and a thick liquid or dry film resist is coated onto the wafer. The resist is then exposed and developed. Electroplating allows to deposit the solder over the top of the resist layer

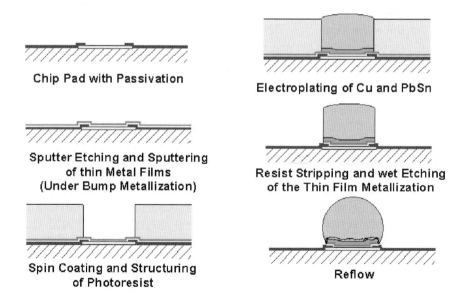

Fig. 16.6 Solder bumping process flow (electroplating) [16]

to form mushroom like structures. Then the resist is stripped and the UBM is etched between the bumps. The reflow process transforms the solder into nearly ball shape and leads to the formation of intermetallic compounds at the UBM/solder interface which is important for a reliable adhesion of the bump to the UBM. The advantage of mushroom plating is that the photo resist layer can be significantly thinner than the final solder ball height and that solder deposition is fast because the solder surface is growing when being plated over the resist edge. A disadvantage is that with mushroom bumping plating control becomes more demanding. Photo resist thickness in mushroom plating typically is between 25 and 60 μm. As a result, mushroom plating becomes incapable for finer bump pitches. Thicker resist (approximately 100 μm) is used for fine pitch bumping where the solder is completely plated in the bump mold.

If micromechanical elements are present on the wafer, care must be taken to avoid possible damaging during thin film processing [20]. Cavity surface contamination can later impair MEMS performance, too. In some cases it will be necessary to protect the MEMS areas by covering with photoresist prior to the sputter step, or by local etching the plating base prior to the electroplating process. Some 3-dimensional structures with cavities like air bridges or acceleration transducers require more than one photoresist layer. In that case a primary coated and patterned resist layer has to fill the hollow space of the complex items and consequently serves as a smooth subsurface for the following UBM deposition.

For solder printing, either a metal or resist stencil (for finer pitches) is used. The deposition of the UBM needs to be done before solder printing. A thin film process based on sputtering (sometimes in combination with plating) can be used for depositing the UBM. A low cost method is the electroless deposition of Ni and immersion deposition of gold (ENIG) on the Al pads. The ENIG process is based on the selective chemical deposition of metal on Al bond pads. Wafers are treated in a sequence of chemical solutions. After each treatment they have to be rinsed carefully in DI water. The principle of the process is shown in Fig. 16.7 [21].

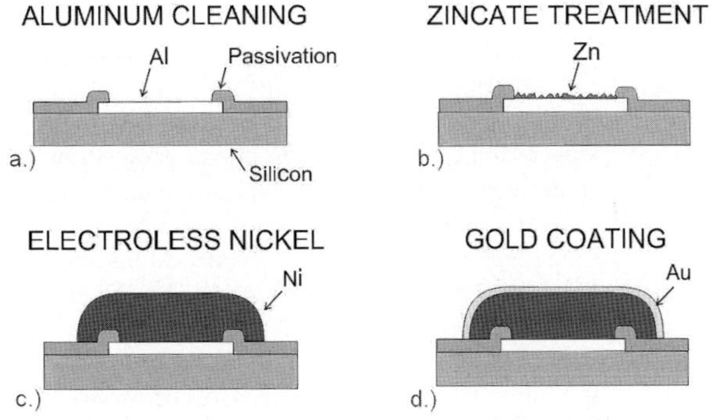

Fig. 16.7 Principle process of electroless Ni bumping: (**a**) bond pad in initial state, (**b**) after zinc deposition, (**c**) after growth of electroless Ni, (**d**) after plating of thin immersion Au [21]

First the surface of Al bond pads is cleaned by immersing the wafers into two Aluminum cleaning baths. The passivation cleaner removes possible residues while the second (Al cleaner) removes thick Al oxides and roughens the surface. In a zincate bath a thin Zn layer is deposited on Al by an exchange reaction in order to activate the surface for subsequent Ni plating. The electroless Ni bath contains mainly Ni ions and hypophosphite. A first Ni layer is deposited on the pads by an exchange reaction between Zn and the Ni ions. On the first layer, additional Ni is plated by a continuous autocatalytic reaction which is necessary to plate more than monolayers without any current. The energy is supplied internally inside the plating bath by the oxidization of adsorbed hypophosphite. The released electrons are able to reduce the Ni^{2+}. The phosphorus converted from hypophosphite is built into the Ni-layer. This can change the mechanical and electrical property of the Ni. The plating rate is 25 µm/hr. In the subsequent immersion Au solution, a thin Au film is deposited by an exchange reaction on the surface of the Ni layer. The Au has a thickness of 0.05–0.08 µm and is required to prevent Ni from oxidation. But the formation of brittle Au–Ni–Sn intermetallic phases at the interface between Ni and solder can impact the long term operation reliability of the interconnections, therefore the Au layer has to be kept as thin as necessary.

The complete process flow is shown in Table 16.1. In addition to the wet chemical treatments mentioned above, the backsides of the wafers have to be protected in order to prevent Ni deposition on Si. This is done by spin-coating of a protective resist on the backside of the wafer. After bump plating, the resist is stripped. All chemicals used in this process are commercially available. They are completely cyanide-free and no organic solvents are used.

For all wet-chemical treatments, up to 25 wafers can be handled together in one carrier. The process requires tanks with seven different chemical baths and additional rinse tanks. The process times are relatively short. They range from 30 s (zincating) to 30 min (immersion Au). By handling the wafer cassettes manually from bath to bath a throughput of 25 wafers per hour can be achieved. In fully automatic systems 100 wafers per hour are possible. The Ni height uniformity is better than ± 5 % over a 200 mm wafer. Within die variation is

Table 16.1 Process steps of electroless Ni bumping and their function (courtesy of Fraunhofer IZM)

Process step	Function
Protective resist coating	Protective resist coating on wafer backside
Passivation cleaning	Removes passivation residues from Al pads
Al cleaning	Removes thick Al oxides and prepares surface for metal deposition
Zincating	Activates Al for Ni deposition
Electroless Ni	Deposition of Ni layer (typ. 5 µm)
Immersion Au	Au finish on Ni (typ. 0.08 µm) to prevent Ni from oxidation
Backside cleaning	Removes protective coating from backside

Table 16.2 Specification of the ENIG process (courtesy of Fraunhofer IZM)

Property	Specification
Wafer material	Si
Bond pad material	AlSi1%, AlSi1%, Cu0.5%, AlCu2%
Pad metal thickness	≥ 1 μm
Passivation	Defect-free Nitride, Oxide, Oxi-Nitride, Polyimide, BCB
Residues on bond pads	
non organic	<5 nm
Organic	not acceptable
Wafer size	100–300 mm
Wafer thickness	>200 μm (> 150 μm)
Bond pad geometry	Any (square, rectangular, round, octagonal)
Passivation opening	>40 μm
Bond pad spacing	>20 μm
Passivation overlap	5 μm
Wafer fabrication process	CMOS, BiCMOS, Bipolar
Ink dots	Acceptable Stability depend on ink
Probe marks	Acceptable
Scribe lines	Must be passivated (thermal oxide) Test structures acceptable
Laser fuses	
AL fuses	Not acceptable
poly Si fuses	Acceptable (with limitations)

correspondingly lower. The specification of the Fraunhofer IZM ENIG process is given in Table 16.2.

Control of the UBM quality is monitored by shear strength which has to be around 150 MPa (min. 100 MPa). The Al etching process has to be restricted to less than 0.5 μm to avoid damages of the Silicon devices. The Ni UBM has been extensively tested. No failures were detected even after 10,000 hrs thermal storage at 300°C, 10,000 cycles AATC ($-55/+125°C$) and 10,000 hrs humidity storage (85°C/85% r.h.) [22].

The solder deposition is then done by printing process shown in Fig. 16.8 [23].

This process was transferred from PCB industry to wafer technology. The solder pastes that are screened into the apertures of the stencil consist of solder particles of 2–150 μm diameter in binders and flux. They are classified according to the particle size of the solder (Table 16.3).

An additional function of the paste is the pre-fixing of the placed components on the substrate before reflow. The viscosity of the pastes should be between 250–550 Pa·s for screen printing and 400–800 Pa·s for stencil printing.

The screens are typically made by laser drilling of metal sheets or by electroforming. The printer aligns the stencil to the substrate or wafer. The paste is pressed within seconds through the stencil and thousands of pads are bumped at a time. One of the key issues is the requirement that all the paste has to be transferred from the stencil to the wafer. Any solder paste residues in the screen

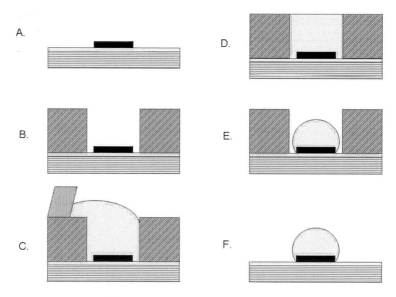

Fig. 16.8 Solder printing using photo-resists (UBM formation [A], stencil or photo-mask [B], printing [C, D], reflow [E] and removal of the mask [F]) [23]

Table 16.3 Solder paste classes

CLASS	1	2	3	4	5	6	7
Solder sphere size [μm]	75–150	45–75	20–45	20–38	15–25	5–15	2–11

will reduce the final bump height uniformity. The metal stencils have to be cleaned by solvents. Mostly water-soluble solder pastes are used in production. To calculate the size of the reflowed bump as a function of pad size and geometry the following equation is used for a reflowed bump as a truncated sphere: $V = (1/2)A \cdot H + (\pi/6) \cdot H^3$ with V as the solder volume, A as the pad area and H for the bump height.

Reflow is necessary to form the bump shape out of the printed solder paste. Due to the presence of flux in the solder paste, the formation of voids during the reflow processes is likely. There is no reliability issue if the voiding is kept under a certain level depending on the design. Solder printing is a fairly simple and inexpensive process step when compared to electroplating or evaporation. An important advantage of stencil printing is the large variety of available solder pastes. This offers flexibility and is of particular importance for selecting lead free solders. Even triple component solders such as SnAgCu (SAC) can be deposited, which is difficult to electroplate.

For fine pitch below 150 μm, a photo-resist mask with a thickness of 70 μm or thicker gains a lot of advantages for the printing process. But it has a higher cost than metal stencils. The process is based on the work done by Flip

Chip International (FCI, former FC Division of K&S, founded as FCT) and is widely used by companies that licensed their Flex-On-Cap-Process (FOC) [23]. As the UBM pad defines the final bump base the molds for the stencil process can have a larger footprint than the final bump in order to have more solder paste being screened into the mold to achieve a larger bump height. Before resist stripping the solder paste has to be heated to be transformed into solid solder. Examples of this process are given in Figs. 16.9 and 16.10 [23].

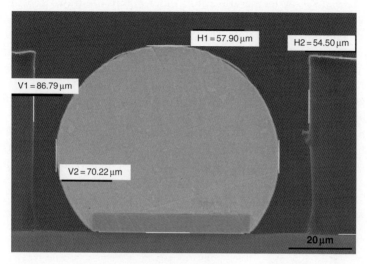

Fig. 16.9 Printed SnAgCu solder into dry-film resist [23]

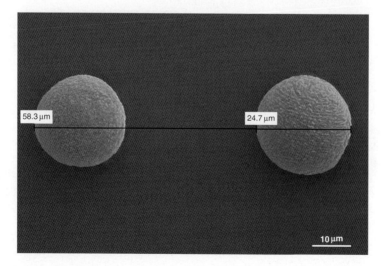

Fig. 16.10 Printed SnAgCu bumps (25 μm with 60 μm pitch) after resist stripping [23]

C4NP is a novel solder bumping technology developed by IBM which addresses the limitations of existing bumping technologies by enabling low-cost, fine pitch bumping using a variety of lead-free solder alloys [24]. It is a solder transfer technology where molten solder is injected into pre-fabricated and reusable glass molds. The basic process flow is given in Fig. 16.11.

The glass mold contains etched cavities which mirror the bump pattern on the wafer. The filled mold is inspected prior to solder transfer to the wafer to ensure high final yields. Filled mold and wafer are brought into close proximity/ soft contact at reflow temperature and solder bumps are transferred onto the entire 300 mm (or smaller) wafer in a single process step without the complexities associated with liquid flux. The C4NP process was transferred to production by a cooperation between IBM and SussMicrotec. Different cost models are proposing low cost but the infrastructure of the glass mold is not established yet.

Other Bumping Technologies: Alternative bump technologies include stud bumping, solder jetting, and ball attach.

Direct ball attach is only used for large solder balls of typical diameters of 300 μm or above. In some cases a vacuum head serves as a template for the ball layout on the wafer and picks up preformed solder spheres from a reservoir. The

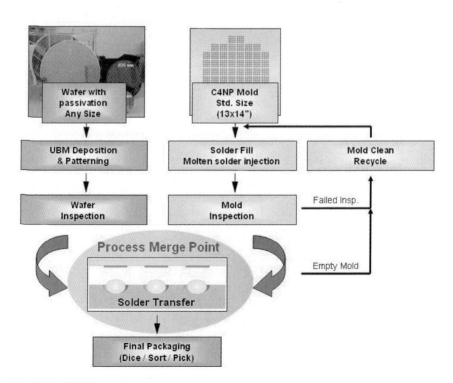

Fig. 16.11 C4NP bumping process sequence

spheres are dipped into flux and are placed on the wafer and are reflowed. The advantage of ball attach is that it is an inexpensive way to provide large solder volumes and that it is easily applicable to any solder type. Ball attach is often used to provide solder balls on redistributed dice where the pitch of the area array pad layout is larger or equal to 500 μm. Equipment makers are trying to push the limits to finer geometries.

Stud bumps are made with wire bonders by cutting of the wire right after bonding it to the IC pad. The bump can be either left with a spike or can be coined creating a flat surface or can be sheared off across the top directly after bonding. This technique is fairly flexible in regards to the desired bump metallurgy and can be used for example for gold bumping and even for solder bumping. Since stud bumping is a serial process it has little importance for mass production but is an important technique for flip chip prototyping and low volume manufacturing. Stud bumping is the major technology applicable to singulated dice.

Solder jetting is a serial and maskless solder deposition technique where solder droplets are ejected from a print head onto the wafer. High ejection frequencies are possible. However, overall process control is difficult and solder jetting has not yet been adoption by the industry. Several attempts have been made in the past without any success.

16.3.1.3 Bump Metallurgy

Bumps based on Gold and AuSn: Gold is a noble metal which does not oxidize or corrode. It has excellent electrical and thermal conductivity. Due to its high melting point gold bumps cannot be reflowed. The interconnection is mainly made to flexible substrates or tapes (TCP: tape carrier packaging) by thermocompression bonding (TAB: tape automatic bonding) or adhesively bonded to glass substrates (COG: chip on glass) using conductive films. Main application is for LCD drivers. The deposition gold bumps is done by electroplating on a sputtered Au layer with an adhesion/diffusion layer of Ti or TiW or other transition metals. A schematic structure is given in Fig. 16.12.

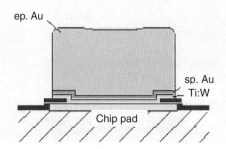

Fig. 16.12 Schematic of a Au bump [16]

Today, minimum gold bump pitch for LCD drivers in mass production is around 30 μm and less with a 10 μm gap between bumps (COG applications allow finer pitch than TAB). These are the tightest pitches of any bumping process in mass production.

For electroplating pure Au, both sulphitic as well as cyanidic electrolytes have been used. The advantages of the sulphitic bath type are its excellent compatibility to the applied photoresist system, its well-known non-toxicity, and the easy softening of the Au bumps during annealing. On the other hand, cyanidic Au electrolytes are easy to handle due to the more stable Au complex and offer deposits with strong adhesion even on Ni and Cu as base metal. Cyanidic solutions tend to underplating, which can become a critical fact for fine-pitch bumping. The metal content of most commercially available Au electrolytes is between 8 and 15 g/l, and the applicable current density of exemplarily 5–20 mA/cm^2 results in a deposition rate of 0.3 up to 1.2 μm/min. All kinds of Au bumping electrolytes work at higher temperatures, mostly between 50 and 70°C. In many cases of thermo compression bonding, the ductility of as-plated Au bumps is not sufficient and has to be improved by subsequent annealing step. During thermal aging at 200°C the initial micro-hardness of around 130 $HV_{0.025}$ is decreasing down to 70–50 $HV_{0.025}$ within a few minutes. The shape of the bump is totally controlled by the resist (thickness range of 30 μm) due to the lack of the reflow. Therefore the combination of resist and electrolyte has to be chosen carefully. Figure 16.13 shows a FIB (Focus Ion Beam) cut of a high structured resist. The resist has vertical sidewalls with a variation in the opening of ± 1 μm for 45.8 μm thickness.

In case of solder bumps, a final reflow is regularly done to homogenize the tin alloy and to form the spherical bump shape. Each solder composition requires

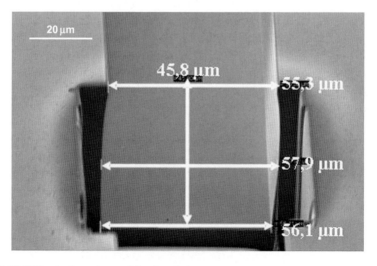

Fig. 16.13 FIB cut of a resist used for Au plating

an adapted temperature-time profile, which depends on the melting point, the ambient medium, and the out-gassing behavior of the solder. This step is not necessary for Au.

The electroplating process can create a lot of stress to the resist which can cause a deformation of the final structure. Figure 16.14 shows a comparison of Au bumps created by a non-optimized combination of plating electrolyte and photo-resist (left hand side) and the optimized combination (right hand side).

The ECD can cause a deformation of the resist to over 20 μm if the wrong combination of materials is chosen.

The **eutectic Au/Sn** system provides a good corrosion resistance and enables a flux-free Flip Chip assembly. Hence, it is the most suitable interconnection material for optical and optoelectronic devices. The basic structure of the AuSn bump is given in Fig. 16.15.

Au and tin are sequentially electroplated. Evaporation in a sandwich-like fashion is also possible.

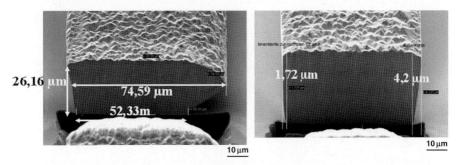

Fig. 16.14. Electroplated Au: non-optimized combination of electrolyte/photo-resist (*left*) and nearly perfect combination (*right*)

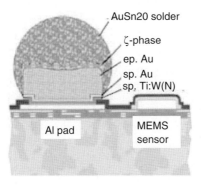

Fig. 16.15 Schematic of a AuSn bump [20]

Bumps based on Solder: An interconnection by soldering is the most common technology for all microelectronic systems [25]. The main advantage is the possibility to connect uneven and/or rough surfaces and allows by the same time the possibility of repair. The basic requirement is a wettable surface on both interconnection sides. To avoid the flow of the solder over the whole surface, a solder mask is needed. A high reliability for the lifetime of the product can be achieved if the solder is highly elastic. If the substrate and the chip has a large CTE mismatch, the solder might be outside its elastic state, and then an underfill has to be used. Underfills are highly filled epoxy polymers. The filler (small particles of SiO_2, for example) is necessary to reduce the CTE of the polymer which is much higher than inorganic materials such as Si or Metals. An overview of the most common solder materials is given in Table 16.4.

Solder interconnects are formed during a reflow process with a reflow temperature which is $10 + °C$ higher than the theoretical melting temperature of the solder.

Fluxes are needed to remove oxides during soldering process. In addition, the tackiness allows to hold the placed components in place before the reflow and the final joining process is completed. Fluxes can be inorganic acids, organic acids, rosin, and no-clean resins. The J-STD classification describes both flux activity and flux residues activity as follows: L = Low or no flux/flux residue activity; M = Moderate flux/flux residue activity and H = High flux/flux residue activity. These classes are further labeled for activity or corrosiveness.

Table 16.4 Selection of solders for flip chip interconnects

Solder	Melting point	Remark
63Pb37Sn	183°C	Eutectic PbSn, low melting point, compatible with organic PCBs, commonly used for most SMDs. Not allowed anymore for ROHS
95Pb5Sn (or similar)	315°C	High-lead, good electromigration behavior, highly reliable thermo-mechanical interconnect, flip chip on ceramic substrate; no reflow of high lead bumps during chip attach on PCB (eutectic PbSn on PCB side), flux free reflow in H_2 atmosphere.
96.5Sn3.5Ag (or similar)	221°C	Currently most common binary lead free solder for flip chip. Typically used in conjunction with electroplating.
97Sn/3Cu	227°C	Difficult to electroplate. Short bath lifetime.
95.5Sn3.9Ag0.6Cu	218°C	Common lead-free solder paste, Cu content reduces Cu consumption from UBM.
80Au20Sn	280°C	Common for flux free opto-electronic assembly on gold finishes, controlled stand off height.
In	157°C	Ideal for temperature sensitive electronic devices due to very low reflow temperatures.
Sn	232°C	Risk of tin whisker formation

Copper Posts (Pillars): Copper has approximately a 10 times better electrical and thermal conductivity than PbSn solders. Copper posts are therefore attractive whenever high currents pass through the interconnect like in the case with power devices. This is particularly important for fine pitch bumping which reduces the size of the bump base and increases the resistance of the bump. Cu pillars are more prone against electromigration and they are therefore a promising candidate for lead-free alternatives especially for high-current applications with small bump size [26]. Electromigration is the transport of material which is caused by the gradual movement of the atoms in a conductor due to the momentum transfer between conducting electrons and diffusing metal atoms. The effect is important in applications where high direct current densities are used, such as high density interconnects and high power applications. As the structure size in electronics such as integrated circuits (ICs) decreases, the practical significance of this effect increases. Cu pillar can be seen a thick UBM structure to alleviate current crowding at the interface between bump and chip pad.

The Cu bumps as well as the Cu routing layers are deposited from a sulfuric acidic bath containing an organic suppressor and an accelerator to achieve a bright and fine-grained Cu crystallization, especially developed for semiconductor application but primary for damascene processing.

Copper posts are not reflowed and therefore can be used to provide high aspect ratio structures for fine pitch bumping without reducing the bump stand-off height. To form interconnection to the substrate, a solder tip is added on top of the post or on the board pad. Copper posts are plated and typical structures are shown in Fig. 16.16.

Unlike solder plating copper post require in-via plating with very thick resist layers of 70 + μm.

Copper deposition will attract more attention in the future by the up-coming trend to 3-D integration. TSV (Through Silicon Vias) are a major step for these technologies. A schematic graph shows the importance of Cu in Fig. 16.17.

The Si-etching by Deep Reactive Ion Etching (Deep-RIE) is a well-know process common in the MEMS industry (Bosch process). Major issues are the passivation of the Si, the deposition of the Cu-seed layer inside the Si via and the Cu filling process.

Plating of Ni: The used Ni electrolyte is chemically based on sulphamic acid and generally origins from electroforming industry, where very thick and low-stress layers are required. The Ni deposits look semi-bright, although a very low codeposition of organic additives happens.

16.3.1.4 Plating of Alloys

For solder bumping, different kinds of stannous alloy electrolytes and a pure Sn bath are used, all of which are based on methane sulphonic acid. Here, a non-porous crystallization and a low co-deposition rate of organics are required to prevent bubble formation inside the bump during the final reflow process. The

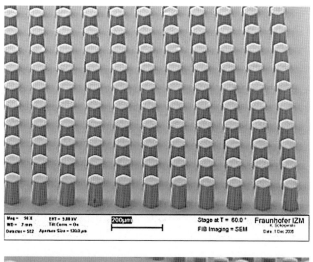

Fig. 16.16 Copper posts after resist stripping (*top*: 80 μm thick Cu plated inside dry film; *bottom*: 60 μm thick Cu plated in liquid type resist)

main issue of plating alloys at the same time is the difference of the electromotoric force (EMF). The difference in EMF is only 10 mV for PbSn. Pb is slightly more noble than Sn. The actual cell potential departs from the equilibrium value by the over-potential or cell polarization which includes all concentration, diffusion and other effects. These can be minimized by bath agitation, high ion content, lower current density etc. In addition the seed layer must be thick enough to avoid voltage drops below a few mV. A deposition rate of over 4 μm/min with a uniformity of less 3% (1 sigma) over 300 mm can be achieved with optimized conditions using automatic plating systems.

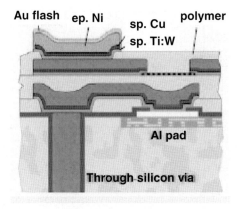

Fig. 16.17 Schematic graph of a TSV filled with Cu [20]

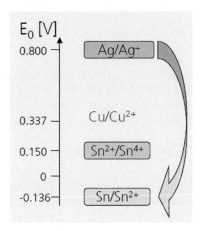

Fig. 16.18 EMF for Ag, Cu and Sn

The difference for the potentials for Sn, Ag and Cu are given in Fig. 16.18. Inhibitors (for example organic additives etc. or complexing agents) are acting differently for each ion partner, resulting in a different change in the deposition potential. For example a complexing agent can change the activity of the given ions. If cyanide ions are added to a Cu-ion solution the following reaction will take place:

$$Cu^+ + 3CN^- \rightleftharpoons [Cu(CN)_3]^{2-}$$

The concentration of Cu^+ ion will be reduced. Free Cu(I) concentrations is in the range of $\sim 0.5 \times 10^{-7}$ mol/l but it will be reduced by ~ 0.25 mol/l NaCN or KCN down to 10^{-26} mol/l. The EMF (Cu^+/Cu): is $+0.35$ V but will change

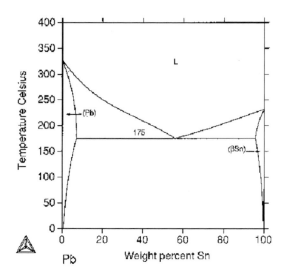

Fig. 16.19 Phase diagram of PbSn [27]

to -1.0 V by adding this cyanide salt. Such complexing agents are necessary to reduce the EMF of Ag/Ag+ to be co-deposited with Sn in case of SnAg ECD.

One advantage of processing PbSn compared to the lead-free alternatives is the melting temperature at the eutectic point shown in the phase diagram in Fig. 16.19 [27].

The eutectic melting temperature of 183°C is changed only slightly by varying the solder composition. This is totally different for SnAg and SnCu as shown in Figs. 20 and 21 [28].

For SnAg3.5 the eutectic melting temperature is 221°C. But the melting point will increase to 285°C by adding another 2.3% Ag. Similar for SnCu going from 227°C melting temperature for the 0.7% Cu eutectic composition to 264°C for 1.1% Cu. This means an exact control of the electroplating process to avoid cold joint in the assembly process. The schematic structure of a final SnAg bump is given in Fig. 16.22.

During ECD processing, several electrochemical reactions are taking place between the solution and both anode and cathode [20]. Permanent changes of the bath chemistry require an individual monitoring of each relevant organic as well as inorganic compound to ensure consistent plating results. The contents of metallic ions, acids, and other anionic compounds are commonly determined by titrimetric or spectrophotometric analysis. The organic additives are consumed during the electroplating process mainly by co-deposition, anodic decomposition, and drag-out loss. All these reactions are nearly in proportion to the current throughput and therefore predictable. Additionally, a time-dependent degradation of organics can often be observed. In the field of microelectroplating, cyclic voltametric stripping (CVS) is the most utilized method for the

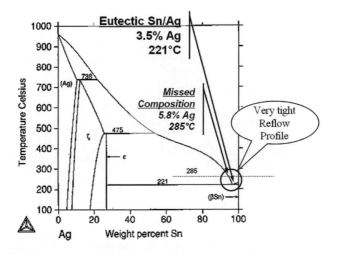

Fig. 16.20 Phase diagram of SnAg [28]

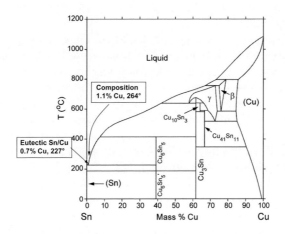

Fig. 16.21 Phase diagram of SnCu [27]

analysis of organic substances. The concentrations of all components mentioned above must be maintained by periodically adding the respective concentrates to the electrolyte. Today, fully automatic control units including the self-dosing function are commercially available for high-volume manufacturer. In the end, the changes of specific gravity, the bleed-out of photoresist and the accumulation of organic breakdown products limit the life-time of the plating bath.

The monitoring of the metal ions inside the lead-free alloy electrolytes is of utmost importance because of the low content of Ag and Cu. Small variations would lead to significant changes of the solder composition and therefore to an

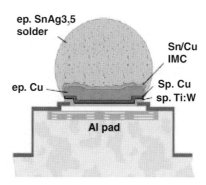

Fig. 16.22 Schematic of a SnAg bump

impact of the reflow manner. Due to the pure tin anodes mainly used for lead-free solder bumping, the Ag and Cu is not continuously dissolving but must be frequently replenished. The composition of the deposited Sn/Ag and Sn/Cu alloys can be determined by differential scanning calorimetry (DSC) and energy-dispersive X-ray analysis (EDX).

A further issue for these high Sn lead-free alternatives is the growth of Whiskers which are fine, hair-like needle mono-crystals. The diameter of these crystals is in the range of ~1 μm and the length can go up to the mm-range, resulting in the risks of electrical shorts between interconnects. One of the possible reasons for whisker formation is the stress from the substrate or by intermetallic layers. Growth rate of Sn whisker is in a range of 3 μm / month to 130 mm/month where tin atoms are diffusing to the bottom of the whisker being highest around 50°C. In general, alloys are reducing the risk of tin whiskers due to limiting the mobility of Sn atoms.

The compositions of selected electrolytes are summarized in Table 16.5. The electrolytes are shipped as ready to use or as a set of concentrates that will be mixed and diluted on site. Depending on the electrolyte, different kinds of analysis have to be made during the operation to monitor the composition of the electrolyte. The chemical suppliers offer application specific solutions to maintain a long lifetime of the electrolyte.

16.3.1.5 Photo Resists

Photo resists are photo-sensitive polymer-based materials that are applied temporarily on the wafer mostly for plating or etching [29]. A summary of the main properties is given in Table 16.6 [1].

The base resin of positive-tone photoresists is typically Novolak whereas negative-tone resists are based on acrylate or an epoxy resin. Positive-tone resists have photo-active compounds that make exposed areas soluble in a diluted alkali (base) solution such as sodium hydroxide (NaOH) or metal ion

Table 16.5 Overview of selected electrolytes [16]

	Cu electrolyte	Ni electrolyte	PbSn electrolyte	Au electrolyte	Sn electrolyte
Contents	$CuSO_4$, sulfonic acis, chloric acid, grain refiner and leveler, wetting agent	$Ni(NH_2SO_3)_2$, boric acid, grain refiner and wetting agent (if necessary)	$Sn(CH_3SO_3)_2$, $Pb(CH_3SO_3)_2$, methane sulfonic acid, grain refiner, wetting agent, oxidation inhibitor	$(NH_4)_3[Au(SO_3)]$, ammonium sulfite, ammonia, organic grain refiner and leveler, complexing agents and stabilizers	$Sn(CH_3SO_3)_2$, methane sulfonic acid, grain refiner, wetting agent, oxidation inhibitor
Metal concentration	20 g/l Cu	45 g/l Ni	Total of 28 g/l	12 g/l Au	20 g/l Sn
Temperature	25°C	50°C	25°C	55°C	25°C
pH Value	<1	4.0	<1	7.0	<1
Current density	10...30 mA/cm^2	10...30 mA/cm^2	20 mA/cm^2	5...10 mA/cm^2	7...15 mA/cm^2
Current efficiency	Nearly 100%	>95%	Nearly 100%	>95%	Nearly 100%
Anode material	Phosphorus alloyed copper	S-activated nickel pellets	Appropriate Pb/Sn alloys	Platinum-covered titanium	Pure tin

Table 16.6 Classification of photo resist and their characteristics [1]

Class	Resin/Dev. Chem	Pre-Bake	Re-Hydr.	Exposure Spectrum	Dose	Side-wall	Resol.	Stripping	Application
Positive liquid	Novolak/ aqueous	Yes	Yes	Broadband (chemically amplified)	High	45°–85° (~90°)	+	Easy	Gold bump, mushroom solder bump, UBM
Negative liquid	Acrylate, epoxy/ aqueous, organic	Yes	No	Broadband	Medium	~90°	+	More difficult	Gold bump, in-via plated solder bump
Negative dry film	Acrylate, epoxy/ aqueous	No	No	Broadband	Low	~90°	−	More difficult	In-via plated solder bump, photo stencil

free TMAH (Tetramethylammonium hydroxide). Negative dry film resist is developed mostly with environmentally friendly aqueous carbonate developer.

An important difference between positive and negative tone resists is that positive tone resist based on Novolak requires a re-hydration step after baking to allow water to penetrate into the resist layer again. The chemical reaction is given in Fig. 16.23 which highlights the importance of the water diffusion for the formation of the acid group which will react with the base during the development [30].

This re-hydration step can take hours for very thick layers. Different examples of openings in Photo-Resists with a thickness of 50 μm are shown in Fig. 16.24.

From the application point of view, the following trends can be observed: Positive tone resist is still very popular because of its ease of use. It is often selected whenever resist thickness is typically less than 50 μm and near 90° side wall angle is not a requirement. Usually positive resist are used for mushroom solder bump plating, gold bumping, rerouting and UBM patterning (in case solder is applied by stencil printing). However, since straight side walls are important in gold bumping, negative tone resists are increasingly used in this application. Resist layers thicker than 50 μm are necessary for in-via solder or Cu pillar plating and photo stencil printing. In most cases negative resists are used. Dry film resists will dominate the market for very thick resists because of the best cost of ownership compared to liquid-based resists. The application process of a dry film is shown schematically in Fig. 16.25. A polymer resist film which is covered by a Mylar™ coversheet is applied onto a wafer through a pressurized hot roll. Mylar™ is a trade-name of Dupont for BOPET (biaxially-oriented polyethylene terephthalate). Still the limited resolution capabilities of dry-films are driving the interest towards negative liquid resist for very fine pitch solder bumping. But further cost pressure of final products could change this trend.

Process technology for electronic packaging and MEMS is being confronted more and more with higher topography on the wafer due to higher complexity of the devices. Especially spin coating of photoresists has severe limitations when dealing with larger three-dimensional features because spin coating tend to cause excessive thickness in the bottom and inadequate thickness at the top of these features. One method to overcome the limitations of spin coating liquid

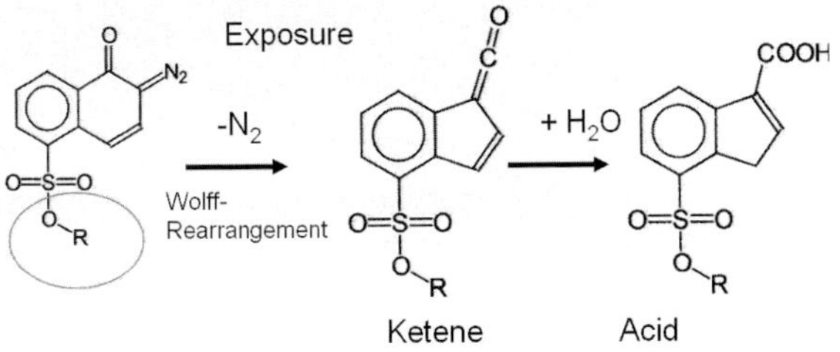

Fig. 16.23 Chemical reaction of the Novolak-resin during exposure and water adsorption

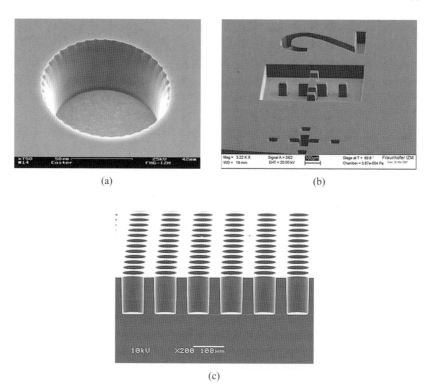

Fig. 16.24 Examples of 50 µm thick resists: Novolak based Clariant AZ 4620 (**a**); negative-type from RHEM (**b**); dry film negative type Dupont WBR 200 with 75 µm opening (**c**)

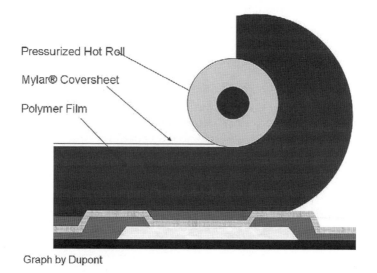

Fig. 16.25 Schematic application process of a dry film (courtesy of Dupont)

photoresists on wafer surfaces with extreme topography is the electrophoretic deposition of photoresists [31].

The electrophoretic resist coating process is based on the electrodeposition of either a negative tone or a positive tone photoresist from an aqueous solution onto a conductive seed layer. In the aqueous resist emulsion, the ionized polymer forms charged micelles comprising solvent, dye, and photoinitiator molecules. When an electric field is applied, the micelles migrate by electrophoresis towards the corresponding electrode and form on the surface a self-limiting, insulating film. This electrode is the wafer that is supposed to be coated. The electrophoretic deposition is finished very quickly, usually after 10 s. The coating can be done in a highly automated resist coater, developed for example by Besi Plating in Drunen/The Netherlands. This coater is based on semiautomatic single wafer operation.

The resulting layer thickness is mainly affected by the applied voltage and the temperature during the deposition. Final resist layer thicknesses between 3 and 20 μm are obtained in dependence of the applied voltage, the bath temperature, and the used resist type. The role of a chemical thickness controller also having an influence on the electrophoretic deposition is discussed. Furthermore, it will be shown how the quality of the resist bath can be controlled and maintained over a long period of time.

Cavities, 400 μm deep, obtained on Si wafers either by wet chemical etching or by dry plasma etching can be conformally coated with electrophoretic resists from Rohm and Haas Electronic Materials (RHEM). The top corners of the cavities are well covered with photoresist even after the full lithographic process (Fig. 16.26).

Using the InterVia 3-D-N resist from RHEM, an aspect ratio of nearly 2 was obtained. The patterned structures which have been defined in the electrode-positable photoresist can be transferred into the underlying plating base by etching the uncovered metal. Furthermore, electrophoretic photoresist has also successfully been used as a plating mold for Cu, Au and Ni.

(a)   (b)

Fig. 16.26 EDPr from RHEM, negative type: (**a**) Overview of Si-cavity; (**b**) Details of the bottom of the cavity which is 400 μm deep

16.3.1.6 Processing of Photo Resist and Photo Polymers

Application of any thin film polymer and photo resist consists of the following processing steps: coating (including resist bake), UV exposure, development and post-bake or cure [32]. There might be additional cures involved like post-exposure bake (PEB) to enhance the photo-process or post development bakes to stop the development process or to reduce crowning effects for PI processing. Essential for all deposition technologies is a clean and void-free surface. This is much more important compared to front-end (FE) processes because packaging is often a process which is done outside a FAB by sub-contractors. In this case the wafers are shipped in sealed boxes which are not a guarantee for a dust-free environment. In addition there might be an electrical test in between which are qualified for wire-bonding interconnection afterwards. WL-CSP is still a smaller market compared to the fast amount of wire-bonded packages worldwide.

The main technologies for the deposition of photo-resist are spin-coating, lamination, spray coating, and electroplating / electrophoretic deposition as shown in Fig. 16.27 [33].

The equipment for deposition of liquid resist / thin film polymers is mainly spin-coaters. Laminators are used to coat dry film resist using heated rolls to enhance the adhesion. The adhesion of the resist on the plating base is of importance to avoid under-plating (un-wanted plating underneath the resist). For reliable further processing i.e. electroplating of the wafer, it is important to precisely clean the wafer at the wafer edge. In the case of liquid film resist, this is done by an edge bead removal and by cutting the laminated dry film. The first step for the spin coating process is dispensing the liquid resist onto the wafer. The dispense pattern can be important for the final film uniformity and especially for the material consumption. In most cases, a spiral dispense from the edge to the center is the optimal method. Then the spin coater rotates the wafer

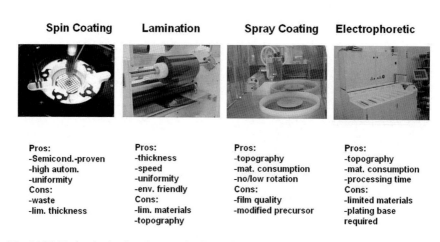

Fig. 16.27 Technologies for photo-resist deposition

at a speed of several hundred to thousand rotations per minute to uniformly spread the resist material over the entire wafer surface. The spin-speed and therefore the centrifugal force is the main parameter for the final film thickness. During this step, most of the solvent in the resist evaporates. In general, the thickness of the final resist film is controlled by resist viscosity, spin speed, surface tension and slightly by the design of the coater module. Using an optimized process and appropriate resist, the thickness for a target of 50 μm can vary less than 1% over 300 mm wafer size.

The polymer thickness h is a function of the angular velocity ω and two parameters K and m:

$$h = K \cdot \omega^{-m}$$

K describes the coater design and the solid content in the resist while the interaction of the polymer and the solvent is given by m [34]. Solvents which evaporate during the spinning process which is for example typical for positive tone Novolak-based resists lead to m = 0.5. Higher values are typical for Polyimides which are dissolved in solvents with lower vapor pressure lime NMP (N-Methylpyrrolidone). The evaporation of solvents can be suppressed by a rotating lid over the spinning chuck (Gyrset^{TM} coating, trademark of Suss MicroTec). This leads to a wider thickness range for a given resist formulation. The surface tension of the resist is responsible for the formation of an edge bead. Depending on the subsequent process steps this edge bead remains on the wafer or has to be removed (full or partial edge bead removal is for example necessary for electroplating). The edge bead is often removed chemically on the spin coater by solvent dispense or by UV exposure and development where the later is the more accurate method.

A bake of the resist is necessary after spin coating to remove the remaining solvent from the resist. Baking temperature depends on the resist type but is typically between 80 and 150°C. The baking process has a big influence on the exposure result observed after resist development. Insufficient baking leads to trapped solvent inside the resist layer. Especially with positive tone resist this is detrimental to side wall control as belly shaped resist profiles appear because the developer becomes more efficient when solvent is still present. In addition, trapped solvent can contaminate a plating bath over time which will deteriorate the plating process because ectrolytes are sensitive to organic additives.

A fully automatic coating system with a handling module, a spin coater with a dispense unit and a hotplate/coolplate is essential for good process control and to obtain repeatable results. Multiple hotplates are important since baking is often the most time-consuming step when processing thick resist.

A special challenge is the resist coating of 3D structures such as trenches or via holes. Electrophoretic resist plating is a common technology to coat severe topography whenever a metal surface is available. The method is based on electroplating. Another deposition technology for severe topography uses spray technology and can be used for metal and even non-metal surfaces. The resist or polymer is modified by dilution and sprayed by a special nozzle over the wafer. Even nearly vertical sidewalls can be coated.

An additional advantage of spray coating is the reduction of material consumption. In spin coating most of the resist is spun over the edge of the wafer. Using spray coating only the wafer surface is covered with resist. For thicker layers multiple spraying steps have to be performed. The film quality (uniformity, flatness) of sprayed resist is not as good as with spin coating because there are no centrifugal forces leveling the film. But for most of the packaging applications, resists are used for etching or as plating molds and polymers for electrical isolation where only a given thickness has to be reached.

Resists and Photo-Polymer for packaging are photo-sensitive at a wavelength spectrum of broad-band UV (i, h and g-line between 365 and 436 nm wavelength) mask aligners and 1X steppers (as opposed to reduction steppers that are used in the front end) both equipped with mercury short arc lamps. An optical resolution of 4 µm which is currently the minimum feature dimension in packaging is achievable with both methods. Overlay accuracy is getting tougher and has to be hardly better than ±1 micron. Mask aligners and steppers are capable of achieving this overlay accuracy.

Mask aligner are proximity printing tools where mask and wafer are separated by an exposure gap of approximately 50 microns. The exposure of the whole wafer is done within one shot using a mask which is slightly larger than the wafer. This is in contrast to steppers which are using reticles with a maximum field size of about 20×40 mm for the exposure. Therefore multiple shots are necessary to expose a full wafer, and it has throughput disadvantage compared to the mask aligner especially for large wafer sizes.

There is a steady trend for most resists and photo-polymer to use aqueous developers which translates into lower material cost than solvent based developers. In addition this is an attribute to "green" production. Puddle or spray development are the processes of choice for these materials. Immersion development is only used for R&D labs or very small productions.

Photo-resists are only temporary materials for plating or etching. Therefore resist stripping is a necessary process step which can affect the final result tremendously. The resist has to be removed completely without causing damages to the structures. Residues would limit the seed-layer etching for example in case of electroplating metal lines or bumps. Most positive-tone resists like Novolak-based materials can be easily stripped in alkaline aqueous solutions. In contrast negative-tone resist are polymerized during exposure. Therefore stronger stripper mostly based on organic solvents like NMP or even amines are necessary. In contrast to the PCB industry which favors a flaking of the resist, a complete dissolution is a must for the equipment used in packaging. Therefore the risk of metal attack is given. Examples of such corrosion effects are shown in Fig. 16.28.

Two different strippers based on organic solvents were evaluated for stripping resists after plating SnCu bumps. The first stripper has attacked strongly the Cu socket which was in contrast to second stripper which results in Sn corrosion.

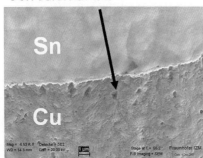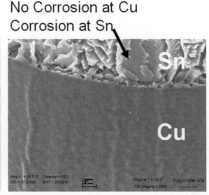

Fig. 16.28 Cu/Sn bumps after striping negative-type resist: *Left*: Corrosion of Cu with stripper A and *right*: Corrosion of Sn using stripper B

16.3.1.7 Polymers for Redistribution Layers (RDL)

Polymeric coatings such as polyurethanes, acrylic, epoxies and silicones have been used for over 40 years to protect printed wiring boards (PCB) from moisture, handling and environmental influences [35]. Special semiconductor grade polymers had to be developed for highly ion sensitive chip passivation layers. Especially for epoxy resins multiple distillation procedures were introduced to remove the sodium and chloride ions which are by-products in the standard epoxy synthesis. Polyimides are the standard passivation layer for memory chips and other devices with the need of surface protection for the handling and testing procedure. Photo-sensitive resins have been developed to reduce processing cost but are mostly higher in price for the base material. Dry-etching requires a masking process with either a hard mask which is a physically deposited and structured metal layer or a thick photo resist coating. The difference in processing for photo-sensitive and dry-etch polymer is given in Fig. 16.29.

Due to cost saving programs, the in-depth characterization of materials that was prevalent in the 1960s through the 1990s has slowed down considerably. In addition, due to mergers, spin-offs and low economic margins, the continuity of polymeric products is not always given. However, polymeric materials became more important with the introduction of new packaging concepts for ICs.

In general thin film polymers have proven to be an integral material basis for many different types of advanced electronic applications. First used as IC stress buffer layers, then established for MCMs, they are now used in various new packages, especially in the field of WLP [36].

Several different redistribution processes have been developed but main process steps are similar to each other. Differences exist mainly in the material selection. As an example, the redistribution technology of Fraunhofer-IZM/ TU Berlin will be described in more detail (Fig. 16.30).

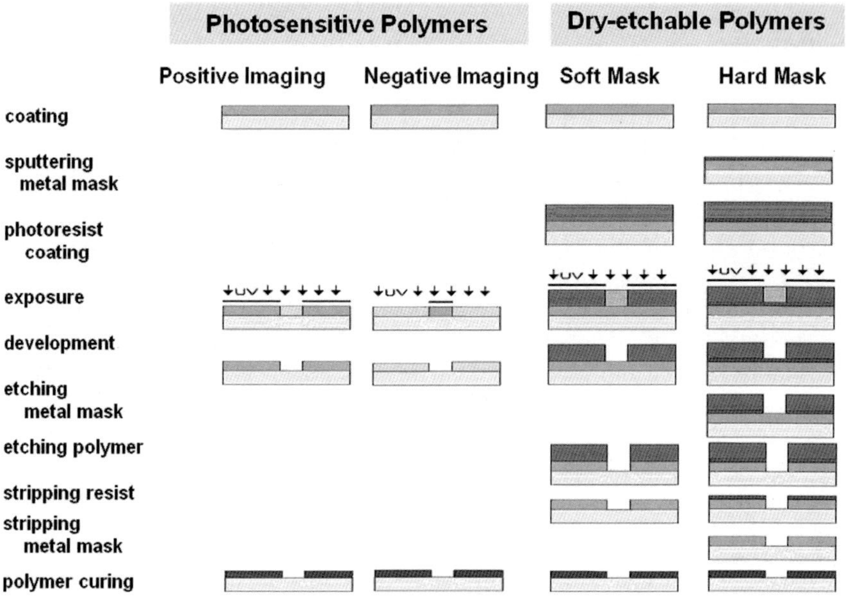

Fig. 16.29 Technologies for structuring thin film polymers

First, a dielectric layer is deposited on the wafer to enhance the passivation layer of the die. Pinholes in an inorganic passivation would give shorts in the rewiring metallization. The polymer layer under the rewiring metallization acts also as a stress buffer layer for the bumping and assembly processes. Using photosensitive polymers requires fewer processing steps for thin film wiring than non-photosensitive materials that have to be dry etched.

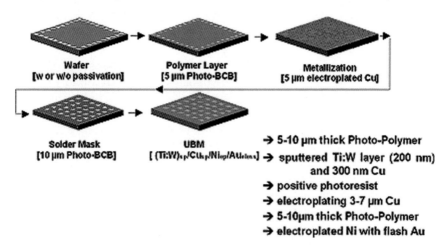

Fig. 16.30 Process flow of an RDL process (Fraunhofer IZM)

Fraunhofer IZM/TU Berlin uses Photo-BCB (Cyclotene). Compared to other polymers BCB, it has a low dielectric constant and dielectric loss, minimal moisture uptake during and after processing, very good planarization, and a low curing temperature. The rewiring metallization consists of electroplated copper traces to achieve a low electrical resistivity. A sputtered layer of Ti:W-Cu (200/300 nm) serves as a diffusion barrier to Al and as a plating base. A positive acting photo resist is used to create the plating mask. After metal deposition, the plating base is removed by a combination of wet and dry etching. The copper process is shown in Fig. 16.31:

A second Photo-BCB layer is deposited to protect the copper and to serve as a solder mask. BCB can be deposited directly over the copper metallization without any additional diffusion barriers. Electroplated Ni/Au is used for the final metallization. Solder balls (high melting or eutectic PbSn) are deposited by solder printing directly on the redistributed wafers. Then the solder paste is reflowed in a convection oven under nitrogen atmosphere and the flux residues are removed in a solvent adapted to the used solder paste (Figs. 16.32 and 16.33).

Mean value of the solder ball diameter can be adapted to the assembly and board requirements between ~100 and 350 μm depending on the ball pitch. Shear testing is the method of choice for a first quality check. Shear stress for these balls should be higher than 130 cN per bump. Dicing the wafer with a standard wafer saw completes the CSP-WL build-up. The reliability of the redistribution layer was evaluated for consumer, medical, automotive and space applications.

The requirements for the selection of a given polymer are quite broad: High decomposition temperature for thermosets or high glass temperature for thermoplastics for the high temperature packaging processes such as solder reflow,

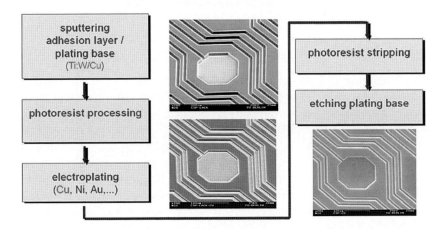

Fig. 16.31 Thin film copper process at Fraunhofer IZM / TUB

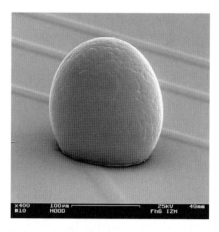

Fig. 16.32 Solder printed PbSn on redistributed wafer (Photo-BCB/Cu)

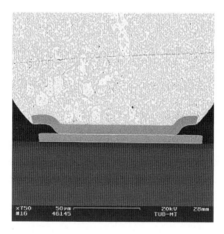

Fig. 16.33 Cross section of screen printed PbSn on redistributed wafer (Photo-BCB/TiW/Cu/ Photo-BCB/TiW/Cu/Ni/Solder)

high adhesion, high mechanical and chemical strength, excellent electrical properties, low water up-take, photo-sensitivity, and high yield manufacturability. An important process difficulty is that these high-end highly crosslinked polymers are nearly insoluble in organic solvents. Therefore, pre-polymers are manufactured which have a molecular weight in the 100 thousands and are dissolved in an organic solvents. These solutions are commonly called precursors and are ready for the spin-on process. The final polymerization is done on the wafer by thermal curing.

The performance of polymers plays a major role in the built-up structure of WLP because it is one of the key layers which act as a buffer between IC and PCB. Polymers with low dielectric constant are preferred because a high capacitance reduces the computing speed between integrated circuits. In addition, the selection of the optimal polymer for a given application depends not only on its physical and chemical properties and processability, but also on its intrinsic interfacial characteristics. Table 16.7 gives a selection of common types of photo-sensitive spin-on dielectric materials listed with some important physical properties.

PI (Polyimide), BCB (Benzocylobutene) and PBO (Polybenzoxazole) are common re-passivation materials. The chemical structures and the chemical reactions are given in Figs. 16.34, 16.35, and 16.36.

BCB and Polyimides (with only a few exceptions) are negative acting materials requiring organic developer while PBO is positive acting. All materials require a bake to remove solvents but no re-hydration step. In some cases a post exposure bake is necessary to enhance the photo initiated polymerization. All materials are exposed by the broadband spectrum and achieve a side wall angle of approximately 40–60°. Resolution is usually not an important requirement for WLP because only vias of 20 microns or larger in diameter have to be opened over larger I/O pads. This is changing for new 3-D integration concepts.

An important feature for thin film polymers is the ability to planarize underlying feature like metal routings [37]. The main parameter for this property is the shrinkage during cure. This shrinkage can very between 40% (PI) and nearly zero for BCB and some of the epoxy-based polymers.

An example of the importance of planarization is given in Fig. 16.37 by an MCM [38].

This multi-chip module concept is a prototype for a pixel detector system for the Large Hadron Collider LHC at CERN, Geneva. The project is part of the ATLAS experiment. The ATLAS experiment is being constructed by 1700 collaborators in 144 institutes around the world. The main purpose of this detector are the search for the Higgs Boson, the last undiscovered particle in the Standard Model of elementary particles and their interactions and the study of the decays of the top quark, which was discovered 1994, with high statistics.

For the pixel detector, a modular system is needed which can be put together to build this large detector system. While the diode-pixel-arrays have an active area of about 10 cm^2, the read-out chips are one order of magnitude smaller because of their higher complexity. In general, the diode-pixel-arrays and the read out chips can be fabricated in wafer size dimensions. Each module is an excellent example showing highest density FC assembly with a 50 μm pitch. A multilayer routing with high planarity is needed for the MCM-D version of this system.

A new class of polymer called Parylene has been introduced for conformal coating. These are deposited from the gaseous phase. Parylene are shipped as a Dimer (two monomers linked together) in solid state [39]. The dimers are vaporized at 150°C and pyrolized into their highly reactive monomer radicals at 680°C. Finally, the monomer enters the deposition chamber at 0.1 torr and room temperature where it simultaneously adsorbs and polymerizes on the parts. The main advantage is the low temperature processing which allows

Table 16.7a Overview of commercially available thin film polymers

		Photosensivity	Developer	Base Chemistry	Curing T (for 1–2hrs) [°C]	Diel Const [1 kHz–1 MHz]	Loss Factor [1 kHz–1 MHz]	Tg/Decompostion T [°C]
Asahi	Pimel G7621	negative	organic	PI	> 350	3.3	0.003	355
Dow Chemical	Cyclotene 4000	negative	organic	BCB	210–250	2.65	0.0008	> 350
Dow Corning	Photoneece PWDC 1000	positive	aqueous	PI	320	2.55 (1 GHz) 2.9	0.002 (1 GHz)	290
Futji Film	WL 5150	negative	organic	Silicone	250	3.2	0.0070	> 350
	Probimide 7000 348	negative	organic	PI	> 350	3.3	0.007	> 350
HDM	PI 2730	negative	organic	PI	> 350	3.2	0.004	> 350
	HD 4000	negative	organic	PI	> 350	2.9	0.003	> 350
	HD 8000 (PBO)	positive	aqueous	PI	> 350	3.2	0.006	350
JSR	WLP 1200	negative	aqueous	nanofilled Novolac	170	3.4	0.0097	300
Nippon Steel	Cardo VPA	negative	aqueous	modified acrylic resin	200	3.8	0.036	> 210
Rohm and Haas	Intervia 8000	negative	aqueous	modified resist	175	3.4	0.03	180
Sumitomo	PBO	positive	aqueous	PBO	320	2.9 (1 GHz)	0.026 (1 GHz)	150–200
Toray	Photoneece BG 2400	negative	organic	PI	> 350	2.9–3.5		380
	Photoneece PW 1000	positive	aqueous	PI	> 350	3.2		255
						2.9		290

Table 16.7b (continued)

		CTE [ppm/K]	Tensile Strength [MPa]	Elongation to Break [%]	Residual Stress [MPa]	Youngs Modulus [GPa]	Water Uptake [%]
Asahi	Pimel G7621	40–50	150	30	40–50		0.8
Dow Chemical	Cyclotene 4000	45	87	8	28	2.9	< 0.2
Dow Corning	Photoneece PWDC 1000	36	130	40	28	3	
	WL 5150	236	6	37	2.6	0.16	
Futji Film	Probimide 7000	27	170	73	30	2.9	1.3
	348	23	123	8		3.2	1.8
HDM	PI 2730	16	170	?	18	4.7	>1.0
	HD 4000	35	200	45	35–37	3.5	
	HD 8000 (PBO)	47	122	11	29	2.5	
JSR	WLP 1200	51	75	7.5		1.6	
Nippon Steel	Cardo VPA	80	7.5	11.5	14–30	2.5	1.6
Rohm and Haas	**Intervia 8000**	58				4.5	<0.38
Sumitomo	PBO	55	100	18		2.5	0.3
Toray	Photoneece BG 2400	25	180	40		3.9	
	Photoneece PW 1000	36	130	40		3	

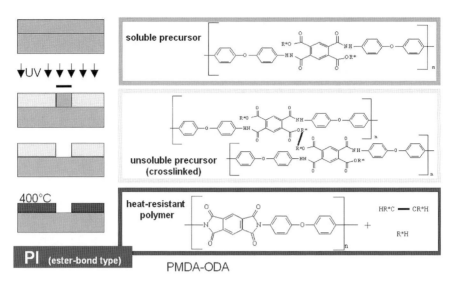

Fig. 16.34 PI (PMDA-ODA) polymer (photo reaction and cure)

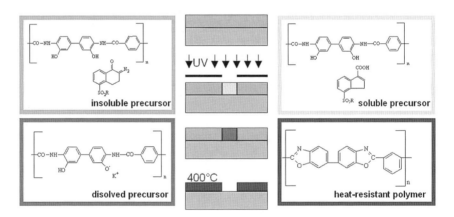

Fig. 16.35 PBO polymer (photo reaction and cure)

coating of highly sensitive parts. Figure 16.38 shows the chemical structures of three types of Parylene. The conformal coating is highly attractive for the upcoming 3-D integration technologies.

16.3.1.8 Adhesion and Copper Diffusion into Polymers

The reliability of a multi-layer thin film structure strongly depends on the adhesion between the different layers [40]. The different interfaces in a Photo-BCB/TiW/Cu/PbSn technology based WLP are:

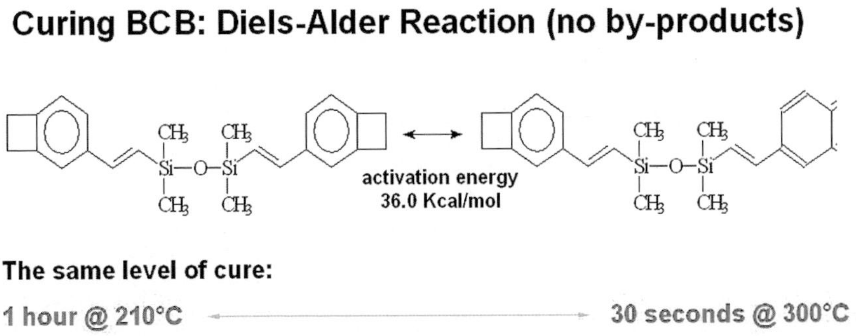

Curing BCB: Diels-Alder Reaction (no by-products)

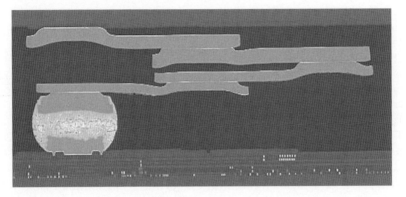

Fig. 16.36 BCB molecule and cure

Fig. 16.37 Multilayer BCB / Cu representing nearly perfect planarization of BCB (pixel detector module)

1. Photo-BCB to the inorganic chip passivation (Silicon-Oxide, Nitride etc.) and to metal (chip pad (Al) or redistribution metallization (Cu))
2. Metal (i.e. redistribution metallization or UBM) to Photo-BCB
3. Photo-BCB to Photo-BCB
4. Solder to UBM

The adhesion in these thin film structures can be described with three mechanisms: roughness, chemical bonding using adhesion promoters, and chemical interlocking/diffusion. Important for a reliable package is the integrity of the interfaces during the lifetime of the microelectronic product. For interface (1), organo-silane based adhesion promoters are used to create essential layers to couple the organic dielectric to the inorganic surfaces. The theoretically ideal structure would be a monomolecular layer, coupling on one side to the inorganic surface and the other to the polymer. Different adhesion promoters for Photo-BCB have been evaluated. High adhesion strengths on several inorganic surfaces were obtained using a Vinylsilane, which is spun directly on

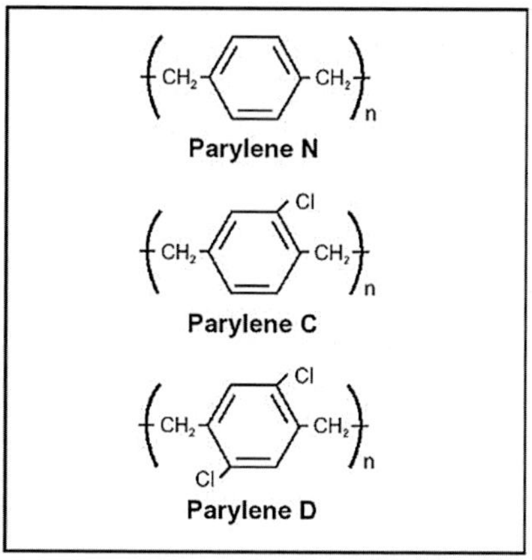

Fig. 16.38 Parylene (3 types)

the wafer before BCB deposition. There is a strong indication that chemical bonding through Si-O bonds is responsible for the adhesion. The thickness of the adhesion promoter layer is in a range of 0.5–5 nm. This adhesion promoter layer improves the adhesion to values over 60 MPa on Al, Cu and different inorganic chip passivations.

The metal to BCB interface (2) depends strongly on the metallization technique [41]. Electroless deposited metals have no adhesion on untreated BCB films with its very smooth surfaces (roughness in the Å range). Sputtering is used as a reliable metallization process for the redistribution of the WLP because the metal atoms have a penetration depth of around 100 nm. This guarantees the strong adhesion of over 80 MPa between the sputtered metal on the Photo-BCB. Paik et al. [42] described the stable interface between Cu and BCB. In addition, there is nearly no influence of the descum process (RIE) using a flourine gas/oxygen mixture, which is necessary to achieve a 100% electrical yield in via holes. Only pure O2-Plasma reduces the adhesion due to an oxidization of the BCB surface [43, 44].

For the Photo-BCB to Photo-BCB interface (3) high adhesion is obtained by performing a partial cure of the underlying BCB layers followed by a final full cure of the whole stack. A correlation between the adhesion strength and the degree of cure was found. The roughness and the surface chemistry of the Photo-BCB layer modified by RIE had no significant effect on adhesion. The mechanisms of chemical interlocking and inter-diffusion are the driving forces for the adhesion between BCB layers. Therefore, the degree of cure is the key to high adhesion.

The solder to UBM interface is based on the formation of intermetallics (IMCs). Due to the brittle nature of those IMCs, there should be minimum growth of these layers during the operation lifetime. Ni is preferred over Cu in the case of Sn-based solders because of the slower growth rates of NiSn IMCs compared to CuSn IMCs.

In conclusion, the surface chemistry and physics should be carefully analyzed for the reliable built-up of thin film structure. Special emphasis has to be made to all modification processes like sputtering, plasma etc.

An additional concern for the high density wiring is the copper diffusion into the polymer. An excellent test structure is a metal insulator semiconductor (MIS) capacitor fabricated on n-type silicon wafers [45]. The polymer under test forms the center of the dielectric in the capacitor. The top and bottom of the polymer layer are passivated by a thermal oxide layer on the bottom and a PECVD oxide layer on the top. Copper ions have been shown to form at the surface of and readily diffuse into SiO_2, providing a source of Cu to potentially diffuse into the polymer under test. The second purpose the oxide layers serve is to isolate the polymer from the metal and silicon surfaces which have been shown to provide additional sources of charged particles that diffuse into the polymer. The gate of the MIS capacitor is fabricated using sputtered / electroplated Cu. An aluminum contact on the back side of the wafer is used for electrical contact for the opposite side of the capacitor. Under BTS (Bias Temperature Stress), the samples are heated to a specified temperature and then a specific electrical bias is applied between the contacts of the capacitor for a specified time [46]. Before and after BTS (CV), measurements are performed. The CV measurement is based on the behavior of a MIS capacitor. For an n-type semiconductor, the voltage on the gate is scanned from a negative voltage to a positive voltage. As the measurement begins with a strong negative bias on the gate the MIS capacitor is in the inversion region. The negative bias on the gate repels the majority carrier electrons and an equivalent positive charge from the fixed ionized dopants remains at/near the semiconductor insulator interface. This depletion region is free from mobile charges and therefore acts as an isolator and forms a capacitor in series with the gate capacitor resulting in the smallest capacitance measured. As the applied bias becomes less negative and the majority carriers begin to return to the area at/near the interface silicon. This is known as the depletion region of the CV curve and the capacitance starts to increase because the depletion capacitance is reduced. As the voltage becomes positive, the majority carriers collect at the semiconductor insulator surface creating a parallel plate capacitor with a value close to the value of the insulator alone. The interesting part of the measurement occurs in depletion region where transition occurs between the inversion and accumulation. For example, if positive charges (Cu^+) have diffused into the polymer the voltage at which depletion begins to occur will be shifted to a more negative voltage because the positive charge in the polymer will compensate for some of the negative charge on the gate. The largest effect happens when charges diffuse through the dielectric layer and collect at the semiconductor surface.

Using the BTS and CV test methods, it was found and verified with SIMS that Copper accumulates at the SiO_2/Si substrate interface for MIS capacitors with Cu/SiO_2/Polymer/SiO_2/n-type Si/Al structure. Copper migration was detected after BTS for an Epoxy Polymer, but not for BCB (Fig. 16.39).

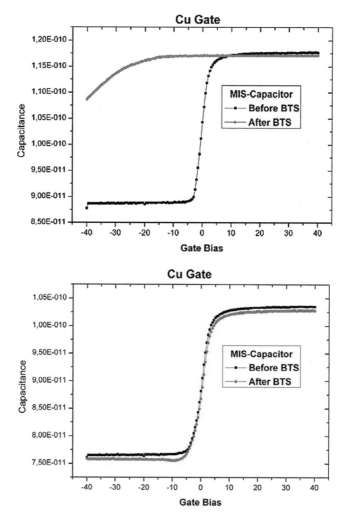

Fig. 16.39 Copper drift using a metal insulator semiconductor (MIS) capacitor (*left*: Epoxy, *right*: BCB)

Using nano-spot EDX, copper was detected in all material systems all over the samples. This is probably caused by strong scattering effects due to the copper electrode. To prepare the electron transparent lamella for TEM, a FIB lift out technique was used. After preparation, the lamella was transferred to a gold specimen holder grid (Au signal in EDX spectra) (Fig. 16.40).

It was proven that Cu diffusion is not an issue for BCB. The behavior of PI depends strongly on the specific type of the polymer chain. Loke et al. reported that BCB has the lowest Cu diffusion for commercially available low k polymers [47].

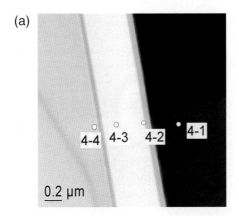

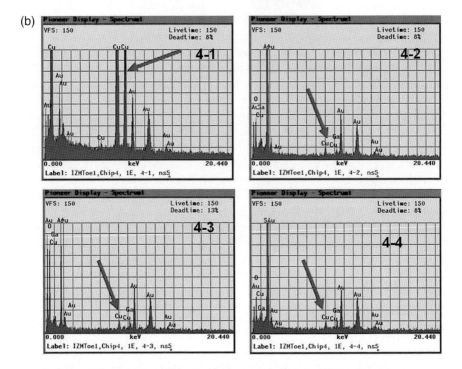

Fig. 16.40 Nano-spot for a /SiO2/BCB/SiO2/n-type Si interface: (**a**) locations for data collections; (**b**) EDX spectra at those four locations

16.4 Materials for Integrated Passives

On-Chip passives have a high potential for achieving advantages in cost, performance, or functionality as compared to discrete passives [1]. Therefore, the integration of passive components on board or wafer level will be an important step towards further performance enhancement and system miniaturization. Three basic material classes are needed for the realization of integrated passives elements: conductors, resistors and dielectrics [48]. These can be made out of metals, polymers, or ceramics. The main difference between polymer and ceramic technologies is the maximum processing temperature which can be up to 300°C for polymers but can reach 700°C and above for the firing of ceramics. Metals like Cu, Au or Al, or metal-filled polymer thick films with a resistivity of less than 0.1 ohm/square are used for the conductors to avoid high parasitic resistance. Alloys like NiCr, CrSi, or TaN, cermets (ceramic-metal composites) or carbon-filled polymers are the materials of choice for resistors having values of 100–10,000 ohm/square. Polymers with a dielectric constant κ of 2–5, amorphous metal oxides ($\kappa = 9$–50) or crystallographic ordered mixed oxides with $\kappa > 1000$ are the central building blocks for capacitors. A wide variety of deposition technologies are used for integrated passives depending on the material types and the structuring process (additive vs. subtractive): sputtering, evaporation, spin-on, lamination, sol-gel, chemical conversion like oxidization, etc. The different processes can be classified according to the MCM types: thin film (MCM-D), polymer thick film (MCM-L) and ceramic thick film (MCM-C). Highest performance is achieved using thin film technology. Although this technology is still struggling with cost, the use of 300 mm substrates (glass, metal sheets or silicon) will allow to reduce the cost in the future. Polymer thick film is closely linked to the PCB infrastructure. Large substrate size or even roll-to-roll processing enables lowest cost processing. In addition, there is a well established PCB infrastructure which has no principal barrier for the integration of passive elements. The main hurdles, on the other hand, are the less controllable tolerances and the limited material spectrum for very high frequency applications. A trade-off has to be made whether the additional cost and complexity to the manufacturing process to add extra internal layers for passives are worth the free board space. Ceramic thick film is cost wise in between. Further cost reduction is achieved by photosensitive materials and improvements in printing but the substrate size is generally limited to 6" square. The high temperature compatibility is an interesting feature for automotive and other applications under extreme conditions.

An integrated passive filter based on thin film technology is given as an example. The build-up process is similar to the redistribution for WLP, therefore the same production line can be used. Copper/BCB technology in conjunction with NiCr sputtering is the core process steps. In Fig. 16.41 the build-up process for inductors, resistors and capacitors are shown.

Up to now the integration of inductor has achieved the highest acceptance. The deposition is done by sputtering and electroplating of copper in combination of a high planarizing polymer like the BCB. A schematic is given in Fig. 16.42.

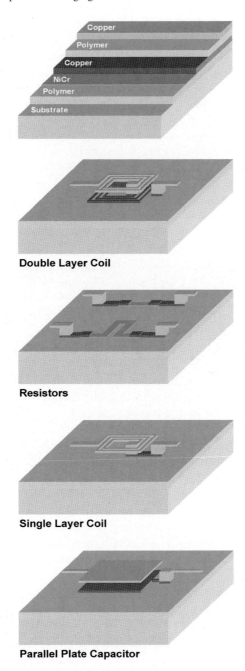

Fig. 16.41 Schematic of thin film build-up process for integrated passive components (Fraunhofer IZM) [49]

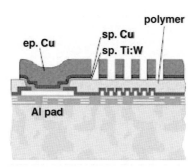

Fig. 16.42 Schematic of inductors as IPD on chip

Inductive sensors or on-chip inductors can be used for angular and linear position sensors, proximity sensors, non-destructive testing, layer thickness measurement, on-chip passive components for RF circuits, DC-DC converters, miniature transformers and position sensor. An example is given in Fig. 16.43.

The operating principle of these microcoil position sensors is based on an oscillator generating a carrier signal at a frequency around 500 kHz. This signal is a current that is sent through the excitation coil and thus generates an AC magnetic field. The magnetic field is proportional to the current and the number of windings. In order to maximize the magnetic field at a given supply voltage (5 V), the resistance must be minimized, which explains the use of copper as the material of choice for the excitation coil.

The AC magnetic field is coupled into the detection coils. Two detection coils are connected in a differential configuration. If the magnetic field is disturbed, e.g. by a tooth of a gear that turns in front of the sensor, a signal will be measured. This signal is the amplitude-modulated carrier signal with the modulation depending on the coupling between the excitation coil and the detection coils. The signal is demodulated using a synchronous demodulator and a low-pass filter. Two detection coil systems are implemented, that have slightly different positions, so that the direction of the movement of the target wheel can be detected. If a target gear rotates in front of the sensor, a sine and cosine signal are measured on the two channels, and the angular position can be calculated with a high precision. Instead of a gear for

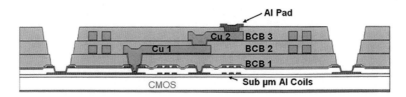

Fig. 16.43 Schematic of inductors on ASIC for a positioning sensor

Fig. 16.44 Positioning sensor based on inductors on chip assembled on FR 4 with SMT passives

rotational applications, a steel band with holes can be used for linear applications. An example of such a sensor assembled by wire-bonding on a PCB is given in Fig. 16.44.

The technology for these microcoils is based on a high-density thin film BCB/Copper process. Due to the ongoing trend of miniaturization, further developments have to be done to increase the wiring density and the height of the copper metallization. A process for 4 µm lines and space in thick photoresist has therefore been developed. Emphasis has been made to reduce underetching of the plated copper structures to guarantee high adhesion of the Cu to the BCB. A cross-cut by FIP (Focus Ion Beam) is shown in Fig. 16.45.

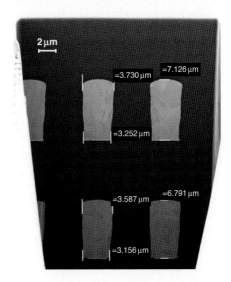

Fig. 16.45 Cross cut (focus ion beam) for a 4 µm lines and space Cu 2-layer wiring isolated and planarized by BCB

BCB is the material of choice due to its ideal planarization capability. The process has been demonstrated on 200 mm ASICs.

References

1. M. Töpper, D. Tönnies, Microelectronic packaging, in: Semiconductor Fabrication Handbook, M. H. Geng (ed), McGraw-Hill, New York, 2005, pp. 21.1–21.54
2. Microelectronic Packaging Handbook, Part 1–3, R. Tummala, E. Rymaszewski, A. Klopfenstein (eds), Chapman & Hall, New York, 1997
3. Multichip Module Technology Handbook, P. Garrou, I. Turlik (eds), McGraw-Hill, New York, 1998
4. A. J. van Roosmalen, "There Is More Than Moore", Proc. 5th Int. Con. on Mech. Sim. and Exp. in Microelectronics and MST, EuroSim2004
5. Ball Grid Array Technology, J. Lau (ed), McGraw-Hill, New York, 1995
6. K. Kosuga, "CSP Technology for Mobile Apparatuses", Proceedings International Symposium on Microelectronics, Philadelphia, Oct. 1997, p. 244
7. Chip Scale Package, J. Lau, S.W. Ricky Lee (eds), McGraw-Hill, New York, 1999
8. K. Iwabuchi, "CSP Mounting Technology", Proc. SEMI Technology Symposium 1996, Chiba/Japan
9. M. Töpper, J. Simon, H. Reichl, "Redistribution Technology for CSP using Photo-BCB", Future Fab International, 1996, p. 363
10. P. Garrou, "Wafer Level Chip Scale Packaging (WL-CSP): An Overview", IEEE Trans. Adv. Packaging, 2000, 23 (2), 200, p. 198
11. P. Garrou, "Wafer Level Packaging has Arrived", Semiconductor International, October 2000, p. 1192
12. M. Töpper, H. Reichl, SECAP, International Advanced Packaging Consortium: Formed to Standardize Process Equipment for Wafer Level Packaging Technologies for 300 mm, Future Fab, 10 July, 2001
13. Principles of Electronic Packaging, D. P Seraphim, R. Lasky, Che-Yu Li (eds), McGraw-Hill, New York, 1989
14. H. Reichl, "Direktmontage", Springer Verlag, Berlin, 1998
15. Area Array Interconnection Handbook, K. Puttlitz, P. Totta (eds), Kluwer Academic Publishers, Dordrecht, 2001
16. L. Dietrich, J. Wolf, O. Ehrmann, H. Reichl, "Wafer Bumping Technologies Using Electroplating for High-Dense Chip Packaging", Proceedings Third International Symposium on Electronic Packaging Technology (ISPT'98), Beijing (China), Aug. 17–20, 1998
17. U. Müller, "Anorganische Strukturchemie", Teubner Verlag, 3. Auflage 1996
18. K. Ruhmer, E. Laine, K. O'Donnell, K. Hauck, D. Manessis, A. Ostmann, M. Töpper, "UBM Structures for Lead Free Solder Bumping using C4NP", EMPC, June 2007
19. T. Kawanobe, K. Miyamoto, Y. Inaba, "Solder Bump Fabrication by Electrochemical Method for FC Interconnection", IEEE Publication CH1671-7/0000, 1981, p. 149
20. L. Dietrich, M. Toepper, O. Ehrmann, H. Reichl, "Conformance of ECD Wafer Bumping to Future Demands on CSP, 3D Integration, and MEMS", Proceedings of the ECTC, San Diego, 2006, p. 1050
21. A. Ostmann, G. Motulla, J. Kloeser, E. Zakel, H. Reichl, "Low Cost Techniques For Flip Chip Soldering", Proc. Surface Mount International Conference, Sept. 1996, San José
22. S. Anhöck, A. Ostmann, H. Oppermann, R. Aschenbrenner, H. Reichl, "Reliability of Electroless Nickel for High Temperature Applications", Intl. Symposium of Advanced Packaging Mat. Conf, Braselton, USA, March 1999, p. 256

23. T. Baumgartner, D. Manessis, M. Töpper, K. Hauck, A. Ostmann, H. Reichl, P. Goncalo C T Jorge, "Printing solder paste in dry film – A low cost fine-pitch bumping technique", Proceedings of EPTC, Singapore, 2007, p. 609
24. P. A. Gruber, L. Belanger, G. P. Brouillette, D. H. Danovitch J.-L. Landreville, D. T. Naugle, V. A. Oberson, D.-Y. Shih, C. L. Tessler, M. R. Turgeon, "Low-cost wafer bumping", IBM J. Res. & Dev. Vol. 49 NO. 4/5 July/September 2005, p. 621
25. Modern Solder Technology for Competitive Electronic Manufacturing, J. Hwang (ed), Mc Graw-Hill, New York, 1996
26. Solder Joint Technology Materials, Properties, and Reliability, King-Ning Tu (ed), Springer, Berlin, 2007
27. Bioh Kim, "Leadfree Solder Deposition for Wafer Level Packaging Applications" 5th Annual SECAP East Asia Seminar Series, Nov. 2004,
28. P-Y Chevalier, Thermochimica Acta, 136 (1988), pp. 45–54
29. G. Messner, I. Turlik, J. W. Balde, P. Garrou, "Thin Film Multichip Modules", ISHM Publication, 1992, p. 5
30. R. Dammel, Diazonaphtoquinoe-Based Resists, SPIE Optical Engineering Press, Bellingham, WA, Vol. TT-11, 1993
31. Th. Fischer, M. Töpper, N. Jürgensen, O Ehrmann, M. Wiemer, H. Reichl, "Conformal coating and patterning of 3D structures on wafer level with electrophoretic photoresists", Proceedings of Smart System Integration Conference, Paris, France, 2007, p. 473
32. Thin Film Technology Handbook, A. Elshabini-Riad, F.D. Barlow (eds), McGraw Hill, New York, 1997
33. M. Töpper, Ch. Lopper, J. Röder, K. Hauck, Th. Fischer, T. Baumgartner, H. Reichl, WLP Photoresists for the 21st century, IWLPC Conference, San José, USA, November 2005
34. W. Daughton, Journal of Electrochemical Society, Vol. 129, No. 1, 1982, p. 173.
35. Coating Materials For Electronic Applications, J. Licari (ed), Noyes Publications, Berkshire, 2003
36. M. Töpper, The Importance of Polymers in WLP, in: Materials for Information Technology, E. Zschech, C. Whelan, T. Mikolajick (eds), Springer, Berlin, 2005
37. P. Chiniwalla, R. Manepalli, K. Farnsworth, M. Boatman, B. Dusch, P. Kohl, S. A. Bidstrup-Allen, "Multilayer Planarization of Polymer Dielectrics", IEEE Transactions on Adv. Packaging, Vol. 24, 2001, p. 41
38. M. Töpper, Th. Fritzsch, V. Glaw, R. Jordan, Ch. Lopper, J. Röder, L. Dietrich, M. Lutz, H. Oppermann, O. Ehrmann, H. Reichl, "Technology Requirements for Chip-On-Chip Packaging Solutions", Proceedings ECTC Conference, Florida, USA, 2005, p. 802
39. A. Hardy, "Protecting Modern Wafer-level Packages", Press release by Specialty Coating Systems, 2008
40. M. Töpper, A. Achen, H. Reichl, "Interfacial Adhesion Analysis of BCB / TiW / Cu / PbSn Technology in Wafer Level Packaging", Proceedings ECTC 2003, p. 1843
41. M. Töpper, Th. Stolle, H. Reichl, "Low Cost Electroless Copper Metallization of BCB for High-Density Wiring Systems", Proceedings of the 5th Intl. Sym. on Advanced Packaging Materials, Braselton, Georgia, USA, March 1999, p. 202
42. K.W. Paik, R.J. Saia, J.J. Chera, "Studies on the Surface Modification of BCB Film", Proceedings MRS, Boston, Nov. 1990
43. F. Krause, K. Halser, K. Scherpinski, M. Töpper, "Surface Modification due to Technological Treatment Evaluated by SPM and XPS Techniques", Proceedings MicroMat, Berlin, April 2000
44. P. Chinoy, "Reactve Ion Etching of Benzocyclobutene Polymer Films", IEEE Trans. Comp. Packaging and Manufact. Tech., Part C, Vol 20, 1997, p. 199
45. A. Loke, "Process Integration Issues of Low-Permittivity Dielectrics with Copper for High-Performance Interconnects", Ph.D. Thesis, Dep. of Electrical Engineering, Stanford University, U.S.A., March 1999, p. 85

46. M. Töpper, K. Hoferling, F. Defo Kamga, H. Reichl, "Copper Migration in Thin Film Polymers", Proceedings of Nano Materials, Berlin, 2007
47. A. Loke et al., "Evaluation of copper penetration in low-k polymer dielectrics by bias-temperature stress", MRS Spring Meeting, Symposium N/O, 1999, p. 1
48. Integrated Passive Component Technology, R. Ulrich, L. Schaper (eds), Wiley Interscience/IEEE Press, New York, 2003
49. K. Zoschke, J. Wolf, M. Töpper, O. Ehrmann, Th. Fritzsch, K. Scherpinski, F.-J. Schmückle, H. Reichl, "Thin Film Integration of Passives – Single Components, Filters, Integrated Passive Devices", Proceedings of ECTC Conference, 2004, p. 294

Chapter 17
Microelectromechanical Systems and Packaging

Y.C. Lee

Abstract Microelectromechanical systems (MEMS) technology enables us to create different sensing and actuating devices integrated with microelectronic, optoelectronic, radio frequency (RF), thermal, and mechanical devices for advanced microsystems. In all these systems that demand low cost and small size, MEMS packaging is usually a major consideration. The relationship between MEMS and packaging, however, is not limited to packaging of MEMS devices. MEMS devices can in fact be used to enhance packaging technologies for microelectronic, optoelectronic and RF systems. In addition, packaging technologies can be applied to fabricate MEMS devices. Therefore, packaging and MEMS technologies are essential to integrate sensors and actuators with other components on a single system platform. MEMS reliability as showstopper can be removed, and there is a great opportunity to apply MEMS and packaging technologies to develop fully integrated micro/nanosystems in the future.

Keywords Microelectromechanical systems · MEMS · Packaging · System integration · Reliability

17.1 Introduction

Microelectromechanical systems (MEMS) technology enables us to create different sensing and actuating devices integrated with microelectronic, optoelectronic, radio frequency (RF), thermal, and mechanical devices for advanced microsystems. Semiconductor fabrication processes allow for cost effective production of these micro-sensing or actuation devices in the 1–100 μm size scale. Hundreds of MEMS-based sensors and actuators and systems have been demonstrated and the number of their applications is growing. A few examples of their diverse applications are pressure sensors, tilt sensors, accelerometers, gyros, chemical micro sensors, lab-on-a-chip, micro biomedical devices,

Y.C. Lee (✉)
University of Colorado, Boulder, CO, 80309-0427, USA
e-mail: leeyc@colorado.edu

resonators, displays, optical switches, radio frequency (RF) switches and passive components, printer heads, energy harvesting and storage, and data storage. The MEMS market is experiencing a period of dramatic growth. For example, in 2007, Tire Pressure Monitoring System (TPMS) became a standard feature offered in tens of millions of automobiles. An accelerometer enables the detection of an iPhone's rotation for changing the contents of the display in its proper landscape or portrait aspect ratio. Nintendo's Wii game console uses accelerometers to detect accelerations in three dimensions of the handheld pointing device. In all these applications that demand low cost and small size, MEMS packaging is usually a major consideration.

Figure 17.1 illustrates some of the differences between MEMS and microelectronics fabrication and packaging [31]. The cantilever beam device shown in the figure represents a simple configuration for pressure sensors, accelerometers, micro-mirrors and radio frequency (RF) switches. MEMS design requires solid modeling and simulation considering electro-thermal-mechanical behaviors. MEMS fabrication involves deposition and removal of micron-thick layers with controlled mechanical and electrical properties [33, 43]. After fabrication, the sacrificial materials are removed in order to release the device for mechanical movements. This release process is usually the first step in MEMS packaging. After release, the devices can be tested on the wafer-level, followed by dicing. The released, diced device is assembled and sealed in a package. These testing, dicing, assembly and sealing steps are very challenging. Without proper protection, the micro-scale, movable features could be damaged easily during these steps [40]. It is always desirable to protect MEMS devices by wafer-level packaging before dicing.

Wafer-level MEMS packaging is very challenging. As a result, most studies have focused on wafer-level capping [39, 14, 48, 30]. As shown in Fig. 17.2,

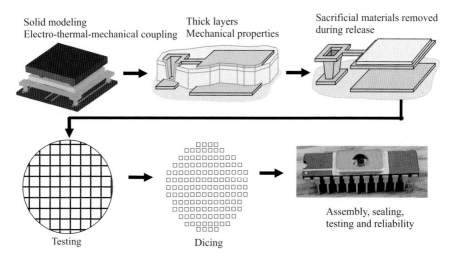

Fig. 17.1 MEMS design, fabrication and packaging [31]

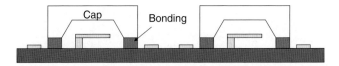

Fig. 17.2 Wafer-level capping for hermetic sealing of MEMS and plastic molding of MEMS devices capped [31]

silicon or glass caps are bonded onto a MEMS wafer for hermetic and/or vacuum sealing. The sealing is accomplished by wafer-to-wafer anodic bonding, soldering, or glass sealing. The capped MEMS devices are diced and packaged through injection plastic molding, which is compatible with that used in microelectronic packaging. Packaging cost and size are reduced substantially. In addition to this example, many other wafer-level capping technologies have been developed. Some examples will be reviewed in the next section on "Packaging of MEMS."

The relationship between MEMS and packaging, however, is not limited to packaging of MEMS devices. MEMS devices can in fact be used to enhance packaging technologies for microelectronic, optoelectronic and RF systems. In addition, packaging technologies can be applied to fabricate MEMS devices. Two more sections of the chapter will discuss a few examples of such "MEMS for Packaging" and "Packaging for MEMS." With these examples, we will be able to appreciate the golden opportunity to fully integrate sensors and actuators with microelectronic, optoelectronic and RF components on a single system platform. Packaging and MEMS technologies are essential to the integration on such a platform. The last section on "Opportunities and Major Challenges" will address this system integration as the opportunity and reliability as the main challenge. Reliability as showstopper can be removed, and we have a great opportunity to apply MEMS and packaging technologies to develop integrated micro/nanosystems in the future.

17.2 Packaging of MEMS

The fundamental problems in packaging of MEMS were illustrated in Figs. 17.1 and 17.2. New concepts have been developed recently to solve these problems. These approaches will be illustrated by three notable examples: a wafer-level capping technology using gold caps stamp-sealed; an encapsulation process developed during device fabrication steps; and a plastic wafer-level capping approach. Of course, MEMS packaging is not limited to wafer-level capping; therefore, we will illustrate another three approaches: environment-resistant packaging, bioMEMS packaging, and flip-chip assembly for hybrid integration. These illustrations will provide an insight into the needs and problems associated with diverse interfaces for MEMS sensors and actuators.

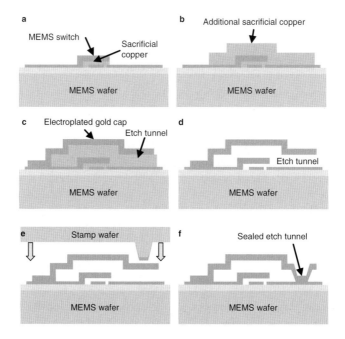

Fig. 17.3 Microshell packaging for RF MEMS using a stamp-sealing process [18]

Figure 17.3 illustrates a recent development of wafer-level capping developed by Heck et al. [18]. Major processing steps are listed below:

a. MEMS switch is fabricated using electroplated gold as the structural material and copper as the sacrificial material.
b. Additional copper is electroplated to enclose the switch.
c. The gold cap layer is formed by electroplating.
d. Copper is removed by etching to release the switch.
e. The device is sealed by a stamp-sealing process.

Figure 17.3-f shows the device sealed. Figure 17.4 shows an image of a stamp-sealed microshell membrane with an emphasis on the etch tunnels sealed. Details of each layer are shown in Fig. 17.5. This microshell approach is different from the typical wafer-level capping approach shown in Fig. 17.2. Gold and copper materials used for the switch device and package are compatible with electroplating; it is a manufacturable, low-cost and low-temperature process. The gold cap also serves as a metal shield to protect the RF MEMS device packaged. In addition, these materials enable the stamp-sealing process, which is very different from the wafer-to-wafer bonding process as illustrated in Fig. 17.2.

Figures 17.6 and 17.7 illustrate another novel approach for wafer-level capping with a MEMS device encapsulated by silicon during the fabrication

17 Microelectromechanical Systems and Packaging

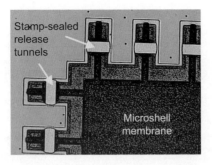

Fig. 17.4 Microshell with stamp-sealed etch tunnels [18]

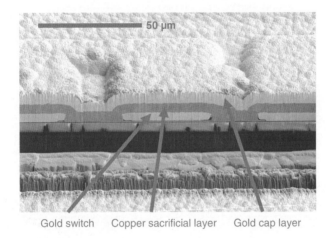

Fig. 17.5 SEM photograph of the cross-section of a capped switch before release [18]

process. No additional capping steps are needed after fabrication. The major processing steps developed by Candler et al. [1] are listed below:

a. Resonator structures are fabricated on a silicon-on-insulator (SOI) wafer. Their cross-sections are shown as squares in Fig. 17.6-a.
b. The resonator trenches (~1 μm wide) are covered by a sacrificial, non-conformal oxide layer with openings for the electrical contacts to the resonator structure.
c. A 2 μm layer of silicon is deposited with vent holes that allow HF vapor access for release.
d. A HF vapor process is used to release the resonators via a timed etch process.
e. The resonators are sealed in silicon encapsulation with silicon deposited at 950°C, followed by chemical-mechanical polishing (CMP). The trenches are

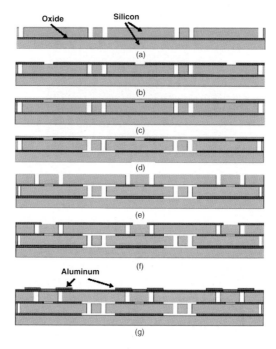

Fig. 17.6 Fabrication and encapsulation for MEMS (Candler et al. [1], ©2006 IEEE)

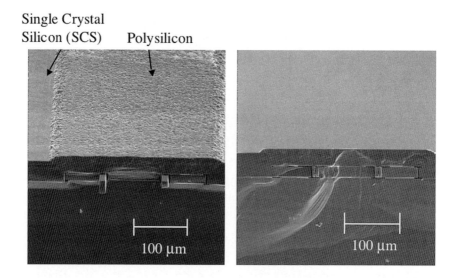

Fig. 17.7 SEM photograph of the cross-section of resonator encapsulated before chemical mechanical polishing (CMP) (Candler et al. [1], ©2006 IEEE)

etched through encapsulation for the electrical contacts. It should be noted that silicon deposition on existing single crystal silicon remains single crystal good for fabricating circuitry beside the resonator.
f. An oxide layer is deposited to cover the trenches with openings for the electrical contacts.
g. Aluminum is deposited and patterned for final contacts.

Figure 17.7 shows MEMS encapsulated before chemical-mechanical polishing (CMP). The image shows the step height between polysilicon and single crystal silicon. This step is caused by the underneath oxide layer and different deposition rates for single and polycrystalline silicon. CMP can remove the step height and polish polysilicon.

The fully integrated device fabrication and encapsulation processes would eliminate the boundary between MEMS fabrication and MEMS capping. After fabrication, the devices encapsulated can be packaged by any microelectronic packaging manufacturers. In addition, the fully integrated process would achieve substantial reduction in area, height and cost. On the other hand, the fully integrated process could limit the number of design options for MEMS devices.

The first two examples illustrate wafer-level capping using metal or polysilicon inorganic materials for hermetic or vacuum sealing. However, wafer-level packaging technologies developed for microelectronic packaging is usually polymer-based; therefore, it is interesting to investigate whether polymer can be applied for MEMS packaging. Figure 17.8 shows a MEMS resonator enclosed by a polymer cap demonstrated by Joseph et al. [24].

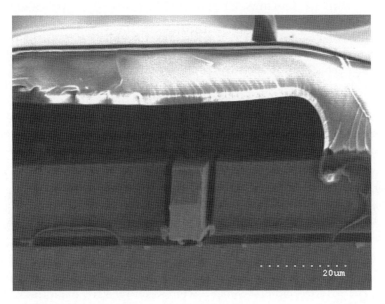

Fig. 17.8 SEM photograph of the cross-section of a microresonator enclosed by a polymer cap (Joseph et al. [24], ©2007 IEEE)

Major processing steps are listed below:

a. A thick photosensitive sacrificial polymer layer (approximately 10 µm) of Unity 2303P (from Promerus LLC) is spin-coated.
b. The wafer is soft-baked on a hotplate at 110°C for 10 min, followed by selective deep UV exposures (248 nm, 1 J/cm^2).
c. The exposed thick film is bake-developed at 110°C for 8–10 min.
d. A thin protection layer (~1 µm SiO_2) is deposited on the thick sacrificial material to provide additional mechanical strength.
e. The sacrificial material is encapsulated by a patterned polymer Avatrel overcoat.
f. The sacrificial polymer covering the beams is decomposed by heating at 170–260°C to form air-cavities for the beam resonators.

Polymer capping is very appealing; however, two issues have to be considered. Outgasing from polymer can generate particles, which can cause MEMS stiction. Hermetic and/or vacuum sealing is essential to most of MEMS devices. The polymer cap has to be covered by additional barrier coating. Such coatings have been developed for organic light emitting diodes (OLED) [35, 16, 2, 3]. One of the technologies is atomic layer deposition (ALD), which is described in the last section on "Opportunities and Major Challenges." It is very important to integrate ALD and other barrier coating technologies with polymer MEMS packaging and improve wafer-level capping to wafer-level packaging technologies.

Wafer-level capping represents major MEMS packaging activities for inertial sensors and RF actuators. However, MEMS has diverse applications, and many of them demand different functional interfaces. Packaging should allow MEMS moving parts to interact with other components through optical, electrical, thermal, mechanical, or chemical interfaces. Challenging problems are usually dependent on specific MEMS functions. The following two examples will illustrate new packages developed with different functional interfaces.

Figure 17.9 shows an environment-resistance package providing excellent vibration and thermal isolation achieved by using crab-leg shaped isolation suspension tethers. The package developed by Lee et al. [29] has three major components listed below:

a. A silicon substrate with signal feedthroughs is the main support of the package. Its shallow recess formed by deep reactive ion etching (DRIE) has a gold layer serving as a shock absorber and a heat radiation shield.
b. A 100-µm thick Pyrex glass isolation platform is anodically bonded on the silicon substrate through suspension tethers.
c. A silicon cap is for vacuum/hermetic sealing.

All these components can be fabricated and assembled in wafer-level processes. The MEMS chip is flip-chip assembled on the glass platform, which is supported by suspension tethers over the shallow recess formed in the supporting silicon substrate. The suspension tethers have electrical interconnect lines and

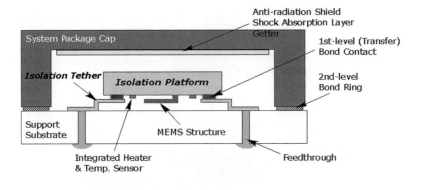

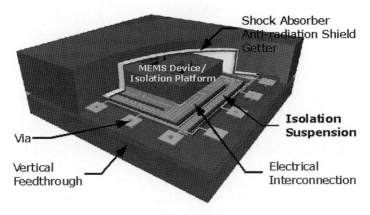

Fig. 17.9 Environment-resistance packaging technology for MEMS (Lee et al. [29], ©2007 IEEE)

vias between pads on the glass platform and vertical feedthroughs through the silicon substrate. The temperature of the MEMS device is oven-controlled by an integrated heater and temperature sensor; it is maintained at a fixed temperature. Vibration isolation is achieved by the thin glass suspension tethers.

Figure 17.10 presents a packaging solution for an optoelectronic biochip reviewed by Velten et al. [50]. This solution is compatible with microelectronic chip packaging as much as possible. In addition, it considers the biochip-compatibility and the measurement of low current. The sensor signal with a resolution better than 1 pA needs to be measured to monitor the decrease of the photocurrent due to analyte binding. A leadless ceramic chip carrier (LCCC) can meet the requirement due to its low leakage current property. The biochip wafer is cut with grooves of 225 μm depth in the 525-μm-thick wafer. After bioaffinity coating, the wafer is broken along the precut grooves. The single chip is glued into LCCC using a conductive epoxy adhesive cured at room temperature.

The major challenge is to attach the microinjection-molded microfluidic module onto the biochip in LCCC. The microfluidic module is placed on the

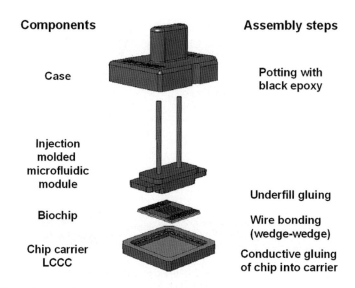

Fig. 17.10 Packaging of an optoelectronic biochip connected to a microfluidic module (Velten et al. [50], ©2005 IEEE)

chip with its microchannel aligned with the optical waveguides. The bioaffinity layer coated on the waveguide should be inside the channel for biosensing. A small amount of low-viscosity UV-curing acrylate glue is applied on the biochip at the fillet; it spreads in the capillary cleft between the chip and the module. Fast UV curing is used to avoid fluid contamination. The biochip is also designed with a large area reserved for adhesion. Inlet and outlet polyetheretherketone (PEEK) tubings are then glued into the microfluidic module. The assembly of LCCC, biochip, and microfluidic module is enclosed by an epoxy potting material.

A comprehensive review of packaging for bioMEMS is beyond the scope of this chapter. This optoelectronic biochip packaged is a good example representing the area. Fluidic coupling is a critical packaging consideration. Optical devices and waveguides have to be integrated with microfluidic channels. More importantly, all these technologies have to be compatible with microelectronic packaging technologies as much as possible.

All the MEMS packages presented so far are for single MEMS devices. For more complex microsystems, multiple MEMS devices are to be integrated with other microelectronic, optoelectronic and microwave devices. For such hybrid integration, MEMS devices should be integrated with other devices on a new, common substrate. A flip-chip assembly process with silicon removal technology has been developed for such transfer and integration [12]. Figure 17.11 presents a MEMS variable capacitor designed for the flip-chip assembly with silicon MEMS device transferred to an alumina substrate. The device developed by Faheem and Lee [13] consists of five components. Tethers are used to

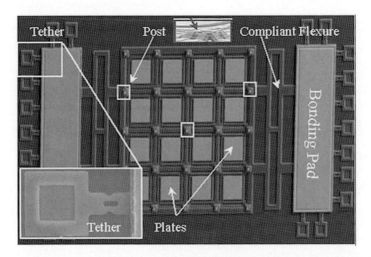

Fig. 17.11 A variable capacitor featured with tethers, bonding pads, compliant flexures, 5×5 "posts", and 4×4 plates on the host silicon substrate (Faheem et al. [12], ©2003 IEEE)

connect the pre-assembly released MEMS device to the silicon substrate. The tethers lightly connect a released MEMS device to its silicon donor substrate. They break naturally after delivering the device to the host substrate during or after the flip-chip assembly.

The bonding pads join the device to the new alumina substrate through solder bumps. Two compliant flexures accommodate the thermal mismatch between the silicon and the alumina substrate after the flip-chip assembly. Arrays of 2×2, 3×3 or 4×4 capacitor plates are designed with each plate surrounded by four "posts" (legs) to support the plate and its flexures. The use of the posts enables a precise gap control, which is critical to the operation of the capacitor plates. Before using posts, the gap is controlled by the solder joints with a large height variation. To reduce such a variation, posts are created by stacking different MEMS layers during device fabrication. When the top plates are pulled down by the electrostatic force, each plate's pull-in voltage is controlled by the precise gap defined by the posts rather than the solder joints. In addition, posts also enable us to design very compliant flexures to reduce thermal mismatch-induced warpage, which can degrade the electrostatic behavior of the MEMS by significantly increasing the pull-in voltage. With tethers and posts, the thermo-mechanical behavior of the variable capacitors becomes controllable after the flip-chip assembly.

17.3 MEMS for Packaging

Packaging of MEMS is usually recognized as the major MEMS packaging activity. However, as mentioned at the beginning of this chapter, there are other MEMS and packaging relationships. One of them is for MEMS devices

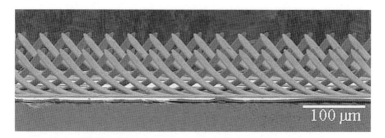

Fig. 17.12 SEM photograph a spring chip (Chow et al. [4], ©2006 IEEE)

developed to support advanced packaging technologies. This section will present the use of MEMS for fine-pitch flip-chip interconnect and testing probes, followed by the use of MEMS for active optical alignment. The third case is about the use of MEMS technology to fabricate co-axial cable-like RF circuits on printed circuit boards.

Figure 17.12 shows a section of 800 springs developed by Chow et al. [4] for interconnect and testing. There are 4 rows with 200 springs in each row. The springs can be fabricated with standard wafer-scale thin-film deposition processes on a BICMOS wafer, Corning 1737 glass or any other substrates. Major processing steps are listed below:

a. A sacrificial layer of titanium is deposited by sputtering.
b. A gold seed layer is sputtered, followed by electroplating of the spring metal through a mask defined for the springs and the electrical interconnect. The spring metal has multiple layers to provide a stress gradient with tensile layers on top. As reported previously, the spring metal layers can be sputtered MoCr layers which are deposited at different pressures. For a low cost approach, electroplated nickel layers with different residual stresses are used by Chow et al. [4].
c. A second mask is used to define release regions around the springs and to protect the signal reroute traces against the release etchant.
d. The springs are released by etching away the sacrificial layer; the springs rise off the wafer to relieve the internal stresses.
e. An alloy is deposited around the spring, followed by electroplated nickel hardened gold. The alloy provides mechanical strength and stiffness, and the gold protects it from oxidation and increases its electrical conductivity.

There are 800 springs on a 3×10 mm chip with each spring 180 μm long, 14 μm wide, and 5 μm thick. The spring tips are 57 μm tall, with $\pm$ 5 μm height variations. The springs are interleaved with a pitch of 40 μm. The pitch of the pads to be probed is 20 μm.

Such micro springs and other similar MEMS springs can be used for testing probes or detachable flip-chip interconnects. Or, they can be used for compliant solder joints [6] or thermal variable resistors for temperature control [25]. These MEMS devices are critical to the advancement of packaging technologies.

17 Microelectromechanical Systems and Packaging

The second example of MEMS for packaging is the use of micro-mirror for optical active alignment, which is commonly used in manufacturing of optoelectronic modules. The current approach uses a precision robot to position the fiber during the alignment and welding processes. Robots with a submicron resolution are expensive, and the precision of fixing the fiber drops significantly during welding, due to thermal shrinkage. As a result, a MEMS-based micromirror is a promising device; it steers a laser beam for active alignment without requiring a precision robot and fixture. Ishikawa et al. [23] demonstrated a micromirror suspended by four thermal actuators for laser-to-fiber coupling. As shown in Fig. 17.13, these structures are lifted up to 45° with mechanical locking mechanisms; thereafter the micromirror can be steered two-dimensionally by the actuators. Each actuator could move out-of-plane due to bending resulting from differential heating. Powering a pair of the actuators could rotate the mirror along a horizontal or vertical axis. Powering a single actuator can rotate the mirror along a diagonal axis. Beam steering with such a large degree of freedom could improve the optical coupling efficiency from 10% to over 80% in an experiment using a vertical-cavity surface-emitting laser (VCSEL) and a multi-mode fiber. Ideally, for laser-to-fiber coupling a micromirror should remain fixed without the need for power after alignment. The device shown in the figure was improved by a new micromirror assembly with a position fixing function achieved by breakable tethers. The micromirror is connected with several bimorphs constrained by tethers. The tethers can be broken one by one through electrical heating. Each broken tether would enable the bimorph to make a movement, which would tilt the mirror surface for beam steering. After such a movement, the position of the mirror is fixed without any power needed to maintain actuation [26].

The third example is about the use of MEMS technology to fabricate coaxial cable-like RF circuits on printed circuit boards as shown in Fig. 17.14.

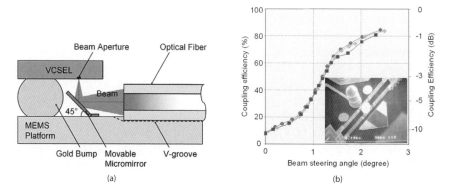

Fig. 17.13 (a) Schematic view of a laser-to-fiber coupling concept and (b) a graph of coupling efficiency / beam steering angle with inset micrograph of the microstructure (Ishikawa et al. [23], ©2002 IEEE)

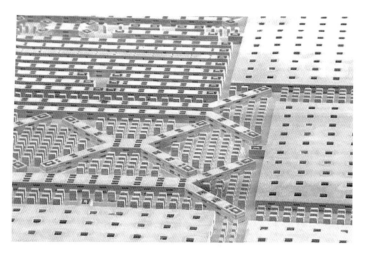

Fig. 17.14 SEM photograph of a recta-coax based cavity resonator (larger areas in the photo) and a number of coaxial lines with various separation distances (Filipovic et al. [15], ©2006 IEEE)

Recta-coax lines and components studied by Filipovic et al. [15] are fabricated using a new sequential microfabrication process. The recta-coax structures are built up layer by layer with a uniform copper stratum as the Layer 1. Layers 2, 4 and 5 are copper strata defined by photo-resist, which is also the sacrificial material. The copper thicknesses for Layers 1, 2, 4, and 5 are 10, 75, 75, and 50 μm, respectively. The inner conductor of the co-axial cable-like circuit has to be supported mechanically without affecting the RF characteristics. This support is achieved by using 15 μm thick polymer deposited and patterned as part of Layer 3 with a total thickness of 100 μm. Once the structure is built, the resist is removed through release holes in the top and side walls. Figure 17.14 shows different recta-coax lines, calibration structures, test sets for isolation, and several different cavity resonators. The resonators use a larger area in the photo. The 250×300 μm feed ports are visible as rectangular coaxial openings at the end of each line/component. The holes in the top and side walls are used for releasing the photoresist. The metal posts seen in the photo do not have any electrical effect. The 50Ω coaxial lines are designed and fabricated with 250 μm tall rectangular and square cross-sections. The inner conductor is supported by periodically (700 μm) spaced 15 μm tall and 100 μm wide dielectric support straps ($\varepsilon = 3.7$, $\tan \delta = 0.05$). The release holes in the side (100×75 μm) and top walls (200×100 μm) are placed in the regions between the straps. These components are built on a high resistivity Si wafer; other substrates can be used, however.

Electrical effects of potential fabrication issues such as strap parameters including the protrusion into the vertical walls, offset layers, under/over etching, etch holes, surface roughness are investigated in other studies [32, 49]. With a proper design, the performance of a line built within the limits of the fabrication tolerances is almost the same as the performance of a perfectly fabricated structure.

17.4 Packaging for MEMS

We have discussed packaging of MEMS devices and the use of MEMS devices for advanced packaging technologies. Another interesting MEMS and packaging relationship is to apply packaging technologies to fabricate MEMS devices. This section will discuss the use of the soldering technology to fabricate three-dimensional surface-micromachined MEMS devices, followed by the use of the flexible circuit technology to fabricate RF MEMS devices. The third case is about the use of the low temperature co-fired ceramics (LTCC) technology to fabricate lab-on-a-chip devices.

One of the most common methods for fabricating MEMS devices is by using surface micro-machining, which is, however, unable to produce highly three-dimensional structures. A common solution is to fabricate flat, 2D hinged components that can be lifted or rotated into assembled structures. These hinged devices need to be assembled after fabrication. Manual assembly usually consists of rotating the plates by hand using high precision micro-manipulators. This form of assembly is not practical for mass assembly and manufacturing though, and is rarely effective. An interesting solution is the use of the surface tension properties of molten solder or glass as the assembly mechanism [17].

The solder method involves using a standard hinged plate with a specific area metallized as solder wet table pads. Once the solder is in place, it is heated to its melting point, and the force produced by the natural tendency of liquid to minimize its surface energy pulls the free plate away from the silicon substrate (Fig. 17.15). Solder is a predominant technology for electronics assembly and

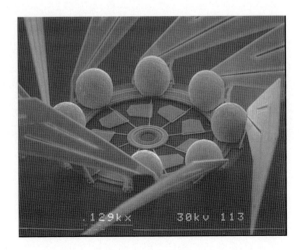

Fig. 17.15 Solder self-assembly of a hinged MEMS plate and a solder assembled three-dimensional MEMS device (Kladitis et al. [27], ©2001 IEEE)

packaging. It is not only used for electrical connections, but also for sub-micron accuracy alignment in many packaging applications such as optoelectronic passive alignment [46]. Using solder, hundreds or thousands of precision alignments can be accomplished with a single batch reflow process. In addition, solder provides high quality mechanical, thermal, and electrical connections. The SEM photograph of the three-dimensional fan developed by Kladitis et al. [27] is an illustration of the excellent capability of solder assembly.

Silicon processing is not the only means to fabricate MEMS devices. In fact, we expect to see more and more MEMS devices to be fabricated using polymer materials. A good example is a flexible circuit-based RF MEMS. Figure 17.16 shows different layers and assembled prototype of X/Ku band switches demonstrated by Ramadoss et al. [36, 37, 38]. Coplanar waveguide (CPW) lines for mounting switches and on-wafer multi-line TRL calibration are patterned on the metallization layer of a Duroid substrate. Photosensitive benzocyclobutene (BCB) dielectric layer is spin-coated and patterned on CPW lines. Adhesive spacer film is milled to create slot-openings. The switch electrode metallization is patterned on Kapton-E polyimide film, which is machined using Excimer laser to create slot-openings. These layers are aligned using a fixture and laminated using a thermo-compression bonding cycle.

These switches are manufacturable using printed circuit board (PCB) facilities, and they can be integrated with PCB-based RF circuits and antennas. We expect them to have an impact on PCB-based applications. However, with the large size, there are concerns about their RF losses. The insertion loss could be less than 0.3 dB and the isolation could reach –50 dB at the designed

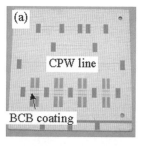
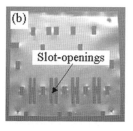
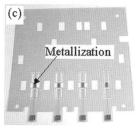
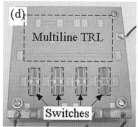

Fig. 17.16 Photographs of (**a**) CPW line with BCB dielectric layer on Duroid substrate; (**b**) adhesive spacer film with milled slot-openings; (**c**) Kapton E polyimide film with switch top electrode metallization and laser machined slot-openings; and (**d**) assembled switch prototype (Ramadoss et al. [36], ©2003 IEEE)

frequencies. Such performance is close to that achieved by thin-film based RF MEMS.

This RF MEMS switch is an example of MEMS devices fabricated using printed circuit boards (PCB) or flexible circuit boards. There are other devices demonstrated for microfluidic, pressure sensor and other sensor and actuator applications. In general, PCB MEMS devices are larger than silicon MEMS devices. Their mechanical responses are slow due to the large size. Barrier coating technologies for polymer capping (see Fig. 17.8) are needed to seal these PCB MEMS devices if hermetic or vacuum sealing is required. On the other hand, embedded capacitors, inductors and resistors are well known to be critical to PCB-based systems demanding small size and low cost. Similarly, there is a good opportunity to laminate PCBMEMS devices with other circuits for small size, low-cost systems. We expect PCB MEMS to become a viable alternative to silicon MEMS in the future.

The third example of packaging for MEMS is the use of ceramics for lab-on-a-chip applications. Figure 17.17 shows a ceramic based polymerase chain reaction (PCR) demonstrated by Sadler et al. [41]. A continuous flow polymerase chain reaction (CPCR) device for DNA amplification and an electronic DNA detection chip are fabricated using a multilayer low temperature co-fired ceramics (LTCC) platform. LTCC allows 3-D integration of microfluidics, heaters, and surface-mount temperature sensors for the device. The device is designed according to the Dupont 951 tape system. Multiple layers of green-sheet ceramic tape are processed, aligned, laminated, and then fired to form the device. Microfluidic channels are formed by mechanical punching or laser machining of the layers designed. Gold is used for sensor electrodes and connection terminals, and silver/palladium is used for the heaters. Both materials are deposited and patterned using thick-film printing techniques. Electrical vias are fabricated by mechanical punching of holes filled with thick-film silver paste. All the layers are aligned, laminated using low-pressure lamination technology, and fired 850°C. Silicon transistors for temperature sensing can be soldered onto the ceramic devices. Finally, a plastic cap is attached to the detection chip using double-sided tape, and inlet and outlet tubing is attached to the CPCR device with epoxy.

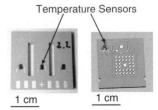

Fig. 17.17 Ceramic-based polymerase chain reaction (PCR) device (Sadler et al. [41], ©2003 IEEE)

17.5 Opportunities and Major Challenges

We have reviewed interesting examples of packaging of MEMS devices, use of MEMS devices to enhance packaging technologies, and use of packaging technologies to fabricate MEMS devices. These examples also indicate a great opportunity to apply packaging and MEMS technologies to fully integrate sensors and actuators with microelectronic, optoelectronic and RF components on a single system platform. This system integration opportunity can be illustrated by the physics subsystem of a chip-scale atomic clock as shown in Fig. 17.18 [28]. This subsystem is designed, fabricated and assembled by engineers at the Teledyne Scientific Company. As shown in Fig. 17.18a, a laser beam emitted from a vertical cavity surface-emitting laser (VCSEL) is passed through a Cesium vapor cell to a photodiode. A local oscillator operating at 4.6 GHz modulates the laser beam. When the modulation frequency reaches 4.59631589 GHz, the atoms are pumped into a coherent dark state. This results in significantly reduced absorption, which is detected by a photodiode. The photodiode signal is fed back to a local oscillator to lock it to this resonance. Thus measuring a very accurate frequency, this is then converted to time.

The cell and VCSEL in this assembly must be maintained at specified temperatures, e.g. 70 ± 1°C, using heaters on either side of the cell. As shown in Fig. 17.18b, the physics assembly is suspended on a specially designed Cirlex suspension with limited conductive heat losses. Low emissivity coating is applied to reduce radiation losses. In addition, all components integrated are sealed in a package to limit convective losses. The physics subsystem of CSAC integrates photonic active and passive devices with vapor cells, thermal and mechanical elements through advanced MEMS and packaging technologies. This subsystem has to be further integrated with microelectronic controllers and RF oscillators for a complete CSAC. CSAC is a good example illustrating how

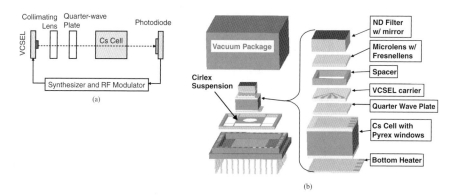

Fig. 17.18 (**a**) Schematic diagram of a chip-scale atomic clock and (**b**) physics subsystem of the clock

MEMS and packaging technologies are used to integrate MEMS devices with microelectronic, photonic and RF devices.

System integration using packaging and MEMS technologies is an exciting opportunity; novel systems such as CSAC will create a new generation of microsystems in the 21st century. More importantly, nanoelectromechanical systems (NEMS) have been demonstrated in various laboratories; they will be integrated with MEMS for significant performance improvement. The integration of MEMS and NEMS is the theme of the research studies conducted at the DARPA Center on Nanoscale Science and Technology for Integrated Micro/Nano-Electromechanical Transducers (iMINT). Here is a short list of possible impacts resulting from some integrated MEMS/NEMS systems being studied at iMINT:

a. a flexible thermal ground plane with a thermal conductivity 100X better than copper's.
b. a nano-scaled barrier coating on polymer with 10,000X enhanced hermeticity.
c. a metamaterial-enabled patch antenna that is 5–10X smaller than current antennas.
d. a biosensor with single molecule sensitivity while reducing the false alarm rate by 100X.
e. a light emitting diode with 3X enhanced efficiency and 100X reduced thermal resistance.
f. a solid state supercapacitor with the capacitor density increased by 25–100X.

MEMS/NEMS and packaging technologies are essential to each system integration. System integration is indeed a great opportunity; however, they are numerous challenging issues for the integration. The systems developed should be manufacturable. For example, the physics subsystem shown in Fig. 17.18 consists of quite a few components vertically stacked. Three-dimensional stacking is not manufacturable for microelectronic packages; therefore, manufacturability of such a physics subsystem is not promising today. How to improve manufacturing processes to accommodate MEMS/NEMS devices integrated with other devices is a major challenge. The approaches shown in Section 17.2: Packaging of MEMS can serve as good examples. In each approach, a wafer-level capping process has been custom-developed for each MEMS device. After this process, MEMS devices capped can be treated as another electronic components going through regular manufacturing processes established for microelectronic packages. It is critical to develop wafer-level capping-like processes to meet the manufacturing challenges for integrated MEMS/NEMS systems.

Another major challenge is reliability. Stiction, fracture, and fatigue, mechanical wear with respect to frequency and humidity, and shock and vibration effects are the major causes of MEMS failures. Their effects on NEMS will be much more influential. During the last 20 years, MEMS products have proven to be reliable [34, 45, 47]. The most reliable MEMS devices are hermetically packaged single-point contact or no-contact devices. Recently, novel MEMS devices with surface contacts have reached impressive reliability levels with billions or hundreds of billions of surface impacts. It is a significant improvement from the early studies on RF MEMS.

Most of the improvements on surface contacts result from design for reliability and material development. For a capacitive RF MEMS switch, the capacitance ratio can be reduced from 100 to 5 after improving RF system designs. With a small ratio, contact area is decreased substantially. Its corresponding stiction force is reduced and the reliability is enhanced significantly. Or, the flexure of the MEMS device can be designed to be very stiff. The high pull-down voltage required for the stiff beam can be adjusted during the snap-down process, so the voltage applied after pull-down can be small. Small voltage can reduce charge accumulation and avoid charge-induced stiction problem. Meanwhile, the stiff beam has a very large spring force, which can overcome surface adhesion even after billions of surface contacts.

For the material development, self-assembled monolayer (SAM) is applied to MEMS devices to avoid moisture-induced stiction failures [7]. Hardened gold alloys are developed to assure reliability of RF contact switches [5]. Tungsten coating can enhance wear resistance of MEMS bearing [42]. In addition, atomic layer deposition (ALD) has also been introduced to provide various nano-scaled surface coatings to enhance MEMS reliability. ALD is one of the best surface coating technologies with precision down to one atomic layer. It is important to assure MEMS reliability affected by the nano-scaled interface between MEMS surfaces. A 25-nm coating can affect the performance of a 100-nm scaled device. More importantly, ALD will be essential to NEMS. The 25-nm coating will be in the same order as the 100-nm scaled devices. Any variations in the nano-scaled coating will have a significant effect on NEMS performance and reliability. ALD provides the best to assure high-quality, precise nano-scaled coatings.

ALD is a thin film growth technique allowing atomic-scale thickness control. ALD utilizes a binary reaction sequence of self-limiting chemical reactions between gas phase precursor molecules and a solid surface [11]. Films deposited by ALD are extremely smooth, pinhole-free and conformal to the underlying substrate surface. This conformality enabled successful coating to cover the entire MEMS device as shown in Fig. 17.19 [21]. Furthermore, ALD is a low temperature process enabling deposition on thermally sensitive materials. For example, we can use photoresist to cover some patterned areas during deposition for selective instead of comprehensive coverage. ALD can be used to grow a variety of materials including oxides, nitrides, and metals.

ALD process consists of two CVD half reactions. One example of this process is the atomic layer deposition of Al_2O_3 consisting of the following binary reaction sequence in which the asterisks designate the surface species:

$$A) \ Al-OH^* + Al(CH_3)_3 \rightarrow Al-O-Al(CH_3)_2^* + CH_4$$

$$B) \ Al-CH_3^* + H_2O \rightarrow Al-OH^* + CH_4$$

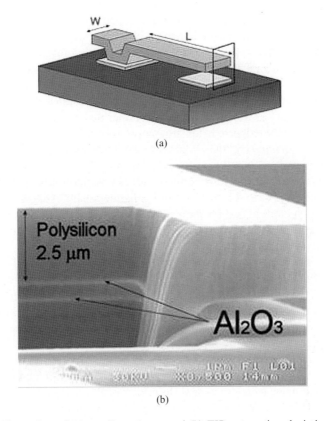

Fig. 17.19 Illustration of (a) cantilever beam and (b) FIB cut section depicting deposited alumina layer ([21], ©2002 IEEE)

In reaction A, the $Al(CH_3)_3$ reacts with the surface hydroxyl groups to deposit a new monolayer of aluminum atoms terminated by methyl groups. In reaction B, the methylated surface reacts with H_2O vapor, thereby replacing the methyl groups with hydroxyl groups. CH_4 is liberated in both the A and B reactions. The net result of one AB cycle is the deposition of one monolayer of Al_2O_3 onto the surface. The ALD Al_2O_3 film growth is extremely linear with the number of AB cycles performed and the growth rate is 1.29 Å/cycle. The deposition rate is about 0.12 nm (one AB cycle) in 6–10 s in a laboratory setup. In a manufacturing setup, the cycle time can be reduced by at least 10 times.

ALD can be used to coat many different nano-scaled, single-layer or multi-layer structures to protect MEMS from different reliability failures. Successful coatings have been developed to solve the following MEMS reliability problems:

- Dielectric Coating to Prevent Electrical Shorts [21, 22]
- Charge Dissipation for Reliable MEMS [8]
- Hydrophobic Coating for Reliable MEMS [19]

In addition, ALD coatings have been developed for nano-scaled devices [20]. These ALD coatings can be further improved by newly developed molecular layer deposition (MLD) [7]. The nano-scaled organic/inorganic multilayer is expected to play a major role in material improved for reliable integrated MEMS/NEMS.

ALD, MLD and many other materials processing technologies are available to remove reliability as the showstopper for integrated MEMS/NEMS systems. However, our knowledge is often not good enough to guide the materials development. Fortunately, the progress to establish basic knowledge in reliability is promising.

Figure 17.20 shows moisture-induced adhesion energy affected by relative humidity with respect to different surface roughnesses [9]. Adhesion results are cantilevers with landing pad roughnesses ranging from 2.6 to 10.3 nm rms. The adhesion energy is extracted by comparing the experimental deflections to finite element method simulations.

Dashed lines report the experimental observations of the cantilevers when the relative humidity (RH) increases. The cantilever is stuck on the surface when RH reaches a critical level; this stiction is a jump in adhesion energy as shown in each dashed line. As the landing pad roughness increases, the RH at which the adhesion initially jumps due to capillary condensation also increases. Once the initial jump occurs, the adhesion energy is independent of RH.

In addition, a detailed model was developed to consider the actual surface topography and the observed surface correlations. The modeling results are shown in solid lines, which represent the effects of RH and surface roughnesses very well. The maximum adhesion energy due to capillary condensation is

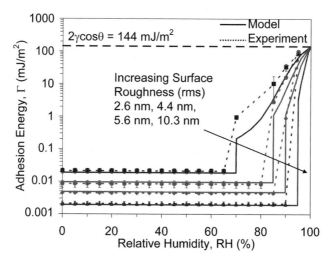

Fig. 17.20 Moisture-induced adhesion energy affected by relative humidity with respect to different surface roughnesses [9]

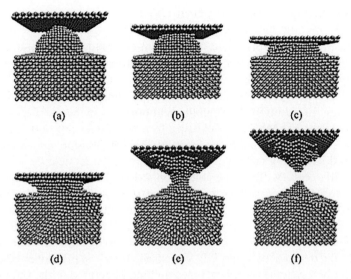

Fig. 17.21 Atomistic view of the asperity evolution as seen along the [001] direction: (**a–c**) increasing load from right after jump to contact to maximum loading, and (**c–f**) unloading from the maximum load to full separation [44]

$2\gamma \cos \theta = 144$ mJ/m^2, which is shown for reference. It is used to include the capillary force in the model.

Figure 17.21 shows the evolution of the contact morphology during a single loading and unloading cycle [44]. Such a contact is commonly observed in RF MEMS contact switches. The molecular dynamics simulations are conducted to study repetitive contact between a hemispherical asperity and a flat. Figure 17.21a shows that the asperity stretches upward to meet the top rigid plate when it is moved towards the asperity. This stretch results in a tensile, elastic strain in the asperity with a contact region that is much wider than a single atom. As the rigid plate continues to move, the system goes into elastic compression that is well characterized by the classic theory for adhesive, elastic contacts. Moving further followed by separation, the system undergoes substantial crystal plasticity (see Fig. 17.21b–f). At larger loads (Fig. 21c and d), the contact has substantially broadened and the asperity has become noticeably thinner.

As demonstrated by Figs. 17.20 and 17.21, knowledge in MEMS reliability is being established. With enough knowledge, yet to be established, we will be able to specify nano-scaled coatings that can be fabricated by atomic layer deposition and molecular layer deposition for reliable MEMS. MEMS devices have proven reliable without contacts with a cyclic life over trillions. Some MEMS devices have proven reliable even with contacts with a cyclic life over 100 billion. With advancement of science and technology, MEMS reliability will not be a showstopper anymore. Of course, we will never be able to solve the MEMS reliability problem. Reliability is always a critical issue in microelectronic packaging, and it is going to be a critical issue in MEMS packaging.

17.6 Conclusions

Several cases have been presented for packaging of MEMS, MEMS for packaging and packaging for MEMS. With movable parts, MEMS possesses a special packaging problem, which is solved by wafer-level capping. With movable parts, MEMS also provides novel solutions to enhance advanced packaging technologies. In addition, packaging technologies can be used to fabricate MEMS devices. MEMS and packaging have an interesting and stimulating relationship. It is essential to understand this relationship and apply both MEMS and packaging technologies properly in various systems.

In the 21st century, there is a golden opportunity to fully integrate sensors and actuators with microelectronic, optoelectronic and RF components on a single system platform. Packaging and MEMS technologies are essential to the integration on such a platform. Reliability is a showstopper for MEMS and NEMS to be integrated. With knowledge and technology to be established, the reliability barrier can be removed. We expect to see many novel integrated MEMS/NEMS systems to be developed and manufactured in the future.

Acknowledgments The author is supported by the DARPA Center on Nanoscale Science and Technology for Integrated Micro/Nano-Electromechanical Transducers (iMINT) through the DARPA S&T Fundamental Program (HR0011-06-1-0048). He is also supported by the DARPA Micro Cryogenic Cooling (MCC) Program (NBCHC060052).

References

1. Candler RN, Hopcroft MA, Kim B, Park WT, Melamud R, Agarwal M, Yama G, Partridge A, Lutz M, and Kenny TW (2006) Long-term and accelerated life testing of a novel single-wafer vacuum encapsulation for MEMS resonators. Journal of Microelectromechanical Systems, Vol. 15, No. 6, pp. 1446–1456
2. Carcia PF, McLean RS, Reilly MH, Groner MD, and George SM (2006) Ca test of Al2O3 gas diffusion barriers grown by atomic layer deposition on polymers. Applied Physics Letters, 89, 031915
3. Chen TN, Wuu DS, Wu CC, Chiang CC, Chen YP, and Horng RH (2007) Improvements of permeation barrier coatings using encapsulated parylene interlayers for flexible electronic applications. Plasma Processes and Polymers, Vol. 4, pp. 180–185
4. Chow EM, Chua C, Hantschel T, Van Schuylenbergh K, and Fork DK (2006) Pressure contact micro-springs in small pitch flip-chip packages. IEEE Transactions on Components and Packaging Technologies, Vol. 29, Issue 4, pp. 796–803
5. Coutu, RA, Reid JR, Cortez R, Strawser RE, and Kladitis PE (2006) Microswitches with sputtered Au, AuPd, Au-on-AuPt, and AuPtCu alloy electric contacts. IEEE Transactions on Components and Packaging Technologies, Vol. 29, June, pp. 341–349
6. Dang B, Bakir MS, Patel CS, Thacker HD, and Meindl JD (2006) Sea-of-leads MEMS I/O interconnects for low-k IC packaging. Journal of Microelectromechanical Systems, Vol. 15, pp. 523–530
7. de Boer MP, Knapp JA, Michalske TA, Srinivasan U, and Maboudian R (2000) Adhesion hysteresis of silane coated microcantilevers. Acta Materialia, Vol. 48, pp. 4531–4541

8. DelRio FW, Herrmann CF, Hoivik N, George SM, Bright VM, Ebel JL, Strawser RE, Cortez R, Leedy KD (2004) Atomic layer deposition of Al2O3/ZnO nano-scale films for gold RF MEMS. IEEE MTT-S International, Vol. 3, pp. 1923–1926
9. DelRio FW, Dunn ML, Phinney LM, Bourdon CJ, and de Boer MP (2007) Rough surface adhesion in the presence of capillary condensation, Applied Physics Letters, Vol. 90, 163104
10. Du Y and George SM (2007) Molecular Layer Deposition of Nylon 66 Films Examined Using in Situ FTIR Spectroscopy, Journal of Physics Chemistry C, Vol. 111, pp. 8509–8517
11. Elam JW and George SM (2003) Growth of ZnO/Al2O3 alloy films using atomic layer deposition techniques. Chemistry of Materials 15, p. 1020
12. Faheem FF, Gupta KC, and Lee YC (2003) Flip-chip assembly and liquid crystal polymer encapsulation for variable MEMS capacitors. IEEE Transactions on Microwave Theory and Techniques, pp. 2562–2567
13. Faheem FF and Lee YC (2004) Tether- and post- enabled flip-chip assembly for manufacturable RF-MEMS. Sensors and Actuators, Vol A-114, No. 2–3, pp. 486–495
14. Felton LE, Hablutzel N, Webster WA, and Harney KP (2004) Chip scale packaging of a MEMS accelerometer. Proc. 54th Electronic Components and Technology Conference, pp. 869–873
15. Filipovic DS, Popovic Z, Vanhille K, Lukic M, Rondineau S, Buck M, Potvin G, Fontaine D, Nichols C, Sherrer D, Zhou S, Houck W, Fleming D, Daniel E, Wilkins W, Sokolov V, and Evans T (2006) Modeling, design, fabrication, and performance of rectangular μ-coaxial lines and components. IEEE MTT-S International Microwave, Symposium Digest, 11–16 June, pp. 1393–1396
16. Groner MD, George SM, McLean RS, and Carcia PF (2006), Gas diffusion barriers on polymers using Al2O3 atomic layer deposition. Applied Physics Letters, Vol. 88, 051907
17. Harsh KF, Bright VM, and Lee YC (1999) Solder self-assembly for three-dimensional micro-electromechanical systems. Sensors and Actuators A, Vol. 77, pp. 237–244
18. Heck J, Bar H, Chou TKA, Tran Q, Ma Q, Weinfeld B, and Rao V (2007) A stamp-sealed microshell package for RF MEMS switches. ASME InterPACK '07, July 8–12, Vancouver, British Columbia, Canada, paper # IPACK2007-33887
19. Herrmann CF, DelRio FW, Bright VM, and George SM (2004) Hydrophobic coatings using atomic layer deposition and non-chlorinated precursors. 17th IEEE International Conference on MEMS, pp. 653–656
20. Herrmann CF, Fabreguette FH, Finch DS, Geiss R, and George SM (2005) Multilayer and functional coatings on carbon nanotubes using atomic layer deposition. Applied Physics Letters, Vol. 87, 123110
21. Hoivik ND, Elam JW, Linderman RJ, Bright VM, George SM, and Lee YC (2003) Atomic layer deposited protective coatings for microelectromechanical systems. Sensors and Actuators A: Physical, Vol. 103, Issue 1/2, pp. 100–108
22. Hoivik ND, Elam JW, Linderman RJ, Bright VM, George SM, and Lee YC (2003) Atomic layer deposited protective coatings for micro-electromechanical systems. Sensors and Actuators, Vol. A-103, pp. 100–108
23. Ishikawa K, Zhang J, Tuantranont A, Bright VM, and Lee YC (2003) An integrated micro-optical system for VCSEL-to-fiber active alignment. Sensors and Actuators, Vol. A-103, pp. 109–115
24. Joseph PJ, Monajemi P, Ayazi F, and Kohl PA (2007) Wafer-level packaging of micromechanical resonators. IEEE Transactions on Advanced Packaging, Vol. 30, Issue 1, Feb. 2007, pp. 19–26
25. Kim HS, Liao HH, Lee BH, and Kenny TW (2006) Design and verification of a low-powered pre-programmable in-package temperature controller. ASME International Mechanical Engineering Congress and Exposition, November 2006, Chicago, Illinois USA, IMECE2006-15136

26. Kitagawa H, Boteler DJ, and Lee YC (2006) Thermo-mechanical behavior of a micromirror for laser-to-fiber active alignment. Proceedings of ASME International Mechanical Engineering Congress and Exposition November 5–10, Chicago
27. Kladitis PE, Linderman RJ, and Bright VM (2001) Solder self-assembled micro axial flow fan driven by a scratch drive actuator rotary motor. The 14th IEEE International Conference on Micro Electro Mechanical Systems, 21–25 Jan, pp. 598–601
28. Laws AD and Lee YC (2007) Thermal and structural analysis of a suspended physics package for a chip-scale atomic clock. ASME InterPACK'07, Vancouver, July 8–13
29. Lee, SH, Lee SW, and Najafi K (2007) A generic environment-resistant packaging technology for MEMS. Solid-State Sensors, Actuators and Microsystems Conference, TRANSDUCERS 2007. 10–14 June, pp. 335–338
30. Lee YC, Parviz BA, Chiou A, and Chen S (2003) Packaging for microelectromechanical and nanoelectromechanical systems. IEEE Transaction on Advanced Packaging, pp. 217–226
31. Lee YC (2007) MEMS packaging and reliability. In: Suhir E, Lee YC and Wong CP (eds) Micro- and Opto-Electronic Materials and Structures: Physics, Mechanics, Design, Reliability, Packaging, Springer, Berlin
32. Lukic M, Rondineau S, Popovic Z, and Filipovic DS (2006) Modeling of realistic rectangular μ-coaxial lines. IEEE Transactions on Microwave Theory and Techniques, vol. 54, pp. 2068–2076
33. Madou MJ (2002) Fundamentals of Microfabrication: The Science of Miniaturization. CRC, Boca Raton
34. MEMS Industry (2004) Report Focus on Reliability, MEMS Industry Group, Pittsburgh, PA, USA
35. Moro L, Krajewski TA, Rutherford NM, Philips O, Visser RJ, and Gross M, Bennett WD, and Graff G (2004) Process and design of a multilayer thin film encapsulation of passive matrix OLED displays. Proc. of SPIE Vol. 5214, pp. 83–93
36. Ramadoss R, Lee S, Lee YC, Bright VM, and Gupta KC (2007) MEMS capacitive series switch fabricated using PCB technology. International J. of RF and Microwave Computer–Aided Engineering, Vol. 17, Issue 4, July, pp. 387–397
37. Ramadoss R, Lee S, Lee YC, Bright VM, and Gupta KC (2006) RF MEMS capacitive switches fabricated using printed circuit processing techniques. IEEE/ASME Journal of Microelectromechanical Systems, pp. 1595–1604
38. Ramadoss R, Lee S, Lee YC, Bright VM, and Gupta KC (2007) MEMS capacitive series switch fabricated using PCB technology. International Journal of RF and Microwave Computer-Aided Engineering, pp. 387–397
39. Riley GA (2004) Wafer-level hermetic cavity packaging. Advanced Packaging, Vol. 3, No. 5, pp. 21–24
40. Roberts, CM, Long LH, and Ruggerio PA (1994) Method for Separating Circuit Dies from a Wafer. US Patent 5362681
41. Sadler DJ, Changrani R, Roberts P, Chou CF, and Zenhausern F (2003) Thermal Management of BioMEMS: Temperature Control for Ceramic-Based PCR and DNA Detection Devices. IEEE Transactions on Components and Packaging Technologies, Vol. 26, No. 2, pp. 309–316
42. Sechrist ZA, Fabreguette FH, Heintz O, Phung TM, Johnson DC, and George SM (2005) Optimization and structural characterization of W/Al2O3 nanolaminates grown using atomic layer deposition techniques. Chemistry of Materials, Vol. 17, pp. 3475–3485
43. Senturia SD (2000) Microsystem Design (Hardcover). Springer, Berlin
44. Song J and Srolovitz DJ (2007) Atomistic simulation of multicycle asperity contact. Acta Materialia Vol. 55, pp. 4759–4768
45. Sontheimer A and Douglass M (1998) Identifying and eliminating digital light processing TM failure modes through accelerated stress testing. TI Technical Journal, July–September 1998, pp. 128–136

46. Tan Q, Lee YC, and Itoh M (2005) Soldering technology for optoelectronic packaging. In: Passive Micro-Optical Alignment Methods, Boudreau R and Boudreau S (eds), CRC Press, Boca Raton
47. Tanner DM (2000) Reliability of surface micromachined MicroElectroMechanical Actuators. 22nd Int. Conf. Microelectronics, Nis, Yugoslavia, pp. 97–104
48. Tseng A, Tang WC, Lee YC, and Allen J (2001) NSF 2000 workshop on manufacturing of micro-electro-mechanical systems. Journal of Materials Processing & Manufacturing Science, Vol. 8, No. 4, pp. 292–360
49. Vanhille KJ, Fontaine DL, Nichols C, Popovic Z, and Filipovic DS (2007) Ka-Band miniaturized quasi-planar high-Q resonators, IEEE Transactions on Microwave Theory and Techniques, Vol. 55, pp. 1272–1279
50. Velten T, Ruf HH, Barrow D, Aspragathos N, Lazarou P, Jung E, Malek CK, Richter M, Kruckow J, and Wackerle M (2005) Packaging of Bio-MEMS: Strategies, technologies, and applications. IEEE Transactions on Advanced Packaging, Vol. 28, No. 4, pp. 533–546

Chapter 18
LED and Optical Device Packaging and Materials

Yuan-Chang Lin, Yan Zhou, Nguyen T. Tran, and Frank G. Shi

Abstract As for integrated circuit (IC) device packaging, the packaging materials are critical to the LED packaging because the device packaging and assembly yield, and the device reliability and lifetime are determined by the quality of packaging and assembly materials as well as their processing. This presents serious challenges to the development of LED packaging materials, which is exactly the objective of this chapter to review those challenges and to point out the direction of further development.

It is proper to point out here that although this chapter will focus on the packaging materials for LEDs, the information provided by this chapter is equally applicable to the packaging of other optical devices including laser diodes, optical sensors, fiber optic devices, optical detectors, optical couplers, etc.

The first section will review materials challenges and some solutions for the packaging of high power LEDs, followed by functions of packaging and materials for advanced optoelectronic device packaging/manufacturing. The advanced encapsulation, lens, chip bonding and PCB materials for high power LED packaging will be presented in these sections. As a conclusion, we will point out the directions of materials for advanced high-power LED and optoelectronic device packaging as well as the requirements and approaches to determine LED performance and reliability.

Keywords Light emitting diode (LED) · optoelectronic · packaging · encapsulant · reliability

---

Y.-C. Lin (✉)
Optoelectronics Packaging and Materials Labs, 916 Engineering Tower, University of California, Irvine, CA 92697-2575, USA

18.1 Background

18.1.1 Introduction

Since the commercialization of blue light emitting diodes (LED) in middle 1990s [1], the applications of LEDs have been rapidly extended. As shown in Table 18.1, the global LED packaging (which has the largest profit margin among LED value chain) alone in 2006 reached the market value of $7.749 billion, which is expected to grow to $11.156 billion in 2009.

The rapid growth and the expected continuing rapid growth in LED and applications is driven by the rapid progress in LED technologies and the global energy supply as well as environmental concerns. For example, 32 tons Hg per year is used to make fluorescent lamps in U.S. alone. In U.S., according to Department of Energy (DOE), 22% of all oil/gas/coal/nuclear energy goes to lighting, and 30%, the largest category, of all electricity is consumed in U.S. by lighting. The efficiency of white LEDs has now exceeded that of the most energy efficient fluorescent lighting, permitting an expected rapid penetration of LED lighting into not only special lighting market, but also the global $70 billion general lighting market. The adoption of LED lighting will make a great impact on our economy, our energy supply, our environment, and thus our life, as shown in the following list:

- LED lighting potential in US (white LEDs @150 lm/W; DOE)
    - ~50% reduction in electricity use for lighting (= 100 power plants; DOE)
    - ~$125B saving on electricity from 2005 to 2025 (DOE)
    - Reduction of 28-40 M tons/yr of carbon emissions
    - Reduction of >1 M tons/yr in nitrous/sulfur dioxide emissions
    - No mercury: 32 tons/yr mercury used for fluorescent lighting
    - Intrinsically safe: no electrocution (vs. 120VAC), no fire
    - Directly compatible with battery and solar technologies (12VDC)

Table 18.1 Global LED chip packaging trend by region (US $, million)

Country	2005	2006	2007*	2008*	2009*	2006**	2009**
China	425	534	695	853	1,089	25.6%	27.7%
Taiwan	1,164	1,313	1,445	1,589	1,732	12.8%	9.0%
Korea	613	750	931	1,111	1,282	22.3%	15.4%
Europe	666	1,106	1,181	1,350	1,546	66.1%	14.5%
US	845	709	851	1,002	1,226	(16.1%)	22.4%
Japan	2,987	3,337	3,637	3,964	4,281	11.7%	8.0%
Total	6,700	7,749	8,740	9,869	11,156	15.7%	13.0%

Source: PIDA, compiled by Digitimes, April 2007
*: forecast; **: year to year growth rate

- The national and local governments have already introduced legislations for adopting more efficient lighting that are best met with LED lighting:
  - 4 States (CA, CT, MD, NJ) energy efficiency act that are best met with LEDs
  - Legislation pending in ~10 other States
  - Austrian, EU, Japan, South Africa, etc. passed legislation for adopting more efficient lighting that are best met with LEDs

This rapid progress in adopting LED lighting can be represented by the England's Palace in converting the center room chandelier (Fig. 18.1): all 32 twenty-five watt tungsten lamps were removed, and a low voltage system controlling 2.8 watt LED lamps were installed initiating an energy saving in excess of 80%. The LED lamps are also used for illuminating the Grand Staircase in Buckingham Palace (Fig. 18.2).

Fig. 18.1 LED chandelier in The England's Palace (Source: LEDtronics Press Releases, November 2007)

Fig. 18.2 LED Illumination on The Grand Staircase in Buckingham Palace. (Source: LEDtronics Press Releases, November 2007)

Although LED lighting is expected to be adopted eventually for general lighting, the largest applications of colored and white LEDs are for (1) automobile interior and exterior lighting; (2) backlighting for mobile devices and small and middle sized liquid crystal displays (LCD); (3) single and traffic lighting; and (4) LED display. In fact, those applications now account for about 90% of the LED needs.

LED packaging, as stated above, has the largest profit margin as well as the market value among the entire LED value chain. Similar to integrated circuit (IC) device packaging, the packaging materials are critical to the LED packaging because the device packaging and assembly yield, and the device reliability and lifetime are determined by the quality of packaging and assembly materials as well as their processing.

Although the basic functions of the packaging materials for LEDs are shared by those of IC devices, i.e., the die bonding, wire bonding, and die encapsulation, there are additional unique requirements of the packaging materials for LED packaging. This can be most clearly illustrated by the encapsulant materials. The encapsulant materials for IC devices are silica filled so that the materials can have low coefficient of thermal expansion (CTE), high flame resistance, and low moisture absorption. But the encapsulant materials for LEDs must not be filled with conventional silica fillers because of optical transmission requirements. This presents serious challenges to the development of LED packaging materials, which is exactly the objective of this chapter to review those challenges and to point out the direction of further development.

It is proper to point out here that although this chapter will focus on the packaging materials for LEDs, the information provided by this chapter is equally applicable to the packaging of other optical devices including laser diodes, optical sensors, fiber optic devices, optical detectors, optical couplers, etc.

The remaining part of this chapter is organized as follows. The first section will review materials challenges and some solutions for the packaging of high power LEDs, followed by functions of packaging and materials for advanced optoelectronic device packaging/manufacturing. The advanced encapsulation, lens, chip bonding and PCB materials for high power LED packaging will be presented in these sections. As a conclusion, we will point out the directions of materials for advanced high-power LED and optoelectronic device packaging as well as the requirements and approaches to determine LED performance and reliability.

18.1.2 Materials Challenges and Solutions for the Packaging of High Power LEDs

In 2006, the optoelectronics market achieved new highs with optoelectronics-enabled devices and components reaching US$565 billion, a 14.5 percent increase over 2005 (**O**ptoelectronics **I**ndustry **D**evelopment **A**ssociation, 10/2007). OIDA forecasts a strong and steady growth over the next decade for the optoelectronics – enabled devices and components with the market revenues expected to surpass US$1.2 trillion by 2017 and a 2007 to 2017 Compound Annual Growth Rates (CAGR) of 7.7 percent. Within the optoelectronics-enabled products, the growth drivers over the next decade will be solar, computing/processing, and consumer display/TVs. These markets will achieve 2007 to 2017 CAGR of 17.3 percent, 5.6 percent and 6.3 percent, respectively. The yearly growth for optoelectronics enabled products and systems in 2006 was led by environment/sensing (43.1 percent) and medical care/welfare (28.6 percent). White High Brightness LEDs (HBLED) will fuel growth of the LED markets to surpass US$14 billion by 2017. The market will be driven by solid state lighting, automotive, and signs/displays. Moreover, solid state lighting devices will grow to more than 30 percent of the lighting market by 2017, thereby giving competition to the incumbent incandescent and fluorescent luminaires. The solid state lighting market is forecasted to grow to more than US$60 billion over the next decade and will compose predominantly of HBLEDs.

LED light sources have considerable advantages over traditional lighting sources, such as extremely long life, high durability, low energy usage, as well as their color-generating abilities, which define them as unique. In addition to producing millions of colors and providing white light in a variety of color temperatures, LED devices are fully dimmable and can be controlled by a simple switch or sophisticated optical feedback driver electronics to balance the Red, Green and Blue (RGB) light output over temperature. Existing lamp technologies simply cannot compete with the huge number of design, control and display possibilities available with LEDs. Additional advantages include the following [2]:

- Long lifetime (20,000 to 100,000+ hours)
- Small form factor for improved design flexibility
- Environmentally friendly – no hazardous materials, i.e., mercury
- Rapid on and restrike times (<100 ns) and digital control with 100% dimming capability
- Highly energy efficient
- Vivid saturated colors without filters
- Dynamic color control – white point tunable
- Cold start capable down to $-40°C$ and high-temperature operation up to $185°C$ junction temperature.

Progress in blue LEDs based on InGaN with respect to light generation efficiency and manufacturability enables many different pathways to new sources of white light for specialty lighting and, in the longer run, for general illumination. As a future lighting source, high power white LEDs should have the following properties: (i) high luminous efficiency, (ii) high power capability, (iii) good color-rendering capabilities, (iv) high reliability, (v) low-cost manufacturability, (vi) environmental friendly, and (vii) unique optical properties, which can be controlled to a degree not possible with conventional lighting sources such as incandescent and fluorescent lamps [1].

High power LED lighting is significant in that it provides decades of lifetime under normal operation and only requires a fraction of the power demanded for traditional lighting solutions. For LED lighting to be a viable lighting source, there are many technical challenges to be resolved. Among them, the light extraction efficiency, the chip overheating, and the light output degradation are the key issues, which turn out to be all related to the packaging materials.

High power LEDs have been limited by efficiency and reliability concerns. Until recently, continuing improvements in light output of high power LEDs are now starting to challenge the packaging materials. In theory, it should be possible to produce more light by driving bigger chips, typically 1 mm $\times$ 1 mm, with higher current. However, most high power LEDs convert only about 15% of the input power into light, with the rest being lost as heat. With high junction temperature caused by high power LEDs, thermal management becomes a critical issue for packaging. Therefore, enhancing heat removal for safe junction temperature operation and minimizing thermal stresses caused by the CTE mismatch of materials are significant for the packaging of high power LEDs. Hence, most LEDs use a traditional epoxy system as encapsulant that tends to degrade quickly upon exposure to high temperature or intense ultraviolet light, causing ingression of destructive moisture and air, and discoloration. These degradations are the main reasons why high power white and blue LEDs have short lifetimes [3, 4].

It is obvious that the above-mentioned critical issues in LEDs packaging are mostly materials dependent. Therefore, the challenges for packaging materials are to increase the light extraction efficiency, minimize the heat generated by chip, conduct more heat out of the package, and withstand heat and UV light. Thermal management issues are critical for lifetime, lumen output, and fixture design of high power LEDs. To improve packaging materials and the lifetime of LEDs, the following are needed: (1) new encapsulant materials with high refractive index to match that of the chips so that light can be significantly extracted from encapsulated chips; (2) new encapsulant materials with better thermal and UV resistance; (3) new encapsulant materials with a good match between the chip's coefficient of thermal expansion and that of the package and high thermal conductive die-bond materials holding the chip in place; and (4) encapsulant materials with good adhesion and low moisture permeability (Table 18.2).

18.1.3 *Thermal and UV stable (Long Lifetime) Encapsulant Materials*

Opto-electronic devices, such as LEDs, have special needs for optically transparent encapsulant materials. The transmission must be stable during

Table 18.2 Materials challenges and solutions for packaging high power LEDs

Challenges	Problems	Packaging Materials Solutions
Light Extraction	Refractive index mismatch between LED die and encapsulant	High refractive index encapslant
		Efficient lens/cup design
Thermal Yellowing	Thermal degradation of encapsulants induced by high junction temperature between LED die and leadframe	Modified epoxy resins or silicone based encapsulant
		Low thermal resistance substrate
UV Yellowing	Photo degradation of encapsulants induced by UV radiation from LED dies and outdoor	UV transparent or silicone based encapsulant
Stress/Delamination	Failure of wire-bond and die attach caused by the CTE mismatch among encapsulant, LED die and leadframe	Low CTE and modulus encapsulants
		Excellent adhesion and CTE matching materials between the bonded surfaces
Lifetime	Only about 20,000 hours compared to the ideal 100,000 hours	Above-mentioned

packaging and assembly, and during the life-time service. For example, the encapsulant materials must be sufficiently tough to resist possible thermal shock produced during the soldering of the chips onto the printed circuit board and the high temperature during the life-time service. For ultraviolet LEDs and outdoor applications, the materials should also be capable of resisting UV-induced yellowing and thus extended lifetime.

The encapsulation of LEDs is a potting or molding process, frequently using thermosetting epoxy resins because of their overall properties and cost [5]. There are several types of epoxy resins that can be used, such as diglycidyl ether of bisphenol-A (BPA) and cycloaliphatic epoxy resins [6]. With the saturated structure, cycloaliphatic epoxy resins are expected to have better UV and weathering resistance, and thus can be used for encapsulating optical devices intended for outdoor applications. BPA epoxy resin is more inexpensive and thermally stable because of the phenyl groups in the main chains. However, its UV resistance is not as good as that of the cycloaliphatic epoxy resins. To take advantages of both types of resins, a new formulation was reported, which blends the two types of resins. It was found that the new system outperforms both of the pure resin systems in terms of thermal aging performance, while the UV aging performance is kept similar to that of the cycloaliphatic epoxy system (Fig. 18.3). To improve the weatherability of the encapsulant, thermal cationic polymerization of hydrogenated bisphenol-A glycidyl ether and its discoloration were studied [7]. Moreover, it is generally well know that the thermal and UV stability of epoxy systems can be further improved by adding the anti-oxidants and UV stabilizers or absorbers [8]. However, what types of additives and their concentrations should be used for the best performance strongly relies on the system.

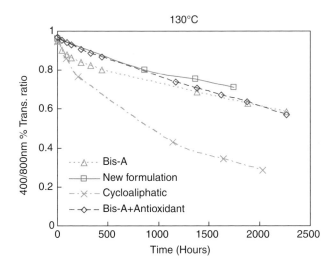

Fig. 18.3 Thermal aging results of different epoxy systems

Silicone is considered as another better choice for high power LEDs and outdoor applications because of its good thermal and UV resistance. But silicone suffers from several issues, such as poor physical properties, poor moisture resistance, dust abstracting, and the need for outer layer protection. To take advantage of the stable siloxane bond for better thermal and UV stability while keeping the benefits of epoxy resins, epoxidized silicone is a possible solution. Investigations were also being conducted into improving thermal and weathering resistance by curing silicone-containing epoxy derivative with anhydride [9]. But this application faces its own challenges.

18.1.4 Stress/Delamination

A CTE mismatch between bonded parts and the bonding solder introduces stresses during temperature cycling in the manufacturing process, which can cause delamination between the bonded surfaces. Sometimes, inappropriate solder and process control can lead to a short circuit in the device. Because of relatively high wettability, solders can overflow a specified region of contacts and create a short.

It is well known that curing of epoxy resins is accompanied by shrinkage and development of internal stress. In fact, the larger the difference between thermal expansion coefficients of the resin and the substrate materials is, the higher the internal stress is, which may cause device failure during processing or reduce LED's reliability. In order to reduce the internal stress, the Young's modulus and the thermal expansion coefficient of the encapsulant must be decreased.

Nano-sized silica fillers can be incorporated into epoxy systems to lower the CTE while keeping the transmission loss at minimum [10]. The addition of the nano filler also increases the toughness and thermal conductivity of the system, resulting in a much better thermal cycling performance.

18.1.5 Reliability and Lifetime

To dissipate the amount of heat that is generated during operation, the LED die needs to be bonded to a heat sink or substrate, often with a solder attach. If voids in the solder attach create an insufficient thermal path, the resulting hot spots will eventually lead to thermal runaway and failure. Whisker growth caused by electromigration, which can come from internal strain, temperature, humidity, and material properties, usually happens near the bonded surface between the solder and the heat sink and can lead to electrical short circuits. In choosing a die attach material, the following should be considered: (1) stress relaxation at the interface; (2) excellent adhesion between the bonded surfaces; (3) effective heat dissipation as well as high thermal conductivity; and (4) CTE matching materials between the bonded surfaces.

Package-related failure can occur in the encapsulant, wire, and phosphor. Wire-bond breakage or detachment and die-attach strength loss are due to overheated epoxy encapsulant. These problems, in turn, cause a delamination between the chip and epoxy. Mechanical stress from lead wires is another failure mechanism, because it can generate open circuits inside the device. Inappropriate pressure, position, and direction applied to lead wire soldering can accumulate the stress at normal operating temperature, bending the leads toward the body of the LED.

Most of white LEDs use yellow or red/green phosphors, which are susceptible to thermal degradation. When two or different phosphors are mixed, each constituent should have compatible lifetime and degradation behavior to keep the status of color. The color temperature and purity level of phosphors also degrade over time.

As a brief summary, new encapsulant materials for future high-power and high-efficiency LED packaging should have the following properties: high refractive index, high thermal and UV resistance, low CTE, low modulus, good adhesion, and low moisture permeability. High refractive index is needed to achieve high light extraction. However, this need can be alleviated by using efficient packaging design, such as the multiple small chip mounting and lens/cup design. To take advantages of both epoxy resins and silicone, developing new epoxidized silicone materials is a possible solution for the packaging of high-power LEDs [11].

18.2 Packaging Function

A typical phosphor-based high power white LED package is a system that combines an LED chip, electrical and thermal connections (wire bond and die attach material), an optional reflector cup, a substrate, a phosphor-containing

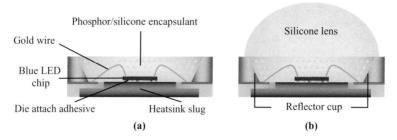

Fig. 18.4 Schematic cross-sectional view of white LED packages

encapsulating material, and an optional optical lens. In such a case, the optical design and encapsulating material with phosphor for LED packaging play very important roles in determining LEDs performance.

As shown in Fig. 18.4(a), a typical high power white LED package structure is comprised of a leadframe with a reflector cup, an LED die, a die attach adhesive, gold wires, and a silicone encapsulant mixed with phosphor material. For the lens-containing type LED package as shown in Fig. 18.4(b), an optically clear lens is added to the flat-top type package. The packaging process is described as follows: (1) The leadframe is cleaned and baked prior to use; (2) The blue LED chip is attached to the center of the reflector cup of the leadframe by using a conductive die attach adhesive; (3) The adhesive is cured at least for 45 minutes at 175°C; (4) Wire-bonding is done to electrically connect the LED die to the leadframe; (5) The silicone encapsulant mixed with a YAG:Ce^{3+} phosphor is applied to fill the reflector cup of the leadframe to form a flat-top surface, and cured by heating; and (6) The optical lens is attached to the top surface of the encapsulant to make the with-lens type package.

18.2.1 Encapsulation and Protection

One of the purposes of using encapsulants in LED packaging is to protect LED chip and gold wire from an adverse environment and increase their long-term reliability. However, the ultimate goal of LED encapsulation is to ensure the device's reliability and increase the production yield with the lowest cost [12].

Moisture and dust in the air are direct causes of semiconductor device defects, in addition to vibration, shock, contaminants, and hostile environmental conditions such as severe thermal cycling. Lighting and magnets can also cause malfunctions. Among them, moisture is the major source of corrosion for LED chip. Electro-oxidation and metal migration are associated with the presence of moisture. The diffusion rate of moisture depends on the encapsulant material and is a function of the diffusive encapsulant thickness and exposure time. Polymer materials, such as epoxies and silicones, which are commonly used in optoelectronic device encapsulation applications are a few orders of magnitude more permeable to moisture than glass and metals. Silicone materials, which have the highest moisture transient penetration rate in most polymers, nevertheless are one of the best device encapsulants. Moreover, the LED chip bond pad areas are etched out for interconnect and need protection as well. That is why LED devices need encapsulating materials to shut out these external influences and serve to protect LED chips, as well as to enhance their reliability and lifetime [13].

18.2.2 Light Extraction

The reliability of an LED package depends on many factors. One of the most important factors is the heat generation inside the package. The generated heat is mainly contributed by low light extraction, which is the primary cause of LED degradation, decreased light output, and lifetime (Fig. 18.5).

There are several techniques to solve the thermal issue. Improving light extraction is a good way to solve this problem. Single large chip mounting enables a compact package size and high light output. However, the quantum efficiencies of chips fall significantly as the die area increases, mainly because less light is emitted from the sidewalls of the chip (Fig. 18.6). In comparison to the use of single large chip, the use of multiple small chips dramatically increases the surface area for light extraction as well as the surface area for heat flow (Fig. 18.7).

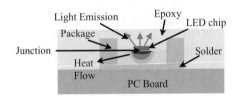

Fig. 18.5 General LED package for heat conduction

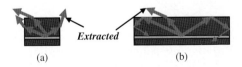

Fig. 18.6 Comparison of light extraction between (**a**) small chip mounting and (**b**) large chip mounting

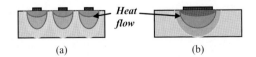

Fig. 18.7 Comparison of heat flow between (**a**) small chip mounting and (**b**) large chip mounting

Therefore, multiple small chip mounting enables a much higher light efficacy and better thermal performance. However, it may increase the complexity of LED packaging, package size, and hence the cost. How to develop multiple-chip based LED packages with a compact size and low cost is still an issue.

Other design parameters of LED package which can also contribute to higher extraction efficiency include a cup at a certain tilted angle and epoxy surface condition (epoxy-air interface) [14]. In Fig. 18.8, a LED chip is placed in a reflective cup filled with epoxy encapsulant. The encapsulant surface is formed with a spherical curvature that expresses as lens height (Fig. 18.6). The epoxy-air interface can be a large single lens, multiple small lenses, or a rough surface.

Light extraction can also be enhanced with epoxy encapsulant of high refractive index and high transparency at wavelengths of interest. Figure 18.9 shows light extraction efficiency as a function of epoxy refractive index and epoxy surface curvature expressed as lens height. At different lens height, the maximal extraction efficiency occurs at a different value of epoxy refractive index, although the light extracted from LED chip ($n_D = 2.4$–3.4) increases as the refractive index of epoxy encapsulant increases (Fig. 18.9). Changing the epoxy curvature greatly changes the light extraction efficiency at the refractive index value of above 1.3.

Absorption coefficient of encapsulant also has a great impact on the extraction efficiency. Figure 18.10 shows that the light extraction efficiency is improved drastically as reducing the absorption coefficient at high refractive index. The light extraction can be further improved when multiple micro-lenses are used in place of a single large lens (Fig. 18.11). Figure 18.10 shows the improvement of light extraction with multi-micro-lenses of 0.5 mm in diameter. The light extraction may be enhanced even more if smaller size of micro-lens is used.

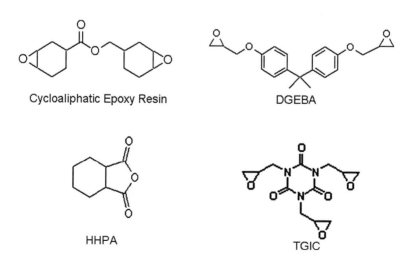

Fig. 18.8 LED lamp with a spherical lens

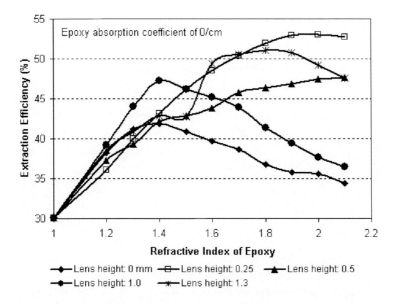

Fig. 18.9 Extraction efficiency as a function of epoxy refractive index and lens curvature expressed as lens height. Absorption coefficient of epoxy is 0 cm^{-1}

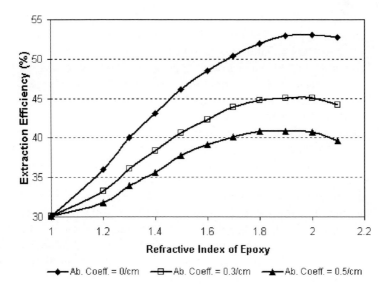

Fig. 18.10 Extraction efficiency as a function of epoxy refractive index and absorption coefficient

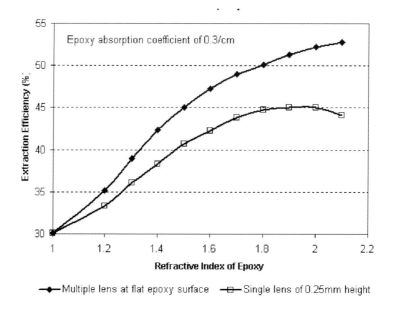

Fig. 18.11 Extraction efficiency is enhanced with multiple hemi-spherical micro-lens of 0.5 mm in diameter. Absorption coefficient of epoxy is 0.3 cm^{-1}

18.2.3 Optics

Many solid state lighting (SSL) luminary applications use a secondary optic to couple light from the LED into a desired beam shape. Many applications use a total internal reflection (TIR) or Fresnel type optic. Those optics at least throw away 10–15% of the total LED output flux. To improve the coupling efficiency of light from an LED to the outside world, the secondary optic can be removed from the package for some applications.

Diffractive optical elements (DOEs) are surface microstructures that are typically used for beam shaping of optical light sources. They use interference, the interaction of waves with each other to "break up" a light wave and "re-arrange" it so that the new wave or waves propagate in the direction of constructive interference. DOEs are not new, but recent advances in design and fabrication technology have made them a cost-effective solution for tackling LED optical design problems such as optical extraction, shrinkage of the overall package size, reduction of assembly costs, and of course beam shaping. DOEs have an important part to play in the future miniaturization of LED packaging. DOEs are advantageous because they usually are planar (flat) and they can be manufactured with lithographic and micromechanical methods. There are three distinct stages of DOE structures integrated with LED packages (Fig. 18.12): (a) Conventional optics: currently, the

a. Conventional Optics
- Dome lenses and mini-reflectors
- Bulky
- Limited beam-shaping possibilities
- Multichip LEDs are very difficult

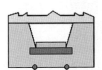

b. Micro-optical elements
- Diffractive and refractive optical microstructures
- Customized beam shaping using diffractive elements
- Flat-small dimensions
- Full optimization for LED chip
- Compatible with IR reflow

c. Monolithic micro-optics
- Advantages of micro-optical elements
- Integration of micro-optics onto LED
- Fabrication on LED wafer
- No assembly required

Fig. 18.12 Evolution of secondary optics for LEDs
(Source: LED Magazine, July, 2005)

traditional approach is to use total internal reflection (TIR) optics or mini-reflectors for optical extraction and beam shaping. The majority of manufacturers provides a primary optic and allows customers to source their own secondary beam shaping optics. (b) Separated micro-optic elements: the first stage in the natural progression towards more compact devices is to replace the bulky, TIR optics with low profile, refractive Fresnel lenses. (c) Integrated monolithic micro-optics: the ultimate goal in the packaging evolution is to fabricate the optical DOE microstructure directly onto the LED at the wafer level. This type of optics offers the simplest packaging solution while maintaining the freedom to homogenize and beam-shape the LED output [15].

18.2.4 Electrical Connections

If LED chips are simply encapsulated in the packaging material to protect them from the external environment, they will be unable to exchange signals with the outside. For the LED packaging, LED die is attached to the metal lead frame by the solder or silver paste (conductive die attach adhesive), so that the lead frame can act as electrical contact to the outside. A separate metal pin is connected to the bond-pad on the top of LED chip by wire bond and acts as second electrode, which allows signals to be sent to LEDs from the outside.

Fig. 18.13 A cross-sectional image of a three-watt Lumileds HBLED (Source: Prismark/Binghamton University)

18.2.5 Thermal Dissipation

Silicon chips heat up during operation. If the temperature of the actual chip becomes too high, the chip will malfunction. Packages need to effectively release this heat. And in the case of semiconductor devices that give off especially high levels of heat, such as the high-power LED, heat sinks or cooling fans can further dissipate the heat.

Standard 5 mm LED packages were originally designed for use in indicator applications, but their design does not allow for sufficient heat dissipation from the LED chip to keep it cool during operation. For the standard 5 mm LED, the maximal admissible package thermal resistance is 7 K/W, which is much lower than the package thermal resistance in newer high-power LEDs designed for illumination applications.

The main approach for enhancing heat dissipation in LED packages is to make the heat-removing path as large-surface as possible and as short as possible. For these purposes, leads in the new packages are changed to contact plates and slugs. Moreover, special requirements exist for leadframes and PCBs. For example, Lumileds proposed the most successful high power LED package from the thermal-management point of view by integrating a heat sink slug into the leadframe. A cross-sectional image of this package is shown in Fig. 18.13.

18.3 Materials for LED and Optoelectronic Device Packaging

The packaging technology is becoming increasingly important for the performance of LEDs in many current and future applications. Some of the older conventional packages today are inadequate for the rapidly improving high brightness AlGaInP or InGaN dice. Novel packages must consider better optical, electrical, and thermal performance. The demand for high reliability puts stringent requirements on the chemical and thermal stability of the packaging, and die attach and encapsulating materials as well as the selected processes. For devices with light converting phosphors such as white LEDs, additional consideration needs to be taken to improve the efficiency and stability of the phosphor materials. The best white LEDs today have achieved the luminous efficacy of 135 lm/W, which is much higher than that of fluorescent light, but still much lower than the possible luminous efficacy of 300 lm/W. Improvement of the luminous efficacy depends not only on the dice but also on the package used for assembly.

18.3.1 Encapsulation Materials for Standard LEDs

LED packages can be divided into two categories: through hole and surface mount. Through hole components like the radial package are loaded to a PC board from one side and soldered from the other. A simple through hole (radial) LED package is shown in Fig. 18.14. It was originally designed for low current indoor applications and has a maximum thermal resistance of 280 K W^{-1}, limiting the electrical input power to a few hundred mW. The LED chip has typically a lateral dimension of 200–300 µm. It is attached to the metal lead frame by the epoxy-based die attach adhesive so that the lead frame can act as electrical contact to the outside. The lead frame is shaped in such a way that it can act as a mirror cup as well as heat sink for the LED chip. A separate metal pin is connected to the bond-pad on the LED chip by gold wire and acts as second electrode. Chip and lead-frame are encapsulated by epoxy to form a dome-shape lens in order to achieve a certain radiation characteristics. The use of epoxy

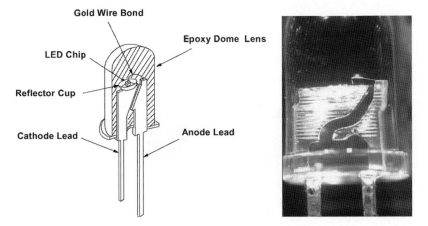

Fig. 18.14 Radial LED lamp
(Photo source: Prismark/Binghamton University)

encapsulant typically doubles the light extraction efficiency due to the enhanced light extraction efficiency at epoxy-chip interface and the dome-shaped epoxy lens. Standard diameters of the epoxy domes are 3 and 5 mm, and the packages are therefore named "3 or 5 mm-LED package" or "lamp type LED".

Surface mount devices (SMDs) are loaded and soldered on the same side, providing several benefits for industrial production such as faster placing in automatic machines, smaller size, less parasitic effects, and lower costs. In particular, in applications where space is limited such as in mobile phones and back lighting unit for laptop, the surface mount technology (SMT) is superior [16].

Figure 18.15 shows a standard SMT-package for LEDs. The die is attached to the lead-frame with the pre-molded plastic housing. A wire bond connects the die's top electrode to the other part of the lead-frame. Then, the cavity of lead-frame is filled with encapsulant. The flat-top of epoxy-air interface of SMT-packages results in a 10% lower extraction efficiency compared to radial packages with epoxy domes. However, for applications with the need for narrower emission profiles or more directionality, a transparent lens can be added.

Depending on the application, SMT packages can be made very small. The smallest devices are 0.5–1 mm wide and tall, which is just a little more than the dimensions of the die. The thermal resistance of SMT packages ranges from 300 to 500 KW^{-1}, which limits the maximum applicable current to 100–150 mA. SMT packages can also house several chips, e.g. the generation of white light or as multiple-color LEDs. An example for a multi-chip SMT package is shown in Fig. 18.16.

High power LEDs are designed for operation currents of 1A or even higher. For this current range, the package has to be capable of dissipating more heat generated from the die with high junction temperature compared to the standard package. Therefore, high power packages, such as the one shown in Fig. 18.17 (with the highest drive currents up to 1500 mA), usually include a heat sink metal base and thermal enhanced substrate (PCB).

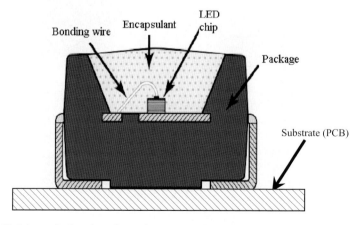

Fig. 18.15 Schematic drawing of a surface mount LED package

Fig. 18.16 Three-chip LED in an SMT package (Source: PLCC-4 Surface Mount LED indicator from Avago Co.)

Fig. 18.17 High power LED package with substrate (Source: Luxeon K2 star from Lumileds Co.)

18.3.1.1 Liquid/Solid Epoxy Encapsulants for Monotonic Color LEDs

Epoxies are one of the most utilized polymeric materials in optoelectronics. Their unique characteristics, such as good mechanical, electrical, and optical properties, excellent chemical and corrosion resistance, good adhesion, and low shrinkage, have made the cost-effective epoxy resins suitable as encapsulating materials for LED applications. Used in low power LEDs, epoxy encapsulant remains transparent and does not degrade over many years for long-wavelength visible-spectrum and IR LEDs [17].

Epoxy based encapsulants fall into two categories based on their physical forms, liquid type and solid type, which is also called molding compound. These two types of encapsulants are used in different packaging processes.

The property parameters of the liquid epoxy encapsulant important for LED applications include: transparency, transparency retention upon thermal and radiation treatment, refractive index, glass transition temperature (T_g), coefficient of thermal expansion (CTE), viscosity, adhesion, flexural strength, modulus, toughness, moisture absorption, and flame resistance. The optical property requirements are certainly unique to the transparent encapsulant. Other property criteria are similar to the epoxy encapsulant for IC industry. However, the traditional approaches to achieve these properties are not suitable for transparent encapsulant systems. For example, while silica fillers are usually added into the epoxy encapsulant to lower the CTE, improve the toughness, as well as flame resistance in IC industry, this approach cannot be taken for transparent encapsulant systems because, the micron-sized fillers would block light impairing the optical properties. Epoxy nanocomposites are reported as a possible solution for LED packages. It was found that, by adding nano-sized silica, the CTE of the epoxy encapsulant was lowered and the toughness was increased, while a good transparency was maintained [10, 18]. Similarly, approaches to increase the toughness and flame resistance need to be carefully selected not to adversely affect the optical properties. In fact, care must be taken for any addition of ingredients into the formulation, because even clear liquid chemicals might have compatibility issue resulting poor transparency after curing. This could be a very challenging task.

Typical liquid epoxy encapsulants for LED applications are thermally curable two-part systems. Part A mainly contains epoxy resin(s), while Part B is made of hardener(s) and catalyst(s). There are other additives, such as anti-UV agent, antioxidant, deformer, and flexibilizer in either Part A or Part B depending upon their chemical nature. When Part A and Part B are mixed and heated at elevated temperature, usually above 120°C, for a few hours, a cross-linked network forms and its properties are highly dependent on the epoxy and hardener selection [6]. Two types of epoxy resins are often used in the epoxy encapsulant formulations, diglycidyl ether of Bisphenol A (DGEBA), an aromatic epoxy resin, and cycloaliphatic epoxy resins. The structures of DGEBA and an example of cycloaliphatic epoxy resin are listed in Fig. 18.18. Among the various hardeners for epoxy resins, organic acid anhydrides are the most often used ones. This is because its low viscosity ensures easy handling, and the resulting network has very good optical and electrical properties. In Fig. 18.17, the structure of hexahydrophthalic anhydride (HHPA) is shown as an example in the anhydride family. DGEBA resin offers good thermal resistance and high strength, but its radiation resistance is not as good as cycloaliophatic resin. Cycloaliphatic resin has high T_g and good radiation resistance but poor moisture resistance.

When different epoxy resins and hardeners are used in the formulation, material performance can be quite different. Therefore it is critical to select the encapsulant based on the specific application requirements. M. Edwards compared three different optically transparent encapsulants and the results are helpful in selecting encapsulants [6]. Among the three systems, DGEBA/anhydride,

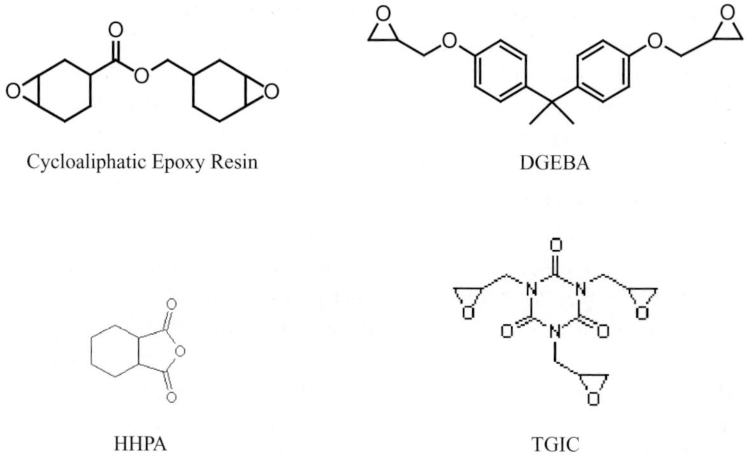

Fig. 18.18 Chemical structures of some key ingredients in epoxy encapsulant formulations

DGEBA/amine, and cycloaliphatic/anhydride, the anhydride systems have significantly better color stability during thermal aging, while the amine system has better adhesion retention after humidity exposure. The anhydride-cured systems have higher T_g, while the amine system is capable of lower temperature curing. As with most materials there are trade-offs in properties that must be considered when selecting materials for a specific application.

Epoxy siloxane hybrid monomer was tried as a new resin for transparent encapsulant systems [19, 20]. The idea is to take advantage of both siloxane resin and epoxy resin. The siloxane resin is stable in response to heat and UV light, while the epoxy resin has excellent adhesion. Use of siloxane alone would avoid discoloration, but its poor adhesion may cause optical delamination. The epoxy siloxane monomers of 1,3-bis[2-(3-{7-oxabicyclo[4.1.0]heptyl}ethyl]-tetramethyldisiloxane (BEPDS) was cured by anhydride. It was found that when catalyst PX-4ET was used, less thermal discoloration was observed. Thermal and UV discoloration varied directly with catalyst concentration, and minimum thermal discoloration was obtained with 0.71–0.35 mol % of PX-4ET. Anhydride concentration affects the physical properties of the materials. Maximum T_g, and minimum CTE and thermal discoloration were achieved when epoxy and anhydride were present in equivalent amounts [19]. The same monomer and other two with different length of dimethyl siloxanes were also cationically cured using thermal cationic initiator CP-77. These monomers showed good reactivity in thermal cationic polymerization. The discoloration of these polymers can be reduced with decreased catalyst concentration. Among the monomers, epoxy tetrasiloxane showed lowest thermal discoloration. Short siloxane groups led to highly rigid glasslike matrix, whereas longer chains produced flexible material with considerable elongation.

Hydrogenated DGEBA is another special epoxy resin studied in attempt to achieve better color stability [7]. Generally when DGEBA is hydrogenated, the UV resistance gets better, but the thermal stability suffers. Alternative epoxy encapsulant that is room temperature stable and was therefore made as one component encapsulant was reported [20]. It was found that the cationically polymerized DGEBA at 0.5 phr (0.5 weight parts per 100 parts of resin) initiator level gives the best compromise of chemical reactivity and optical properties, and therefore is suitable for LED applications as an optically transparent encapsulant. This encapsulant, on one hand, is room temperature stable for at least six months, yet cures fast at low temperature on the other hand. Moreover, the encapsulant has a high refractive index of 1.6 resulting from the high density of aromatic structure in the cross-linked network. Packaged with this encapsulant, the LED device was shown to exhibit an increased light output as demonstrated by both simulation and experimental measurement. The introduced system represents an example of a one-component encapsulant that is optically transparent, cures fast at low temperature with no sacrifice of room temperature stability, and has high refractive index at the same time. There are also few UV cure epoxy encapsulants offered in the market, which are one component and room temperature stable.

Epoxy resin based compositions have been widely used in the formation of molding compound for use as electronic packaging materials and encapsulants for semiconductor elements and electronic circuits. While molding compounds for semiconductor devices are application specific, there are a number of important materials property criteria for package performance. These key properties include the coefficient of thermal expansion (CTE), glass transition temperature (T_g), room and high temperature modulus, the dependence of melt viscosity on time and temperature, adhesion characteristics (i.e. to leadframe and other package metallizations and die passivation), moisture absorption rate, warpage control and wire sweep performance, etc. It is important to understand that a change in any package material including die attach adhesive can affect mold compound reliability and, as such, compatible material selection is imperative.

Solid clear molding compounds (CMC), which are partially cured or "B-staged" compounds in a form of "pellet", are well known for encapsulating optoelectronic devices by transfer molding process. Compared with the traditional molding compound, optical properties, such as transparency in the desired wavelength range, and the transparency retention during service, need to be considered besides the other properties important for a molding compound.

Katsumi Shimada disclosed a process producing an epoxy molding compound for photosemiconductor element encapsulation [21]. The formulation contains an epoxy resin, a hardener, and a catalyst. The process comprises a first step of melt-mixing the ingredients together and a second step of regulating viscosity of the molten mixture obtained in the first step at a given temperature. Preferred epoxy resins are bisphenol A, bisphenol F, novolac, alicyclic epoxy resins, and triglycidyl isocyanurate (TGIC), which is excellent in transparency and resistance to discoloration. Although such an epoxy resin may be liquid at room temperature, a solid epoxy resin is preferred. Examples of the hardener used in the formulation include acid anhydrides and phoenolic hardeners, which are colorless or light yellow. Preferred hardening accelerators are the tertiary amines, imidazole derivatives, and phosphorus compounds [21].

To overcome the poor photo resistance at short wavelength and increased linear expansion coefficient of epoxy-siloxane composition modified from the alicyclic epoxy resin, Hisataka Ito disclosed an epoxy resin composition for photosemiconductor element encapsulation, which comprises the following components: (a) an epoxy resin, i.e., cycloaliphatic epoxy resin and triglycidyl isocyanurate, (b) an acid anhydride curing agent, i.e., mixture of 4-methylhexahydrophthalic anhydride and hexahydrophthalic anhydride, (c) a silicone resin having the constituent siloxane unit and having at least one hydroxyl group or alkoxy group bonded to a silicon atom per molecule, wherein substituted or unsubstituted aromatic hydrocarbon groups occupy 10% by mole or greater among the monovalent hydrocarbon groups bonded to silicon atoms, and (d) a curing accelerator, i.e., tertiary amines, imidazole derivatives, and phosphorus compounds [22].

To provide a molding compound useful as an optoelectronic encapsulant for LEDs with improved resistance to heat and UV light, Dale Starkey disclosed the molding compound including a partially cured epoxy composition, an antioxidant, and optionally, a phosphor material substantially uniformly distributed throughout the epoxy composition. As the epoxy component, triglycidyl isocyanurate sold under the trade name TEPIC by Nissan Chemical Industries, Ltd. is particularly desirable. The cyclic anhydride component used in the formulation is desirably a cycloaliphatic anhydride, such as hexahydrophthalic anhydride. In order to promote reaction of cyclic anhydride component and the epoxy component, the anhydride ring must be opened. Such ring opening can be accomplished, for example, by active hydrogens. In desirable applications, a polyol is incorporated into the epoxy composition to assist in the ring opening of the anhydride and promote curing of the epoxy composition. In addition to the epoxy composition, the molding compound can include one or more antioxidant materials, i.e., IRGANOX 1035, 1010 and 1076 from Ciba Specialty Chemicals. Among the mold release agents, higher fatty acids of 12 to 20 carbon atoms or lower alcohol (1 to 3 carbon atoms) esters, preferably saturated and most preferably stearic acid or methyl stearate, are utilized because they are very effective to facilitate the releasing of the product from the mold [23].

Casting/Dispensing/Potting Processes for LED Packaging

In casting/dispensing/potting processes, liquid epoxy encapsulants are used. Casting process is commonly used for encapsulating the lamp type LEDs. During the process, the semi-packaged LED (chip/gold wire/leadframe) is

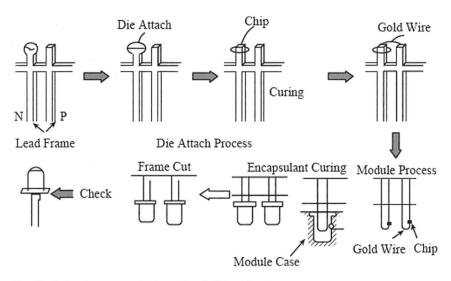

Fig. 18.19 Casting process for lamp type LED packaging

placed inside the module case, which is filled with well-mixed liquid encapsulant. The LED is released from the module case after the encapsulant is completely cured (Fig. 18.19).

Molding Process for LED Packaging

Transfer, compression, and injecting moldings are some of the current molding processes for LED packaging. In these processes, solid form encapsulants are commonly used. For high-volume manufacturing, encapsulation via transfer molding offers the cost-effective and high-performance process for LED packaging. In the transfer molding process, leadframes or laminate arrays are loaded – manually or automatically – into the cavity of the bottom mold and then the mold is closed and clamped under high pressure. Pellets of pre-heated (softened) mold compound material are then transferred under pressure via plunger movement into the mold cavity where the thermosetting material fills the available mold volumes and cures, yielding an encapsulated device [24].

The typical transfer molding process is described as follows (Fig. 18.20): (1) Load semi-packaged LED (chip/leadframe/PCB) into the cavity of the mold; (2) Preheat the molding compound pellet and mold; (3) The mold is closed up and the molding compound pellet is placed into a portion of the mold called the "pot". The plunger (on the top-most part of the mold) fits snugly into the "pot"; (4) When the mold reaches the desired temperature (150~175°C), the melted molding compound is forced through the gate into the cavity by the plunger with hydraulic pressure (500~1000 psi). The

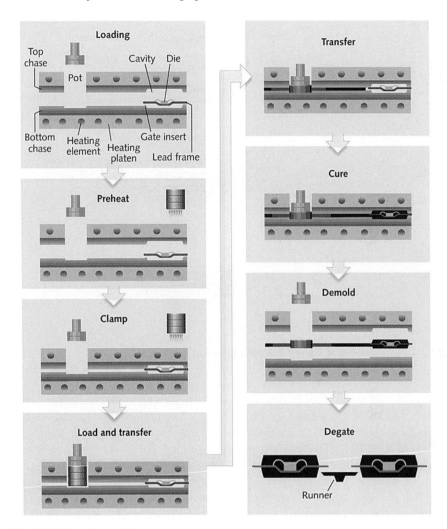

Fig. 18.20 Transfer molding process

mold is held closed while the compound cures; (5) The plunger is raised up and mold is opened. The LED part can be removed and the "transfer pad" material may be removed and thrown away; (6) A post-mold cure is needed for some types of molding compounds; and (7) The flash and the gate may need to be trimmed.

In some specific applications, compression molding is still the best and simplest way for high-volume manufacturing. For LED encapsulation, compression molding is being evaluated due to some advantages in comparison to transfer molding, such as using either liquid or solid compound and reduced materials wasting during the molding process (Fig. 18.21).

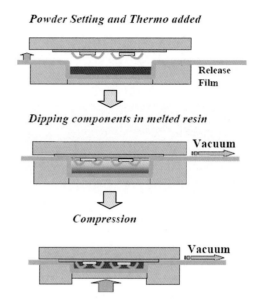

Fig. 18.21 Compression molding process

18.3.1.2 Liquid/Solid Epoxy Encapsulants for White LEDs

White LED Packages

Currently, there are three general approaches to generate white light from high power LEDs, illustrated in Fig. 18.22 [25, 26]. The first method directly mixes light from three (or more) monochromatic sources, i.e., red, green, and blue LEDs to produce a white source matching with the RGB sensors in human eyes. These trichromatic LED-based white-light sources offer a high luminous efficacy of radiation, a broad range of color temperatures, and excellent color-rendering index (CRI) exceeding 85. However, as the device temperature increases, the chromaticity locus of the trichromatic source shifts toward a lower xy chromaticity coordinate that corresponds to a higher correlated color temperature (CCT). This can be due to the stronger temperature

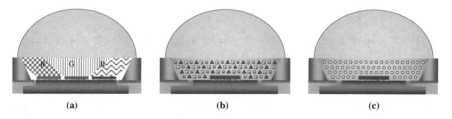

Fig. 18.22 Three methods for fabricating white light from LEDs (**a**) red + green + blue LEDs. (**b**) UV LED + RGB phosphors. (**c**) blue LED + yellow phosphor

dependence of the red LED emission power, which decreases faster at high temperature than that of green and blue LEDs [27]. Moreover, the driving electric circuit for this method is too complex to use. It requires the use of sophisticated optical feedback driver electronics to balance the RGB output over temperature and over operational lifetime. For general illumination usage, white LEDs need to have simple driving circuits and broader spectrum [28]. The second method uses an ultraviolet LED to excite a combination of red, green, and blue phosphors and simultaneously generate three different colors in such a way that none of the UV LED light is allowed to escape. This approach is advantageous because the emission color would be very reproducible and stable. Moreover, high CRI, similar to fluorescent lamps, can be realized. However, there is a significant loss of efficiency due to photon down-conversion from UV to blue, green, and red. In addition, red phosphors that can be efficiently pumped by UV LEDs have yet to be developed [29]. The third method uses a blue LED to excite a yellow emitting phosphor integrated into the LED package. This method is the most common approach of making high-power white LEDs for general illumination, because it is much easier to fabricate compared to the previous two methods, while providing much higher efficiency, acceptable CRI, and lifetime [14, 28].

For phosphor based white LEDs, the phosphor absorbs the short-wavelength emitted from the primary LED chip and down-converts it to a longer-wavelength. For example, a white LED can be achieved by combining an LED that emits blue light and a yellow emitting phosphor such as cerium activated yttrium aluminum garnet ($Y_3Al_5O_{12}:Ce^{3+}$). The blue LED emits a first light typically with peak wavelength of 440 to 480 nm as an excitation light. The yellow phosphor absorbs a portion of blue light and emits a second broadband light with peak wavelength of 560 to 580 nm, generally in the yellow light portion of the spectrum. The combination of the yellow light with the unabsorbed blue light is perceived as white light by human eyes. The optical properties of dichromatic white light is dependent on the balancing of the blue light power with the yellow light power that is affected by the amount of phosphor (the thickness of the phosphor-containing encapsulant), concentration, density, particle size, distribution, and geometry of phosphor inside LED packages [30–32].

Phosphor Placement in White LED Packages

For the conventional phosphor dispensing method, the phosphor is typically mixed in either the epoxy or silicone gel encapsulant that surrounds the LED chip (Fig. 18.23a). Due to the difficulty in consistently dispensing small amounts of phosphor-containing encapsulants and the tendency for the phosphor particles to settle during the curing process, there has been a significant variation in the correlated color temperature (CCT) of commercially produced white LEDs. In addition, even within one LED, the thickness of the phosphor-containing encapsulant through which light travels can also vary significantly

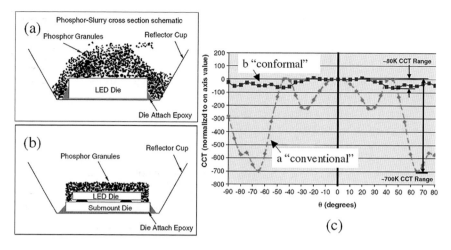

Fig. 18.23 (**a**) Cross-section of a white LED made by dispensing a drop of phosphor-containing encapsulant onto a blue LED chip. (**b**) Cross-section of a white LED made by depositing a conformal phosphor coating directly on a blue flip-chip. (**c**) Variation in CCT as a function of viewing angle for the devices shown in parts (a) and (b)

depending upon the emission angle of the light and whether it is emitted from the top or the sides of the chip. Recently, high-power white LEDs have been released by Lumileds where a conformal phosphor coating on the LED chip was employed (Fig. 18.23b). The flip-chip construction of the LED is essential for such a coating process because the top of the LED chip must be free of wire bond pads which would be covered by such a phosphor coating. The resulting improvement in color uniformity of such a coating process is shown in Fig. 18.23c. A ten-fold reduction in CCT variation as a function of viewing angle was observed and the CCT variation was brought down to the point where it was undetectable by the human eye [29].

Moreover, white LEDs produced by traditional phosphor dispensing techniques frequently appear to be yellow on the edges and blue in the center of the beam because of a lack of uniformity in phosphor coverage of the die. This color non-uniformity results in a varying mix of emissions from the underlying blue LED and the yellow phosphor-converted photons across the LED package (Fig. 18.24b).

Lumileds' patented conformal coating technology utilizes special tools and procedures to prevent these color irregularities by distributing the phosphor at a regulated and consistent thickness over the entire Luxeon chip. The superior color uniformity of white Luxeon LEDs compared to other white solid-state light sources is noticeable to the naked eye (Fig. 18.24a). It can also be quantified by measuring the correlated color temperature (CCT) over the entire 180-degree LED viewing angle, since the color variation in conventional LEDs typically manifests itself as a cooler white in the middle with a warmer white or yellowish ring off axis in the light beam [33].

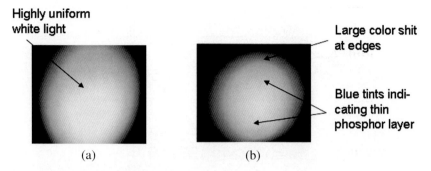

Fig. 18.24 (a) Lumileds Luxeon white LEDs using conformal phosphor coating. (b) Conventional white LEDs using traditional phosphor dispensing method

Remote Phosphor Package

In most commercial LEDs, the phosphor is either deposited directly onto the chip surface (conformal coating), or is dispersed in the encapsulation material, such as silicone, that is placed over the LED chip. In these conventional LED packages, a significant portion of the light emitted by the phosphor could be absorbed by the reflector cup, substrate, fillet of die attach adhesive, and especially LED chip. This absorption loss adds additional thermal load on the LED package, reduces the white LED's overall light extraction efficiency, and is detrimental to its lifetime.

In order to minimize this absorption loss, the phosphor is dispensed remotely from the LED chip. It is shown in Fig. 18.25 by Tran [34] that the luminous efficacy was greatly improved as the phosphor was placed far away from the LED chip.

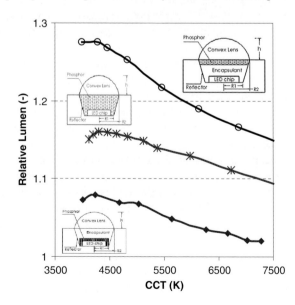

Fig. 18.25 The dependence of luminous efficacy on phosphor placement

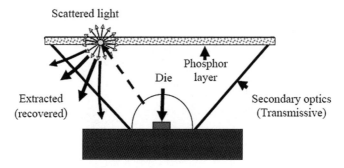

Fig. 18.26 White LED using SPE concept

The LRC at RPI in April 2005 reported a remote phosphor technique known as scattered photon extraction (SPE) to significantly improve extraction efficiency and alleviate the associated thermal loading (Fig. 18.26). The new SPE method works by moving the phosphor away from the die and by shaping the primary optic surrounding the die to extract a significant proportion of the back-scattered light before it is absorbed by the package [35–37].

Phosphor Dispersion

For phosphor based white LEDs, a phosphor can be defined as a material, which absorbs the short-wavelength emitted from the primary LED chip and down-converts it to a longer-wavelength. For example, a white LED can be achieved by combining an LED that emits blue light and a yellow emitting phosphor such as cerium activated yttrium aluminum garnet ($Y_3Al_5O_{12}:Ce^{3+}$). The combination of the yellow light with the unabsorbed blue light is perceived as white light by human eyes. Typical phosphor particle size is 2–20 microns in diameter with a specific gravity of 4.5. These conventional phosphors have a refractive index of 1.7–2.3 for visible light. Phosphors typically come in powder form and are dispersed into a liquid encapsulant, i.e., epoxy or silicone (specific gravity of 1–1.3), by different weight ratios. The resulting well-mixed phosphor/encapsulant mixture is used to encapsulate the LED die. However, the phosphor particles tend to settle inside the phosphor/liquid encapsulant mixture during the curing process, thereby providing a non-uniform distribution of phosphor throughout the cured encapsulant, which exhibits a yellow white or bluish white color of white LEDs. Moreover, for the mass production, phosphor particles settled inside the phosphor/encapsulant mixture before applying to fill the reflector cup of the leadframe can cause the inconsistent CCT for the LEDs of the same batch.

To overcome the settling issue of phosphor particles inside the phosphor/liquid encapsulant mixture, the anti-settling agent (fumed silica) can be used to prevent the settling of phosphor during the curing process. The suspending

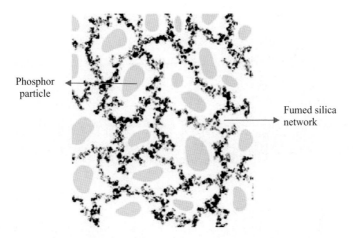

Fig. 18.27 Diagram illustrating the suspending action of the fumed silica network [42]

action of the fumed silica network is illustrating in the Fig. 18.27. The particles of phosphor materials are suspended or isolated in the three-dimensional network and prevented from coalescing with each other or settling. The fumed silica has a refractive index of 1.4–1.5, while the liquid encapsulants have a refractive index of 1.4–1.6. Therefore, more radiation from the primary LED chip is absorbed by phosphor materials due to the scattering effect of radiation inside the network when the fumed silica is present. Using this technique, the phosphor amount of white LED can be reduced by using the package with fumed silica compared to the use of the package without fumed silica at the same CCT, and the color uniformity of mixed light can also be achieved. For the white LED, CCT (K) can further be reduced by utilizing fumed silica, which can be attributed to more long-wavelength converting emission from phosphor materials caused by the scattering effect of radiation [38–41].

Figure 18.28 shows images of both cured sample and the packaged white LED with fumed silica added in the phosphor/encapsulant mixture. It is found that the settling of phosphor particles can be efficiently prevented when fumed silica is added, providing a uniform distribution of phosphor throughout the cured encapsulant. In comparison to the white LED package without fumed silica, the one with fumed silica has the advantages in anti-settling of phosphor particles, better color uniformity and reduced phosphor usage.

18.3.2 Encapsulation Materials for High Power LEDs

Epoxy encapsulants are commonly used in standard LED applications. However, for the high power LEDs (input power >1 W), the epoxy encapsulants show photo-thermal induced degradation caused by both high junction

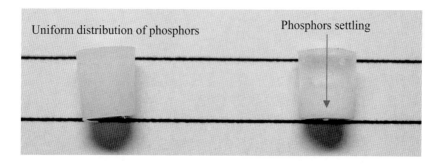
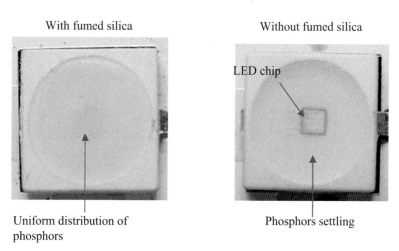

Fig. 18.28 Images of both cured samples and the packaged white LEDs with and without fumed silica added in the phosphor/encapsulant mixture

temperature from these LED diodes and photo-generated transmission losses from blue/UV LEDs. The degradation causes severe discoloration of epoxy encapsulants at encapsulant-die interface and thus the reduction in lifetime. In contrast to traditional epoxy encapsulants, silicone material is expected to have high resistance to discoloration caused by both the high temperature and blue/UV emissions from high power LED light sources, as well as wide operation temperature, stress-relieving, and better flame resistance.

18.3.2.1 Silicone Encapsulant for UV, Blue and Other Color LEDs

Introduction

Silicones are finding wide applicability as packaging materials for HBLEDs. Silicones have many advantages over other materials due to its thermal stability, low modulus, low shrinkage, low moisture absorption, good optical clarity, and adjustable refractive index.

Silicones are highly transparent in the UV-visible wavelength region. With minimal to no absorption losses, the light produced by the LED chip is transmitted efficiently through the silicone material.

Silicones can also be formulated to achieve a wide range of cured modulus values. The hardness can range from soft gels, to harder flexible elastomers, and even up to very hard resinous materials. The cured modulus is dictated by two factors: the crosslink density and the ratio of linear to branched silicon species in the polymer.

When compliant gels and soft elastomers are used to encapsulate devices, they provide a soft stress-relieving characteristic that can cushion the devices from internal and external stresses. A critical characteristic of a good encapsulant is adhesion, and silicones can be designed to have good adhesion to the various substrates and components used to build LEDs.

Another key attribute of silicones is their unique cure chemistry. Silicones developed for the LED market are thermoset materials and as such are cured with a thermal process. Such a system has several advantages. It can be offered in either one-part or two-part compositions, can be accelerated with heat, shows low cure shrinkage, and is free of cure by-products [42].

Silicone Chemistry

Silicones can be considered a "molecular hybrid" between glass and organic linear polymers. As shown in Fig. 18.28, when there is an absence of R groups, only oxygen attached to the silicon atom, the structure is essentially an inorganic glass (called a Q-type Si). If one oxygen is substituted for an R group (i.e. methyl, ethyl, phenyl, etc.) a resin or silsequioxane (T-type Si) material is formed. The silsequioxanes are more flexible than the Q-type materials. Finally, if two oxygen atoms are replaced by organic groups, a very flexible linear polymer (D-type Si) is obtained. The last structure shown (M-type Si) has three oxygen atoms replaced by R groups, resulting in a polymer chain terminating group. By varying the nature and number of R substituents in the molecular structure shown in Fig. 18.29, it is possible to control and tailor the optical and mechanical properties of the cured network [43].

The most common siloxane polymer is polydimethylsiloxane. In this structure the two methyl groups are attached to each silicone atom and the oxygen atoms join the silicone atoms in a chain. The diagram below shows their typical structure (Fig. 18.30).

Different polysiloxanes can provide a variety of excellent elastomeric properties that can be chosen according to the specific application, i.e., temperature stability (-115 to $260°C$), fuel resistance, optical clarity (with refractive indexes as high as 1.60), low shrinkage (<2%), and low shear stress. For example, the hardness of silicone encapsulants can be varied from Shore D>60 (hard resins) to Shore A (elastomers) to Shore 00 (soft gels). In HBLED packaging, hard resins and elastomers are preferred for optical lenses, whereas soft gels were used for encapsulating the chip and the wire-bonds inside the lenses that are

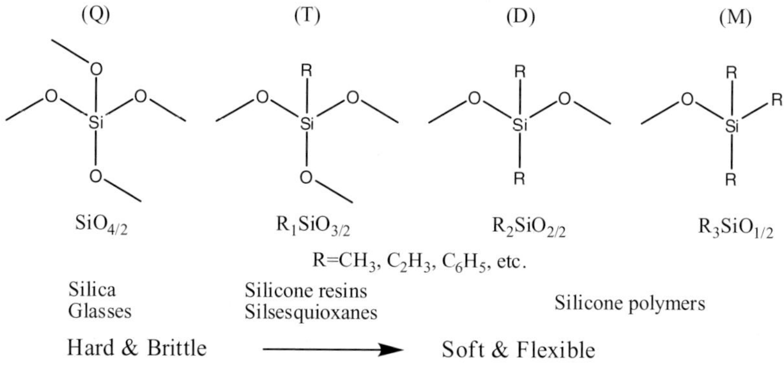

(Q)   (T)   (D)   (M)

$SiO_{4/2}$ $R_1SiO_{3/2}$ $R_2SiO_{2/2}$ $R_3SiO_{1/2}$

$R = CH_3, C_2H_3, C_6H_5$, etc.

Silica Glasses   Silicone resins Silsesquioxanes   Silicone polymers

Hard & Brittle ⟶ Soft & Flexible

Fig. 18.29 Generic structural units in the silicone polymer family

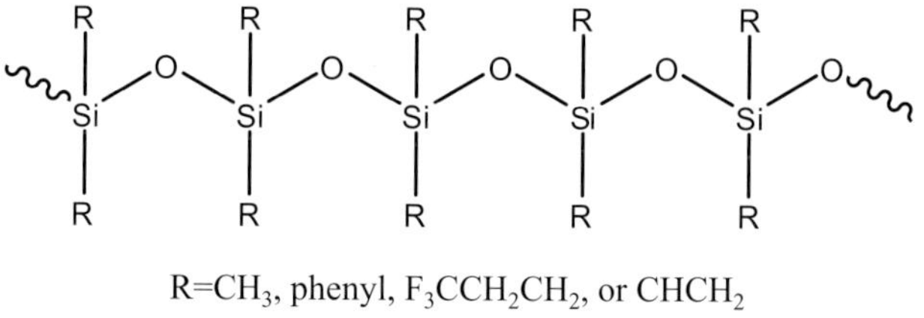

$R = CH_3$, phenyl, $F_3CCH_2CH_2$, or $CHCH_2$

Fig. 18.30 Chemical structure of polysiloxanes

more sensitive to stress. Dimethyl silicones, or dimethylpolysiloxanes, are the most common used silicone polymers. These types of polymers are typically the most cost effective to produce and generally yield good physical properties in silicone elastomers and gels.

The chemical structure of dimethylpolysiloxanes shown below (Fig. 18.31) contains vinyl groups, which are commonly used in a platinum catalyzed addition reaction. All dimethylpolysiloxanes have a refractive index of 1.40, 25 °C at 598 nm.

Methyl phenyl silicone systems contain diphenyldimethylpolysiloxane copolymers. The phenyl functionality boosts the refractive index of silicone systems from 1.4 upwards to 1.6, depending on the concentration of phenyl groups

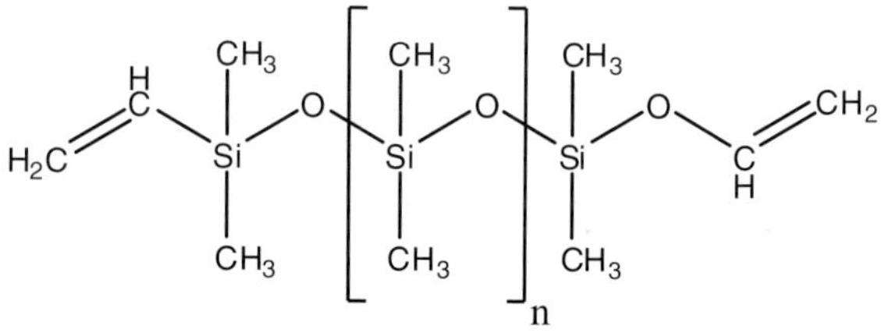

Fig. 18.31 Chemical structure of dimethylpolysiloxanes, vinyl-terminated

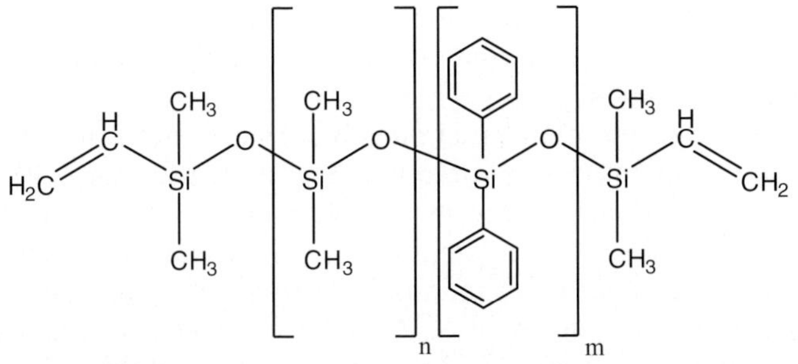

Fig. 18.32 Chemical structure of diphenyldimethylpolysiloxane, vinyl-terminated

inside the structure. Silicone polymers with diphenyl functionality are useful in biophotonic applications (e.g., intraocular lenses) where higher refractive index materials can be useful in creating a thin lens. The diagram below (Fig. 18.32) shows a typical structure for a methyl phenyl silicone.

Cure Chemistry

Silicones can be cured by platinum catalyzed addition, tin condensation, peroxide, or oxime. For various cure systems as described above, platinum systems are the most appropriate for HBLED applications. Addition cure chemistry provides an extremely flexible basis for formulating silicone elastomers. An important feature of this cure system is that no byproducts are formed, allowing fabrication of parts with good dimensional stability. Platinum catalyzed silicones utilize a platinum complex to participate in a reaction between a hydride functional siloxane polymer and a vinyl functional siloxane polymer. The result is an ethyl bridge between the two polymers. The diagram below shows their reaction mechanism (Fig. 18.33).

For the basic formulation, vinyl-terminated polydimethylsiloxanes with viscosities greater than 200 cps generally have less than 2% volatiles and form the base polymers for these systems. The crosslinking polymer is generally a methylhydrosiloxane-dimethylsiloxane copolymer with 15–50 mole % methylhydrosiloxane. The catalyst is usually a complex of platinum in alcohol, xylene, divinylsiloxanes or cyclic vinylsiloxanes. The system is usually

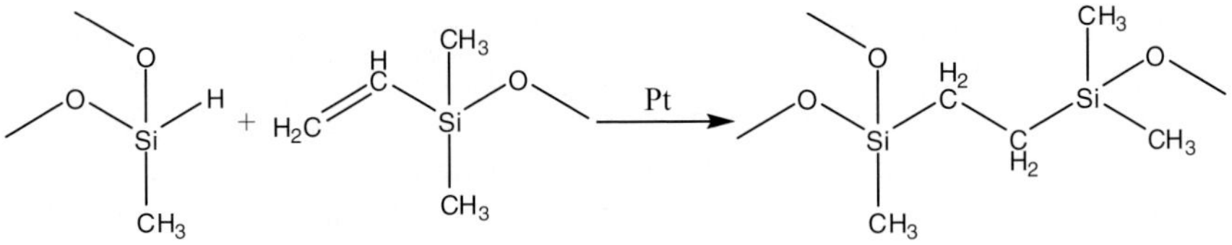

Fig. 18.33 Addition cure of silicones- the platinum catalyzed hydrosilylation reaction

prepared in two parts. By convention, the A part usually contains the platinum at a level of 5–10 ppm, and the B part usually contains the hydride functional siloxane [44–47].

Platinum systems are often cured quickly with heat but can be formulated to cure at low temperatures or room temperature if necessary. The possibility of inhibiting the cure is the main disadvantage of platinum systems. Inhibition is defined as either temporarily or permanently preventing the system from curing. Some types of inhibitors are purposefully added to these systems to control the rate of cure. However, contact with tin, sulfur, and some amine containing compounds may permanently inhibit the cure. Compounds that inhibit the cure can be identified easily by attempting to cure a platinum catalyzed system in contact with the compound, as inhibition results in uncatalyzed regions of elastomer systems or inconsistency in cure over time.

Effect of Refractive Index of Silicone on Optical Output

Refractive index is an import optical property for LED encapsulating materials. Due to the refractive index difference between the low refractive index of encapsulant ($n = 1.4$–1.6) and the high refraction index ($n = 2.4$) of GaN, light is trapped and reflects many times between top and bottom of the GaN layer, decreasing the light extraction efficiency. It is well-known that the higher refractive index of the encapsulant can be obtained, the more light extraction of the LED device can be achieved. However, with enhanced technologies, any way of destroying parallelism of the GaN layer – etching, lapping, polishing or texturing – will help to extract the light before it is absorbed in GaN (Fig. 18.34). Therefore, the refractive index of encapsulant is less significant for extracting light from GaN layer compared to that of GaN layer without any surface roughening.

Moreover, for some typical applications, planar-surface LEDs are frequently used under circumstances where the intended viewing angle is close to normal incidence or the LED is intended to blend in with a planar surface. For the flat-surface package, light inside the encapsulant cannot be extracted if it is totally internal-reflected at the encapsulant-air interface. Under such circumstance, light output of high power LEDs encapsulated by the encapsulant with low refractive index could be higher than the encapsulant with high refractive index.

To verify the above mentioned mechanism and get a better understanding the effect of refractive index of encapsulant on optical output, two types of packaged blue LEDs, in which one has a flat-top (FT) emitting surface and the other has a flat-top-with-lens (FTWL) type, are fabricated by using the same leadframe and investigated on their optical powers (mW) before and after the encapsulation with different refractive indices of silicones. It is found that the optical power increasing ratio decreases as the refractive index of silicone increases for both types of packages after the encapsulation, while the FTWL

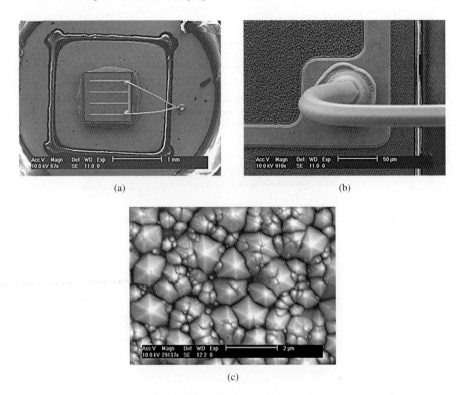

Fig. 18.34 SEM images of (**a**) SemiLED 460 nm high power 1 mm × 1 mm blue LED chip, (**b**) gold wire bonding, (**c**) Surface roughening structure on LED chip

package has a higher optical output than that of FT package due to less total internal reflection effect inside the FTWL package (Fig. 18.35).

18.3.2.2 Silicone Encapsulant for White LEDs

As mentioned earlier, a 2-part platinum cure system will most likely be chosen to mix with phosphors for white LEDs encapsulation. By adding phosphor, most of the physical properties of cured silicone including adhesion will decrease. Although the mixed silicone usually has a higher viscosity than mixed epoxy system, the settling issue of phosphor particles inside the phosphor/silicone mixture still presents during the cure process. Therefore, the fumed silica (surface treated), which has a good compatibility with silicone, can further be used to prevent the settling of phosphor particles. For silicone curing, cure inhibitors such as sulphur, amines, and tins should be recognized in the HB LED package. Often, fluxes and die attach adhesives contain such materials. If these materials cannot be altered, custom silicone dispersion

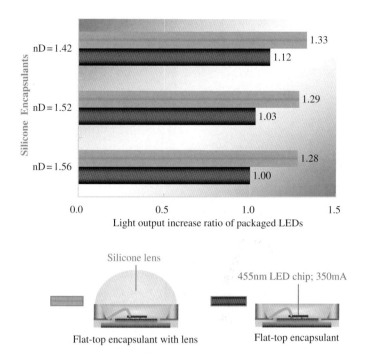

Fig. 18.35 Light output increasing ratio after silicone encapsulation with and without silicone lens ($n_D = 1.42$)

formulations may need to be developed to compensate for this inhibition. The package design and stress at the die level will be the primary factors in choosing the hardness and adhesion strength of the silicone dispersion when cured.

Most high refractive index silicone systems have the best thermal stability of any available silicones. The addition of the phenyl functional groups in silicone systems not only increases refractive index but also decreases the effects of temperature on the chemical structure of the silicone polymer due to steric hindrance. However, recent high power white LEDs utilizing UV or blue LEDs as primary sources have shown photo-generated transmission losses caused by phenyl functional groups in the silicone, resulting in the discoloration of high refractive index silicones and thus the reduction in lifetime of LEDs. In contrast to high refractive index silicones, low refractive index silicones without phenyl groups are expected to have better resistance to discoloration caused by high radiation flux from blue/UV LEDs light sources in high power white LEDs.

Fumed silica is the most common used filler in silicone systems to increase the strength, and is the main ingredient that produces many silicone rubbers used in multiple industries today. These particles are nonporous and allow a better interaction with the silicone into which they are dispersed. To improve the solubility and wet-out of the fumed silica even more into the silicone, the fumed silica is often treated with a chemical to reduce the attractive forces called hydrogen bonding.

Consistency is the key in making a dispersion of any powder. The objective is to uniformly mix and disperse the powder particles throughout the entire liquid. The ideal dispersion would be to have the phosphor particles all broken down into primary particles of uniform size and separated from each other uniformly with each particle covered by a uniform layer of silicone. Non-uniform mixing will cause clumping and settling and not provide the desired effects required from the dispersion, in our case, optimum light output. Consistency is dependent on the shearing capacity of the dispersion equipment, the length of shearing time, the viscosity of the liquid, and the particle size and density of the powder.

Typically, the maximum shear time is determined when the viscosity has plateaued or slightly decreased. The higher the surface area of the particle, the more difficult the powder is to disperse and the more energy is needed to wet-out the surface areas. Another key property affecting dispersions is the powder particle structure. The more highly structured the particle means that it is harder to get into the space around the particles and more difficult to disperse.

18.3.3 Optical Lens Materials

The choice of the optical lens material for LED packaging normally depends on the optical, mechanical, and environmental requirements set to the end-product, total costs and manufacturing possibilities. The quality of a lens material is characterized by transmission and dispersion factor (so called Abbe factor), as well as their thermal and UV resistance. Glass is the best lens materials, but its cost is higher than other materials. Plastic materials can also be used, when developing a system with fewer requirements for optical precision, and when there is a need for good integration of the optical system to the rest of the mechanical structure. Suitable standard thermoplastic materials for lenses are acrylics (PMMA), polycarbonate (PC), polystyrene (PS), poly(styrene-co-acrylonitrile) (SAN), and polyamide 12 (PA12). Among them, optical grade PMMA and high quality UL rated PC (Fig. 18.36) are commonly used for LED applications to ensure excellent optical properties and efficiency as well as long-term material stability and durability. When using thermoplastics, special attention must be paid to material specific properties, material behavior in the molding process, and the manufacturing and design of molds.

As discussed in earlier sections, silicone-based materials offer many advantages and are ideally suited for the HBLED market. Silicone lens fabrication can vary from a soft gel to a resin-based composition. Some small complex lenses have been fabricated using resin-based compositions for precision-molding applications. In this fabrication process, a lens mould-cup clamps down on the

Fig. 18.36 LED lenses made from high quality UL rated Polycarbonate

base LED package, into which the high-flow silicone material is injected and then thermally cured. Meanwhile, optical parts have been created using a variety of techniques such as casting, compression molding, and injection molding. Molding processes using multi-cavity moulds are ideal for high-volume, low-cost production of optical parts such as LED lenses. Molding with silicones also allows for designing unique optical elements and features for improved extraction of light from LED devices. Commercial fabrication of LED lenses has begun by various suppliers using the two-part addition curing silicone compositions [48].

18.3.4 Optical Chip Bonding Materials

With the emergence of high power LEDs, the thermal management of LED package is demanded. Despite of package design, all packaging materials, especially die attach materials, need to have high thermal conductivities in order to improve heat conduction from LED dice. The equivalent thermal circuit from the LED die to the board is shown in Fig. 18.37.

However, the die attach materials have the lowest thermal conductivities compared to other packaging materials. The die attach materials act as a bottleneck to control the heat flow path and thus affect the heat flow rate. In other words, the performance of LEDs can be decisively determined by die attach materials in terms of electrical and thermal conductivities as well as stability under harsh environment. Many efforts have been put in the selection and development of die attach materials. A detailed review article has been published to discuss the materials, applications, and recent advances of electrically conductive adhesives as an environmental friendly solder replacement in the electronic packaging industry [49].

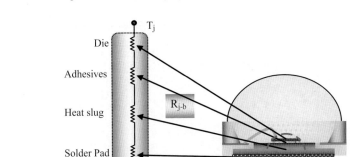

Fig. 18.37 Schematic diagram of equivalent thermal circuit from the junction to the board

Generally, die attach materials can be mainly divided into two groups: organic metal adhesives and solder paste. Organic adhesives are suspensions of metal particles in a polymer carrier. The particles generally are several microns in size, usually in the form of thin flakes of silver. The carrier which consists of resin (such as an epoxy, a silicone, or a polyimide) provides adhesion and cohesion to make a bond with the correct mechanical strength, while the metal particles provide electrical and thermal conductivity. The minimum curing temperature of polymer adhesives is around 150–170 °C [50].

Lead-free solders normally have higher reflow temperature than tin-lead solders [51]. The processing temperature of solders is higher than that of polymer based metal adhesives. Moreover, the use of solder pastes for relatively large dies have led to the problem of voiding underneath the die, which leads to the thermal conduction and stress issues [52]. On the other hand, solder pastes often contain a lot of flux in order to remove native oxide and increase wettability of the surface. The use of silicone makes the use of solders difficult since silicone is extremely sensitive to many chemicals including solder flux [53]. However, the high power LEDs are characterized by the relatively large dice and silicone encapsulant have to be used because of its high thermal stability and excellent radiation resistance. Thus, the development of organic metal adhesives is now being actively investigated for replacing solder based die attach materials for high power LEDs.

18.3.5 PCB Materials for High Power LED

Due to the development of high power LED packaging and materials, high-flux LED modules also grow rapidly. As shown in Fig. 18.38, flux per package has doubled every 24 months for more than 30 years. High-power chip and

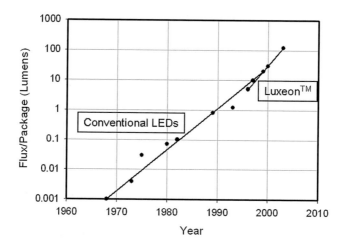

Fig. 18.38 The evolution of flux per package (Source: Lumileds)

packaging technology has introduced an inflection point. The LED development has also been challenged by the heat dissipation. For LED light emitting efficiency, only fifteen to twenty percent of input power has been transferred to light, and the rest of it is converted to heat. Therefore, if the heat can't be dissipated into environment, LED reliability and lifetime will be a big issue. Figure 18.39 shows the effect of junction temperature to the relative light intensity, and typical life expectancy of LEDs. At high junction temperature, the light output and lifetime are decreased linearly. Thus, reducing LED junction temperature is very important. Thermal management at package is to dissipate heat to the environment. A schematic diagram of typical single-chip LED package is shown in Fig. 18.40. It includes optical lens, LED chip, transparent encapsulant, phosphor, and heat sink slug. The process is to attach LED chip on a heat sink by the solder or the silver paste. Reducing thermal resistance by the heat sink slug is the most popular LED packaging module. This LED packaging module is used by Lumileds, Osram, Cree, and Nicha. The LED module can be packaged on a substrate to form a light bar, light matrix, and light circle. For the applications such as mini projector and car head light, they need several thousand lumens and always use multi-chip LEDS package and chip on board (COB). By using COB package, thermal resistance of substrate material plays an important role in heat dissipation issue [54–58].

18.3.5.1 LED Substrate Materials

In LED package, single or multi LED chips are attached on heat sink slug by solder or adhesive, and encapsulated by an epoxy resin. Finally a lens is attached on it, as shown in Fig. 18.38. When the LED input power increases, substrate has to be capable of dissipating more heat generated from the LED die

Fig. 18.39 Relative light intensity (**a**) and typical life expectancy (**b**) of LED as a function of junction temperature

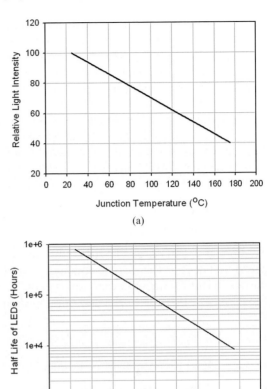

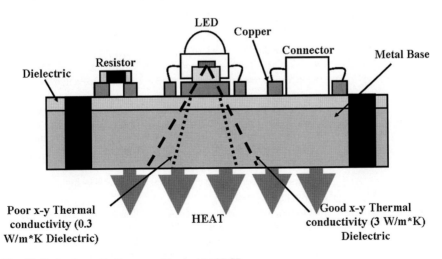

Fig. 18.40 A schematic diagram of typical MCPCB

to the environment. Hence, substrate materials with structure strength and heat dissipation are suitable for packaging application. Traditional LED with low heat generation is packaged on the copper-foil PCB. However, for the high power LED, the copper-foil PCB cannot meet the requirement of heat dissipation. Therefore, it needs to attach copper-foil PCB on an aluminum substrate, which is so-called Metal Core PCB. Another way is to coat insulated layer or dielectric layer on aluminum substrate and then the circuit can be printed on dielectric layer. The LED module can be wire-bonded on the PCB. In order to reduce the thermal resistance of dielectric layer, through holes will be applied in the bottom of the heat slug at LED module, which is so-called direct die attach. Table 18.3 summarizes commonly used substrates and their properties for comparison.

18.3.5.2 Printed Circuit Board (PCB)

Traditionally, LED circuit board is FR4. It can be made of a single or multiple copper foil design. The thermal conductivity of FR4 is around 0.36 W/mK and the coefficient of thermal expansion is about 13~17 ppm/K. PCB is advantageous because it is less costly, well developed, and suitable for the use of large size of board.

18.3.5.3 Metal Core Printed Circuit Board (MCPCB)

Because of low thermal conductivity and poor heat dissipation capability of PCB, metal core PCB is developed; i.e. PCB is attached on aluminum plate.

Table 18.3 The comparison of different LED substrate materials

Substrate	Substrate Property
PCB	• Low performance & cost (CTE 13–17 ppm/°C, $K = 0.36$ W/m.K)
	• Large panel sizes, up to 0.004" (100 μm) Cu thickness
MCPCB (Metal Core PCB)	• Moderate to high cost, high CTE (17–23 ppm/°C)
	• Low thermal conductivity through dielectric ($K = 1$–2.2 W/m·K)
	• Operating temperature limited to ~140°C, process temperature limited to 250–300°C
	• Large panel sizes (18 × 24"), thick Cu available for heat spreading (1–20 mil)
	• Medium to high cost, low CTE (4.9–8 ppm), medium to high thermal conductivity ($K = 24$–170 W/m·K)
Ceramic (Al2O3/ AlN)	• Small panel sizes (< 4.5" sq)
	• Very high operating temperatures, easily handles high power
	• Medium to high cost, low CTE (5.3–7.5 ppm)
DBC (Direct Bond Cu)	• High thermal performance (24–170 W/m.K) with thick Cu for excellent heat spreading (5–24 mils)
	• Very high process & operating temperatures (up to 800°C)
	• Easily handles high power & current

MCPCB (Fig. 18.38) is suitable for the use of high power LEDs. However, the thermal conductivity of dielectric layer is still low. The following research will focus on reducing the thermal resistance of dielectric layer.

18.3.5.4 Ceramic Substrate (Al_2O_3/AlN)

Ceramic substrates are electrical insulators without extra dielectric layers. They have good thermal conductivities and compatible coefficient of thermal expansion (CTE) with LED chips, Si substrate, or sapphire. For example, AlN and SiC have thermal conductivities about 170~230 W/mK and CTE around 3.5~5 ppm/K (Table 18.4). Ceramic substrates are suitable for high power LEDs and use under high temperatures. However, due to their high cost, ceramic substrates often apply in the small scale of substrates.

18.3.5.5 Direct Bond Copper (DBC)

Metal substrate bonded with ceramic can achieve high thermal conductivity and low coefficient of thermal expansion as well as good electrical insulation. For example, pure copper will form CuO with O_2 and decrease its melting point from 1083°C to 1065°C, which is the eutectic point of CuO. CuO and Al_2O_3 or AlN will form the compound after increasing to high temperature (Fig. 18.41). This copper bond ceramic substrate has better heat dissipation properties and is suitable for the use in high power LEDs.

Recent advances in high power LEDs have resulted in high input power and junction temperature. Thus, to accommodate the different applications, substrates with reduced weight, high thermal conductivity, and low CTE have been developed. Table 18.5 shows some advanced metal composite substrates and their thermal properties. With the progress in materials and package methods of LED, light conversion efficiency increases continuously. The products include in backlighting of LCD, display, automobile, and lighting. The most important challenge in LEDs is thermal management, since the heat dissipation adversely affects the LEDs lifetime and light output. The improvement in thermal

Table 18.4 The comparison of AlN ceramic substrate and other materials

Materials	Thermal Conductivity (W/m·K)	CTE (ppm/K)	Specific Gravity	Specific Thermal Conductivity (W/m·K)
Silicon	150	4.1	2.3	65
Gallium Arsenide	54	6.5	5.3	8
Gallium Nitride	130	6	6.1	21
Alumina	20	6.7	3.9	5.1
Aluminum Nitride	170–230	3.5–5.7	3.3	51–70
Aluminum	150–230	23	2.7	50–70
Copper	400	17	8.9	45
AlSiC	200	8.4	3	67

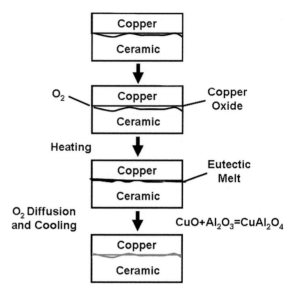

Fig. 18.41 The process of fabricating copper bond ceramic substrate

Table 18.5 Advanced LED substrate materials and their properties

Materials	K (W/m·K)	CTE (ppm/K)	Density(r) (g/cc)	K/ρ (W/m·K)
Cu-Mo-Cu	184	7.0	10	18.4
Cu-Invar-Cu	164	6.02	8.45	19.4
SiC/Al	170~220	8.75~11.5	2.9~3.0	57~73
Cont.CF/Al	300~800	3.2~11	2.3~2.5	120~315
Disc.CF/Al	218~290	4~7	2.3~2.7	92~100
Cont.CF/Cu	330~800	6.5~9.5	4.2~6.8	50~200
Disc.CF/Cu	300~400	7~10.9	4.5~6.6	50~100
Gr Flake/Al	400~420	6~7	2.3~2.7	195~200
Diamond/Al	400~600	4.5~5.0	3.4	174~260
Diamond + SiC/Al	550~600	7.0~7.5	3.1	177~194
Diamond/Cu	600~1200	5.8	5.9	330~670
CVD Diamond	1100~1800	1~2	3.5	310~510
HOPG	1500~1700	−1	2.3	650~740

conductivity of dielectric layer is critical. One of possible solutions is to develop high thermal conductivity and low CTE metal base substrate as well as the improvement in thermal conductivity of dielectric layer [59].

18.4 Materials and LED Performance and Reliability

Lumen maintenance is simply the amount of light emitted from a source at any given time relative to the light output when the source was first measured. It is usually expressed as a percentage. The steady decline over time is known as lumen maintenance. LEDs experience lumen depreciation but it happens over a much longer period of time, usually tens of thousands of hours. Not all LEDs deliver the same lumen maintenance. An LED is a complex package of materials that must all work together to deliver long lifetimes. Everything from the design of the chip, thermal management, optic material, phosphors, and even the assembly of the entire package will affect lumen maintenance. A power LED industry group, the Alliance for Solid-State Illumination Systems and Technologies (ASSIST), found that 70% lumen maintenance is close to the threshold at which the human eye can detect a reduction in light output. The ASSIST research shows a 30% reduction in light output is acceptable to the majority of users for general lighting applications [60].

Many applications for encapsulants require high strength and durability at low temperatures. Many of these same applications also require resistance to thermal cycling between high and low operating temperatures. These application were affected by more than just diurnal (day/night) or seasonal variations. Outdoor equipments, such as automotive parts, light fixtures, transformers, etc., were subjected to the normal temperature variations that occur during the day or season. Also, they are exposed to the temperature variations that occur when energizing and deenergizing equipments.

Due to CTE mismatch of packaging materials within the LED package, exposure to high internal temperatures beyond the maximum ratings or repeated thermal cycling can potentially cause different types of catastrophic failures. The excessive internal temperatures can arise either due to excessive ambient temperature as well as excessive junction temperature of LED chip, which could be caused by either excessive forward currents or excessive thermal resistance.

Depending on the applications, the packaged LED encapsulated by different encapsulants should withstand a couple of hundred non-operating temperature cycles from –40°C to 120°C. However, catastrophic failures can occur more quickly at higher/lower temperature excursions. The most common type of failure due to thermal overstress is broken gold wires. The broken wires are a normal wearout mechanism for LEDs. However, the number of cycles to failure is accelerated by the magnitude of the temperature excursion. While wire failures are rare, the most common type of bond wire failure caused by thermal overstress is a broken stitch bond, where the wire breaks immediately above the stitch.

Excessive temperatures can cause delamination between the LED die and the encapsulant. Figure 18.42 shows a sketch of delamination between the LED die and the silicone encapsulant. Generally, this delamination problem does not cause a catastrophic failure but can cause a permanent reduction in light output. In white LEDs, delamination can either occur between the phosphor coating and the silicone encapsulant or between the InGaN die and the phosphor coating [61].

The LED performance and reliability are determined by the amount and the quality of light output. However, these parameters are all related to the thermal

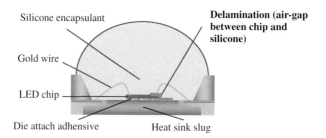

Fig. 18.42 Delamination between the LED chip and encapsulant

management of LED package. In high power LEDs, since the light conversion efficiency is still lower than 50% and thus the conduction of redundant heat in LED packages is demanding. The most important parameter, which can monitor the performance and reliability of LED, is the junction temperature (T_j). The increased junction temperature adversely decreases the light output and life span of LEDs [62].

Various indirect techniques have been utilized to measure junction temperature of LEDs including micro Raman spectroscopy [63], thermal resistance [64], electroluminescence [65], and photoluminescence [66]. However, the most straightforward method is to measure the relationship between diode forward voltage and junction temperature of LEDs. The diode forward voltage method consists of two series of measurements, a calibration measurement and real junction temperature measurement. In calibration measurement, the device-under-test (DUT) is placed in a temperature controlled oven and connected to the drive and measurement equipment. After the junction has come to thermal equilibrium with temperature controlled oven, a pulse current is sourced into the DUT to prevent self-heating and voltage drop is measured. From theoretical evolution, the temperature coefficient of forward voltage is [67]:

$$\frac{dV_f}{dT} \approx \frac{k}{e} \ln\left(\frac{N_D N_A}{N_C N_V}\right) - \frac{\alpha T(T + 2\beta)}{e(T + \beta)^2} - \frac{3\,k}{e} \qquad (18.1)$$

where V_f is forward voltage; N_C and N_V are effective densities of states at the conduction-band and valence-band edges, respectively; α and β are the Varshni parameters; and N_A and N_D are dopant concentrations.

In real application, a linear equation is fitted with the corresponding voltage drop with various set points of oven temperature in order to get a calibration curve. The calibration curve serves as the reference for the deduction of the junction temperature from DC measurement and establishes the relationship between the forward voltage and junction temperature.

The accuracy of forward voltage method was confirmed by monitoring the board temperature in different ambient temperatures. Suppose the thermal resistance from the junction to the board keeps constant under different

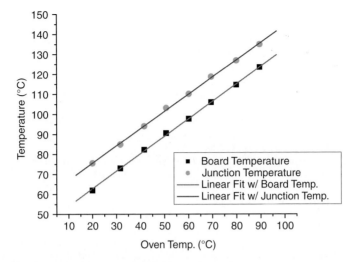

Fig. 18.43 Calculated junction temperatures and corresponding measured board temperatures in different ambient

ambient temperatures. Both stable voltage drop and board temperature were simultaneously measured with different ambient temperatures. The corresponding junction temperatures were calculated with the forward voltages by plugging into the calibration curve. Figure 18.43 shows the calculated junction temperature and measured board temperature have almost parallel linear tracks; i.e. the difference in slope is as small as 0.012.

After determining the junction temperature of LEDs, the reliability test is meaningful. Moreover, the junction temperature varies with different operation current of LEDs. This situation is similar to the LEDs operated in different ambient temperatures. So, in LED reliability test, it should contain the light output along different drive current with time.

Acknowledgment The authors would like to thank Dr. Yongzhi He for providing technical advices through the course of the research in this chapter. In addition, the authors would like to thank Jiun Pyng You and Yeongher Lin for valuable discussions and their endless help on the figures and tables.

References

1. E. Fred Schubert, "Light-emitting diodes", Cambridge: Cambridge University Press, 2nd ed., 2006
2. "LED lighting technology: lessons from the USA", Report of a DTI global watch mission, March 2006
3. C. W. Wessner, "Partnership for solid-state lighting: Report of a workshop", Washington, DC: National Academy Press, pp. 58–61, 2002

4. F. M. Steranka et al., "High power LEDs – Technology status and market applications", *Physica Status Solidi (A)*, Vol. 194, No. 2, pp. 380–388, 2002
5. R. J. M. Zwiers et al., "Development of a new low-stress hyperred LED encapsulant," *IEEE Transactions on Components, Hybrids, and Manufacturing Technology*, Vol. 12, No. 3, pp. 387–392, 1989
6. M. Edwards et al., "Comparative properties of optically clear epoxy encapsulants", *Proceeding of the SPIE, International Society for Optical Engineering*, Vol. 4436, pp. 190–197, 2004
7. Y. Morita, "Cationic polymerization of hydrogenated bisphenol-A glycidyl ether with cycloaliphatic epoxy resin and its thermal discoloration", *Journal of Applied Polymer Science*, Vol. 97, pp. 1395–1400, 2005
8. J. C. Huang, "Comparison of epoxy resins for applications in light-emitting diodes", *Advances in Polymer Technology*, Vol. 23, No. 4, pp. 298–306, 2004
9. Y. Morita, "Curing of epoxy siloxane monomer with anhydride," *Journal of Applied Polymer Science*, Vol. 97, No. 3, pp. 946–951, 2005
10. Y. Zhou et al., "Optical nanocomposite materials for photonic packaging," *Proceedings of PhoPack*, pp. 34–36, 2003
11. Y.C. Lin et al., "Materials challenges and solutions for the packaging of high power LEDs", *International Microsystems, Packaging, Assembly Conference Taiwan*, pp. 1–4, 2006
12. C.P. Wong, "Thermal-mechanical enhanced high-performance silicone gels and elastomeric encapsulants in microelectronic packaging", *IEEE Transactions on Components Packaging and Manufacturing Technology (PART A)*, Vol. 18, No. 2, pp. 270–273, 1995
13. J. H. Lau et al., "Electronic packaging: design, materials, process, and reliability", New York : McGraw-Hill, 1998
14. H. Luo et al., "Analysis of high-power packages for phosphor-based white-light-emitting diodes", *Applied Physics Letters*, Vol. 86, No. 24, 243505, 2005
15. M. Rossi et al., "Micro-optics promote use of LEDs in consumer goods", *LED Magazine*, 2005
16. J. P. Dakin et al., "Handbook of Optoelectronics", New York: Taylor & Francis, 2006
17. R. N. Kumar et al., "Ultraviolet radiation curable epoxy resin encapsulant for light emitting diodes", *Journal of Applied Polymer Science*, Vol. 100, No. 2, pp. 1048–1056, 2006
18. Y. Zhou et al., "Epoxy-based optically transparent nanocomposites for photonic packaging", *Proceedings of Advanced Packaging Materials: Process, Properties and Interfaces, 9th international symposium*, pp. 100–102, 2004
19. Y. Morita et al., "Thermal initiated cationic polymerization and properties of epoxy siloxane", *Journal of Applied Polymer Science*, Vol. 100, pp. 2010–2019, 2006
20. Y. Zhou et al., "One component, low temperature and fast cure epoxy encapsulant with high refractive index for LED applications", *IEEE Transactions on Advanced Packaging*, Vol. 31, No. 3, pp. 484–489, 2008
21. US Patent 6,713,571, Mar.30, 2004
22. US. 20060204760, Sep.14, 2006
23. US patent 7,125,917, Oct.24, 2006
24. L. Rector et al., "Molding transforms to meet advanced market requirements", *Advanced Packaging*, 2007
25. D.A. Steigerwald et al., "Illumination with solid state lighting technology", *IEEE Journal of Selected Topics in Quantum Electronics*, Vol. 8, No. 2, pp. 310–320, 2002
26. A. Bergh et al., "The promise and challenge of solid-state lighting", *Physics Today*, Vol. 54, No. 12, pp. 42–47, 2001
27. S. Chhajed et al., "Influence of junction temperature on chromaticity and color-rendering properties of trichromatic white-light sources based on light-emitting diodes", *Journal of Applied Physics*, Vol. 97, No. 5, Art. No. 054506, 2005

28. M. Yamada et al., "A methodological study of the best solution for generating white light using nitride-based light-emitting diodes", *IEICE Transactions on Electronics*, E88C, No. 9, pp. 1860–1871, 2005
29. F. M. Steranka et al., "High power LEDs – Technology status and market applications", *Physica Status Solidi (A), Applied Research*, Vol. 194, No. 2, pp. 380–388, 2002
30. E. F. Schubert et al., "Solid-state lighting – a benevolent technology", *Reports on Progress in Physics*, Vol. 69, No. 12, pp. 3069–3099, 2006
31. E. F. Schubert et al., "Solid-state light sources getting smart", *Science*, pp. 1274–1278, 2005
32. M. S. Shur et al., "Solid-state lighting: Toward superior illumination", *Proceedings of the IEEE*, Vol. 93, No. 10, pp. 1691–1703, 2005
33. "Lumileds delivers white LEDs with patented conformal coating ", *LED magazine*, 2005
34. N. T. Tran, "Simulation studies for the design and manufacturing of optical sensors and white light emitting devices", PhD Dissertation.
35. N. Narendran, Y. Gu, J. P. Freyssinier-Nova, and Y. Zhu, "Extracting phosphor-scattered photons to improve white LED efficiency", *Physica Status Solidi (A)*, Vol. 202, No. 6, pp. R60–R62, 2005
36. N. Narendran, "Improved performance white LED", *Fifth International Conference on Solid State Lighting, Proceedings of SPIE*, Vol. 5941, pp. 45–50, 2005
37. "Cyberlux acquires rights to remote phosphor technology", *LED magazine*, 2006
38. US. 20060181192, Aug. 17, 2006
39. US. 7045956, May. 16, 2006
40. US. 6791259, Sep. 14, 2004
41. "CAB-O-SIL® Untreated Fumed Silica Properties and Functions", Cabot Corp, www.cabot-corp.com
42. A. Norris, M. Bahadur, and M. Yoshitake, "Silicone materials development for LED packaging", Dow Corning Corporation
43. A. W. Norris et al., "Novel silicone materials for LED packaging", *Proceedings of SPIE*, Vol. 5941, 594115, Sep. 14, 2005
44. W. Noll, "Chemistry and technology of silicones", New York: Academic Press, 1968
45. R. Jones et al., "Silicon-containing polymers", Dordrecht: Kluwer Academic Publishers, 2000
46. S. Clarkson et al., "Siloxane polymers", New Jersey: PTR Prentice Hall, 1993
47. W. Lynch, "Handbook of silicone rubber fabrication", New York: Van Nostrand Reinhold Company, 1978
48. A. Norris et al., "Silicone materials development for LED packaging", *LED Magazine*, 2006
49. Y. Li et al., "Recent advances of conductive adhesives as a lead-free alternative in electronic packaging: Materials, processing, reliability and applications", *Materials Science and Engineering R*, Vol. 51, pp. 1–35, 2006
50. K. Gilleo, in: J.S. Hwang (Ed.), "Environment-friendly electronics: Lead-free technology, electrochemical publications Ltd., Port Erin, UK, 2001 (Chapter 24)
51. M. Abtew et al., "Lead-free solders in microelectronics", *Materials. Science and Engineering R*, Vol. 27, pp. 95–141, 2000
52. W. C. Wu et al., "Investigation on the long term reliability of power IGBT modules", *Proceedings of the International Symposium Power Semiconductor Devices and ICs*, Yokohama, Japan, 1995
53. L. H. U. Andersson et al., "Silicone elastomers for electronic application. I. Analyses of the noncrosslinked fractions", *Journal of Applied Polymer Science*, Vol. 88, pp. 2073–2081, 2003
54. J. Petroski, "Cooling High Brightness LEDs: Developments, Issues, and Challenges" *Next Generation Thermal Management Materials and Systems Conference*, June 15–17, 2005.

55. F. Wall, "Bringing it all together-the basics of building an LED module/assembly", *IMAPS Advanced Technology Workshop on Power LED Packaging and Assembly*, Palo Alto CA, USA, Oct. 26–28, 2005
56. J. Kolbe, "Benefits of insulated metal substrates in high power LED application", *IMAPS Advanced Technology Workshop on Power LED Packaging and Assembly*, Palo Alto CA, USA, Oct. 26–28, 2005
57. D. Saums, "Developments in CTE-Matched, High Thermal Conductivity Composite Baseplate Materials for Power LED Packaging", *IMAPS Advanced Technology Workshop on Power LED Packaging and Assembly*, Palo Alto CA, USA, Oct. 26–28, 2005
58. A. Roth et al., "Direct copper bonded substrates for use with Power LEDs", *IMAPS Advanced Technology Workshop on Power LED Packaging and Assembly*, Palo Alto CA, USA, Oct. 26–28, 2005
59. J. D. Hwang, "The Status and Development of LED Thermal Enhanced Substrate", www.materialsnet.com.tw, Vol. 231, 2006
60. "Understanding Power LED Lifetime Analysis", white paper, Philips Lumileds Lighting, San Jose, CA, USA
61. "Luxeon Reliability", Application Brief, Philips Lumileds Lighting, San Jose, CA, USA
62. N. Narendran et al., "Solid-state lighting: failure analysis of white LEDs", *Journal of Crystal Growth*, Vol. 268, No. (3–4), pp. 449–456, 2004
63. S. Todoroki et al., "Temperature distribution along the striped active region in high-power GaAlAs visible lasers", *Journal of Applied Physics*, Vol. 58, pp. 1124–1128, 1985
64. S. Murata et al., "Adding a heat bypass improves the thermal characteristics of a 50 μm spaced 8-beam laser diode array", *Journal of Applied Physics*, Vol. 72, pp. 2514–2516, 1992
65. P. W. Epperlein et al., "Influence of the vertical structure on the mirror facet temperatures of visible GaInP quantum well lasers", *Applied Physics Letters*, Vol. 62, pp. 3074–3076, 1993
66. D. C. Hall et al., "Technique for lateral temperature profiling in optoelectronic devices using a photoluminescence microprobe", *Applied Physics Letters*, Vol. 61, pp. 384–386, 1992
67. Y. Xi et al., "Junction and carrier temperature measurements in deep-ultraviolet light-emitting diodes using three different methods", *Applied Physics Letters*, Vol. 86, pp. 031907–031909, 2005

Chapter 19
Digital Health and Bio-Medical Packaging

Lei Mercado, James K. Carney, Michael J. Ebert, Scott A. Hareland, and Rashid Bashir

Abstract This chapter reviews the healthcare trends and implications, as well as electronic packaging applications in implantable devices, pacing leads, bio-medical sensors, and point-of-care sensors. Each presents unique opportunities and challenges for electronic packaging and materials.

Keywords Implantable devices · bio-medical sensors · leads · point of care · packaging

19.1 Introduction

The revolution in personal computers and cell phones that has driven the telecommunication and semiconductor industries has resulted in ever more powerful consumer-friendly products. At the same time, however, the products have become commodities with low margins and ever higher investments required to stay competitive. This has left manufacturers looking for new markets to drive growth and profits.

At the same time, the population is aging and there is an increasing demand for better medical therapies. A variety of medical devices have been proposed to meet this need and, since medical devices have typically commanded high margins, many of the biggest players in the electronics field have set up digital health divisions. Medical devices, however, pose new challenges in terms of electronic packaging and materials. In addition, the industry presents high barrier of entry for new entrants, such as long regulatory approval time, high quality and reliability demands, liability and patient safety considerations. Medical devices are usually manufactured in small volume, which reduces the cost effectiveness of high-volume manufacturing processes that provides significant leverage for the large electronic companies. There are also significant

---

L. Mercado (✉)
Neuromodulation, Medtronic, Inc, 4000 Lexington Ave N, Shoreview, MN, 55126, USA
e-mail: lei.l.mercado@medtronic.com

market development challenges to raise awareness of both patients and physicians about medical devices due to the many alternatives available, such as medications and minimally invasive therapies.

19.2 Healthcare Trends – Opportunities/Challenges on Medical Devices and Electronic Packaging

19.2.1 Healthcare Trend and Key Drivers

Healthcare will see significant changes in the next decade due to a variety of drivers, including economic pressure, demographic changes, patient-centered care, IT and technology advances. Healthcare is facing enormous and unprecedented cost pressure. Medicare faces bankruptcy in 10 years, even if the reimbursement is reduced by 10% each year. The population is aging with the baby boomers moving into their golden ages. While improvements in medical care have allowed people to live longer, that longevity also means people will develop and live with additional chronic conditions requiring medical management. It is also expected that the delivery of care will change from hospital-centered care to patient-centered care. Patients will take a more active role in managing their own health and move the point of care to their homes.

19.2.2 Implications of Healthcare Trends on the Opportunities/ Challenges of Electronic Packaging

Each of the key drivers in healthcare trends presents opportunities and challenges for the electronic packaging and materials.

19.2.2.1 Economic Pressure

The reimbursements for medical devices have been on a steady decline in an attempt to reduce healthcare cost. This puts downward pricing pressure on device manufacturers who must reduce costs and improve efficiency. As patients start to cover more of the cost of their medical care out of their own pockets, they also demand therapies and devices with better quality and lower cost.

19.2.2.2 Demographic Changes

Many patients are facing co-morbidities as they age. Some are managed by a number of different physicians each with a specialty such as general practitioners, internal medicine, cardiology, nephrology, and endocrinology. Treating one condition, however, may prove ineffective unless the other conditions (co-morbidities) are also monitored and treated. Unfortunately, there may be little to no communication among the various specialists. New technologies must be developed that can measure all of the information necessary to manage these patients and communicate it to the managing physicians.

19.2.2.3 Patient-Centered Care

Patients will be more involved and empowered in managing their own health. The slow shift to consumer-driven, patient-centered healthcare will make meeting patient needs a critical success factor. Miniaturization is a top requirement for patients who would prefer minimum interference to their daily life. For the same reason, patients do not want to be concerned with the interaction of their devices with their environment, therefore demanding the devices to be MRI (Magnetic Resonance Imaging) safe and electromagnetic compatible. The globalization and diversification of the patient population also made customization increasingly important to satisfy the needs of various cultures and geographies.

Most implantable devices are battery-powered. Battery longevity is an important concern for the patients. When the battery is depleted, not only do patients have to pay for a replacement device, they often have to undergo the surgical procedures again to take out the existing device and put in the new one. This increases patient cost, inconvenience, and potential risks of infection. Therefore, an ongoing challenge is to increase battery capacity, reduce energy consumption, while simultaneously decreasing the size of the implantable device.

The aging baby boomer population will put high demand in hospital availability and compete for the limited clinician time. Patients are also less tolerant of driving for hours for a routine follow-up. Therefore remote patient management is being embraced by both clinicians and patients alike.

19.2.2.4 Information Technology Advances

The advance in information technology is driving the paradigm shift in medical information management. Seventy percentage of hospitals are making progress in establishing Electronic Health Records. The technology advance will allow increase in patient data collection. On the other hand, information overload and medical staff shortage lead clinicians to demand actionable information. This put increasing demands on data storage and processing capabilities.

19.3 External Packaging of Implantable Medical Devices

19.3.1 Biological Hermeticity

External packaging of implantable medical devices serves as a biological barrier between the body and sensitive electronics, helps absorb mechanical forces applied to the device, and also may serve some key electronic functionality for a variety of therapies. Any object placed inside the body for medical purposes must meet strict controls and undergo rigorous testing to ensure that the packaging is biocompatible. The accepted definition of biocompatibility was stated by David Williams in 1987: Biocompatibility is the ability of a material to perform with an appropriate host response in a specific application [1]. Examples of undesirable responses include cytotoxicity (toxic to cells), mutagenicity and/or chromosomal aberrations, sensitization (allergic response), pyrogenicity (fever producing), or hemolysis (red blood cell damage). Fortunately, in spite of this list of potentially adverse bio-responses to implants, there are a number of widely tested and approved materials used extensively throughout the device industry for the external packaging of implantable medical devices. The external construction of most implantable pulse generators (IPGs) such pacemakers, neurostimulators, drug pumps, implantable cardioverter defibrillators (ICDs) used in chronic medical implants are constructed of relatively common, non-exotic materials such as titanium metal for the "cans" and polyurethane or silicone compounds and adhesives for interface headers. These materials have been the primary components in external device packaging for decades and are backed by literally billions of patient hours in the field with a high degree of reliability and demonstrated biocompatibility. Biocompatibility of implantable leads have additional challenges which will be described in Section 19.4.

Occasionally, patients with sensitivities or allergies to various metals can be provided with custom devices (usually plated with gold) in order to reduce or eliminate any allergic type reactions to devices. While general metal allergies are not uncommon, it is rare for any implantable medical device to require such special coatings.

Some medical applications, such as device leads, focus a great deal of energy on pursuing new materials that are more robust and tolerant of the biological interaction, but most device "cans" are relatively happy with the current state of affairs. Some new technologies on the horizon that will impact external packaging include investigations into different surface coatings to reduce the chance of infection. New implantable medical electronics that rely on novel materials as sensor components will have to demonstrate adequate safety and biocompatibility before they will be approved for use.

19.3.2 Electrical Compatibility

Another primary function of the external packaging is to work with the device's circuitry to keep the sensitive electronics safe from a myriad of external electrical and magnetic sources of interference. Several decades ago, the first few generations of internally implanted pacemakers had the electronics encapsulated in a polymeric material without the benefit of a metal can to act as a Faraday shield around the device. This could possibly lead to device interference as those who remember once ubiquitous signs warning pacemaker patients of the presence of microwave ovens can attest. These warnings of microwaves are now relics of the past, due primarily to the metallic can and the design of electronic input circuitry that serves as a gateway for both bio-signal sensing and transmission of therapies to patients. Several standards exist that prescribe the level of immunity to electrical and magnetic fields that an implantable medical device, especially one providing life-sustaining therapy, is required to exhibit. These include EN45502-2-1 (for low power devices), EN45502-2-2 (for high power devices), and CD ANSI/AAMI PC69 standards. These standards define test methods, criteria for device performance during and after exposure, and rationale for testing devices to certain frequency ranges of electro-magnetic radiation at certain power levels.

Low frequency ($f \leq 450$ MHz) emitters and power level requirements are typically those encountered in radio and television transmission, electronic article surveillance gates, RFID systems, some wireless services, and some medical procedures (e.g. diathermy, RF ablation, etc.). Testing at intermediate frequencies (450 MHz $\leq f \leq 3$ GHz) is centered around technologies that include cell phones and some radio systems. This range of frequencies has seen explosive growth in the last decade or so and will continue to evolve new modulation schemes and applications that devices will need to withstand. At very high frequencies ($f > 3$ GHz) such as microwave radiation, there are few requirements due to the understanding that both the limited sources and the natural protection of the device electronics afforded by both the device can and the body provide ample immunity to these radiators.

New technologies and gadgets are constantly being introduced into the marketplace, so the effort required to characterize and catalog these emitters is rapidly changing. Medical device manufacturers are constantly asked about potential device interference due to new sources such as hybrid car engines, portable music systems (e.g. iPods), video game systems with wireless transmitters, etc. The device construction (packaging and input circuit design) requires that the external interference does not change the therapeutic behavior of the device or adversely interact with the device in a way that places a patient at risk. This includes safe device operation, maintenance of device settings and programming, and safe therapy delivery to the patient during the specified interference exposure levels.

19.3.3 Mechanical Requirements

Mechanical requirements for device reliability are also strongly dependent upon the external construction of the device and the packaging used inside the can to protect the sensitive electronics. Requirements exist that outline the use conditions (temperature, vibration, shock, etc.) and performance criteria during device transportation, storage, and handling. Once the device is implanted, the external packaging is expected to protect the device from scenarios such as cyclic loadingconditions (repeated muscular motion forces on the device), atmospheric pressure changes (from high altitude commercial aircraft to scuba diving), and mechanical shock (blunt trauma). Again, existing standards such as EN45502-2-1 and -2 prescribe minimum requirements, but do not force adherence to any particular design methods or practice.

Additional mechanical requirements exist for the connector block that acts as the interface between the implantable device and leads that connect the device to the appropriate organ or tissue. It is important for the connector to maintain mechanical integrity under both implant (e.g. lead insertion) and chronic implant conditions (e.g. cyclic loading) in order to keep a viable pathway between the leads and the electronics that drive them.

19.3.4 Electrical Pathway

Electrically, the external packaging of the device may become part of the electrical circuit formed between the device and the human body. In many pacing applications, the device can be programmed to act as an electrode that completes a circuit with one or more lead electrodes. This configuration can be found in some pacing applications (so called unipolar pacing) or bio-impedance measurements such as those made across the thoracic cavity between the lead tips and the device implanted in the pectoral region of the chest (see Fig. 19.1).

In high power defibrillation therapies found on ICDs, the device may become an integral portion of the circuit that permits more efficient delivery of high

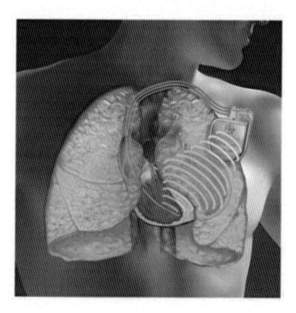

Fig. 19.1 Electrical pathway formed between the device can and a lead

energy from the lead defibrillation electrodes across the cardiac muscle into the can. Without the can in the defibrillation circuit, it would require a great deal more energy and/or different, and possibly less comfortable, device and lead configurations within the body in order to shunt the energy across the heart to stop a life-threatening arrhythmia.

19.3.5 Internal Packaging

Inside an implantable medical device, the electronics and packaging look surprisingly similar to off-the-shelf consumer electronic devices. While the earliest pacemakers were extremely simple in their design, requiring only a power source and a few transistors to provide a stable series of electrical output pulses to stimulate cardiac tissue, modern pacemakers provide a wide array of functions including on-board microprocessors for signal processing and therapy optimization, wireless telemetry for communication to the outside world, and diagnostic data storage. These additional features support the crucial sensing and pacing functions of the device as well as additional high voltage therapies included in ICD products. Figure 19.2 shows an implantable pacemaker

Fig. 19.2 Internal view of a pacemaker (Medtronic EnRhythm Model P1501DR)

(Medtronic EnRhythm Model P1501DR) with one half of the shield removed to show internal components including ICs, discrete components, electrical feedthroughs, and a battery.

The range of signals required to perform all of these functions is quite varied and provides a source of challenges for both the electronic circuitry and the packaging that pulls all that functionality together. Figure 19.3 illustrates a view of signal amplitude and pulse width (proxy for frequency) characteristics for several basic functions of an ICD product. Sensing, important to the optimal operation of a device, must be sensitive to cardiac signals with typical amplitudes between 0.5 and 30 mV and frequencies <100 Hz. Similar sensing requirements are also required for non-cardiac devices such as neurostimulators. This crucial sensing must be performed in the presence of pacing signals (0.5–5 V amplitudes with 0.1–1 ms pulse widths are typical).

Occasionally, high voltage defibrillation therapies are needed to terminate life threatening tachyarrhythmias. These therapies can generate high energy (8–35 J) output pulses with voltage amplitudes of several hundreds of volts and ~100 ms pulse durations. Because of this wide variation of applications: from low amplitude sensing requirements, pacing therapies, high voltage defibrillation therapies, on-board microprocessor and memory functions, and telemetry frequencies in the 100's of kHz for close (order cm) range up to 100's of MHz for distance (order meters) telemetry applications all coupled with aggressive power management techniques, there is not really any single integrated circuit (IC) technology that adequately and simultaneously addresses these requirements while maintaining sufficient noise immunity for proper sensing. These conflicting requirements and technological capabilities limit the ability of system design to accommodate all of the functions on a single piece of silicon, and typical designs will select the best IC technology for the given performance, power, and reliability requirements. While many circuit techniques, including highly tuned filtering and blanking periods, are applied in order to minimize the impact of these signals on the sensing capabilities of the device, packaging plays a critical role in permitting these various functions to be interconnected with each other

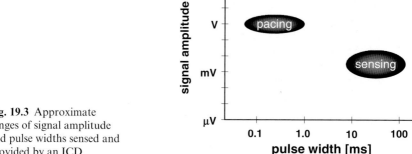

Fig. 19.3 Approximate ranges of signal amplitude and pulse widths sensed and provided by an ICD

while also maintaining proper noise immunity, especially to the delicate sensing function of the device.

One important aspect of electronic packaging that is worth mentioning is the relatively benign thermal environment encountered in implantable medical products once placed inside the body. The device is held at a relatively constant 37°C body temperature throughout its implant lifetime. In addition, designs that place a premium on extremely low power consumption (<100 microWatts dissipation typical) keeps component and package heating to an absolute minimum throughout the vast majority of a product's lifetime. Thermal heating that occurs during high voltage defibrillation therapy is kept under design control and only constitutes a very small fraction of time in a typical device application. Both baseline temperatures and thermal gradients in the electronic packaging and not typically high reliability risk items for ICs in these products.

The primary components of an implantable medical device include a battery, an electronic assembly with a populated circuit board, large capacitors (for high voltage therapy applications), telemetry antennae, sensors (such as motion and magnetic field), and additional connectors. On the electronic assembly, the circuit board is comprised of both discrete electronic components and ICs. The ICs themselves are packaged in a variety of form factors, including well known flip chip and stacked die assembly processes in modern devices. One drive in the industry is the continued reduction in device size for cosmetic purposes, patient comfort, and optimal device implant location selection. These requirements place continuous pressure on both the external and internal packaging considerations. Figure 19.4 shows the evolution in ICD device volumes over the last decade and a half.

Pacemaker products do not require the larger capacitors for high energy defibrillation therapy nor the high rate battery designs to support it, so they are significantly smaller in volume than their ICD cousins by roughly a factor of three. Reduction in ICD device volume is clearly evident since the first models

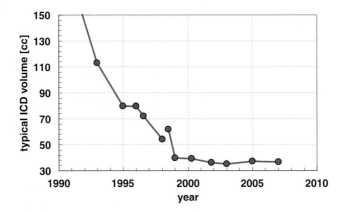

Fig. 19.4 Evolution of typical ICD volume vs. year

introduced circa 1990, but has not been a consistent device requirement. Typically, increases in device volume or periods of relative constant device size (from the late 1990s to day) are driven by addition of new features and technologies into devices that consume some of the natural decrease in product volume that would occur if device performance stayed constant. Some of these new features over the past few years include a trend towards higher defibrillation energies ($\sim$35 J today) which may require modifications to batteries and high voltage capacitors, addition of distance telemetry to enable longer range device to receiver communications, greater computational functionality, or larger batteries to support longer device lifetimes between replacements. In addition to traditional device configurations, new implantable monitors and therapeutic devices are continuously driving device size reduction in order to reach new locations within the body and simplify implantation procedures away from surgery towards more benign approaches such as direct injection through a needle. Research into new and improved design techniques, material advances, therapy optimization, and packaging technologies will all contribute to reductions in device volumes in the upcoming years.

19.3.6 Soft Errors and Single Event Upsets

Because of the aggressive power management and moderate computational burden used in implantable medical devices, there is a strong push towards running ICs at extremely low voltages in order to save power. These operating modes can lead to exacerbated sensitivity to soft-errors from alpha particles, thermal neutrons, and energetic neutrons from cosmic radiation. Our understanding of the physics of soft errors and their impact on integrated circuit technology is certainly not new, but very high reliability requirements in life-sustaining devices lead to a variety of safety features incorporated into them. In order to minimize the power, area, and computational burden of these features (such as error correcting codes, redundancy, data integrity checking, etc.) it is highly desirable to utilize packaging materials and technologies that minimize the potential impact of alpha particle induced upsets by maintaining high levels of purity in materials and cleanliness in processing and manufacturing. Cleaner end product packaging reduces the burden and necessity of additional safety features in the product.

19.4 Leads in Medical Devices

19.4.1 Overview of Leads

Pacing leads are the "wires" that carry the electricity from the pacemaker or ICD to the heart. Pacemaker leads are implanted through a vein in the chest and fixated inside the heart. There is usually one lead put in the right ventricle, another positioned in the right atrium and in heart failure patients another lead may be inserted into the coronary sinus and positioned over the left ventricle. Defibrillator leads are typically inserted into the right ventricle.

Once inside the heart the lead must be fixated to the muscle. Fixation is either active (traumatic fixation), such as an extendible/retractable helical electrode [2] or passive (atraumatic) such as a tine [3] shown in Fig. 19.5. The lead bodies are compliant and flex with each beat of the heart. A heart rate of 60 beats per minute corresponds to flexing the lead approximately 32 million times per year.

Like implantable pulse generators, the leads must be both biocompatible and biostable. Factors that affect the biocompatibility of the pacing lead would include: materials, lead design or shape, implant location, skill of the implanter and the ability of material/device to resist degradation within the body (biostability). The biocompatibility of the materials and the lead must be assessed prior to use in people. Testing is performed per guidelines and test methods outlined in ISO 10997. These tests evaluate hemocompatibility, pyrogenicity, acute and chronic toxicity, sensitization and carcinogenicity. The leads are also implanted in animals to assess both long term biocompatibility and biostability.

The lead and the materials that comprise the lead must withstand the chemical and mechanical environment within the body. The humoral (bulk) environment within the body consists of water, electrolytes (e.g., Na^+, K^+, Ca^{+2}, Mg^{+2}, Cl^-, HPO_4^{-2}, SO_4^{-2}, HCO_3^-), proteins, fatty acids, lactic acid, uric acid, creatinine, bilirubin, bile salts, glucose, urea and many more. Additionally a pacing lead or any implanted device must deal with the inflammatory or foreign body response. Once implanted a cascade of reactions start that result in the pacing lead being covered with a layer or layers of foreign body giant cells and/or macrophages and a fibrotic capsule composed primarily of collagen containing phagocytic cells and fibroblasts [4]. The cellular component of the foreign body reaction can have a significant effect on the biostability of the materials in the pacing lead. These cells can release number of enzymes and oxidants to destroy the foreign body (pacing lead). The compounds that seem to have the greatest effect on biostability appear to be the oxidants (H_2O_2, O_2^-, OH) and hydrolytic enzymes.

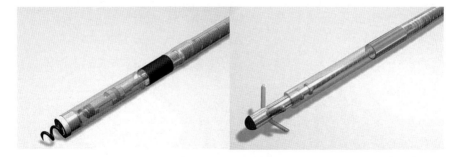

Fig. 19.5 Lead with active fixation extendible/retractable helix and a tined lead

Pacing leads are comprised of a connector that plugs into a pulse generator, electrical conductor or conductors that carry charge to the pacing site, insulation that isolates the conductors and electrodes.

19.4.2 Lead Connector

Lead connection is often looked at as trivial; however, improper insertion of the lead into the pulse generator connector is one of the leading causes for reoperation. Lead connectors were not always standardized between the manufacturers. Several unique connector designs were introduced in the early 1980s. This meant the lead from one manufacturer could not be directly inserted into the pulse generator from another manufacturer without an adapter. In the mid 1980s, a joint IEC/ISO International Pacemaker Standards Working Group defined a formal international standard for lead connectors, IS-1 shown in Fig. 19.6.

However, this standard only defines connections for unipolar and bipolar brady pacing lead designs. The development of the implantable cardioverter defibrillators (ICD's) brought about the standardized DF-1 connection for high voltage lead connectors. Today, the pacing companies are working toward a new connector standard, IS-4, that would allow a reduced size lead connector with multiple connections. As one looks toward the future, there is opportunity to design connectors that eliminate the problems associated with improper insertion and allow the multiple connections that will be needed for multiple electrodes and sensors on the lead.

19.4.3 Conductors

There are two types of conductors used in pacing leads: coils and cables shown in Fig. 19.7. The coil was first suggested in 1961 by Dr. William Chardack and greatly reduced conductor fracture [5]. The original wire used in coils was

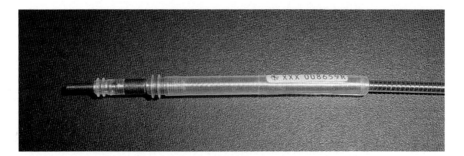

Fig. 19.6 IS-1 Lead Connector

Fig. 19.7 Cross section of a defibrillator lead body showing cable and coil conductors, tubing insulating the conductors, multi-lumen tubing separating the conductors and a protective tube over the multi-lumen tubing

stainless steel which occasionally corroded. Platinum and platinum alloys were used to reduce concerns with corrosion but were very expensive and still fractured. These materials were eventually replaced by the super alloy, MP35N. MP35N has both excellent corrosion properties and mechanical properties making it ideal for use in coils. MP35N coils were improved by using multiple smaller diameter wires. The use of multiple smaller diameter wires allowed the electrical resistance to drop and improved the flex life of the coil. To further reduce the electrical resistance, the drawn filled tube (DFT) was used. In the DFT wire the core of the MP35N wire was replaced with silver which dramatically reduced resistance. The corrosion resistant MP35N on the outside of the wire protected the silver from corroding.

Cables were implemented to further reduce electrical resistance in defibrillator leads. Cables are comprised of many strands of pure MP35N or silver cored MP35N wire. However, both coils and cables are still susceptible to fatigue and fracture within the body. Additionally, it is known that cobalt corrosion byproducts from the MP35N can catalyze the degradation of polyurethanes and other materials used for pacing lead insulation.

19.4.4 Insulation

Teflon, or polyethylene was used as insulation in early leads. Bonding concerns made Teflon difficult to use in manufacturing, so it was abandoned. The use of polyethylene insulation was stopped because it makes stiff leads which increases the possibility of perforating the heart, and is not biostable [6, 7]. Polyester polyurethanes were tried because of their excellent mechanical properties, but were stopped because they are subject to rapid degradation in water. Silicone rubber became the material of choice for insulation because it was nontoxic, chemically inert and biostable. Silicone rubber has low tear strength and had to be used with thicker walls to minimize mechanical damage. Silicone rubber also has a high coefficient of friction in blood which made it difficult to pass two leads in the same vein. As a result, dual chamber pacing did not realize its full potential in the 1970s. In the early 1980s polyether polyurethane began to be used as lead insulation. Polyether polyurethane is hydrolytically stable unlike the polyester polyurethane, stronger mechanically than silicone and is slippery when wetted with blood. The increased mechanical properties allowed the insulation thickness to be downsized. The smaller size combined with lubricious surface in blood made it easy to place two leads in one vein making dual chamber pacing a practical therapy.

Unfortunately, the softer polyether polyurethanes were discovered to be subject to two previously unknown failure mechanisms, metal ion oxidation (MIO) and environmental stress cracking (ESC) [8, 9, 10] shown in Fig 19.8 (a) and (b). Pacemaker lead manufacturers have learned how to design around these failure mechanisms to produce excellent longevities [11].

High performance silicone rubbers have replaced the early silicone rubbers and allowed the insulation wall thickness to be reduced. Additionally, surface treatments were developed to make silicone rubber more lubricious and easier to implant. Thus, today, silicone rubber leads can be made substantially smaller and easier to use, but still not as small and tough as polyurethane leads. Silicone rubber, while chemically inert, still has failure mechanisms, including susceptibility to mechanical damage. Silicone rubber insulation failure due to compressive creep (cold flow) or wear is a concern in multiple lead implants. Thus, it is probably correct to say that at the present time, there is no optimum insulation material. Manufacturers continue to research new, biostable polymers for insulation.

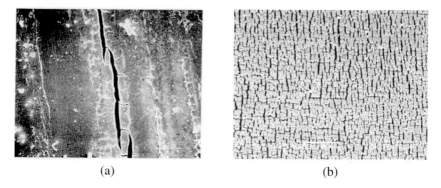

Fig. 19.8 (a) MIO breach in inner insulation and (b) ESC in outer insulation

19.4.5 Electrodes

Early transvenous leads were relatively large diameter (12 French, 0.156") as were the electrodes. The large electrode size had low pacing impedance which resulted in high current drain and shorter pacemaker longevity. Early work by Irnich showed that the theoretically optimum (spherical) electrode for stimulation was about 0.7 to 1 mm in radius, corresponding to the thickness of the connective tissue that forms around it [12]. The next generation of pacing electrodes was smaller which increased impedance and reduced battery current drain, however, the smaller size also increased the impedance associated with sensing (source impedance) [13]. Mismatch between the input impedance of the pulse generator's sensing circuit (too low) and the source impedance (too high) can result in signal attenuation and sensing failure. This drove electrode size to the 6 to 12 mm^2 size range in order to optimize impedance and minimize signal attenuation.

In the late 1970s, the totally porous and porous surface electrode were introduced [14, 15]. These structures produced high pacing impedance because of their small size (defined by the electrode's radius), but their increased surface area from the porosity resulted in much lower source impedance. Thus, porous electrodes provided better sensing than polished electrodes. An added benefit was that the pores facilitated tissue ingrowth, which aided fixation (Fig. 19.9a, b, and c).

In 1979, the carbon electrode was introduced with microporous surface structure [16]. The microporosity further improved the performance of the electrodes and different coating began to be added to increase the interfacial surface area. These coatings are used today and include platinum black, titanium nitride [17] and iridium oxide [18].

The steroid-eluting electrode was introduced by Stokes in 1982 [19]. The steroid was combined with silicone to form a plug which was positioned inside the electrode shank behind the porous tip (Fig. 19.10). This electrode technology combined porosity and microporosity with a glucocorticosteroid resulting

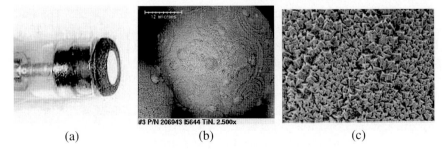

Fig. 19.9 Sintered, porous, titanium nitride coated electrode (**a**) and scanning electron microscope photographs at 2500X (**b**) and 20,000X (**c**) magnification, respectfully

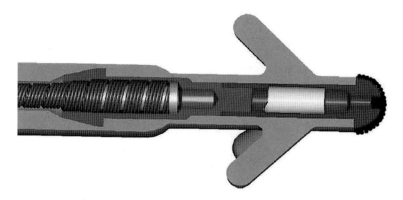

Fig. 19.10 Cross section of steroid eluting electrode

in minimal to no threshold rise as a function of implant time [20, 21]. The steroid mitigates the foreign body response at the electrode tip, preventing threshold rise that would occur in its absence [22]. The addition of the steroid not only prevented threshold rise it significantly reduced exit block. Exit block is a phenomenon where pacing thresholds continue to rise beyond the ability of the pacemaker to capture the heart.

Electrode materials used in early transthoracic temporary pacemakers included tantalum, silver plated copper and stainless steels [23]. Stainless steel was used in the early 1960s for implantable electrodes, but gave way to more corrosion-resistant materials. Platinum, platinum alloys, and Elgiloy became the materials of choice for the vast majority of permanent leads. However the current density, which governs corrosion, was high enough to cause the Elgiloy to corrode. The issues with current flow were partially solved by coupling a capacitor between the output terminals of constant voltage generators. "Capacitively coupled" generators limit current output below that required for significant corrosion.

Other materials with better corrosion resistance have been studied extensively. Both titanium and tantalum are excellent electrode materials [24, 25]. Under controlled conditions, oxides of varying structures are grown on their surfaces. These oxides are stable in the body, even when charged. Titanium oxide electrodes have been used successfully for over 25 years in Europe [26, 27]. The only negative aspect of titanium is its poor radiopacity, which makes implant more difficult because the tip does not show up on X-ray.

Leads are available with a bewildering array of electrode designs, fixation mechanisms, conductor configurations, insulation materials, etc. As technology progresses over the sixth decade of cardiac pacing, lead reliability will improve and the evolution will continue at a remarkable pace. New conductor and insulation technologies are needed that will eliminate the major device failure mechanisms (such as ESC, MIO, creep, wear, crush, fracture). New electrode materials will be developed that allow drug elution as well as improve the electrical performance of the lead. Meanwhile, the lead may look like a simple wire, but in many ways, it is far more complex than the pulse generator.

19.5 Implantable Biomedical Sensors

19.5.1 Overview of Implantable Sensors

The previous two sections contained overviews of the advances and opportunities for chronically implanted pacemakers, defibrillators and leads that are used to correct problems with the rhythm of the heart. Recently, medical devices have been demonstrated that are intended to manage or correct numerous other medical conditions either through the use of closed-loop implanted systems or by providing important information to the managing physician who can change drug therapies or order procedures. In these systems, one of the key components is a sensor that measures the physiologic variable of interest. The specific requirements of the sensor are determined by the variable to be measured, the environment in which the sensor is to be used, the accuracy of the sensed signal, and the lifetime of the device.

The applications for these sensors are extremely varied. For example, sensors that are intended to help diagnose problems in the digestive system may be swallowed or placed in a specific position that is prone to extremely high levels of moisture and acidity. These sensors are only expected to last a few days before they are passed through the system. Sensors that are placed under the skin or into the head, cardiovascular system, or eyes of people in order to manage or correct the effects of chronic diseases are expected to operate flawlessly for years. Some applications for these sensors include the management of diabetes, hypertension, heart failure and blindness. Chronically implanted sensors have also been used as a method to diagnose an infrequent but problematic condition or provide an early warning for the onset of a condition for which a patient is at risk. Diagnosing the cause of fainting is an example of the former condition while providing an early warning for the start of a stroke or heart attack is an example of the latter. Another application for implanted sensors consists of devices that are intended to monitor the status of another implanted medical device such as an artificial cervical disc or a prosthetic hip or knee joint. These sensors are used to measure the stresses on the device or determine if excessive wear is occurring.

In the remainder of this section, different sensors will be described including their key requirements.

19.5.2 Sensors for Gastrointestinal Diagnosis

The Bravo^{TM} pH Monitoring System from Medtronic (Minneapolis, MN) is shown in Fig. 19.11. This system is used to measure the pH level in the esophagus

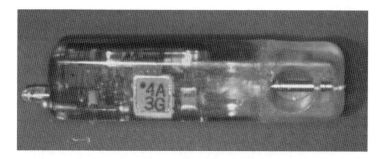

Fig. 19.11 Photograph of the Bravo esophageal pH sensor for the diagnosis of gastroesophageal reflux disease

of patients who are suspected of having gastroesophageal reflux disease [28]. The Bravo sensing capsule is approximately 2.5 × 0.6 × 0.5 cc and contains a pH sensor, electronic circuitry, a radio transmitter, and a battery. Unlike pacemakers and defibrillators that have a titanium can for the external packaging, the Bravo circuitry is potted in epoxy, leaving only the electrodes for the pH sensor exposed. The choice of epoxy rather than titanium permits the telemetry to operate over a few meters of distance and keeps the cost of the device low.

The Bravo sensor capsule is positioned in the esophagus just above the entry into the stomach using a specially designed catheter. The sensor is attached to the wall of the esophagus using a metal pin and the catheter is removed. The sensor then transmits the pH sensor reading to a nearby receiver every few seconds for 24 to 48 hours.

The sensor must survive in an extremely hostile environment of high humidity and acidity. The selected epoxy encapsulation is thick enough to prevent moisture from reaching the circuitry while also providing a durable, low-cost protective layer for the circuits.

A second device with a similar operating environment is the PillCam from Given Imaging, Ltd (Yoqneam, Israel). The device is 11 mm × 26 mm and weigh less than 4 grams. It is swallowed and, while moving through the stomach and small intestines, takes and transmits two images per second for approximately eight hours. The result is more than 50,000 images. The application is to detect polyps, cancer, or causes of bleeding and anemia inside of the small intestine. The capsule contains a camera, lights, RF transmitter and batteries. A unique requirement for the PillCam packaging is that it must accommodate the optical functions of the device. The capsule has a clear end that allows the internal lights to illuminate the lining of the small intestine and the image to be focused on the camera.

19.5.3 Implantable Pressure Sensors

The measurement of pressure is used to diagnose and manage a number of serious medical conditions including heart failure, glaucoma, and hydrocephalous.

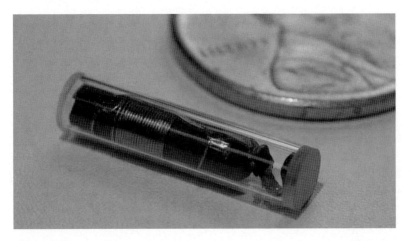

Fig. 19.12 Photograph of the wireless pressure sensor capsule for the measurement of intra-cardiac pressures (Courtesy of ISSYS)

A number of companies are developing pressure sensors for chronic implantation to manage these conditions. ISSYS (Ypsilanti, MI USA) is developing a two-part system consisting of an implantable, batteryless, sensor shown in Fig. 19.12 and a companion hand-held reader. The implantable sensor module contains a silicon MEMS (MicroElectroMechanical System) pressure sensor along with custom electronics and a telemetry antenna. The pressure sensor is powered and interrogated using an RF signal from the external reader. The reader transmits power to the sensor and the sensed pressure is in turn transmitted back using inductive magnetic telemetry.

The sensor is intended to be implanted in the left atrium of the patient's heart for the management of heart failure [29]. This also drives a number of requirements for the packaging of the device. First, the device is expected to last more than 10 years. Therefore, the package must be impervious to moisture. Second, the device must be safe. Therefore it must be made using biocompatible materials and cannot cause clots to form that might break off and cause a stroke. Finally, the package must allow the RF energy to easily couple to the internal antenna. ISSYS has chosen to use a glass capsule that is sealed to a silicon MEMS pressure sensor. A non-thrombogenic coating is applied to glass to prevent the formation of clots.

Mesotec (Hanover Germany) is developing an implantable pressure sensor to measure Intraocular Pressure (IOP) in the eye of a patient with glaucoma [30]. The pressure sensor is a single silicon IC that contains an integrated pressure sensor as well as the recovery circuitry and RF telemetry. The sensor IC is sealed inside an annular silicone ring, called the mesoRing, as shown in Fig. 19.13. The selection of the polymer and the sealing process are critical to the proper operation of this device. Not only must the polymer protect the circuitry but it must also not

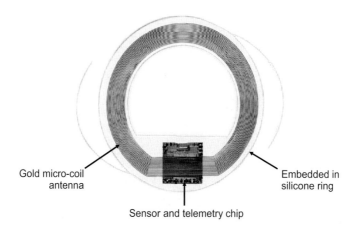

Fig. 19.13 Photograph of the mesoRing telemetric Intraocular Pressure Sensor implant (courtesy of Mesotec)

swell or cause a force to be applied to the surface of the pressure sensor that would produce an offset.

The ring also contains a foldable gold micro coil connected to the IC. Similar to the ISSYS design, power delivery and communication are accomplished through RF inductive coupling to an external reader. The mesoRing is approximately 1 cm across and is implanted during the eye surgery to replace a diseased lens.

Pressure sensors are also in development to monitor chronic diseases like heart failure. One such device is the Chronicle ® Implantable Hemodynamic Monitor from Medtronic (Minneapolis, MN USA). As shown in Fig. 19.14 the Chronicle system consists of an implanted pacemaker-sized monitor and a pressure-sensing lead placed in the right side of the heart. The cardiac pressures are measured continuously and stored in the device. The data are then intermittently transmitted by the patient to a secure website accessible by treating clinicians [31].

The pressure sensor is the small capsule that is approximately 3 cm from the distal tip of the lead. The Chronicle pressure sensor capsule consists of a titanium housing that has been machined to have a thin diaphragm in one section. This diaphragm is one plate of a capacitor and deflects due to cardiac pressure to produce a change in the capacitance. This pressure-dependant capacitance controls the output of an IC inside of the capsule. Therefore, the titanium housing provides not only protection for the internal circuitry, but the machined diaphragm is the pressure sensitive element. Because the pressure sensor is on a lead that is placed in the heart, the construction must be very robust to repeated loading. It should be expected that the lead will flex tens of millions of times during the life of the patient.

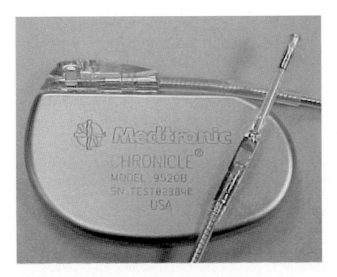

Fig. 19.14 Photograph of the Chronicle Implantable Hemodynamic Monitor and pressure sensor on the lead

St. Jude Medical (St. Paul, MN) and Transoma Medical (St. Paul, MN) are each developing pressure sensors that are to be placed in the heart. Although there are differences between these devices and the Chronicle in terms of sensor design and position of implant, the requirements for the package to be robust, hermetic, and mechanically stable are common.

19.5.4 Implantable Sensors for the Blind

An exciting application of sensors in ophthalmic applications is the design and manufacture of implantable medical devices to aid in vision improvement. Optobionics (Naperville, IL USA) is developing the Artificial Silicon Retina^{TM} (ASR) microchip to restore some sight to patients who have been blinded by retinitis pigmentosa or age-related macular degeneration [32]. In these patients, the retinal cells are damaged but the underlying nerves are still intact. The system is designed to stimulate the damaged retinal cells, allowing them to send visual signals again to the brain. The ASR microchip is a silicon chip 2 mm in diameter and 25 microns thick, which is surgically implanted under the retina. It contains an array of approximately 5,000 "microphotodiodes," each with its own stimulating electrode. These microphotodiodes are designed to convert the light energy focused on them by the lens of the eye into impulses that stimulate the remaining functional cells of the retina.

Alternative approaches to returning sight to the blind are being investigated at other institutions. The Doheny Eye Institute and the Intraocular Retinal

Prosthesis Group at the University of Southern California [33] is using a small external camera to generate an image which is transmitted to a system surgically implanted in the eye. The implanted components contains electronic circuitry that receives the transmitted signal, converts it into a set of electrical impulses and an array of electrodes that deliver the impulses to the nerves of the eye. Ultimately, the desire is to place the camera in the eye as well.

In all of the vision applications, the materials and packaging for the implanted portion of the system must simultaneously meet a number of requirements. Because there is a need to keep the devices thin, the outer coating of the device must be thin as well. These layers must protect the underlying electronic components from the fluids in the eye but must also be biocompatible and prevent irritation, even after years of implantation. Polymer coatings such as silicone or deposited layers such as diamond-like coatings are being investigated.

19.5.5 Implanted Sensors for Orthopedics and Spine

Sensors implanted in orthopedic and spinal devices are new being investigated to provide accurate information on the mechanical stresses and wear of the implanted systems. In these application, great care must be taken to protect the sensors from being crushed by the large forced generated by the motion of the joints as well as to design a package that will not leak after millions of cycles. MicroStrain (Williston, VT USA) has built an investigational, full artificial knee replacement that can wirelessly report digital, 3 dimensional torque and force data back to computers [34, 35]. The sensing system was enclosed in a custom titanium alloy total knee replacement. The unit includes all of the electronics and is hermetically sealed using laser welding. Piezoelectric transducers sense the local strain in the titanium, and the wireless sensor sends the data to an external antenna. The system itself must be mechanically robust enough to withstand millions of knee movements. The information gained from this device can be used to develop design improvements, refine surgical instrumentation, guide postoperative physical therapy and potentially detect the individual activities that would overload the implant.

19.5.6 Implanted Glucose Sensors

Biochemical sensors are an area of active research. The search for a long-lived, continuous glucose sensor for the management of diabetes has been underway for decades [36] but the magic device still remains elusive. At the time of this writing, there are two companies, Medtronic, Inc. [37] and Dexcom, Inc. (San Diego, CA) that have FDA-approved continuous glucose sensors. These are not the dream design of a small sensor capsule that is inserted under the skin and accurately measures glucose levels for years. Instead, these are devices for which the sensing element contains electrochemical electrodes coated with a glucose converting enzyme. The electrodes are printed onto a long, thin polymer film. The electrodes are inserted through the skin and remain in place for three to seven days before they are replaced. A small transmitter is connected to the external end of the sensing electrode. The electronics in the transmitter measures the glucose value regularly and sends the information to a nearby receiver.

Dexcom has also reported on their progress to build a fully implanted monitor [38]. The first version of the sensor that was implanted in humans consisted of an electrochemical sensor packaged in a small, cylindrical device about the size and shape of an AA battery. The package also contains a battery, circuit board, microprocessor, and radio-transmitter. One further feature of the sensor was a multilayered membrane that coated the sensor and was intended to control the body's natural response to form a fibrous tissue layer around the sensor. This layer must be avoided because it would prevent the diffusion of glucose to the sensor surface, resulting in erroneous glucose readings. These sensors have been implanted in patients for up to three months.

Sensors for Medicine and Science (Germantown, MD) is developing an implantable sensor which relies on fluorescence to measure glucose concentration under the skin. A drawing of the sensor is shown in Fig. 19.15. The sensor capsule contains an LED to excite the fluorophores in the indicator layer, optical detectors to measure both the LED emission and the fluorescence, and circuitry to drive and control the optoelectronic components. The implanted capsule also has circuitry to harvest energy from an RF signal from an external monitor.

Implanted biomedical sensors are an area of active research and technical development although only a handful have been commercially released to date due to the difficulty of simultaneously achieving accuracy, stability, and biocompatibility in an extremely small package. The requirement for better information

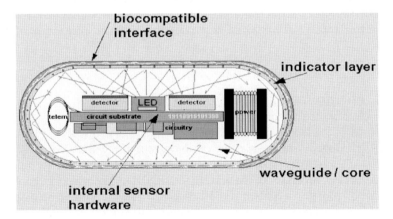

Fig. 19.15 Drawing of a fluorescence-based implantable glucose sensor capsule (courtesy of Sensors for Medicine and Science)

on patient health as an enabler of lower cost healthcare combined with the number of organizations that are now moving into this area will result in a number of devices being released in the near future.

19.6 Chip-Based Point of Care Sensors: Opportunities and Challenges

19.6.1 Introduction

Point-of-care biosensors can have a wide and profound impact on diagnosis and management of disease. Point-of-care biosensors can be used at the bedside, at the doctor's office, or at home. The need for such sensors and devices has been on the rise due to an increased need to detect targets of disease for a more efficient management of disease and also for the increased need to manage the disease at the individual level. The fact that we are witnessing the largest ever aging population in the US in the coming decades, very high costs of health care, and widespread AIDs epidemic in much of the underdeveloped part of the world, point of care biosensors are urgently needed to help tackle these complex issues. These sensors need to be cheap, disposable and one-time-use and sensitive and be able to detect the target biological entities such as cells, bacteria, viruses, proteins, DNA, or small molecules.

Currently, only a few examples of point of care sensors are available in the market. The pregnancy test and the blood glucose monitoring are perhaps the two most common examples of tests which can be purchased over the counter, are easy to use, and provide critical information needed by the user. Two additional examples include the cholesterol test and the iSTAT/Abbott cartridges for detection of blood gases, ions, and most recently cardiac markers. Many more opportunities exist in making such point-of-care sensors more pervasive, especially if microfluidics, micro and nanotechnology and lab-on-a-chip technologies are employed in the development and realization of these sensors.

19.6.2 Microsystems, BioMEMS, and BioChips

The point-of-care biosensors available over the counter today are limited to proteins, enzymes, or small molecules (as in the case of the iSTAT or the Cholesterol tests). Many more opportunities exist for the detection of cells, microorganisms, viruses, proteins, DNA, and small molecules. The recent technological advances in top–down silicon nanotechnology and development of microfluidic devices present themselves with new opportunities for small, sensitive, one time use, point of care diagnostic biochip sensors capable of rapid and highly accurate analysis of samples of body fluid. Micro and nanofabrication techniques have enabled researchers to produce very small-scale sensors. In general, the small size of sensors not only improves their sensitivity and also enables formation of arrays. Moreover, micro-electro-mechanical-systems (MEMS), micro-fluidics and nanotechnology also offer fabrication of sensors that have the capability of detecting biological entities, possibly without using any labels.

The use of micro and nano-scale detection technologies is justified by, (i) reducing the sensor element to the scale of the target species and hence providing a higher sensitivity, (ii) reduced reagent volumes and associated costs, (iii) reduced time to result due to small volumes resulting in higher effective concentrations, and (iv) amenability of portability and miniaturization of the entire system. Monitoring of cancer biomarkers from serum or blood, detection of viruses or bacteria from blood or urine, detection and counting of CD4+ white blood cells, and detection of presence of live cells from drinking water, are only some of the examples of possible applications. The actual sensing modalities could include electrical methods and all its derivatives, mechanical sensing, or optical sensing (Bashir, et al. 2004). The sources of sample for diagnostic applications can include saliva, blood, or urine, with blood being the richest repository of information about the individual's state of health (Toner and Irimia, 2005). In addition to body fluids, other samples such as water and other fluids used industrial microbiology and pharmaceutical manufacturing, and fluids from extracts of food samples also compose a large market segment for such point-of-care or point-of-test sensors. Figure 19.16 shows the steps to be performed for point-of-care sensors sample analysis. The target fluid is to be metered and injected into a device. Depending on the target entity to be detected, the sample might have to be processed, i.e. target entities separated from other entities in the sample. Such sample preparation and analyte extraction or concentration is needed to increase the signal to noise ratio and to be able to detect minute quantities of the target. Then the target entity is detected and the results need to be displayed in the reader or the data analysis system. The model here is that of a disposable cartridge/sensor and a reusable reader/system.

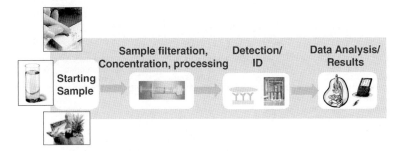

Fig. 19.16 Overview of steps to be performed in a point of care biochip sensor

19.6.3 Sensor Technology Platform

Microfluidic devices can handle fluids in the range of few tens of hundreds of microlitres depending on the device architecture and the application. Depending on the target analyte, the possible technology elements in such a biochip sensor are shown in Fig. 19.17. Some means to dispense and transport the sample are needed. The sample might have to be filtered to separate cells from serum, for example, or separate viruses and bacteria from blood cells. Further sorting or separation might be done using charge or pH based separation, or by the use of electrophoresis or dielectrophoresis. Then the target entities might have to be captured using surface receptors that bind to antibodies to achieve specific capture. Once the target cells are captured, in some cases, the cells might have to be grown and cultured to either increase their number, e.g. for the case of detection of live bacteria, or for detection of proteins secreted by growing cells. Then the cells might need to be lysed to analyze the cellular contents using bio-molecular identification techniques. These techniques could be label free, for example, using electrical (impedance, electrochemical, or field effect) or mechanical sensors (micro or nano-cantilever). Optical detection with fluorescence labels or other techniques such chemiluminescence or bioluminescence can be used with integrated photo-detectors (in the system/reader). Not all modules have to be used for all applications and the order could also be switched around depending on the target analyte and entity. Figure 19.18 shows images from literature of BioMEMS and microfluidic devices that are being developed in academia or industry.

Mechanical detection for biochemical entities and reactions has more recently been used through the use of micro- and nano-scale cantilever sensors on a chip. These cantilever sensors (diving board type structures) can be used in

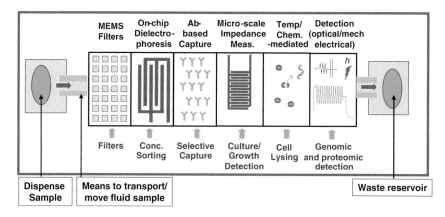

Fig. 19.17 Overview of a possible technology platform for on-chip detection of microorganisms or cells from fluids. Not all modules might be needed and the sequence of modules can be rearranged based on the type of assay needed

19 Digital Health and Bio-Medical Packaging

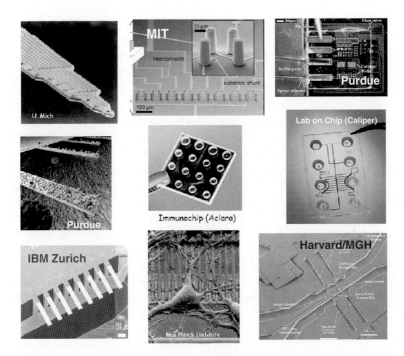

Fig. 19.18 Various biochip sensors and lab on chip devices from literature

two modes, namely stress sensing and mass sensing. In the stress sensing mode, the bio-chemical reaction is performed selectively on one side of the cantilever. A change in surface free energy results in a change in surface stress, which results in measurable bending of the cantilever. Thus, label-free detection of biomolecular binding can be performed. The bending of the cantilever can then be measured using optical means (laser reflecting from the cantilever surface into a quad position detector, like in an AFM) or electrical means (piezo-resistor incorporated at the fixed edge of the cantilever). In the mass sensing mode, the cantilever is excited mechanically so that it vibrates at its resonant frequency (using external drive or the ambient noise, for example). The resonant frequency is measured using electrical or optical means, and compared to the resonant frequency of the cantilever once a biological entity is captured. The change in mass can be detected by detection of shift in resonant frequency, assuming the spring constant does not change. The quality factor is decreased with increased damping, for example in a fluid, and hence the minimum detectable mass is much higher in damped mediums as compared to low-damped mediums.

Electrical or electrochemical detection techniques have also been used quite commonly in Biochips and BioMEMS sensors. These techniques can be amenable to portability and miniaturization, when compared to optical detection techniques, however, recent advances in integration optical components on a chip can also produce smaller integrated devices. Electrochemical biosensors include three basic types; (i) amperometric biosensors, which involves the electric current associated with the electrons involved in redox processes, (ii) potentiometric biosensors, which measure a change in potential at electrodes due to ions or chemical reactions at an electrode (such as an ion Sensitive FET), and (iii) conductometric biosensors, which measure conductance changes associated with changes in the overall ionic medium between the two electrodes. There are more reports on potentiometric and amperometric sensors, specially, due to the established field of electrochemistry, and many of these sensors have been used as the micro and nano-scale.

Optical detection techniques are perhaps the most common due to their prevalent use in biology and life-sciences. There is a very significant amount of literature on BioMEMS devices with optical detection. Optical detection techniques can be generally based on Fluorescence or Chemiluminescence. Fluorescence detection techniques are based on fluorescent markers that emit light at specific wavelengths and the presence and enhancement, or reduction (as in Fluorescence Resonance Energy Transfer) in optical signal can indicate a binding reaction. The detectors include photodiodes or CCD elements which are integrated in the readers so that the cost of the disposables are minimized as much as possible.

19.6.4 Bio-Chip Packaging Issues and Challenges

The packaging of biochips and lab-on-chips poses significant challenges as compared to micro-electronic device packaging. Such devices need to possibly have all or some of these interfaces [41]; (i) electrical interface, for the case of electromechanical detection, (ii) optical interfaces, for fluorescence or other means of optical detection to reaction regions in the chip, (iii) fluidic interfaces, for possibly transferring fluid from the cartridge to the chip itself, and (iv) mechanical interfaces, for applying pressure to move and pushing the fluids in the chip, for example. The co-existence of these multiple interfaces makes the packaging issues more complex than other device or sensor packages. In addition, the cost needs to be low enough for these devices to be used in throw-away one-time-use applications. Figure 19.19 shows drawn schematic of such sensor layouts, and Fig. 19.20 shows a concept schematic of packaging system for such micro-fluidic biochip sensors.

The design criteria and the performance specifications for any of these devices could include any or all of these considerations below [41]. (i) The package should eliminate manual handling and connection of individual fluidic ports, (ii) The package and sensor itself should be designed such that a failure more in one does not compromise the other, (iii) If the devices is to be used a one-time-use sensor, the package must also protect the sensor from the environment till the device is used, and subsequent to use, the package and device must hold all fluids used during the test and prevent any contamination to the environment, (iv) The package should be able to handle the desired pressure

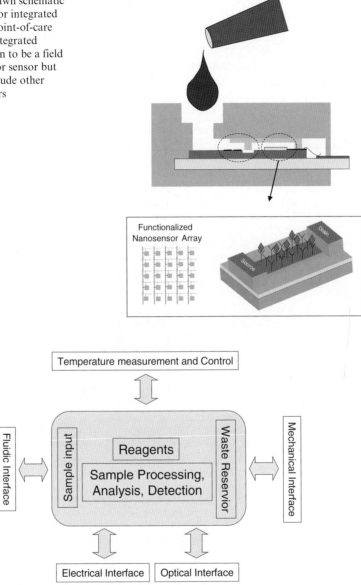

Fig. 19.19 Drawn schematic of cartridges for integrated biochips for point-of-care testing. The integrated sensor is shown to be a field effect transistor sensor but could also include other types of sensors

Fig. 19.20 Schematic of packaging requirements for integrated lab-on-chip and microfluidic biochip sensors

ranges for movement of fluids inside the biochip sensor, (v) Any or all of the interface schemes mentioned above must be designed in the package, (vi) The package and the biochip must not contaminate or degrade the biological entity of interest under test, (vii) The package and biochip must not be permeable to gases or fluids for them to leak to the environment, (viii) The fluidic interface must be reliable and robust, (ix) The materials used for building the package needs to be sterilized before the introduction of the sample to be tested. These are only some of the requirements, and depending on the specific test and assay under development, the requirements might be increased or decreased from the above list.

References

1. Williams DF, *The Williams Dictionary of Biomaterials* (Liverpool, UK: Liverpool University Press, 1999), 40
2. Bisping HJ and Rupp H. A new permanent transvenous electrode for fixation in the atrium. In Watanabe Y (ed), *Proceedings of the Vth International Symposium on Cardiac Pacing*. Amsterdam, Exerpta Medica, pp. 543–547, 1977
3. Citron P and Dickhudt E. US Patent No. 3959502, Endocardial electrode. 1976
4. Anderson JA. Inflammatory response to implants. *ASAIO* II(2):101–107, 1988
5. Chardack W, Gage A and Greatbatch W. Correction of complete heart block by a self contained and subcutaneously implanted pacemaker. *J Thorac Cardiovasc Surg* 42:814, 1961
6. Dolezel B, Adamirova L, Naprstek A and Vondracek P. *In vivo* degradation of polymers. I. Change of mechanical properties in polyethylene pacemaker lead insulations during long-term implantation in the human body. *Biomaterials*, 10(2):96–100, 1989
7. Wasserbauer R, Beranova M, Vancurova D and Dolezel B. Biodegradation of polyethylene foils by bacterial and liver homogenates. *Biomaterials*, 11(1):36–40, 1990
8. Byrd CL, McArthur W, Stokes K, Sivina M, Yahr WZ and Greenberg J. Implant experience with unipolar polyurethane pacing leads. *PACE,* 6(5):868–882, 1983
9. Zhao Q, Topham N, Anderson JM, Hiltner A, Loden G and Payet CR. Foreign-body giant cells and polyurethane biostability: *In vivo* correlation of cell adhesion and surface cracking. *J Biomed Mater Res* 25:177–183, 1991
10. Zhao Q, McNally AK, Rubin KR, Reiner M, Wu Y, Rose-Caprara V, Anderson JM, Hiltner A, Urbanski P and Stokes K. Human plasma α_2-macroglobulin promotes in vitro stress cracking of Pellethane 2363-80A: *In vivo* and *in vitro* correlations. *J Biomed Mater Res* In Press
11. Beanlands DS, Akyurekli Y and Keon WJ. Prednisone in the management of exit block. In Meere C (ed), *Proceedings of the VIth World Symposium on Cardiac Pacing*, Montreal, PACESYMP, 1979, Chapter 80-3
12. Irnich W. Physikalische Uberlegungen zur elektrostimulation. *Biomedizin Technik* 3:97–104, 1973
13. Barold SS, Ong IS and Heile RA. Matching characteristics of pulse generator and electrodes. A clinicians concept of input and source impedance and their effect on demand function. In Meere C (ed), *Proceedings of the VIth World Symposium on Cardiac Pacing*, Montreal, PACESYMP, 1979, Chapter 34-3
14. Amundson D, McArthur W, MacCarter D and Mosharrafa M. Porous electrode-tissue interface. In C Meere (ed), *Proceedings of the VIth World Symposium on Cardiac Pacing* Montreal, PACESYMP, Chapter 29-16
15. Wilson GJ, MacGregor DC, Bobyn JD, Lixfeld W, Pillar RM, Miller SL and Silver MD. Tissue response to porous-surfaced electrodes: Basis for a new atrial lead design. In C Meere (ed), *Proceedings of the VIth World Symposium on Cardiac Pacing*, Montreal, PACESYMP, Chapter 29-12

16. Beck-Jansen P, Schuller H and Winther-Rasmussen S. Vitreous carbon electrodes in endocardial pacing. In C Meere (ed), *Proceedings of the VIth World Symposium on Cardiac Pacing*, Montreal, PACESYMP, Chapter 29-9
17. Schaldach M, Bolz A, Breme J, Hubmann M and Hardt R. Acute and long-term sensing and pacing performance of pacemaker leads having titanium nitride electrode tips. In Antonioli E, Aubert AE and Ector H (eds), *Pacemaker Leads 1991*, Amsterdam, Elsevier, 1991, pp. 441–450
18. Del Bufalo AGA, Schlaepfer J, Fromer M and Kappenberger L. Acute and long-term ventricular stimulation with a new iridium oxide-coated electrode. *PACE* 16(6):1240–1244, 1993
19. Stokes KB, Graf JE and Wiebusch WA. Drug-eluting electrodes-improved pacemaker performance. In Potvin AR and Potvin JH (eds), *Frontiers of Engineering in Health Care-1982*. Proceedings, Fourth Annual Conference IEEE Engineering in Medicine and Biology Society, pp. 499–502, 1982
20. Stokes KB, Bornzin GA. and Wiebusch, WA. A steroid-eluting, low-threshold, low-polarizing electrode. In Steinbach K (ed), *Cardiac Pacing*, Darmstadt, Steinkopff Verlag, pp. 369–376, 1983
21. Mond H and Stokes K. The Electrode-Tissue Interface: The Revolutionary Role of Steroid Elution. *PACE* 15(l):95–107, 1992
22. Stokes K and Anderson J. Low Threshold Leads: The Effect Of Steroid Elution. In Antonioli GE (ed), *Pacemaker Leads*, Amsterdam, Elsevier, 537–542, 1991
23. Thevenet A, Hodges PC and Lillehei CW. Use of myocardial electrode inserted percutaneously for control of complete atrioventricular block by artificial pacemaker. *Dis Chest* 34:621–631, 1958
24. Dawson WW (ed). Electrode materials study, contract number NIH-71-2286, Tenth Quarterly Report, Nov. 1973–Jan. 1974
25. Johnson PF, Bernstein JJ, Hunter G, Dawson WW and Hench LL. An *in vitro* and *in vivo* analysis of anodized tantalum capacitive electrodes: corrosion response, physiology and histology. *J Biomed Mater Res* 11:637–656, 1977
26. Maiolino P, Del Bene P, Cecci A, Cappelletti F, Pauletti M, Al Bunni M and Audoglio R. Titanium oxide electrode: 60 Months clinical experience of low energy pacing. In Antonioli GE, Aubert AE and Ector H (eds), *Pacemaker Leads 1991*, Amsterdam, Elsevier, pp. 491–496, 1991
27. Audoglio R and Gatti AM. Non-stoichiometric titanium oxide: Why is it a so effective material for low energy pacing? In Antonioli GE, Aubert AE and Ector H (eds), *Pacemaker Leads 1991*, Amsterdam, Elsevier, pp. 491–496, 1991
28. Pandolfino JE, Richter JE, Ours T, Guardino JM, Chapman J, and Kahrilas PJ. Ambulatory esophageal pH monitoring using a wireless system. *Am J Gastroenterol* 98:740–749, 2003
29. Najafi N and Ludomirsky A. Initial Animal Studies of a Wireless, Batteryless, MEMS Implant for Cardiovascular Applications. *Biomedical Microdevices* 6:61–65, 2004
30. Stangel K, Kolnsberg S, Hammerschmidt D, Hosticka BJ, Trieu HK, and Mokwa W. A programmable intraocular CMOS pressure sensor system implant. *IEEE J Solid-State Circuits* 36:1094–1100, 2001
31. Steinhaus D, Reynolds DW, Gadler F, Kay GN, Hess MF, and Bennett T. Implant experience with an implantable hemodynamic monitor for the management of symptomatic heart failure. *Pacing Clin Electrophysiol* 28:747–53, 2005
32. Chow AY, Chow VY, Packo K, Pollack J, Peyman G, and Schuchard R, The artificial silicon retina microchip for the treatment of vision loss from retinitis pigmentosa. *Arch Ophthalmol* 122:460–469, 2004
33. Liu W, McGucken E, Cavin R, Clements M, Vichienchom K, Demarco C, Humayun M, de Juan E, Weiland J, and Greenberg R. A retinal prosthesis to benefit the visually

impaired. In Teodorescu N (ed), *Intelligent System and Techniques in Rehabilitation Engineering*, CRC Press, 99, pp. 31–87, 2000
34. D'Lima DD, Townsend CP, Arms SW, Morris BA, and Colwell CW. An Implantable Telemetry Device to Measure Intra-Articular Tibial Forces. *J Bio-Mechanics* 38:299–304, 2005.
35. Kirking B, Krevolin J, Townsend C, Colwell CW Jr, and D'Lima DD. A multiaxial force-sensing implantable tibial prosthesis. *J Biomechanics* 39:1744–1751, 2006
36. Klonoff DC. Continuous glucose monitoring: roadmap for 21st century diabetes therapy. *Diabetes Care* 28:1231–1239, 2005
37. Gross TM, Bode BW, Einhorn D, Kayne DM, Reed JH, White NH, and Mastrototaro JJ. Performance evaluation of the MiniMed continuous glucose monitoring system during patient home use. *Diabetes Technol Ther* 2:49–56, 2000
38. Garg SK, Schwartz S, and Edelman SV. Improved Glucose excursions using an implantable real-time continuous glucose sensor in adults with type 1 diabetes. *Diabetes Care* 27:734–738, 2004
39. Bashir R. BioMEMS: State of the art in detection and future prospects, *Adv Drug Delivery Rev* 56:1565–1586, 2004
40. Toner M and Irimia D. Blood-on-a-Chip. *Annu Rev Biomed Eng* 7:77–103, 2005
41. Lee K. The development of highly functional cartridge for rapid detection of microbial contaminants. MS Thesis, Purdue University, 2006

Subject Index

A

ACAs/ACFs, 365
Adhesion, 35, 118, 135, 161, 164, 170, 171, 173, 230, 256, 263, 276, 279, 283, 287, 289, 291, 301, 309, 311, 314, 316, 341, 345, 350, 353, 355–356, 367, 376, 380, 381, 383, 384, 394, 398, 421, 424, 425, 431, 433, 481, 493, 506, 515, 553, 556, 564, 578, 584, 588, 590, 597, 610, 620, 624, 634, 635, 637, 647, 650, 661
Adhesives, 24, 28, 38, 52, 54, 127, 151, 256, 274, 275, 279, 365–368, 374, 376, 377, 379–394, 397, 407, 409, 421, 442, 505, 665, 668, 684
Advanced BGA / CSP, 407, 415–416
Ajinomoto Build-up Film (ABF), 243, 247
Anodic bonding, 51, 52, 56, 57–58
Aramid paper, 274, 284

B

Back side treatment, 219
Ball grid array (BGA), 3, 4, 365, 408, 548
BCB, 16, 24, 25, 28, 30, 31, 34, 42, 160, 161, 314, 495, 497, 549, 550, 556, 559, 583–594, 597, 616
BGA, 3, 4, 38, 39, 40, 47, 88, 162, 204, 206, 208, 211, 254, 261, 264, 326, 340, 349, 351, 365, 376, 378, 408, 415, 420, 434, 548
Bio-medical sensors, 681
Blind via (BV), 243, 247
Bonding, 16–28, 39, 51–72, 96–99, 102, 113–174, 222, 376, 409, 410, 414, 603, 668–669, 693
Bonding wire, 113, 120, 130–134, 151, 160, 168, 170, 409, 414, 417, 632

Bumping, 131, 168, 170, 265, 330, 331, 373, 377, 384, 385, 386, 388, 389, 547, 550, 551–593

C

Carbon nanotube, 453, 454, 503, 512, 514, 517
Ce materials (TIMs), 437
Ceramic filler, 274, 299, 481
Chip scale package (CSP), 4, 365, 378
Chip warpage, 407, 412, 420, 428, 430, 431, 432, 434
Coefficient of thermal expansion, 55, 79, 132, 195, 243, 350, 357, 368, 391, 411, 420, 428, 440, 496, 506, 548, 632, 634, 650, 672, 673, 307, 343
Compliant I/O, 77, 88–91
Composites, 280, 285, 286, 294, 297, 421, 443, 453, 454, 455, 459, 481, 483, 485, 487, 594
Conductive particles, 168, 365, 366, 367, 368, 370, 371, 373, 375, 378, 379, 380, 396, 441, 443, 483, 485
Conductivity, 44, 62, 85, 118, 132, 248, 285, 286, 287, 291, 344, 365, 366, 368, 369, 371, 376, 378, 380, 382, 383, 384, 390, 393, 394, 397, 410
Conductor Loss, 80, 274, 289, 291, 294, 301, 302
Contact resistance, 99, 292, 367, 368, 370, 371, 376, 378, 379, 382, 390, 392, 394, 395, 396, 397, 398, 438, 448, 449, 453, 468, 515, 552
Copper clad laminate (CCL), 243, 248, 249, 250, 252, 256, 257, 258, 259
Copper Foil, 54, 274, 275, 287, 288, 289, 290, 291, 292, 294, 301, 302, 493, 511, 672
Copper interconnects, 77, 95–101
Coreless, 243, 264, 265–267
Corrosion inhibitor, 365, 396, 397

713

D

Damage, 22, 31, 64, 115, 123, 134, 135, 140, 141, 145, 154, 159, 165, 172, 219, 221, 224, 225, 233, 234, 238, 239, 276, 305, 316, 348, 358, 367, 419, 421, 535, 684, 694

Density factor, 437, 438, 441

Die attach, 38, 52, 64, 135, 151, 159, 160, 267, 268, 307, 351, 352, 357, 407, 409, 411, 412, 413, 415, 416, 418, 419, 420, 421, 422, 425, 427, 430, 432, 506, 635, 637, 638, 643, 645, 650, 657, 665, 669, 672

Dielectric constant, 7, 53, 55, 79, 80, 89, 118, 161, 243, 245, 267–269, 273, 274, 278, 280, 281, 286, 294, 295–298, 299, 300, 309, 315, 375, 481, 482, 484, 486, 488, 505, 513, 519, 583, 584, 594

Die stacking, 1, 39

3D integration, 1, 2, 3, 4, 5, 6, 9, 10, 11, 13, 46, 47

Direct bonding, 16, 17, 52, 56, 61–62, 69, 162, 170

E

Electrically conductive adhesives (ECAs), 365–398, 505, 668

Electrodeposited copper, 100, 274

Electronic packaging, 52, 58, 60, 61, 62, 308, 323, 407–409, 411, 415, 547, 575, 650, 668, 681, 682–683, 689

Embedded capacitor, 459, 480, 481, 484, 488, 497

Embedded resistor, 274

Encapsulant, 87, 89, 309, 629, 632, 634–638, 640, 646–652, 655, 658–661, 664, 666, 670, 675

Epoxies, 52, 53, 196, 275, 276, 277, 298, 376, 377, 398, 442, 581, 638

Epoxy molding compounds (EMCs), 325, 339, 341–346, 408, 418

Epoxy resin, 52, 255, 269, 277, 278, 294, 310, 344, 345, 348, 353, 354, 355, 357, 368, 382, 394, 396, 407, 412, 413, 421, 422, 423, 424, 426, 433, 515, 572, 635, 648, 649, 650, 651

F

Films, 52, 53, 54, 96, 114, 136, 138, 154, 160, 274, 275, 276, 287, 377, 409, 415, 418, 421, 422, 425, 430, 431, 432, 433, 479, 484, 489, 490, 493, 494, 505, 508, 514, 563, 575, 590, 594

First-level interconnect, 113, 114, 307

Flame retardant, 193, 277, 339, 341, 345–346

Flip-chip, 77, 78, 307, 311, 318, 332

Fluxless soldering, 51, 52, 65–71

FR-4, 94, 95, 97, 102, 162, 274, 297, 316

G

Glass fabric, 248, 250, 274, 279, 280–283, 295, 296, 297, 300

Glass reinforcement, 274, 294, 295, 296, 303

Grinding, 28, 31, 32, 222–229, 231, 233, 236, 254, 255

H

High density interconnect (HDI), 243, 247, 253–256, 567

High temperature and high frequency bonding, 113

I

ICAs, 365, 366, 379–398

Impact performance, 365, 390, 391, 397–398

Implantable devices, 681, 683

Input/Output, 77, 122, 308

Interconnect, 1, 7, 9, 28, 31, 41, 53, 61, 71, 79, 80, 83, 85, 88, 89, 92–95, 100–106, 113, 114, 115, 117, 118, 119, 127, 132, 168, 174, 243, 245, 247, 307, 308, 312, 316, 322, 324, 390, 393, 492, 506, 513, 552, 566, 567, 612, 639

Interconnection of stacked and thinned ICs, 113

L

Laser, 11, 58, 65, 114, 128, 145, 146, 236, 237, 238, 239, 247, 253, 254, 263, 266, 267, 268, 269, 270, 273, 366, 386, 512, 559, 613, 616, 617, 618, 632, 707

LCP, 55, 256, 274, 275, 276, 279, 294, 297, 299, 300, 496–497

Lead-free, 92, 182–185, 188, 189, 192, 195, 197, 199, 200, 201–205, 208, 210, 211, 212, 259–261, 265, 277, 279, 286, 293, 315, 316, 317, 319, 321, 332, 341, 349, 354, 368, 372, 379, 390, 411, 412, 506, 507, 508, 562, 567, 570, 572

Leads, 59, 60, 89, 90, 141, 159, 172, 222, 238, 250, 255, 278, 293, 321, 322, 328, 355, 357, 441, 442, 461, 475, 477, 484, 507, 526, 551, 557, 579, 637, 644, 669, 684, 686, 690, 691, 692

Subject Index

Light emitting diode (LED), 608, 619, 629, 630, 675–677, 703
Low dielectric constant, 53, 55, 161, 243, 267, 309, 375, 583, 584
Low-K ILD, 317
Low stress, 35, 317, 346, 358, 360, 361, 407, 411, 414, 415, 420, 434

M

Magnetic inductor(s), 459, 462, 463, 464–470, 472–480
Materials, 16, 18, 44, 52, 53, 54, 58, 60, 61, 80, 85, 92, 95, 97, 98, 101, 104, 115, 118, 125, 128, 129, 130, 132, 138, 139, 142, 160, 161, 167, 171, 181, 190, 193, 195, 201, 203, 225, 243, 245, 246, 248, 250, 251, 256, 257, 258, 259, 263, 265, 266, 267, 269, 274, 276–289, 294–304, 314–317, 321, 323, 325, 328, 330, 332, 341, 345, 358, 359, 365–371, 374, 376, 383, 388, 390, 392, 398, 407, 409, 410, 411, 420, 421, 464, 480, 489, 493, 508, 532, 555, 581
Mechanical performance, 108, 221, 230, 277, 304, 480
Microelectromechanical Systems (MEMS), 33, 55, 56, 141, 390, 517, 548, 557, 565, 567, 601–623
Micro-system, 1, 4
Moisture Absorption, 274, 278, 284, 300, 309, 315, 343, 351, 355, 382–383, 396, 414, 418, 420, 632, 648, 650, 660
Moisturized reflow resistance, 339, 349–356, 357
Moldability, 339

N

Nanogenerator, 503, 523–527, 529, 533, 535, 536, 537
Nanopackaging, 503–537
Nanoparticle, 483–488, 504–506, 508–511, 515, 527, 528
Nanosolder, 503
Nanotechnology, 441, 453–454, 483, 503, 504, 518, 704, 705
Nanowire, 503, 506, 518–520, 527–535

O

Optoelectronic, 507, 565, 601, 603, 609, 610, 613, 618, 624, 629, 632, 633, 638, 644–673
Organic substrates, 95, 97, 98, 100, 107, 243, 244, 245, 246, 308, 309, 507

P

Package crack resistance, 407, 411, 412, 415, 418–434
Packaging, 3–4, 11, 38, 39, 40, 45, 46, 47, 52, 55, 56, 58, 60, 61, 62, 77, 78, 85, 87, 89, 91, 97, 101, 129, 156, 168, 174, 235, 240, 243, 245, 247, 308, 315, 317, 323, 325, 340, 346, 366, 374, 407, 408, 409, 411, 415, 418, 439, 455, 481, 497, 504–507, 517, 535–538, 547, 549, 551, 563, 575, 578, 580, 610, 611, 618, 623, 632, 634, 640, 643, 645, 647, 652, 660, 667, 670
Passives, 4, 5, 28, 243, 244, 269, 460, 505, 550, 593, 594, 597
Pb-free, 63, 181, 192, 193, 194, 200, 204, 209, 418, 419
Peel strength, 305, 418, 420, 422, 423, 424, 427, 431, 434
Photo-Resists, 547, 552, 560, 575, 580
PI, 16, 24, 25, 34, 53, 256, 300, 427, 429, 431, 433, 547, 549, 556, 578, 585, 587, 588, 592
Piezotronics, 503
Point of care, 681, 682, 704–710
Polyimide, 30, 41, 53, 54, 92, 118, 138, 139, 160, 161, 244, 256, 275, 276, 278, 279, 315, 345, 378, 381, 383, 407, 411, 412, 415, 419, 421, 427, 433, 434
Polymers, 450, 537, 581–588, 594

Q

Quality factor, 392, 459, 460, 461, 464, 469, 473, 474, 475, 476, 477, 707

R

Redistribution, 87–88, 89, 162, 245, 263, 269, 301, 313, 317, 325, 327, 377, 547–552, 581, 583, 589, 590, 594
Reinforcement, 274
Reliability, 119, 128, 150–155, 181, 201–213, 303–306, 312–315, 376, 379, 415, 418–434, 603, 623, 674–677
Rheology, 196, 393, 437
Rolled annealed copper, 274, 289

S

SAC, 181, 183, 206, 207, 210, 211, 212, 316, 560
Semiconductor nanomaterial, 503
Singulation, 3, 32, 219, 228, 230, 231, 235–240, 328, 331

SnAgCu, 181, 182, 183, 184, 190, 207, 507, 560, 561
Solder, 52, 54, 56, 60, 61, 62–66, 69, 71, 77, 78, 79, 80, 84, 85, 86, 87, 92, 94–104, 107, 115, 118, 136, 158, 162, 181–196, 199, 200–213, 246, 249, 250, 251, 258, 260–270, 276, 278, 291, 293, 304, 305, 308, 309, 312–332, 340, 341, 349, 352, 353, 354, 357, 369, 371, 376, 377, 389, 390, 394, 397, 409, 410, 411, 412, 415, 418, 419, 427, 433, 452, 554
Solder-free, 77, 92, 94–104
Soldering, 54, 61, 62, 63, 64, 71, 115, 181–196, 200, 204, 206, 210, 278, 291, 293, 304, 316, 341, 349, 371, 390, 411, 412, 415, 418, 419, 427, 554
Surface finish, 192, 193, 202, 210, 212, 224, 230, 247, 259, 260, 262, 264, 266, 291, 293, 294, 304, 305, 321, 371, 372
Surface mount technology (SMT), 187, 318, 321, 324, 325, 327, 379, 390, 489, 547, 548, 549, 550, 646
System Integration, 601, 603, 618, 619

T

Thermal interface, 438–450, 454, 514, 515
Thermal resistance, 437, 438, 440, 441, 447, 450–455, 515, 619, 635, 644, 645, 646, 648, 670, 672, 673, 675, 676
Thermoplastic, 54, 55, 275, 276, 279, 330, 366, 367, 370, 376, 381, 386, 387, 388, 389, 390, 421, 424, 583, 667
Thermoset, 274, 275, 276, 278, 296, 323, 368, 370, 377, 381, 386, 388, 416, 426, 635, 661
Thin Film, 11, 42, 114, 136, 143, 160, 161, 284, 291, 388, 459, 462, 496–498, 505, 517, 550, 556, 557, 581, 582, 586, 588, 594, 597, 617, 620

Thin film resistor, 459, 489, 490, 491, 495, 496, 497
Thinning, 14, 16, 28, 31, 32, 33, 39, 40, 219, 220, 222, 228, 229, 230, 233, 234–235, 239, 241, 255
Thru-silicon vias (TSVs), 1
Tin-silver-copper, 181, 205

U

UBM, 66, 72, 314, 385, 516, 547, 552, 553, 554, 556, 557, 559, 560, 566, 574, 575, 589, 590
Underfill, 24, 77, 80, 90, 91, 92, 94, 102, 103, 106, 109, 119, 158, 168, 258, 261, 309–332, 369, 385, 387, 388, 389, 548, 566

W

Wafer bonding, 16, 18, 96, 97, 99, 101, 102, 108, 604
Wafer-bow, 219
Wafer Level Packaging, 58, 89, 244, 547, 549, 551, 552, 602, 608
Warpage, 194, 195, 267, 316, 339, 349, 356–357, 358, 407, 412, 415, 420, 428, 430, 431, 611, 650
Water absorption, 53, 55, 248, 251, 257, 279, 343, 349, 350, 353–354, 355, 357, 360, 375, 407, 420, 421, 422, 425, 426, 427, 434, 496
Wirebonding, 159, 160, 162, 163, 164, 174, 377
WL-CSP, 547, 550, 578

Z

Zinc oxide, 503, 518

About the Editors

Dr. Daoqiang Daniel Lu currently is the head of Product Development department of Electronics Materials Division of Henkel Corporation in Yantai, China. He received his MS and PhD degrees on Polymer Science and Engineering from Georgia Institute of Technology in 1996 and 2000, respectively. Prior to joining Henkel, Dr. Lu worked for the R&D Department of Intel Corporation as a Sr. Scientist and program manager for 7 years. He also had worked for Lucent Technologies, Amoco Electronic Materials Division, and the Electronics Materials Group of National Starch and Chemical Company before. Dr. Lu has extensive experience in electronic packaging materials and processing.

Dr. Lu received many awards including the IEEE/CPMT Outstanding Young Engineer Award in 2004, the IEEE ECTC best poster paper in 2007, Intel's most patent filing in 2003–2007, Intel Divisional Recognition Awards in 2002, 2003, and 2007, Intel most patent granting of the year for 2006 and 2007, Best Graduate Student of the Year of Georgia Tech Packaging Research Center in 2000, and Best Paper of the Session of International Symposium and Exhibition on Advanced Packaging Materials in 2000.

Dr. Lu has published more than 40 technical papers, wrote chapters for four books, holds 40 US patents, and has more than 40 pending patent applications. He has been serving key roles in organizing international conferences and teaching professional development short courses in these conferences.

Dr. Lu is a senior member of IEEE, an associate editor of the *IEEE Transactions on Advanced Packaging*, and a Guest Professor of Shanghai University (Shanghai, China).

Prof. C. P. Wong is a Regents' Professor and holder of the Charles Smithgall Institute Endowed Chair (one of the two Institute Endowed Chairs at GT) at the School of Materials Science and Engineering at Georgia Institute of technology. He received his B.S. degree from Purdue University, and his Ph.D. degree from the Pennsylvania State University. After his doctoral study, he was awarded a two-year postdoctoral fellowship with Nobel Laureate Professor Henry Taube at Stanford University.

He joined AT&T Bell Laboratories as a member of the technical staff and has been involved with the research and development of the polymeric materials (inorganic and organic) for electronic and photonic applications. He became an AT&T Bell Laboratories Fellow (the most prestigious award bestowed by Bell Labs) for his fundamental contributions to low-cost high performance plastic packaging of semiconductors. Since 1996, he is a Professor at the School of Materials Science and Engineering at the Georgia Institute of Technology. He was named a Regents Professor (highest rank professor) in July 2000, elected the Class of 1935 Distinguished Professor in 2004 for his outstanding and substained contributions in research, teaching and services, and named holder of the Georgia Tech Institute Endowed chair in 2005.

His research interests lie in the fields of polymeric materials, materials reaction mechanism, IC encapsulation, in particular, hermetic equivalent plastic packaging, electronic manufacturing packaging processes, interfacial adhesions and nano functional material syntheses and characterizations. He is one of the pioneers who demonstrated the use of organic polymers as device encapsulant to achieve reliability without hermeticity in plastic IC packaging, elucidated the fundamental conductivity fatigue of conductive adhesives for lead-free interconnects, syntheses and characterizations of nano composites, lead-free alloys, developed the first no flow underfill for low cost flip-chip, ultra high k capacitor composites and novel lotus effect coating materials.

He received many awards, among those, the AT&T Bell Labs Fellow Award in 1992, the IEEE CPMT Society Outstanding and Best Paper Awards in 1990, 1991, 1994, 1996, 1998, 2002, the IEEE CPMT Society Outstanding Sustained Technical Contributions Award in 1995, the Georgia Tech Sigma Xi Faculty Best Research Paper Award in 1999, Best MS, PhD and undergraduate Thesis Awards in 2002 and 2004, respectively, the University Press (London) Award of Excellence, the IEEE Third Millennium Medal in 2000, the IEEE EAB Education Award in 2001, the IEEE CPMT Society Exceptional Technical

Contributions Award in 2002, elected as holder of the Charles Smithgall Institute Endowed Chair at Georgia Tech in 2005, the Georgia Tech Outstanding PhD Thesis Advisor Award , the IEEE Components, Packaging and Manufacturing Technology Field Award in 2006, the Sigma Xi's Monie Ferst Award in 2007 and the ASM's TEEM Award in 2008.

He holds over 50 U.S. patents, numerous international patents, has published over 500 technical papers. Dr. Wong is a Fellow of the IEEE, AIC, and AT&T Bell Labs, and was the technical vice president (1990 and 1991), and the president of the IEEE-CPMT Society (1992 and 1993). He was elected a member of the National Academy of Engineering in 2000.

Printed in the United States of America

DISCARDED
CONCORDIA UNIV. LIBRARY

CONCORDIA UNIVERSITY LIBRARIES
MONTREAL